INTERNATIONAL ATOMIC WEIGHTS*

Name	Symbol	Atomic Number	Atomic Weight	Name	Symbol	Atomic Number	Atomic Weight
Actinium	Ac	89	(227)†	Mercury	Hg	80	200.6
Aluminium	Al	13	27.0	Molybdenum	Mo	42	95.9
Americium	Am	95	(243)	Neodymium	Nd	60	144.2
Antimony	Sb	51	121.8	Neon	Ne	10	20.2
Argon	Ar	18	39.9	Neptunium	Np	93	(237)
Arsenic	As	33	74.9	Nickel	Ni	28	58.7
Astatine	At	85	(210)	Niobium	Nb	41	92.9
Barium	Ba	56	137.3	Nitrogen	N	7	14.01
Berkelium	Bk	97	245	Nobelium	No	102	(254)
Beryllium	Be	4	9.01	Osmium	Os	76	190.2
Bismuth	Bi	83	209.0	Oxygen	O	8	16.00
Boron	B	5	10.8	Palladium	Pd	46	106.4
Bromine	Br	35	79.9	Phosphorus	P	15	31.0
Cadmium	Cd	48	112.4	Platinum	Pt	78	195.1
Calcium	Ca	20	40.1	Plutonium	Pu	94	(242)
Californium	Cf	98	(251)	Polonium	Po	84	210
Carbon	C	6	12.01	Potassium	K	19	39.1
Cerium	Ce	58	140.1	Praseodymium	Pr	59	140.9
Cesium	Cs	55	132.9	Promethium	Pm	61	(147)
Chlorine	Cl	17	35.5	Protactinium	Pa	91	(231)
Chromium	Cr	24	52.0	Radium	Ra	88	(226)
Cobalt	Co	27	58.9	Radon	Rn	86	(222)
Copper	Cu	29	63.5	Rhenium	Re	75	186.2
Curium	Cm	96	(247)	Rhodium	Rh	45	102.9
Dysprosium	Dy	66	162.5	Rubidium	Rb	37	85.5
Einsteinium	Es	99	(254)	Ruthenium	Ru	44	101.1
Erbium	Er	68	167.3	Samarium	Sm	62	150.4
Europium	Eu	63	152.0	Scandium	Sc	21	45.0
Fermium	Fm	100	(253)	Selenium	Se	34	79.0
Fluorine	F	9	19.0	Silicon	Si	14	28.1
Francium	Fr	87	(223)	Silver	Ag	47	107.9
Gadolinium	Gd	64	157.3	Sodium	Na	11	23.0
Gallium	Ga	31	69.7	Strontium	Sr	38	87.6
Germanium	Ge	32	72.6	Sulfur	S	16	32.1
Gold	Au	79	197.0	Tantalum	Ta	73	180.9
Hafnium	Hf	72	178.5	Technetium	Tc	43	(99)
Helium	He	2	4.00	Tellurium	Te	52	127.6
Holmium	Ho	67	164.9	Terbium	Tb	65	158.9
Hydrogen	H	1	1.008	Thallium	Tl	81	204.4
Indium	In	49	114.8	Thorium	Th	90	232.0
Iodine	I	53	126.9	Thulium	Tm	69	168.9
Iridium	Ir	77	192.2	Tin	Sn	50	118.7
Iron	Fe	26	55.8	Titanium	Ti	22	47.9
Krypton	Kr	36	83.8	Tungsten	W	74	183.9
Lanthanum	La	57	138.9	Uranium	U	92	238.0
Lawrencium	Lr	103	(257)	Vanadium	V	23	50.9
Lead	Pb	82	207.2	Xenon	Xe	54	131.3
Lithium	Li	3	6.94	Ytterbium	Yb	70	173.0
Lutetium	Lu	71	175.0	Yttrium	Y	39	88.9
Magnesium	Mg	12	24.3	Zinc	Zn	30	65.4
Manganese	Mn	25	54.9	Zirconium	Zr	40	91.2
Mendelevium	Md	101	(256)				

*More precise values of the atomic weights based on the latest IUPAC reports are given in Appendix 9.
†For those elements all of whose isotopes are radioactive, the parentheses indicate the isotope with the longest half-life. (For a definition of *half-life*, see Section 7-4.2.)

chemistry
experimental foundations

chemistry
experimental foundations

ROBERT W. PARRY
Professor of Chemistry
University of Utah

LUKE E. STEINER
Professor of Chemistry Emeritus
Oberlin College

ROBERT L. TELLEFSEN
Chemistry Teacher
Napa High School
Napa, California

PHYLLIS M. DIETZ
Chemistry Teacher
Fountain Valley High School
Huntington Beach, California

PRENTICE-HALL, INC., ENGLEWOOD CLIFFS, NEW JERSEY

The cover photograph depicts a lighted candle silhouetted against the multicolored interference pattern produced by a Michelson interferometer using white light. The brilliant light of the interferometer has washed out nearly all the light of the flame, but the hot gases from the flame have sharply deformed the interference pattern and produced a large discontinuity on the edge of the flame. This unique view of the candle symbolizes our revision, which takes a new look at chemistry as a dynamic and distinctly contemporary science.

Photographed expressly for *Chemistry: Experimental Foundations* by Drs. Cagnet, Françon, and Mallick of the Laboratoire d'Optique of the Faculté des Sciences de Paris in France.

CHEMISTRY: Experimental Foundations
Supplementary Materials
Laboratory Manual, Laboratory Notebook,
Achievement Tests, Teachers Guide.

© 1970 by Prentice-Hall, Inc., Englewood Cliffs, New Jersey. All rights reserved. No part of this book may be reproduced in any form or by any means without permission in writing from the publisher. Printed in the United States of America.
13-128975-6   3 4 5 6 7 8 9 10

Design by Arthur F. Soares; Art Production Jerrold J. Stefl; Assistant John A. Brandes; Drawings by George Bakacs.

Prentice-Hall International, Inc., *London*
Prentice-Hall of Australia, Pty. Ltd., *Sydney*
Prentice-Hall of Canada, Ltd., *Toronto*
Prentice-Hall of India Private Ltd., *New Delhi*
Prentice-Hall of Japan, Inc., *Tokyo*

Preface

CHEMISTRY: EXPERIMENTAL FOUNDATIONS is an authorized revision of the original CHEM Study textbook. As in the original book, experimentation is the vehicle for presenting chemistry as it is today. Unifying principles are developed from experimental observation. Through the use of such principles, you avoid seemingly endless memorization and chemistry emerges as a science rather than a mass of information. The cornerstone of modern science—the development of principle from observation—is the cornerstone of this revision.

This book is a chemistry book, not a book *about* chemistry. It asks you to participate in scientific activity and to share the excitement of discovery. As a consequence we hope that you will have a better appreciation of the power and limitations of scientific methods. At the end of the course you *will not* know all of chemistry, but we hope that you will know how a chemist works. We hope, further, that you will have formed the habit of questioning. We hope that you will seek true understanding rather than being satisfied with dogmatic assertions. In these hopes lie many of the most important, yet hidden, benefits of a study of science.

We live in a scientific age, one which science and its product, technology, dominate. Successful technology demands very careful application of fundamental scientific principles. But before technology can succeed, science must advance. We hope this book can contribute to that advancement.

In preparing this revision the authors have been assisted by many people. Our debt to the Prentice-Hall staff is unusually large. The assistance rendered by editors Marion Cahill McDanield and Kelvin Kean literally made this book possible. Without their enthusiasm and help, completion of the manuscript would have been doubtful. To Edgar P. Thomas and James M. Guiher we are grateful for sound administrative guidance, and to Aletta Biersack, Anthony Caruso, and the many other members of the Prentice-Hall production staff who worked effectively in producing the book, we express our appreciation.

We are particularly appreciative of the constructive comments and help rendered by Harold Pratt, Jefferson County Schools, Lakewood, Colo.; Kermit Waln, Thomas Jefferson H. S., Denver, Colo.; Raymond T. Byrne, Batavia H. S., Batavia, N. Y.; and Charles Tier, St. Joseph's H. S., Metuchen, N. J., who evaluated the manuscript at various stages. We also appreciate the contributions of the following people who lent us their valuable advice and experience:

Margaret B. Anderson, Westfield H. S., Westfield, Mass.; Prof. W. A. Andrews, University of Toronto; Joseph E. Anthony, Pennsylvania Bureau of General and Academic Education; Curtis Carlson, Saratoga H. S., Saratoga, Calif.; Don Chaney, Los Gatos H. S., Los Gatos, Calif.; V. L. Chapman, Lord Byng Secondary School, Vancouver, B. C., Canada; Harry E. Choulett, Berkeley H. S., Berkeley, Calif.; Prof. H. R. Crane, University of Michigan; Jerry L. Craven, Lyons Twp. H. S., La Grange, Ill.; Calvin F. Delano, Westside H. S., Omaha, Nebr.; Robert W. Fatzinger, Arvada H. S., Arvada, Colo.; Dan Feller, Sunset H. S., Beaverton, Oreg.; John Friesen, Mennonite Collegiate Institute, Gretna, Man., Canada; Glen Hampshire, Jefferson H. S., Portland,

Oreg.; Wallace Helber, Culver Military Academy, Culver, Ind.; Richard L. Hutchens, Clackamas H. S., Milwaukie, Oreg.; Gerard A. Kass, Northport H. S., Northport, N. Y.; William F. Kieffer, College of Wooster, Wooster, Ohio; Thomas K. Lewis, Newport Catholic H. S., Newport, Ky.; Lowell Lockridge, North H. S., Des Moines, Ia.; Lloyd A. Meech, Jr., South H. S., Denver, Colo.; Margaret Nicholson, Acalanes H. S., Lafayette, Calif.; Michael Saltzman, Bronxville Public Schools, Bronxville, N. Y.; Sister Clarice Lolich, Dominican Sister, Mission San Jose, Pasadena, Calif.; Sister Marie Berge, O.P., St. Thomas Aquinas H. S., Ft. Lauderdale, Fla.; R. Lane Trantham, South Carolina State Department of Education; Prof. A. B. Van Cleave, University of Saskatchewan, Regina, Sask., Canada; Prof. Helen L. Whidden, Randolph-Macon Woman's College, Lynchburg, Va.; John D. Wood, Aloha H. S., Beaverton, Oreg.

We also wish to thank Ella Marie Kean and Patty Freeman for their assistance in preparing the manuscript.

Finally, we thank the teachers and students who have used the CHEM Study material and who have guided us in this revision. To them goes our deepest appreciation.

RWP
LES
PMD
RLT

Contents

1. THE OBSERVATIONAL BASIS OF CHEMISTRY 2
1-1 What Is Chemistry? 3 1-2 The Activities of Science 3 1-3 Observation and Description 4 1-4 The Search for Regularities 6 1-5 Wondering Why 10 1-6 Conclusions and the Accuracy of Observations 13 1-7 Communicating Scientific Information 17 1-8 Highlights 18 Questions and Problems 19

2. THE ATOMIC THEORY: ONE OF OUR BEST SCIENTIFIC MODELS 20
2-1 Growth of a Model 21 2-2 Molecular Weights, the Mole, and the Molar Volume 30 2-3 Atoms 35 2-4 Elements and Compounds 42 2-5 Highlights 47 Questions and Problems 48

3. PUTTING IDEAS TO WORK: THE CONSERVATION LAWS 50
3-1 Determining the Number of Moles From Laboratory Quantities 51 3-2 The Conservation of Mass 54 3-3 Calculations Based Upon Chemical Equations 59 3-4 Energy Change in Chemical Reactions: The Conservation of Energy 61 3-5 Highlights 64 Questions and Problems 64

4. MORE ABOUT GASES: THE KINETIC THEORY 66
4-1 Molar Volumes and the Distance Between Particles in the Gas Phase 67 4-2 Measurements on Gases 71 4-3 The Kinetic Theory 76 4-4 A General Gas Equation 82 4-5 The Consequences of Theory 86 4-6 Highlights 89 Questions and Problems 89

5. LIQUIDS AND SOLIDS: CONDENSED PHASES OF MATTER 92
5-1 Phase Changes 93 5-2 Solutions 105 5-3 Highlights 112 Questions and Problems 112

6. WHY WE BELIEVE IN ATOMS 114
6-1 Observation and Conclusion 115 6-2 Observation and Belief in Atoms 117 6-3 The Electrical Nature of Matter and the Makeup of Atoms 119 6-4 The Electrical Properties of Condensed Phases 123 6-5 The Electrical Conductivity of Solutions and an Electrical Model for Atoms 125 6-6 Other Consequences of the Existence of Ions 134 6-7 Highlights 135 Questions and Problems 136

7. THE STRUCTURE OF THE ATOM 138
7-1 Elementary Atomic Architecture 139 7-2 "Seeing" Parts of Atoms 147 7-3 Highlights of Elementary Atomic Architecture (7-1) and "Seeing" Parts of Atoms (7-2) 154 7-4 Properties of Nuclei—Radioactivity and Nuclear Chemistry 155 Questions and Problems 158

8 ATOMIC STRUCTURE AND THE PERIODIC TABLE 160

8-1 The Loss of Electrons From an Atom—The Ionization Energy of Atoms 161 8-2 A Model for Electron Arrangement 166 8-3 The Chemical Significance of Ionization Energies 168 8-4 The Alkali Metals 171 8-5 The Halogens 174 8-6 Hydrogen—A Family by Itself 178 8-7 The Elements From Three Through Ten—The Second Row of the Periodic Table 179 8-8 The Periodic Table 185 8-9 Models and Physical Properties—Bonding Patterns Across the Second Row 186 8-10 Highlights 187 Questions and Problems 188

9 ENERGY CHANGES IN CHEMICAL AND NUCLEAR REACTIONS 190

9-1 Energy Changes and Chemical Reactions 191 9-2 The Conservation of Energy 198 9-3 The Energy Stored in a Molecule 201 9-4 The Energy Stored in a Nucleus 203 9-5 Highlights 208 Questions and Problems 208

10 THE RATES OF REACTIONS 210

10-1 Factors Affecting Reaction Rates 212 10-2 The Role of Energy in Reaction Rates 222 10-3 The Rate of Nuclear Processes 230 10-4 Highlights 231 Questions and Problems 232

11 EQUILIBRIUM IN PHASE CHANGES AND IN CHEMICAL REACTIONS 234

11-1 Equilibrium in Phase Changes 235 11-2 Equilibrium in Chemical Systems 238 11-3 Altering the Equilibrium Condition or the State of Equilibrium 244 11-4 Attainment of Equilibrium 245 11-5 Predicting New Equilibrium Concentrations—Le Chatelier's Principle 246 11-6 Application of Equilibrium Principles—The Haber Process 249 11-7 Quantitative Aspects of Equilibrium 250 11-8 The Factors Determining Equilibrium 256 11-9 Highlights 262 Questions and Problems 263

12 SOLUBILITY EQUILIBRIUM 266

12-1 Solubility: A Case of Equilibrium 267 12-2 The Factors that Fix the Solubility of a Solid 271 12-3 The Factors That Fix the Solubility of a Gas in a Liquid 273 12-4 Aqueous Solutions 274 12-5 An Application of Equilibrium Concepts to Ionic Solutions 280 12-6 Highlights 285 Questions and Problems 285

13 IONIC EQUILIBRIUM: ACIDS AND BASES 288

13-1 Electrolytes—Strong and Weak 289 13-2 Experimental Introduction to Acid and Base Systems 297 13-3 Conceptual and Operational Definitions 301 13-4 Acid-Base Titrations 302 13-5 The pH Scale 305 13-6 Strengths of Acids 305 13-7 An Expansion of Acid-Base Concepts 312 13-8 Highlights 315 Questions and Problems 316

14 OXIDATION-REDUCTION REACTIONS 318
14-1 Electrochemical Cells 319   14-2 Electron Transfer and Predicting Reactions 328   14-3 Balancing Oxidation-Reduction Equations 339   14-4 Electrolysis 348   14-5 Highlights 353   Questions and Problems 354

15 ELECTROMAGNETIC RADIATION AND ATOMIC STRUCTURE 356
15-1 Light and Color 357   15-2 Energy and an Interpretation of the Hydrogen Spectrum 369   15-3 Mechanical Atomic Models 376   15-4 Wave Models for the Hydrogen Atom 380   15-5 The Quantum Theory 382   15-6 Electronic Structure and the Periodic Table 386   15-7 Ionization Energies, Energy Levels, and the Periodic Table 393   15-8 Highlights 399   Questions and Problems 400

16 MOLECULAR ARCHITECTURE: GASEOUS MOLECULES 402
16-1 The Covalent Bond 403   16-2 The Bonding Capacity of the Second-Row Elements 409   16-3 Trend in Bond Type Among the Second-Row Fluorides 415   16-4 Molecular Geometry 417   16-5 Molecular Shape and Electric Dipoles 423   16-6 Multiple Bonds 426   16-7 Methods of Structure Determination 430   16-8 Highlights 432   Questions and Problems 433

17 MOLECULAR ARCHITECTURE: LIQUIDS AND SOLIDS 434
17-1 Molecular Solids—van der Waals Forces 435   17-2 Covalent Bonds and Network Solids 438   17-3 The Metallic State 440   17-4 Boron—The Element Where Metals and Network Solids Meet 448   17-5 Ionic Structures—The Ionic Bond and Its Consequences 452   17-6 Hydrogen Bonds 454   17-7 Highlights 457   Questions and Problems 457

18 THE CHEMISTRY OF CARBON COMPOUNDS 460
18-1 Sources of Carbon Compounds 461   18-2 Molecular Structure of Carbon Compounds 462   18-3 Some Chemistry of Organic Compounds 473   18-4 Nomenclature 483   18-5 Hydrocarbons 485   18-6 Polymers 492   18-7 Highlights 496   Questions and Problems 496

19 THE THIRD ROW OF THE PERIODIC TABLE 498
19-1 The Elements of the Third Row—Their Physical Properties 499   19-2 The Elements of the Third Row as Oxidizing and Reducing Agents 505   19-3 The Acidic and Basic Character of the Hydroxides of the Elements of Row Three 510   19-4 Highlights 516   Questions and Problems 516

20 VERTICAL GROUPS OF THE PERIODIC TABLE 518
20-1 Some Important Properties in Two Columns of the Periodic Table—the Alkaline Earth Metals and the Halogens 519   20-2 Sizes of the Atoms and Ions in the Second and Seventh Columns 522   20-3 Highlights 526   Questions and Problems 526

21 THE TRANSITION ELEMENTS: THE FOURTH ROW OF THE PERIODIC TABLE 528
21-1 Identification and Electronic Structure of Transition Elements 529   21-2 Complex Ions 535   21-3 Specific Properties of Fourth-Row Transition Elements 542   21-4 Some Chemistry of Chromium—a Typical Transition Metal 546   21-5 Some Chemistry of Iron—Our Most Widely Used Transition Metal 547   21-6 Highlights 548   Questions and Problems 549

22 SOME ASPECTS OF BIOCHEMISTRY 550
22-1 Molecular Composition of Living Systems 551   22-2 Molecular Structures in Biochemistry 557   22-3 The "Nucleic Acids," DNA and RNA, Nature's Messengers 561   22-4 Highlights 567

23 ASTROCHEMISTRY 568
23-1 The Chemistry of Our Planet Earth 569   23-2 The Chemistry of the Moon 573   23-3 The Chemistry of the Planets 576   23-4 The Sun 580   23-5 Stellar Atmospheres 581   23-6 Interstellar Space 581   23-7 Epilogue 582

APPENDICES 1–11 583

ACKNOWLEDGEMENTS 607

INDEX 609

chemistry
experimental foundations

. . . I wish you not only the joy of great discovery; I wish for you a world of confidence in man and man's humanity, a world of confidence in reason, so that as you work you may be inspired by the hope that what you find will make men freer and better . . . .

J. ROBERT OPPENHEIMER (1904–1967)

CHAPTER 1 The Observational Basis of Chemistry

Which image is real and which is a reflection? Only careful observation will tell.

What do scientists really do? What does it mean when someone says he is a chemist? How does he spend his time? Are the TV commercials honest when they show someone in a white coat frowning through horn-rimmed glasses at dramatically bubbling concoctions which may be tooth paste or infernal machines? How can you tell which you've discovered so they will know what to award you the Nobel prize for?

We don't promise to answer *all* of these questions in one easy chapter, but at least we'll clue you in on SOME OF THE SUCCESS SECRETS OF THE MEN IN THE INDEPENDENT TESTING LABORATORIES.

1-1 WHAT IS CHEMISTRY?

What does the word *chemistry* suggest to you? To many people chemistry means miracle fibers such as nylon and dacron, medicines from aspirin to antibiotics, polymers from floor tiles to bread wrappers, fuels from heating oil to high-test gasoline, and thousands of other products which make our surroundings more comfortable and attractive. This is the chemistry that shows above the surface and which benefits us all in a very material way. But chemistry to a chemist is more like an iceberg: most of its activity is not visible on casual inspection. What then is chemistry?

As we all know, a dictionary definition of baseball does not carry with it the intense emotional thrill of a ball driven out of the park or of a close decision at home plate. To a baseball player or a baseball fan, baseball is not easily defined. You have to play the game! Certainly no dictionary definition of music can describe the wild excitement or the dreamy, romantic mood which can be generated by musicians. Chemistry, like baseball and music, is not defined with any ease. It is a living subject, requiring mental and physical activity. Like baseball, you have to participate in the activity to understand it properly. Let us put the chemical ball into play by taking a brief look at the activities of science.

1-2 THE ACTIVITIES OF SCIENCE

Have you ever noticed earthworms on top of the ground after a heavy rain? Hungry birds are quick to spot these worms and eat their fill. On days when it does not rain, the birds frequently gather around the lawn sprinkler. Our feathered friends, primitive scientists that they are, seem to make a generalization: water in the ground means worms on top. The life of the bird is better because of this simple generalization.

Man is different in degree but not in kind. He observes and studies his surroundings more carefully than does any other creature. He organizes the information he has gathered. Then he examines the organized information for regularities; like the bird, man also notes that water in the ground means worms on top. Many fishermen have profited from this generalization.

Unlike the bird, man is not content just to find and use the simple regularity. He wonders why the regularity exists. Does the water in the ground prevent the worm from getting air? This is a reasonable hypothesis and suggests that earthworms will drown if we place them in a glass of water. "Wondering why" has suggested a possible characteristic of earthworms. More sophisticated questions are possible. We might ask: How much water is needed in the ground to drive the worms to the surface? Why does the worm need air? How does he use it? Man differs from the bird because he wonders about these things.

Man differs in another very important way. Communication between birds is primitive and inefficient. Man, on the other hand, has developed an effective communication system which permits

The ACTIVITIES of SCIENCE

Fig. 1-1 **Water in the ground means worms on top.**

him to pass his observations and thoughts on to men throughout the world. Communication lies at the heart of human advancement in all fields.

In summary, we can now list the basic activities of science. These are:

(1) to accumulate information through observation;
(2) to organize this information and to seek regularities in it;
(3) to wonder why the regularities exist; and
(4) to communicate the findings and their probable explanations to others.

There is no fixed order in which these activities must be carried out; no rigid "scientific method" requires that the steps be done in the order given above; in fact, "wondering why" usually suggests the need for more carefully controlled observations. A carefully controlled sequence of observations is frequently called an **experiment.** Conditions for experiments in chemistry are often controlled most easily in a laboratory, but the study of nature should not be confined to a single room. Science is all around us!

1-3 OBSERVATION AND DESCRIPTION

The power of observation is in no way peculiar to science. The ingenious detective who astounds everyone with his ability to draw conclusions from what he sees has fascinated mystery story readers from the days of Sherlock Holmes to the days of James Bond, Agent 007. Many of you know of James Bond's amazing skills in many areas, but the uncanny powers of observation possessed by Sherlock Holmes are less familiar. If you do not know Mr. Holmes, an introduction is in order; if you do know him, you will probably enjoy seeing him in action again.

THE POWER OF OBSERVATION

Dr. Watson is describing an encounter between the great detective, Sherlock Holmes, his client, Mr. Wilson, and himself, Dr. Watson. (From Sir Arthur Conan Doyle, "The Red-Headed League.")

> The portly client puffed out his chest with an appearance of some little pride and pulled a dirty and wrinkled newspaper from the inside pocket of his greatcoat. As he glanced down the advertisement column, with his head thrust forward and the paper flattened out upon his knee, I took a good look at the man and endeavoured, after the fashion of my companion, to read the indications which might be presented by his dress or appearance.
>
> I did not gain very much, however, by my inspection. Our visitor bore every mark of being an average commonplace British tradesman, obese, pompous, and slow. He wore rather baggy gray shepherd's check trousers, a not over-clean black frock-coat, unbuttoned in the front, and a drab waistcoat with a heavy brassy Albert chain, and a square pierced bit of metal dangling down as an ornament. A frayed top-hat and a faded

brown overcoat with a wrinkled velvet collar lay upon a chair beside him. Altogether, look as I would, there was nothing remarkable about the man save his blazing red head and the expression of extreme chagrin and discontent upon his features.

Sherlock Holmes's quick eye took in my occupation, and he shook his head with a smile as he noticed my questioning glances. "Beyond the obvious facts that he has at some time done manual labour, that he takes snuff, that he is a Freemason, that he has been in China, and that he has done a considerable amount of writing lately, I can deduce nothing else."

Mr. Jabez Wilson started up in his chair, with his forefinger upon the paper, but his eyes upon my companion.

"How, in the name of good-fortune, did you know all that, Mr. Holmes?" he asked. "How did you know, for example, that I did manual labour? It's as true as gospel, for I began as a ship's carpenter."

"Your hands, my dear sir. Your right hand is quite a size larger than your left. You have worked with it, and the muscles are more developed."

"Well, the snuff, then, and the Freemasonry?"

"I won't insult your intelligence by telling you how I read that, especially as, rather against the strict rules of your order, you use an arc-and-compass breastpin."

"Ah, of course, I forgot that. But the writing?"

"What else can be indicated by that right cuff so very shiny for five inches, and the left one with the smooth patch near the elbow where you rest it upon the desk?"

"Well, but China?"

"The fish which you have tattooed immediately above your right wrist could only have been done in China. I have made a small study of tattoo marks and have even contributed to the literature of the subject. That trick of staining the fishes' scales of a delicate pink is quite peculiar to China. When, in addition, I see a Chinese coin hanging from your watch-chain, the matter becomes even more simple."

Mr. Jabez Wilson laughed heavily. "Well, I never!" said he. "I thought at first that you had done something clever, but I see that there was nothing in it, after all."*

Scientists particularly appreciate Mr. Wilson's last comment. Much of science seems routine and easy after the answer is known, but the problem usually looks different at the start. Every activity has Monday-morning quarterbacks like Wilson. Mr. Holmes was continually disturbed by Dr. Watson's poor powers of observation. He once told him ("A Scandal in Bohemia"): "You SEE, but you do NOT OBSERVE." Good scientists, like good detectives, cannot afford the luxury of *seeing without observing*. Science and clever detective work have much in common. Subtle clues lead to an explanation if one *observes*. If one only *sees*, the answer may pass by unnoticed.

Good observation takes concentration, alertness to detail, patience, and practice. We must identify and control the important variables governing a process. Consider an example from your own experience—the burning candle. How do we sort out the conditions which need to be controlled? Be ready for surprises here; conditions which seem important at first may have little effect on the process, whereas conditions that do not seem important may be crucial. For

* Reprinted by Permission of the Estate of Sir Arthur Conan Doyle.

The POWER of OBSERVATION

Fig. 1-2 **Good observation requires concentration, alertness to detail, patience, and practice.**

example, the nature of the torch or match used to light the candle is of no real significance once the process is underway, but the location of the candle in the room may be crucial in determining its burning characteristics. Why is the location important? Because a candle is strongly influenced by air currents or "drafts" and the air currents differ in varying parts of the room. One must *observe* and not just *see* to be a Sherlock Holmes or a good experimentalist.

Review your own description of a burning candle and compare your essay with the one in Appendix 1. How many of your observations are listed there? How many observations have you made which are not listed in Appendix 1? (A count of four means you are as good as Sherlock Holmes.)

1-4 THE SEARCH FOR REGULARITIES

Observation always leads to questions. One of the first questions to concern us is: What regularities appear? The discovery of regularities permits a simplification of the observations. In our earlier discussion the regularity "water in the ground means worms on top" summarized thousands of separate observations by both birds and men. Instead of each observation standing alone, many are classed together and can be considered more effectively.

The search for regularities is usually fun, but it is not always easy. In any process of exploration, wrong turns may lead to blind alleys. Not every step is an advance, and yet there is no way to advance other than by taking steps. Some move us ahead; some move us back. We hope that more lead us ahead than take us back. How the search proceeds can be seen in the following fable from science fiction.

1-4.1 Martin the Martian

Martin the Martian brought his spacecraft down in a cold, earthly woods. Martin was an unusually brilliant being, but he was operating in a new environment: Mars has no oxygen and therefore no fire as we know it. It was cold as Martin stepped into the dark night. He looked around quickly and saw in the distance some creatures whom he believed to be less intelligent beings. These creatures were huddled around a glowing pile of cylindrical objects and were making strange sounds. Martin approached the group as closely as he dared and found to his surprise that much heat was coming from the glowing pile of cylinders (a campfire). If only Martin could get his own pile of hot cylinders, life would be much more comfortable for him. How to do it? That was the problem. Surely if those backward creatures on Earth could make a heat source, the brilliant Martin could do as well.

His first step was to wait until the Earthlings disappeared into their tents. Then Martin quietly approached the pile of cylinders (a campfire), pulled out several logs by their cool ends, and carried the glowing pieces over to a spot near his spaceship. Soon he, too, had a pile of glowing cylinders. He warmed himself and felt pleased with the high level of Martian science.

Fig. 1-3 **Martians in a new environment.**

His happiness was short-lived. His cylinders began to disappear as the evening wore on. The Earthlings had piled all kinds of materials on their fire and the flames had leaped higher. What should Martin get to put on his fire? He went out into the woods as the Earthlings had done before him; logs were scarce and his Martian science did not tell him what to bring back.

Experimentation provided the only answer. He collected all kinds of materials and piled each object on his fire. In each case he wrote down what he saw. After a few trips the information shown in the table below appeared as a page in his notebook. (We have used our description of the object since his code designation was too complex for our backward minds.)

TABLE 1-1 "BURNABILITY" OF OBJECTS

English Designation of Object	Observation
Tree limb	Burns fairly well if not soaked in water.
Fence post	Burns well.
Rubber hose	Burns, but causes air pollution. Can't stand this!
Dynamite cap	Catastrophe! Wise to avoid these in future! Scattered hot cylinders.
Large rock	Didn't burn.
Large glass marble	Didn't burn; seemed to crack open.
Wooden pole	Burns well; get lots of these.

As Martin surveyed the limited data in his notebook, his mathematically trained soul was stirred. He suddenly noticed that everything which burned was cylindrical. He had run into problems with the dynamite cap and the rubber hose, but he thought he could recognize these objects in the future. They were special cases of burning. He put forth a *hypothesis:*

Cylindrical Objects Burn

In the future he would only collect cylindrical objects.

Using his hypothesis as a guide, Martin brought to his campsite an old cane, a baseball bat, and more tree limbs. All burned brightly and Martin was proud of his rapid mastery of the Earthly environment. He noted with pride that he had passed up a large wooden door, a large box of newspapers, and a piece of chain. Clearly, only cylindrical objects burn!

Do you feel sorry for our Martian visitor? Clearly, his generalization is not true; he is being misled. But wait: the generalization states a regularity discovered among all of the observations available to him—namely, the items on *his* list. If we do not go beyond *his* list, the generalization is true. *A generalization is reliable within the bounds defined by the experiments that lead to the rule!*

On his next trip to gather fuel, Martin staggered into camp dragging three pieces of iron pipe, two ginger ale bottles, and the axle from an old car. He passed up the wooden door, a long wooden 4″ × 4″, and a sack of wooden tent stakes. His fire dwindled and died. Martin was astonished!

MARTIN the MARTIAN

Fig. 1-4 Clearly, all cylindrical objects do *not* burn.

During the long cold night that followed, Martin the Martian became Martin the Miserable Martian. His hypothesis was in trouble. He had to draw some new and disturbing conclusions, and he was *very cold*. His new conclusions were:

(1) All cylindrical objects do *not* burn.
(2) Tree limbs, fence posts, and other cylindrical objects listed in the table still burn.
(3) The list is still useful.

In the bright, warm sun of the day following the cold, miserable night, Martin looked at his list and at the objects which burned. Stimulated by the warm sunshine and the appearance of the things which did burn, he proposed a new hypothesis:

Wooden Objects Burn

Martin is on the right track. His miserable nights are behind him.

If at this point you are inclined to wonder about the low state of Martian science, be careful. Think about how you might fare in the Martian environment? At least Martin has used good scientific practice. All of us, beginning students and experienced scientists alike, make observations, organize them, and seek regularities to help us in the effective use of our knowledge. Preliminary regularities are frequently stated as hypotheses; good hypotheses grow into theories as confidence in the generalization grows. A theory is retained as long as it is consistent with the known facts of nature or as long as it is an aid in systematizing knowledge.

Recent scientific history reads much like our Martian fable. In 1960 chemists around the world were teaching students in high schools and colleges that gases with eight electrons in the outer shell form no compounds. Such gases were said to be inert. A CHEM Study film was originally made emphasizing that compounds (we shall meet this word again shortly) of the inert gases *do not* form. Remember: All Cylindrical Objects Burn! In 1962, Professor Neil Bartlett, a scientist working in Canada, made the "inert" gas xenon combine with a new compound of platinum and fluorine, platinum fluoride (PtF_6). Professor Bartlett had figuratively thrown the cylindrical iron pipe into the fire. Clearly, all cylindrical objects do not burn. And, just as clearly, some atoms with eight electrons in the outer shell form compounds. Parts of the film and most of our ideas on "inert" gases were changed seriously as a result of Professor Bartlett's work. An old and accepted generalization was destroyed by clever observations that created a whole new field of chemistry. (See Section 8-3.2.)

Science has its limitations and scientists make mistakes, but such mistakes are unavoidable if we are to progress.

1-4.2 Regularities in the Melting of Solids

Like Martin the Martian, we too have been accumulating information. Some information on the melting of solids has been gathered in the laboratory and in class, and we are now in a position to propose a generalization.

REGULARITIES in the
MELTING of SOLIDS

Hypothesis: A solid melts to a liquid when the temperature is raised sufficiently. *The temperature at which a solid melts is characteristic.* When the warm liquid is recooled, it solidifies at this same temperature.

Further experiments will show that this hypothesis is true for thousands and thousands of solids, but some further restriction is needed. (Even the postulate All Wooden Objects Burn needs some further restriction. Ebony burns poorly.) This need for further restriction for some substances can be turned to our advantage in trying to classify materials. For example, people have found or prepared hundreds of thousands of substances which will melt at a sharp and characteristic temperature when the solid is heated, and which will freeze at this same temperature when the liquid is cooled. Other materials will soften without sharp melting or may even give irregular behavior, as did the dynamite cap in the Martian's fire. A few materials like carbon will melt only at extreme temperatures under very high pressures. Some characterization of materials on the basis of melting behavior would appear to be reasonable. Indeed, chemists have found that the *melting temperature and melting behavior are useful pieces of information in the identification of a substance.* One of the characteristic properties recorded for use in identifying a new pure material is its **melting point.** Ice melts in water exposed to laboratory air at 0.00°C. We take advantage of this fact when we fill our soft drink cooler with ice in the summertime.

A special set of words has been found useful in dealing with processes such as melting and freezing. We say that ice is the **solid phase** of water and that water is the **liquid phase.** Ice and water are different **phases** of the same substance. The change that occurs when a solid melts or a liquid freezes is called a **phase change.** We shall consider later other phase changes, such as the change of a liquid to a gas or a gas to a liquid.

Fig. 1-5 **The temperature at which ice melts is characteristic and never changes.**

EXERCISE 1-1

You are given a liquid which may be any *one* of three pure substances—benzene, toluene, or xylene. How could you tell which one of these three liquids had been given to you? Use the following characteristics:

Substance	Formula	Melting Point (°C)	Boiling Point (°C)	Color	Density (g/ml)
Benzene	C_6H_6	5.5	80.10	colorless	0.879
Toluene	C_7H_8	−95	110.63	colorless	0.867
Xylene	C_8H_{10}	−29	144.41	colorless	0.880

1-5 WONDERING WHY

We have already experienced some of the activities of science. First comes careful observation under controlled conditions, then organization of the information and the search for regularities of behavior. There is one more activity that, like dessert, fittingly comes last. This activity may be called "wondering why," and it arises from an irresistible urge to know more than merely What happens? We must also seek the answer to *Why* does it happen? This activity is probably the most creative and the most rewarding part of a science, and it differentiates man's science from that of the bird's "observation." What does it mean to answer a question beginning Why?

1-5.1 Explanations: A Model for Gases

Let us see what it means to search for an explanation to a familiar problem. What happens, for example, when someone blows up a balloon? As a person forces his breath (a gas) into the balloon, the balloon grows larger, the rubber is stretched more tightly, and the balloon becomes harder. Can we "explain" *what we see?* Can our "explanation" suggest new experiments? The answer to these questions begins with "wondering why." Why does the balloon grow larger? Why does it become harder?

There are two ways to proceed in trying to answer these questions. We have already examined one of these ways—to look more closely at the balloon, to record carefully what we see, and to seek regularities in what we observe. The second is to look *away* from the balloon and to seek similar behavior in another situation that we understand better. Sometimes the well-understood situation or **model** is very useful in helping us to understand the problem at hand. If we have little data, many different models will frequently offer an explanation for what is known; on the other hand, as we learn more and more about a problem, the number of models which will fit the facts becomes smaller and smaller. Let us try to formulate a model to represent the gas in the balloon by using a system about which we all know something.

Most of us have seen small balls made of "super-rubber." When such balls are dropped to the floor, they rebound upward, rising *almost* to the height from which they were dropped. When thrown directly against the wall in a small room, the ball bounces back and forth between the walls many times before gradually slowing down and stopping. Could there be a connection between the motion of the rubber balls and the air in the balloon?

A collection of bouncing balls is a *relatively simple system* for experimental study and can be described in quantitative terms. As such, it might be a good model system for representing the gas in the balloon. Suppose we picture air, or any other gas, as a collection of miniature balls bouncing around and colliding with the walls of the container. When the ball strikes the container, it pushes against it. But the wall pushes back with an equal force, and the ball leaves the wall going in a different direction. If there were an enormous number of particles, there would be many such collisions per second. Such a model could account for the "*push*" of the gas on the balloon wall. The "push" acting on a given area of wall surface (say, a square this big: ☐) is called the **gas pressure.** We could say that the collisions of the balls with the balloon wall account for the observed **pressure** of the gas. If more gas is added to the balloon, there will be more particles, hence more wall collisions per second, hence more "push" per unit of wall area, hence higher pressure. As the pressure (*push per unit of area*) on the balloon wall increases, the rubber walls will stretch outward exposing *more wall area*. This process continues until the push per unit of area is almost the same as it was before. In short, the balloon grows bigger if more gas is added and the pressure of gas in the balloon remains essentially constant.

Our model, based on bouncing rubber balls, has passed its first test: It reproduces several of the observations made when we blew up the balloon. In this sense it is a good model and deserves more careful examination. It leads to many predictions which will be explored more carefully in Chapter 4.

EXERCISE 1-2

Suppose that we force gas into a steel tank with a pressure gauge attached. The gauge will read 1 atmosphere when we have a given quantity of gas present. What would the pressure be if we doubled the amount of gas in the tank?   *2 atmosphere.*

1-5.2 Differences Between Model and System

The model for gas behavior describes the observations we have made so far. Unfortunately, this success can often lead into trouble. When a model is a very good one, we may start to think that the model *is* the system; then confusion results. Let us see how such a problem can arise.

When our "super-rubber" ball was dropped from a given height above the floor, it rebounded to a level *almost* as high as that from which it fell originally. *Note that it did not bounce up quite as high as its original position.* If the ball, after being dropped, had bounced

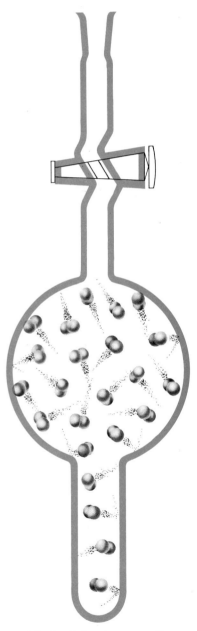

Fig. 1-6 **Could there be a connection between the motion of the "super-rubber" balls and the air in the balloon?**

up exactly to its original position, we would have said that it was **perfectly elastic.** *A perfectly elastic ball would bounce forever once it got started.* On the other hand, collisions of the "super-rubber" ball with the walls or floor are *not* perfectly elastic; we know the ball will slow down a little after each collision and finally stop—no matter how hard we throw it originally. How would we expect the gas to behave if we assumed that the little balls in our gas were exactly like "super-rubber" balls?

Let us consider the following experiment. A sample of argon gas is sealed into a heavy steel tank containing a gauge to tell us the pressure of the gas inside the tank. (Remember that pressure is a measure of the push of the gas particles against a given area of the tank's surface.) We read and record the pressure shown on the pressure gauge and then let the tank stand in the room at a constant temperature for five years. The model now permits us to make a prediction. If each small particle of gas were *exactly* like a "super-rubber" ball, it would slow up a little after each collision with the wall. After five years the particles of gas would have stopped bouncing and would have fallen to the bottom of the tank. The pressure in the tank (due to the push exerted by gas particles when they hit the walls) should gradually fall to zero and should be zero after a five-year interval.

The model has made a prediction which can be tested experimentally. Repeated observation will show that, as long as the tank is stored in a room of reasonably constant temperature, the pressure will *not* drop if the tank does not leak. The observation is in direct conflict with our prediction. Clearly, the small particles of the gas in the steel tank do *not* slow down after each collision as the "super-rubber" balls do. We must then conclude that if the moving-particle model for a gas is to be correct, *the collisions of the gas particles with the walls and with each other must be perfectly elastic.* The particles do not quit bouncing as time passes.

We see now that the model system ("super-rubber" balls) cannot be *exactly* like the gas in the balloon—but the resemblance is fairly close. Further, we know where at least one of the differences between a real gas and a model of a gas lies: Gas particles must undergo *perfectly elastic* collisions; they must be *more elastic* than the "super-rubber" balls.

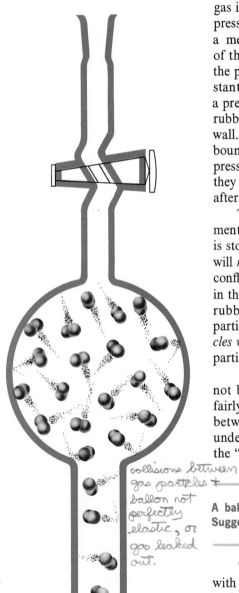

Fig. 1-7 If each small particle of gas were exactly like a "super-rubber" ball, it would slow up a little after each collision with the wall.

EXERCISE 1-3

A balloon filled with gas gradually shrinks on standing for some time. Suggest a model to explain this simple observation.

This is the characteristic pattern of an explanation. It begins with a Why? question that asks about a process that is *not* well understood (Why does a balloon expand as we blow into it?). An answer is suggested in terms of a process that is better understood ("super-rubber" balls bouncing off the walls). The model process may suggest new experiments or observations which can be performed. The results may show that our unknown system is very much like the model system or differs from it in certain ways (inelastic versus elastic collisions). If differences are extreme, we may abandon the model system and search for a new one. If differences are not extreme, the model will be retained or modified while we remember that the model and the system are not exactly the same. Since it is difficult to visualize the particles that make up the air, we attempt to explain the properties of the gas (air) in terms of the behavior of "super-rubber" balls, which are easy to see, to handle, to measure, and to study. Their behavior is well understood.

The search for explanation is, then, the search for likenesses that connect the system under study with a model system already studied. The explanation is considered to be "good" when:

(1) the model system is well understood—that is, when the regularities in the behavior of the model system have been thoroughly explored; and
(2) there are close similarities between the system being studied and the model system.

Our "super-rubber" ball model constitutes a good explanation because:

(1) how a "super-rubber" ball rebounds is well understood—we can calculate in mathematical detail just how much push the ball exerts on the wall at each bounce; and
(2) there seems to be a close similarity between the random motion of "super-rubber" balls and the motion of gas particles—exactly the same mathematics describes the pressure behavior if the gas is pictured as a collection of many small particles, endlessly in motion, bouncing elastically against the walls of the container.

The particle explanation of the gas phase is therefore a good one. Now, perhaps, you can see that answering the question Why? is merely a highly sophisticated form of seeking regularities. It is indeed a regularity of nature that gases and "super-rubber" balls have properties in common. The special creativity shown in the discovery of this regularity is that the likeness is not readily apparent; only with considerable thought can it be established. Fittingly, there is a special reward for the discovery of such hidden likenesses. The discoverer can bring to bear on the system being studied all the experience and knowledge accumulated from the well-understood system.

1-6 CONCLUSIONS AND THE ACCURACY OF OBSERVATIONS

Let us put some of our earlier generalizations to work in solving a practical problem. Assume that you are mixing chemicals and that a reaction takes place. One of the products which you separate from the reaction mixture is a white crystalline solid. Chemical theory (which we shall talk about later) tells us that the product is probably *one* of the ten materials listed in Table 1-2. Can the

TABLE 1-2

Material	Melting Point (°C)	Boiling Point (°C)
A	32.0	212
B	10.1	110
C	7.5	78
D	6.0	150
E	12.0	93
F	18.0	298
G	5.5	95
H	40.0	321
I	0.0	120
J	−25.0	55

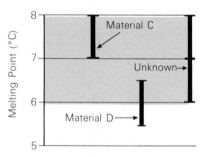

Fig. 1-8 **Within the limits defined by uncertainty the unknown could be either material C or material D.**

product be identified using a few experiments and the numbers in the table. The answer is: It depends! Let us see how.

Procedures first considered in Exercise 1-1 suggest that it would be helpful to measure the melting point of the solid product. We do so, and find a value of 7°C, which is then recorded as the melting point of the unknown solid. Armed with this information about our material, we now look at the list of possible compounds. We find a value of 7.5°C listed as the melting point of material C. This seems to be the value closest to the value of 7°C recorded for our unknown, so we tentatively identify the unknown solid as material C.

But wait! Are we sure that material D, with a melting point of 6°C, is ruled out? It clearly depends upon how accurately the value of 7°C is known and how accurately the values in the table are known. Suppose the values in the table could be off as much as 0.5°C either way. Then it is possible that the value recorded for material C (7.5°C) should really have been as high as $7.5 + 0.5°C = 8.0°C$, or else perhaps as low as $7.5 - 0.5°C = 7.0°C$. It would be convenient for us to record this value as $7.5 \pm 0.5°C$. Both the high and low possibilities are indicated easily in this way.

What about our own observation? If our own observation were off as much as one degree in either direction, allowed values for the melting point of the unknown solid could be 8°C, 7°C, or 6°C. This sequence might be conveniently recorded as $7 \pm 1°C$.

Let us now look at values for material D in the table. Using the uncertainty of $\pm 0.5°C$ for D, values ranging from 6.5°C to 5.5°C are acceptable.

It is now clear that the lowest of the possibilities for our measured value (6°C) falls well within the range of possible values for material D (6.5°C to 5.5°C). That is, material D could melt at a temperature as high as 6.5°C, while the unknown could have a melting point as low as 6°C. We see now that the unknown can be either C or D if the limits of error used above are valid.

EXERCISE 1-4

Suppose our laboratory measurement of the melting point could be in error by as much as 2°C either way. What possibilities for the unknown solid now exist? Can you suggest other experiments which might help you decide between the possibilities for the unknown solid?

1-6.1 Uncertainty in Science

The ability to use melting points in identifying material isolated from the reaction depends, then, upon how accurately we know the melting points. One is tempted to say: Why do we not do our work properly? Why do we not determine the melting points of everything exactly? Then we would not have these annoying questions of uncertainty to bother us!

Unfortunately, the very nature of scientific observation is such that *we have some uncertainty in every measurement we make.* Suppose, as an example, that in determining a melting point we must read the thermometer shown in expanded form in Figure 1-9. It is clear on even casual inspection that the temperature is *about* 31°C. Yet a report of 31°C might be considered unsatisfactory for many uses, since it does not convey *all* the available information. More careful inspection of the figure shows that a value of 30.9°C can easily be read. Probably all observers would agree that the value is 30.9 and not 30.8 or 31.0. We feel reasonably certain about the three numbers—three, zero, and nine.

Are we then in a position to read the value more closely than 30.9? Very careful examination of the figure shows that the top of the mercury column is a little above 30.9. Is it 30.92? Can we be sure it is not 30.91 or 30.95? In all honesty, we cannot say! The fourth number is quite uncertain, but an *estimate* of its value tells us that the total reading is a little *above* 30.9. Certainly, an overly optimistic report of 30.92251 would be misleading at best, and might even be dishonest in a sense, since it implies information we do not have. Other people reading the same instrument could perhaps read 30.91°C or 30.94°C; but any objective observer would have to conclude that we really did not read the last three figures in 30.92251, but rather that we manufactured them without any observational basis.

What, then, would be a meaningful way to record what we see? Since we all agree on the value 30.9°C, these numbers (which *are* certain) should be recorded. Since the next uncertain number adds a little extra information, we can justify recording this as well. We could thus make a case for writing down four figures, say 30.92, all of which are *significant!* The numbers 251 in 30.92251 are *not* significant; they convey no meaningful information and are not recorded. Scientists have formalized this notation somewhat. They say that the number 30.92 has four **significant figures.** *All of the certain digits plus the next uncertain one are classed as significant figures.* This is a convenient and simple way to indicate something about the **reproducibility** or **precision** of the measurement. It tells us that all observers would agree on the value 30.9 but that the last number might vary from one observer to another. We shall use the method of significant figures frequently to indicate the precision of our measurements.

One other point about significant figures is worth mentioning. Suppose we have a number which is read as 200, but may actually be as high as 300 or as low as 100. How many significant figures do we have? Clearly, even the first number is uncertain, so we cannot have more than *one* significant figure. The zeros recorded are useful in locating the decimal point, but they certainly *do not* tell us anything about the reproducibility of the measurement. We can then state that *the number of significant figures has nothing to do with the location of the decimal point. Zeros that merely indicate the location of the decimal point are not significant.*

For example, each of the numbers 36.09, 3.609, .003609, and 360.9 has *four* significant figures. The number 36.090 has *five* significant figures. The number in the thousandths place was read and has the value zero; otherwise we would have written 36.09 to indicate that we did not read beyond the hundredths place. But what about the number 36090? In this case, the last 0 may be significant, or it

UNCERTAINTY in SCIENCE

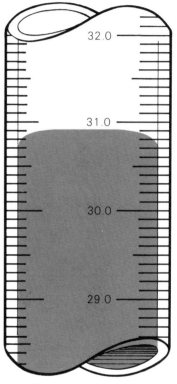

Fig. 1-9

may merely mark the decimal place. How do we indicate significance? To make it clear that the last number is significant, the number should be written with five significant figures as 3.6090×10^4. If the last number is not significant, 3.609×10^4 should be used.

There is one very serious limitation in using significant figures alone to indicate uncertainty. While we know that the last number is uncertain when we write four significant figures, we do not know *how* uncertain. The number with *four significant figures,* 30.92, could be $30.92 \pm .03$ or it could be $30.92 \pm .01$. Clearly, $30.92 \pm .03$ conveys more information than just 30.92. The method of indicating uncertainty by plus-or-minus values has much to recommend it if the size of the uncertainty in the last place is known. Another useful way to indicate uncertainty is to indicate the percentage of uncertainty in the number. For example, $30.92 \pm .03$ could also be written $30.92 \pm 0.1\%$. The method selected to indicate uncertainty will frequently depend on the system studied. For the problem involving the melting point of solids, the $\pm 0.5°$ or $\pm 1.0°$ was very useful in arriving at our decision. Significant figures alone would not have been as helpful.

1-6.2 Accuracy and Precision

So far, this treatment of uncertainty seems to center on the question: How closely can different observers read the same instrument? Answers to this question tell us something about the *precision of the measurement.* A measurement of melting point by a given thermometer may be extremely reproducible, giving the same result out to about $\pm .03°C$ each time. Such a result is **precise,** but it does not necessarily mean that we know the melting point with great **accuracy**! You say why not? Different observers can use *this* thermometer and get the same result each time. Why is the result not *accurate*?

The answer to your question may lie in the nature of the measuring instruments used. For example, let us suppose that there is a small error in placing the reading marks on our thermometer stem. Suppose that, through carelessness in thermometer manufacture, the value marked as 30.60 on a Bureau of Standards calibration-thermometer is marked as 30.90 on our thermometer. The error would never show when we used the thermometer to determine the melting point of our solid, yet the value we read would be $0.30°C$ away from all of the comparable tabulated values. (Most values in published tables have been taken with a Bureau of Standards thermometer. This is a very *accurate* instrument.) We would have to say that the measurement of melting point using our thermometer was of high **precision** (highly reproducible using our thermometer), but of limited **accuracy** because of errors in the thermometer calibration. *Accuracy represents how closely the measurement approaches to the true or accepted value. Precision describes the measurement's reproducibility using a given measuring instrument or experimental set-up.*

Any measurement has some uncertainty resulting from limitations in the measuring device and in the experimenter's ability to use the device. Estimation of the uncertainty associated with the use of

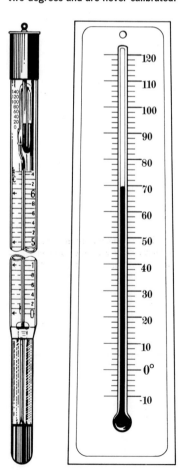

Fig. 1-10 **Left:** Beckmann differential thermometers certified by the National Bureau of Standards are both precise and accurate. Each is calibrated against an NBS standard. (Courtesy of Lux Scientific Instrument Corporation.) **Right:** Household thermometers are rarely precise or accurate. They can be read within one or two degrees and are never calibrated.

each different piece of your laboratory apparatus is given in the laboratory manual and will be helpful in determining the degree of confidence you can have in a laboratory measurement taken with any given instrument.

It is important to indicate the uncertainty in a measurement. This is done most easily by using the proper number of significant figures. It is, however, more meaningful to indicate plus-or-minus values after the result. Uncertainty in measurement influences quantities calculated from these measurements. How such uncertainty makes the calculated result uncertain will be indicated in the laboratory work. (See Appendix 4 of the Lab Manual.)

1-7 COMMUNICATING SCIENTIFIC INFORMATION

One of the most important reasons for man's progress is his ability to communicate information to his fellow man. It is not necessary for each of us to invent the atomic description of matter. That was done for us by others and passed on to us by means of books and papers. The *way* in which ideas are communicated to others is of extreme importance. In fact, one can almost say that *a scientific advance is important only if it is told to others in a manner which they can understand.* This statement is beautifully illustrated by the following episode from the early history of the atomic theory:

In 1789 William Higgins, then in Britain, published a book entitled *A Comparative View of the Phlogiston and Antiphlogiston Theories.* In this book many of the arguments which were to form the basis for the atomic theory were introduced; however, Higgins's presentation was difficult to follow and his views were not widely read or accepted. Some nineteen years later John Dalton, a Manchester schoolteacher, published independently a book entitled *A New System of Chemical Philosophy.* Dalton, though a man of limited scholarly attainment and background, made a masterful presentation of his ideas. He used pictures and symbols to make

Fig. 1-11 **A scientific advance is important only if it is told to others in a manner which they can understand.**

the concept of atoms into a working hypothesis for all chemists of the era. Dalton's book attracted much attention and led ultimately to the acceptance of the atomic theory.

In 1814 Higgins formally claimed that his book, published in 1789, had contained a description of the atomic theory. Indeed, Higgins's claim to priority was supported by Humphrey Davy, who had written in 1811: "It is not a little curious that the first views of the Atomic Chemistry, which has been so expanded by Dalton, are to be found in a work published in 1789 by William Higgins." Many historians of science have since examined Higgins's book and few have recognized the atomic theory in its pages. For this reason, Dalton is usually given sole credit for the atomic theory; Higgins is seldom mentioned. This latter fact is unfortunate since it does appear that Higgins did indeed anticipate some of the concepts of the atomic theory. But because his presentation failed to communicate his ideas in a manner which could be understood and applied by most scientists, his contribution attracted little attention and was not considered to be of great value.

We repeat: A scientific advance is important *only* if it is told to others *in a manner which they can understand*. As we go along, we will encounter many different ways to present an idea. The best method is always decided by the *reader,* not the *author.*

Today ideas in science come at us from all directions. We have books, review and research journals, direct laboratory observation, and television. Review journals carry articles written by competent scientists to summarize all the available information in a field. Such articles go out of date very rapidly. The American Chemical Society publishes a review journal with the name *Chemical Reviews*. Research journals record the results of original experiments and describe new and developing ideas in science. The American Chemical Society publishes such research journals as the *Journal of the American Chemical Society, Journal of Organic Chemistry, Journal of Physical Chemistry, Inorganic Chemistry,* and *Biochemistry*. These journals present ideas on the frontiers of scientific thought.

1-8 HIGHLIGHTS

Man's study of his surroundings involves a number of activities such as:

(1) *Accumulating information through observation.*
(2) *Organizing information and seeking regularities.*
(3) *Wondering why these regularities exist.* In giving an explanation it may be necessary to "build" or formulate a scientific model. A good model will lead to new information and to new insight into the system.
(4) *Communicating information.*

There is no fixed order for these activities. But, when properly done, they lead to scientific advances such as the theory of relativity, a vaccine for polio, rockets capable of traveling to other planets, and the synthesis of a virus (life?) in the laboratory (see Chapter 22). Where we go from here is anyone's guess—and your responsibility!

QUESTIONS AND PROBLEMS

1. How many significant figures would be appropriate in reading your watch to get the time? Estimate the plus-or-minus uncertainty in reading your watch.

2. Let us assume that the "super-rubber" ball model for a gas is an accurate one. What would happen to the *pressure* if the balls moved more rapidly in a metal tank of fixed volume?

3. Blow up a balloon and then release it so that the air can escape. Offer an "explanation" for what you see.

4. The weight of a piece of metal may be as much as 264 g (grams), or as little as 260 g. How many significant figures are used to express the weight of the metal? How would the weight be given using the plus-or-minus designation? What is the percentage uncertainty in the measurement?

5. The piece of metal used in question 4 is cut in two. One piece weighs 230 ± 2 g. What is the *largest* weight that the other piece could have? What is the *smallest* weight that the other piece could have? Express the weight of the other piece using a plus-or-minus estimate to indicate the uncertainty in your measurement.

6. The burning of a candle is a relatively complex process which raises many questions. We have listed five questions here which are worthy of answers, but we are *not* in a position to answer them yet. We shall, however, keep them in mind as we progress.

   (a) Why is energy liberated in the burning of a candle?
   (b) Why did the candle not burn while it was stored in your drawer?
   (c) What was the role of the match you used to light the candle?
   (d) What is the role of the wick of the candle?
   (e) Why does a flame emit colored light?

   Can you add questions to this list based on your own work with the candle? There are surely many more questions about a phenomenon as complex as the burning of a candle!

... hypotheses ought to be fitted merely to explain the properties of things and not attempt to predetermine them except insofar as they can be an aid to experiments ....

SIR ISAAC NEWTON (1642–1727)

CHAPTER 2
The Atomic Theory: One of Our Best Scientific Models

When hydrogen chloride and ammonia mix, a white smoke appears. Can this tell us anything about atomic theory?

In Chapter 1 we found that the behavior of gases became less mysterious when we related it to the model of "super-rubber" balls in constant motion. Building on this model, we find that we can count the particles, find out how much each kind weighs, and determine what smaller particles make up each kind of gaseous "super-rubber" ball. We find that it is the kind and arrangement of these smaller particles, called atoms, which determine what kind of substance we are talking about—whether it will be a deadly poison or a new flavor for chewing gum. In all chemical reactions from the burning of a candle to the explosion of TNT, atoms break apart and recombine. Our elastic "super-rubber" ball model apparently can be stretched and stretched! We even develop a shorthand for naming chemical particles and describing the changes which happen to them. We're off and running!

One of the activities of science is the search for regularities. Valid regularities permit us to make reasonable predictions about the results of future experiments. If the apparent regularity is false (for example, All Cylindrical Objects Burn), it soon runs into conflict with experiment and is either forgotten or modified so it is valid. If the apparent regularity is true (for example, All Wooden Objects Burn), it will predict with reasonable accuracy the results of many experiments and can be used to replace long lists of data. A good regularity saves lots of wear and tear on the memory.

As more and more of the predictions of a regularity are verified, our faith in its value grows; finally, we dignify the regularity by calling it a **law.** *A regularity that directly correlates experimental results is generally called a* **rule** *or* **law.** As we shall see, a law is most useful if it is a quantitative statement. We shall often use a law as the starting point when we "wonder why." In seeking an explanation for a regularity, it is customary to reason by analogy and to attempt to explain the law in terms of things which are better understood. For example, in Chapter 1 it was suggested that a gas is made up of little particles which are like constantly moving "super-rubber" balls. In using the analogy, we constructed a mental **model** for the gas. *The explanation of the behavior of the gas in terms of the motion of these particles is called a* **theory** *or* **principle.**

As you can see, rules, laws, models, theories, and principles all have a common aim. They systematize our experimental knowledge. They all state regularities among known facts. Usually when the explanation involves a real physical system (such as bouncing balls), we call it a **model;** when it involves more abstract ideas linking model and system (for example, a mathematical equation), we call it a **theory.** This differentiation is not of great importance, however, and we may even use the words *model* and *theory* interchangeably.

When seeking an explanation, we sometimes find more than one acceptable model. If this happens, the model or theory which is most useful is usually preferred. A useful model points to new directions of thought, suggests new experiments, and can be expanded or modified to account for the results of many new studies. A useful model is the heart of scientific thought.

2-1 GROWTH OF A MODEL

In order to watch a theory grow, let us examine in more detail the model which was used to describe a gas confined in a balloon. The gas was pictured as a collection of small, rapidly moving particles which rebounded from the walls of the balloon in a series of perfectly elastic collisions. As the particles rebounded from the balloon wall, they pushed on it and pressure resulted. This description of the system reminded us of a collection of rapidly moving "super-rubber" balls confined in a box. Although the analogy was close, there was at least one very obvious and significant point of difference. We know that "super-rubber" balls in a box would gradually slow down and stop as the box was allowed to stand. Collisions between "super-rubber" balls and the walls (and with each other) are *not* perfectly elastic. As a result of imperfect elasticity, each ball slows down slightly after every collision and finally stops. On the other hand, collisions of air particles with the walls of the balloon must be regarded as being perfectly elastic on the average or we run into a serious conflict with experiment. If the particles of a gas were to slow down slightly after every collision, the pressure in a closed steel tank would gradually diminish as the tank stood in a classroom, held at a constant temperature. *Experience tells us that this does not happen.* Despite this known and understandable difference between gas particles and "super-rubber" balls, the "super-rubber" ball model for gases is a very good one. Its development is worthwhile.

Correlations in science are most useful when they are quantitative, when they answer the question "How much?" as well as "What happens?" *A numerical relationship between the applied pressure and the volume would be called a* **quantitative** *correlation.* Let us try some simple, but significant experiments in an effort to establish a *quantitative relationship* between pressure and volume.

2-1.1 Pressure-Volume Measurements for Air

The equipment for this study is simple. We need (1) a plastic syringe with a rubber seal, similar to the one found in your laboratory; (2) a rubber stopper bored part way through so that it can close off the end of the syringe; (3) a clamp to hold the syringe upright; and (4) five or six bricks, like those found on a construction site.

Fig. 2-1.1   **Apparatus for studying change in volume as pressure on a gas is increased.**

Fig. 2-1.2   **Pressure × volume = a constant. As pressure increases, volume decreases.**

To do the experiment, the plunger of the syringe is pulled out, the outlet end is closed off with the rubber stopper, the syringe is clamped in an upright position, and a brick is balanced on top of the plunger. Read the volume of air trapped in the syringe under a load of one brick. See Figure 2-1.1. The weight of one brick is pushing down and exerting extra pressure on the gas in the syringe. In Table 2-1 we record the pressure as "one brick" (a convenient temporary unit). Two bricks are added and the volume is read again. Actual values for air obtained by the authors in five trials are shown in Table 2-1. The apparatus and steps in the experiment are shown in Figure 2-1.2. Values of this type are easily obtained in your own laboratory or classroom.

TABLE 2-1 PRELIMINARY PRESSURE-VOLUME MEASUREMENTS FOR AIR

Pressure* (bricks)	Volume (ml)†					Average Volume (ml)	$\frac{1}{\text{volume}}$ (1/ml)
	Trial 1	Trial 2	Trial 3	Trial 4	Trial 5		
1.0	27.0	29.0	28.0	28.0	29.5	28.5	3.50×10^{-2}
2.0	23.0	22.0	23.0	22.0	22.0	22.5	4.45×10^{-2}
3.0	18.0	17.0	17.0	17.5	18.0	17.5	5.70×10^{-2}
4.0	14.0	14.5	13.5	14.0	13.5	14.0	7.15×10^{-2}
5.0	12.0	12.0	12.0	12.5	11.5	12.0	8.30×10^{-2}

* Only whole bricks were used.   † Read to closest 0.5.

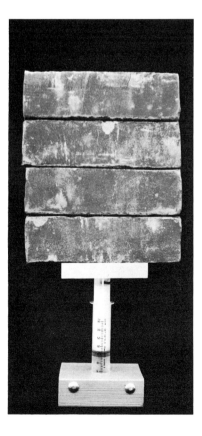
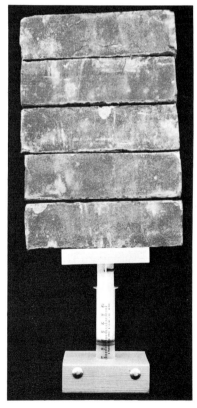

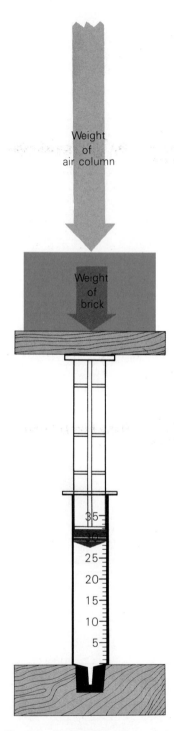

Fig. 2-2 An invisible column of air also pushes down (increases pressure) and confines the gas (decreases volume) in the cylinder.

2-1.2 Pressure Due to the Atmosphere—Total Pressure

The numbers in Table 2-1 show clearly that as the pressure (number of bricks) increases, the volume decreases; a simple **qualitative** relationship is clear. Before we try to obtain a *quantitative relationship* between pressure and volume, we must know what the *total pressure* on the gas really is.

What is pushing down on the plunger? Several bricks are exerting pressure. In addition, an invisible column of air extends from the top of the plunger up into space. (See Figure 2-2.) This air column has weight, too, and must be considered in determining the total weight on the plunger. By applying relatively simple algebra to the numbers of Table 2-1 (and to the numbers of Appendix 2), we can show that the air column has a weight equivalent to that of about 1.7* bricks. Let us add the weight of the atmosphere to each pressure reading of Table 2-1. The result is seen in the third column of Table 2-2. The total pressure is the pressure due to the weight of the bricks plus the pressure due to the weight of the atmosphere. Table 2-2 now contains two columns (3 and 4) which tell us *quantitatively* how *volume* of a gas changes with the *total pressure* applied to that gas.

TABLE 2-2 CORRECTED PRESSURE-VOLUME MEASUREMENTS FOR AIR

Pressure Due to Bricks	Pressure Due to Air (bricks)	Total Pressure (bricks)	Average Volume (ml)	Pressure × Volume (bricks × ml)
1	1.7	2.7	28.5	77
2	1.7	3.7	22.5	83
3	1.7	4.7	17.5	82
4	1.7	5.7	14.0	80
5	1.7	6.7	12.0	80

2-1.3 Presentation of Data

We now have numbers in a table. What is the best way to present these numbers so that we can obtain a maximum amount of new information? Let us retrace our steps. First, we blew up a balloon and noticed that its sides were forced out as air was forced into the balloon. This qualitative observation suggested a model for a gas. Next, we used the plastic syringe to obtain some rather crude quantitative measurements of pressure versus volume for a gas. The pressure value due to the weights of the bricks had to be added to the pressure value due to the weight of the air column in order to obtain the value for the total pressure pushing down on the plunger. The corrected pressure-volume values were summarized in Table 2-2. What more can the values of Table 2-2 tell us? To answer this question, let us present our data in pictorial form.

* We can also calculate this value independently if we know the weight of the brick, the pressure of the air in grams per square centimeter, and the size of the piston in the cylinder. The value of 1.7 is again obtained. See Appendix 2.

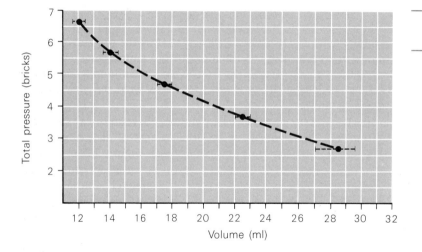

Fig. 2-3 **Pressure versus volume of a gas.**

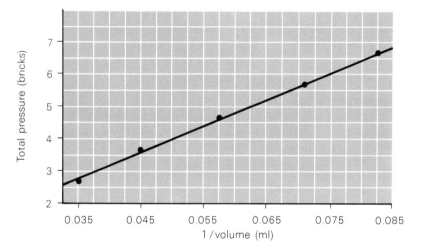

Fig. 2-4 **Pressure versus 1/volume of a gas.**

In Figure 2-3 we have plotted pressure versus volume and an interesting curve has resulted—a short section of a hyperbola. In Figure 2-4 we have plotted pressure versus 1/volume and a straight line has been obtained. This is a particularly significant and useful* relationship because it shows simply that pressure is directly proportional to 1 divided by the volume ($1/V$). The statement suggests a simple mathematical relationship:

$$P \text{ is proportional to } \frac{1}{V}$$

or

$$P = \text{proportionality constant} \times (1/V)$$

If both sides of this equation are multiplied by V, we obtain the mathematical expression

$$PV = \text{a constant}$$

* A plot of this type in Appendix 2 helped us to decide that the pressure of the air is equal to the pressure of 1.7 bricks.

This expression leads us to multiply pressure by volume in Table 2-2; the result is shown in the last column of Table 2-2 and indicates that pressure × volume is a constant. The $P \times V$ values shown in Table 2-2 are as constant as one could expect from our relatively crude measurements; more precise measurements demonstrate clearly that for a gas such as air, $P \times V$ is indeed a constant under the conditions used here.

In summary, we have presented our data on pressure and volume of a gas sample in four ways:

(1) in a *qualitative* statement which asserts that volume decreases as pressure increases;
(2) in a table which summarizes *quantitatively* values of volume as pressure increases;
(3) in two graphs, the first showing *pressure* versus *volume* and giving a curve, and the second showing *pressure* versus *1 over volume* and giving a straight line; and
(4) in the form of a mathematical statement, $P \times V =$ a constant, which was suggested by the plot of P versus $1/V$.

EXERCISE 2-1

Look at the numbers in Table 2-2. Suppose the value of 14.0 ml shown as the volume under a pressure of four bricks were really 14.5 milliliters. What would the value of pressure (P) × volume (V) be? Suppose it were 13.5 ml, what would the value of $P \times V$ be? What range in values of $P \times V$ is indicated just by the precision with which the volume was read? Repeat for the volume value at two bricks.

2-1.4 Quantitative Pressure-Volume Data and the Growth of the Model

We have summarized significant information about the relationship between pressure and volume for air. How does the particle model correlate with the data which have been accumulated? We picture the small particles which make up air as rebounding back and forth between the container walls. In such a model the pressure is determined by

(1) the size of the push each colliding particle gives to the walls of the container and
(2) the number of collisions made with a unit area of the wall in any one second.

If the volume were cut to half its former value and the number of particles held constant, we would expect *twice* as many particles per liter. With twice as many particles per liter, we would expect twice as many collisions with a unit area of the wall in any one second (the frequency of collisions is doubled). The model suggests that doubling the number of collisions made with the wall should double the "push" per unit area, hence double the total pressure. Experiment and model agree. We have seen that when the amount of gas and the temperature are held constant and the total pressure is doubled, the volume will be cut to half its original value (Table 2-2).

The model is supported by the quantitative data.

2-1.5 Pressure-Volume Measurements for Other Gases

All observations made up until now on air agree with the "super-rubber" ball model. Do gases other than air behave in the same way? Many gases with sharply contrasting properties are available for our study. Tanks labeled oxygen, hydrogen, carbon dioxide, carbon tetrafluoride, ammonia, hydrogen chloride, or hydrogen bromide can be purchased and their gases tested. Each of these pure gases can be put into the pressure-volume syringe and studied. The results of several of these studies are shown in Table 2-3. We were careful to start with the same volume of each gas at the atmospheric pressure in the room. In each case, the syringe was filled to the same volume with a given gas; then bricks were added

TABLE 2-3 PRESSURE-VOLUME MEASUREMENTS FOR SELECTED GASES AT ROOM TEMPERATURE (25°C)

	Carbon Dioxide				Nitrogen (Trial 1)		
Pressure Due to Bricks	Total Pressure (bricks)	Volume (ml)	Pressure × Volume (bricks × ml)	Pressure Due to Bricks	Total Pressure (bricks)	Volume (ml)	Pressure × Volume (bricks × ml)
1	2.7	29.2	79	1	2.7	29.2	79
2	3.7	21.5	80	2	3.7	22.8	84
3	4.7	17.0	80	3	4.7	17.8	84
4	5.7	13.5	77	4	5.7	15.0	85
5	6.7	11.7	78	5	6.7	12.5	84

	Ammonia				Nitrogen (Trial 2)		
Pressure Due to Bricks	Total Pressure (bricks)	Volume (ml)	Pressure × Volume (bricks × ml)	Pressure Due to Bricks	Total Pressure (bricks)	Volume (ml)	Pressure × Volume (bricks × ml)
1	2.7	27.0	73	1	2.7	29.2	79
2	3.7	22.1	82	2	3.7	22.5	83
3	4.7	18.0	84	3	4.7	18.0	85
4	5.7	13.6	78	4	5.7	14.7	84
5	6.7	12.0	80	5	6.7	12.1	81

Hydrogen Bromide			
Pressure Due to Bricks	Total Pressure (bricks)	Volume (ml)	Pressure × Volume (bricks × ml)
1	2.7	29.2	79
2	3.7	21.6	80
3	4.7	17.0	80
4	5.7	13.8	79
5	6.7	11.9	80

one by one to increase the pressure and the volume was recorded for each pressure reading. Since all observations were made in the same room over a short period of time, the temperature was very close to being constant for all observations. The data in Table 2-3 indicate an amazing and fortunate fact. *Not only does the relationship "P × V = a constant" seem to apply to all gases studied, but, within the precision of our measurements, the numerical constant is the same for the four new gases as well as for air.*

Why should this be? What property or properties of gases are implied by this observation? For one thing this observation says that each gas exhibits behavior consistent with the particle model we used successfully for air. Each gas behaves like air. Perhaps, then, *all gases are made up of particles?* Furthermore, the fact that all gases show the same value for pressure × volume suggests that, if all gases *are* made up of particles, the bombardment by particles must produce about the same pressure regardless of the kind of gas used. Perhaps all gas particles are the same and we have equal numbers of particles of similar type in the syringe each time.

2-1.6 Properties of Gases

If all gas particles were the same, all gases would have the same properties. Experiment quickly reveals that this is not so. The gases we worked with show striking differences in properties outside the syringe. Ammonia dissolves readily in water. If a drop of an organic dye, phenolphthalein, is added to the water after ammonia has dissolved, the solution changes from colorless to red. The water containing the dissolved ammonia has a bitter taste and an obnoxious odor characteristic of gaseous ammonia. The gas has the same odor. Hydrogen bromide also dissolves readily in water, but a drop of phenolphthalein added to the water containing hydrogen bromide shows no color change. The water solution has a sour taste and an odor which is completely different from that of ammonia. Nitrogen, in contrast, will not dissolve in water to an appreciable degree and has no odor.

Gas particles, then, are *not* the same. Different gases must be made up of *different* kinds of particles. If this were not so, the differences between gases would be very difficult to understand. A name for these particles of gases would be helpful. *The particles of gases are called* **molecules.** Molecules of different gases differ. But we are still left with the question: Why do different gases give the same value of pressure × volume in our syringe experiments? To find the answer we need more experimental information.

MIXTURES OF GASES

What happens when gases are mixed? We have already established that when more gas molecules are forced into a given volume, the pressure due to the gas rises. Thus, if the volume occupied by a given quantity of gas at constant temperature is halved, the pressure doubles. Forcing one gas into another should bring about an increase in the number of molecules per unit volume, hence an increase in pressure.

To test this hypothesis, a syringeful of nitrogen is forced into a syringeful of hydrogen bromide. We find that the pressure *must be doubled* if the combined gas mixture is to be retained in one syringe. In a similar way a syringeful of nitrogen is forced into a syringeful of ammonia. Again, the pressure must be doubled if all the gas is to be retained in one syringe. The behavior of these gas mixtures is in agreement with the particle model.

Let us now check the result by forcing a syringeful of ammonia into a syringeful of hydrogen bromide. Our initial observation surprises us. A white smoke and powder appear in the tube and the plungers in each syringe move to the bottom of the tubes. *A solid of trivial volume is formed from the two syringefuls of gas:*

$$1 \begin{Bmatrix} \text{syringeful of} \\ \text{hydrogen bromide} \end{Bmatrix} \text{ plus } 1 \begin{Bmatrix} \text{syringeful of} \\ \text{ammonia} \end{Bmatrix} \text{ gives } \begin{Bmatrix} \text{white solid} \\ \text{of trivial volume} \\ \text{compared to gases} \end{Bmatrix}$$

What happens if we take one syringeful of ammonia and *half* a syringeful of hydrogen bromide? Experiment shows that white powder and smoke again appear, but the volume does not drop to zero. Instead, half a syringeful of gas remains. This remaining gas has the properties of ammonia. Our experiments show that only *half the ammonia is used!*

$$\tfrac{1}{2} \begin{Bmatrix} \text{syringeful} \\ \text{of} \\ \text{hydrogen} \\ \text{bromide} \end{Bmatrix} \text{ plus } 1 \begin{Bmatrix} \text{syringeful} \\ \text{of} \\ \text{ammonia} \end{Bmatrix} \text{ gives } \begin{Bmatrix} \text{white} \\ \text{solid} \end{Bmatrix} \text{ plus } \tfrac{1}{2} \begin{Bmatrix} \text{syringeful of} \\ \text{ammonia} \end{Bmatrix}$$

or

$$\tfrac{1}{2} \begin{Bmatrix} \text{syringeful of} \\ \text{hydrogen bromide} \end{Bmatrix} \text{ plus } \tfrac{1}{2} \begin{Bmatrix} \text{syringeful of} \\ \text{ammonia} \end{Bmatrix} \text{ gives } \begin{Bmatrix} \text{white} \\ \text{solid} \end{Bmatrix}$$

If we repeat the experiment using one syringeful of hydrogen bromide but only half a syringeful of ammonia, the white solid again appears, but only *half* the hydrogen bromide is used.

$$1 \begin{Bmatrix} \text{syringeful} \\ \text{of} \\ \text{hydrogen} \\ \text{bromide} \end{Bmatrix} \text{ plus } \tfrac{1}{2} \begin{Bmatrix} \text{syringeful} \\ \text{of} \\ \text{ammonia} \end{Bmatrix} \text{ gives } \begin{Bmatrix} \text{white solid} \\ \text{of trivial} \\ \text{volume} \\ \text{compared} \\ \text{to gases} \end{Bmatrix} \text{ plus } \tfrac{1}{2} \begin{Bmatrix} \text{syringeful} \\ \text{of} \\ \text{hydrogen} \\ \text{bromide} \end{Bmatrix}$$

The result of these three experiments can be summarized by the statement

$$1 \begin{Bmatrix} \text{volume of} \\ \text{hydrogen bromide gas} \end{Bmatrix} \text{ plus } 1 \begin{Bmatrix} \text{volume of} \\ \text{ammonia gas} \end{Bmatrix} \text{ gives } \begin{Bmatrix} \text{white solid of} \\ \text{trivial volume} \end{Bmatrix}$$

How does this result fit in with the particle model for gases? The simplest interpretation is that one molecule of hydrogen bromide combines with one molecule of ammonia to give a particle which helps to form the white solid.

$$1\begin{Bmatrix} \text{molecule of} \\ \text{hydrogen bromide} \end{Bmatrix} \text{plus } 1 \begin{Bmatrix} \text{molecule of} \\ \text{ammonia} \end{Bmatrix} \text{gives} \begin{Bmatrix} \text{particle which} \\ \text{makes up the} \\ \text{white solid} \end{Bmatrix}$$

Such a hypothesis would be consistent with our observations if one syringeful of hydrogen bromide gas contained the same number of hydrogen bromide molecules as one syringeful of ammonia gas. Remember that one *volume* of ammonia reacts with one *volume* of hydrogen bromide. This is a bold suggestion. It could explain the pressure-volume constant if one molecule of ammonia were as effective as one molecule of hydrogen bromide in generating pressure.

2-1.7 Avogadro's Hypothesis

The simplicity of the equation relating volume and pressure suggested to the French physicist André Ampère (1775–1836) around 1810 that equal volumes of gases at the same temperature and pressure contain an equal number of molecules. This thought was little more than speculation until the Italian chemist Amedeo Avogadro (1776–1856), working independently, combined the information on combining volumes of gases with the pressure-volume data and proposed what is usually called **Avogadro's hypothesis:** *Equal volumes of gases, measured at the same temperature and pressure, contain an equal number of molecules.*

You may object, saying that the evidence which we presented is consistent with Avogadro's hypothesis, but it certainly does not prove the hypothesis. There may be many other explanations which would offer equally good interpretations of the facts. Your position would be perfectly reasonable and justifiable. In fact, you would be in rather distinguished company. John Dalton (founder of the atomic theory) and J. J. Berzelius (one of the great chemists of all time) took this position and for almost 50 years Avogadro's hypothesis was not accepted. *No single experiment or series of experiments ever established the validity of Avogadro's hypothesis.* Rather, the slow accumulation of facts, all of which could be explained simply and directly by Avogadro's hypothesis, finally brought about its general acceptance. It stands today as one of the cornerstones of modern science. The controversy over Avogadro's views and the vigorous interplay of personalities during the period 1800 to 1860 make that part of the history of science fascinating reading.*

We shall tentatively accept Avogadro's hypothesis and see how it can explain the many observations we shall make throughout this course. This, in fact, is precisely how the hypothesis was accepted into the structure of modern chemistry.

2-2 MOLECULAR WEIGHTS, THE MOLE, AND THE MOLAR VOLUME

Avogadro's hypothesis, like the statement "All Wooden Objects Burn," has far-reaching consequences. We already know that gases

---

* See, for example, E. Farber, "The Evolution of Chemistry," Chapter 10, The Ronald Press, New York, 1952.

differ one from another. Properties of gases vary widely. It is logical to assume that molecules of different gases have different weights. One way in which we might determine *relative* weights of some gas molecules would be to weigh 1 liter* of each of several different gases. If, then, Avogadro's hypothesis is true and molecules of two different gases have different weights, the 1-liter samples of each gas (measured at the same temperature and pressure) will have different weights. The ratio of the weight of *1 liter* of gas A to *1 liter* of gas B should be the same as the ratio of the weight of *one molecule* of gas A to *one molecule* of gas B. This follows because we are assuming that equal volumes of gases at the same temperature and pressure have the same number of molecules. Let us investigate the weights of equal volumes of gases measured at the same temperature and pressure.

2-2.1 The Relative Weights of Gases

We are going to fill a 1-liter bulb with gas (see Figure 2-5), weigh the bulb plus gas, then pump out the gas and weigh the empty bulb. The difference between these two weights will give us the weight of the gas in the bulb. By selecting a pressure to confine the gas which is equal to the average pressure of the earth's atmosphere at sea level (a pressure of 1.7 bricks), and by selecting a cold temperature of 23°F (or −5°C, or degrees Celsius†) at which to measure the gas volume, the weights will be close to whole numbers. This will simplify the arithmetic involved in obtaining weight ratios. For that reason alone, we shall work under these cold and unusual conditions for our initial experiments.

One liter of *carbon dioxide* weighs 2.0 grams (g) at 23°F (−5°C) and 1 atmosphere (atm) pressure. One liter of *carbon tetrafluoride* weighs 4.0 g under these same conditions. If 1 liter of carbon dioxide contains the same number of molecules as does 1 liter of carbon tetrafluoride, then *one molecule* of carbon tetrafluoride must be 4.0 g/2.0 g or twice as heavy as *one molecule* of carbon dioxide. If the *relative weight* of a carbon dioxide molecule were 1.0, then the relative weight of a carbon tetrafluoride molecule would be 2.0. If we knew the weight of a carbon dioxide molecule, we could determine the weight of a carbon tetrafluoride molecule.

This procedure for obtaining the *relative* weights of gas particles using carbon dioxide as a standard can be applied to many gases. As an example, let us weigh hydrogen molecules. One liter of hydrogen at 1 atm pressure and −5°C weighs 0.092 g. If the relative weight of the carbon dioxide molecule is taken as 1.0, then the relative weight of the hydrogen molecule is

$$\frac{\text{weight of hydrogen sample}}{\text{weight of carbon dioxide sample}} = \frac{0.092 \text{ g}}{2.0 \text{ g}} = 0.046 = \frac{1}{22}$$

$$\frac{\text{weight of hydrogen molecule}}{\text{weight of carbon dioxide molecule}} = \frac{1/22}{1} = \frac{1}{22}$$

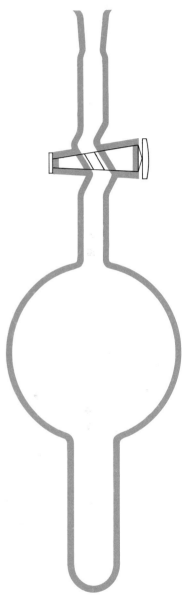

Fig. 2-5 **One-liter bulb for weighing a gas.**

---

* A **liter** is the unit of volume used in science; it is equal to 1.06 quarts.

† Temperatures in chemistry are measured in degrees Celsius. The scale is defined more precisely in Chapter 4.

If the carbon dioxide molecule is 1.0, the hydrogen molecule is 1/22. But *fractional* numbers for particle weights are inconvenient. Since we can use any particle as a reference standard and can arbitrarily assign any weight to it, we could call the relative weight of hydrogen particles 1.0; then a carbon dioxide particle would have a *relative* weight of 22 and a carbon tetrafluoride particle would have a *relative* weight of 44 $\left(\frac{4.0 \text{ g}}{0.092 \text{ g}} = 44\right)$. This standard is fine if we do not find a gas lighter than hydrogen. If we do, we shall be back to fractional weights. Fortunately, no gas available to us in the laboratory in weighable amounts is lighter than hydrogen. It is convenient, therefore, to adopt hydrogen as our temporary standard. Then all other gases will have relative weights greater than the standard.

Since we are concerned only with *relative* weights, the actual number assigned to our temporary standard is arbitrary. It is most convenient to assign gaseous hydrogen molecules a relative weight of 2.0. As you may know, each hydrogen molecule is composed of two separate units called **atoms.** By assigning a value of 2.0 to the hydrogen *molecule,* a value of 1.0 can be assigned to the lightest atom, the hydrogen *atom.* Arbitrarily, then, a hydrogen molecule is assigned a relative weight of 2.0. On this scale a carbon dioxide molecule will have a relative weight of 2.0 × 22 or 44 and a carbon tetrafluoride molecule a relative weight of 2.0 × 44 or 88.

EXERCISE 2-2

One liter of the gas methane (a component of fuel gas) has a weight of 0.736 g at −5°C and 1 atm pressure. What is the relative weight of a methane molecule if hydrogen molecules are taken as 2.0? One liter of hydrogen at 1 atm and −5°C weighs 0.092 g.

2-2.2 Molecular Weights

We have just used Avogadro's hypothesis to calculate the relative weights of molecules of different gases. It is appropriate to call these relative weights* **molecular weights.** As you will recall from the preceding section, any gas can be selected as a standard (first we used carbon dioxide, then hydrogen) and any value can be assigned to the standard (we assigned a value of 1.0 to carbon dioxide, then found it more convenient to assign a value of 2.0 to hydrogen).

If the relative weight of an oxygen molecule is determined using hydrogen with a value of 2.0 as a standard, we find that the molecular weight of oxygen is 32. Early chemists could obtain oxygen in pure form simply by heating an orange powder called mercuric oxide. Furthermore, when oxygen was assigned the molecular weight of 32.00, molecular weights of a large number of other gases turned out

---

*In any one spot *on earth,* **mass** and **weight** are frequently used as equivalent terms; however, in *space* the terms have different meaning. Weight refers to the gravitational attraction between an object and the earth. Mass refers to the actual *quantity* of material present. We shall frequently use the more scientifically precise term mass instead of weight, but we are still saddled with the archaic terms molecular weight and atomic weight. The world standard for mass is a kilogram platinum cylinder in Paris.

to be whole numbers. These two facts made oxygen a very convenient primary standard for early chemists. To this day, when we determine the molecular weight of gases by weighing a sample of the gas and a sample of a standard, oxygen is usually selected as the primary standard and a value of 32.00 is *assigned* to it.

Let us now briefly summarize the procedure we can use to obtain molecular weights of gaseous substances:

(1) We weigh a given volume of a standard gas (usually oxygen) at a given temperature and pressure.
(2) We assign a molecular weight to the standard. Chemists use oxygen and assign it a value of 32.00.
(3) We weigh an equal volume of the unknown gas using the same temperature and pressure as we used for the standard gas.
(4) We calculate the molecular weight of the unknown gas by means of the relationship

$$\frac{\text{molecular weight of unknown}}{\text{molecular weight of standard}} = \frac{\text{weight of unknown gas sample}}{\text{weight of standard gas sample}}$$

If oxygen is the standard, the value for the molecular weight of the standard is 32.00. If another gas is used as a standard, its molecular weight must be known at least as accurately as we desire the molecular weight of the unknown. Finally, neither the temperature nor pressure need be known if the temperature and pressure used when weighing the unknown are the same as were used when weighing the standard.

EXERCISE 2-3

A sample of the interesting gas diborane is weighed in a bulb. The gas is found to weigh 0.600 g. The diborane is removed from the bulb and replaced with an equal volume of oxygen gas measured under the same conditions of temperature and pressure. The sample of oxygen weighs 0.696 g. What is the molecular weight of the diborane?

2-2.3 The Mole

An experimental procedure for determining molecular weights (the relative weights of molecules) has been outlined. Notice that in using this procedure, we made two arbitrary choices. First, oxygen was selected as the standard gas; and second, a molecular weight of 32.00 was assigned to it. After we made these choices, we could determine the molecular weight of any other gas.

An interesting question now arises: What unit should be assigned to molecular weight values? In choosing a weight unit, it is convenient to select a unit appropriate to the usual weight dimensions of the item being weighed. For example, we use grams to measure out laboratory reagents, tons to express the weight of ocean liners, and pounds to buy hamburger meat. Because an individual molecule is so small, we need a very small, new unit to express the mass of a single molecule. We unknowingly defined this new unit when we selected the oxygen molecule as a standard and assigned it a value of

32.00. This new unit, the **atomic mass unit (amu),** is so small that 32.00 amu are present in the mass of a single oxygen molecule.*

By definition, then, the oxygen molecule has a mass of 32 amu. One hydrogen molecule has a mass of 2 amu; one carbon dioxide molecule, a mass of 44 amu; and one carbon tetrafluoride molecule, a mass of 88 amu. Note that these are *relative* masses which are based on our earlier experiments. Each value gives the relative mass of a molecule as compared to the standard oxygen molecule.

While the atomic mass unit is convenient for expressing the mass of a single molecule, it is not convenient in laboratory operations. For example, it is impossible to weigh out 32 atomic mass units of oxygen. But it is very easy to weigh out 32 *grams* (g) of oxygen. We need two mass scales then—one for individual molecules (amu) and one for weighing reagents in laboratory operations (g).

In fact, a quantity equal to the molecular weight of a substance in grams (*e.g.,* 32.0 g of oxygen) is so easy to handle in the laboratory that it has been given a special name. It is called a **mole.** By using methods which are a little beyond us right now, it can be shown that 1 mole of oxygen (32.0 g) contains 6.02×10^{23} individual oxygen molecules. The same type of experiment shows that 1 mole of carbon dioxide (44.0 g) contains 6.02×10^{23} carbon dioxide molecules. Similarly, 1 mole of carbon tetrafluoride (88.0 g) contains 6.02×10^{23} carbon tetrafluoride molecules. This value—6.02×10^{23}—keeps appearing. It is known as **Avogadro's number.** *One mole of anything always contains 6.02×10^{23} units.*

The mole is to the chemist as the dozen is to the grocer. The chemist handles chemicals in moles (6.02×10^{23} units) just as the grocer handles eggs in dozens (12 units). (We shall use this analogy again in Chapter 3). In summary, 1 mole of a gas is equal to its molecular weight expressed in grams and 1 mole contains 6.02×10^{23} molecules.

More precise weight and volume relationships show that if the molecular weight of oxygen is arbitrarily established as 32.00, the molecular weight of hydrogen is 2.02. A 2.00-g sample of hydrogen is not quite 1 mole; a 2.02-g sample is exactly 1 mole and it contains 6.02×10^{23} molecules.

2-2.4 The Volume Occupied by One Mole of a Gas— The Molar Volume

What volume does 1 mole of a gas such as oxygen occupy? Our previous study of pressure-volume relationships (Section 2-1.1) showed that the volume of a gas is dependent upon the pressure on the gas. If we want to determine the volume occupied by 1 mole of gas, we must specify the pressure. Let us adopt the average pressure of

---

*How many grams are there in 1 amu? To answer this question, we must know the weight of an oxygen molecule in grams. Using experimental methods a little beyond us here, it can be shown that one oxygen molecule has a mass of 5.31×10^{-23} g. One atomic mass unit, then, is

$$\frac{5.31 \times 10^{-23} \text{ g/oxygen molecule}}{32 \text{ amu/oxygen molecule}} = 1.66 \times 10^{-24} \frac{\text{g}}{\text{amu}}$$

the air at sea level as our standard of pressure for this measurement and call it 1 atmosphere (atm). Similarly, we must specify a temperature. A convenient and easily reproduced temperature is that of ice melting slowly under 1 atm of pressure. This temperature, defined as zero degrees Celsius (0°C), is the standard which we shall use. Under these conditions (1 atm and 0°C) 1 mole of oxygen (32.00 g) has a volume of 22.4 liters and 1 mole of nitrogen (28.00 g) has a volume of 22.4 liters. Values for two other gases are shown in Table 2-4. Notice that these values are also close to 22.4 liters per mole. Similar measurements made on many other gases show that 1 mole of a gas occupies approximately 22.4 liters at 0°C and 1 atm of pressure.

TABLE 2-4 LABORATORY DATA FOR DETERMINING THE VOLUME OCCUPIED BY 1 MOLE* OF A GAS AT 0°C AND 1 ATM PRESSURE

Gas	Mass of Flask (1.00 Liter) Empty (g)	Mass of Flask Plus Gas (g)	Mass of 1.00 Liter of Gas (g/liter)	Mass of 1 Mole of Gas (g/mole)	Molar† Volume (liters/mole)
Oxygen	157.35	158.78	1.43	32.0	22.4
Nitrogen	157.35	158.60	1.25	28.0	22.4
Carbon monoxide	157.35	158.59	1.24	28.0	22.5
Carbon dioxide	157.35	159.32	1.97	44.0	22.3

* Remember that the mole is the quantity equivalent to one molecular weight in grams.
† The value for the volume of 1 mole of a gas can be obtained from laboratory data by dividing the mass of 1 mole of gas by the mass of 1 liter of the gas, measured in the laboratory at 0°C and 1 atm pressure. We shall use this relationship later. A unit check shows

$$\frac{g/mole}{g/liter} = \frac{liters}{mole}$$

We have seen in Section 2-1.3 that for a given quantity of gas pressure × volume = a constant in a room at constant temperature. If we take 1 mole of a gas at 0°C and 1 atm pressure, we can write

$$\text{volume} = \frac{22.4 \text{ liters}}{\text{mole}} \quad \text{if pressure is 1 atm}$$

We can then write

$$\text{pressure} \times \text{volume} = 1 \text{ atm} \times \frac{22.4 \text{ liters}}{\text{mole}}$$

or

$$P \times V = 22.4 \text{ atm} \times \frac{\text{liters}}{\text{mole}}$$

2-3 ATOMS

In the foregoing discussion we saw that molecules of different substances have different weights, but one gas molecule is just as effective as another in generating pressure on the walls of a container. Molecules have some similarities and some differences. Let us focus attention on the differences between collections of molecules in gases and possible explanations for these differences.

The gas molecule seems to be the unit particle in our "super-rubber" ball model for gases; however, it would be logical to ask: Can molecules be broken into simpler particles? Most of us already know the answer to this question. *Molecules can be broken into atoms.* The word *atom* is a firmly established part of our culture. We speak of the Atomic Energy Commission, an atomic power plant, the atomic bomb, the atomic age, and even of the atomic theory. What is the scientific background for our belief in atoms? Let us look again at some experiments.

2-3.1 The Combining Volumes of Gases— Chemical Change

If in a darkened room we mix in a bulb 1 liter of gaseous hydrogen and 1 liter of gaseous chlorine and if we then shine an ultraviolet sunlamp on the bulb, we see a flash and hear a resounding bang. A sizable amount of energy has been released. Further, if we examine the contents of the bulb after the explosion, we find no remaining hydrogen and no chlorine. A new material called hydrogen chloride is present in the bulb. Its properties are completely different from those of the hydrogen and chlorine we used in the beginning. Chlorine, for example, is a green gas, but hydrogen chloride is colorless; these two gases can be easily distinguished by color. Both hydrogen and hydrogen chloride are colorless; but their behavior with water is strikingly different. When hydrogen is fed into an inverted bottle filled with water, hydrogen bubbles collect above the water surface; but when hydrogen chloride is similarly fed into water, most of the gas disappears into the water—we say that most of the hydrogen chloride dissolves and only a small portion of the gas appears above the water surface. Many other differences could be cited. The new substance, hydrogen chloride, is completely different from the mixture of hydrogen and chlorine with which we started.

We say that a change resulting from a reaction such as that of hydrogen gas with chlorine gas is a **chemical change.** *In a chemical change the products are very different from the reactants and a large energy change is usually observed.* The flash, bang, and heat indicate energy released.

EXERCISE 2-4

When a spark is passed through a mixture of gaseous hydrogen and gaseous oxygen, water is produced. Is this a chemical change? Defend your answer by comparing reactants and products as to their melting points, role in combustion, and role in supporting life.

What are the molecular implications of this chemical reaction? It would be helpful to compare volumes so we could apply Avogadro's hypothesis and count molecules. Experiment shows that

$$1 \left\{ \begin{array}{c} \text{liter of} \\ \text{hydrogen} \end{array} \right\} \text{ plus } 1 \left\{ \begin{array}{c} \text{liter of} \\ \text{chlorine} \end{array} \right\} \text{ gives } 2 \left\{ \begin{array}{c} \text{liters of} \\ \text{hydrogen chloride} \end{array} \right\}$$

All gas volumes are measured at the same temperature and pressure. It is not important for us to know exactly how many molecules are present in 1 liter of hydrogen; we can arbitrarily represent this number as n. Avogadro's hypothesis tells us that we must also have n *molecules* of chlorine and $2n$ *molecules* of hydrogen chloride, since equal volumes of gases at the same temperature and pressure contain equal numbers of molecules. We can now write

$$n \begin{Bmatrix} \text{molecules of} \\ \text{hydrogen} \end{Bmatrix} + n \begin{Bmatrix} \text{molecules of} \\ \text{chlorine} \end{Bmatrix} \longrightarrow 2n \begin{Bmatrix} \text{molecules of} \\ \text{hydrogen chloride} \end{Bmatrix}$$

The symbol + now replaces "plus" and the arrow replaces "gives" in the earlier expression. If we were to let n equal 1 (a sample too small to work with), we could write

$$1 \begin{Bmatrix} \text{molecule of} \\ \text{hydrogen} \end{Bmatrix} + 1 \begin{Bmatrix} \text{molecule of} \\ \text{chlorine} \end{Bmatrix} \longrightarrow 2 \begin{Bmatrix} \text{molecules of} \\ \text{hydrogen chloride} \end{Bmatrix}$$

We can now make a most significant observation: each simple *molecule* of hydrogen chloride was formed from *half a molecule of hydrogen* plus *half a molecule of chlorine.* Clearly, each single molecule of hydrogen was broken into two identical pieces in the reaction, and each piece was then used to make a molecule of hydrogen chloride. Similarly, each chlorine molecule had to be broken into two identical parts in the reaction. This conclusion is not altered if we let n, the number of molecules, equal 2, 20, 51, 3,137, or 3,721,256,372, because the value of n can be eliminated by dividing both sides of the equation by n. The number of hydrogen chloride molecules is always twice the number of molecules of hydrogen or chlorine regardless of the value of n.

Since the hydrogen molecule must be broken into two pieces in this reaction, it is convenient to identify the pieces. These smaller fragments or building blocks for molecules are the **atoms.** Experiments of the type just described provide some of the evidence for belief in the existence of atoms of hydrogen and chlorine. It is apparent that each hydrogen molecule must contain at least two atoms, since each molecule is broken into two identical parts in the formation of hydrogen chloride. The observations could also be explained if each hydrogen molecule contained four, six, eight, ten, or any even number of atoms. The observations would not be consistent with the existence of three, five, seven, or any odd number of hydrogen atoms in the hydrogen molecule, since a molecule containing an odd number of atoms could not be broken into two equal parts without splitting at least one atom into two pieces. According to the definitions and models we are using, an atom *cannot* be split in a *normal chemical process*. We conclude that a molecule of hydrogen must contain an even number of atoms. We can apply the same argument to the chlorine molecule. Each chlorine molecule must contain an even number of atoms. A similar study of the reaction between nitric oxide and oxygen shows that each oxygen molecule also must contain an even number of atoms:

$$2n \begin{Bmatrix} \text{molecules} \\ \text{nitric oxide} \end{Bmatrix} + n \begin{Bmatrix} \text{molecules} \\ \text{oxygen} \end{Bmatrix} \longrightarrow 2n \begin{Bmatrix} \text{molecules} \\ \text{nitrogen dioxide} \end{Bmatrix}$$

Scientists always seek the simplest explanation of the known facts. The simplest interpretation is that each molecule of hydrogen contains *two* atoms; each molecule of chlorine contains *two* atoms. and each molecule of oxygen contains *two* atoms. No experiment performed to the present is inconsistent with the simplest assumption. We conclude, then, that each molecule of hydrogen contains two hydrogen atoms, each molecule of chlorine contains two chlorine atoms, and each molecule of oxygen contains two oxygen atoms. Such molecules are said to be **diatomic.** They contain two atoms.

Some molecules contain many more than two atoms. For example, the following volume relationships describe the formation of the gas ammonia from nitrogen and hydrogen:

$$1 \left\{ \begin{array}{l} \text{liter} \\ \text{nitrogen} \end{array} \right\} + 3 \left\{ \begin{array}{l} \text{liters} \\ \text{hydrogen} \end{array} \right\} \longrightarrow 2 \left\{ \begin{array}{l} \text{liters} \\ \text{ammonia} \end{array} \right\}$$

If we divide both sides of the equation by 2, we see that each ammonia molecule must contain half of a nitrogen molecule and three halves of a hydrogen molecule:

$$\tfrac{1}{2} \left\{ \begin{array}{l} \text{liter} \\ \text{nitrogen} \end{array} \right\} + \tfrac{3}{2} \left\{ \begin{array}{l} \text{liters} \\ \text{hydrogen} \end{array} \right\} \longrightarrow 1 \left\{ \begin{array}{l} \text{liter} \\ \text{ammonia} \end{array} \right\}$$

$$\tfrac{1}{2} \left\{ \begin{array}{l} \text{molecule} \\ \text{nitrogen} \end{array} \right\} + \tfrac{3}{2} \left\{ \begin{array}{l} \text{molecules} \\ \text{hydrogen} \end{array} \right\} \longrightarrow 1 \left\{ \begin{array}{l} \text{molecule} \\ \text{ammonia} \end{array} \right\}$$

Since half of a nitrogen molecule is one nitrogen atom and half of a hydrogen molecule is one hydrogen atom, the gas ammonia must contain one nitrogen atom and three hydrogen atoms. The ammonia molecule, then, contains four atoms.

While we needed the concept of molecules to understand or "explain" the *pressure-volume* relationship of gases, we need the concept of atoms to help us understand the *chemical changes* gases undergo.

2-3.2 Formulas and Equations

In trying to *describe* the chemical changes which occur when hydrogen gas and chlorine gas explode, we wrote

$$n \left\{ \begin{array}{l} \text{molecules} \\ \text{hydrogen} \end{array} \right\} + n \left\{ \begin{array}{l} \text{molecules} \\ \text{chlorine} \end{array} \right\} \longrightarrow 2n \left\{ \begin{array}{l} \text{molecules} \\ \text{hydrogen chloride} \end{array} \right\}$$

While this statement as written is accurate and contains much information, it can be condensed. A form of scientific shorthand is desirable and available.

Suppose we represent one hydrogen atom by the symbol H. Since a hydrogen molecule contains two hydrogen atoms, the hydrogen can be represented as H—H where the line between the H's suggests some kind of bond holding the two atoms together. If a chlorine atom is represented as Cl, a chlorine molecule would be Cl—Cl, and a hydrogen chloride molecule would be H—Cl. The chemical process can now be summarized by writing a **chemical equation:**

$$\text{H—H} + \text{Cl—Cl} \longrightarrow 2 \text{ H—Cl} \tag{1}$$

In this notation it is clear that one molecule of hydrogen, containing two hydrogen atoms, combines with one molecule of chlorine, containing two chlorine atoms, to give two molecules of hydrogen chloride, each containing one hydrogen atom and one chlorine atom. Further, bonds between the hydrogen atoms in the hydrogen molecule are broken during reaction. Bonds between the chlorine atoms in the chlorine molecule are also broken during reaction. (As you will learn later, the ultraviolet light used to start the reaction breaks the bonds between the chlorine atoms. That is what initiates the explosion.) Finally, new bonds between hydrogen and chlorine must be formed to give hydrogen chloride. This breaking of bonds and formation of new bonds is one of the important characteristics of a chemical process.

EXERCISE 2-5

(1) Write an *equation* using the notation H—H for hydrogen, and so on, to represent the reaction of nitrogen and hydrogen to give ammonia (see page 38).

(2) *Two liters of hydrogen gas combine with 1 liter of oxygen gas to give 2 liters of water vapor.* Write an equation using the notation above to represent the process.

While we have simplified the notation tremendously by using H—H in place of the words "molecule of hydrogen," we can simplify the notation even further. A hydrogen molecule can be written as H_2. The subscript 2 indicates that we have two hydrogen atoms in each molecule. Similarly, the chlorine molecule can be written as Cl_2. Since the hydrogen chloride molecule contains one hydrogen atom and one chlorine atom, it is written as HCl and the bond between atoms is implied. The **chemical equation** which represents the process observed is then

$$H_2 + Cl_2 \longrightarrow 2\ HCl \qquad (2)$$

The equation is simply a representation in an abbreviated form of laboratory observations. The notations H_2 and HCl are called the **formulas** of the molecules they represent.

EXERCISE 2-6

Use the following symbols: carbon = C, hydrogen = H, oxygen = O, nitrogen = N, and chlorine = Cl.

(1) Two liters of carbon monoxide combine with 1 liter of oxygen to give 2 liters of carbon dioxide. (All volumes are measured at the same temperature and pressure.) If carbon monoxide has the formula CO, what is the formula for carbon dioxide? Write the equation for the burning of CO to give carbon dioxide. Burning is the process in which CO combines with molecular oxygen.

(2) Two liters of nitric oxide decompose to form 1 liter of nitrogen and 1 liter of oxygen. (All volumes are measured at the same temperature and pressure.) Write the word equation for the process. Write the symbols for the *products* formed, indicating the relative numbers of molecules of each product. Now write the formula for nitric oxide.

FORMULAS and EQUATIONS

2-3.3 Chemical Implications of Atoms

The concept of atoms brings many other benefits. By assuming the existence of atoms we can answer the question: Why do molecules differ? We can explain differences between various molecules in terms of the kinds and arrangement of atoms in each molecule. For example, water is different from hydrogen chloride or ammonia because it contains a different combination of atoms arranged in a different way. Throughout the course we shall investigate properties of substances found in nature or prepared in the laboratory and we shall seek explanations in terms of the numbers, types, and arrangements of the atoms present. Frequently, the detailed arrangement of atoms in a molecule is extremely important in defining the properties of that molecule. Many illustrations will appear as we progress.

2-3.4 Structural Formulas and Molecular Models

You have just been told that the way in which atoms are bonded is very important in determining the properties of a substance. It would be helpful to represent the *arrangement* of atoms in the molecule as well as the number and kinds of atoms present. For this purpose **structural formulas** and **models** are used. Thus, the structural formula for water is

$$\overset{O}{\underset{H\quad\quad H}{\diagup\diagdown}}$$

As in the case of the H—H and H—Cl molecules, the dashes indicate the chemical bonds or the connections between atoms. The angular arrangement indicates that the two hydrogen atoms and the oxygen atom do not lie in a straight line but make an H—OH angle of about 104 degrees.* Sometimes, for more complicated molecules, structural details such as bond angles cannot be represented easily on a piece of paper. A structural formula is a representation on a piece of paper of the three-dimensional molecule. For example, the molecule methane (CH$_4$) is represented by the planar structural formula

$$\begin{array}{c} H \\ | \\ H-C-H \\ | \\ H \end{array}$$

even though the actual molecule has four hydrogens arranged around carbon at the corners of a regular tetrahedron (see Figure 2-6).

The structural formula for methane and the actual molecular model (see Figure 2-6) illustrate that no written formula is quite as effective as a molecular model in helping to visualize molecular shapes. Since molecular shape is so important in establishing properties of substances, a number of physical models can be used to represent molecules. Some of the more commonly used models are seen in Figure 2-7. Selection of the particular kind depends upon

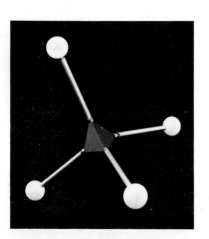

Fig. 2-6 Model of a methane (CH$_4$) molecule.

---

*Details of molecular geometry are obtained using experiments of the type described at the end of Chapter 16.

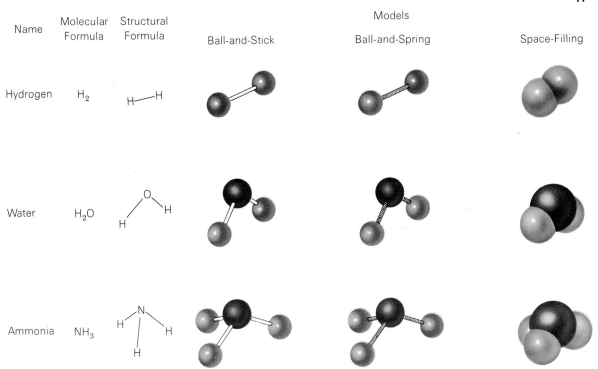

Fig. 2-7 **Different representations of hydrogen (H_2), water (H_2O), and ammonia (NH_3) molecules.**

what must be emphasized. The **ball-and-stick model** and **ball-and-spring model** indicate bond orientations. Ball-and-spring models are useful to indicate molecular flexibility and molecular vibration. The **space-filling model** provides a more realistic view of the spacial relationships and the crowding among *nonbonded atoms*. See Figure 2-8.

2-3.5 Atoms and Molecules in Liquids and Solids

So far, we have accumulated evidence to indicate that gases are made up of molecules and gas molecules are made up of atoms. We have also observed the burning of a candle in the laboratory. Gases such as carbon dioxide and water vapor (steam) were recognized as products of the burning of the candle.

We might now raise the question: What is the candle made of? Since a candle burns to produce gases, collections of molecules containing atoms, the simplest assumption that we can make is that the *candle itself contains molecules made up of atoms*. These "candle molecules" would then combine with the *diatomic* oxygen molecules (remember: diatomic means two atoms per molecule) to give gaseous products. This simple assumption, that *liquids and solids contain atoms and that such atoms are frequently combined into molecules in the liquid or solid,* is consistent with all we know about both liquids and solids. *All matter is made up of atoms.* This is the fundamental postulate of the atomic theory. The atomic theory is amazingly powerful and can well be considered the cornerstone of all modern science.

Fig. 2-8 **Model showing the crowding of nonbonded atoms.**

2-4 ELEMENTS AND COMPOUNDS

In giving the pictorial equation for the reaction between gaseous chlorine and hydrogen we wrote

$$\text{H—H} + \text{Cl—Cl} \longrightarrow 2\,\text{H—Cl} \qquad (1)$$

If we look at this equation, we immediately see two distinct classes of molecules. In the first class *identical atoms* are bound together to make the molecules; we see H—H and Cl—Cl. A substance made up of one kind of atom is called an **element.** The second class of molecule observed in our equation is hydrogen chloride. It contains one hydrogen atom and one chlorine atom. Each molecule of hydrogen chloride contains more than one kind of atom. Pure hydrogen chloride contains *more than one kind of atom.* Pure hydrogen chloride has *identical molecules,* each of which contains one hydrogen *atom* and one chlorine *atom* in chemical combination. Hydrogen chloride, a substance which contains more than one kind of atom in chemical combination, is a **compound.**

2-4.1 The Elements

An element is a substance that contains only one kind of atom. A few of the elements such as gold, silver, and iron were known to the ancient Greeks and Romans. At the beginning of the nineteenth century only about 26 elements were known. One hundred years later 81 elements were known. Figure 2-9 shows the number of known elements as a function of time. This graph shows that the rate of discovery of new elements is declining; further, it shows that we have about 103 known elements today. But the graph does not show that today all new elements are made by nuclear reactions (Chapter 7); they are not discovered in nature. An extrapolation of the graph (an extension beyond the last known point) suggests that there may be a limit to the number of elements which can exist. However, many distinguished nuclear scientists now question this limit. They suggest that relatively stable elements of much higher mass may well be made in the future. If such is the case, the curve would bend sharply upward.

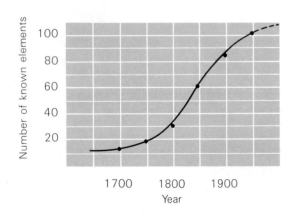

Fig. 2-9.1 **(Left)** The discovery of the elements: total number of known elements as a function of time.

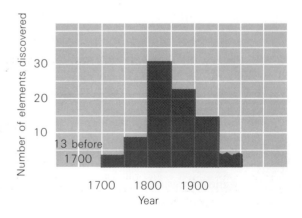

Fig. 2-9.2 **(Right)** The discovery of the elements: number of elements discovered in each half-century since 1700.

2-4.2 Atomic Weights and Molecular Weights

If each mole of oxygen contains 6.02×10^{23} molecules, and if each molecule of oxygen contains two atoms, then 1 *mole* of oxygen *molecules* contains 12.04×10^{23} atoms. We may ask the question: What is the mass of 1 mole (6.02×10^{23}) of oxygen *atoms?* Clearly it will be half the mass of 1 mole of oxygen *molecules,* or 16.00 g. We can then state that 1 mole of oxygen *atoms* (6.02×10^{23} atoms) will weigh 16.00 g. The molecular weight of a substance was defined earlier as the *relative mass* of a molecule of the substance expressed on a scale in which the oxygen molecule is assigned a value of 32.00 amu. It is reasonable to define an **atomic weight** as *the mass of an atom on this same scale.* The oxygen atom has an *atomic weight* of 16.00, since the *diatomic* oxygen molecule is assigned a molecular weight of 32.00. *The atomic weight of an atom is the relative mass of that atom expressed on a scale in which an oxygen atom is assigned a value of 16.00.** One mole of atoms of an element is 6.02×10^{23} atoms. This mole of atoms has a mass equal to the atomic weight of the element expressed in grams. For oxygen the atomic weight is 16.00. One mole of oxygen *atoms* has a mass of 16.00 g. For hydrogen the atomic weight is 1.008; 1 mole of hydrogen *atoms* has a mass of 1.008 g. For carbon the atomic weight is 12.01, and 1 mole of carbon *atoms* has a mass of 12.01 g. Atomic weight values are used frequently. Table 2-5 on page 44 lists working values.

Common sense tells us that the weight of an object should be equal to the sum of the weights of its parts. This fact, summarized as the **Law of Conservation of Mass,** is true within our ability to weigh on balances. If we want to know the *molecular weight* of HCl, we simply add the atomic weight of one hydrogen atom and the atomic weight of one chlorine atom:

$$\text{mol wt HCl} = \text{at. wt H} + \text{at. wt Cl} = 1.0 + 35.5 = 36.5$$

EXERCISE 2-7

(1) Show that the weight of 1 mole of CO_2 is 44.0 g.
(2) Show that the weight of 1 mole of SO_2 is 64.1 g.
(3) Calculate the weight in grams of 6.02×10^{23} molecules of CO. Calculate the weight in grams of 3.01×10^{23} molecules of fluorine.

EXERCISE 2-8

What is the molecular weight of hydrogen cyanide (HCN)?

2-4.3 More on Symbols, Formulas, and Names

SYMBOLS OF THE ELEMENTS

In Section 2-3.2 we explored ways for effectively representing chemical processes. Word equations and equations using symbols were

---

* Actually, atomic weight is now based on a specific carbon atom as a standard. This standard is difficult to describe at this time, and the change from oxygen to a specific carbon atom is unimportant in most chemistry. It is only -0.0045 percent change. The difference may be important in nuclear processes. A table of precise atomic weights appears in Appendix 9.

TABLE 2-5 INTERNATIONAL ATOMIC WEIGHTS*

Name	Symbol	Atomic Number	Atomic Weight	Name	Symbol	Atomic Number	Atomic Weight
Actinium	Ac	89	(227)†	Mercury	Hg	80	200.6
Aluminium	Al	13	27.0	Molybdenum	Mo	42	95.9
Americium	Am	95	(243)	Neodymium	Nd	60	144.2
Antimony	Sb	51	121.8	Neon	Ne	10	20.2
Argon	Ar	18	39.9	Neptunium	Np	93	(237)
Arsenic	As	33	74.9	Nickel	Ni	28	58.7
Astatine	At	85	(210)	Niobium	Nb	41	92.9
Barium	Ba	56	137.3	Nitrogen	N	7	14.01
Berkelium	Bk	97	245	Nobelium	No	102	(254)
Beryllium	Be	4	9.01	Osmium	Os	76	190.2
Bismuth	Bi	83	209.0	Oxygen	O	8	16.00
Boron	B	5	10.8	Palladium	Pd	46	106.4
Bromine	Br	35	79.9	Phosphorus	P	15	31.0
Cadmium	Cd	48	112.4	Platinum	Pt	78	195.1
Calcium	Ca	20	40.1	Plutonium	Pu	94	(242)
Californium	Cf	98	(251)	Polonium	Po	84	210
Carbon	C	6	12.01	Potassium	K	19	39.1
Cerium	Ce	58	140.1	Praseodymium	Pr	59	140.9
Cesium	Cs	55	132.9	Promethium	Pm	61	(147)
Chlorine	Cl	17	35.5	Protactinium	Pa	91	(231)
Chromium	Cr	24	52.0	Radium	Ra	88	(226)
Cobalt	Co	27	58.9	Radon	Rn	86	(222)
Copper	Cu	29	63.5	Rhenium	Re	75	186.2
Curium	Cm	96	(247)	Rhodium	Rh	45	102.9
Dysprosium	Dy	66	162.5	Rubidium	Rb	37	85.5
Einsteinium	Es	99	(254)	Ruthenium	Ru	44	101.1
Erbium	Er	68	167.3	Samarium	Sm	62	150.4
Europium	Eu	63	152.0	Scandium	Sc	21	45.0
Fermium	Fm	100	(253)	Selenium	Se	34	79.0
Fluorine	F	9	19.0	Silicon	Si	14	28.1
Francium	Fr	87	(223)	Silver	Ag	47	107.9
Gadolinium	Gd	64	157.3	Sodium	Na	11	23.0
Gallium	Ga	31	69.7	Strontium	Sr	38	87.6
Germanium	Ge	32	72.6	Sulfur	S	16	32.1
Gold	Au	79	197.0	Tantalum	Ta	73	180.9
Hafnium	Hf	72	178.5	Technetium	Tc	43	(99)
Helium	He	2	4.00	Tellurium	Te	52	127.6
Holmium	Ho	67	164.9	Terbium	Tb	65	158.9
Hydrogen	H	1	1.008	Thallium	Tl	81	204.4
Indium	In	49	114.8	Thorium	Th	90	232.0
Iodine	I	53	126.9	Thulium	Tm	69	168.9
Iridium	Ir	77	192.2	Tin	Sn	50	118.7
Iron	Fe	26	55.8	Titanium	Ti	22	47.9
Krypton	Kr	36	83.8	Tungsten	W	74	183.9
Lanthanum	La	57	138.9	Uranium	U	92	238.0
Lawrencium	Lr	103	(257)	Vanadium	V	23	50.9
Lead	Pb	82	207.2	Xenon	Xe	54	131.3
Lithium	Li	3	6.94	Ytterbium	Yb	70	173.0
Lutetium	Lu	71	175.0	Yttrium	Y	39	88.9
Magnesium	Mg	12	24.3	Zinc	Zn	30	65.4
Manganese	Mn	25	54.9	Zirconium	Zr	40	91.2
Mendelevium	Md	101	(256)				

* More precise values of the atomic weights based on the latest IUPAC reports are given in Appendix 9.
† For those elements all of whose isotopes are radioactive, the parentheses indicate the isotope with the longest half-life. (For a definition of *half-life*, see Section 7-4.2.)

given. The obvious choice of the symbol to represent an element was the first letter of the name of the element. We chose O for oxygen, H for hydrogen, N for nitrogen, and C for carbon. Eight other elements* are named by this simple and obvious convention.

Since there are 103 elements to name but only 26 letters to use, it is obvious that a second letter must be used to identify some elements. What choices for the second letter are most reasonable? Since chemistry is an international subject, international agreement on symbols is desirable but difficult to get. Such agreement has been obtained. The symbols listed in Table 2-5 have been adopted by the International Union of Pure and Applied Chemistry, an international organization of chemists. After using just the first letter of the name for a symbol, the most obvious choice is the first *two* letters of the name. Many elements are named by this convention: helium is He, lithium is Li, beryllium is Be, neon is Ne, aluminum is Al, silicon is Si, argon is Ar, calcium is Ca, and titanium is Ti. Some additional elements have been named using this convention. How many of the group of 34 can you identify? (See Table 2-5.)

In a few cases it is convenient to use the first letter followed by a letter *other than* the second letter of the name. Frequently the choice of the second letter is obvious from the sound of the name. For example, magnesium is Mg, chlorine is Cl, chromium is Cr, manganese is Mn, and zinc is Zn. Many additional elements have been named in this way.

Finally, ten rather common elements have symbols derived from the Latin name of the element rather than the English name. Indeed, many elements in this group are easily obtained from their ores and were therefore known to the alchemists, hence the Latin name. For example, gold is Au from the Latin *aurum* and silver is Ag from the Latin *argentum*. Because the elements in this group are rather common and their symbols are frequently needed, these are collected in Table 2-6. You are urged to look over these eleven symbols.

* These elements are boron, fluorine, sulfur, phosphorus, iodine, vanadium, uranium, and yttrium.

MORE on SYMBOLS, FORMULAS, and NAMES

**EDWARD WICHERS
(1892–   )**

Edward Wichers is an American chemist who spent most of his professional life with the United States Bureau of Standards in Washington, D.C. Although a distinguished chemist in his own right, Dr. Wichers is given special recognition here for his role as a scientific statesman. As a result of his leadership, professional competence, and dedication, separate atomic weight scales used by chemists and physicists were abandoned in 1961 and replaced by a new scale. Carbon-12 was selected as the new atomic weight standard and assigned a value of 12.000000. This careful choice resulted in a minimum number of changes for chemists and physicists. Dr. Wichers' resolution of one of the more difficult organizational problems of science has simplified life for future generations of scientists.

Edward Wichers played a key role in the International Union of Pure and Applied Chemistry for many years, serving as its vice-president from 1952–1959. He retired as Associate Director of the Bureau of Standards in 1962.

TABLE 2-6 CHEMICAL SYMBOLS THAT ARE NOT DERIVED FROM THE COMMON ENGLISH NAME OF THE ELEMENT

Common Name	Symbol	Symbol Source*
Antimony	Sb	stibnum
Copper	Cu	cuprum
Gold	Au	aurum
Iron	Fe	ferrum
Lead	Pb	plumbum
Mercury	Hg	hydrargyrum
Potassium	K	kalium
Silver	Ag	argentum
Sodium	Na	natrium
Tin	Sn	stannum
Tungsten	W	wolfram

* All entries in this column are Latin, except for *wolfram*, which is German.

FORMULAS AND NAMES

We indicated earlier that the symbols for the elements can be used in writing formulas for molecules. The hydrogen molecule is H_2, hydrogen chloride is HCl, and water is H_2O. We repeat: The symbols have a *quantitative* significance. In the formula of a molecule containing hydrogen, H stands for *one* atom. A molecule of ordinary table sugar has the formula $C_{12}H_{22}O_{11}$ because the sugar molecule contains 12 carbon atoms, 22 hydrogen atoms, and 11 oxygen atoms. Note that the hydrogen and oxygen are present in a 2 to 1 ratio, just as they are in water. This formula suggested to earlier chemists that sugar was made from carbon and water, hence was a **hydrate** of carbon. Traces of the ancient idea are still found in our listing of sugar as a carbo*hydrate* (carbon-hydrate).

NOMENCLATURE

We have indicated both names and symbols for many *elements*. What about the names of *compounds*? Is there logic to the system naming them? The answer is a qualified "Yes." There are some rules, but they—like the rules of grammar—are not always obeyed in practice.

The simplest kind of compound is one containing only two *kinds* of atoms. This is called a **binary compound.** The name of a binary compound is formed by taking the name of the element most like a metal, joining it to the stem of the name of the element least like a metal, and adding *-ide.* Let us see how this works. Common table salt is NaCl; the name is sodium chloride. *Chlor-* is the stem of chlorine. We add *-ide* for the second part of the name of the substance. Similarly, the compound $CaCl_2$ is named calcium chloride. It is used on the roads in the summer to keep down dust. We have already called the compound HCl hydrogen chloride. The gas H_2S is hydrogen sulfide.

EXERCISE 2-9

Name: Na_2S, CaF_2, MgO, Al_2O_3, and $ScCl_3$? The stem of oxygen is *ox-*, of fluorine *fluor-*, of sulfur *sulf-*.

At times *more than one kind of molecule containing the same two kinds of atoms* can be formed. For example, carbon forms at least two compounds with oxygen, CO and CO_2. If we apply the rule for naming binary compounds, we would say there are two carbon *ox*ides. In order to differentiate these oxides, one is called carbon *mon*oxide and the other is called carbon *di*oxide. The prefix *mono-* means one, *di-* means two, *tri-* means three, and *tetra-* means four. Thus, carbon monoxide is an oxide containing one carbon atom and one oxygen atom. (When the number of carbon atoms is not specified, one atom is assumed.) Carbon dioxide contains one carbon atom and two oxygen atoms. Sulfur trioxide is SO_3. Carbon tetrafluoride is CF_4. Nitrogen trifluoride is NF_3, while N_2F_2 is dinitrogen difluoride.

2-5 HIGHLIGHTS

The model advanced in Chapter 1 to explain the behavior of a gas in a balloon has grown rapidly. Quantitative pressure-volume measurements for many gases showed that

$$\text{pressure} \times \text{volume} = \text{a constant}$$

as long as we started with the same initial volume of gas measured at the same temperature and pressure. This fact was combined with information on combining volumes of gases to provide initial evidence for **Avogadro's hypothesis:** Equal volumes of gases at the same temperature and pressure contain the same number of molecules.

Using Avogadro's hypothesis as a starting point, we obtained **relative masses** of gaseous molecules. A standard gas was needed to establish a **molecular weight** scale. Oxygen was selected as the standard gas and assigned a value of 32.00 for its molecular weight. The relative mass of any other gas on this scale is the molecular weight of the gas. A 32.00-g sample of oxygen is a **mole** of oxygen. A mole is 6.02×10^{23} units. One mole of oxygen gas contains **6.02×10^{23} molecules.** One mole of a gas occupies 22.4 liters if the gas is measured at 0°C and 1 atm pressure.*

We then noted that molecules of the same substance are the same, while molecules of different substances vary. Reasons for the differences were sought in the nature of the molecules. Consideration of a **chemical change** led us to the view that molecules contain smaller building blocks called **atoms.**

Having proposed the existence and ways of representing atoms, we represented molecules through **structural formulas** and **models.** These models and formulas picture what is known about the number, types, and arrangement of atoms in the molecules. Atoms are present in all matter.

Atomic weights give relative weights of *atoms,* just as molecular weights give relative weights of molecules. Molecular weights of molecules can be obtained by adding up the atomic weights of the atoms which make up the molecule.

Thus, a model (theory) grows. As it is tested in an ever-widening range of experience, the model often becomes more complex. In some cases the development of the model is a slow and logical process; in other cases an incisive and brilliant suggestion may open the way for spectacular correlations and progress (Avogadro's hypothesis). Increasing complexity of a model is frequently offset by the advantage of developing interrelationships among many different parts (that is, by discovering hidden likenesses). The atomic theory, as developed to correlate chemical behavior, is much more complicated than is needed to explain the simple gas behavior first mentioned in Chapter 1. Nevertheless, correlations developed between "super-rubber" balls and the air in a balloon have provided us with a substantial start in understanding chemistry.

---

* Deviations from this statement are considered in Chapter 5.

QUESTIONS AND PROBLEMS

1. Suppose you took a 1.0-g sample of hydrogen and obtained pressure-volume data. What would be the value of the constant in the expression $P \times V = a$ constant if $T = 0°C$? What would be the value of the constant if you took 1.0 g of NH_3? Watch the significant figures. Remember that for 1 mole of hydrogen (2.0 g), the constant is 22.

2. What do you *observe* in the laboratory when you take a bottle containing a solution of ammonia gas (NH_3) and open it next to a bottle containing a solution of hydrogen chloride gas (HCl)? Only one single, solid product forms. Can you construct a mental *model* for what you see and write an *equation* to describe the observed process?

3. Gaseous uranium hexafluoride is important in the preparation of uranium as a source of "nuclear energy." A flask filled with this gas is weighed under certain laboratory conditions (temperature and pressure), and the weight of the gas is found to be 3.52 g. The same flask is filled with oxygen gas and is weighed under the same laboratory conditions. The weight of the oxygen in the flask is found to be 0.32 g. What is the ratio of the weight of one uranium hexafluoride molecule to the weight of one oxygen molecule? State any guiding principles needed in answering this question.

4. There are a number of gas mixtures that combine in the general pattern

$$1 \begin{Bmatrix} \text{volume} \\ \text{gas A} \end{Bmatrix} + 1 \begin{Bmatrix} \text{volume} \\ \text{gas B} \end{Bmatrix} \longrightarrow 2 \begin{Bmatrix} \text{volumes} \\ \text{gas C} \end{Bmatrix}$$

(All are measured at the same temperature and pressure.) The combination of hydrogen and chlorine to give HCl is one example. Why do these data show that both gas A and gas B must have an *even* number of atoms in the molecules? What explicit assumption is used in your proof?

5. One liter of a substance containing only hydrogen and oxygen decomposes on exposure to light to give 1 liter of water vapor (H_2O) and ½ liter of oxygen gas (O_2). (All were measured at the same temperature and pressure.) How many atoms of hydrogen were present in each original molecule? How many atoms of oxygen were present in each original molecule? Write the formula for the original compound.

6. A glass bulb from which all air has been pumped out weighs 150.300 g. Pure oxygen is admitted to the bulb and the bulb plus gas is reweighed. A weight of 151.050 g is obtained. The oxygen is pumped out and an unknown gas is added to the bulb so that it is at the same temperature and pressure as the oxygen was earlier. The bulb plus the new gas weighs 152.360 g. What is the molecular weight of the new gas? (*Answer*: 88.)

7. What would the molecular weight of the new gas in question 6 have been if we had assigned the molecular weight of 20.00 to oxygen instead of 32.00? What would the molecular weight of hydrogen have been if we had assigned a molecular weight of 20.00 to oxygen as our standard?

8. The weight of 0.500 liter of gas A measured at 0°C and 1 atm pressure is 0.625 g. What is the weight of 1 liter of gas A? What is the weight of 22.4 liters of this gas? What is the molecular weight of the gas? Which one of the following gases could A be: N_2, CO_2, O_2, Cl_2?

9. A flask of gaseous CCl_4 was weighed at a measured temperature and pressure. The flask was flushed and then filled with oxygen at the same temperature and pressure. The weight of the CCl_4 vapor will be about (a) the same as that of oxygen, (b) one fifth as heavy as the oxygen, (c) five times as heavy as the oxygen, (d) twice as heavy as the oxygen, or (e) one half as heavy as the oxygen.

10. How many molecules are contained in 1.01 g of hydrogen gas?

11. How many particles are there in a mole?

12. The earth has about 3,000,000,000 or 3×10^9 people. If we had 1 mole of dollars to divide among all the people in the world, how much would each person receive?

13. A stone about the size of a softball weighs roughly a kilogram (1,000 g). How many moles of such stones would be needed to account for the entire mass of the Earth, which is about 6×10^{27} g?

14. Determine the weight, in grams, of one silver atom. (*Answer*: 1.79×10^{-22} g.)

15. If 1½ moles of hydrogen gas (H_2) react in a given experiment, how many grams of H_2 does this represent?

16. (a) What does the molecular formula CBr_4 mean? (b) What information is added by the following structural formula?

$$Br-\underset{\underset{Br}{|}}{\overset{\overset{Br}{|}}{C}}-Br$$

17. A pure white substance, on heating, forms a colorless gas and a purple solid. Is the substance an element or a compound? Explain.

18. What do the following symbols represent: K, Ca, Co, CO, Pb?

19 Write formulas for silicon dioxide (common sand, or silica); sulfur dichloride; nitrogen trifluoride; aluminum trifluoride; and dinitrogen difluoride.

20 (a) Write formulas for hydrogen chloride, hydrogen bromide, hydrogen iodide, boron trichloride, carbon tetrachloride, nitrogen trichloride, and oxygen dichloride. (b) Locate in Table 2-5 the symbol for each element involved in these compounds.

21 For each of the following substances give the name of each kind of atom present and the total number of atoms represented in the formula shown:

Name	Formula
(a) graphite (pencil lead)	C
(b) diamond	C
(c) sodium chloride (table salt)	NaCl
(d) sodium hydroxide	NaOH
(e) calcium hydroxide	$Ca(OH)_2$
(f) potassium nitrate	KNO_3
(g) magnesium nitrate	$Mg(NO_3)_2$
(h) sodium sulfate	Na_2SO_4
(i) calcium sulfate	$CaSO_4$

22 All of the following substances are called acids. What element do they have in common?

Name	Formula
(a) nitric acid	HNO_3
(b) hydrochloric acid (or hydrogen chloride)	HCl
(c) hydrofluoric acid (or hydrogen fluoride)	HF
(d) sulfuric acid	H_2SO_4
(e) phosphoric acid	H_3PO_4

23 Here are the names of some common chemicals and their formulas. Name the elements in each material:

Name	Formula
(a) hydrogen peroxide	H_2O_2
(b) jeweler's rouge	Fe_2O_3
(c) light bulb filament	W
(d) tetraethyl lead	$Pb(C_2H_5)_4$
(e) baking soda	$NaHCO_3$
(f) octane	C_8H_{18}
(g) household gas	CH_4

QUESTIONS and PROBLEMS

24 Write the formulas for the following compounds and give the weight of 1 mole of each: carbon disulfide, sulfur hexafluoride, nitrogen trichloride, osmium tetroxide.

25 Consider the following data:

Element	Atomic Weight
A	12.01
B	35.5

A and B combine to form a new substance, X. If 4 moles of B atoms combine with 1 mole of A atoms to give 1 mole of X molecules, then the weight of 1 mole of X is (a) 47.5 g, (b) 74.0 g, (c) 83.0 g, (d) 154.0 g, or (e) 166.0 g.

26 Calculate the molecular weight for each of the following: SiF_4, HF, Cl_2, Xe, NO_2.

27 What is the *molecular weight* of $C_2H_2F_4$?

28 Two volumes of hydrogen fluoride gas combine with one volume of the gas dinitrogen difluoride to form two volumes of a gas G.

(a) According to Avogadro's hypothesis, how many molecules of G are produced from one molecule of dinitrogen difluoride?

(b) If X = number of atoms in a molecule of hydrogen fluoride, Y = number of atoms in a molecule of dinitrogen difluoride, and Z = number of atoms in a molecule of G, write the relation among X, Y, and Z appropriate to the combining volumes given.

(c) For each of the following possible values of X and Y, calculate the required value of Z.

If X is	and Y is,	then Z must be
1	2	
1	4	
2	2	
2	4	

(d) No odd value of Y is suggested in part (c). Prove that Y must be an even integer.

Because of its practical use, and for its own intrinsic interest, the principle of the conservation of energy may be regarded as one of the great achievements of the human mind.

SIR WILLIAM CECIL DAMPIER (1867–1952)

CHAPTER 3

Putting Ideas to Work: The Conservation Laws

Only a few chemical reactions involve as great an energy change as this, but all involve some energy change.

Anyone who has ever followed a recipe—either successfully or unsuccessfully—knows from experience that *quantities* are important. It is not enough to specify that one needs eggs, flour, baking powder, and milk to bake a cake, nor is it enough to say that the mixture has to be put into a heated oven. One needs to know how many eggs and how much flour, baking powder, and milk. We must specify the temperature of the oven and the time of baking. If you have only one egg, you can't bake a very large cake, and if the oven isn't working you can't bake one. So it is with chemical reactions. You have to have the right proportions of the right molecules and the right energy conditions or the desired result won't be obtained. While it is individual molecules which react with each other, we can estimate numbers of molecules conveniently either by weighing them or by measuring their volume. Learning to weigh and measure the right quantities of chemicals for successful reactions is the subject considered in this chapter.

Our growing model for matter has been expanded until it now provides a classification scheme for everything from muddy water to gases obtained in cylinders from the storeroom. Despite these successes, one of the most important sides of the atomic theory remains undeveloped—namely, its ability to give numerical (quantitative) relationships involving the weights of reactants and products in chemical processes. Let us examine these numerical or quantitative relationships using the mole as the measuring unit. In doing this, we shall use several similar numerical relationships.

3-1 DETERMINING THE NUMBER OF MOLES FROM LABORATORY QUANTITIES

Since laboratory measurements of reagents are usually made in grams or made in units of volume such as liters, the laboratory values must be converted to moles if we are to use the mole as the unit for measuring chemicals. Remember that the mole is to chemistry as the dozen is to the merchant selling eggs or oranges. Though the term *mole* may look strange at first, you can learn to use it just as easily as you now use the word *dozen*. Let us run some parallel calculations using both moles and dozens. To begin, we shall work with dozens.

Example (1a) How many dozens of eggs are contained in 36 eggs?

You know that there are 12 eggs per dozen. This is written 12 eggs/doz. We then divide 36 eggs by 12 eggs/doz:

$$\frac{36 \text{ eggs}}{12 \frac{\text{eggs}}{\text{doz eggs}}} \tag{1}$$

The units can be handled just like numbers here. The unit "eggs/doz eggs" is really just like a numerical fraction. Reducing the fraction to lowest terms, we have

$$36 \text{ eggs} \times \frac{1 \text{ doz eggs}}{12 \text{ eggs}} = 3.0 \text{ doz eggs} \tag{2}$$

The unit "eggs" in numerator and denominator reduces to 1, giving 3.0 as the proper number and "doz eggs" as the proper unit.

Working with units in this way is called **simple unit analysis** and is most helpful in providing a check on calculations. If the units of the answer are incorrect, a mistake has been made somewhere in the procedure.

Example (1b) How many moles of carbon dioxide (CO_2) will be found in 18.06×10^{23} molecules of carbon dioxide?

If we remember that there are 6.02×10^{23} particles per mole, the problem can be written as a straightforward division process:

Chapter 3 / PUTTING IDEAS to WORK: The CONSERVATION LAWS

$$\frac{18.06 \times 10^{23} \text{ molecules CO}_2}{\frac{6.02 \times 10^{23} \text{ molecules CO}_2}{\text{mole CO}_2}}$$

$$= 18.06 \times 10^{23} \text{ molecules CO}_2 \times \frac{1 \text{ mole CO}_2}{6.02 \times 10^{23} \text{ molecules CO}_2}$$

$$= 3.00 \text{ moles CO}_2 \tag{3}$$

Note that 6.02×10^{23} units per mole is used just like 12 eggs per dozen. The unit "molecules CO_2" reduces to 1 and the unit "mole CO_2" remains after we divide.

We can pursue the analogy further. Let us assume that a dozen hen eggs weighs 1.5 pounds. Then we ask:

Example (2a) How many dozens of hen eggs will be found in a crate containing 45 pounds of eggs?

The answer is obtained by a straightforward division process:

$$\frac{45 \text{ lb eggs}}{1.5 \frac{\text{lb eggs}}{\text{doz eggs}}} = 30 \text{ doz eggs} \tag{4}$$

Note that unit analysis again indicates the correct units for the answer.

─────── **EXERCISE 3-1** ───────

Invert the fraction

$$\frac{\text{lb eggs}}{\frac{\text{lb eggs}}{\text{doz eggs}}}$$

and multiply to show that the final unit is doz eggs.

─────────────────────────

The weight relationships involving moles are identical. Only the units differ.

Example (2b) How many *moles* of hydrogen gas will be found in a tank containing 60 g of hydrogen?

We know that each *mole* of hydrogen gas weighs 2.0 g (compare with 1.5 lb per doz eggs). We then write

$$\frac{60 \text{ g hydrogen}}{2.0 \frac{\text{g hydrogen}}{\text{mole hydrogen}}} = 30 \text{ moles hydrogen} \tag{5}$$

A related question can now be answered using the answers from Examples (2a) and (2b).

Example (3a) How many individual eggs are contained in 45 lb of eggs?

The answer is clearly

$$\frac{45 \text{ lb eggs}}{1.5 \frac{\text{lb eggs}}{\text{doz eggs}}} \times 12 \frac{\text{eggs}}{\text{doz eggs}} = 360 \text{ eggs} \qquad (6)$$

Note that units reduce properly. You know intuitively that 30 doz eggs [answer to Example (2a)] will be 360 eggs. For molecules the logic is identical.

Example (3b) How many individual hydrogen *molecules* are contained in 60 grams of hydrogen?

We write

$$\left(\frac{60 \text{ g H}_2}{2.0 \frac{\text{g H}_2}{\text{mole H}_2}}\right) \times \left(6.02 \times 10^{23} \frac{\text{molecules H}_2}{\text{mole H}_2}\right) = 1.8 \times 10^{25} \text{ molecules H}_2 \quad (7)$$

$$30 \text{ moles H}_2 \times \left(6.02 \times 10^{23} \frac{\text{molecules H}_2}{\text{mole H}_2}\right) = 1.8 \times 10^{25} \text{ molecules H}_2$$

These illustrations indicate clearly that in converting weight of a substance to *moles* of that substance, we divide the weight by the weight of 1 mole.

$$\text{number moles} = \frac{\text{weight substance (g)}}{\text{weight 1 mole (g/mole)}} \qquad (8)$$

To find the number of molecules, we multiply the number of moles by Avogadro's number:

$$\text{number molecules} = \text{number moles} \times \left(6.02 \times 10^{23} \frac{\text{molecules}}{\text{mole}}\right) \qquad (9)$$

EXERCISE 3-2

How many moles of CO_2 are in a tank containing 24 g of CO_2? (Obtain the molecular weight of CO_2 from the atomic weights of its components.)

EXERCISE 3-3

(1) Calculate the molecular weight of $C_2F_2H_4$. 66 g/mole
(2) How many *moles* of $C_2F_2H_4$ would be present in 132 g of $C_2F_2H_4$? 2 moles
(3) How many *molecules* of $C_2F_2H_4$ would be present in 132 g of $C_2F_2H_4$? 12.04×10^{23}
(4) How many *fluorine atoms* would be present in 132 g of $C_2F_2H_4$? 24.08×10^{23}

We can also convert volume of a gas to number of moles of a gas if we divide through by volume of 1 mole (for example, 22.4 liters at 0°C and 1 atm).

Example (4) How many moles of chlorine would be contained in 89.6 liters of Cl_2 measured at 0°C and 1 atm?

The answer is

$$\text{moles Cl}_2 = \frac{89.6 \text{ liters Cl}_2}{22.4 \frac{\text{liters Cl}_2}{\text{mole Cl}_2}} = 4.00 \text{ moles Cl}_2 \qquad (10)$$

EXERCISE 3-4

(1) How many *moles* of nitrogen are contained in 33.6 liters of N_2 measured at 0°C and 1 atm?

(2) How many molecules of nitrogen are there in this sample of gas?

3-2 THE CONSERVATION OF MASS

We noted briefly in Chapter 2 that if a chemical equation is properly written, it must ultimately show that *the total number of atoms of each kind does not change as a result of the reaction.* This idea was one of the basic postulates of John Dalton when he proposed the atomic theory. He wrote: "We might as well attempt to introduce a new planet into the solar system, or to annihilate one already in existence, as to create or destroy a particle of hydrogen." We observed this principle when we wrote equations based on combining volume relationships. For example, when hydrogen burns, our observations can be summarized by writing

$$2 \begin{Bmatrix} \text{volume} \\ \text{hydrogen} \\ \text{gas} \end{Bmatrix} + 1 \begin{Bmatrix} \text{volume} \\ \text{oxygen} \\ \text{gas} \end{Bmatrix} \longrightarrow 2 \begin{Bmatrix} \text{volume} \\ \text{water} \\ \text{vapor} \end{Bmatrix} + \begin{Bmatrix} \text{energy} \\ \text{as heat} \end{Bmatrix} \quad (11)$$

Remembering that an *equation is simply a representation in an abbreviated form of laboratory observations,* we can now reduce this description to the following equation:

$$2 \text{ H}_2 + \text{O}_2 \longrightarrow 2 \text{ H}_2\text{O} + \text{energy as heat} \quad (12)$$

There are four hydrogen atoms on the left and four hydrogen atoms on the right. Similarly, there are two oxygen atoms on both the left and the right. Clearly, atoms are neither gained nor lost in the process. In chemical reactions atoms are *conserved.* Belief in the conservation of atoms is based upon a generalization tested by many decades of experimentation. *Matter can be neither created nor destroyed.* Since we often measure a quantity of matter in terms of its mass, we say that *mass is conserved.* The conservation of mass is fundamental to our use of arithmetic in chemical equations. A chemical equation is **balanced** when it shows conservation of numbers of atoms and, consequently, conservation of mass.

3-2.1 Writing Balanced Equations for Reactions

How can we write a balanced equation for a reaction? We must:

(1) Know what reactants are consumed and what products are formed. (This information is obtained in the laboratory.)
(2) Know the correct formula of each reactant and each product. (This information is calculated from laboratory observations and will be discussed in detail in a later chapter.)
(3) Satisfy the Law of Conservation of Atoms.

The foregoing steps were applied one at a time when we wrote the balanced equation for the burning of hydrogen gas in oxygen to give water vapor or the burning of hydrogen gas in chlorine to give hydrogen chloride. These steps can be applied equally well to reactions involving both solids and gases. For example, metallic magnesium burns in air to give magnesium oxide, heat, and a dazzling white light. Careful experiments have shown that gaseous oxygen from the air is used in this burning operation. Further experiments show that the product formed (magnesium oxide) has the formula MgO. Let us now carry out steps (1) and (2) by writing down the correct formulas for all reactants and products. The information comes from laboratory experiments.

$$Mg + O_2 \longrightarrow MgO + \text{energy as heat and light} \qquad (13)$$

The above expression does not yet conserve atoms. To conserve atoms we must place appropriate numbers before each formula so that the same numbers of atoms of each element appear on both the left- and right-hand sides of the equation. *We must not change the formulas in this process since formulas came from separate laboratory studies.*

Let us start the balancing operation by arbitrarily considering one molecule of oxygen. Since one molecule of oxygen contains *two atoms* of oxygen, it will form *two* molecules of MgO.* (Each molecule of MgO contains one atom of oxygen, as the absence of a subscript tells us.) We then write

$$? \, Mg + 1 \, O_2 \longrightarrow 2 \, MgO + \text{energy} \qquad (14)$$

The question mark reminds us that the number of atoms of Mg has not been decided yet. Clearly, two Mg atoms are now needed on the left-hand side of the equation; hence we write

$$2 \, Mg + 1 \, O_2 \longrightarrow 2 \, MgO + \text{energy} \qquad (15)$$

The equation (*15*) is now balanced: we have the same number of magnesium and oxygen atoms on both the left- and right-hand sides of the equation.

3-2.2 Balanced Equations and the Mole

Suppose we were to start with *two* oxygen molecules instead of one. We would then have four oxygen atoms, which would produce four units of MgO. Clearly, four Mg atoms would be required. We could then write

$$(2 \times 2 \, Mg) + (2 \times 1 \, O_2) = 2 \times 2 \, MgO + 2 \times \text{energy} \qquad (16)$$

The original numbers in front of each substance on both left- and right-hand sides of the balanced equation (*15*) can be multiplied by

---

* As you will see later, solid MgO does not contain distinct molecules of MgO, but the idea of an MgO molecule to define the mole is convenient and acceptable.

2 *or any other number,* just as long as every coefficient is multiplied by the same number. We can write for the general case

$$(n \times 2 \text{ Mg}) + (n \times 1 \text{ O}_2) \longrightarrow (n \times 2 \text{ MgO}) + (n \times \text{energy}) \qquad (17)$$

===== EXERCISE 3-5 =====

Show that atoms are conserved when $n = 8$ and when $n = 44$ in equation (17).

===== EXERCISE 3-6 =====

Suppose ten hydrogen molecules and ten oxygen molecules are mixed. How many molecules of water could be formed? What would be left over?

===== EXERCISE 3-7 =====

One million oxygen molecules react with sufficient hydrogen molecules to form water molecules. How many water molecules are formed? How many hydrogen molecules are consumed?

If n were 6.02×10^{23}, we could write

$$(6.02 \times 10^{23} \times 2 \text{ Mg}) + (6.02 \times 10^{23} \times 1 \text{ O}_2) \longrightarrow$$
$$(6.02 \times 10^{23} \times 2 \text{ MgO}) + (6.02 \times 10^{23} \text{ energy}) \qquad (18)$$

or

$$2 \text{ moles magnesium atoms} + 1 \text{ mole oxygen gas} \longrightarrow$$
$$2 \text{ moles magnesium oxide} + \text{much more energy} \qquad (18a)$$

Numbers of atoms are conserved as before; the equation is balanced. *The balanced equation can be interpreted in terms of moles as well as individual atoms and molecules.* We can always multiply the coefficients by a common factor or divide by a common factor and obtain equally valid equations. In all equations the coefficient 1 may be dropped, but it is never wrong to retain it.

3-2.3 Balanced Equations and Mass Balance

As in the burning of hydrogen, our balanced equation shows the conservation of mass. Two moles of magnesium metal weigh 48 g. (See the table on page 44.) One mole of oxygen weighs 32 g. These combine to give $48 + 32 = 80$ g of MgO. Let us summarize the information given to us by a balanced equation.

$$2 \text{ Mg} + \text{O}_2 \longrightarrow 2 \text{ MgO} + \text{energy} \qquad (19)$$

$$2 \begin{Bmatrix} \text{magnesium} \\ \text{atoms} \end{Bmatrix} + 1 \begin{Bmatrix} \text{oxygen} \\ \text{molecule} \end{Bmatrix} \longrightarrow 2 \begin{Bmatrix} \text{magnesium} \\ \text{oxide} \\ \text{molecules} \end{Bmatrix} + \{\text{energy}\} \qquad (19a)$$

$$2 \begin{Bmatrix} \text{moles} \\ \text{magnesium} \\ \text{atoms} \end{Bmatrix} + 1 \begin{Bmatrix} \text{mole} \\ \text{oxygen} \\ \text{molecules} \end{Bmatrix} \longrightarrow 2 \begin{Bmatrix} \text{moles} \\ \text{magnesium} \\ \text{oxide} \\ \text{molecules} \end{Bmatrix} + \{\text{energy}\} \qquad (19b)$$

$$48 \begin{Bmatrix} \text{grams} \\ \text{magnesium} \\ \text{metal} \end{Bmatrix} + 32 \begin{Bmatrix} \text{grams} \\ \text{oxygen} \\ \text{gas} \end{Bmatrix} \longrightarrow 80 \begin{Bmatrix} \text{grams} \\ \text{magnesium} \\ \text{oxide solid} \end{Bmatrix} + \{\text{energy}\} \quad (19c)$$

If equation (*19*) can imply moles [equation (*19b*) above], we can just as well begin by choosing 1 mole of magnesium metal (24 g). This will form 1 mole of magnesium oxide. One mole of magnesium oxide contains 1 mole of oxygen atoms. This is the number of oxygen atoms contained in ½ mole of oxygen molecules. Thus, we can write

$$1 \text{ Mg} + \tfrac{1}{2} \text{ O}_2 \longrightarrow 1 \text{ MgO} + \text{energy} \quad (20)$$

This is also a balanced equation. It is just as correct as the original balanced equation involving whole numbers and can be converted to the original if we multiply both sides of the equation by 2.

3-2.4 Other Examples

Let us examine a somewhat more complex equation associated with the burning of natural gas. Natural gas is largely methane, which has the formula CH_4. When methane burns, it produces carbon dioxide (CO_2) and water (H_2O). Oxygen from the air is also used. The equation summarizing these experimental facts is

$$? \text{ CH}_4 + ? \text{ O}_2 \longrightarrow ? \text{ CO}_2 + ? \text{ H}_2\text{O} + \text{energy} \quad (21)$$

Let us arbitrarily select 1 mole of methane as our starting quantity. Since 1 mole of methane contains 1 mole of carbon atoms, *1 mole* of CO_2 will be generated. We now can write

$$1 \text{ CH}_4 + ? \text{ O}_2 \longrightarrow 1 \text{ CO}_2 + ? \text{ H}_2\text{O} + \text{energy} \quad (21a)$$

The question marks remind us that we have not yet determined the numbers for oxygen and water. Since 1 mole of methane (CH_4) contains 4 moles of hydrogen atoms, 2 moles of water (H_2O) will be generated:

$$1 \text{ CH}_4 + ? \text{ O}_2 \longrightarrow 1 \text{ CO}_2 + 2 \text{ H}_2\text{O} + \text{energy} \quad (21b)$$

The products now contain 2 moles of oxygen atoms in 1 mole of CO_2 (1×2) and 2 moles of oxygen atoms in 2 moles of H_2O (2×1). A total of 4 moles of oxygen atoms—or 2 moles of O_2—is required:

$$1 \text{ CH}_4 + 2 \text{ O}_2 \longrightarrow 1 \text{ CO}_2 + 2 \text{ H}_2\text{O} + \text{energy} \quad (22)$$

Check the atoms on both sides of the equation to be sure that everything balances.

EXERCISE 3-8

Balance the equation for the combustion of CH_4 starting with 1 mole of O_2 rather than 1 mole of CH_4; then show that the equation is balanced.

EXERCISE 3-9

Ammonia gas, NH_3, can be burned with oxygen gas, O_2, to give nitrogen gas, N_2, and water, H_2O. See if you can follow the logic of the following steps in balancing this reaction.

$$NH_3 + O_2 \longrightarrow N_2 + H_2O$$
$$NH_3 + O_2 \longrightarrow 1\ N_2 + H_2O$$
$$2\ NH_3 + O_2 \longrightarrow 1\ N_2 + H_2O$$
$$2\ NH_3 + O_2 \longrightarrow 1\ N_2 + 3\ H_2O$$
$$2\ NH_3 + \tfrac{3}{2}\ O_2 \longrightarrow 1\ N_2 + 3\ H_2O$$

State briefly what was done in each step.

The same logic is used in writing a balanced equation for the burning of candle wax or paraffin. Paraffin (candle wax) is made up of molecules of several sizes. We shall use the molecular formula $C_{25}H_{52}$ as representative of its molecules. Since the numbers are larger, the system appears to be more complex, but the technique is just the same. One mole of candle wax contains the Avogadro number (6.02×10^{23}) of these molecules ($C_{25}H_{52}$). The starting expression is then

$$C_{25}H_{52} + O_2 \longrightarrow CO_2 + H_2O \tag{23}$$

Let us start with 1 mole of paraffin since this mole contains the largest number of atoms in the above equation. Since there are 25 moles of carbon atoms in 1 mole of paraffin, we must form 25 moles of CO_2.

$$1\ C_{25}H_{52} + ?\ O_2 \longrightarrow 25\ CO_2 + ?\ H_2O \tag{24}$$

Since there are 52 moles of hydrogen atoms in 1 mole of paraffin, we must form 52/2 or 26 moles of H_2O.

$$1\ C_{25}H_{52} + ?\ O_2 \longrightarrow 25\ CO_2 + 26\ H_2O \tag{25}$$

Again a question mark is used for numbers which are not yet established. Note that we need 25 moles of oxygen molecules to form the CO_2 and 26/2 or 13 moles of oxygen molecules to form the 26 moles of H_2O. Then, $25 + 13 = 38$ moles of oxygen. We can now write the balanced equation

$$1\ C_{25}H_{52} + 38\ O_2 \longrightarrow 25\ CO_2 + 26\ H_2O \tag{26}$$

EXERCISE 3-10

Check the number of atoms on each side of the equation for paraffin combustion to show that it is balanced.

EXERCISE 3-11

Balance the equation for the combustion of sugar ($C_{12}H_{22}O_{11}$) in O_2 to give CO_2 and H_2O.

$1\ C_{12}H_{22}O_{11} + 12\ O_2 \longrightarrow 12\ CO_2 + 11\ H_2O$

3-3 CALCULATIONS BASED UPON CHEMICAL EQUATIONS

In Section 3-1 we learned to convert grams of a pure substance to moles of substance and liters of gas to moles of gas. In Section 3-2 we learned to balance chemical equations. These two skills can now be combined to help us determine the quantities of reactants and products involved in a chemical process. Let us see how this works. We shall start with a really simple example to illustrate a general procedure which is applicable to many chemical problems.

Example (1) What weight of magnesium oxide would be produced by burning 3.00 moles of magnesium atoms?

The balanced equation is

$$2 \, Mg + O_2 \longrightarrow 2 \, MgO + energy \qquad (27)$$

The equation tells us that 2 moles of Mg atoms produce 2 moles of MgO. If we divide both sides by 2, then the new equation shows that 1 mole of magnesium atoms will produce 1 mole of MgO.

$$\tfrac{2}{2} \, Mg + \tfrac{1}{2} \, O_2 \longrightarrow \tfrac{2}{2} \, MgO \qquad (28)$$

or

$$Mg + \tfrac{1}{2} \, O_2 \longrightarrow MgO$$

Then 3.00 moles of magnesium atoms would give 3.00 moles of MgO. Our unit representation is

$$\underbrace{\frac{1.00 \text{ mole MgO}}{\text{mole Mg atoms}}}_{\text{FROM EQUATION}} \times \underbrace{3.00 \text{ moles Mg atoms}}_{\text{FROM PROBLEM}} = 3.00 \text{ moles MgO} \qquad (29)$$

$$3.00 \text{ moles MgO} \times 40.3 \, \frac{\text{g MgO}}{\text{mole MgO}} = 121 \text{ g MgO}$$

Example (2) Ammonia gas can be burned under appropriate conditions to give water and nitric oxide. The balanced equation is

$$4 \, NH_3 + 5 \, O_2 \longrightarrow 4 \, NO + 6 \, H_2O \qquad (30)$$

(a) How many moles of oxygen are required to burn 68 g of ammonia?

First we might ask: How many moles of ammonia are found in 68 g of NH_3? Using the information from Section 3-1, we can write

$$\frac{68 \text{ g NH}_3}{17 \, \frac{\text{g NH}_3}{\text{mole NH}_3}} = 68 \text{ g NH}_3 \times \frac{1 \text{ mole NH}_3}{17 \text{ g NH}_3}$$

$$= \frac{68}{17} \text{ moles NH}_3 = 4.0 \text{ moles NH}_3 \qquad (31)$$

The balanced equation now tells us that 5.0 moles O_2 will be needed to burn 68 g of ammonia.

(b) How many grams of nitric oxide will be produced?

The balanced equation tells us that 4.0 moles of NH_3 will yield 4.0 moles of NO. If 1 mole of NO weighs 30 g (14 + 16 = 30), then 4.0 moles weigh

$$4.0 \text{ moles NO} \times 30 \frac{\text{g NO}}{\text{mole NO}} = 120 \text{ g NO} \qquad (32)$$

(c) What weight of water will be produced if we burn 10 g of NH_3?

The first step involves conversion of grams of ammonia to moles of NH_3 using the methods of Section 3–1. We write

$$\frac{10 \text{ g } NH_3}{17 \frac{\text{g } NH_3}{\text{mole } NH_3}} = 10 \text{ g } NH_3 \times \frac{1 \text{ mole } NH_3}{17 \text{ g } NH_3} = 0.59 \text{ mole } NH_3 \qquad (33)$$

The second step involves determining the number of moles of H_2O produced by 0.59 mole of NH_3. The balanced equation is

$$4 NH_3 + 5 O_2 \longrightarrow 4 NO + 6 H_2O \qquad (34)$$

If we determine how many moles of water are produced from 1 mole of NH_3, we can then use simple multiplication to determine how many moles of water are produced from any number of moles of NH_3. To determine the products from 1 mole of NH_3, let us divide both sides of the equation by 4:

$$\tfrac{4}{4} NH_3 + \tfrac{5}{4} O_2 \longrightarrow \tfrac{4}{4} NO + \tfrac{6}{4} H_2O \qquad (35)$$

Then we write

$$NH_3 + \tfrac{5}{4} O_2 \longrightarrow NO + \tfrac{3}{2} H_2O \qquad (36)$$

Since $\tfrac{3}{2}$ moles of H_2O are produced per mole of NH_3, we can write

$$0.59 \text{ mole } NH_3 \times \tfrac{3}{2} \frac{\text{moles } H_2O}{\text{mole } NH_3} = 0.89 \text{ mole } H_2O \qquad (37)$$

The third and final step involves converting moles of water to grams of water. Simple multiplication is required:

$$0.89 \text{ mole } H_2O \times \frac{18 \text{ g } H_2O}{\text{mole } H_2O} = 16 \text{ g } H_2O \qquad (38)$$

In actual practice it is not necessary to work out the arithmetic of each step before proceeding to the next step. The entire problem can be formulated as follows:

$$10 \text{ g } NH_3 \times \frac{1 \text{ mole } NH_3}{17 \text{ g } NH_3} \times \tfrac{3}{2} \frac{\text{moles } H_2O}{\text{mole } NH_3} \times \frac{18 \text{ g } H_2O}{\text{mole } H_2O} = 16 \text{ g } H_2O \qquad (39)$$

Note that units reduce until only grams of H_2O remain.

61

CHEMICAL REACTIONS:
The CONSERVATION of ENERGY

EXERCISE 3-12

Show that 3.80 moles of oxygen are needed to burn 35.3 g of paraffin by the reaction considered in Exercise 3-10.

$.1\, C_{25}H_{52} + 3.8\, O_2 \rightarrow 2.5\, CO_2 + 2.6\, H_2O$

EXERCISE 3-13

How many moles of oxygen (O_2) are required to produce 242 g of magnesium oxide (MgO) by the equation

$$2\, Mg + 1\, O_2 \longrightarrow 2\, MgO$$

$6\, Mg + 3\, O_2 \rightarrow 6\, MgO$

EXERCISE 3-14

(1) Write the equation for the reaction which took place in Experiment 8, Part III. What residue did you obtain on evaporating the solution in beaker 2?

$AgNO_3 + NaCl \rightarrow AgCl + NaNO_3$

(2) In Experiment 8 you determined the number of silver chloride moles formed in the reaction of some sodium chloride with a known amount of silver nitrate. How many moles of sodium chloride reacted with the silver nitrate? Compare this with the number of moles of sodium chloride you added.

3-4 ENERGY CHANGE IN CHEMICAL REACTIONS: THE CONSERVATION OF ENERGY

Up to the present time, we have been careful to use numbers to identify the number of moles of each reactant and the number of moles of each product involved in a given chemical equation, but we have not tried to be quantitative about the energy changes associated with any specific chemical process. On the other hand, laboratory work has clearly indicated that the burning of a given weight of candle gives a definite and identifiable quantity of heat energy. The **calorie** is the unit with which we measure heat energy. It is the quantity of heat energy required to raise the temperature of 1 g of water 1°C.

If we burn twice as much candle, twice as much heat will be generated. Our equation should show this fact. Similar relationships are established in burning magnesium to give magnesium oxide and in burning hydrogen to give water. If we allow 1 mole of magnesium atoms to react with ½ mole of gaseous oxygen molecules, forming 1 mole of magnesium oxide, 146,000 calories of energy (heat) will be evolved.* This fact can be expressed by adding a quantitative energy term to our balanced equation:

$$Mg(solid) + \tfrac{1}{2}\, O_2(gas) \longrightarrow MgO(solid) + 146{,}000\ cal \qquad (40)$$

The quantity of energy liberated when 1 mole of magnesium is burned to give 1 mole of magnesium oxide is called the **molar heat of combustion** of magnesium.

---

* It is assumed in this measurement that reactants were originally at 25°C and products were cooled to 25°C before final measurement of the heat evolved.

Obviously, if 2 moles of magnesium are burned to give 2 moles of magnesium oxide, 1 mole of gaseous molecular oxygen will be needed and $2 \times 146{,}000$ cal of energy will be liberated. Similarly, if only 0.025 mole of magnesium is burned to give 0.025 mole of magnesium oxide, $0.025 \times 146{,}000$ cal of heat or about 3,700 cal will be released.

A comparable set of relationships can be used to describe the burning of hydrogen gas to give water. The **molar heat of combustion** of gaseous H_2 to give liquid H_2O at 25°C is 68,000 cal:

$$H_2 + \tfrac{1}{2} O_2 \longrightarrow H_2O + 68{,}000 \text{ cal} \qquad (41)$$

$$2\, H_2 + O_2 \longrightarrow 2\, H_2O + 136{,}000 \text{ cal} \qquad (41a)$$

$$0.025\, H_2 + \tfrac{0.025}{2} O_2 \longrightarrow 0.025\, H_2O + 1{,}700 \text{ cal} \qquad (41b)$$

Energy can be treated as a regular product of these chemical reactions. A reaction in which energy appears as a product is called an **exothermic reaction** (heat is released).

EXERCISE 3-15

(1) How much heat is released when 2 moles of hydrogen burn to form water? when ½ mole burns to form water? 136,000 cal. 34,000 cal.
(2) How much heat energy is released when 36 g of magnesium metal is burned in air to give solid MgO? 131,400 cal.

3-4.1 The Source of Energy in Chemical Processes

Where does the energy in the burning of hydrogen come from? This is an appropriate "wondering why" question. Let us look for the answer by examining a bond representation for the burning of hydrogen in oxygen to give water:

$$H{-}H + H{-}H + O{-}O \longrightarrow H{-}O{\overset{H}{}} + H{-}O{\overset{H}{}} \qquad (42)$$

In these representations it is clear that two hydrogen molecules are involved. For hydrogen molecules to react with oxygen to form water molecules, the bonds between hydrogen atoms in each molecule must be broken at some point in the process. This requires energy. Similarly, the bond between the two oxygen atoms in the oxygen molecule must be broken at some point in the process. This also uses up energy.

The steps considered so far have been absorbing energy, not *releasing* it! Where then does the energy which is released come from? The only possible source of energy left to be considered is the formation of bonds between two hydrogen atoms and an oxygen atom to give a water molecule. Clearly, large amounts of energy are released when the H—O bonds are formed. The energy released in this process supplies all of the energy needed to (1) break the H—H bonds in hydrogen molecules, (2) break the O—O bonds in oxygen molecules, and (3) release all of the heat observed as a product of the reaction. This breaking of chemical bonds and formation of new bonds is the *defining characteristic* of a chemical process.

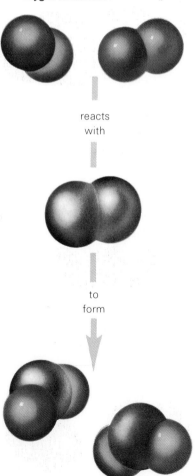

Fig. 3-1 Formation of water molecules from hydrogen molecules and oxygen molecules.

reacts with

to form

Because breaking and forming bonds generally involves relatively large amounts of energy, chemical processes are usually accompanied by sizable energy changes. Let us review our laboratory observations to see how this works.

When we melted wax, the solid wax could be recovered merely by cooling. No chemical bonds were broken. On the other hand, when we burned the candle, the products of combustion could not be converted back into wax or a shiny new candle by cooling! Instead, carbon dioxide (CO_2) and liquid water (H_2O) were identified as products: *in the burning of the wax, new products were created—chemical bonds were broken, and new bonds were formed.* It is not surprising then that the energy released in the combustion of wax was well over two hundred times greater than the energy released when the same amount of wax froze to a solid. Sizable amounts of energy are usually associated with processes involving the breaking of bonds and formation of bonds.

3-4.2 The Conservation of Energy

The balancing of chemical equations is based on the **Law of Conservation of Mass.** Mass can be neither created nor destroyed. Energy has also been handled in our chemical equations and treated like a product of the reaction. Is another conservation law implied? Perhaps: *Energy* can neither be created nor destroyed. A very large collection of experimental evidence indicates that this statement is true for all chemical processes. Like the Law of Conservation of Mass, the **Law of Conservation of Energy** provides one of the cornerstones of modern science.

3-4.3 The Decomposition of Water

We have just observed that when 1 mole of gaseous H_2 and ½ mole of gaseous O_2 combine to give 1 mole of liquid water, 68,000 cal of energy are released. According to the Law of Conservation of Energy, 68,000 cal must be added to convert 1 mole of water back to 1 mole of H_2 and ½ mole of O_2.

Let us examine the experiments to test this prediction. If we place water with a few drops of sulfuric acid in the electrolysis apparatus shown in Figure 3-3 on the next page, and pass a direct electric current through the solution, water will be decomposed. In the apparatus shown, two electrical conductors (called electrodes) are immersed in the liquid. When the electrodes are connected to a source of electrical energy, hydrogen gas appears at one electrode and oxygen gas appears at the other. If we operate the apparatus until 1 mole of water has decomposed (18 g of H_2O), 1 mole of hydrogen gas and ½ mole of oxygen gas are produced. We observe also that electrical energy equivalent to 68,000 cal of heat has been used. The equation for this process is

$$68{,}000 \text{ cal} + H_2O \longrightarrow \tfrac{1}{2} O_2 + H_2 \qquad (43)$$

The equation is just the reverse of the equation written for the combustion of 1 mole of hydrogen gas. Such a process in which energy is a reactant is called an **endothermic reaction.**

The DECOMPOSITION of WATER

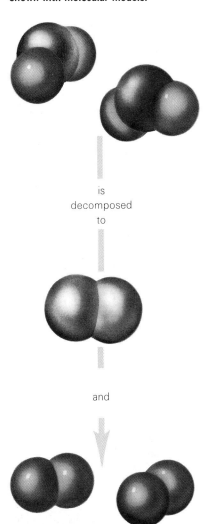

Fig. 3-2 **The decomposition of water shown with molecular models.**

is decomposed to

and

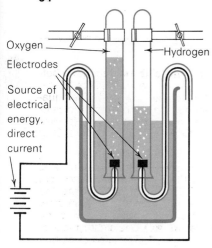

Fig. 3-3 **The apparatus for the electrolytic decomposition of water.**

3-5 HIGHLIGHTS

Laboratory measurements of substances are frequently made in units of grams or liters of gas. Grams are converted to moles through division by the weight of 1 mole. Liters of gas are converted to moles through division by the volume of 1 mole in liters (22.4 liters at 0°C and 1 atm). A procedure for balancing equations was developed based on the **Law of Conservation of Mass:** Mass can be neither created nor destroyed. These two skills—the conversion of laboratory measurements to moles and the balancing of equations—have been used to determine quantities of products, or reactants, involved in chemical reactions. Another conservation law, the **Law of Conservation of Energy,** was suggested by the appearance of balanced equations involving energy terms. This law—Energy can be neither created nor destroyed—is supported by a large body of human experience. Processes in which energy appears as a product are called **exothermic reactions.** Processes in which energy appears as a reactant are called **endothermic reactions.**

QUESTIONS AND PROBLEMS

1. A chemist weighs out 10.0 g of water (H_2O), 10.0 g of ammonia (NH_3), and 10.0 g of hydrogen chloride (HCl). How many moles of each substance does he have? Show that your calculation gives the correct units in each case.

2. How many moles are contained in 49 g of pure H_2SO_4? Show that the procedure gives the correct set of units.

3. (a) How many moles of atoms are present in 9.0 g of aluminum (Al)? (b) How many individual atoms of Al are present in 9.0 g of aluminum?

4. If hydrogen gas (H_2) and fluorine gas (F_2) are allowed to mix, a violent explosion usually results. One liter of hydrogen combines with 1 liter of fluorine to give 2 liters of hydrogen fluoride (HF) at high temperatures. (a) Write the equation. (b) How many *molecules* of HF will form from one molecule of H_2? (c) How many *moles* of HF will be formed from 1 mole of H_2? (d) How many *moles* of F_2 are required to produce 8 moles of HF? (e) How many *molecules* of F_2 are required to produce 8 *moles* of HF?

5. Nitric oxide (NO) combines with oxygen of the air (O_2) to give the brown gas nitrogen dioxide (NO_2), and no other products. (a) Write an equation for the process described. (b) Balance the equation. (c) How many moles of oxygen atoms are there in 2 moles of NO? (d) How many moles of oxygen atoms are there in 1 mole of O_2? (e) How many moles of oxygen atoms are there in 2 moles of NO_2? (f) Use the answers to parts (c), (d), and (e) to verify that the reaction is written so as to conserve oxygen atoms. (g) Show that nitrogen atoms are conserved in the balanced equation.

6. The equation for the preparation of ammonia (NH_3) is

$$N_2 + H_2 \longrightarrow NH_3$$

(a) Balance this equation. (b) Show that both nitrogen atoms and hydrogen atoms are conserved in the balanced equation. (c) How many moles of NH_3 will be produced from 1 mole of N_2? (d) What *weight* of NH_3 will be produced from 1 mole of N_2?

7. In the manufacture of nitric acid (HNO_3), nitrogen dioxide reacts with water to form HNO_3 and nitric oxide (NO):

$$3 NO_2 + H_2O \longrightarrow 2 HNO_3 + NO$$

(a) Verify that the equation conserves oxygen atoms. (b) How many molecules of nitrogen dioxide are required to form 25 molecules of nitric oxide? (c) How many moles of nitric oxide will be formed from 0.60 mole of nitrogen dioxide?

8. If 3 g of substance *A* combine with 4 g of substance *B* to make 5 g of substance *C* and some *D*, how many grams of *D* will be produced?

9. One step in the manufacture of sulfuric acid is to burn sulfur (S_8) in air to form a colorless gas with a choking odor. The name of the gas is sulfur dioxide and it has the molecular formula SO_2. On the basis of this information: (a) Write the balanced equation for this reaction. (b) Interpret the equation in terms of molecules. (c) Interpret the equation in terms of moles. (d) Two moles of sulfur (S_8) will produce how many moles of sulfur dioxide (SO_2)?

10 Balance the equations for each of the following reactions. Begin on the basis of 1 mole of the substance underscored.

(a) $\underline{Li} + Cl_2 \longrightarrow LiCl$
(b) $Na + \underline{Cl_2} \longrightarrow NaCl$
(c) $Na + F_2 \longrightarrow \underline{NaF}$
(d) $\underline{Na} + Br_2 \longrightarrow NaBr$
(e) $O_2 + \underline{Cl_2} \longrightarrow Cl_2O$

11 Balance the equations for each of the following reactions involving oxygen. Begin on the basis of 1 mole of the substance underscored.

(a) With metallic nickel: $\underline{Ni} + O_2 \longrightarrow NiO$
(b) With metallic lithium: $Li + \underline{O_2} \longrightarrow Li_2O$
(c) With the rocket fuel hydrazine (N_2H_4): $\underline{N_2H_4} + O_2 \longrightarrow N_2 + H_2O$
(d) With acetylene (C_2H_2) in an acetylene torch flame: $\underline{C_2H_2} + O_2 \longrightarrow CO_2 + H_2O$ (Answer: $C_2H_2 + \frac{5}{2} O_2 \longrightarrow 2 CO_2 + H_2O$.)
(e) With the important copper ore chalcocite (Cu_2S) (the process called "roasting" the ore): $\underline{Cu_2S} + O_2 \longrightarrow Cu_2O + SO_2$
(f) With the important iron ore iron pyrites (FeS_2) (again, "roasting" the ore): $\underline{FeS_2} + O_2 \longrightarrow Fe_2O_3 + SO_2$

12 (a) Balance the equations for the decomposition (to elements) of ammonia (NH_3), nitrogen trifluoride (NF_3), and nitrogen trichloride (NCl_3). Base each equation upon the production of 1 mole of N_2.

$NH_3 \longrightarrow 1 N_2 + H_2$
$NF_3 \longrightarrow 1 N_2 + F_2$
$NCl_3 \longrightarrow 1 N_2 + Cl_2$

(b) Rewrite the equations to include the information that the decomposition of ammonia is *endothermic*, absorbing 22.1 kilocalories/mole of N_2 formed; that the decomposition of NF_3 is *endothermic*, absorbing 54.4 kcal/mole of N_2; and that the decomposition of NCl_3 is *exothermic*, releasing 109.4 kcal/mole of N_2.
(c) One of the three compounds NH_3, NF_3, and NCl_3 is dangerously explosive. Which would you expect to be the explosive substance? Why?

13 Balance the following equations.

(a) When KNO_3 is heated to high temperatures, oxygen is given off. (Start with $O_2 = 1$.)
___ $KNO_3 \longrightarrow$ ___ $KNO_2 +$ ___ O_2
(b) Black iron oxide can be converted to metallic iron and water if it is heated in a hydrogen stream. Balance starting with underlined substance equal to 1.
___ $\underline{Fe_3O_4} +$ ___ $H_2 \longrightarrow$ ___ $Fe +$ ___ H_2O
(c) Metallic zinc gives off H_2 and is converted to $ZnCl_2$ if it is dropped into a solution of HCl in water.
___ $\underline{Zn} +$ ___ $HCl \longrightarrow$ ___ $H_2 +$ ___ $ZnCl_2$
(d) ___ $P_2O_5 +$ _3_ $H_2O \longrightarrow$ _2_ H_3PO_4
(e) _2_ $Al +$ _3_ $H_2SO_4 \longrightarrow$ _3_ $H_2 +$ _1_ $Al_2(SO_4)_3$

QUESTIONS and PROBLEMS

14 If a piece of sodium metal is lowered into a bottle of chlorine gas, a reaction takes place. Table salt (NaCl) is formed. (a) Write the equation for the reaction. (b) How many moles of NaCl will be formed from 1 mole of Na? (c) How many moles of NaCl will be formed from 2.30 g of Na?

15 Methane, the main constituent of natural gas, has the formula CH_4. Its combustion products are carbon dioxide and water. (a) Write the equation for the combustion of methane. (b) One mole of methane produces how many moles of water vapor? (c) One-eighth mole of methane produces how many moles of carbon dioxide? (d) How many moles of water vapor are produced by 4.0 g of methane?

16 If potassium chlorate ($KClO_3$) is heated gently, the crystals will melt. Further heating will decompose it to give oxygen gas and potassium chloride (KCl). (a) Write the equation for the decomposition. (b) How many moles of $KClO_3$ are needed to give 1.5 moles of oxygen gas? (c) How many moles of KCl will be given by 0.33 mole of $KClO_3$? (d) How many moles of oxygen gas will be produced by 122.6 g of $KClO_3$?

17 One gallon of gasoline can be considered as about 25 moles of octane (C_8H_{18}). (a) How many moles of oxygen must be used to burn this gasoline, assuming that the only products are carbon dioxide and water? (b) How many moles of carbon dioxide are formed? (c) How much does this carbon dioxide weigh? (Express your answer in kilograms.) (d) What weight of carbon dioxide is released into the atmosphere when your automobile consumes 10 gallons of gasoline? Express this answer in pounds (1 kg = 2.2 lb).

18 Iron (Fe) burns in air to form a black, solid oxide (Fe_3O_4). (a) Write the equation for the reaction. (b) How many moles of oxygen gas are needed to burn 1 mole of iron? (c) How many grams of oxygen gas are needed? (d) Can a piece of iron weighing 5.6 g burn completely to Fe_3O_4 in a vessel containing 0.05 mole of O_2?

19 (a) According to the equation given in question 7, how many grams of nitric acid will be formed from 1 mole of nitrogen dioxide? (b) How many *more* grams of nitric acid could be formed if the nitric oxide formed could be completely converted into nitric acid? (Assume 1 mole of nitric oxide gives 1 mole of nitric acid.)

20 Hydrazine (N_2H_4) can be burned with oxygen to provide energy for rocket propulsion. The energy released is 150 kcal/mole of hydrazine burned. (a) How much energy will be released if 10.0 kg of hydrazine fuel are burned? (b) Compare the energy that will be released if the same weight of hydrogen (10.0 kg) is burned as a fuel instead. (See Section 3-4.)

I am never content until I have constructed a mechanical model of the object that I am studying. If I succeed in making one, I understand; otherwise, I do not.

LORD WILLIAM THOMSON KELVIN (1824–1907)

CHAPTER 4 More About Gases: The Kinetic Theory

Inflating a balloon is a simple demonstration of kinetic theory.

This chapter could well be titled "Much Ado About Almost Nothing," since gases are mostly empty space. It is the empty space which permits a gas to be easily compressed. It is the endless, rapid motion of the particles banging against each other and the walls of the container which causes a gas to fill space and exert pressure. Heating of gas particles causes them to move more rapidly and bang more vigorously against the walls of the container. Either the pressure or the volume, or both must get larger if the temperature is raised.

In order to describe an amount of gas, then, we must specify not only the volume it occupies, but the temperature and pressure at which it was measured. In this chapter we learn of these measurements and the interrelationships between them.

MOLAR VOLUMES and the
DISTANCE BETWEEN PARTICLES

The model that we suggested in Chapter 1 to explain the behavior of air in a balloon has grown rapidly. In Chapter 2 we made observations on the way in which the volume of a given quantity of gas changes as pressure on the gas is increased. Our model explained these observations on gases in both a qualitative and quantitative way. Next we assumed that Avogadro's hypothesis was true and we used this hypothesis, together with our model for gases and some chemical observations, to develop the concepts of molecules, atoms, molecular weights, and atomic weights. These concepts then provided a quantitative procedure for examining weights of chemically reacting materials, and gave us a substantial basis for further development of the atomic theory. The study of gases, therefore, is fundamental to our understanding of chemical concepts.

The gaseous state is the simplest form of matter; yet because most gases are difficult or impossible to see directly, their study requires a good imagination. A good scientific imagination will be useful in much of our later work, particularly in the study of atomic structure. We found in Chapter 2 that the regularities gases exhibited could be expressed as mathematical statements such as "pressure × volume = a constant." We shall examine these regularities in more detail in this chapter. We shall find that their detailed interpretation, called the **kinetic theory,** provides us with an understanding of temperature on the molecular level. The kinetic theory is a very powerful tool in science.

4-1 MOLAR VOLUMES AND THE DISTANCE BETWEEN PARTICLES IN THE GAS PHASE

With our model of a gas it was necessary to postulate the existence of particles (molecules) which have a rapid, random motion and which collide with each other and with the walls of the container. One of the interesting questions which we have not yet answered is: How far apart are these particles? Relative to the size of the molecules, are they close together or far apart?

The answer can be found by looking at the volume occupied by 1 mole (6.02×10^{23} molecules) of a substance when it exists as a solid, liquid, and gas. In Chapter 2 we calculated the volume of 1 mole of a *gas* at melting ice temperature (0°C) and at the pressure exerted by the air at sea level (1 atmosphere). This quantity was called the **molar volume** of the gas. We found that all gases have a molar volume which is fairly close to the value 22.4 liters at 0°C and 1 atm pressure.

One mole of a pure material in the solid state has a definite and reproducible volume, too, if we specify the temperature and pressure. In contrast to gases, however, volumes for different solids are *not* the same, even at the same temperature and pressure. Similarly, 1 mole of a pure material in the liquid state has a definite volume at a given temperature and pressure, but different liquids have different molar volumes. Let us examine molar volumes for a relatively simple material and see what this study tells us about the distance between molecules in the gas phase.

4-1.1 The Molar Volume of Nitrogen and Intermolecular Distances

In Chapter 2 we accumulated quite a lot of information about nitrogen. We know that gaseous nitrogen is composed of nitrogen molecules having the formula N_2; 1 mole of nitrogen weighs 28.0 grams (molecular weight = 2 × 14.0) and occupies a volume of 22.4 liters or 22,400 milliliters at 0°C and 1 atm pressure (the average pressure of the air at sea level).

Let us investigate nitrogen in its solid and liquid states. If a 28.0-g sample of gaseous nitrogen were cooled somewhat below −210°C, a white solid would form. If we now measure the volume occupied by this 28.0-g sample of white solid, a value very close to 27.2 milliliters is obtained.* Above −210°C the solid melts to give liquid nitrogen. The volume occupied by 28.0 g of this liquid at a temperature slightly above −210°C is 34.6 ml.

These significant numbers on nitrogen are easily summarized:

$$\text{molar volume gaseous } N_2 = 22{,}400 \text{ ml}$$
$$\text{molar volume liquid } N_2 = 34.6 \text{ ml}$$
$$\text{molar volume solid } N_2 = 27.2 \text{ ml}$$

Note that the volume occupied by 1 mole of gas is 650 times larger than the volume occupied by 1 mole of liquid nitrogen (22,400 ml/34.6 ml) and 825 times larger than the volume occupied by 1 mole of solid nitrogen (22,400 ml/27.2 ml). *If we assume that the size of a molecule itself does not change significantly in going from liquid to gas, then we must conclude that the molecules have become separated from each other in the gas phase.* The free space between gaseous molecules must be about 650 times larger than the space between the liquid molecules. Experiments with other materials lead to similar conclusions about the gaseous state. The actual volume of the molecules in a gas is trivial in comparison to the volume occupied by the gas itself.

Gases are compressed relatively easily because it is not difficult to force more molecules into the empty space between molecules. On the other hand, liquids and solids are compressed with great difficulty because there is very little space between molecules.

EXERCISE 4-1

How many molecules of nitrogen are present in 1 liter of gas at 0°C and 1 atm pressure? (*Answer:* 2.69×10^{22} molecules.)

EXERCISE 4-2

(1) Calculate the volume (in ml) occupied by one nitrogen molecule in the solid phase. (*Answer:* 4.52×10^{-23} ml/molecule.)

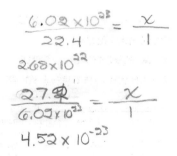

* The molar volume of the solid is obtained from its density by dividing 28.0 g/mole by the density 1.03 g/ml.

$$\text{molar volume solid } N_2 = \frac{28.0 \text{ g/mole}}{1.03 \text{ g/ml}} = 27.2 \text{ ml/mole}$$

(2) Recognizing that 1 ml is 1.00 cm³, estimate the length of the edge (in centimeters) of a cube that has the volume calculated in (1). Use one significant figure. Now express your answer in ångströms (1 Å = 10^{-8} cm). (*Answer:* 4 Å.)

EFFECT of a CHANGE in TEMPERATURE on the MOLAR VOLUME of a GAS

$$\frac{1 \text{ ml}}{1 \text{ cm}^3} = \frac{4.52 \times 10^{-23} \text{ ml}}{}$$

4-1.2 The Effect of a Change in Temperature on the Molar Volume of a Gas

We have seen in Section 4-1.1 that one of the very significant relationships of Chapter 2 is contained in the statement: *One mole* of gas occupies 22.4 liters at 0°C and 1 atm pressure. This statement was true for all of the gases which we have considered. If we did not want to hold the pressure at 1 atm, we found that the volume could be calculated from the simple relationship

$$P \times V = 22.4 \left(\frac{\text{liters} \times \text{atm}}{\text{mole}}\right) \quad (1)$$

$4 \text{ atm} \cdot V = 22.4 \frac{\ell \cdot \text{atm}}{\text{mole}}$

$V = 5.6 \, \ell/\text{mole}$

provided that the temperature was held constant at 0°C. For example, if the pressure was raised to 2 atm, the volume for 1 mole could be calculated from the relationship

$$2 \text{ atm} \times V = 22.4 \left(\frac{\text{liters} \times \text{atm}}{\text{mole}}\right) \quad (2)$$

or

$$V = 11.2 \text{ liters/mole at 2 atm and 0°C}$$

$4 \text{ atm} \cdot V = 44.8 \frac{\ell \cdot \text{atm}}{2 \text{ mole}}$

$V = 11.1 \, \ell/2 \text{ mole}$

EXERCISE 4-3

(1) What volume would 1 mole of gas occupy at 4 atm pressure? 5.6
(2) What volume would 2 moles of gas occupy at 4 atm pressure? 11.1

It is relatively easy to calculate a new gas volume after the pressure is changed *if we hold both the quantity of gas and the temperature constant;* but what happens to the gas volume as we change the temperature? An experiment provides the answer.

Table 4-1 shows the results of some pressure-volume measurements made on 1 mole of ammonia gas at 25°C (at approximately room temperature). Allowing for the expected amount of experimental uncertainty, it is clear that within the pressure range 0.2–2.0 atm, the pressure-volume product for 1 mole of ammonia at 25°C is 24.5 ± 0.7. This value is significantly larger than the value 22.4 ± 0.7 found at 0°C for ammonia in measurements of similar accuracy.

If we examine the precision of the measurements which gave us the values 22.4 and 24.5, it is apparent that even though the values are uncertain to ± 0.7, the difference between the two numbers is real and not a result of poor laboratory measurements or crude instrumentation. A more accurate study of the ammonia system at 25°C confirms our earlier conclusion. The data are in Table 4-2. We note that the average value is 24.40 ± 0.06. On the other hand, we see

TABLE 4-1 PRESSURE AND VOLUME OF 1 MOLE OF AMMONIA GAS (NH_3) AT 25°C

Pressure (atm)	Volume (liters)	P × V (atm × liters)
0.200	123	24.6
0.400	60.0	24.0
0.600	43.0	25.8
0.800	29.3	23.4
1.00	25.7	25.7
1.50	15.9	23.9
2.00	12.1	24.2
		Average 24.5 ± 0.7

TABLE 4-2 ACCURATE PRESSURE-VOLUME MEASUREMENTS OF 1 MOLE OF AMMONIA GAS (NH_3) AT 25°C

Pressure (atm)	Volume (liters)	P × V (atm × liters)
0.1000	244.5	24.45
0.2000	122.2	24.44
0.4000	61.02	24.41
0.8000	30.44	24.35
2.000	12.17	24.34
		Average 24.40

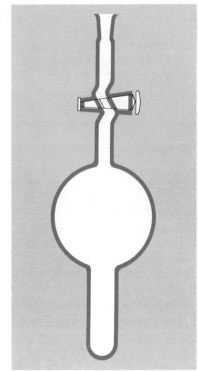

Fig. 4-1 A 1-liter gas bulb used to weigh and handle gases in the laboratory.

that the individual values are not randomly distributed around 24.40. Instead, there is a steady decline from 24.45 at 0.1 atm to 24.34 at 2.0 atm. This is significant and will be examined more carefully in Chapter 5.

Our observations on 1 mole of ammonia can now be summarized by two expressions:

$$P \times V = 22.4 \text{ liters} \times \text{atm} \quad at\ 0°C \qquad (3)$$
$$P \times V = 24.4 \text{ liters} \times \text{atm} \quad at\ 25°C \qquad (4)$$

This result does not surprise us since most of us know from earlier experience that a gas expands when heated at constant pressure.

We were pleased earlier to note that the expression

$$P \times V = 22.4 \text{ liters} \times \text{atm} \quad \text{at } 0°C \qquad (3)$$

applied to many gases, not just to O_2, NH_3, or N_2. We now wonder how general is the relationship

$$P \times V = 24.4 \text{ liters} \times \text{atm} \quad \text{at } 25°C \qquad (4)$$

Consider the following experiment. The air is pumped out of the liter flask shown in Figure 4-1. The flask is weighed empty (without air), then weighed again after it has been filled with a new gas at 1 atm pressure and 25°C. The difference in weight is the weight of 1 liter of the new gas that was added. From this information and the molecular weight of the gas, we can calculate the volume of 1 mole of that gas at 25°C and 1 atm. Table 4-3 shows the results. We find

TABLE 4-3 THE VOLUME OF 1 MOLE OF GAS AT 25°C AND 1 ATM PRESSURE

Gas	Weight of Flask Empty W_1 (g)	Weight of Flask + Gas W_2 (g)	Weight of 1 Liter of Gas $W_2 - W_1$ (g/liter)	Molar Weight MW (g/mole)	Volume $MW/(W_2 - W_1)$ (liter/mole)
Oxygen (O_2)	157.35	158.66	1.31	32.0	24.5
Nitrogen (N_2)	157.35	158.50	1.15	28.0	24.3
Carbon monoxide (CO)	157.35	158.50	1.15	28.0	24.4
Carbon dioxide (CO_2)	157.35	159.16	1.81	44.0	24.3

that all gases have about the same molar volume at 25°C and 1 atm. More precise measurements show that whether the gas is O_2, N_2, CO, or CO_2, the same volume—24.4 ± 0.2 liters—contains 6.02×10^{23} molecules at 25°C and 1 atm.

4-2 MEASUREMENTS ON GASES

Our model for gases requires increasingly precise measurements of gas properties. It is therefore appropriate to pause and consider in more detail just how measurements on gases are made. Let us first examine the logic and the physics behind the measurement of gas pressure.

4-2.1 Measuring the Pressure on a Gas

We have been using the word *pressure* for some time. The original definition referred to pressure as the "push" on unit area of the wall of the containing vessel. Our model for gases attributed this "push" to collisions between rapidly moving molecules and the walls of the container. This model was attractive because it could explain a very important observation about pressure: *A gas exerts pressure equally on all the walls of its container*. This fact is important in the operation of all pressure-measuring devices.

Consider the device shown in Figure 4-2.1. Two bulbs, each of the same volume and containing the same amount of a given gas, are connected by a U-tube half filled with liquid mercury. Molecules in the bulbs strike all walls of the container randomly, including the surface of the mercury in the U-tube. Since each bulb has the same volume and each contains the same number of molecules of a given gas, an equal number of collisions generating comparable force should be expected on each square centimeter of mercury surface in bulbs A and B. In short the "push" per unit area on the two surfaces of mercury should be the same and the surfaces should stand at the same height. This "push" per unit of area is called **pressure**. *Pressure is defined as force per unit area of surface.*

Now add a sizable quantity of the given gas to the bulb on the right (bulb B) in Figure 4-2.2. The number of collisions with a unit

MEASURING the PRESSURE on a GAS

molecular weight g/mole
1 liter gas weight g/l

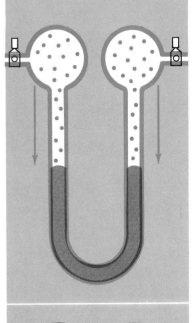

Fig. 4-2.1 Pressure of gas in side A equals pressure of gas in side B.

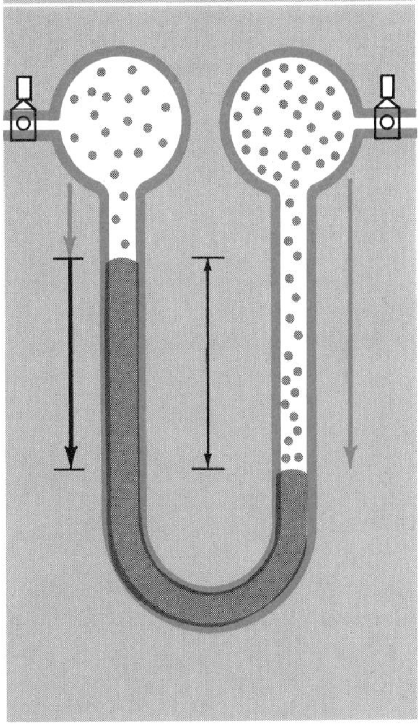

Fig. 4-2.2 Pressure of gas in A plus pressure due to weight of mercury column of height h equals pressure of gas in B.

Chapter 4 / MORE ABOUT GASES: The KINETIC THEORY

area of the mercury surface on the right side (B) should now be larger than the number of collisions with the same area of mercury surface on the left side (A). More collisions mean that the push per unit area (or pressure) will be higher on side B. The mercury column then moves until the force per unit area due to the weight of the mercury column (height h) just balances the extra force per unit area generated by additional molecular collisions on side B. Our model suggests that the height of the mercury column (h) might be a good measure of the difference in pressure (or push on the surfaces) in bulbs A and B.

Such a U-tube, known as a **manometer,** permits us to measure the *difference* in pressure exerted on the surfaces of the two arms of the manometer. Two types of manometers are shown in Figure 4-3. Figure 4-3.1 shows a closed-end manometer. Here all the gas above the mercury surface (except for a trivial amount of mercury vapor, $P = 10^{-3}$ mm Hg) has been removed from arm B. Thus, the pressure in the bulb is balanced by the force or "push" per unit area resulting from the weight of the mercury column; the gas pressure is 105 millimeters of mercury (expressed as 105 mm Hg).

The apparatus shown in Figure 4-3.2 differs in that the right-hand tube is open. In this type of manometer, *atmospheric pressure* is exerted on the right-hand mercury column. Hence the pressure of the gas in the flask (force per unit area of surface) plus the pressure exerted by the mercury column (force per unit area of surface) equals atmospheric pressure (force per unit area of surface).

$$\begin{Bmatrix} \text{atmospheric} \\ \text{pressure} \end{Bmatrix} = \begin{Bmatrix} \text{pressure} \\ \text{in flask} \end{Bmatrix} + \begin{Bmatrix} \text{pressure exerted} \\ \text{by mercury column} \end{Bmatrix} \quad (5)$$

$(760 \text{ mm Hg}) = (105 \text{ mm Hg}) + \quad (655 \text{ mm Hg})$

Fig. 4-3.1 **Closed-end manometer.** Gas pressure in arm A equals pressure due to weight of mercury column in arm B. (Pressure of gas = 105 mm.)

Fig. 4-3.2 **Open-end manometer.** Gas pressure (105 mm) equals atmospheric pressure (760 mm) minus pressure of mercury column (655 mm).

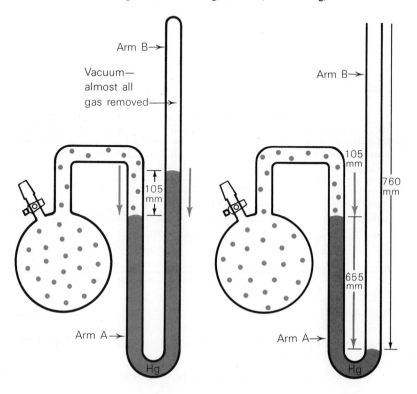

The pressure indicated by our open-end manometer can now be calculated as 760 mm − 655 mm = 105 mm. The value indicated for the pressure in the flask is the same as that indicated by the closed-end manometer. This open-end manometer has the disadvantage that an absolute measurement of pressure in the flask *cannot* be made until the atmospheric pressure is known. It is, however, easier to build than is the closed-end manometer.

The atmospheric pressure is usually measured by means of a **barometer** (see Figure 4-4). A barometer can be made by filling a long tube (closed at one end) with mercury and inverting it in a dish of mercury. If the tube is long enough, mercury will flow from the tube until the column of mercury exerts a downward pressure which is exactly balanced by the pressure of the air. In Figure 4-4 the pressure of the air is 760 mm. The air pressure pushes down on the surface of the mercury in the dish, thus holding up mercury in the tube.*

MEASURING the PRESSURE on a GAS

Fig. 4-4 **(a) Barometer and (b) open-end manometer. A barometer is really an open-end manometer with a vacuum in the closed end above the mercury column. Compare with Figure 4-3.2.**

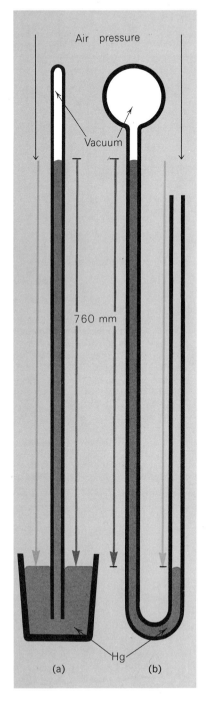

EXERCISE 4-4

A flask of gas is attached to an open-end manometer. For the following three cases decide whether the gas pressure is equal to, greater than, or less than atmospheric pressure.

(1) The mercury level on the flask side is 20 mm lower than the level on the atmosphere side.   g press. > atm press.
(2) The mercury levels are equal.   g = atm
(3) The mercury level on the flask side is 20 mm higher than the level on the atmosphere side.   g press. < atm press.

Explain your reasoning.   atm P = g P + Hg P

EXERCISE 4-5

If the atmospheric pressure is 748 mm, what is the pressure of the gas sample for the three cases in Exercise 4-4?   1) 768 mm Hg   2) 748 mm Hg   3) 728 mm Hg

---

* Since pressure is defined as force per unit area, and since weight is the force exerted downward by a body as a result of gravitational attraction, the downward pressure exerted by the mercury column is the weight of the column divided by the area over which the weight is distributed. In absolute units this is

pressure due to Hg column = $\left\{\begin{array}{l}\text{mass of Hg}\\\text{per unit of volume}\end{array}\right\} \times \left\{\begin{array}{l}\text{acceleration}\\\text{due to gravity}\end{array}\right\} \times$

area of tube × height of column × $\dfrac{1}{\text{area of tube}}$

In absolute units (dynes/cm²) this is

$\left\{\begin{array}{l}\text{pressure}\\\text{due to}\\\text{Hg column}\end{array}\right\} = \dfrac{13.6 \text{ g Hg}}{\text{cm}^3 \text{ Hg}} \times \dfrac{980 \text{ cm}}{\text{sec}^2} \times$ area Hg column × height column × $\dfrac{1}{\text{area Hg column}}$

All quantities are either constant or cancel, so the pressure due to the mercury column is proportional to the height of the column.

$$P = (13.6 \times 980) \times \text{height of column}$$

Chapter 4 / MORE ABOUT GASES: The KINETIC THEORY

4-2.2 Temperature and Its Measurement

What is temperature? What does the word imply to you? Among other things it implies the following: When the thermometer outside your window reads $-20°F$, you know that you will be cold if you go out to sunbathe. When the thermometer outside reads $110°F$, you know that you will be very hot, particularly if you are foolish enough to go for a walk in the sun. Temperature tells you something about your personal comfort; it does this because temperature is a number which tells you which way thermal energy or heat will flow. If the temperature is significantly lower than the temperature of your body, thermal energy (heat) will pass from your body to the surroundings. If your body cannot replace this energy (heat) fast enough, you feel cold. On the other hand, if energy (heat) is flowing into your body from its surroundings at a rate which is greater than the rate at which energy is removed, you feel hot!

Let us consider another example. If a piece of hot metal is dropped into a glass of water at room temperature, the temperature of the water rises (the water becomes hotter) and the temperature of the metal falls (the metal becomes colder). Ultimately, the metal and water will reach a point where standardized thermometers will show identical readings for each (Figure 4-5). No net exchange of energy can be observed between water and metal. The water and metal are in **thermal equilibrium.** We say that the water and the metal have the same temperature. In an experimental sense, *temperature can be considered as a number which tells us which way thermal energy (or heat) flows between two bodies.*

So far we have been concerned only with temperatures of liquids and solids; yet gas temperatures are far easier to understand on a molecular level. Let us examine the temperature of gases.

To measure the temperature of a gas, we immerse a thermometer in it. We now know that if the thermometer is colder than the system, thermal energy flows into the thermometer until the gas and the thermometer are at the same temperature. If the thermometer were hotter than the gas, thermal energy would flow from the thermometer to the gas. When thermal equilibrium is reached, the gas and the thermometer have the same temperature. We then read the temperature of the gas on the thermometer.

There are many kinds of thermometers. Any material can be fashioned into a thermometer if it has a readily measured property that changes with temperature. Almost all solids and liquids change volume when the temperature is changed. We use this property of mercury and glass to construct the ordinary mercury thermometer. Both the glass bulb and glass stem of the thermometer expand *slightly* as the thermometer becomes hotter. On the other hand, the mercury contained inside the bulb and glass stem expands much more with a given change in temperature than does the glass which holds it. As the mercury occupies more volume, its level rises in the stem and we read a number which indicates this new and higher volume and thus a new and higher temperature.

Gases can also be used to indicate changes in temperature. A gas held at constant volume will exert a higher pressure as the temperature is increased. One can then transform the higher pressure

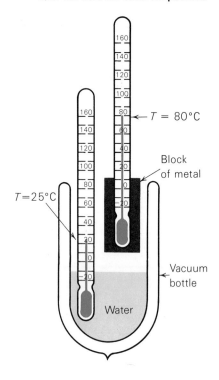

Fig. 4-5 The meaning of temperature. When two objects at different temperatures are brought together, they will reach thermal equilibrium; both will have the same temperature.

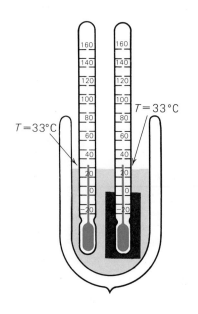

reading into a suitable higher temperature reading by means of a table or mathematical relationship.

The way in which we express temperature—the definition of the temperature scale (or degree)—is rather arbitrary. In general we need two fixed points; for the Celsius (or centigrade) scale these points are the freezing point of water (which is called zero, 0.00) and the boiling point of water under a pressure of 760.00 mm (which is called 100.00). If the property (such as expansion of mercury up the stem) changes regularly between these two fixed points, we just mark the fixed points, label them, and divide the measured difference between them (*i.e.*, distance between boiling and freezing marks) into 100 equal parts called degrees. The most troublesome point arises if the property does not change regularly with temperature; then more involved methods of standardization must be used. This detail will not concern us here, but it is important in making accurate thermometers.

4-2.3 Standard Temperature and Pressure (STP)

All the discussion until now has emphasized the importance of temperature and pressure. It should be apparent that chemists usually control and measure these conditions during experiments. Further, as you probably noticed when we described the molecular volume of 22.4 liters, a temperature of 0.00°C and a pressure of 1.00 atm were repeatedly specified. It would be useful to refer many other experimental results to *a standard and generally accepted set of temperature and pressure conditions.*

Such a set of standard conditions has been adopted on an international basis. *The standard temperature is 0.00°C.* Ice water slush gives a steady temperature of 0.00°C on standing in an insulated vessel such as a thermos bottle. Such a bath is readily made in the laboratory and can be maintained for long periods of time. Its temperature is so reproducible that it is used to set 0° on the Celsius temperature scale and 32° on the Fahrenheit scale. By general agreement the temperature 0.00°C (equilibrium temperature of ice water) is accepted as a standard temperature for reporting data involving gases.

Air pressure fluctuates from day to day and from place to place and decreases as the altitude increases, so it is not possible to describe accurately a "standard atmospheric pressure." Instead, an *arbitrary* standard pressure has been chosen which is relatively close to an average pressure at sea level. By international agreement *a standard pressure, 1.00 standard atmosphere, is defined as that pressure which will support a pure mercury column 760.00 mm high, measured at 0.00°C.* This pressure is frequently referred to as 1.00 atmosphere (1 atm).

To summarize: the conditions 0.00°C and 760.00 mm pressure are called standard temperature and pressure and are abbreviated STP.

EXERCISE 4-6

Express the pressures in Figure 4-3 in atmospheres rather than millimeters of mercury.

4-2.4 A Commentary on Avogadro's Hypothesis

One of the most important relationships in all our former arguments is Avogadro's hypothesis: *Equal volumes of gases measured at the same temperature and pressure contain equal numbers of molecules.* This relationship permits us to determine the relative weights of gaseous molecules and places molecular and atomic weights on a sound experimental footing. Its truth was established not by a single experiment but by many separate observations such as those we have been considering. Since we are investigating questions of measurement, some commentary on the reliability of Avogadro's hypothesis seems appropriate. Avogadro's hypothesis is important not because it is exact, but because it applies to all gases regardless of whether their molecules are large or small. The molecules of different gases actually have different sizes and slightly different attractions for each other. As a result, different gases do *not* have *exactly* the same number of molecules in a given volume. Such variations are, however, usually small and generally can be reduced to less than 1 percent by controlling experimental conditions. These deviations do not necessarily impair the usefulness of Avogadro's hypothesis as a method for determining the molecular weight of a gas and for "counting" molecules. This is a very fortunate fact of nature!

Examination of Table 4-3 (see page 71) substantiates this fact. Molecular weights of gases can be determined independently by an instrument known as a mass spectrometer (to be discussed in Appendix 4). If the molecular weights of gases, determined in a mass spectrometer, are used as a basis for weighing out 1 mole of each of several gases, we can make a direct measurement of the molar volumes for these gases. Data for oxygen, nitrogen, carbon monoxide, and carbon dioxide are shown in Table 4-3. We see that these gases follow Avogadro's hypothesis to three significant figures.

4-3 THE KINETIC THEORY

The model of a gas as a collection of particles in endless motion requires that each particle possess *energy of motion*. This energy of motion is called **kinetic energy.** We shall not be surprised to learn, then, that the model for gases is called the **kinetic theory of gases.** An application of the quantitative mathematical concepts of the kinetic theory permits us to be a little more specific in our description of molecular motion.

Considerable evidence, direct and indirect, indicates that gas molecules travel in straight lines until they meet other gas molecules or the walls of the container. Then they bounce off; some hit head-on and some at an angle; the net result is a helter-skelter movement of molecules in all directions and at all speeds. Since an individual molecule will change both speed and direction of motion even in a single second to give a zig-zag path, it is pointless to speak of the velocity of a given molecule. We can, however, speak of the *average speed* of a collection of molecules at any instant. The average speed of a collection of gas molecules does not change with time unless the temperature is changed. It is a measurable and reproducible quantity. At room temperature the average speed of a nitrogen molecule is determined to be about ¼ mile per second or about 900 miles per hour. Although the average distance between molecules is small in an absolute sense, it is large in comparison to the size of the molecules themselves (Section 4-1.1). This means that molecules can travel relatively long distances without colliding. On the average, at room temperature and 1 atm pressure a molecule will travel about 15 times the average distance between molecules before colliding with another molecule.

The molecules making up solids and liquids are also in motion but their motion will be restricted. Such motion is harder to describe because the particles are closer together.

4-3.1 Pressure Changes Resulting from a Change in the Number of Molecules

Earlier we noted that gas pressure is a result of collisions between molecules and the container walls. Such molecular collisions exert a force on the wall. If we remember (Section 4-2.1) that pressure is defined as force per unit area, it is easy to predict that twice as many molecules in the same volume will give twice as many collisions per unit of time and area, hence twice as much force per unit area. The pressure will be twice the original value. In brief, the number of collisions per unit area per second is proportional to the number of molecules per unit of volume (temperature assumed constant). Experimentally this means: *If volume and temperature remain constant, the pressure is directly proportional to the number of moles of gas per unit volume.*

EXERCISE 4-7

What happens when an air hose is fastened to an automobile tire in the service station? Can you think of any case where the pressure of the tire could go down rather than up?

EXERCISE 4-8

A container of fixed volume contains 2 moles of gas at room temperature. The pressure in the container is 4 atm. Three moles of gas are added to the container at the same temperature. Use the result just stated to show that the pressure is now 10 atm. $\frac{4}{2} = \frac{x}{5}$ $x = 10$

4-3.2 Partial Pressures

We have talked about the pressure in vessels filled with pure gases such as O_2, N_2, and NH_3. Is the description of gases changed seriously if we work with a gas mixture? This question can be answered by using the kinetic model for a gas to interpret a rather simple set of experiments.

Consider the box illustrated in Figure 4-6. This box has a total volume of 5 liters. It is divided into two sections—one of a 4-liter volume and one of a 1-liter volume—by a breakable diaphragm or wall. The 4-liter section is filled with N_2 at *1 atm pressure* and the 1-liter section is filled with O_2 at *1 atm pressure*. The entire box is held at

PARTIAL PRESSURES

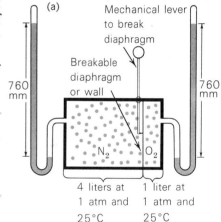

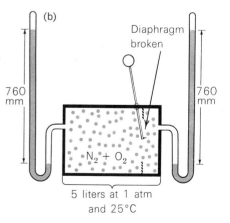

Fig. 4-6 **Partial pressure.** Manometers read the same (a) before and (b) after the diaphragm is broken.

25°C or room temperature. Note that each section of the box contains gas at 1 atm.

Let us now pull the handle and break the diaphragm. The N_2 and O_2 molecules mix. We observe the two manometers carefully after the diaphragm is broken and find that the pressure is *the same* as it was before the diaphragm was broken.

Can we rationalize this observation in terms of the kinetic theory? In Section 4-1.2 we found that the pressure-volume relationship for gases can be expressed as

$$P \times V = \text{a constant at } 25°C \tag{6}$$

Thus we can write for N_2 before the diaphragm is broken

$$1 \text{ atm} \times 4 \text{ liters} = 4 \text{ liters} \times \text{atm at } 25°C$$

(The value of the constant here is 4, not 24.5, because we have taken only 4 liters, not 1 mole or 24.5 liters, at 25°C.) After the diaphragm is broken, the volume through which N_2 molecules can roam is 5 liters. Qualitatively, we can see that the pressure due to N_2 must fall. How much does it fall? Since we know $P \times V = 4$ for our sample of N_2 at 25°C, we can calculate the value of P when the volume goes to 5 rather than 4 liters.

$$P \times 5 \text{ liters} = 4 \text{ liters} \times \text{atm}$$
$$P = \tfrac{4}{5} \text{ atm} \tag{7}$$

We find that N_2 in the total container exerts a pressure of $\tfrac{4}{5}$ atm. Similar calculations can be made for oxygen. The original pressure is 1 atm and the original volume is 1 liter; therefore,

$$P \times V = 1 \text{ liter} \times 1 \text{ atm} = 1 \text{ (liter} \times \text{atm)}$$

for O_2 at 25°C. (Note that the constant here is 1, not 24.5, because we took only 1 liter, *not* 1 mole or 24.5 liters, of O_2 at 25°C.) If the volume available to O_2 molecules is expanded to 5 liters, we get

$$P \times 5 \text{ liters} = 1 \text{ liter} \times \text{atm}$$
$$P = \tfrac{1}{5} \text{ atm} \tag{8}$$

The oxygen in the container after the diaphragm is broken exerts a pressure of $\tfrac{1}{5}$ atm.

We can then write

$$\begin{array}{ll} P \text{ due to } N_2 = & \tfrac{4}{5} \text{ atm} \\ \underline{P \text{ due to } O_2 = } & \underline{\tfrac{1}{5} \text{ atm}} \\ \text{total } P & = 1 \text{ atm} \end{array}$$

One atm total pressure was observed in our experiment after the diaphragm was broken. *The model agrees with the experimental observation.*

We call the pressure due to nitrogen the *partial pressure of nitrogen*. This is sometimes indicated as $P_{N_2} = \tfrac{4}{5}$ atm. We call the pressure due to oxygen the *partial pressure of oxygen*, or $P_{O_2} = \tfrac{1}{5}$ atm. The pressure exerted by each of the gases in a gas mixture is called the **partial pressure** of that gas. The partial pressure is the pressure that the gas would exert if it were alone in the container. The **total pressure** is the sum of the partial pressures.

$$P_{\text{Total}} = P_{N_2} + P_{O_2} = \tfrac{4}{5} \text{ atm} + \tfrac{1}{5} \text{ atm} = 1 \text{ atm} \tag{9}$$

The above simple additive relationship is a result of the makeup of a gaseous phase. *There is so much space between the molecules that each molecule behaves almost independently.* Each molecule contributes its share to the total pressure through its occasional collisions with the container walls. The total pressure felt will simply be the *sum* of these individual molecular collisions. The nitrogen molecules, taken together, undergo wall collisions that generate ⅘ atm. The oxygen molecules taken together undergo collisions that generate ⅕ atm. The sum total of these is the 1 atm observed.

EXERCISE 4-9

If 21 percent of the molecules in the air are oxygen molecules, what is the partial pressure of oxygen gas in the air when the barometer reads 740 mm Hg?

[handwritten: $.21\,O_2 + .79\,G = 740\text{ mmHg}$]

[handwritten answer: 155 mm Hg]

4-3.3 Temperature and the Kinetic Theory

What does temperature imply on a molecular scale, particularly as it applies to gases? We have agreed that the pressure exerted by gas molecules *at a given temperature* will be dependent upon the number of molecules hitting a unit of surface in a given time. But the masses of the molecules and their velocities will also be important. For example, a baseball exerts more force than a tennis ball thrown with the same velocity. A baseball exerts more push if it is a "fast ball" than if it is a "slow ball." To understand how the kinetic theory deals with these factors we must consider temperature.

We defined temperature as a number that tells us which way thermal energy (heat) will flow when two systems are brought into close contact with each other. Suppose we have two gases, nitrogen (N_2) and carbon monoxide (CO), in separate containers. The nitrogen is at a much higher temperature originally than is the carbon monoxide (see Figure 4-7.1). According to our definition of temperature, if N_2 is mixed with CO (Figure 4-7.2), some of the energy of the nitrogen must be transferred to the carbon monoxide because energy flows from the body at higher temperature (N_2) to the body at lower temperature (CO). The energy of the nitrogen is stored in the energy of motion of its molecules (kinetic energy of molecules). Thus the molecules of N_2 originally had more energy of motion on the average than did the molecules of CO. Since molecules of CO and N_2 both have the same molecular weight (28), the hotter nitrogen molecules must have a higher average speed than the cooler CO molecules.* Collisions of the rapidly moving molecules of N_2

---

* The kinetic energy of a body is determined both by its mass and its velocity. The quantitative definition of kinetic energy is

$$\text{kinetic energy} = \tfrac{1}{2} \text{mass} \times (\text{velocity})^2$$

[handwritten: $KE = \tfrac{1}{2} m v^2$]

TEMPERATURE and the KINETIC THEORY

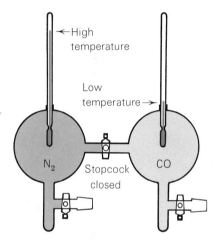

Fig. 4-7.1 When two gases at different temperatures are kept in separate containers they will stay at different temperatures.

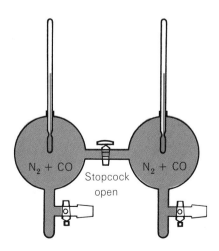

Fig. 4-7.2 When mixed the two gases will exchange energy and reach thermal equilibrium.

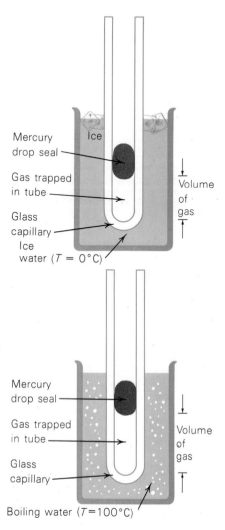

Mercury drop seal
Gas trapped in tube
Glass capillary
Ice water ($T = 0°C$)
Volume of gas

Mercury drop seal
Gas trapped in tube
Glass capillary
Boiling water ($T=100°C$)
Volume of gas

Fig. 4-8 **The volume of a given quantity of gas will increase when the gas is heated.**

TABLE 4-4 CHANGE OF VOLUME OF A GAS WITH CHANGE IN TEMPERATURE

Temperature (°C)	Relative Volume As Measured by Length of Sample
0	1.00
50	1.18
100	1.37
200	1.73

with the somewhat more slowly moving molecules of CO slow down the N_2 molecules *on the average* and speed up the carbon monoxide molecules *on the average*. In this way kinetic energy is transferred from N_2 molecules to CO molecules until the average speed of the CO molecules equals the average speed of the N_2 molecules. The gases then have equal kinetic energy and no further net transfer of energy from N_2 to CO is expected. When kinetic energy is no longer transferred from N_2 to CO, the gases are in thermal equilibrium: *the two gases are at the same temperature.*

The above specific example illustrates a more general and basic postulate of the kinetic theory. *When gases are at the same temperature, the molecules of the gases have the same average kinetic energy.**

4-3.4 Absolute Temperature and the Dependence of Volume and Pressure on Temperature

ABSOLUTE TEMPERATURE

The quantitative relationships between gas temperature and gas pressure or gas volume were first studied by Jacques Charles in 1787. The relationships he developed are referred to frequently as **Charles' law**. A few simple experiments will show what is involved.

In a small-bore glass tube, ½ meter in length and closed at one end, we place a drop of mercury. This mercury falls and finally traps a sample of air in the bottom of the tube (see Figure 4-8). Since the tube has a uniform bore, we can use the length of the air sample as a measure of its volume. The mercury plug moves up or down and maintains a constant pressure.

We may place the tube in ice water (0°C) and measure the relative volume of the air sample. If the tube is immersed in water boiling at 1 atm pressure (100°C), the relative volume has a higher value. From these data and from similar measurements at other temperatures, we collect data such as those in Table 4-4.

When we plot these results with relative volumes on the vertical axis (ordinate) and temperatures on the horizontal axis (abscissa), we obtain the graph shown in Figure 4-9. The straight line passes through the experimental points. When extended upward, it shows that the volume at 273°C is double that at 0°C. Extrapolated, or extended downward, the line shows that the volume would become zero at −273°C. The volume change per degree Celsius is ½₇₃ of the volume at 0°C. Actually, all gases liquefy before their temperature reaches −273°C.

If gases are heated or cooled at constant volume, the pressure changes per degree Celsius also at the rate of ½₇₃ of its value at 0°C. Then the pressure of a gas would become zero at −273°C. In terms of the kinetic theory, the motion of the molecules would cease at this temperature.† The kinetic energy would become zero.

---

* Gases at the same temperature do not have the same average molecular velocity unless the gases have the same molecular weight. This was true for N_2 and CO. For equal kinetic energy,

$$\tfrac{1}{2}m_1(v_1)^2 = \tfrac{1}{2}m_2(v_2)^2$$

where m is the mass of the particle and v is its velocity.

† The motion referred to here is the motion of the molecules with respect to the container, not the vibrations and rotations of the atoms in the molecule.

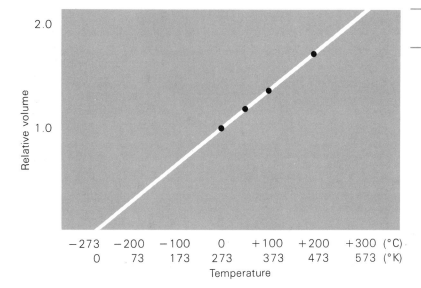

Fig. 4-9 **A plot of temperature versus volume for gases.**

There are great advantages to an *absolute* temperature scale that has its zero point at −273°C. As noted earlier, the "zero" of temperature in the Celsius scale is fixed at an arbitrary temperature selected because it is easily measured. In contrast, the zero point of the absolute scale has a definite meaning in the kinetic theory. Furthermore, we find that *the volume of a fixed amount of gas (at constant pressure) varies directly with temperature expressed on an absolute temperature scale.* The pressure of a fixed amount of gas (at constant volume) varies directly with temperature.* According to the kinetic theory, the kinetic energy of the molecules varies directly with the absolute temperature. For these reasons, in dealing with gas relations, we shall usually express temperature on an absolute temperature scale.

This temperature scale, with the same size degree as the Celsius scale, is called the Kelvin scale and values on this scale are expressed in degrees Kelvin (°K). Both Kelvin and Celsius temperatures are shown in Figure 4-9. Notice that all numerical values on the Kelvin scale are 273 degrees higher than the corresponding temperatures on the Celsius scale (more precisely, 273.1 degrees higher).

EXERCISE 4-10

(1) Express the following temperatures in degrees Kelvin:
  Boiling point of water   100°C   373°K
  Freezing point of mercury   −38.9°C 234.1°K
  Boiling point of liquid nitrogen −196°C 77°K
(2) Express the following temperatures in degrees Celsius:
  Melting point of lead   600°K   327°C
  A normal room temperature   298°K   25°C
  Boiling point of liquid helium 4°K   −269°C

---

* This direct relation between volume and temperature (at constant pressure) is called Charles' law.
$$V = k \cdot T$$

EXERCISE 4-11

In the laboratory a student obtained the following result: 2.00×10^{-3} mole of magnesium produces a volume of hydrogen that occupies 49.0 ml at 25°C and 1 atm pressure.

(1) If 1 mole of magnesium produces 1 mole of hydrogen, use these data to calculate the volume of 1 mole of hydrogen.
(2) Calculate the volume 1 mole of hydrogen will occupy at 0°C (273°K) and 1 atm.

We have remarked that at a temperature of zero on the absolute temperature scale there would be no motion of molecules relative to the container. The kinetic energy of translation (motion with respect to the container) would become zero. Very interesting phenomena occur at temperatures near 0°K. (The superconductivity of many metals and the superfluidity of liquid helium are two examples.) Hence, scientists are extremely interested in methods of reaching temperatures as close to absolute zero as possible. Two low-temperature coolants commonly used are liquid hydrogen (which boils at 20°K) and liquid helium (which boils at 4°K). Helium, under reduced pressure, boils at even lower temperatures and provides a means of reaching temperatures near 1°K. More unusual techniques have been developed to produce still lower temperatures (as low as 0.001°K), but even measuring these temperatures becomes a severe problem.

4-4 A GENERAL GAS EQUATION

In considering gases we have identified four general variables which are interrelated. These variables—pressure, volume, temperature, and number of moles of gas—have been considered in pairs until now. We have held two variables constant and then studied variations in the remaining two. In our very first investigation of the relationship between pressure and volume, for example, we considered a fixed quantity or number of moles of gas at constant temperature. Number of moles and temperature were therefore held constant while pressure and volume were varied. In studying the relationship of temperature and volume in Section 4-3.4, the number of moles of gas and the pressure were held constant while volume and temperature were varied. Is this pattern always necessary or desirable? Let us see if a more general equation can be developed.

4-4.1 Development of a General Gas Equation

In studying the general relationship between pressure and volume we found that pressure × volume = a constant. The absolute size of the constant is dependent upon the quantity of gas and the temperature. If 1 liter of gas is taken and a pressure of 1 atm is used, then

$$P \times V = 1 \text{ liter} \times 1 \text{ atm} = 1 \text{ (liter} \times \text{atm)} \qquad (6b)$$

Values of $P \times V$ for six different starting volumes at 0°C and 1 atm are shown in Table 4-5. It is clear that the size of the constant is determined by the quantity of gas taken. We found that there were a number of advantages in selecting 22.4 liters. We recognize 22.4 liters at 1.00 atm and 0.00°C as 1 mole of gas. It has 6.02×10^{23} molecules, or 1 mole of molecules. It is then convenient to express the constant as $n \times 22.4$, where n is the number of moles of gas taken.

$$P \times V = n \times \left(\frac{22.4 \text{ liters} \times \text{atm}}{\text{mole}}\right)^* \text{ at } 0°C \qquad (10)$$

We now have three variables related through a single expression. Let us include temperature as well. When the temperature was varied from 0°C to 25°C, the value of the PV constant increased from 22.4 to 24.4 (see Section 4-1.2). At 25°C we can write

$$P \cdot V = n \cdot \left(\frac{24.5 \text{ liters} \times \text{atm}}{\text{mole}}\right) \text{ at } 25°C \qquad (11)$$

If we remember how temperature alters the volume of a gas (Section 4-3.4), we can write

$$V = \frac{22.4 \text{ liters}}{\text{mole}} \times \frac{(273+25)°K}{(273)°K} = \frac{22.4 \text{ liters}}{\text{mole}} \times \frac{298}{273} = \frac{24.4 \text{ liters}}{\text{mole}} \text{ at } 25°C \quad (12)$$

In place then of 24.5 in our PV expression at 25°C, we can write (if P is still 1 atm)

$$P \cdot V = n \left[22.4 \times 1 \frac{(\text{liters} \times \text{atm})}{\text{mole}}\right] \times \frac{(273+25)°K}{273°K} \qquad (13)$$

If we now remember that $(273+25)°K = 298°K$, the temperature of the gas on the absolute or Kelvin scale, we can write

$$P \cdot V = n \left[\frac{22.4}{273}\left(\frac{\text{liters} \times \text{atm}}{\text{mole} \times °K}\right)\right] T \qquad (14)$$

where T is the absolute temperature of the gas.

We now have a general relationship involving P, V, n, and T. The numerical quantity

$$\frac{22.4}{273}\left(\frac{\text{liters} \times \text{atm}}{\text{moles} \times °K}\right)$$

does not change as we change P, V, n, and T. It simply reflects our earlier arbitrary choices for molar volume, standard temperature, and standard pressure. Since this number does not change, it is worthwhile doing the division needed to evaluate the constant.

$$\frac{0.081 \text{ liters} \times \text{atm}}{\text{mole} \times °K}$$

* The unit mole must go into the denominator of the constant because the 22.4 value is for a single mole or is 22.4 liters × atm/mole.

DEVELOPMENT of a
GENERAL GAS EQUATION

TABLE 4-5 PRESSURE-VOLUME PRODUCT FOR GASES AT 0°C

Starting Volume (liters)	Starting Pressure (atm)	P × V (liters × atm)
1	1	1
2	1	2
4	1	4
11.2	1	11.2
22.4	1	22.4
44.8	1	44.8

This quantity is of great importance in physical chemistry. It is assigned the symbol R and is known as the **ideal gas constant.** The general gas equation can then be written as

$$PV = nRT \qquad (15)$$

If P is given in mm Hg and V is given in ml, R has a different numerical value, but the same overall significance.

$$\frac{22{,}400 \text{ ml} \times 760 \text{ mm Hg}}{\text{mole} \times 273°K} = \frac{62{,}400 \text{ ml} \times \text{mm Hg}}{\text{mole} \times °K}$$

The expression $PV = nRT$ is known as the **ideal gas law.** Chemists find it a most convenient and powerful equation. The numerical value selected for R depends on the units which are being used in the problem. This fact is summarized as follows:

IF PROBLEM HAS:

Pressure in:	Volume in:	Value of R is:
atm	liters	$0.0821 \left(\dfrac{\text{liters} \times \text{atm}}{\text{mole} \times \text{degree}} \right)$
mm Hg	liters	$62.4 \left(\dfrac{\text{liters} \times \text{mm Hg}}{\text{mole} \times \text{degree}} \right)$
mm Hg	ml	$62{,}400 \left(\dfrac{\text{mm Hg} \times \text{ml}}{\text{mole} \times \text{degree}} \right)$
atm	ml	$82.1 \left(\dfrac{\text{atm} \times \text{ml}}{\text{mole} \times \text{degree}} \right)$

4-4.2 Some Examples of Gas Law Calculations

Some sample problems are included here to indicate how gas law calculations can be made.

Example (1) A 1.00-liter gas sample at 0°C and 1 atm pressure weighs (0.715 ± 0.005) g. Which one of the following gases could be indicated: CO_2, CO, N_2, O_2, CH_4, NH_3, HCl?

The information given permits the calculation of a molecular weight since the molar weight will be the weight of 22.4 liters of gas measured at STP. We then can write

$$\text{molar weight gas} = \frac{(0.715 \pm 0.005) \text{ g}}{\text{liter}} \times \frac{22.4 \text{ liters}}{\text{mole}} = \frac{16.0 \text{ g}}{\text{mole}}$$

molecular weight gas = 16.0

To decide which of the gases is indicated, we must calculate the molecular weights for each of the gases shown. For CO_2 the value is equal to the atomic weight of carbon plus two times the atomic weight of oxygen or $12.00 + (2 \times 16.0) = 44.00$. Similar calculations for all other gases give these molecular weights: $CO = 28$, $N_2 = 28$, $O_2 = 32$, $CH_4 = 16$, $NH_3 = 17$, and $HCl = 36.5$. The answer appears to be CH_4, which has a molecular weight of 16. To make certain, however, we must make sure that the error in measurement was not large enough to permit the true measured value to be 17. If we assume an error of ±0.005 in the weight and ±0.2 in molar volume, the largest possible values will be 0.715 + 0.005 = 0.720 and 22.4 + 0.2 = 22.6. The molecular weight calculated using these maximum values is 16.3. The value for NH_3 would seem to be just outside our limit of error. Usually additional types of chemical tests would be desirable to check our conclusions.

Example (2) What pressure would be required to compress a 2-mole sample of gas into 10 liters at 0°C?

We start with the knowledge that *1 mole* of gas at STP (0°C and 1 atm) occupies 22.4 liters. Two moles of gas will then occupy

$$2 \text{ moles} \times \frac{22.4 \text{ liters}}{\text{mole}} = 44.8 \text{ liters at } 0°C \text{ and } 1 \text{ atm}$$

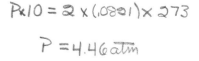

We have shown that

$$P \times V = \text{a constant at } 0°C$$

For 2 moles of gas $V = 44.8$ liters (see above). Hence, under 1 atm (given) we can write

$$1 \text{ atm} \times 44.8 \text{ liters} = 44.8 \text{ (atm} \times \text{liters)}$$

The constant in the equation above is 44.8; hence, for 2 moles of gas at 0°C we can write

$$P \times V = 44.8 \text{ (atm} \times \text{liters)}$$

If $V = 10$ liters (given), then

$$P \text{ (atm)} \times 10 \text{ liters} = 44.8 \text{ (atm} \times \text{liters)}$$

$$P = \frac{44.8 \text{ atm} \times \text{liters}}{10 \text{ liters}}$$

$$P = 4.48 \text{ atm}$$

Example (3) Barometer A (a barometer is used to measure atmospheric pressure) has some gas on top of the column. It reads 742 mm Hg. A new barometer, B, from the storeroom reads 757 mm Hg. What is the gas pressure above the mercury in barometer A? (See Figure 4-10.)

Barometer A is now serving as a manometer which measures the *difference* between the pressure above the mercury column and the pressure of the atmosphere outside. Pressure reading by barometer A = atmospheric pressure − pressure inside. Hence,

$$742 \text{ mm Hg} = 757 \text{ mm Hg} - \text{pressure inside}$$
$$\text{pressure inside} = 15 \text{ mm Hg}$$

Example (4a) What is the partial pressure of each gas in a mixture containing by volume: 20 percent He, 30 percent CO, 10 percent H_2, and 40 percent CH_4 if the total pressure is 800 mm Hg?

Fig. 4-10

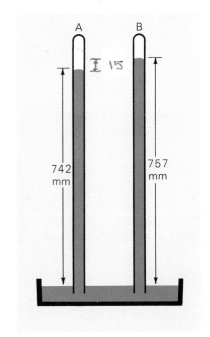

Twenty percent He by volume means that 20 percent of the molecules in the container are He. Thus,

$$800 \text{ mm Hg} \times 0.20 = 160 \text{ mm He}$$
$$800 \text{ mm Hg} \times 0.30 = 240 \text{ mm CO}$$
$$800 \text{ mm Hg} \times 0.10 = 80 \text{ mm H}_2$$
$$800 \text{ mm Hg} \times 0.40 = \underline{320 \text{ mm CH}_4}$$
$$800 \text{ mm total}$$

Example (4b) What would be the *total pressure* if water vapor at a pressure of 18 mm were added to the dry gases with no change in total volume?

$$\text{total pressure} = 800 + 18 = 818 \text{ mm Hg}$$

Example (5) (Use of the Ideal Gas Law.) What volume would be occupied by 3 moles of nitrogen under a pressure of 120 atm and a temperature of 50°C?

Given: $PV = nRT$

$$P = 120 \text{ atm}$$
$$V = \text{volume (liters)}$$
$$T = 50°C + 273 = 323°K$$
$$n = 3 \text{ moles}$$

Since P is given in atm and V in liters, we shall use R in these units:

$$R = 0.082 \left(\frac{\text{atm} \times \text{liters}}{\text{mole} \times °K} \right)$$

$$120 \text{ atm} \times V \text{ (liters)} = 3 \text{ moles} \times 0.082 \left(\frac{\text{atm} \times \text{liters}}{\text{mole} \times °K} \right) \times 323°K$$

$$V = \frac{3 \times 0.082 \times 323}{120} \text{ liters} = 0.66 \text{ liter}$$

4-5 THE CONSEQUENCES OF THEORY

4-5.1 Avogadro's Hypothesis and the Kinetic Theory

One of the fundamental postulates of the kinetic theory is: If two gases are at the same temperature, the molecules of the gases have the same average kinetic energy. The ability of this postulate to explain Avogadro's hypothesis is one of its most important achievements.

We may state Avogadro's hypothesis in this form: If two gases at the same temperature have the same number of particles in a given volume, they must exert the same pressure. Yet, as remarked in Section 4-3.3, the mass of a molecule, as well as its velocity, should influence the pressure exerted. If the molecules of our two gas samples have different masses, they must have different speeds in order to have the same kinetic energies. The lighter molecules must travel faster, so they will strike the container walls more times per second. The effect of the more frequent collisions exactly counteracts the lower "push" per collision from these molecules of lower mass. The result is in perfect accord with Avogadro's hypothesis: Two gases at the same concentration and at the same temperature exert the same pressure, even though their molecules have different masses.

Avogadro's hypothesis can be shown quite readily in an approximate way. The kinetic energy of a moving particle is expressed by the equation

$$KE = \frac{1}{2}mv^2 \quad (16)$$

where m is the mass of the particle and v is the velocity. Therefore, for gas A and gas B at the same temperature, we have

$$(KE)_A = (KE)_B \quad (17)$$

$$\frac{1}{2}m_A v_A^2 = \frac{1}{2}m_B v_B^2$$

or

$$m_A v_A^2 = m_B v_B^2 \quad (18)$$

Now suppose we place n molecules in a cubical box with a side of length d. The pressure is fixed by

(1) the number of wall collisions per second on each square centimeter, and
(2) the momentum transferred per collision:

$$\text{pressure} = \left(\frac{\text{collisions}}{\text{second}}\right)\left(\frac{1}{\text{area}}\right)\left(\frac{\text{momentum}}{\text{collision}}\right) \quad (19)$$

Momentum depends upon mass and velocity. Let us consider the momentum change in a head-on collision. The particle approaches the wall with momentum mv and leaves with this same momentum *in the opposite direction*. The faster the molecule is traveling, the harder it hits. The momentum transferred to the wall is then

$$\text{momentum change} = 2mv \quad (20)$$

The collisions per second with the wall, on the other hand, depend upon the container dimension and the molecular velocity. Let us assume that one third of the molecules bounce back and forth in a given direction between two opposite walls. This assumption follows because we have three *pairs* of walls in the cube and n molecules. Thus, there are $n/3$ molecules bouncing between any one pair of walls. One of these walls receives a collision each time one of the molecules travels the box dimension d and back, a distance of $2d$. The time required is $2d/v$. Then we write

$$\frac{\text{collisions}}{\text{second}} = \left(\frac{\text{number of particles bouncing back and forth}}{\text{time for a particle to travel distance } 2d}\right)$$

$$= \left(\frac{n/3}{2d/v}\right) = \left(\frac{n}{3}\right)\left(\frac{v}{2d}\right) = \frac{nv}{6d} \quad (21)$$

Combining (19), (20), and (21), we find

$$\text{pressure} = \left(\frac{\text{collisions}}{\text{second}}\right)\left(\frac{1}{\text{area}}\right)\left(\frac{\text{momentum change}}{\text{collision}}\right)$$

$$= \left(\frac{nv}{6d}\right)\left(\frac{1}{d^2}\right)(2mv)$$

$$= \frac{1}{3}\left(\frac{n}{d^3}\right)(mv^2) \quad (22)$$

Applying equation (22) to each of the gases A and B,

$$P_A = \frac{1}{3}\left(\frac{n_A}{d^3}\right)(m_A v_A^2) \qquad (23)$$

$$P_B = \frac{1}{3}\left(\frac{n_B}{d^3}\right)(m_B v_B^2) \qquad (24)$$

If the gases have the same pressure, $P_A = P_B$, we can equate (23) and (24) so that

$$\frac{1}{3}\left(\frac{n_A}{d^3}\right)(m_A v_A^2) = \frac{1}{3}\left(\frac{n_B}{d^3}\right)(m_B v_B^2) \qquad (25)$$

If the temperatures of the gases are the same, (18) is applicable and equation (25) becomes

$$\frac{n_A}{d^3} = \frac{n_B}{d^3} \qquad (26)$$

Thus we see that at the same temperature and pressure, the two gases have the same number of molecules per unit volume. This is Avogadro's hypothesis.

4-5.2 The Ideal Gas Law from the Kinetic Theory

Equation (22) states

$$P = \frac{1}{3}\left(\frac{n}{d^3}\right)mv^2$$

If we remember that the volume of the cubic box is d^3, we can write

$$d^3 = V$$

Then

$$P \cdot V = \frac{1}{3}(nmv^2) = \left(\frac{n}{3}\right)mv^2 \qquad (27)$$

If we also remember that the value of the temperature is proportional to the kinetic energy, we can write

$$\frac{mv^2}{2} \propto \text{temperature}$$

or

$$(T \cdot \text{constant}) = mv^2 \qquad (28)$$

If the number of molecules is now taken as proportional to the number of moles of gas, we write

$$\text{number moles} \cdot (6.02 \times 10^{23}) = \text{number molecules}$$

If n' is taken as the number of moles and n as the number of molecules, then

$$n' = \frac{n}{6.02 \times 10^{23}} \quad \text{or} \quad n = 6.02 \times 10^{23} \times n'$$

Then

$$\frac{n}{3} = \frac{6.02 \times 10^{23} \times n'}{3}$$

We can then write

$$P \cdot V = \underbrace{n' \cdot \frac{6.02 \times 10^{23}}{3}}_{\frac{n}{3}} \times \underbrace{\text{constant} \times T}_{mv^2}$$

If we collect all of the constants into a single constant, R, we arrive at

$$\frac{6.02 \times 10^{23} \cdot \text{constant}}{3} = R \qquad (29)$$

We can then write

$$PV = n'RT \qquad (15)$$

which is identical to the ideal gas law, except that we have used n' for number of moles of gas.

4-6 HIGHLIGHTS

Regularities observed in the behavior of gases have contributed much to our understanding of the structure of matter. One of the most important regularities is Avogadro's hypothesis: Equal volumes of gases contain equal numbers of particles (at the same pressure and temperature). This relationship is valuable in the determination of molecular formulas, and these formulas must be known before we can understand chemical bonding.

We have explored the meaning of temperature. According to the **kinetic theory,** when two gases are at the same temperature, the molecules of the two gases have the same average kinetic energies. Changing the temperature of a sample of gas at constant pressure reveals that the volume is directly proportional to the temperature if the temperature is expressed in terms of a new, absolute scale. The melting point of ice (0°C) on this new scale, the **Kelvin scale,** is 273°K. The boiling point of water at 1 atm (100°C) is 373°K. The zero temperature on the Kelvin scale corresponds to the hypothetical loss of all molecular motion.

This progress substantiates our confidence in the usefulness of the atomic theory and it encourages us to develop the model further. We shall see that the concepts we have developed in our consideration of gases are also useful in considering the behavior of condensed phases—liquids and solids.

QUESTIONS AND PROBLEMS

Review Chapters 2 and 3.

1 How many molecules are there in a molar volume of a gas at 100°C? at 0°C?

2 What is the molecular weight of a gas if at 0°C and 1 atm pressure, 1.00 liter of the gas weighs 2.00 g? (Answer: 44.8.)

3 The gas sulfur dioxide combines with oxygen to form the gas sulfur trioxide:

$$2SO_2(g) + O_2(g) \longrightarrow 2SO_3(g)$$

What ratio would you expect for the following?

(a) $\dfrac{\text{number of } SO_3 \text{ molecules produced}}{\text{number of } O_2 \text{ molecules consumed}}$

(b) $\dfrac{\text{volume of } SO_3 \text{ gas produced}}{\text{volume of } O_2 \text{ gas consumed}}$

All volume measurements are made at STP.

4 A glass bulb weighs 108.11 g after all the gas has been removed from it. When filled with oxygen gas at atmospheric pressure and room temperature, the bulb weighs 109.56 g. When filled at atmospheric pressure and room temperature with a gas sample obtained from the mouth of a volcano, the bulb weighs 111.01 g. Which of the following molecular formulas for the volcano gas could account for the data: CO_2, OCS, Si_2H_6, SO_2, NF_3, SO_3, S_8, a gas mixture half CO_2 and half Kr?

5 A carbon dioxide fire extinguisher contains about 10 lb (4.4 kg) of CO_2. What volume of gas could this extinguisher deliver at room conditions?

6 Hydrogen for weather balloons is often supplied by the reaction between solid calcium hydride (CaH_2) and water to form solid calcium hydroxide [$Ca(OH)_2$] and hydrogen gas (H_2). (a) Balance the equation for the reaction and decide how many moles of CaH_2 are required to fill a weather balloon with 250 liters of H_2 at STP. (b) What weight of water is consumed in forming the hydrogen? (*Answer:* 0.20 kg.)

7 Gas is slowly added to the empty chamber of a closed-end manometer (see Figure 4-3.1). Draw a picture of the manometer mercury levels, showing in millimeters the difference in heights of the two mercury levels (a) before any gas has been added to the empty gas chamber, (b) when the gas pressure in the chamber is 300 mm, (c) when the gas pressure in the chamber is 760 mm, and (d) when the gas pressure in the chamber is 865 mm.

8 Repeat question 7 but with an open-end manometer (see Figure 4-3.2). Atmospheric pressure is 760 mm.

9 The balloons that are used for weather study are quite large. When they are released at the surface of the earth, they contain a relatively small volume of gas compared to the volume they acquire when aloft. Explain.

10 A 1.50-liter sample of dry air in a cylinder exerts a pressure of 3.00 atm at a temperature of 25°C. Without change in temperature, a piston is moved in the cylinder until the pressure in the cylinder is reduced to 1.00 atm. What is the volume of the gas in the cylinder now?

11 Suppose the total pressure in an automobile tire is 30 psi (pounds per square inch) and we want to increase the pressure to 40 psi. What change in the amount of air in the tire must take place? Assume that the temperature and volume of the tire remain constant.

12 The density of liquid carbon dioxide at room temperature is 0.80 g/ml. How large a cartridge of liquid CO_2 must be provided to inflate a life jacket of a 4.0-liter capacity at STP? (*Answer:* 9.8 ml.)

13 A student collects a volume of hydrogen over water. He determines that there are 2.00 × 10^{-3} mole of hydrogen and 6.0×10^{-5} mole of water vapor present. If the total pressure inside the collecting tube is 760 mm, what is the partial pressure of each gas? (*Answers:* partial pressure H_2 = 738 mm; partial pressure H_2O = 22 mm.)

14 A candle burns under a beaker until the flame dies. A sample of the gaseous mixture in the beaker now contains 6.08×10^{20} molecules of nitrogen, 0.76×10^{20} molecules of oxygen, and 0.50×10^{20} molecules of carbon dioxide. The total pressure is 764 mm. What is the partial pressure of each gas?

15 Consider two closed glass containers of the same volume. One is filled with hydrogen gas, the other with carbon dioxide gas, both at room temperature and pressure. (a) How do the number of moles of the two gases compare? (b) How do the number of molecules of the two gases compare? (c) How do the number of grams of the two gases compare? (d) If the temperature of the hydrogen container is now raised, how do the two gases now compare in pressure, volume, number of moles, and average molecular kinetic energy?

16 The boiling points and freezing points in degrees Celsius of certain liquids are listed below. Express these temperatures on the absolute temperature (degrees Kelvin) scale.

(a) liquid helium: boiling point = −269

(b) liquid hydrogen: freezing point = −259 (*Answer:* 14°K)

(c) liquid hydrogen: boiling point = −253 (*Answer:* 20°K)

(d) liquid nitrogen: freezing point = −210

(e) liquid nitrogen: boiling point = −196

(f) liquid oxygen: freezing point = −219

(g) liquid oxygen: boiling point = −183

17 If exactly 100 ml of a gas at 10°C is heated to 20°C (pressure and number of molecules remaining constant), the resulting volume of the gas will be which of the following? (a) 50 ml (b) 1,000 ml (c) 100 ml (d) 375 ml (e) 103 ml.

18 Why is it desirable to express all temperatures in degrees Kelvin when working with problems dealing with gas relationships?

19 A gaseous reaction between methane (CH_4) and oxygen (O_2) is carried out in a sealed container. Under the conditions used, the products are hydrogen (H_2) and carbon dioxide (CO_2). Energy is released, so the temperature rises during the reaction. (a) Write the balanced equation for the reaction. (b) Will the final pressure be larger or smaller than the original pressure if the temperature does not change? (c) By what factor does the pressure change if 1 mole of methane and 1 mole of oxygen are mixed and allowed to react? Assume that temperature changes from 25°C to 200°C. (*Answer:* 2.38.)

QUESTIONS and PROBLEMS

20 Automobiles are propelled by burning gasoline (C_8H_{18}). Oxygen reacts with the gasoline to form carbon dioxide and water, releasing enough energy to heat the gas from about 300°K to about 1500°K.

Balance the equation for the reaction and decide whether the work done by the gas in the cylinder is mainly due to pressure rise resulting from change in number of moles of gas or pressure rise resulting from heating.

21 Why does the pressure build up in a tire on a hot day? (Answer in terms of the kinetic theory.)

22 A vessel contains equal numbers of oxygen and of hydrogen molecules. The pressure is 760 mm Hg when the volume is 50 liters. Which of the following statements is *false*? (a) On the average, the hydrogen molecules are traveling faster than the oxygen molecules. (b) On the average, more hydrogen molecules strike the walls per second than oxygen molecules. (c) If the oxygen were removed from the system, the pressure would drop to 190 mm Hg. (d) Equal numbers of moles of each gas are present. (e) The average kinetic energies of oxygen and hydrogen are the same.

23 The vapor pressure of a molten metal can be measured with a device called a Knudsen cell. This is a container closed across the top by a thin foil pierced by a small, measured hole. The cell is heated in a vacuum, until the vapor above the metal streams from the small hole (it effuses).

Two identical Knudsen cells are heated to 1000°C, one containing lead and the other containing magnesium. (a) Contrast the average kinetic energies of the lead and magnesium atoms within each cell. (b) Contrast the average speeds of the lead and magnesium atoms leaving each cell. (c) At this fixed temperature the number of atoms which leave the cell per unit of time is determined by two factors: (1) the pressure exerted by metal vapor in the cell and (2) the mass of the gaseous particles. Explain.

24 Compressed oxygen gas is sold in 40-liter steel cylinders. The pressure at 25°C is 130 atm. (a) How many moles of oxygen does such a filled cylinder contain? (b) How many kilograms of oxygen are in the cylinder? The ideal gas law is useful here. (*Answer to* (b): 6.8 kg.)

25 What is the volume of 0.18 mole of CO_2 at a temperature of 177°C and a pressure of 312 mm Hg?

26 Show how you would calculate R from the following data:

(a) pressure = 760 mm, volume = 12.2 liters, temperature = 298°K, and number of moles = 0.5

(b) pressure = 760 mm, molar weight of gas = 64.1 g/mole, density of gas = 2.97 g/liter, temperature = 273°K

(c) pressure = 108 cm, volume = 33.6 liters, number of grams sample = 3.03 g, molar weight = 2.02 g/mole, temperature = 38.6°K

27 Suppose standard conditions had been selected as 25°C and 2.0 atm instead of at 0.00°C and 1.00 atm. What would the new value of R have been?

From a drop of water, a logician could infer the possibility of an Atlantic Ocean or a Niagara Falls without having seen or heard of one or the other.

SIR ARTHUR CONAN DOYLE (1859–1930)

CHAPTER 5 Liquids and Solids: Condensed Phases of Matter

Water in its two condensed phases—at 0°C water can exist as both a liquid and a solid.

The "ideal gas" has thus far been considered as a collection of rebounding particles which are so far apart that attraction between particles is negligible. We found that no gas is truly "ideal," that the attraction between particles *is* of consequence, and that there is great variation in the strength of these attractive forces. It is as a result of this attraction between real particles that solids and liquids—the crowded or condensed phases of matter—exist.

As gas particles are found to be less than "perfect" on crowding, their differences become more interesting than their similarities. The magnitude of the attractive forces between particles determines many of the physical properties of any substance—boiling point and freezing point, the amount of energy necessary to produce a phase change, and behavior when mixed with other substances. While solids and liquids are not as "simple" to study as free and unattached gas particles, there is a certain relief in dealing with substances in a form that we can see, feel, measure, and weigh more readily than we can gases.

Our model for gases has moved from a purely qualitative description of gas behavior to a quantitative and mathematical description. (1) Experimentally we found that the expression $PV = $ a constant describes the behavior of a number of gases if a constant quantity of gas at a constant temperature is considered. (2) We were then able to use the "super-rubber" ball model for gases to derive the expression $PV = $ a constant (again, quantity of gas and temperature assumed constant). (3) It was then possible* to expand the "super-rubber" ball model to obtain the more general expression $PV = nRT$. The same general expression is also obtainable from experiment.

So far the successes of our model have been spectacular; but before we become too self-satisfied, let us try a few more experiments. Trouble lies ahead—but trouble which may mean progress! The late Peter Debye, Nobel Laureate in chemistry, once said, "I love to see unexpected experimental results; I am bound to learn something." Here is the experiment; we hope we learn something.

5-1 PHASE CHANGES

5-1.1 The Pressure-Volume Behavior of Ammonia at High Pressure—Non-Ideal Gases

In Table 4-2 of Chapter 4 we summarized results obtained when 1 mole of ammonia gas (NH_3) was compressed at 25°C. The volume of the gas was carefully measured as the pressure was increased by stages. The relationship $PV = 24.4 \pm 0.1$ was obeyed very well for pressures from 0.1 to 2.0 atm, but we noticed (pages 69–70) even then that the values were not randomly distributed around 24.4. Instead, a gradual decrease of the "constant" from 24.45 at 0.1 atm to 24.34 at 2.0 atm was seen. Why? What happens if we go beyond 2 atm? Is the relationship $PV = 24.4 \pm 0.1$ still true?

The data shown in Table 5-1 indicate that the relationship $PV = $ a constant is not valid at pressures greater than 2 atm. The

TABLE 5-1 ACCURATE PRESSURE-VOLUME MEASUREMENTS FOR 17.00 GRAMS OF AMMONIA GAS AT 25°C

Pressure (atm)	Volume (liters)	P × V (atm × liters)
0.1000	244.5	24.45
0.2000	122.2	24.44
0.4000	61.02	24.41
0.8000	30.44	24.35
2.000	12.17	24.34
4.000	5.975	23.90
8.000	2.925	23.40
9.800	2.360	23.10 (condensation beginning)
9.800	0.020	0.20 (no gas left; liquid only)
20.00	0.020	0.40 (only liquid present)
50.00	0.020	1.0 (only liquid present)

---

* You saw that it was possible to derive $PV = nRT$ from the gas laws even though you may not have followed the detailed development.

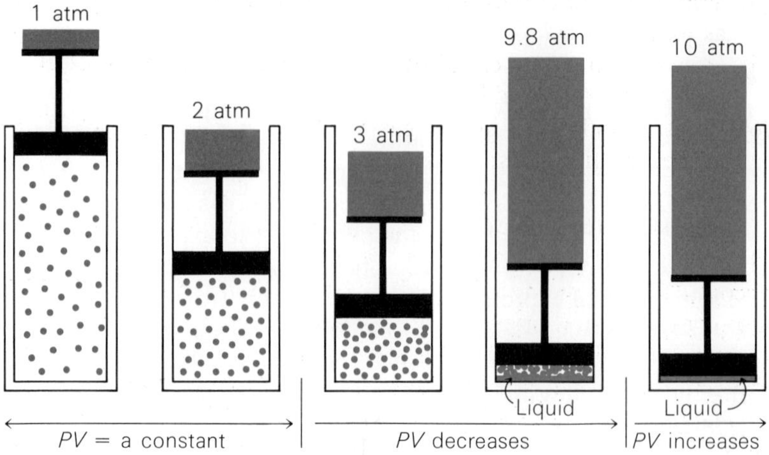

Chapter 5 / LIQUIDS and SOLIDS: CONDENSED PHASES

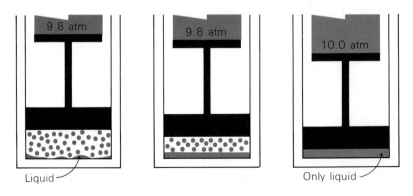

Fig. 5-1.1 **Behavior of ammonia as pressure is increased while temperature remains at 25°C.**

value of the "constant" (originally 24.4) becomes noticeably smaller above 2.0 atm until, at a pressure of 9.8 atm, liquid droplets begin to appear in the gas container and the value of the product obtained by multiplying pressure and volume drops from 23.10 to 0.20. As we try to increase the gas pressure above 9.8 atm by pushing a piston into the container, *we find that more gas is converted to liquid and the pressure returns to 9.8 atm.* We are surprised to learn that *the pressure on the ammonia cannot be raised above 9.8 atm at 25°C until all the gas has been converted to liquid* (Figure 5-1.2). As soon as the last

Fig. 5-1.2 **Enlarged view of the liquefaction of ammonia at 25°C.**

of the gas has disappeared, pressure on the liquid can be increased without any obvious volume change. For a liquid the PV product increases again as pressure increases but without a significant volume change. It is clear that the simple relationship $PV = $ a constant is not a good description of the liquid phase, and our simple gas law fails most rapidly during the change from gas to liquid. These experiments are illustrated in Figure 5-1.

These observations give us much to "wonder why" about. For example:

(1) Why did the value of the constant 24.4 begin to get smaller above about 2 atm pressure?
(2) What is so special about a pressure of 9.8 atm for ammonia at 25°C?
(3) What is the relation between liquid and gaseous phases?

(4) Why does the expression $PV =$ a constant not apply to liquids as well as gases?
(5) How does our "super-rubber" ball model for gases account for the behavior of liquids?
(6) What modification in the gas model is needed to explain the properties of liquids?

Question (6) is the most significant. It can also be phrased: How must the model for gases be altered in order to explain or account for the nature of liquids? Let us suppose that a liquid is also composed of "super-rubber" balls, collected in the lower part of the container. Each ball still has kinetic energy since the liquid is at the same temperature as the gas from which it was formed. This means that the balls (or molecules) can vibrate back and forth and move around pushing the other balls about, but *some forces must be acting among the balls as a group to hold them together.*

Clearly, this is an important point. Our development of the kinetic theory for gases assumed that gas particles exert no real forces on each other. In contrast, we cannot understand the existence of the liquid phase unless we assume that real molecules *do exert* forces of attraction on each other. These forces are very small when the molecules are far apart (that is, when the gas is at low pressure). As we developed the model for gases, agreement between theory and experiment was good at low pressures, even though we did not consider intermolecular attractions. This is because when forces are very small, we can ignore them. But as molecules are pushed closer and closer together at higher pressures, the forces of attraction between molecules become important. At intermediate pressures molecules are attracted to each other just a little and the volume is a little smaller than it would be if there were no attraction between molecules. Because the *volume is less* than it would be without molecular attraction, the PV product for NH_3 at 25°C drops *below* 24.4 at pressures a little higher than 2 atm. This is a reasonable answer to question (1).

When we tried to push the molecules close enough together to give a gas pressure higher than 9.8 atm at 25°C, the forces acting between molecules became strong enough—because of the shorter distance—to pull the molecules together into liquid droplets. Condensation of the gas into liquid continued until the pressure of the remaining gas dropped again to 9.8 atm [question (2)]. The model describes the difference between liquid and gaseous phases in terms of intermolecular forces holding liquid molecules together [questions (3) and (5)]. Finally, we find that in explaining the relationship $PV =$ a constant for gases using the kinetic theory, we assumed that the molecules of gas were far apart and moved rapidly through open space. This is not so for liquids. Liquid molecules are actually held together in a movable mass [question (6)]. It really is not strange that PV is *not* a constant for a liquid [question (4)].

Let us summarize what has been said. The kinetic theory for gases is based on an "ideal" gas—one in which the molecules exert no force on each other. Every gas approaches such "ideal" behavior when its pressure is low enough and temperature high enough. At very low pressures molecules are, on the average, so far apart that their attractive forces are negligible. An *ideal* or *perfect gas* obeys the ideal gas law ($PV = nRT$). Attractions between molecules first cause deviations from the ideal gas law and finally the formation of liquids. Most gases show some deviation from ideality. The larger the deviation from ideal behavior, the smaller the **molar volume** of the gas at STP and the easier the gas is to liquefy. (See data for ammonia in Table 5-1 on page 93.)

EXERCISE 5-1

The following table indicates the boiling points and the molar volumes (at 0°C and 1 atm) of some common gases:

Gas	Formula	Boiling Point (°C)	Molar Volume (liters)
Helium	He	−269	22.426
Nitrogen	N_2	−196	22.402
Carbon monoxide	CO	−190	22.402
Oxygen	O_2	−183	22.393
Methane	CH_4	−161	22.360
Hydrogen chloride	HCl	−84.0	22.248
Ammonia	NH_3	−33.3	22.094
Chlorine	Cl_2	−34.6	22.063
Sulfur dioxide	SO_2	−10.0	21.888

(1) What regularity is suggested in the relationship between the boiling points and molar volumes? Plot the values. Choose a scale so that range 21.500–22.500 will occupy a full page in your notebooks.

(2) Account for this regularity. (*Hint:* What does a high boiling point suggest about intermolecular forces?)

5-1.2 Liquid, Solid, or Gaseous States?— The Factors Involved

Although our description of pressure-volume relationships for ideal gases is fairly good, it is not very widely applicable. Only a handful of substances are gases under normal conditions of temperature and pressure. Of the hundred or so elements, eleven are gases* and two (bromine and mercury) are liquids at STP (0°C and 1 atm). All the rest are solids. As for compounds, more than a million have been prepared. Yet more than 99 percent of these are liquids or solids at STP. Gases clearly are in the minority.

Look again at the data for ammonia in Table 5-1. Here we see that the classification of ammonia as a gas or a liquid does not really mean much unless we give the temperature and the pressure for the system. Ammonia is a gas at 25°C and 1 atm; it is a liquid at 25°C and 10 atm. Pure substances can be converted from gas to liquid in a process which we call a **phase change.** Phase changes are familiar to you: ice melts to water and water boils to give steam (water vapor); solid iron melts to liquid iron; naphthalene melts to a liquid; gaseous ammonia condenses to give liquid ammonia; candle wax

* The eleven gaseous elements at STP are hydrogen (H_2), helium (He), nitrogen (N_2), oxygen (O_2), fluorine (F_2), neon (Ne), chlorine (Cl_2), argon (Ar), krypton (Kr), xenon (Xe), and radon (Rn).

melts to give liquid wax. These are only a few of the phase changes with which you are familiar.

Phase changes are important in the identification of pure substances and in extending the usefulness of the kinetic theory. The temperature at which a pure solid under a given pressure changes to a liquid is its **melting point.** The melting point of a pure substance* is a characteristic property of that substance and can be used to identify the material. For example, ice changes to water at 0.00°C; methylsalicylate, ordinary oil of wintergreen, melts at −8.6°C; and trinitrotoluene, the widely known explosive TNT, melts without explosion at 80.7°C. Chemists use melting points frequently to identify substances.

5-1.3 Energy Required for Phase Changes— The Heat of Vaporization

When a pan of water is warmed, the input of energy causes the water temperature to rise. At a certain temperature the water begins to boil. Gas bubbles form in the liquid and rise to the surface where they burst. Liquid water is being converted to gaseous water. If we place a thermometer in the boiling water, we find that the *temperature remains constant as long as any liquid water remains* (see Figure 5-2). *The energy added by the gas flame is being used to convert water in the liquid phase to water in the gas phase at constant temperature.* If we stop adding energy (turn off the gas flame), the boiling stops immediately. In summary we can say: *When water changes from the liquid phase to the gaseous phase at constant pressure, energy must be supplied; yet the boiling temperature remains constant.*

Clearly, water in the gas phase has more energy than water in the liquid phase. The equation representing the change is

$$H_2O(liquid) + \text{thermal energy} \longrightarrow H_2O(gas) \text{ at } 100°C \text{ and } 1 \text{ atm} \qquad (1)$$

How much energy is involved in the process? Using a calorimeter like the one you used in the laboratory, we discover that it takes 9.70 kilocalories of energy to vaporize 1 mole of water (6.02×10^{23} mole-

---

* The melting point of a pure substance will change slightly with pressure; hence, for precise work pressure must be specified. Changes in melting point due to changes in pressure are usually very small, however, and will not concern us in the laboratory work for this course.

Fig. 5-2 **Temperature of water above a flame as a function of time. (Pressure of atmosphere = 760 mm Hg.)**

Fig. 5-3 **Schematic representation of the vaporization of 1 mole of liquid water.**

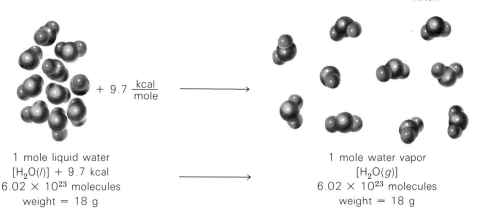

1 mole liquid water
[$H_2O(l)$] + 9.7 kcal
6.02×10^{23} molecules
weight = 18 g

+ 9.7 $\dfrac{\text{kcal}}{\text{mole}}$ ⟶

1 mole water vapor
[$H_2O(g)$]
6.02×10^{23} molecules
weight = 18 g

cules or 18.0 g). The equation can be written in somewhat abbreviated but more precise form as

$$H_2O(l) + (9.70 \times 10^3 \text{ cal}) \longrightarrow H_2O(g) \text{ at } 100.00°C \text{ and } 1 \text{ atm} \quad (2)$$

The value 9.70×10^3 cal is called the **molar heat of vaporization** of water. This is the energy required to separate 6.02×10^{23} molecules of water one from another, as pictured in Figure 5-3.

EXERCISE 5-2

How much energy is required to evaporate 2 moles of water? ½ mole of water? 1.94×10^4

When water vapor condenses to liquid water, the molecules release the energy which was absorbed in separating them. One mole of gaseous water will release 9.70×10^3 cal of energy (heat) when condensed to liquid water at the same temperature. The amount of heat energy released is numerically equal to the molar heat of vaporization.

The amount of energy required to evaporate 1 mole of pure liquid to give 1 mole of pure vapor (gas) varies over a wide range for different pure substances. Table 5-2 shows the boiling points and heats of vaporization of a variety of liquids. In each case, energy is absorbed by the liquid molecules as they separate into gaseous molecules.

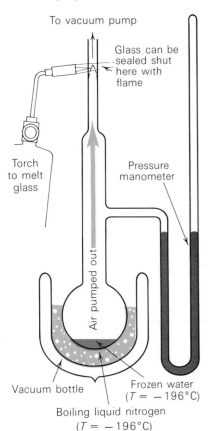

Fig. 5-4.1 **Preparing a flask to study the vapor pressure of water.**

TABLE 5-2 THE NORMAL BOILING POINTS AND MOLAR HEATS OF VAPORIZATION OF SOME PURE SUBSTANCES

Substance	Phase Change (liquid) $\longrightarrow$ (gas)	Boiling Point (°K)	Boiling Point (°C)	Molar Heat of Vaporization (kcal/mole)
Neon	$Ne(l) \longrightarrow Ne(g)$	27.2	−245.8	0.405
Chlorine	$Cl_2(l) \longrightarrow Cl_2(g)$	238.9	−34.1	4.88
Water	$H_2O(l) \longrightarrow H_2O(g)$	373	100	9.7
Sodium	$Na(l) \longrightarrow Na(g)$	1162	889	24.1
Sodium chloride	$NaCl(l) \longrightarrow NaCl(g)$	1738	1465	40.8
Copper	$Cu(l) \longrightarrow Cu(g)$	2855	2582	72.8

We shall see in Chapter 8 that the extreme range of heats of vaporization shown in this table can be explained using rather simple principles. These principles provide a basis for qualitative predictions of boiling point, heat of vaporization, and other properties.

5-1.4 Liquid-Vapor Equilibrium—Vapor Pressure

We have been considering vaporization of a liquid at its usual boiling point. But liquids can vaporize at other temperatures. Let us consider this process, again beginning with liquid water.

A 50-ml sample of liquid water is placed in the flask shown in Figure 5-4.1. The liquid is frozen with a very cold refrigerant such as liquid nitrogen ($-196°C$), and the air in the flask is pumped out using a vacuum pump. When the air has been removed as completely as possible, the glass tube is melted shut at the point so indicated in the figure. We can now place the sealed flask in a series of constant temperature baths maintained at $0°C$, $25°C$, $50°C$, and $100°C$. The results are seen in Figure 5-4.2.

Our observations indicate that

(1) *there is a pressure above the liquid even after the air has been removed,* and
(2) *the pressure above the liquid increases as the temperature is raised.*

The pressure reading in the flask will change when the flask is first put into the thermostat, but ultimately *many different sealed flasks of pure water give the same reading at the same temperature.* When the pressure reading does not change with time, we say that the liquid and its vapor are in **equilibrium.** The pressure reading is then the **equilibrium pressure.** The **equilibrium pressure reading** in

LIQUID-VAPOR
EQUILIBRIUM—VAPOR PRESSURE

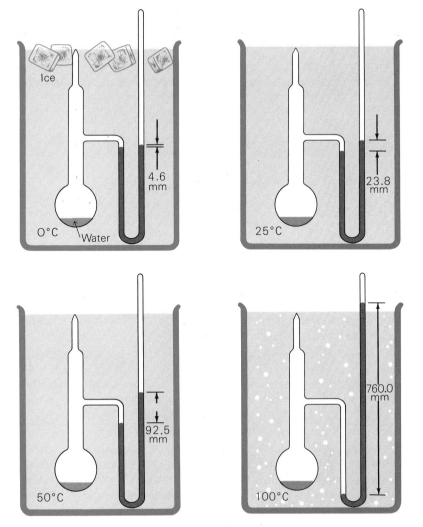

Fig. 5-4.2 **Vapor pressure of water at four different temperatures.**

Chapter 5 / LIQUIDS and SOLIDS: CONDENSED PHASES

TABLE 5-3 VAPOR PRESSURE OF WATER

Temperature (°C)	Vapor Pressure (mm Hg)
0	4.6
25	23.8
50	92.5
100	760.0

TABLE 5-4 VAPOR PRESSURE OF BENZENE

Temperature (°C)	Vapor Pressure (mm Hg)
0	27
25	94
50	271
100	1,360

the flask gives the **vapor pressure** of water at the temperature of the thermostat. In a more formal sense we can say: *The pressure exerted by a vapor in equilibrium with its pure liquid phase is the vapor pressure of the liquid at the temperature used.* At equilibrium no net evaporation or condensation is taking place. We are unable to detect measurable changes.

From the experiment just described, we can now tabulate the vapor pressure of water at four different temperatures in Table 5-3 at the left. The experiment can be repeated using many liquids other than water. In this way we learn a third fact: *Different liquids at the same temperature give different vapor pressures.* For example, if liquid benzene is used in the experiment instead of water, the vapor-pressure values at 0°C, 25°C, 50°C, and 100°C are those in Table 5-4 at the lower left. A repetition of this type of experiment for many liquids indicates another generalization: *The vapor pressure of every liquid increases as the temperature is raised.*

Our observations on vapor pressure can now be summarized:

(1) The pressure exerted by a vapor in equilibrium with its liquid phase at constant temperature is the vapor pressure of the liquid at that temperature.
(2) The vapor pressure of a liquid is dependent only upon the nature of the liquid and the temperature.
(3) Different liquids at any one temperature have different vapor pressures.
(4) The vapor pressure of every liquid increases as the temperature is raised.

=============== EXERCISE 5-3 ===============

Watch water boil in a pan or beaker. What is inside the bubbles that rise from the bottom and sides during boiling?

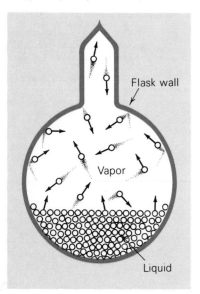

Fig. 5-5 **Kinetic theory view of liquid-vapor equilibrium.**

5-1.5 The Kinetic Theory for the Vapor Pressure of a Liquid—Partial Pressures

Does the **kinetic theory** (or "super-rubber" ball model) for liquids and gases explain what we have seen? Let us put it to the test.

We suggested earlier that molecules attract each other as they are pushed closer together. At some point this attraction is great enough to pull the molecules together into droplets or aggregates which we call liquids. We noted that molecules in liquids have kinetic energy too; they are in motion. At any instant some move rapidly, some move slowly. If a given molecule near the surface begins a rapid movement away from the surface of the liquid, it may have enough kinetic energy to overcome the forces of attraction which hold it to the liquid. It then leaves the liquid and enters the vapor (gas) phase (Figure 5-5). In the vapor phase it behaves like any other gas molecule; its collisions with the walls produce pressure. The model tells us that if a liquid has strong forces of attraction between molecules, then only a few molecules will have enough energy to overcome the forces of attraction and escape. Only a relatively few molecules will be able to evaporate. *Thus, liquids, at the same temperature, with strong attractive forces between molecules will have a lower vapor pressure than will liquids with weak intermolecular forces (temperature assumed constant).*

EXERCISE 5-4

Which liquid has the stronger intermolecular forces, water or benzene?

Clearly, different liquids at the same temperature should have different values for their vapor pressure. (See Table 5-5.) So far, so good. The model is in agreement with observation. What about temperature effects? An increase in temperature of liquid or gas is associated with an increase in the average kinetic energy of the molecules. As the average kinetic energy increases, a larger number of molecules will have enough kinetic energy (i.e., are moving fast enough) to overcome the forces of attraction in the liquid and escape into the vapor (Figure 5-5). At higher temperatures there will be more molecules in the vapor phase. Hence, vapor pressure will increase with temperature.

TABLE 5-5 VAPOR PRESSURES OF LIQUIDS

Temp (°C)	Water (mm Hg)	Ethyl Alcohol (mm Hg)	Carbon Tetrachloride (mm Hg)	Methyl Salicylate (mm Hg)	Benzene (mm Hg)
−10	2.1	5.6	19		15
−5	3.2	8.3	25		20
0	4.6	12.2	33		27
5	6.5	17.3	43		35
10	9.2	23.6	56		45
15	12.8	32.2	71		58
20	17.5	43.9	91		74
25	23.8	59.0	114		94
30	31.8	78.8	143		118
35	42.2	103.7	176		147
40	55.3	135.3	216		182
45	71.9	174.0	263		225
50	92.5	222.2	317		271
55	118.0	280.6	379		325
60	149.4	352.7	451	1.41	389
65	190.0	448.8	531	1.90	462
70	233.7	542.5	622	2.52	547
75	289.1	666.1	820	3.40	643
80	355.1	812.6	843	4.41	753
85	433.6	986.7	968	5.90	877
90	525.8	1187	1122	7.63	1020
95	633.9	1420	1270	9.93	1280
100	760.0	1693.3	1463	12.8	1360

What happens as time passes? Some molecules leave the liquid and go into the vapor. Some molecules in the vapor collide with the liquid surface and return to the liquid phase. Soon a condition will exist where the number of molecules leaving the liquid in any one second is just equal to the number of molecules returning to the liquid in any one second. To the observer on the outside, no measurable changes are taking place. We say that the system is at **equilibrium**: the *rate* at which molecules leave the liquid phase is equal to the *rate* at which they return to the liquid phase.

So far we have been concerned only with the vapor pressure produced by a pure liquid in a flask from which all other gases have been removed. What will happen if other gases are present? Surely, other gas molecules above the surface will collide with molecules escaping from the liquid surface (see point A in Figure 5-6). After the collision, the escaping molecule may well bounce back into the liquid. The *rate* at which molecules escape from the surface is *reduced* by the presence of another gas above the liquid. If this were the only consequence of other gases being above the liquid, the vapor pressure would fall; but *we find that the vapor pressure does not fall*. The explanation lies in the fact that molecules *returning to the surface* are also blocked to some degree by collisions with the inert* gas molecules (see point B in Figure 5-6). In short, *the "inert" gas has just as much effect on the rate of return of molecules as it does on their rate of escape.* The presence of an "inert" gas above the liquid does not alter the *partial pressure* of the vapor (gas from the liquid) above that liquid. *The equilibrium partial pressure due to gaseous water molecules above a liquid water surface is independent of the presence of air or other gas above that surface.* The measured pressure due to air and water above the water surface will be the *sum* of the *partial pressures* of the gaseous water and the other gases which make up the air. For example, if a flask originally contained no measurable amount of air (it was evacuated as in the experiment), liquid water placed in the flask at 20°C would evaporate until the pressure rose from 0.0 mm to 17.5 mm. If the flask originally contained dry air at 750.0 mm, liquid would evaporate until the total pressure rose from 750.0 mm to 767.5 mm (the partial pressure of water vapor changing from 0.0 to 17.5 mm in both cases). Experiment fully confirms this prediction.

5-1.6 The Kinetic Theory and the Heat of Vaporization

We are all familiar with the cooling effect a breeze has when it blows over a person wearing a wet bathing suit. A person can feel cold even on a very hot day. Why? It is convenient to say that the evaporation of water has a "cooling effect" and that the breeze increases the rate of evaporation of water, but this does not really provide a fundamental answer. What does the kinetic theory model tell us about this question? According to the kinetic theory, those molecules in the liquid surface with the highest kinetic energy will have the greatest *probability* of overcoming the intermolecular forces of attraction and escaping at any instant. In short, the more energetic or more rapidly moving molecules will escape, while the less energetic, more slowly moving molecules will remain. The molecules remaining will thus have lower kinetic energy on the average than did mole-

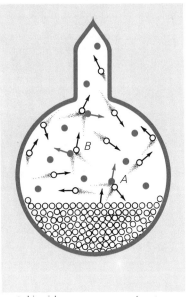

○ Liquid or vapor molecules   ● Inert gas molecules

Fig. 5-6 **Effects of an "inert" gas on liquid-vapor equilibrium.**

---

* "Inert gas" as used here refers to any gas which does not react with the liquid being vaporized.

cules of the original liquid. Since temperature is directly proportional to the average kinetic energy of the molecules, the temperature of the remaining liquid is lower. Stated in a simpler but less accurate way, we can say: The faster moving, "hot" molecules go to the vapor; the slower moving, "cold" molecules remain in the liquid. The liquid cools. If we want the temperature to remain constant, external energy—the **heat of vaporization**—must be added to supply the energy lost in overcoming the intermolecular forces of attraction between molecules in the liquid phase.

5-1.7 The Boiling Point

The model tells us that the vapor pressure of a liquid increases as temperature rises (Section 5-1.5). If a liquid in an open dish is heated, its temperature will go up and its vapor pressure will rise until bubbles appear in the liquid and boiling begins. The temperature and vapor pressure will then remain constant as the liquid boils (Figure 5-2).

Separate experiments show that when the liquid boils, the vapor pressure of the liquid is just equal to the pressure of the atmosphere. Why is the pressure of the atmosphere important to boiling? The kinetic theory provides the answer. Consider the bubble just below the liquid surface in Figure 5-7. Vapor molecules from vaporization of the liquid are present inside the bubble. These vapor molecules bombard the bubble walls and tend to push them outward. At the same time the atmosphere pushes downward on the surface of the liquid. The push of the atmosphere is transmitted equally in all directions throughout the liquid and tends to collapse the bubble. If the atmospheric pressure is greater than the vapor pressure of the liquid, the bubble collapses. If the pressure inside the bubble (due to vapor pressure of liquid at that temperature) is slightly higher than atmospheric pressure, the bubble grows and rises to the surface where it breaks. Bubble formation is characteristic of the boiling process.

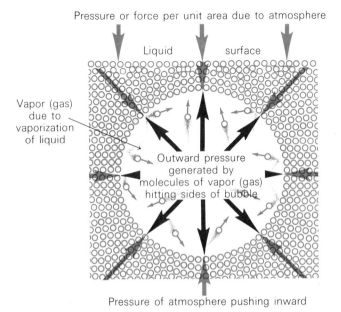

Fig. 5-7 **Bubble in a boiling liquid.**

Chapter 5 / LIQUIDS and SOLIDS: CONDENSED PHASES

The boiling point of a liquid is the temperature at which the vapor pressure of the liquid is equal to the pressure of the atmosphere above the liquid. If atmospheric pressure is 760.0 mm Hg, water boils at 100.0°C. At 100.0°C the vapor pressure of water is 760.0 mm Hg. If the atmospheric pressure is 750 mm Hg, water boils at 99.6°C, because the vapor pressure of water is 750 mm Hg at 99.6°C.

The **normal boiling point** of a liquid is defined as the temperature at which the vapor pressure of the liquid is exactly 1 standard atm or 760.0 mm Hg.

=== EXERCISE 5-5 ===

What is the normal boiling point of ethyl alcohol (see Table 5-5)?

75-80°C

=== EXERCISE 5-6 ===

Suppose a closed flask containing liquid water is connected to a vacuum pump and the pressure over the liquid is gradually lowered. If the water temperature is kept at 20°C, at what pressure will the water boil?

17.5 mm Hg

=== EXERCISE 5-7 ===

Answer Exercise 5-6 substituting ethyl alcohol for water. Repeat for carbon tetrachloride at 40°C.

43.9 mm   *216 mm*

=== EXERCISE 5-8 ===

760.0
633.9 mm
126.1

A thermometer in a pot of boiling water on a mountain reads 95.0°C. What is the atmospheric pressure on the mountain? The rule of thumb for altitude-pressure correlations is that the pressure falls about 25 mm for every 1,000 feet of elevation. Estimate the height of this mountain.

5,000 ft.

5-1.8 Solid-Liquid Phase Changes

According to the model just developed, molecules in solid and liquid phases (called *condensed phases*) are held together by intermolecular attraction. In liquids, molecules are irregularly spaced, randomly oriented, and reasonably free to move over each other. In crystalline solids, the molecules occupy regular positions and are held together more firmly (see Figure 5-8). One mole of a pure solid has lower energy than 1 mole of its liquid phase at the same temperature.

The difference between the energy of a substance in liquid form

Fig. 5-8 **Schematic representation of the melting of 1 mole of ice.**

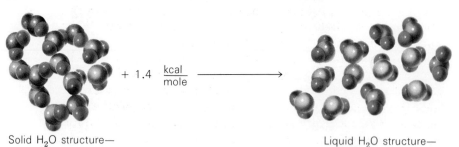

Solid H_2O structure—
strongly ordered

$+ 1.4 \frac{kcal}{mole}$

Liquid H_2O structure—
strongly disordered

and its energy in solid form is usually much smaller than the difference between the energies of the liquid and gaseous forms. For ice we can write

$$H_2O(solid) + 1,440 \text{ cal} \longrightarrow H_2O(liquid) \text{ at } 0°C \qquad (3)$$

or, 1.44×10^3 cal of energy would be required to convert 1 mole of ice (18.0 g) to 1 mole of liquid water at $0°C$. This is much less than the 9.7×10^3 cal required to convert 1 mole of water to vapor at $100°C$. Table 5-6 contrasts the melting points and the heats of melting per mole of the pure substances listed in Table 5-2. The energy of melting per mole is known as the **molar heat of melting** or the **molar heat of fusion**.

TABLE 5-6 THE MELTING POINTS AND HEATS OF MELTING OF SOME PURE SUBSTANCES

Substance	Phase Change (solid) $\longrightarrow$ (liquid)	Melting Point (°K)	Melting Point (°C)	Molar Heat of Melting (kcal/mole)
Neon	Ne(s) $\longrightarrow$ Ne(l)	24.6	−248.4	0.080
Chlorine	Cl_2(s) $\longrightarrow$ Cl_2(l)	172	−101	1.53
Water	H_2O(s) $\longrightarrow$ H_2O(l)	273	0	1.44
Sodium	Na(s) $\longrightarrow$ Na(l)	371	98	0.63
Sodium chloride	NaCl(s) $\longrightarrow$ NaCl(l)	1081	808	6.8
Copper	Cu(s) $\longrightarrow$ Cu(l)	1356	1083	3.11

We find an extreme range for the heats of melting of these substances. The molar heats of melting may vary from 0.080 kcal/mole for neon to 6.8 kcal/mole for sodium chloride (common table salt). The value for sodium chloride is 85 times that for neon. As you may well suspect, there are very great differences in the forces that bind solids together. These forces affect many properties other than melting point and heat of melting. We shall consider them in more detail later.

5-2 SOLUTIONS

Sodium chloride, sugar, ethyl alcohol, and water are four **pure substances**. Each is characterized by definite properties such as vapor pressure, melting point, molar heat of melting, boiling point at a given pressure, and density at a given temperature. Suppose we mix some of these pure substances. Sodium chloride dissolves when placed in contact with water. The solid disappears, becoming part of the liquid. Similarly, sugar in contact with water dissolves. When ethyl alcohol is added to water, the two substances mix to give a liquid similar in appearance to each of the original liquids.

The salt-water mixture, the sugar-water mixture, and the alcohol-water mixture are all _homogeneous._ Each mixture has only a single liquid phase. Such homogeneous mixtures are called **solutions.** In the laboratory we shall handle solutions frequently. Solutions of many types surround us. Filtered seawater is a solution of salts in water; the air we breathe (sans smog) is oxygen and other gases dissolved in nitrogen; soft drinks are solutions made by dissolving sugar, flavoring, carbon dioxide, and occasionally other ingredients in water; even the gasoline we burn in our cars is a solution made by mixing organic compounds called *hydrocarbons* with various anti-knock agents, and other soluble special purpose ingredients.

Solutions differ from pure substances in that their properties vary, depending upon the relative amounts of the pure substances present. The behavior of a solution during a phase change is dramatically different from the behavior of a pure substance. For this reason it is useful to make a distinction between pure substances and solutions. These differences provide a basis for deciding whether a given material is a pure substance or solution. For example, suppose we compare salt water and distilled water. (Each sample is a homogeneous system.) How do we know which one, if either, is a pure substance? We cannot tell just by looking at them. True, there are differences such as taste, density, and so on, but these differences considered separately do not tell us which sample is a pure substance and which a mixture. More sophisticated observation is needed. Let us look further.

5-2.1   Solutions and Phase Changes

A sample of water freezes at 0°C (at 760 mm Hg). If we freeze 10 ml out of a 100-ml sample of water, a 10-ml ice cube and 90 ml of pure water will remain. If we continue this process, separating ice and water each time, ten ice cubes will result—all the same size and composition, all pure water. In a similar way we can boil the original sample of water (at 100°C) until 10 ml has vaporized and condensed (see Figure 5-9). A 90-ml sample of hot water and a 10-ml sample of freshly distilled water result. If the boiling process is continued, nine more 10-ml quantities of freshly distilled water will be obtained. The properties of the different samples of a pure substance, separated through freezing or boiling, remain unchanged. A pure substance has a definite and reproducible melting point and

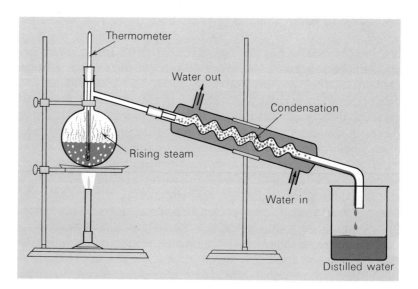

Fig. 5-9 **A distillation unit.**

boiling point at a given pressure. In fact, chemists use melting and boiling points of pure substances to identify compounds.

Solutions behave differently. Suppose that we start with a 100-ml sample of a salt solution. We have made the solution by dissolving a convenient amount (say, 1 g) of sodium chloride in the water. This solution is not very different from the pure water, but the difference is important. We are not surprised to find that it begins to boil close to, but not exactly at, 100.0°C. If we collect the steam (gaseous water) from the salt solution and condense it in a separate vessel (see Figure 5-9), we find that the resulting liquid behaves like pure water rather than like the solution from which it came. Furthermore, after a 10-ml sample of water has been vaporized, we are left with 90 ml of a solution whose composition is not the same as the original. It is saltier: the same amount of salt is present in the solution containing less water. The first solution had 1 g of salt dissolved in 100 ml of solution; the second solution had 1 g of salt dissolved in 90 ml of solution. There is a definite difference in the relative amounts of the components.

This difference results in different properties for the two solutions. Figure 5-10 contrasts the behavior of pure water and a concentrated salt-water solution during boiling. As the amount of water

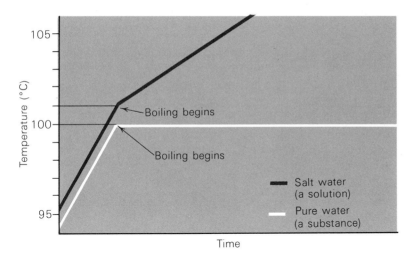

Fig. 5-10 **Behavior on boiling.**

per gram of salt decreases, the boiling point of the salt solution increases. The salt solution starts to boil at a temperature which is somewhat higher than the boiling point of pure water. As the boiling continues, the temperature of the pure water remains constant whereas the temperature of the salt solution keeps rising. If we boil off all the water, solid salt remains.

Thus, by **distilling**—that is, by *evaporating and recondensing the resulting vapor in a separate vessel*—we can separate a pure liquid from a solution; and by **crystallizing**—that is, by *forming a crystalline solid*—we can frequently obtain a pure solid from a solution. We usually call the pure substances making up the solution (such as salt and water) the **components** of the solution. It should be obvious that crystallization or distillation will *not* separate a pure substance into components. *A pure substance consists of only one component.* The more closely alike the components of a solution are, the harder it is to separate them.

In nature, solutions are much more common than pure substances, and heterogeneous systems (systems containing more than one phase, such as muddy water) are much more common than solutions. When we want pure substances, we often must prepare them from solutions through successive phase changes.

Although our main interest is in liquid solutions, gas and solid solutions are also important for certain situations. We shall consider them briefly and then return to liquid solutions, which are the most important of the three from a chemist's point of view.

5-2.2 Gaseous Solutions

All gas mixtures are homogeneous; hence, all gas mixtures are solutions. Air is an example. There is only one phase—the gas phase—and all the molecules behave as gas molecules. The molecules themselves may have come from gaseous, liquid, or solid substances. Whatever the source of the components, this gaseous solution—air—is a single, homogeneous phase. As with other solutions, the components of air can be separated by phase changes. Through an exotic low-temperature distillation process, liquid air can be separated into liquid oxygen (boiling point $-182°C$), liquid nitrogen (boiling point $-196°C$), and other substances such as argon.

5-2.3 Solid Solutions

Solid solutions are not particularly common in an elementary laboratory study in chemistry, but they are very common in nature. Many of our common minerals and gem stones, such as emeralds, sapphires, rubies, and amethysts, are solid solutions which have always fascinated men. Steel is also a solid solution in which carbon atoms replace some of the iron atoms of the iron crystal. Similarly, gold atoms can replace some of the copper atoms in a copper crystal to give a gold-copper alloy. The alloy is a solid solution. In the same manner, copper atoms can replace gold atoms in a gold crystal to give another solid solution, a copper-gold alloy. Many other metal alloys are solid solutions. Solid solutions are extremely important today in transistors and in so-called solid-state radios, amplifiers, record players, and television units.

5-2.4 Liquid Solutions

In your laboratory work you will deal mostly with liquid solutions. Liquid solutions can be made by mixing two liquids (e.g., alcohol and water), by dissolving a gas in a liquid (e.g., carbon dioxide in water to give "soda water"), or by dissolving a solid in a liquid (e.g., sugar in water). The result is a homogeneous, transparent system containing more than one substance. By definition, the result is a solution.

In such a solution, each component is diluted by the other. The effects of this dilution process are usually fairly easy to see. Consider for a moment the vapor pressure of a salt solution. Since the solution contains salt, which does not vaporize, and water, which does, there will be fewer vaporizable water molecules per unit of volume than in pure water. Furthermore, water molecules that are present in solution may well be restricted in their movement by their attraction for the dissolved salt. The net effect is that the "escaping tendency" of the water in the solution will be *lower* than the "escaping tendency" of the pure water. As a result the *vapor pressure* due to gaseous water above the solution will be lower than the vapor pressure due to gaseous water above pure water at the same temperature. We know that pure water has a vapor pressure of 760 mm Hg at 100°C. It is necessary to heat a salt solution *above* 100°C to reach this same vapor pressure. For this reason the boiling point of salt water at a given pressure is higher than the boiling point of pure water at the same pressure. The amount by which the boiling point is raised is dependent upon the relative amounts of water and salt present in the solution. The more salt in a given volume of solution, the less pure water and the higher the boiling point.

Correspondingly, a lower temperature is required to crystallize ice from salt water or from an alcohol-water solution than is required to crystallize ice from pure water. "Antifreeze" substances added to an automobile radiator act on this principle. They dilute the water in the radiator and lower the temperature at which ice can crystallize from the solution. Again, the amount that the freezing temperature is lowered depends on the relative amounts of water and antifreeze.

In general, the properties of a solution depend upon the relative amounts of the components. It is important to be able to specify quantitatively what is present in a solution; that is, to specify its **composition**—the relative amounts of the components, or pure substances, that were mixed to form the solution. If there are two components, one is called the **solvent** and the other is called the **solute.** *These are merely terms of convenience.* Since both components must intermingle to form the final solution, we cannot make any important distinction between them. When a liquid solution is made from a pure liquid and a solid, it is usually convenient to call the liquid component the solvent.

5-2.5 Expressing the Composition of Solutions

To indicate the composition of a particular solution we must know *relative amounts,* as well as the *kinds,* of components used. When we specify the relative amounts of each component present in the solution, we have specified the **concentration** of the solution. Chemists have many different ways of expressing concentration. The choice of method depends upon the way in which the solution will be used. We shall need only one method of expressing solution concentration in this course. That method will be described below.

In your laboratory work you have frequently found it convenient to measure out a given quantity of a reagent by using a *measured volume of a solution of known concentration.* Rapid measurement of reagents represents one of the most important uses we shall have for solutions in our work. Further, reagents dissolved in solution will frequently react more rapidly and completely than will the pure reagents. This is a welcome bonus resulting from using solutions.

How can a solution be used to measure out a given quantity of reagent? Let us see how this works. Suppose we weigh out carefully 1 mole of table salt. (From the formula this is 23.0 g Na + 35.5 g Cl or 58.5 g of NaCl per mole.) The salt is dissolved in about 500 ml of water; then the resulting solution is transferred completely to the 1,000-ml volumetric flask shown in Figure 5-11. Water is added, the flask is shaken, and more water is added until the volume of the solution is 1 liter (at the mark on the flask) at 25°C. The solution is now shaken to be sure that liquid and solid are thoroughly mixed. In the final solution we have 1 mole of solute (salt) per liter of solution. A solution containing 1 mole of solute per liter of solution is called a *1-molar* solution. **Molarity** *is determined by dividing moles of solute by liters of solution.* Notice that the concentration of water is not specified, though we must add definite amounts of water to make the solutions.

Solutions of many molarities can be prepared. For example, a solution containing 2 moles of solute per liter of solution is called a 2-molar ($2M$) solution. A solution containing 0.750 mole of solute in 2.00 liters of solution would have a molarity of 0.375. This would be written as 0.375 M.

How is a solution of known molarity used to measure out a given quantity of reagent? If we pour a ½-liter quantity (500 ml) of a 1.00-molar solution into a beaker, the beaker contains ½ mole of the reagent (NaCl). If we measure out 100 ml or a 0.100-liter sample of solution, we have measured out 0.100 mole of the dissolved reagent. In general the number of moles of reagent contained in a given quantity of a solution of known molarity is given by the expression

$$\left\{ \begin{array}{c} \text{number of moles} \\ \text{of solute} \end{array} \right\} = \left\{ \begin{array}{c} \text{molarity of} \\ \text{solution} \end{array} \right\} \times \left\{ \begin{array}{c} \text{volume of solution} \\ \text{taken in liters} \end{array} \right\} \quad (4)$$

$$\left\{ \begin{array}{c} \text{number of moles} \\ \text{of solute} \end{array} \right\} = \frac{1.000 \text{ mole}}{\text{liter}} \times 0.500 \text{ liter} = 0.500 \text{ mole}$$

=== EXERCISE 5-9 ===

(1) A 10-mole quantity of sugar is dissolved in 5 liters of water. What is the molarity of the sugar solution? 2 M
(2) In the laboratory (Experiment 8) you weighed out 2.34 g of NaCl and put it in enough water so that the final volume of the solution was 15.0 ml (*Note: 15 milliliters, not liters*). What was the molar concentration of the salt solution formed? (Review Section 3-1.)

5-2.6 Solubility

When sugar is added to water in a beaker, it dissolves. As more sugar is added, the concentration increases until the sugar begins to accumulate on the bottom of the beaker. From this point on, so long as the temperature remains constant, *the concentration of sugar in the solution will not change.* This solution, containing all of the sugar that it will hold at equilibrium at a given temperature, is called a **saturated solution.** A solution in equilibrium at a given temperature with excess solid is said to be saturated. The solution and

Fig. 5-11 Volumetric flask.

$\dfrac{.04 \text{ mole}}{.015 \ell} = \dfrac{2.34 g}{15 \text{ ml}} = 2.66 \text{ M}$

solid are in equilibrium (solution is saturated) if a small crystal added to the solution neither increases or decreases in volume. In a saturated solution the net volume of the added crystal will remain constant.

When a fixed amount of liquid has dissolved all the solid it can hold at equilibrium at a given temperature, the concentration reached is called the **solubility** of that solid in the liquid. The solubilities of solids in liquids range from very low to very high values. For example, a saturated solution of sodium chloride in water at 20°C has a concentration of about 6 moles NaCl per liter of water solution (6 M). In contrast, a saturated solution of NaCl in ethyl alcohol at 20°C has a concentration of only 0.009 moles NaCl per liter of alcohol solution (0.009 M). The solubility of silver nitrate ($AgNO_3$) *in water* is greater than 5 moles per liter (5 M). In contrast, silver chloride (AgCl) has a solubility *in water* of only 10^{-5} mole per liter. You used these facts in the laboratory (Experiments 7 and 8). The $AgNO_3$, having good solubility in water, dissolved easily to form a solution which reacted with the copper coil. In Experiment 8 you were able to take advantage of the low solubility of AgCl to separate it as a solid from the remaining solution.

The solubilities of solids in any solvent will change with temperature. An increase in temperature will sometimes increase solubility, sometimes reduce it.

Because of this range of solubilities, the word *soluble* does not have a precise meaning. There is usually an upper limit to the solubility of even the most soluble solid, while even the least soluble solid yields a few dissolved particles per liter of solution. If a compound has a solubility of more than $\frac{1}{10}$ mole per liter (0.1 M), most chemists usually say it is *soluble*. When the solubility is below 0.1 M (10^{-1} M), most chemists usually say the compound is *slightly soluble*. Compounds with solubility below about 10^{-3} M are sometimes said to be *very slightly soluble*. With solubility much below about 10^{-4} M, a substance is described as *insoluble*. We use glass containers for pure water because glass has a negligible solubility in water.

EXERCISE 5-10

Classify the solubility in water of the following substances that you have used in your experiments: copper, magnesium, silver chloride, hydrogen chloride (g), and silver nitrate.

5-2.7 Variations in the Properties of Solutions

Though many solutions are colorless and closely resemble pure water in appearance, the differences between solutions are great. Furthermore, the ability of apparently similar solvents to dissolve different solids varies widely. For example, pure sugar dissolves in both pure water and in pure (ethyl) alcohol. Sodium chloride or table salt dissolves in water but not noticeably in ethyl alcohol. Iodine, however, does not dissolve in water to any extent but is quite soluble in ethyl alcohol. Clearly, solubility is determined by the nature of both solute and solvent.

The examples just described give us four solutions containing a substantial amount of solute:

(1) sugar in water
(2) sugar in ethyl alcohol
(3) sodium chloride in water
(4) iodine in ethyl alcohol

Solution (4) is readily distinguishable by its dark brown color. The other three are colorless, but they can be identified easily by taste. Chemists, however, have safer and more meaningful ways of distinguishing them: *these solutions differ markedly in their ability to conduct an electric current.* Solution (3) conducts an electric current much more readily than does pure water. In contrast, solutions (1) and (2) conduct the electric current no better than does pure water.

Differences in solubility and electrical conductivity are very important in chemistry. We shall investigate electrical conductivity and the electrical nature of matter in the next chapter.

HIGHLIGHTS

Liquids and solids (condensed phases) exist because molecules attract each other when they are close together. These forces of attraction can be ignored in a study of gases because gas molecules are separated by large distances; but in liquids and solids molecules are close together and the forces of attraction between them, therefore, are important.

In changing a liquid to a gas at constant temperature, energy from the outside is needed to separate the liquid molecules one from another. The energy required to convert 1 mole of liquid to 1 mole of vapor at constant temperature is called the **molar heat of vaporization** of the liquid. The model for liquids can be used to rationalize the fact that all liquids show a **vapor pressure** which is dependent only on the nature of the liquid and the temperature of the liquid. A liquid boils when its vapor pressure is equal to the confining or atmospheric pressure.

The energy required to convert 1 mole of solid to 1 mole of liquid is the **molar heat of melting** or the **molar heat of fusion.** As we might expect, it is less than the molar heat of vaporization since molecules do not have to be separated so far. **Solutions** are homogeneous mixtures. The concentration of a solute in a liquid solution is described most easily in terms of **molarity** (the number of moles of solute per liter of solution).

Solutions have properties that differ from the properties of the pure components which made up the solutions. Some solutions conduct electricity; others do not.

QUESTIONS AND PROBLEMS

1. An assumption is made in the kinetic theory of gases which fails in considering the liquid and solid states? What is the assumption?

2. Why does the product of pressure "times" volume become smaller as one compresses water vapor at 110°C? (Remember that water boils at 100°C.)

3. Can liquid water exist at temperatures above 100°C? Explain.

4. A liquid is heated at its boiling point. Although energy is used to heat the liquid, its temperature does not rise. Explain.

5 What is the maximum amount of heat that you can lose as 1 g of water evaporates from your skin?

6 Which of the following will require more energy: (a) changing 1 mole of liquid water into gaseous water; or (b) decomposing, by electrolysis, 1 mole of water? Explain. (Review Chapter 3.)

7 Choose the liquid having the higher vapor pressure from each of the following pairs (assume all substances are at room temperature): (a) mercury, water; (b) Vaseline, motor oil; (c) a perfume, honey. Describe in each case the basis for your choice.

8 Explain in terms of the kinetic theory each of the following observations: (a) If you have on a wet bathing suit, a breeze will produce a very noticeable cooling effect. (b) Evaporation increases as the temperature increases. (c) A liquid evaporates more rapidly if the vapor is pumped away by a vacuum pump. (d) Heat is required to change water to steam at 100°C.

9 Explain why the boiling point of water is lower in Denver, Colorado (altitude, 5,280 ft), than in Boston, Massachusetts (at sea level).

10 Both carbon tetrachloride (CCl_4)—used in dry cleaning and in some fire extinguishers—and mercury (Hg) are liquids whose vapors are poisonous to breathe. If CCl_4 is spilled, the danger can be removed merely by airing the room overnight; but if mercury is spilled, it is necessary to pick up the liquid droplets with a "vacuum cleaner" device. Explain.

11 A sample of nitrogen is collected over water at 18.5°C. The vapor pressure of water at 18.5°C is 16 mm. (a) When the pressure on the sample has been equalized against atmospheric pressure (756 mm) what is the partial pressure of nitrogen? (b) What will be the partial pressure of nitrogen if the volume is reduced by a factor 740/760? [*Answer to* (b): 760 mm.]

12 A cylinder contains nitrogen gas and a small amount of liquid water at a temperature of 25°C. (The vapor pressure of water at 25°C is 23.8 mm.) The total pressure is 600.0 mm Hg. A piston is pushed into the cylinder until the volume is halved. What is the final total pressure? (*Answer:* 1176 mm.)

13 Because of its excellent heat conductivity, liquid sodium has been proposed as a cooling liquid for use in nuclear power plants. (a) Over what temperature range could sodium be used in a cooling system built to operate at 1 atm pressure or lower? (b) How much heat would be absorbed per kg of sodium to melt the solid when the cooling system is put in operation? (c) How much heat would be absorbed per kg of sodium if the temperature rose too high and the sodium vaporized? Use the data in Tables 5-2 (page 98) and 5-6 (page 105).

14 Water is a commonly used cooling agent in power plants. Repeat question 13 considering 1 kg of water instead of sodium. Contrast the results for these two coolants.

15 How much heat must be removed to freeze an ice tray full of water at 0°C if the ice tray holds 500 g of water?

16 Which of the following statements about seawater is *false*? (a) It boils at a higher temperature than pure water. (b) It melts at a lower temperature than pure water. (c) The boiling point rises as the liquid boils away. (d) The melting point falls as the liquid freezes. (e) The density is the same as that of pure water.

17 How many grams of methanol (CH_3OH) must be added to 2.00 moles of H_2O to make a solution containing equal numbers of H_2O and CH_3OH molecules? How many molecules (of all kinds) does the resulting solution contain?

18 How many moles of ammonium chloride (NH_4Cl) are present in 0.30 liter of a 0.40-*M* NH_4Cl solution? How many grams? (*Answer:* 6.4 g.)

19 Write directions for preparing the following aqueous solutions: (a) 1.0 liter of 1.0-*M* lead nitrate [$Pb(NO_3)_2$] solution, (b) 2.0 liters of 0.50-*M* ammonium chloride (NH_4Cl) solution, (c) 0.50 liter of 2.0-*M* potassium chromate (K_2CrO_4) solution.

20 How many liters of a 0.250-*M* K_2CrO_4 solution contain 38.8 g of K_2CrO_4? (*Hint:* How many moles of K_2CrO_4 in 38.8 g?)

21 List three properties of a solution you would expect to vary as the concentration of the solute varies. *boiling point*

22 Explain why a pressure cooker reaches a higher temperature than water boiled in an open beaker. What determines the temperature in a pressure cooker?

23 A steel cylinder containing the element chlorine has a pressure gauge on it which reads 6.6 atm. When some gaseous chlorine is let out into the hood, the pressure gauge reading drops, then gradually builds back up to 6.6 atm. This happens many times as the tank is used. Explain why.

It is the glory of a good bit of work that it opens the way for better things and thus rapidly leads to its own eclipse.

SIR ALEXANDER FLEMING (1881–1955)

CHAPTER 6 Why We Believe in Atoms

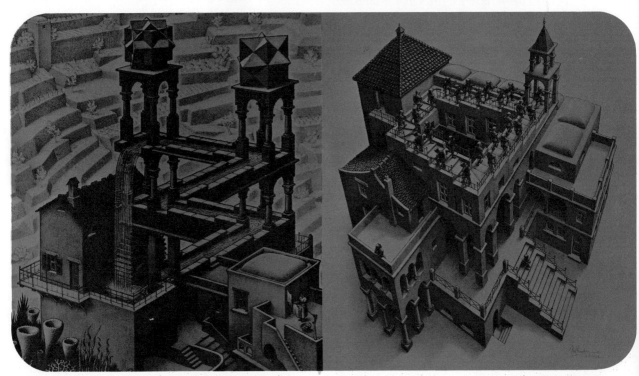

Some things, we see clearly, like these two pictures, are not believable. Other things we cannot see, like atoms, are quite believable.

M. C. Escher; Escher Association

"Great fleas have little fleas upon their backs to bite 'em, and little fleas have lesser fleas, and so *ad infinitum*."

To explain what we encountered in Chapter 2 we found it necessary to subdivide molecules into smaller particles called atoms. Now we find that in order to explain the fact that some substances in solution conduct an electric current we must think of them as comprised of still smaller subdivisions which are electrically charged. Electric charge is familiar to anyone who has ever combed his hair, scuffed his feet across deep carpet and touched his little brother, or who nonchalantly stuck a bobby pin in an electrical outlet. The relationship between atoms and electric charge is explored in this chapter.

So far we have found it convenient to postulate the existence of rapidly moving molecules in order to explain the behavior of gases. An expansion of the model to include intermolecular attraction helped us to understand the behavior of liquids and solids. An application of the model to chemical processes required that we recognize the presence of atoms in molecules. Chemical evidence suggested that many molecules divide in a number of chemical processes, hence they must consist of at least two or more parts called atoms. By postulating the existence of such atoms, we could explain the law of combining volumes for gases if we were willing to assume the validity of Avogadro's hypothesis: Equal volumes of gases measured at the same temperature and pressure contain the same number of molecules. So far then our belief in the atomic theory seems to hinge on the validity of Avogadro's hypothesis. Indeed, if this were the only reason for believing in atoms, modern science would be built on a shaky foundation. There must be more evidence!

6-1 OBSERVATION AND CONCLUSION

The atom is so incredibly small that direct observation of it in a conventional sense is out of the question. Still, as we decided earlier, every theory, even the atomic theory, should be thought about and criticized; the evidence upon which it is based should be examined and understood. It is one thing to ask, Do we believe in atoms? and quite another to ask, Why do we believe in atoms? In this chapter we shall begin to consider the second and harder question: Why do we believe in atoms? Strangely enough, most of the evidence upon which our belief in atoms is based arose from *many* simple chemical observations rather than from one conclusive experiment.

The atomic theory was a topic of conversation among the ancient Greek philosophers. The observations upon which the original atomic theory was based were made in the eighteenth and nineteenth centuries by investigators who used equipment less sophisticated than that found in the average high school laboratory today.

6-1.1 Why We Believe in Garbage Collectors

Let us approach our study of atoms using a rather transparent—even ridiculous—example of how we arrive at conclusions in day-to-day living. This example parallels very closely the scientific account which follows it. Look for the connections.

A new tenant is told by his neighbor that the garbage collector comes every Thursday, early in the morning. Later, in answer to a question from his wife about the matter, the tenant says, "I have been told that there is a garbage collector and that he comes early Thursday morning. We shall see if this is true." The tenant, if he behaves like a true scientist, accepts the statement of the neighbor (who has had opportunity to make observations on the subject). However, he accepts it *tentatively* until he himself knows the evidence for the conclusion.

After a few weeks, the new tenant has made a number of observations consistent with the existence of a Thursday-morning garbage collector. Most importantly, the garbage does disappear every Thursday morning. Secondly, the tenant receives a bill from the city once a month for municipal services. There are several supplementary observations that could corroborate the garbage collector's existence. The tenant is sometimes awakened at 5:00 A.M. on Thursdays by a loud banging and the sound of a truck. Occasionally the banging is accompanied by happy whistling, sometimes by a dog's bark.

The tenant now has many reasons to believe in the existence of the garbage collector, *yet he has never seen him.* Being a curious man, he sets his alarm clock on Wednesday night for 5:00 A.M. Looking out the window Thursday morning, his first observation is that it is surprisingly dark and things are difficult to see (an observation made frequently in scientific research also). However, he sees a shadowy form pass by, a form that looks like a man carrying a large object.

Seeing is believing—or is it? Which of these pieces of evidence really constitutes "seeing" the garbage collector? The answer is *all* of the evidence taken together furnishes the basis for accepting the Garbage Collector Theory of Garbage Disappearance. The direct vision of a shadowy form at 5:00 A.M. would not constitute "seeing" a garbage collector if the garbage did not disappear at that time; the form might have been the paper boy or the milkman. Neither would the garbage disappearance *alone* constitute "seeing" the garbage collector; perhaps a possum or a large dog eats the garbage and drags the cans away. (Remember, a dog's bark was heard.) No, the tenant is convinced that there is a garbage collector because the conviction is consistent with many observations and is inconsistent with none. Other explanations fit the observations too, but not as well. (The tenant has never heard a dog or a possum whistle gaily.) The garbage collector theory passes the test of a good theory—that is, it is useful in explaining a large number of experimental observations. This was true even before the tenant set eyes on the shadowy form at 5:00 A.M.

Yet, we must agree, there are advantages to the "direct vision" type of experiment. Often more detailed information can be obtained in this way. Is the garbage collector tall? Does he have a mustache? Could the garbage collector be a woman? This type of information is less easily obtained from indirect methods of observation. If we want answers to these questions, it may be worthwhile to set the alarm clock and go outside at 5:00 A.M. on a Thursday morning, even if we are absolutely convinced that there is a garbage collector.

Earlier in our study you were a new tenant. You were told that chemists believe in atoms because it helps to explain the chemical combination of gases. You were asked to accept that explanation *tentatively* and you were promised, indirectly, that other evidence would be made available to you.

Now we are going to review additional types of evidence that form the basis for belief in the atomic theory. Our belief will be further strengthened by successful application of the model throughout the remainder of the course.

But belief in atoms is not enough. We shall find many experiments which can be most successfully explained by assuming that atoms themselves have substructures, that atoms can be broken into smaller particles. We shall therefore expand the model to include the structure of the atom.

6-2 OBSERVATION AND BELIEF IN ATOMS

It was noted earlier that Higgins and Dalton were guided by the chemical evidence of the late eighteenth century in their formulation of the atomic theory. What kind of evidence for atoms is provided by chemistry today? We shall consider in turn the definite composition of compounds, the simple weight relations among compounds, and, in review, the combining volumes of gases. Such information provides a chemical basis for believing in atoms.

6-2.1 The Law of Definite Composition

Compounds are found to have definite composition, no matter how they are prepared. For example a 2.016-g sample of hydrogen gas combines with 16.000 g of oxygen gas to produce 18.016 g of liquid water. If a sample of water from the Pacific Ocean is distilled several times, the resulting product will contain 2.016 g of hydrogen for each 16.000 g of oxygen. Water of exactly the same composition will be obtained by distilling water samples from the Great Salt Lake, from the English Channel, or from a snow drift in Antarctica. The percent composition by weight for pure water is always the same.* This is a general observation which applies to all pure compounds. The percent composition by weight for a pure compound is always the same.

The atomic theory provides a ready explanation for the definite composition of chemical compounds. It says that compounds are composed of atoms, and every sample of a given compound must contain the same relative number of atoms of each of its elements. Since the atoms of each element have a characteristic weight, the weight composition of a compound is always the same. Thus the definite composition of compounds provides experimental support for the atomic theory and, further, provides us with the key used today in determining the formula of a compound from laboratory analytical data.

6-2.2 The Law of Simple Multiple Proportions

In many cases, two elements brought together under different conditions can form two or more different compounds. For example, hydrogen and oxygen can form both water and another compound called hydrogen peroxide in which the weight of hydrogen is one-sixteenth the weight of oxygen. If the hydrogen weight is 1.008, the oxygen weight is 16.000. *One mole of hydrogen atoms* is combining with *1 mole of oxygen atoms* to give a compound. What is the formula of

---

* This statement applies if the naturally occurring distribution of isotopes is not disturbed. More will be said about isotopes and their influence on chemical composition.

the compound? It could be HO, H$_2$O$_2$, H$_3$O$_3$, H$_4$O$_4$, and so on, and still be consistent with the information on composition. Since the molecular weight of the compound is found to be 34, the molecular formula must be H$_2$O$_2$. Remembering that the weight ratio of oxygen to hydrogen in water is 8 to 1, we note that the ratio of oxygen to hydrogen in hydrogen peroxide is exactly twice this ratio (16 to 1). What a simple relationship! The factor of 2 is the ultimate in simplicity. Such simple numerical weight relationships are almost always found when two or more elements form more than one pure *compound*. As you can easily see from the formulas H$_2$O and H$_2$O$_2$, the law of multiple proportions is explained very easily by the atomic theory (see Figure 6-1). The ratio of hydrogen to oxygen atoms in water is 2:1 and in hydrogen peroxide is 1:1. Is it surprising that water contains twice as much hydrogen per gram of oxygen as does hydrogen peroxide? This same type of observation is repeated again and again whenever two elements form more than a single pure compound.

Indeed, the atomic theory was first deduced from evidence of this very type. Chemists had found that the law of definite composition was obeyed whenever compounds were formed. In searching for an explanation for this and related facts, John Dalton proposed the atomic theory. Its value in explaining many additional phenomena such as the data leading to the law of multiple proportions soon led to general acceptance of the theory in chemistry.

Fig. 6-1 Simple multiple proportions of oxygen to hydrogen in H$_2$O and H$_2$O$_2$.

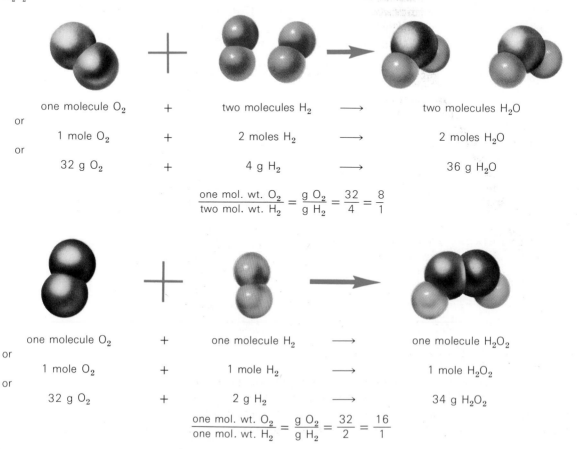

EXERCISE 6-1

Two compounds are known that contain only nitrogen and fluorine. Careful analysis shows that 23.67 g of compound I contain 19.00 g of fluorine and that 26.00 g of compound II contain 19.00 g of fluorine.

(1) For each compound, calculate the weight of nitrogen combined with 19.00 g of fluorine.

(2) What is the ratio of the calculated weight of nitrogen in compound II to that in I?

(3) Compound I is NF_3. This compound has one atom of nitrogen per three atoms of fluorine. How many atoms of nitrogen are there per three atoms of fluorine for each of the molecular formulas N_2F_2 and N_2F_4? Compare these atom ratios to the weight ratio obtained in part (2) and convince yourself that compound II could have the formula N_2F_4 but not N_2F_2.

6-2.3 The Law of Combining Volumes

In Chapter 2 the fundamental argument for believing in atoms centered on the fact that 1 liter of hydrogen will combine with 1 liter of chlorine to give 2 liters of hydrogen chloride. Invoking Avogadro's hypothesis (that: Equal volumes of gases measured at the same temperature and pressure contain equal numbers of molecules) then led to the conclusion that each molecule of hydrogen chloride must contain one half of a molecule of chlorine and one half of a molecule of hydrogen. Clearly each molecule of hydrogen and each molecule of chlorine is broken into two pieces in forming hydrogen chloride. This suggestion is consistent only if chlorine and hydrogen molecules contain sub-units or atoms. Similar observations were made for many other reactions involving gases. These observations on the combining volumes of gases then necessitated postulating the existence of atoms of many different kinds.

6-3 THE ELECTRICAL NATURE OF MATTER AND THE MAKEUP OF ATOMS

In Chapter 5 it was noted that some solutions conduct electricity. This raises an interesting "wondering why" question. What is there about a salt solution which makes it conduct electricity? In fact one can ask an even more fundamental question: *How does the salt solution conduct electricity?* We are led clearly and irrevocably to a study of electrical phenomena in general. After learning a little about electrical phenomena, we shall be able to explore the connection between electrical conductivity and the nature of atoms. As in the garbage collector analogy, we hope to "see" into the atom and identify its parts by collecting many separate bits of information.

6-3.1 Some Simple Electrical Phenomena

Electrical phenomena are so common that we often take them for granted. Some, like the electric charge we generate when walking across a wool rug in winter, have been known for centuries. Others are comparatively new to us. Try to name five you have seen.

Does your list include the following:

(1) the attraction of a comb for your hair on a dry day,
(2) the flash of a bolt of lightning,
(3) the shock you get if you touch a bare wire in a radio set or other electrical appliance,
(4) the heat generated by an electric current passing through the heating element of an electric stove,
(5) the light emitted by the filament of a light bulb as electric current is passed through it,
(6) the magnetic field generated by a current passing through a coil of wire,
(7) the work done by an electric motor when a current passes through its coils, or
(8) the emission of "radio waves" by the antenna of a radio or television station?

How are the foregoing facts interconnected? What is the relationship between the attraction of a comb for your hair and the current passing through the electric motor? What does it mean to say that an electric current "passes through" a coil of wire? Even more basically, what is an electric current?

6-3.2 Detection of Electric Charge

To begin answering these questions, we must work with an electrometer, a device for detecting and measuring electric charge. Figure 6-2 shows a simple electrometer. It consists of two spheres of very light weight, each coated with a thin film of metal. The spheres are suspended near each other by fine metal threads in a box closed to exclude air drafts. Each suspending thread is connected to a brass terminal. Next to the box is a "battery," a collection of electrochemical cells. There are two terminal posts on the battery. We shall call these posts P_1 and P_2. If post P_1 is connected by a copper wire to the left terminal of the electrometer and post P_2 is connected to the right terminal, we observe that the two spheres move toward each other. Evidently the wires have transmitted to the spheres the property of exerting force on each other—an attractive force. The force is still present when the air in the electrometer is removed with a vacuum pump. The spheres react to each other "across space." They "feel" force at a distance.

If the wires are now disconnected, the attractive force remains. However, if the two electrometer terminals are connected by a copper wire, the spheres return to their original positions and hang vertically again. The attraction is lost.

We see that the battery transfers to the electrometer spheres the property of attracting each other. It is natural to imagine that something has been transferred from the battery to the spheres. This "something" is called **electric charge.** The movement of this electric charge from the battery through the metal threads to the spheres is called an **electric current.** This electric charge is lost when the two electrometer terminals are connected by a copper wire.

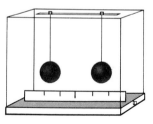

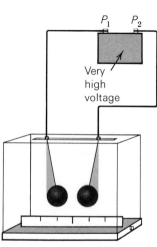

Fig. 6-2 **A simple electrometer.**

We can learn more about electric charge by another use of the electrometer. Connect one wire from the battery (say, post P_1) to the base of the electrometer and the other wire (from post P_2) to both terminals, as in Figure 6-3. This time the two spheres move apart—they repel each other! When both spheres are given electric charge from the battery post labeled P_2, they repel instead of attract. In Figure 6-2 we saw that when one sphere received charge from post P_1 and the other sphere received charge from post P_2, the two spheres attracted. There must be at least two kinds of charge!

Now let us reverse the wires so that both spheres are charged from battery post P_1. This time P_2 is connected to the base of the electrometer. Again we observe that the spheres move apart. When both spheres are connected to the same battery post, the two spheres repel each other.

This represents a large gain in our knowledge of electric charge. One kind of charge comes from battery post P_1. Until we have reason to do otherwise, we shall call this kind of charge C_1. The other kind of charge comes from battery post P_2; we shall call this kind of charge C_2. The spheres attract or repel each other when they carry charge according to the following pattern:

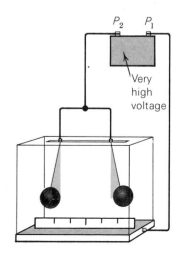

Fig. 6-3 **The electrometer with both spheres connected to battery post P_2.**

$$\begin{array}{ll} C_1 \text{ attracts } C_2 & \textit{Unlike charges attract} \\ C_1 \text{ repels } C_1 & \textit{Like charges repel} \\ C_2 \text{ repels } C_2 & \textit{Like charges repel} \end{array} \quad (1)$$

We have one more observation to formulate. When the spheres were given different charges (as in Figure 6-2), the charges could be removed by connecting the two terminals by a copper wire. Then the spheres lost all attraction for each other. We interpret this behavior to mean that either C_1 or C_2 (or both) has moved through the wire so as to join the other kind of charge. When C_1 and C_2 are united, no charge remains. Symbolically we can say

$$C_1 + C_2 = \text{no charge} \quad (2)$$

Our accumulated evidence shows that there are at least two kinds of electric charge, which we have symbolized as C_1 and C_2. These two kinds of charge possess the properties summarized as statements (1) and (2). We may wonder if there are other kinds of charge. We can answer this by investigating other ways of producing electric charges. A variety of electrochemical cells are known which show the electrometer behavior just described. Some frictional processes leave charges on two surfaces rubbed together. The attraction of a comb for your hair, for instance, is caused by charges left on the comb as it rubs against your hair. The properties of electric charges produced by rubbing a hard rubber rod with cat fur were investigated many decades ago. The hard rubber was found to carry charge C_1 and the cat fur to carry charge C_2. If a glass rod is rubbed with silk, the glass rod is left with charge C_2 and the silk with charge C_1.

No matter how an electric charge is produced, we always find these same two types, C_1 and C_2, and *only* these two. Any method of producing C_1 also produces an equivalent amount of C_2. We conclude: *There are two and only two types of electric charge.*

6-3.3 The Effect of Distance

Figure 6-4 shows two electrometers with different distances between the two pairs of spheres. Though the charges on the spheres come from the same battery, there is more deflection of the spheres when they are closely spaced (the top of Figure 6-4) than when they are widely spaced. *When the spheres are closer together, the deflection is larger.* We might conclude that the force of attraction varies with distance and is stronger when the charges are closer together. Careful quantitative studies show that the force is inversely proportional to the square of the distance d between the two spheres:

$$\text{electric force} \propto \frac{1}{d \times d} \text{ or } \frac{1}{d^2} \tag{3}$$

where d is the distance between the centers of the two spheres.

6-3.4 Electric Force—a Fundamental Property of Matter

We have learned that a battery can transfer to the spheres of an electrometer a property called electric charge. When this happens, the spheres exert force on each other. The discussion brings up two "wondering why" questions.

The first is: Why do the electric charges appear; and what causes the battery to transfer to the electrometer the property called electric charge? We shall examine this question carefully later in the course because the operation of an electrochemical cell is extremely important in chemistry. It is the topic of Chapter 14 in this book. For the moment, all we can say is that electric charge does come from the battery of electrochemical cells, thus indicating that the matter within the cells contains electric charges.

The second question probes deeper: Why do the two electrometer spheres, when charged, exert force on each other; how can we explain this phenomenon? We say that the spheres have an excess of charge C_1 or C_2 and that these charges exert force on each other. This does not really explain electric force at a distance. We are left with the equivalent questions: Why do two like charges repel each other? Why do two opposite charges attract each other? Without an answer, we say: It is a fundamental property of matter that matter can acquire electric charge, and, when it does, it will then exert force on other charged bodies. Such a statement may be taken as a definition of the term *fundamental property*—a property which is generally observed but for which diligent search has failed to yield a useful model. Without an explanation of a property, we call the property *fundamental*. It is a curious fact that when a property has resisted explanation for quite a time and it becomes classified as a fundamental property, an explanation no longer seems to be necessary.

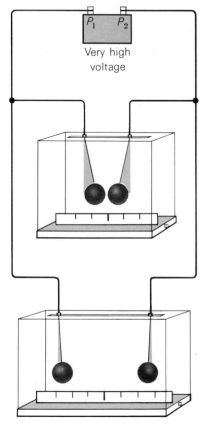

Fig. 6-4 **Contrast of deflections in two electrometers with different distances between spheres.**

6-4 THE ELECTRICAL PROPERTIES OF CONDENSED PHASES

We have mentioned very briefly that salt solutions conduct electric current and that sugar solutions do not. An electric current has been defined as the movement of electric charge. Hence, *when we say that an electric current flows through a salt solution, we mean that there is a movement of electric charge through the solution.* We shall be concerned here with the manner in which this charge moves and with the relationship of this question to the nature of atoms.

6-4.1 The Electrical Conductivity of Water Solutions

Water is a very poor conductor of electricity. Yet when sodium chloride dissolves in water, the solution conducts readily. The dissolved sodium chloride must be responsible. How does the dissolved salt permit charge to move through the liquid? One hypothesis consistent with our observations is that when salt dissolves in water, particles with electric charge are produced. The movement of these charged particles through the solution would account for the current. Since sodium chloride is neutral and has the formula NaCl, and since sodium chloride is responsible for carrying the current in the solution, perhaps charged particles of two different types appear in the solution when sodium chloride dissolves in water. A logical possibility would be that sodium chloride breaks into an atom of sodium which carries one kind of charge and an atom of chlorine which carries a different kind of charge. These two charges of opposite type, when taken together, would give a neutral molecule (see page 121). It would be convenient to indicate the type of charge on the sodium atom (C_2) by the symbol + or positive. Similarly, it would be convenient to indicate the type of charge on the chlorine atom (C_1) by the symbol − or negative. These are symbols of convenience only. The sodium atom with a positive or + charge would then be symbolized as Na^+; the chlorine atom with a negative or − charge would be symbolized as Cl^-. Atoms or groups of atoms that carry electric charge are called **ions.**

With these symbols, we can write the equation for the reaction that occurs when sodium chloride dissolves in water:

$$\text{NaCl } (solid) + \text{water} \longrightarrow \text{Na}^+ \text{ } (in \text{ } water) + \text{Cl}^- \text{ } (in \text{ } water) \quad (4)$$

This equation shows that when solid sodium chloride dissolves in water, Na^+ ions (sodium ions) and Cl^- ions (chloride ions) are present in the solution. It is convenient to abbreviate equations as much as possible while still keeping all the essential information. On the left side of the equation, the term *water* is usually not written since its presence is implied by the symbols on the right.

$$\text{NaCl } (solid) \longrightarrow \text{Na}^+ \text{ } (in \text{ } water) + \text{Cl}^- \text{ } (in \text{ } water) \quad (5)$$

We have already used the abbreviations for solid (s), liquid (l), and gas (g). Now we need an expression for *in water* when we deal with solutions. The term used is *aqueous,* abbreviated *aq.* Thus the usual form of the above equation is

$$NaCl(s) \longrightarrow Na^+(aq) + Cl^-(aq) \tag{6}$$

Here is a model of a salt solution that will aid in discussing electrical conductivity. The solid dissolves, forming the charged particles $Na^+(aq)$ and $Cl^-(aq)$; the charges can move about in the solution. An electric current can pass through the solution by the movement of these ions. The $Cl^-(aq)$ ions move in one direction, causing negative charge to move that way. The $Na^+(aq)$ ions move in the opposite direction, causing positive charge to move that way. As we shall show in more detail later, these movements carry charge through the solution and the current flows.

Sugar dissolves in water, but the resulting solution conducts an electric current no better than does pure water. We conclude that when sugar dissolves, no charged particles result, that no ions form. Sugar must be quite different from sodium chloride!

Calcium chloride ($CaCl_2$) is another crystalline solid that dissolves readily in water. A one molar $CaCl_2$ solution conducts an electric current even better than a one molar sodium chloride solution. Calcium chloride is, in this regard, like sodium chloride and unlike sugar. The equation for the reaction is

$$CaCl_2(s) \longrightarrow Ca^{2+}(aq) + 2Cl^-(aq) \tag{7}$$

The equation shows that when calcium chloride dissolves, ions are present—$Ca^{2+}(aq)$ and $Cl^-(aq)$ ions. In this case, each calcium ion has twice the positive charge held by a sodium ion, $Na^+(aq)$; hence it is written with the arabic numeral 2 and the charge symbol as $Ca^{2+}(aq)$. The chloride ion that forms, $Cl^-(aq)$, is the same negative ion that is present in the sodium chloride solution, though it comes from the calcium chloride instead of the sodium chloride solid. Because both $CaCl_2(s)$ and $NaCl(s)$ dissolve in water to form aqueous ions, they are considered to have similar conductivity.

Silver nitrate ($AgNO_3$) is a third solid substance that dissolves in water to give a conducting solution. The reaction is

$$AgNO_3(s) \longrightarrow Ag^+(aq) + NO_3^-(aq) \tag{8}$$

This time the ions formed are silver ions, $Ag^+(aq)$, and nitrate ions, $NO_3^-(aq)$. The silver ion is a silver atom* with a positive charge; it carries the same charge as an aqueous sodium ion. The aqueous nitrate ion carries a negative charge, the same charge as the aqueous chloride ion. This time, however, the negative charge is carried by four atoms, one nitrogen and three oxygen atoms, that remain together. Since this group, NO_3^-, remains together and acts as a unit, it has a distinctive name, the **nitrate ion.**

These three solids—sodium chloride, calcium chloride, and silver nitrate—are similar, and therefore classified together. They all dis-

---

* As you shall see shortly, the positive charge arises because a *neutral* silver atom has lost an electron to give an *ion* with a positive charge.

solve in water to form aqueous ions and to give conducting solutions. These solids are called **ionic solids.**

EXERCISE 6-2

Each of the following ionic solids dissolves in water to form conducting solutions. Write equations for each reaction.

(1) potassium chloride (KCl)
(2) sodium nitrate ($NaNO_3$)
(3) calcium bromide ($CaBr_2$)
(4) lithium iodide (LiI)
(5) copper chloride ($CuCl_2$)

1) $KCl(s) \rightarrow K^+(aq) + Cl^-(aq)$
2) $NaNO_3(s) \rightarrow Na^+(aq) + NO_3^-(aq)$
3) $CaBr_2(s) \rightarrow Ca^{2+}(aq) + 2Br^-(aq)$

6-4.2 Effect of Concentration

$M = \frac{\text{moles solute}}{\text{liters solution}}$

A solution containing 0.1 mole per liter of NaCl conducts much better than a solution containing 0.01 mole per liter of NaCl. Conductivity depends not only on the presence of ions but also upon the number of ions per liter of solution.

Silver chloride is a solid that shows the concentration effect rather dramatically. This solid does *not* dissolve appreciably in water. In fact, it is frequently classed as insoluble. When solid silver chloride is placed in water, very little solid enters the solution and there is only a *very slight* increase in the conductivity of the solution over that of pure water. Yet there is a real and measurable increase —a few ions are formed. Careful measurements show that even though silver chloride is much less soluble in water than sodium chloride, it is like sodium chloride in that all the solid that does dissolve forms ions in solution—aqueous ions. The reaction is

$$AgCl(s) \longrightarrow Ag^+(aq) + Cl^-(aq) \qquad (9)$$

Silver chloride, like sodium chloride, is an ionic solid.

6-5 THE ELECTRICAL CONDUCTIVITY OF SOLUTIONS AND AN ELECTRICAL MODEL FOR ATOMS

A number of separate observations have been made covering

(1) chemical observations that suggest atoms,
(2) the behavior of spheres carrying electric charge, and
(3) the ability of some aqueous solutions to conduct the electric current.

We recall the fable:

(1) The garbage disappears every Thursday morning,
(2) a loud banging is heard along with the barking of a dog, and
(3) a bill from the city is received.

In each of the two cases cited above, the problem is: How can we interpret the facts? In one case we made an effort to "see" the garbage collector. Let us concentrate on "seeing" atoms. Our "first glance" will suggest many other experiments.

6-5.1 Electrons and Protons

In discussing the ability of a solution to conduct electricity, a new idea was suggested. Particles exist which carry the property called electric charge. Two kinds of particles and two kinds of charge exist; one kind of particle is represented by the symbol +, the second by the symbol —. It would be convenient to give names as well as symbols to these charges. The following statements concerning names and properties are proposed:

(1) Matter includes particles, each of which carries a unit of electric charge. These particles are called **electrons** and **protons.**
(2) Each *electron* carries one unit of charge C_1. This has the symbol —.
(3) Each *proton* carries one unit of charge C_2. This has the symbol +.
(4) These particles exert force at a distance on each other in accordance with the electrical behavior we have observed.
   (a) Since C_1 repels C_1, minus repels minus, or electrons repel electrons.
   (b) Since C_2 repels C_2, plus repels plus, or protons repel protons.
   (c) Since C_1 attracts C_2, minus attracts plus, or electrons attract protons.
   (d) Since $C_1 + C_2 =$ no charge, $(+) + (-) = 0$, or one electron + one proton = no charge, or one unit C_1 + one unit $C_2 =$ no charge.

With this expanded version of the particle model of matter, it is much easier to talk about electrical phenomena.

If a piece of matter (such as one of the electrometer spheres) has the same number of electrons and protons, there are just as many units of charge of type C_1 and of type C_2. Since $C_1 + C_2 =$ no charge, the sphere will have no charge. A body with no net charge (with equal numbers of protons and electrons) is said to be **electrically neutral.** If we remove some of the electrons from the sphere, it will then have an excess of protons, hence a net charge of type C_2. If we add an excess of electrons to the sphere, it will have a net charge of type C_1. The amount of **net charge** is the difference between the amount of charge C_1 and charge C_2.

It is a mathematical convenience to express the net charge in terms of algebraic symbols. For this reason, we identified the type of charge called C_1 as "negative charge" and the type called C_2 as "positive charge" in our earlier discussion. Notice the advantages. The combination of five units of C_1 and three units of C_2 leaves a net of two units of C_1. This can now be expressed

$$5(-1) + 3(+1) = -5 + 3 = -2$$

EXERCISE 6-3

There was a time when atoms were said to be the fundamental particles of which matter was composed. Now we describe the structure of the atom in terms of the fundamental particles we have just named, protons and electrons, plus another kind of particle called the neutron. Why are atoms no longer said to be fundamental particles (see Section 6-3.4)?

EXERCISE 6-4

Suppose ten protons and eleven electrons are brought together. These charges, grouped together, have the same net charge as how many electrons? Remember that one proton plus one electron gives no charge. Write an algebraic expression to show the answer.

6-5.2 An Electrical Model for Electrode Processes

The model for an atom has been allowed to "grow" to include gain or loss of positive or negative charge—ions result (see page 123). The electrical conductivity of solutions has been interpreted in terms of the movement of these ions through the solution. The picture is really only partially completed. If we continue to "wonder why" about electrical conductivity, several other questions seem very important. For example, we might ask:

(1) What happens at the surface where the wire meets the liquid?
(2) Does electricity move through the metal wires as it does through the solution—by movement of ions?
(3) Since ions have + and − charges, can we really separate them so that the positive ions gather around one **electrode** (the wire dipping into solution) and the negative ions around the other **electrode** (the other wire in the solution)? (See Figure 6-5.)

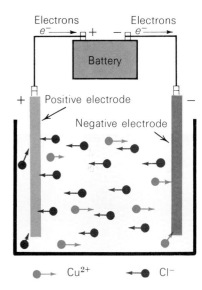

Fig. 6-5 Schematic representation of passage of electric current through copper chloride solution.

A more detailed model for electrical conductance will help us to find an answer for all of these questions. Question (1) is the easiest since it can be answered by careful experimental observation. Let us start with the question: What happens when the wire meets the liquid?

Consider for the moment a dilute water solution of copper(II)* chloride ($CuCl_2$). Since the solution is known to be an electrical conductor, analogy to the case of NaCl and $CaCl_2$ suggests the following equation for the solution process:

$$CuCl_2(s) \longrightarrow Cu^{2+}(aq) + 2Cl^-(aq) \qquad (10)$$

The ions present in the solution are $Cu^{2+}(aq)$ and $Cl^-(aq)$. According to our earlier model, the $Cu^{2+}(aq)$ ions should move toward the **negative electrode** and the $Cl^-(aq)$ ions toward the **positive electrode**. Let us watch the platinum wire hooked up to the negative terminal of the battery. (This is the negative electrode.) We shall pay particular attention to the wire under the solution. *A coating of neutral, red copper metal forms over that surface of the silvery platinum wire which is under the surface of the liquid.* A chemical process seems to be taking place on the wire. Copper ions, $Cu^{2+}(aq)$, are being converted to copper metal (see Figure 6-5).

---

* The symbol (II) following the word copper indicates $CuCl_2$. Copper(I) chloride is CuCl.

How can the model be expanded to accommodate this new set of observations? What we have seen can be summarized by the equation

$$Cu^{2+}(aq) + ? \longrightarrow Cu(\text{solid metal})$$

Simple algebra tells us that the question mark in the above equation must be replaced by two negative charges* if uncharged copper is to be obtained. Two electrons, $2e^-$, would do the job:

$$Cu^{2+}(aq) + 2e^- \longrightarrow Cu(\text{solid metal}) \qquad (11)$$

If we postulate an electron as a reactant, the observed transformation of copper ions to copper metal is easily explained.

At the positive electrode (the wire in the solution attached to the positive terminal of the battery), bubbles of chlorine gas grow, break away from the metal, and rise to the surface. Again a chemical process must be taking place; negatively charged chloride ions, $Cl^-(aq)$, are being converted to neutral chlorine gas (Cl_2):

$$2Cl^-(aq) \longrightarrow Cl_2(g) + ?$$

The equation is unbalanced. What is missing? The equation could be balanced if we used electrons, e^-, as a product instead of as a reactant:†

$$2Cl^-(aq) \longrightarrow Cl_2(g) + 2e^- \qquad (12)$$

The model for conductivity of solutions can now be made more detailed. Postulate: Electrons given up by the chloride ions flow through the wire to the battery (which really serves as an "electron pump"), then over to the negative electrode where copper ions are converted to copper metal (see Figure 6-5). This postulate has many consequences. It suggests that electric current moves in the wire by a net flow of electrons, whereas electric current moves in the solution by migration of ions. The suggestion is consistent with the fact that a mechanical generator can set up a current in a wire and there is *no* evidence of material transport as a result of current flow in a wire. On the other hand, the concentration of ions around the electrode falls in a solution as electrolysis continues. It falls around each electrode first, then the depleted volumes around each electrode grow bigger and bigger (see Figure 6-6). Ions, really bulky material such as $Na^+(aq)$ and $Cl^-(aq)$, are displaced in the electrolysis of a solution! Conductivity in a wire differs from that in solution.

---

* Loss of two positive charges would also give a neutral system; but this possibility is tentatively rejected since later arguments (Chapter 7) will show that gain of the electrons is the only acceptable process. Since the positively charged protons are present *in the nucleus,* they are held firmly and are not easily lost. Protons also have a large mass compared to the electron, so the loss of two protons from the copper would give a nucleus of lower mass and different properties.

† Again, a gain of two positive charges would meet all algebraic requirements for giving us neutral chlorine atoms, but this possibility is tentatively rejected. Evidence considered in Chapter 7 will justify the choice used here.

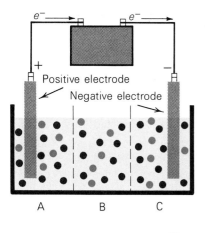
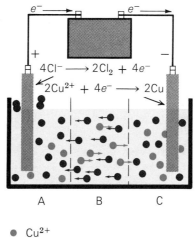

Fig. 6-6 The movement of ions in electrolysis. (The ion balances for compartments A, B, and C are given below.)

● Cl^- ● Cu^{2+}

Compartment A—Ion Balance

CHLORIDE BALANCE
10 Cl^- orig.
−4 Cl^- converted to Cl_2 at electrode.
+2 Cl^- diffuse in from Cell B.

Net remaining = 8 Cl^-

COPPER(II) BALANCE
5 Cu^{2+} orig.
−1 Cu^{2+} diffuses into B.

Net remaining = 4 Cu^{2+}

Net loss = 1 $CuCl_2$

Compartment B—Ion Balance

CHLORIDE BALANCE
10 Cl^- orig.
−2 Cl^- diffuse out to A.
+2 Cl^- diffuse in from C.

Net remaining = 10 Cl^-

COPPER(II) BALANCE
5 Cu^{2+} orig.
−1 Cu^{2+} diffuses out to C.
+1 Cu^{2+} diffuses in from A.

Net remaining = 5 Cu^{2+}

No change

Compartment C—Ion Balance

CHLORIDE BALANCE
10 Cl^- orig.
2 Cl^- diffuse into B.

Net remaining = 8 Cl^-

COPPER(II) BALANCE
5 Cu^{2+} orig.
−2 Cu^{2+} converted to Cu at electrode.
1 Cu^{2+} diffuses in from B.

Net remaining = 4 Cu^{2+}

Net loss = 1 $CuCl_2$

The postulate says that as ions carrying one kind of charge are removed at the electrode, the companion ion of opposite charge moves out toward the opposite electrode (see Figure 6-6). As a result, though ions of opposite charge move in opposite directions, we do not build up a large excess of either positive or negative ions in any part of the solution or around the electrode. This is comforting since accumulation of an excess of ions of the same charge at any point would be very difficult and would require a very high electrical voltage.

6-5.3 Electrolysis and the Particulate Nature of Electric Charge

To understand the full impact of the foregoing experiments on the development of the atomic theory, we must turn back the scientific clock to the views held in the nineteenth century. When Michael Faraday first performed his electrolysis experiments in the early 1830's, the atomic theory had been proposed, but no one had yet suggested the existence of electrons. There was no reason to suspect that electricity consisted of individual units. Faraday observed that the quantity of electricity necessary to deposit a given weight of an element from solutions of its different compounds was always equal to a constant or some simple multiple of this constant. For example, the amount of electricity that will deposit 6.03 g of metallic mercury from a solution of mercuric perchlorate [$Hg(ClO_4)_2$] will deposit the same number of grams of mercury from a solution of mercuric nitrate [$Hg(NO_3)_2$]. On the other hand, this same amount of electricity will deposit exactly twice as much mercury ($2 \times 6.03 = 12.06$ g) from a solution of mercurous perchlorate [$Hg_2(ClO_4)_2$].

If we restate Faraday's experimental finding in terms of the atomic theory, we see that the number of atoms of mercury, hence weight of mercury, deposited by a certain quantity of electricity is a constant [6.03 g from $Hg(ClO_4)_2$] or a simple multiple of this constant [2×6.03 g from $Hg_2(ClO_4)_2$]. Apparently, by measuring the electricity passed through the system, we can count atoms. A simple interpretation is that there are "packages" of electricity. During electrolysis, these packages are parcelled out—one, two, or three to an atom.

The second of Faraday's observations was that the *weights* of *different elements* that were deposited by the same amount of electricity gave simple whole-number ratios when divided by the atomic weights of these elements. For example, suppose an electric current were passed through the three electrolysis cells pictured in Figure 6-7. The two ammeters have the same reading showing that the current entering the cell at the right is identical to that leaving the cell at the left. Thus, the electric circuit guarantees that the same amount of electricity passes through each of the three cells.

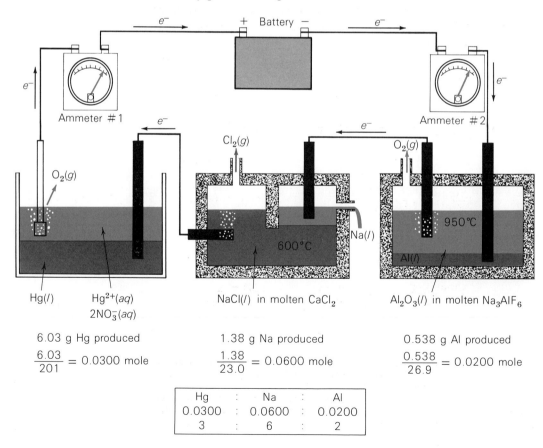

Fig. 6-7 **Weights of three different elements deposited by a given amount of electricity.**

In the first cell the net reaction is the production of metallic mercury and gaseous oxygen through electrolysis of aqueous mercuric nitrate:

$$2Hg^{2+}(aq) + 4NO_3^-(aq) + 2H_2O \xrightarrow[\text{energy}]{\text{electrical}} 2Hg(l) + O_2(g) + 4H^+(aq) + 4NO_3^-(aq) \quad (13)$$

After current has passed through the cell for a definite time, the weight of the mercury produced is found to be 6.03 g.

In the second cell molten sodium chloride is electrolyzed.* The net reaction is

$$NaCl(l) \xrightarrow[\text{energy}]{\text{electrical}} Na(l) + \tfrac{1}{2}Cl_2(g) \quad (14)$$

The same current that produced 6.03 g of mercury is found to produce 1.38 g of molten sodium.

The third cell represents another industrial process, the electrolytic process for manufacturing aluminum. Here Al_2O_3 is electrolyzed† and the net reaction in the cell is

$$Al_2O_3(l) \xrightarrow[\text{energy}]{\text{electrical}} 2Al(l) + \tfrac{3}{2}O_2(g) \quad (15)$$

We find that the same current that produced 6.03 g of mercury produces 0.538 g of aluminum.

Thus, after the same amount of electricity is passed through the three cells, the weights of metals produced are found to be 6.03 g Hg, 1.38 g Na, and 0.538 g Al.

How are these weights related? Faraday realized that simple numbers will result if each weight is divided by the appropriate atomic weight:

$$\frac{\text{wt Hg}}{\text{at. wt Hg}} = \frac{6.03 \text{ g Hg}}{201 \text{ g Hg/mole Hg}} = 0.0300 \text{ mole Hg atoms}$$

$$\frac{\text{wt Na}}{\text{at. wt Na}} = \frac{1.38 \text{ g Na}}{23.0 \text{ g Na/mole Na}} = 0.0600 \text{ mole Na atoms}$$

$$\frac{\text{wt Al}}{\text{at. wt Al}} = \frac{0.538 \text{ g Al}}{26.9 \text{ g Al/mole Al}} = 0.0200 \text{ mole Al atoms}$$

These numbers are related to each other as shown by the ratios

Hg	Na	Al
0.0300 mole	: 0.0600 mole	: 0.0200 mole
3 moles Hg atoms	: 6 moles Na atoms	: 2 moles Al atoms

---

* In practice, calcium chloride must be added to such a cell to lower the melting point of the salt and, even then, the temperature must be high (600°C). This is the commercial method for manufacturing metallic sodium.

† This is the basis for the commercial manufacture of aluminum. Another salt is added as solvent to lower the melting point. A mixture of Al_2O_3 and Na_3AlF_6 (cryolite) can be electrolyzed at 950°C.

A given quantity of electric current deposited six atoms of Na for every three atoms of Hg and two atoms of Al. Our results are quite clear: *A certain amount of electricity will deposit a fixed number of atoms, or some simple multiple of this number, regardless of which element is chosen.* For example, if a given amount of current deposits 6 moles of sodium atoms, the same amount of current will deposit 3 moles of mercury atoms or 2 moles of aluminum atoms. Stated another way, twice as much current is required to deposit 1 mole of mercury as 1 mole of sodium, and three times as much current is required to deposit 1 mole of aluminum as to deposit 1 mole of sodium.

Thus in both of Faraday's experiments, we find that an atom can carry only a fixed quantity of charge, or some simple multiple of this quantity. Notice the far-reaching significance of this observation! Electric charge must come in packages! An atom can carry one, two, or three packages of charge, but it *cannot* carry 1.5872 or 1.4301 packages. Whatever the package of charge is, it is the same for all atoms.

Historically, the realization that electric charge comes in packages led to the proposal that electricity is composed of particles. Since atoms carry electric charges, atoms must contain electrons and their counterparts, protons. In the following chapter we shall consider something of the general arrangement of these particles in atoms, and the simpler chemical implications of such an arrangement. In Chapter 15 we shall consider the deeper implications of the wave nature of electrons in atoms.

EXERCISE 6-5

Remember that in the electrolysis just described 0.0600 mole of sodium, 0.0300 mole of mercury, and 0.0200 mole of aluminum were produced.

(1) How many moles of Cl_2 will be generated in the NaCl(*l*) cell for each mole of sodium [see equation (14)]?
(2) How many moles of chlorine will be given off in liberating 0.0600 mole of sodium?
(3) What volume will this chlorine occupy at STP?
[*Answers*: (1) ½ mole Cl_2/mole Na, (2) 0.0300, (3) 0.672 liters Cl_2 at STP.]

EXERCISE 6-6

If the overall reaction for the deposition of mercury is given by the equation

$$2Hg^{2+}(aq) + 4NO_3^-(aq) + 2H_2O \xrightarrow{\text{electrical energy}} 2Hg(l) + O_2(g) + 4H^+(aq) + 4NO_3^-(aq)$$

what volume of oxygen (at STP) will be generated in the electrolysis experiment of Exercise 6-5? (*Answer*: 0.336 liter oxygen at STP.)

EXERCISE 6-7

Applying Avogadro's hypothesis, explain how the results of Exercises 6-5 and 6-6 show that atoms are "counted" by the electric current.

6-5.4 Measuring a Mole of Electrons

In the electrolysis operation just described, we made no effort to "count" electrons by use of conventional electric meters. Usually the electrician measures the flow of electricity in **amperes**. If our model is correct, *the number of amperes tells us how many electrons flow past a given point per second*. Additional experiments show that if 6.25×10^{18} electrons flow past per second, we have 1 ampere.* This number of electrons, 6.25×10^{18}, is called the **coulomb**—it is a quantity of electrons just like a gallon is a quantity of water. We find that 96,500 coulombs of electricity are the same as 1 mole of electrons:

$$96{,}500 \text{ coulombs electricity} \times \left(6.25 \times 10^{18} \frac{\text{electrons}}{\text{coulomb electricity}}\right) = 6.02 \times 10^{23} \text{ electrons}$$

We thus can summarize the above discussion in several concise mathematical statements:

$$1 \text{ ampere (or amp)} = 6.25 \times 10^{18} \frac{\text{electrons}}{\text{sec}}$$

Thus, 1 amp flowing for 1 sec gives 6.25×10^{18} electrons = 1 coulomb of electricity. Therefore, **amps × sec = coulombs.**

$$96{,}500 \text{ coulombs} = 1 \text{ mole of electrons}$$

We can then write

$$\frac{\text{coulombs}}{96{,}500 \text{ coulombs/mole of electrons}} = \text{moles of electrons}$$

Thus,

$$\frac{\text{amps} \times \text{sec}}{96{,}500 \text{ coulombs/mole of electrons}} = \text{moles of electrons} \qquad (16)$$

This is a useful relationship. It can be used, for example, to answer the question: How many moles of silver atoms will be deposited by a current of 2 amps flowing for 30 minutes? Since our equation is

$$Ag^+(aq) + e^- \longrightarrow Ag(metal) \qquad (17)$$

1 mole of electrons will deposit 1 mole of silver atoms. Our question could then be rephrased by asking how many moles of electrons flow in a current of 2 amps flowing for 30 minutes. The answer is

$$\frac{2 \text{ amps} \times 30 \text{ min} \times 60 \text{ sec/min}}{96{,}500 \text{ amps} \times \text{sec/mole of electrons}} = 0.0373 \text{ mole of electrons} \qquad (18)$$

or 0.0373 mole of silver atoms.

---

*An ampere is actually defined in terms of deposition of silver. One ampere is that steady-state flow of direct current which will deposit 0.001118 g of silver metal per second from a silver nitrate solution.

6-6 OTHER CONSEQUENCES OF THE EXISTENCE OF IONS

The concept of ions appears in many segments of chemistry, not just in a study of the passage of electric current through solutions. Just a few places where the concept of ions is also needed are described below.

6-6.1 Precipitation Reactions in Aqueous Solutions

Though both silver nitrate and sodium chloride have high solubility in water, silver chloride is almost insoluble. What will happen if we mix a solution of silver nitrate and sodium chloride? The solution obtained immediately after mixing will contain both $Ag^+(aq)$ and $Cl^-(aq)$ in high concentration! The $Ag^+(aq)$ came from the reaction

$$AgNO_3(s) \longrightarrow Ag^+(aq) + NO_3^-(aq) \qquad (8)$$

and the $Cl^-(aq)$ came from the reaction

$$NaCl(s) \longrightarrow Na^+(aq) + Cl^-(aq) \qquad (6)$$

The concentrations of $Ag^+(aq)$ and $Cl^-(aq)$ far exceed the solubility of silver chloride. The result is that solid will be formed (remember Experiment 8).

The formation of solid from a solution is called **precipitation:**

$$Ag^+(aq) + Cl^-(aq) \longrightarrow AgCl(s) \qquad (19)$$

Notice that the equation indicates the change that takes place when silver nitrate solution and sodium chloride solution are mixed. We could have written a more complete equation:

$$Ag^+(aq) + NO_3^-(aq) + Na^+(aq) + Cl^-(aq) \longrightarrow$$
$$AgCl(s) + NO_3^-(aq) + Na^+(aq) \qquad (20)$$

However, the two ions $NO_3^-(aq)$ and $Na^+(aq)$ do not play an active role in the reaction, nor do they influence the reaction that does occur. Consequently, they may be omitted from the equation for the reaction. It is convenient to have the balanced chemical equation show only the *species* which actually participate in the reaction. These species are called the **predominant reacting species.**

=== EXERCISE 6-8 ===

(1) Write the complete equation for the reaction of Experiment 12, the precipitation of PbI_2. Include all reactants and products.
(2) Rewrite the above equation eliminating those ions which are not active in the reaction. That is, write the equation showing only the predominant reacting species.

6-6.2 Balancing Equations Involving Ions

The above equations involve charged species called ions. When we considered how to balance equations for chemical reactions (Section 3-2.1), we dealt with reactions involving electrically neutral particles. We were guided by the rule that atoms are conserved. This principle is still applicable to reactions involving ions. In addition, we must consider the balance of charges. *A chemical reaction does not change the total electric charge.* Consequently, the sum of the electric charges on the reactants must be the same as the sum of the electric charges on the products. Calcium chloride dissolves to give aqueous Ca^{2+} and Cl^- ions. The balanced equation tells us that the neutral solid, calcium chloride, dissolves to give one Ca^{2+} ion for every two Cl^- ions. Summing these electric charges,

$$\text{(charge on } CaCl_2 \text{ solid)} = \text{(charge on } Ca^{2+} \text{ ion)} + 2\text{(charge on } Cl^- \text{ ion)}$$
$$0 = (+2) + 2(-1)$$
$$\text{SUM} \quad 0 = 0$$

In a balanced equation for a chemical reaction, charge is conserved.

EXERCISE 6-9

Balance the equations for the reactions given below. For each of the balanced equations, sum up the charges of the reactants and compare to the sum of the charges of the products.

(1) $PbCl_2(s) \longrightarrow Pb^{2+}(aq) + 2Cl^-(aq)$
(2) $K_2Cr_2O_7(s) \longrightarrow 2K^+(aq) + Cr_2O_7^{2-}(aq)$
(3) $Cr_2O_7^{2-}(aq) + H_2O \longrightarrow 2CrO_4^{2-}(aq) + 2H^+(aq)$

6-7 HIGHLIGHTS

Many different facts justify our belief in atoms. Though no one fact is convincing in itself, the entire array of facts provides a most convincing case.

Among the facts reviewed here are

(1) the law of constant composition, ✓
(2) the law of multiple proportions, ✓
(3) the law of combining volumes for gases, ✓
(4) Faraday's laws of electrolysis, and ✓
(5) the behavior of ions. ✓

To understand ions and the electric current, we took a side road and examined the behavior of an electrometer and the nature of electric charge. We found that *ions* are important in many aspects of chemistry, including precipitation processes. Our efforts have placed additional experimental supports around the model for matter. As the model continues to "grow," it becomes more detailed and more powerful. Further development promises to be worthwhile.

Chapter 6 / WHY WE BELIEVE in ATOMS

QUESTIONS AND PROBLEMS

1. A compound of carbon and hydrogen is known that contains 1.0 g of hydrogen for every 3.0 g of carbon. What is the atomic ratio of hydrogen to carbon in this substance? (Review Chapter 3.)

2. There are two known compounds containing only tungsten and carbon. One is the very hard alloy tungsten carbide, used for the edges of cutting tools. Analysis of the two compounds gives, for one, 1.82 g and, for the other, 3.70 g of tungsten per 0.12 g of carbon. Determine the simplest formula for each. (*Hint*: Determine the grams of tungsten per 12 g of carbon in each case. How many moles of tungsten atoms do these weights represent?) (Review Chapter 3.)

3. John Dalton thought the formula for water was HO. (Half a century passed before the present formula for water was generally accepted.) What relative weights did he then obtain for oxygen and hydrogen atoms?

4. Nitrogen forms five compounds with oxygen in which 1.00 g of nitrogen is combined with 0.572, 1.14, 1.73, 2.29, and 2.86 g of oxygen, respectively. Show that the relative weights of the elements in these compounds are in the ratio of small whole numbers. Explain these data using the atomic theory.

5. Three liters of gaseous carbon monoxide (CO) combine with 1 liter of a *new form* of gaseous oxygen to give 3 liters of gaseous CO_2. How many atoms must be present in each molecule of the new form of oxygen?

6. Name two forces other than electric force that are felt at a distance.

7. What would you expect to observe if one electrometer sphere were charged by your hair and the other by the comb used to comb your hair?

8. Why do scientists claim that there are only two kinds of electric charge?

9. It is known that electric charges attract or repel each other with a force inversely proportional to the square of the distance between them. If two spheres like those in the electrometer (see Figure 6-2) are negatively charged, what will be the change in the force of repulsion if the distance between them is increased to four times the original distance?

10. Why do two electrically neutral objects with large mass attract each other?

11. The salt ammonium sulfate [$(NH_4)_2SO_4$] dissolves in water to form a conducting solution containing ammonium ions (NH_4^+) and sulfate ions (SO_4^{2-}). (a) Write the balanced equation for the reaction when this ionic solid dissolves in water. (b) Verify the conservation of charge by comparing the charge of the reactant to the sum of the charges of the products. (c) Suppose 1.32 g of ammonium sulfate is dissolved in water and diluted to 0.500 liter. Calculate the concentrations of $NH_4^+(aq)$ and $SO_4^{2-}(aq)$.

12. Each of the following ionic solids dissolves in water to form conducting solutions. Write equations for each reaction. (a) rubidium nitrate ($RbNO_3$), (b) silver perchlorate ($AgClO_4$), (c) sodium iodide (NaI), (d) strontium iodide (SrI_2), (e) sodium sulfate (Na_2SO_4) (f) copper sulfate ($CuSO_4$).

13. A chloride of iron called ferric chloride ($FeCl_3$) dissolves in water to form a conducting solution containing ferric ions (Fe^{3+}), and chloride ions (Cl^-). (a) Write the equation for this reaction. (b) If 0.10 mole of $FeCl_3$ is dissolved in 1.0 liter of water, what is the molar concentration of ferric chloride in the solution? (c) How many moles of Cl^- ions are produced for each mole of ferric chloride? [Check your equation in (a).] (d) How many moles of chloride *ions* are present in 1 liter of solution? (e) What is the *molar* concentration of chloride ions in the final solution? (f) What is the *molar* concentration of Fe^{3+} ions in the final solution? [Answer to (e): 0.30 *M*; to (*f*): 0.10 *M*.]

14. One liter of solution contains 0.100 mole of ferric chloride ($FeCl_3$) and 0.100 mole of ammonium chloride (NH_4Cl). Calculate the concentrations of Fe^{3+}, Cl^-, and NH_4^+ ions. (*Answers*: concentration $Fe^{3+} = 0.100$ *M*, concentration $NH_4^+ = 0.100$ *M*, concentration $Cl^- = 0.400$ *M*.)

15. Which of the following solutions, if used to replace the $Hg(NO_3)_2$ solution in the electrolysis experiment (see Figure 6-7), would give the same number of moles of metal as the mercuric nitrate solution gave of mercury? More than one choice may be required. (a) $CuSO_4$ (b) $AgNO_3$ (c) $Ga(NO_3)_3$ (d) $Al_2(SO_4)_3$ (e) $CdCl_2$.

16. If n coulombs will deposit 0.119 g of tin from a solution of $SnSO_4$, how many coulombs are needed to deposit 0.119 g of tin from a solution of $Sn(SO_4)_2$? $2n$

17. Suppose two more cells were attached to the three in Figure 6-7. In the one cell, at one of the electrodes, copper is being plated from $CuSO_4$ solution; and at one of the electrodes in the other cell, bromine [$Br_2(g)$] is being converted to bromide ion (Br^-). How many grams of Cu and Br^- would be formed during the same operation discussed in the figure?

18. Write equations for the reactions between aqueous bromide ions and (a) aqueous lead ions and (b) aqueous silver ions. Both lead bromide ($PbBr_2$) and silver bromide (AgBr) are only slightly soluble.

19. When solutions of barium chloride ($BaCl_2$) and

potassium chromate (K_2CrO_4) are mixed, the following reaction occurs:

$2K^+(aq) + CrO_4^{2-}(aq) + Ba^{2+}(aq) + 2Cl^-(aq)$
$\longrightarrow BaCrO_4(s) + 2K^+(aq) + 2Cl^-(aq)$

(a) Show how charge is conserved. (b) Rewrite the equation showing predominant reacting species only. (c) Suppose 1.00 liter of 0.500 M $BaCl_2$ is mixed with 1.00 liter of 0.200 M K_2CrO_4. Assuming $BaCrO_4$ has negligible solubility, calculate the concentrations of all ions present when precipitation stops. (*Answers*: concentration $K^+ = 0.200\ M$, concentration $Cl^- = 0.500\ M$, concentration CrO_4^{2-} = negligible, concentration $Ba^{2+} = 0.150\ M$.)

20 Electric current flowing through a solution of silver nitrate deposits metallic silver at one electrode and releases oxygen gas at the other electrode. To which battery terminal is the electrode connected at which silver is deposited? If a 5.0-g deposit of silver is obtained after 10 min, what was the average reading of the ammeter during the electrolysis?

QUESTIONS and PROBLEMS

21 Two ions, A and B, one carrying two positive charges and the other carrying two negative charges, are placed 2 Å (2×10^{-8} cm) apart. Two other ions, C and D, one carrying one positive charge and the other carrying one negative charge, are placed 1 Å apart (1×10^{-8} cm). The force between ions A and B is (a) larger than, (b) less than, or (c) the same as the force between ions C and D. (*Optional*)

22 Electric current was passed through a solution containing silver nitrate ($AgNO_3$), then into a solution containing an ion, B, of atomic weight 58.7. A 10.8-g sample of silver metal was deposited and a 2.93-g sample of the other metal, B, was deposited by the same current. How many electric charges did each ion of metal B carry in the original solution?

. . . if I have seen further it is by standing on the shoulders of giants.

JOHN OF SALISBURY (?–1180)

CHAPTER 7 The Structure of the Atom

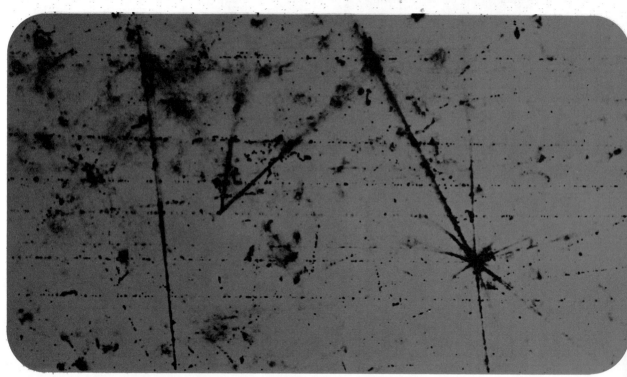

This photographic record of an atomic disintegration is consistent with our ideas of atomic structure.

Subdivision of atoms into positively and negatively charged particles leads us into a number of housekeeping questions. Are there any other kinds of particles in the atom? How big are these particles? How much do they weigh? How are they arranged in the atom? Can subatomic particles themselves undergo changes?

Although investigating something as tiny as an atom and its sub-particles has many of the same difficulties as knitting with boxing gloves on, experiments have been devised which at least partially answer these questions about atomic architecture. In hindsight, these classic experiments are beautifully simple. All that was needed was someone to think them up.

May we remind you that there are still many unanswered questions waiting for a beautifully simple experiment?

Observations summarized in Chapter 6 provide additional reasons for believing in atoms. In addition, charged particles called ions were needed to explain electrical conductivity, precipitation reactions, and electrochemical processes at electrodes. Electrometer experiments provided evidence for positive and negative charges which seemed to be very much involved in the makeup of ions and atoms. The positive particles were called protons; the negative particles, electrons. How are protons and electrons used in the construction of the atom?

A closer approach to a "direct-vision" type of experiment is needed to obtain such detailed information. In the earlier garbage-collector analogy we were almost convinced that there was a garbage collector even before any attempt was made to see him. Details about him were obtained as we progressively "saw" him. The more clearly we saw him, the more clearly the details were defined.

We now believe in atoms, but we lack detail. It is now time to set our scientific alarm clock for five o'clock A.M. and look closely at matter to try to "see" the atom. More and more of our questions seem to point to the details of atomic structure. For example: Why does salt give ions in solution while sugar does not? Why does sodium give a positive ion while chlorine gives a negative ion? Why do hydrogen gas and chlorine gas combine explosively when exposed to bright light? Why is diamond hard while white phosphorus is relatively soft? These questions are not easily answered, and in some cases the answers remain incomplete despite the best efforts of brilliant men. Still, progress has been made. The road to understanding these "wondering why" questions leads us into the world of atomic and nuclear structure.

In contrast to the experiments supporting our belief in atoms (review Section 6-2), many of the key experiments on atomic and nuclear structure involve complex instrumentation and reasonably sophisticated physical concepts. For these reasons, only an abbreviated presentation of experiments supporting our belief in the nuclear atom is given in the next section. Additional detail is given in Chapter 15 and in the appendices. As you will see in Chapter 8 and later, a large amount of chemistry can be correlated by a relatively simple model of the nuclear atom. These useful correlations save much wear and tear on the memory and strengthen our belief in the nuclear model for atomic structure. We shall use the nuclear model to correlate some chemical facts and then proceed with a more careful examination of the details of electron arrangements in atoms in Chapter 15.

7-1 ELEMENTARY ATOMIC ARCHITECTURE

7-1.1 The Components of the Atom

Experiments on the electrical conductivity of solutions suggested that molecules can be broken into charged particles or ions. Ions seemed to involve protons and electrons. Combination of equal numbers of protons and electrons would give a neutral particle (review Section 6-5.1). Perhaps atoms are made up of equal numbers of protons and electrons. Let us investigate this possibility.

By means of an instrument known as a mass spectrograph (described in Appendix 4), it is possible to determine the mass of the proton on the atomic weight scale. The value to two significant figures is 1.0. The mass of an electron determined on this same scale to two significant figures is 0.00055 or 5.5×10^{-4} (see Section 7-2.3). Can protons and electrons combine to make gold, the alchemist's dream? Let us add up the parts.

Rather sophisticated experiments of the type done by Rutherford (see Section 7-2.5) and Mosley (see Appendix 5 for greater detail) indicate that each gold atom contains 79 protons. Since gold atoms are electrically neutral, each gold atom must also contain 79 electrons. Adding up the parts for a gold atom, we find $(79 \times 1.0) + [79 \times (5.5 \times 10^{-4})]$. The second number $[79 \times (5.5 \times 10^{-4})]$ is only 0.04. This value is beyond the limit of significance in our first measurement; hence, the total weight of the gold atom should be 79. Measurement of the actual mass of a gold atom by chemical methods or by a mass spectrometer gives a value of about 197 on the atomic weight scale. If we compare 197 with 79, it is clear that each gold atom must contain something *in addition* to the protons and electrons counted. Experiments beyond those we are concerned with here identify that extra something as the **neutron.** *The neutron has zero charge and a mass of 1.0* on the atomic weight scale. Thus, a neutral gold atom must contain 79 protons, 118 neutrons, and 79 electrons.

The properties of the particles used to construct this picture of the gold atom are summarized in Table 7-1. These three particles—electron, proton, and neutron—are arranged in different combinations to give the more than 100 different elements known to man.

TABLE 7-1 APPROXIMATE CHARGE AND MASS OF SOME FUNDAMENTAL PARTICLES

Particle	Charge (Relative to the Electron Charge)	Approximate Mass (Relative to the Mass of a Proton)
Electron	1−	$1/1836$
Proton	1+	1
Neutron	0	1

7-1.2 A Model for an Atom: The Nuclear Atom

How are these fundamental particles arranged to give an atom? This question was a perplexing one at the turn of the century. The answer to it came from unexpected sources. The English physicist J. J. Thomson (1856–1940) was studying the passage of electricity through gases held in a tube at very low pressures (described in Section 7-2.1). His experiments showed that positive and negative particles were always produced in the tube as electricity passed through. The negative particles were identified as electrons regardless of the gas in the tube, while the positive fragment varied according to the gas in the tube. These experiments suggested to Thomson that a neutral gas atom might be a sphere of positive electricity in which separate negative electrons were imbedded like plums in a pudding. Thomson's "plum-pudding model" seemed reasonable. Then in 1909 Ernest Rutherford,* H. Geiger, and E. Marsden began a series of experiments which led to an amazing conclusion: Dalton's atoms (Section 6-2.2) are *not* like plum puddings, nor are they hard little spheres. Instead, each atom contains an unbelievably small and positively charged nucleus, wherein most of the atomic mass is concentrated. This nucleus occupies approximately only one part in 10^{13} of the total volume of a typical atom such as copper. In the relatively large volume of space outside this nucleus, electrons of low relative mass move about. These are referred to as **extranuclear electrons.** Strange as it may seem, *Dalton's atoms are mostly empty space, not hard little spheres.* (The experiments are described in Section 7-2.5.)

A hydrogen atom, the lightest atom known, contains one proton in the nucleus and one electron outside the nucleus. The nuclei of all other known elements have nuclear charges that are whole-number multiples of the positive charge on the proton. For example, the helium nucleus contains two protons and has a total positive charge of 2. Outside this nucleus there are two electrons. Hence, the overall atom is neutral. The neutral lithium atom has three protons in the nucleus (charge of 3+) and three electrons outside. Neutral beryllium has four protons in the nucleus and four electrons outside. As the above examples show, the number of protons in the nucleus determines the nuclear charge and the nuclear charge determines the identity of the atom. The number of protons in the nucleus gives the **atomic number** of the element. Thus, hydrogen has an atomic number of 1; helium, an atomic number of 2; lithium, an atomic number of 3; and carbon, an atomic number of 6. The number of protons in the nucleus determines the identity of the atom.

Section 7-1.1 showed that protons and electrons alone do not account for the mass of a gold atom. Neutrons in the nucleus make up the additional mass required. Thus, a helium atom with an atomic number of 2 and an atomic mass of 4 has two protons plus two neutrons in the nucleus and two electrons outside the nucleus. We say it has a mass number of 4 because it contains two protons (mass 1 each) and two neutrons (mass 1 each). The electron mass is negligible. Lithium, with an atomic number of 3 and an atomic mass of 7, has three protons plus four neutrons in the nucleus and three electrons outside the nucleus. Again, the mass of the electrons is negligible. Carbon, with an atomic number of 6 and an atomic mass of 12, has six protons plus six neutrons in the nucleus and six electrons outside the nucleus.

A MODEL for an ATOM:
The NUCLEAR ATOM

EXERCISE 7-1

(1) What is the atomic number of boron, the element just before carbon? of beryllium, the element right after lithium?
(2) What is in the nucleus of the nitrogen atom, atomic mass 14? What is in the nucleus of the boron atom, atomic mass 11?

---

* Ernest Rutherford (1871–1937) was a distinguished British physicist and former student of Thomson. Although he originally subscribed to the "plum-pudding" model, facts led him to the nuclear model.

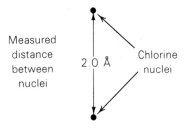

Fig. 7-1.1 **Results of experimental measurement of distance between two chlorine nuclei in a gaseous chlorine molecule.**

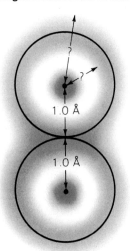

Fig. 7-1.2 **The chlorine radius is half the distance between nuclei, or 1.0 Å. The fact that there is no definite "edge" to the atom prevents a meaningful measurement of size.**

7-1.3 The Size of an Atom and of a Nucleus

How large is an atom? This would be an easy question to answer if atoms were like billiard balls and had well-defined outer surfaces. Unfortunately, atoms do not have well-defined boundaries; they have very fuzzy, indistinct boundaries so that it is impossible to tell where the surface of the atom really is. This strange property is a result of the wave nature of the electron and will be considered in more detail in Chapter 15.

For the present, let us define very carefully what we mean by atomic size. It is possible to devise experiments permitting accurate measurement of the distance between each of the nuclei in a molecule. Because these experiments involve rather complicated concepts and ideas, we shall postpone a description of them and use only the results of such experiments here. One important result gives the distance between the two chlorine nuclei in Cl_2. The value is 2.0×10^{-8} cm. [Because this distance is so small, scientists use a much smaller unit of length to express it. The unit is the **ångström**. One ångström (Å) is equal to 10^{-8} cm.] The distance between the two chlorine nuclei in Cl_2 is 2.0 Å. If the chlorine molecule consists of two spheres (atoms), then each sphere has a radius equal to half the distance between the two nuclei (see Figure 7-1). We could then assign a radius of 1.0 Å to each chlorine atom. This is a well-defined experimental quantity based on specific measurements.

EXERCISE 7-2
---

(1) The measured distance between the carbon nucleus and the chlorine nucleus in carbon tetrachloride is 1.76 Å. Using this value and information given in Figure 7-1.1, calculate the radius of a carbon atom. .76 Å
(2) If the distance between a carbon nucleus and a bromine nucleus in CBr_4 is 1.94 Å, what is the radius of a bromine atom? 1.18 Å
(3) What is the distance between nuclei in Br_2? 2.36 Å

An atomic radius, as we have defined the term above, varies from approximately 4×10^{-9} cm (0.000 000 004) to approximately 2.5×10^{-8} cm (0.000 000 025). A nucleus is much smaller. A typical nuclear diameter is about 10^{-12} cm or about 1/10,000 of the diameter of an atom. Suppose we represent a nucleus by a dot this big: (•). The atomic diameter as defined above would then be a circle 32 feet (10 meters) in diameter! Another analogy may be helpful. Suppose a chlorine atom were expanded until the diameter of the atom were as large as a very large university football stadium, The University of Michigan stadium, for example, which seats 101,000 people. From the outer row of seats on the east side to the outer row of seats on the west side is about 187 meters or about 600 feet. On this scale the nucleus of the chlorine atom would be a sphere on the 50-yard line which is as large as the one shown in the margin. The electrons could be pictured as birds flying constantly in the stadium. The nucleus is truly very small compared to the size of the atom.

7-1.4 Mass Numbers and Isotopes

All the atoms of a given element have the same number of protons in the nucleus, hence the *same nuclear charge* and the same atomic number. Do all the atoms of an element have the same mass number (or same atomic weight)?* Almost all hydrogen atoms do have the same mass. Their mass is equal to the mass of the single nuclear proton (1.0 on the atomic weight scale) plus that of the single extranuclear electron (0.00055 on the atomic weight scale).

EXERCISE 7-3

What atomic weight for a hydrogen atom is indicated if the precise mass of a proton is 1.0075 units and the electron is 5.5×10^{-4}? Does this check the experimental atomic weight of hydrogen as shown in Appendix 9?

[handwritten: 1.0075 + .00055 = 1.00805]

A few hydrogen atoms are different, however. About 0.016 percent of all hydrogen atoms in ordinary hydrogen has a nucleus whose mass is approximately twice as great as that of ordinary hydrogen atoms. Each such heavy hydrogen nucleus is composed of one proton (charge 1, mass 1) plus one neutron (charge 0, mass 1). The total mass of each such nucleus on the atomic weight scale is 2! This heavy hydrogen atom is called hydrogen-2 or, alternatively, **deuterium.** It has a **mass number**† of 2 (one proton plus one neutron). The two kinds of hydrogen atoms, having the same atomic number but different mass numbers, are called **isotopes.** All the elements have a number of isotopes. For example, *ordinary oxygen* contains three kinds of oxygen atoms. *All* have eight protons in the nucleus, making them oxygen atoms. The most common isotope in ordinary oxygen is oxygen-16. Oxygen-16 makes up 99.76 percent of ordinary oxygen. Its atoms have eight neutrons in the nucleus in addition to the eight protons. The total mass of such oxygen atoms is $8 + 8$ or 16. A second isotopic species with eight protons and nine neutrons is known as oxygen-17 (about 0.04 percent of all oxygen). The third isotopic variety (about 0.20 percent of all oxygen) has a nucleus containing eight protons and ten neutrons and is, therefore, oxygen-18. The makeup of a number of the isotopes of different elements is summarized in Table 7-2 on the next page. In short, *atoms which have the same atomic number but different numbers of neutrons in the nucleus are known as isotopes of the same element.*

EXERCISE 7-4

(1) Fluorine-19 (atomic number 9) makes up 100 percent of natural fluorine. Give information like that in Table 7-2 for fluorine-19.
(2) Gold-197 (atomic number 79) makes up 100 percent of natural gold. Give information like that in Table 7-2 for gold-197. *[handwritten: 79p 118n, mass no. = 197]*
(3) Uranium-235 (atomic number 92) makes up 0.71 percent of natural uranium; the remainder is uranium-238. Give information like that in Table 7-2 for uranium-235 and uranium-238. *[handwritten: 235 - 92p 143n, 238 - 92p 146n, mass no. = 235, = 238]*

---

* While chemists for historical reasons still speak of atomic weight, the mass of the atom is really implied. Remember, *mass* refers to the actual amount of material, while *weight* represents the Earth's attraction for this quantity of material.
† The mass number is the mass to the nearest *whole* number.

7-1.5 Ion Formation

How do ions fit into this more detailed view of the atom? Earlier studies with the electrometer verified that positive charges attract negative charges. The picture just outlined indicates that the positive charge on the nucleus due to protons is just neutralized by electrons outside the nucleus. The result is a neutral atom. Consider the helium atom:

Symbolically we can write $(2+) + (2-) = 0$. A neutral atom is formed. If one electron is removed from the helium atom we can write

Symbolically this gives $(2+) + (1-) = (1+)$. *A helium atom which lacks one electron gives a helium ion with a charge of $1+$*. If two electrons are removed from a helium atom, a helium ion with a charge of $2+$ is formed.

TABLE 7-2 COMPOSITION OF ISOTOPES

Isotope of Element	Abundance in Nature (%)	Symbol	Protons in Nucleus	Neutrons in Nucleus	Electrons in Neutral Atom	Atomic Number	Atomic Mass or Mass Number
Hydrogen-1	99.884	^{1}H	1	0	1	1	1
Hydrogen-2	0.016	^{2}D	1	1	1	1	2
Helium-3	1.34×10^{-4}	^{3}He	2	1	2	2	3
Helium-4	100	^{4}He	2	2	2	2	4
Lithium-6	7.4	^{6}Li	3	3	3	3	6
Lithium-7	92.6	^{7}Li	3	4	3	3	7
Beryllium-9	100	^{9}Be	4	5	4	4	9
Boron-10	18.83	^{10}B	5	5	5	5	10
Boron-11	81.17	^{11}B	5	6	5	5	11
Carbon-12	98.892	^{12}C	6	6	6	6	12
Carbon-13	1.108	^{13}C	6	7	6	6	13
Nitrogen-14	99.64	^{14}N	7	7	7	7	14
Nitrogen-15	0.36	^{15}N	7	8	7	7	15
Oxygen-16	99.76	^{16}O	8	8	8	8	16
Oxygen-17	0.04	^{17}O	8	9	8	8	17
Oxygen-18	0.20	^{18}O	8	10	8	8	18
F-19	100	^{19}F	9	10	9	9	19

Negative ions are formed by adding electrons to neutral atoms. Consider a neutral fluorine-19 atom:

$$\begin{pmatrix}9\,p^+\\10\,n\end{pmatrix} \qquad\qquad 9\,e^-$$

NUCLEUS　　　　NINE EXTRANUCLEAR ELECTRONS

If an extra electron is added to a fluorine atom we have

$$\left[\begin{pmatrix}9\,p^+\\10\,n\end{pmatrix} \qquad 10\,e^-\right]^- \qquad \text{One electron added}$$

NUCLEUS　　　　TEN EXTRANUCLEAR ELECTRONS

Symbolically this gives $(9+) + (10-) = (1-)$. A fluorine atom containing one extra electron is a negatively charged fluoride ion. *Positively charged ions are formed by removing electrons from neutral atoms. Negatively charged ions are formed by adding electrons to neutral atoms.* Groups of atoms can also lose electrons to give positively charged ions. Notice the ammonium ion (NH_4^+). Similarly, *groups of atoms* can pick up extra electrons to give negatively charged ions such as nitrate (NO_3^-) and sulfate (SO_4^{2-}).

7-1.6　Energy Relations in Ion Formation

Under what conditions are ions produced? In general, positive ions are produced under conditions in which large amounts of energy are available. For example, atoms in a flame will lose electrons as will atoms in the path of an electrical discharge. A lightning flash or a beam of high-energy electrons will produce ions. Equations representing this process for Na and Ca atoms can be written as

$$Na(g) + \text{energy} \longrightarrow Na^+(g) + e^- \qquad (1)$$

$$Ca(g) + \text{energy} \longrightarrow Ca^+(g) + e^- \qquad (2)$$

$$Ca^+(g) + \text{energy} \longrightarrow Ca^{2+}(g) + e^- \qquad (3)$$

Note the energy terms in the above equations. Why are they necessary? Since a positively charged nucleus attracts a negatively charged electron, work must be done or energy must be used to *pull* the electron away from the atom. A positive ion and a free electron result. *Work* and *energy* are synonymous here, and they indicate that an external agent must exert force on the electron to pull it away from the neutral atom.

In a few cases neutral atoms release energy when they pick up an electron to form a negative ion. Thus, when fluorine is converted to fluoride ion,

$$F(g) + e^- \longrightarrow F^-(g) + \text{energy} \qquad (4)$$

energy is released. This is not a very common situation, but is observed when certain negative ions are formed. On the other hand,

**GLENN T. SEABORG
(1912–)**

Glenn T. Seaborg began life humbly in a small mining town, Ishpeming, Michigan. His Swedish-American parents moved to southern California, where, to finance his undergraduate education at the University of California, he worked as a stevedore, an apricot picker, a laboratory assistant in a rubber company, and a linotype apprentice. He received his Ph.D. from the Berkeley campus in 1937.

Along with Professor Edwin M. McMillan, Seaborg produced neptunium (atomic number 93) and plutonium (no. 94), for which they were awarded the Nobel Prize in 1951. During World War II Seaborg developed a method for extracting and purifying plutonium from uranium. From 1944 to 1953 he and research teams which he trained and supervised established the existence of americium (no. 95), curium (no. 96), berkelium (no. 97), californium (no. 98), einsteinium (no. 99), fermium (no. 100), mendelevium (no. 101), and nobelium (no. 102).

From 1958–1961 Seaborg was Chancellor of the University of California at Berkeley. In 1961 he was appointed Chairman of the Atomic Energy Commission.

energy is *always* absorbed when a neutral atom is converted to a *positive ion* and an *electron*.

In Chapter 6 we wrote ion-electron equations to describe reactions occurring at each electrode when electricity is passed through an electrochemical cell. We arbitrarily rejected a model in which a neutral copper atom would be obtained by loss of positive charges from Cu^{2+}, as in the following:

$$Cu^{2+} \longrightarrow Cu + \text{two positive charges} \qquad (5)$$

The reason becomes clearer now. Protons are very heavy particles compared to electrons and are bound in the small central nucleus. Electrons, on the other hand, are held on the outside of the nucleus. Furthermore, the *addition or subtraction of a whole proton from the nucleus would change the identity of the atom*.

7-1.7 A Summary of Information on Atomic Architecture

Let us summarize the present state of our knowledge of the atom.

(1) The *nucleus* of the atom consists of very tiny and very dense,* positively charged protons plus equally tiny and equally dense, neutral neutrons confined in a very small volume.
(2) The *number of protons in the nucleus* gives the *atomic number* of the element and thus the *positive charge* on the nucleus.
(3) The positive charge on the nucleus is compensated completely in a neutral atom, partially in a positive ion, and excessively in a negative ion by very light electrons which move in a relatively large volume outside the nucleus.
(4) Since electrons are *attracted* to the positively charged nucleus, work is *always* required to pull an electron away from a neutral atom or positive ion.
(5) *The atomic number of an element is an indication of its chemistry.* The nuclear charge (atomic number) and the electrons which it attracts are of major importance in determining the way in which an atom will behave toward another atom.
(6) Atoms having the same atomic number but different numbers of neutrons in the nucleus are known as *isotopes* of that element. Mass differences cause only minor chemical differences in atoms. Only the most precise measurements will indicate the very slight chemical differences between different isotopes of the same element. Thus, we can speak of the chemistry of oxygen without specifying which isotopes are being used.
(7) The mass of a given isotope is given to an accuracy of better than ±0.1 of a mass unit by adding up the number of protons plus the number of neutrons in the nucleus. The mass of the isotope is called its *mass number*. *The mass number for an isotope is equal to the number of protons plus the number of neutrons.*

---

* "Dense" means large mass per unit of volume.

7-2 "SEEING" PARTS OF ATOMS

In the preceding section (Section 7-1) many assertions about the details of atomic structure were made. In general the evidence presented was meager or nonexistent. Some of the classic experiments supporting the earlier statements are described here. Most require some knowledge of physics. It is not necessary for you to understand all these experiments in detail in order to use the nuclear model of the atom; however, a study of this section should be of assistance in learning how a more detailed view of the atom was obtained.

7-2.1 "Seeing" Electrons

Passing electricity through solutions of salts led us to the conclusion that an electric current consists of moving electrons. Passing electricity through gases gives a more detailed view of these electrons. We are all familiar with the red glow of a "neon sign." Closer examination of one of the segments of a neon sign shows that it has much in common with the apparatus shown in Figure 7-2. This is a glass tube fitted with electrodes so that a potential of about 10,000 volts can be applied across the space between the electrodes. Let us fill the original tube with neon, then gradually begin to pump the gas out. When the pressure reaches 0.01 atm,

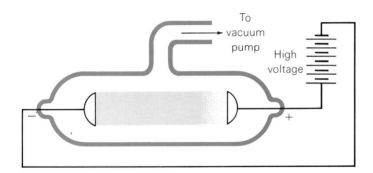

Fig. 7-2 **An electric discharge through a gas.**

the familiar red glow of the neon sign appears. The color depends on the gas selected. Different gases give different colors. If the vacuum pump continues to operate, the color will gradually disappear when the pressure reaches 10^{-6} atm and a fluorescent glow will appear on parts of the tube wall. This glow is of particular interest to us. To study it more carefully, a new tube is made which has a metal disk placed in front of the negative electrode. A triangular hole has been cut out of this metal disk (Figure 7-3, next page). When the tube operates, a sharp triangular image appears on the tube wall opposite the disk (see A on diagram). It appears that the radiation is traveling in straight lines from the negative electrode to the opposite wall. This behavior is characteristic of light. If we now bring an ordinary magnet near the tube, the beam of "light" can be bent and moved around. No one ever saw a light beam which could be moved around by a magnet of this type. The beam is clearly *not* ordinary light. Let us now bring the battery circuit of our apparatus into action by closing switch S_1. Plate P_1 becomes negatively charged (hooked to the negative pole of the battery or electrical generator) and P_2 positively charged (hooked to the positive pole of the battery or generator). Immediately the beam bends toward the positive electrode. Further, *the behavior of the beam is independent of the kind of gas in the tube.* Something in the beam is negatively charged! It is reasonable to suggest that the beam is a stream of the same kind of negative particles (or electrons) which we dealt with in our electrolysis experiments. Such a suggestion has been adequately verified by many other observations.

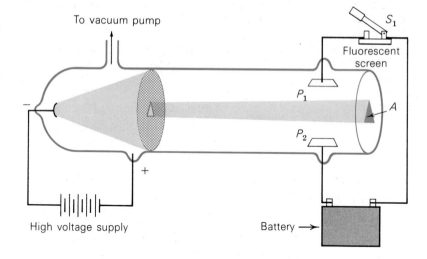

Fig. 7-3.1 **An electric discharge tube, very low pressure. Some electrons leaving the negative electrode pass through the triangular hole to produce a triangular spot on the screen.**

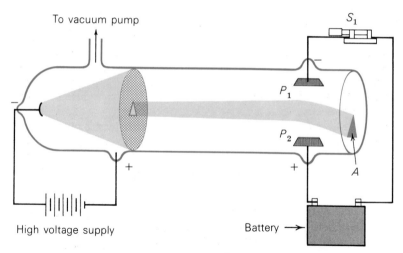

Fig. 7-3.2 **Deflection of beam of electrons by electric charge on plates P_1 and P_2.**

Have we really "seen" an electron? No, not really; but the experiment does have some features of a direct-vision experiment in that it gives us direct and detailed information about the beam. We know, for example, that it is negatively charged; a more careful study of the bending of the beam by a magnet would permit us to measure the ratio of electron charge to electron mass. (The value of e/m for the electron is 1.759×10^8 coulombs*/g.) This measurement, though important, is omitted here and described in some detail in Appendix 3.

What have we really seen? We have *not* seen the electron directly; rather we have seen a burst of light on the fluorescent screen—damage resulting from the collision of the electron with the zinc sulfide on the

---

* The **coulomb** is a unit of electric charge. Its magnitude can be evaluated by its relation to the **ampere**. One coulomb of charge passing a point in a wire every second is a current of 1 ampere. One mole of electrons has, then, 96,500 coulombs of charge. In a wire carrying 10 amp, it takes about 2½ hours for 1 mole of electrons to pass any point.

glass screen. We have seen, to use our analogy, the footprints in the garden rather than the garbage collector. We assumed that the footprints were caused by the garbage collector. From certain properties of the footprints—size, shape, depth, and spacing—we decided that the garbage collector was a man. Perhaps we might even have developed a detailed description of him, including his height, weight, and stride (remember Sherlock Holmes in Chapter 1).

You will find that this is typical of most of the experiments that allow us to "see" atoms and their components. What we actually see is their "footprints"—bursts of light on a screen, dark spots or lines on a photographic plate, and noise from a Geiger counter. It is from such footprints that the characteristics of particles in the atom have been established with great certainty.

7-2.2 The Charge on the Electron

In 1909 Robert Millikan (1868–1953), an American physicist, determined the charge on an electron by using an apparatus similar to that shown schematically in Figure 7-4. Tiny droplets of oil were sprayed into the space above the metal plates in the figure. Now and then an oil droplet would fall through the tiny hole in the upper plate into the space between the plates. The rate at which this oil droplet fell was determined by watching it through a telescope. When the rate of fall was established, the upper plate of the apparatus was connected to the positive terminal of a high-voltage battery and the lower plate was connected to the negative terminal of this battery. Then a beam of X rays was passed through the chamber to ionize gases and provide a ready source of free electrons. After the passage of the X rays, the rate of fall of the droplet changed in sudden jumps. Millikan interpreted this as evidence for gain or loss of one or more electrons by the oil droplet. If the oil droplet carried enough electrons and the charge on the plate were large enough, the fall of the droplet could be completely stopped. It could even be made to rise. By balancing the electrical force which made the droplet rise against the known force of gravity, which made it fall, the amount of electricity on the droplet could be estimated:

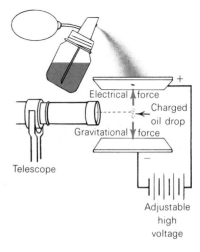

Fig. 7-4 **Millikan's oil-drop apparatus for determining the electron charge.**

$$\left\{\begin{array}{l}\text{electrical}\\ \text{force}\end{array}\right\} = k \left\{\begin{array}{l}\text{voltage}\\ \text{on plates}\\ \overline{\text{distance}}\\ \text{between}\\ \text{plates}\end{array}\right\} \times \left\{\begin{array}{l}\text{charge on}\\ \text{electron}\end{array}\right\} \times \left\{\begin{array}{l}\text{number of}\\ \text{electrons}\\ \text{on drops}\end{array}\right\} \quad (6)$$

$$\underbrace{}_{\text{ELECTRIC FIELD}} \quad \underbrace{}_{\text{AMOUNT OF ELECTRICITY ON DROP}}$$

The equation can be represented symbolically as

$$\text{electrical force} = E \times e \times n \quad (7)$$

where E is the electric field. This is directly proportional to the voltage on the plates and is inversely proportional to the distance between the plates. The symbol e stands for the charge on the electron and n represents the number of excess electron charges on the drop. When the falling action of the drop is suspended, the electrical force pulling up is just equal to the gravitational force pulling down. We can write

$$\text{electrical force pulling up} = \text{gravitational force pulling down} \quad (8)$$
$$E \times n \times e = \text{gravitational force}$$

Since the gravitational force pulling down is known from other experiments and since E can be determined from the voltage on the plates and the distance between them, the charge on the drop ($n \times e$) can be determined.

Millikan made thousands of determinations of the charges on drops of oil, glycerine, and mercury. The charge on the drop was sometimes positive, as a result of the loss of an electron, or sometimes negative, as a result of the gain of an electron. Still, in every case the amount of charge was some whole number times a value which he assigned to the charge on the electron ($n \times e$). Today the accepted value for this charge, e, is 1.602×10^{-19} coulomb/electron (see Section 6-5.4). These experiments could then be combined with the electrolysis experiments, the "neon" electrical discharge-tube experiments, and with experiments giving e/m (Appendix 3) to give us a more detailed picture of the electron as a particle.

EXERCISE 7-5

Suppose five measurements of oil droplet charges give the values listed below:

-4.83×10^{-19} coulomb
3.24×10^{-19}
-9.62×10^{-19}
-6.44×10^{-19}
-4.80×10^{-19}

(1) Divide each charge by the smallest value to investigate the relative magnitudes of these charges.
(2) Assuming each measurement has an uncertainty of $\pm 0.04 \times 10^{-19}$, decide what electron charge is indicated by these experiments alone.

7-2.3   The Mass of the Electron

It was mentioned earlier that the ratio of electron charge to electron mass ($e/m = 1.759 \times 10^8$ coulombs/g) can be determined by an analysis of the deflection of a beam of electrons under the influence of a magnet and a pair of charged plates. We have also considered Millikan's experiment giving the charge on the electron as 1.602×10^{-19} coulomb/electron. The results of these two experiments can now be combined to give the *mass* of the electron.

$$\frac{e}{m} = 1.759 \times 10^{-8} \text{ coulombs/g} \quad (9)$$

$$e = 1.602 \times 10^{-19} \text{ coulomb/electron} \quad (10)$$

$$m = \frac{1.602 \times 10^{-19} \text{ coulomb/electron}}{1.759 \times 10^8 \text{ coulombs/g}} = 9.11 \times 10^{-28} \frac{\text{g}}{\text{electron}} \quad (11)$$

7-2.4   "Seeing" Positive Ions

Experiments conducted in the evacuated neon gas discharge tube demonstrated that electrons are present and that all are the same regardless of the gas in the tube. On the other hand, a small amount of gas (10^{-6} atm) must be present in the tube for electrons to appear. If the gas is removed as completely as possible, no electron beam is seen. The electrons appear to come from the gas in the tube. We can write

$$\text{neon atom} = \text{neon ion}^+ + e^- \quad (12)$$

We expect that a positively charged neon ion would be produced. It is possible to isolate a beam of these positive ions by constructing an apparatus similar to that shown in Figure 7-5. The beam of positive ions can be deflected in magnetic and electrostatic fields just as the electron beam could. Again, such deflection permits us to determine the ratio e/m for the positive ion just as was done for the electron. If we assume that the charge on the positive ion results from the loss of an electron from the atom, we can use the value for the charge on the electron and the ratio of e/m for the ion to determine the *mass* of the positive ion. Very precise mass determination for particles can be achieved in this way. An instrument to determine the mass of positive ions in this fashion is known as the **mass spectrograph** or **mass spectrometer**. It is described in more detail in Appendix 4.

The results of measurements with the mass spectrometer reveal several very important points:

(1) The mass of the *positive ions* in a gas discharge tube changes if we change the gas in the tube. In contrast, the mass of the *electrons* in this same tube is independent of the type of gas present!
(2) Positive ions have a much higher mass than electrons. Thus, the observations support the model, which stated that the positive ions are fragments of gas which remain *after* electrons have been knocked off.
(3) Even a pure gas such as neon will give positive ions which differ somewhat in mass.

This last point is important. Let us examine it in more detail. When neon gas is put in the mass spectrometer, the bending of the positive ion beam shows that neon consists of atoms with three different masses— neon-20, neon-21, neon-22. *Neon is a mixture of three different isotopes.* The relative abundance of these isotopes in normal neon can be determined by measuring the intensity of the spots caused by each of the ion beams (see Appendix 4). This is a most important bit of information in support of the atomic model of the atom, since it permits independent determination of the apparent atomic weight for a mixture of isotopes.

EXERCISE 7-6

In a certain sample of neon in the mass spectrograph, 90.0 percent of the atoms have a mass of 20.0 and 10.0 percent have a mass of 22.0.
(1) Calculate the apparent atomic weight of this mixture. $\frac{90 \times 20 + 10 \times 22}{10} = 20.2$
(2) What would be the weight of 1 liter of this gas at 0°C and 1 atm?
[Answer to (1): 20.2 is the atomic weight.] •9013 grams

7-2.5 "Seeing" the Nucleus

We have described a nucleus which is about 10^{-12} cm in diameter. This number is so incredibly small that it is difficult to imagine, let alone see. Remember that if an *atom* were as big as a large football stadium, the nucleus would be about as big as a marble. How did Rutherford, Geiger, and Marsden go about "seeing" such a very small nucleus?

Let us consider a similar problem. Imagine that you are placed at the edge of a field covered with a large tent (Figure 7-6.1, next page). The tent is empty except for a nonliving object of unknown size, shape, and composition placed near the center of the tent. You are to investigate this object using only a rifle and ammunition. You will fire bullets into the

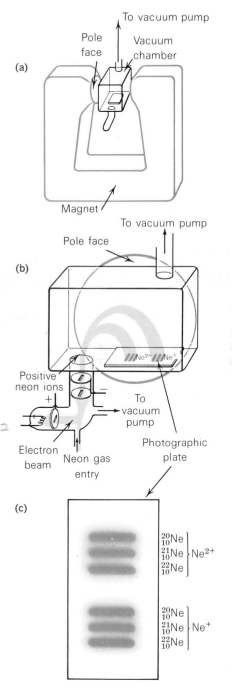

Fig. 7-5 **A mass spectrograph and the mass spectrum of neon: (a) a schematic drawing of a mass spectrograph, (b) an enlargement of the evacuated unit between the poles of the magnet, and (c) the photographic plate from (b).**

tent. By examining holes in the tent, you will be able to tell where each shot came out. If you stand at one point and fire a volley of shots, the bullets not hitting the object will pass right through the tent and emerge on the other side. Bullets hitting the unknown object will be deflected and will not arrive in the expected place at the opposite side of the tent. You could judge the size of the object by identifying the area from which bullets were deflected or scattered. Further, if the unknown object scatters a few of the bullets right back *toward* you, you would judge that it is heavy and made of a hard substance. A light or soft object might cause some deflection, but it could not exert the force needed to throw the bullet back in the direction from which it came.

In place of bullets the physicists Rutherford, Geiger, and Marsden used the newly discovered **alpha particles,** which are given off spontaneously from the radioactive decay of samples of radium (see Section 7-4). These alpha particles are nuclei of helium atoms with a mass number of 4 and a charge of $2+$. They are emitted with a velocity of about 2×10^9 cm per second. A narrow beam of particles is obtained by placing the radium in a deep lead box with a small hole in the top; this is truly a nuclear rifle (see Figure 7-6.2). This nuclear rifle was fired at a thin sheet of gold foil (about 10,000 atoms thick) with this beam of alpha particles. The alpha particles were detected by the light they produced when they collided with plates covered with zinc sulfide. These so-called "scintillation screens" were much like the picture tube in a television set; such a screen gives off a burst of light when an alpha particle strikes it. (The screen in a television set gives off a burst of light when an electron strikes it. These bursts give the image.) The scintillation screen could be moved in a circle so that particles coming off in any direction could be counted. The apparatus operated in a vacuum chamber so that no deflections would be caused by the impact of the alpha particles upon gaseous molecules.

The first observation made with this apparatus was that *apparently* all the alpha particles passed through the foil undeflected. Let us see if this result is consistent with the model of the atom proposed by Thomson. What would happen to the alpha particles if they were shot into a solid made up of closely packed Thomson atoms? At first we might think that they would be stopped or deflected back upon colliding with the atoms. But since it was observed that the alpha particles went straight through the metal foil, we must reconsider the problem. When we shoot at a

Fig. 7-6.1 **Rutherford, Geiger, and Marsdens experiment. Gaining information on size, shape, and composition of a small object inside a very large tent using an air rifle as a probe.**

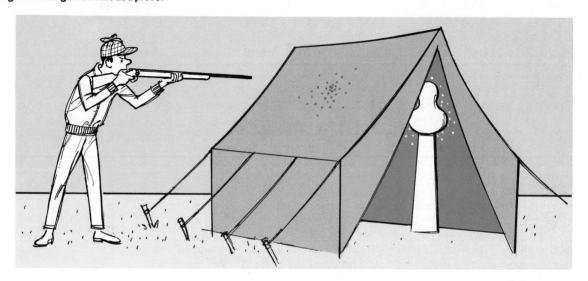

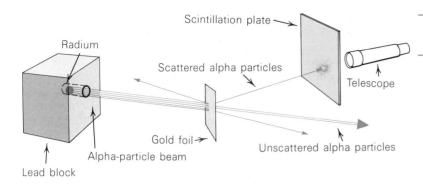

Fig. 7-6.2 **Rutherford's apparatus for observing the scattering of alpha particles by a metal foil.** (The entire apparatus is enclosed in a vacuum chamber.)

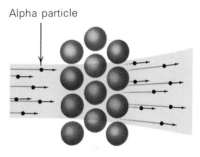

Fig. 7-7 **The scattering of alpha particles by a metal foil made of Thomson atoms.**

paper target with a high-powered rifle, the projectile forces its way through the paper. The alpha particles produced by radium have very high kinetic energy and are very much like bullets from a high-powered rifle. Perhaps the very high kinetic energy allows an alpha particle to force its way right through the atoms of the metal foil. Since a rifle bullet fired into paper passes through undeflected, it seems reasonable to conclude that the alpha particle would also pass through the metal foil undeflected.

According to the Thomson model a metal foil is considered to have essentially uniform density. If this is true there is no way for bombarding alpha particles to be deflected through large angles. At best, the alpha particles might suffer slight deflections from many collisions with many atoms. The model predicts the scattering distribution shown in Figure 7-7.

The first result of Rutherford's experiments seemed to be quite consistent with the Thomson picture of the atom. On more careful examination, however, an astounding discovery was made. By moving the screen around the metal foil, Rutherford and his co-workers were able to observe that a very few scintillations occurred at many different angles; some of these angles were nearly as large as 180°. It was as if some alpha particles had rebounded from a head-on collision with an immovable object. In describing his experiment Rutherford said: *"It is about as incredible as if you had fired a 15-inch shell at a piece of tissue paper and it came back and hit you."* It was impossible to explain the simultaneous observation of large-angle and small-angle deflections by using the Thomson atom.

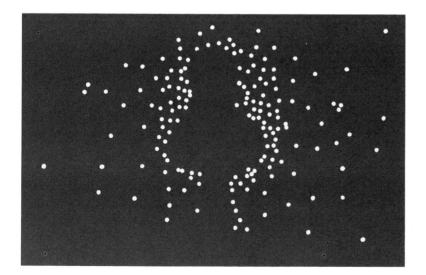

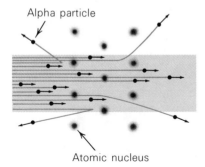

Fig. 7-8 **The scattering of alpha particles by a foil made of Rutherford nuclear atoms.**

To explain his experimental results, Rutherford designed a new picture of the atom which we have described in Section 7-1. Rutherford's model requires that most of the volume of the atom be a region of very low density. However, the atom must contain a tiny, very high-density and positively charged nucleus (see Figure 7-8).

If we allowed alpha particles to impinge upon a metal foil made up of Rutherford atoms, only a few of the particles would be appreciably deflected by the foil. The heavy, fast-moving alpha particles would brush past the lighter electrons without being deflected. Because most of the volume of the metal foil is relatively empty space, the greatest number of alpha particles would pass through the metal undeflected. It is possible, however, for a few particles to be scattered through very large angles. Since both the alpha particles and the nucleus of the atom are positively charged, they repel each other. This repulsion becomes large only when the alpha particles come quite close to the nucleus. Since the nucleus is much heavier than the alpha particles, the alpha particle bounces off, just as a steel post deflects a rifle bullet.

Besides providing a qualitative picture of the atom, Rutherford's experiments provided a method of measuring the charge of the nucleus. The repulsive force that a nucleus exerts on an alpha particle depends upon the magnitude of the charge on the nucleus. Rutherford was able to estimate nuclear charge from the pattern made by scattered alpha particles. The first measurements of the nuclear charge by this method were not very accurate. But by 1920 the alpha-particle scattering experiments were so refined that they could be used to determine nuclear charge accurately.

7-3 HIGHLIGHTS OF ELEMENTARY ATOMIC ARCHITECTURE (7-1) AND "SEEING" PARTS OF ATOMS (7-2)

The mass of the proton and electron can be determined on the atomic weight scale. The mass of atoms is *not* obtained by adding up the masses of protons and electrons; additional particles with a charge of 0 and a mass of 1 (neutrons) are also required as a component of the **atomic nucleus**. Addition of, or removal of, electrons gives **ions**. Removal of electrons from a neutral atom or positive ion requires work. Defining the size of an atom is difficult. Approximate atomic diameters of 1×10^{-8} to 5×10^{-8} cm are given by experiments measuring the distance between nuclei of atoms in molecules. The nucleus of the atom has a diameter of about 10^{-12} cm or about 1/10,000 the size of the atom. Most of the **mass** of the atom is concentrated in the nucleus and is equal to the sum of the masses of protons and neutrons which make up the nucleus. The **atomic number** gives the number of protons in the nucleus and the number of electrons outside the nucleus in the neutral atom. **Isotopes** of a given element are atoms of the same atomic number but of different mass. The atomic number determines the chemistry of the atom. Isotopes of a given element differ in mass; differences in their chemistry are trivial.

Experiments which come closer to describing a "direct view" of an atom and its components are described herein. A more detailed view is given in Appendices 3, 4, and 5. All observations made to date are consistent with our model for the nuclear atom.

7-4 PROPERTIES OF NUCLEI—
RADIOACTIVITY AND NUCLEAR CHEMISTRY

NUCLEAR STABILITY and
NUCLEAR DECAY PROCESSES

You have been told that isotopes of the same element have almost identical chemical properties. We may then properly ask: How do different isotopes of the same element differ? Unless we have some way of measuring differences, a classification scheme for isotopes is neither useful nor philosophically appropriate. The nuclear model suggests the answer: *Isotopes of the same element differ in those properties that are really properties of the nucleus itself.* For example, isotopes of the same element differ in mass because mass is essentially a nuclear property. The difference shows up in the mass spectrograph (see Appendix 4). On the other hand, since all isotopes of the same element have the same nuclear charge and the same number of electrons, their chemical properties are almost exactly the same. Probably the most dramatic difference between isotopes involves differences in **nuclear stability.** Some nuclei undergo spontaneous change to give new nuclei. Let us explore this process more carefully.

7-4.1 Nuclear Stability and Nuclear Decay Processes

If the compound zinc sulfide (ZnS) is mixed with a very small amount of radium salt, the mixture will glow in a dark room. Such mixtures are sometimes used in paints to make, for example, the hands and numbers of watch dials luminous. Look carefully at a luminous watch dial in a dark room. You will probably see some individual flashes of light on the paint surface. The flashes of light come at random and remind one of raindrops hitting the surface of a pond.

Indeed, the analogy is a happy one since the flashes result from alpha particles hitting the zinc sulfide surface. The alpha particles result from the spontaneous destruction of radium atoms. Occasionally a radium atom will explode and eject an alpha particle from its nucleus at high speed. Measurements of the mass and charge of such an alpha particle show that it has a mass number of 4 and a charge of 2. The *fragment* of radium atom remaining after loss of the alpha particle must have a mass number that is four units *less than* the mass number of the original radium. Also, it must have a nuclear charge that is two units less than radium.

It would be convenient to have concise symbols that could be used to write an equation summarizing the process. Because each isotope is characterized by two important numbers—atomic number and mass number—the symbol for an atom of a given isotope should indicate what these numbers are. *A certain isotope is represented by the chemical symbol of the element with the atomic number at its lower left and the mass number at its upper left.* Radium of atomic number 88 and mass number 226 is written as

$$\text{MASS NUMBER} \longrightarrow {}^{226}_{88}\text{Ra} \longleftarrow \text{ATOMIC NUMBER}$$

Nitrogen-13, of atomic number 7 and mass number 13, is ${}^{13}_{7}\text{N}$. Plutonium of atomic number 94 and mass number 239 is ${}^{239}_{94}\text{Pu}$.

EXERCISE 7-7

How many protons and how many neutrons are contained in the nucleus of $^{239}_{94}$Pu? 94 p + 145 n

The nuclear decay process for radium-226 can now be indicated by the equation

$$^{226}_{88}\text{Ra} \longrightarrow {}^{222}_{86}\text{Rn} + {}^{4}_{2}\text{He} \tag{13}$$

Notice that both *electric charge* and *mass number* are conserved in the equation. The number of protons and neutrons (nucleons) is the same at the start and finish of the process. In nuclear processes, two elementary laws must be obeyed:

(1) *The total number of electric charges must be the same on both sides of the equation.*
(2) *Total mass numbers must be the same on both sides of the equation.*

Conservation of charge is indicated in the equation by the fact that 86 protons in radon plus 2 protons in the alpha particle ($^{4}_{2}$He) give the original 88 protons found in $^{226}_{88}$Ra. That is, the sums of numbers in the lower left-hand corners must be the same on both sides of the equation: $88 = 86 + 2$. Similarly, conservation of neutrons (conservation of mass) is indicated by the fact that the sum of the numbers in the upper left-hand corners is the same on opposite sides of the equation. That is, $226 = 222 + 4$. Mass is conserved to a good approximation.* The fate of extranuclear electrons is not specifically indicated in the equation. Still, the conservation of these electrons is indicated by the fact that when nuclear charge is conserved, electrons will be conserved since electrons counterbalance nuclear charge in neutral particles.

EXERCISE 7-8

$^{238}_{94}\text{Pu} \rightarrow {}^{4}_{2}\text{He} + {}^{234}_{92}\text{U}$

Plutonium-238 ($^{238}_{94}$Pu) decays to give an α-particle and a new atom. What is the atomic number of the new atom? What is its mass number?

Alpha-particle loss is not restricted to radium-226. Isotopes of many of the elements with an atomic number above 82 undergo alpha decay. Thus, alpha-particle loss is observed with bismuth-211 ($^{211}_{83}$Bi), polonium-208 ($^{208}_{84}$Po), radon-222 ($^{222}_{86}$Rn), uranium-238 ($^{238}_{92}$U), and many others.

None of the lighter elements (elements having atomic numbers below 82) show alpha activity, but some isotopes of the lighter elements have unstable nuclei. These nuclei emit **beta particles** rather than alpha particles. Mass and charge measurements on beta particles show that they are *electrons* traveling at high speed. With this information the beta decay process for strontium-90 can be represented by the equation

$$^{90}_{38}\text{Sr} \longrightarrow {}^{90}_{39}\text{Y} + {}^{0}_{-1}e \tag{14}$$

* This question will be considered with a higher degree of precision in Chapter 9.

Notice that the laws for both conservation of charge [38 = 39 + (−1)] and conservation of mass (90 = 90 + 0) are obeyed in this process too.*

Beta decay is not limited to the light elements. Even a few of the isotopes of the heavier elements are beta-active. Thus, americium-242 undergoes beta decay to give curium-242:

$$^{242}_{95}\text{Am} \longrightarrow \,^{242}_{96}\text{Cm} + \,^{0}_{-1}e \tag{15}$$

EXERCISE 7-9

Potassium-43, $^{43}_{19}\text{K}$, decays by beta-particle loss. Write the nuclear equation for the process. $^{43}_{19}\text{K} \rightarrow \,^{43}_{20}\text{K} + \,^{0}_{-1}e$

Let us look a little more carefully at the process of β-particle loss. Notice that when strontium-90 is converted to yttrium-90, one of the neutrons in the nucleus is converted to a proton and an electron. The proton stays in the nucleus, but the electron is ejected. The equation for the process is

$$^{1}_{0}n \longrightarrow \,^{1}_{1}p + \,^{0}_{-1}e \tag{16}$$

It is significant to note that *even the fundamental particles of the atom undergo change in nuclear processes*. This is never so in chemical processes. Thus, if a given nucleus has more neutrons and fewer protons than a stable isotope of the same element, β-particle loss is a logical way of increasing stability. Accordingly, those isotopes of a given element with the highest mass number have the highest probability of being beta-active. While carbon-12 is stable, carbon-14, of a higher mass number, decays by β-particle loss. Fluorine-19 is stable, but fluorine-20 is beta-active. The nucleus of highest mass number exhibits beta activity in each case.

Another type of nuclear decay is associated with isotopes which have *fewer* neutrons than do stable nuclei of that element. Such nuclear processes also involve changes in the fundamental particles in the nucleus. The nitrogen-14 nucleus is stable; it has seven protons and seven neutrons. The nitrogen-13 nucleus is radioactive; it has seven protons and six neutrons. A process which changes a proton to a neutron would help move the system toward stability. We might expect a special process resulting in loss of a positively charged particle of zero mass from a proton. Such a particle is called a **positron.** *The positron is a positively charged electron.* On the atomic scale it has a charge of 1+ and a mass of approximately zero. The equation for the decay of nitrogen-13 is

$$^{13}_{7}\text{N} \longrightarrow \,^{13}_{6}\text{C} + \,^{0}_{1}e \tag{17}$$

Notice that the conservation laws are obeyed here as well: Conservation of charge—7 = 6 + 1; conservation of mass—13 = 13 + 0. Positrons are never seen in chemical processes. They occur only as

---
* Because strontium-90 resembles calcium in its chemical properties, it is incorporated into bone tissue, where its radioactive nature creates a hazard for many years.

a result of nuclear disintegration; a new element is always generated by positron-loss.

7-4.2 Rate of Nuclear Decay—Half-Life

Study of the rate of nuclear disintegration shows that in any period of time, a *constant fraction* of the nuclei will undergo decomposition. This observation allows us to characterize or describe the rate of nuclear decay in a very simple manner. It is customary to specify the length of time it takes for *half* the nuclei to decay; this length of time is known as the **half-life** of the nucleus. For example, measurements show that after 4.5×10^9 years, half the atoms in any sample of $^{238}_{92}U$ will decay to $^{234}_{90}Th$. The half-life of $^{238}_{92}U$ is 4.5×10^9 years. A nucleus is considered to be stable if its half-life is much longer than the age of the earth, which is now judged to be about 5×10^9 years. Nuclei that are very unstable are characterized by half-lives which are short, in some cases only a fraction of a second.

7-4.3 Factors in Nuclear Stability

Our discussion on nuclear stability now leads us to one of the most significant "wondering why" questions in all science: What holds the nucleus together? Relatively simple calculations with Coulomb's law

$$\text{force} = \frac{\text{charge}_1 \times \text{charge}_2}{\text{distance}^2} \qquad (18)$$

show that the *repulsive* force between two protons in a helium nucleus is about ten billion times as great as the repulsive force between two protons in the hydrogen molecule (H_2), yet the helium nucleus itself is very stable. It is clear that *fantastically strong forces* must be overcoming the forces of repulsion between protons and must be binding the protons and neutrons into a very stable unit, one which lasts indefinitely without undergoing a decay process. The nature of these forces is not understood and remains one of the most exciting problems in modern-day physics. Very large nuclear accelerators are being used and extremely large ones are being built to study this question.

QUESTIONS AND PROBLEMS

1 A carbon atom contains six protons and six electrons. What other particles (name and number) are needed to account for the known mass of a carbon-12 atom?

2 List the number and kind of fundamental particles in a neutral lithium atom having a nuclear charge of 3 and a mass number of 7.

3 Which of the following statements is *false*? The atoms of oxygen differ from the atoms of every other element in the following ways: (a) The nuclei of oxygen atoms have a different number of protons than the nuclei of any other element. (b) Atoms of oxygen have a higher ratio of neutrons to protons than the atoms of any other element. (c) Neutral atoms of oxygen have a different number of electrons than neutral atoms of any other element. (d) Atoms of oxygen have different chemical behavior than atoms of any other element.

4 How many electrons would be required to weigh 1 g? What would be the weight of a mole of electrons?

5 The nucleus of an aluminum atom has a diameter of about 2×10^{-13} cm. The atom has an average diameter of about 3×10^{-8} cm. Calculate the ratio of the diameters. Calculate the ratio of the volumes.

6 Suppose a copper atom is thought of as occupying a sphere 2.6×10^{-8} cm in diameter. If a spherical model of the copper atom is made with a 5.2-cm diameter, how much of an enlargement is this?

7 An average dimension for the radius of a nucleus is 1×10^{-12} cm and for the radius of an atom, 1×10^{-8} cm. Determine the ratio of atomic volume to nuclear volume.

8 The distance between the nuclei of the atoms in elementary sulfur molecules (S_8) is 2.08 Å. Using this information and the information in Section 7-1.3, estimate the distance between sulfur and chlorine nuclei in S_2Cl_2.

9 Using the information in Section 7-1.3, estimate the distance between chlorine and bromine nuclei in the compound ClBr.

10 Assume that the nucleus of the fluorine atom is a sphere with a radius of 5×10^{-13} cm. Calculate the density of matter in the fluorine nucleus.

11 Suppose an atom is likened to bees flying around their beehive. The beehive would be compared to the nucleus and the bees roving about the countryside would be compared to the electrons of the atom. (a) If the radius of the beehive is 25 cm, what is the average radius of the flight of the bees if they are to maintain proper scale with the atom? Express your answer in kilometers. (b) Describe qualitatively the distribution of bees around the hive as a function of direction and of distance. (c) At any instant, where is the concentration of bees apt to be highest? (Use a value of about 10^{-8} cm for the radius of the atom.)

12 Helium, as found in nature, consists of two isotopes. Most of the atoms have a mass number 4 but a few have a mass number 3. For each isotope, indicate the (a) atomic number, (b) number of protons, (c) number of neutrons, (d) mass number, and (e) nuclear charge.

13 How do isotopes of one element differ from each other? How are they the same?

14 For which of the following processes will energy be absorbed? (a) separating an electron from an electron (b) separating an electron from a proton (c) separating a proton from a proton (d) removing an electron from a neutral atom.

15 How can a potassium atom (K) be converted to a potassium ion (K$^+$)? What would be needed to convert the gas sulfur trioxide (review nomenclature) to the sulfite ion (SO$_3{}^{2-}$)?

16 For *every* atom, less energy is needed to remove one electron from the neutral atom than is needed to remove another electron from the resulting ion. Explain.

17 About how many molecules would there be in each cubic centimeter of the tube shown in Fig-

QUESTIONS and PROBLEMS

ure 7-2 when the glow appears? when the glow disappears again because the pressure is too low?

18 When several oil drops enter the observation chamber of the Millikan apparatus, the voltage is turned on and adjusted. One drop may be made to remain stationary, but others will move up while still others will continue to fall. Explain these observations.

19 Dust particles may be removed from air by passing the air through an electrical discharge and then between a pair of oppositely charged metal plates. Explain how this removes the dust.

20 Describe the spectrum produced on a photographic plate in a mass spectrograph if a mixture of the isotopes of oxygen (^{16}O, ^{17}O, and ^{18}O) is analyzed. Consider only the record for $1+$ and $2+$ ions. (See Appendix 4.)

21 How much would 0.754 mole of chlorine-35 atoms weigh? How much would 0.246 mole of chlorine-37 atoms weigh? What is the weight of a mole of "average" atoms in a mixture of the above samples? What is the atomic weight of the naturally occurring mixture of these isotopes of chlorine?

22 (*Extra Credit*) Hydroxylamine (NH$_2$OH) is subjected to electron bombardment. The products are passed through a mass spectrograph. The two pairs of lines formed indicate charge/mass ratios of 0.0625, 0.0588 and 0.1250, 0.1176. How can this be interpreted?

23 Platinum and zinc have the same number of atoms per cubic centimeter. Would thin sheets of these elements differ in the way they scatter α-particles? Explain.

24 How does the argon atom (atomic number 18, atomic weight 39) differ from the potassium atom (atomic number 19, atomic weight 39)?

25 What changes in mass and atomic number result when an atom emits an α-particle? a β-particle?

26 Write the equations for the nuclear reactions which occur when each of the following nuclei emit α-particles: $^{200}_{84}$Po, $^{203}_{85}$At, $^{222}_{89}$Ac, and $^{244}_{98}$Cf. Write equations for the reactions which occur when each of the following emit β-particles ($_{-1}^{0}e$): $^{238}_{93}$Np, $^{244}_{95}$Am, $^{185}_{66}$Dy, and $^{3}_{1}$H.

27 A radioactive element A with mass 220 and atomic number 85 emits an α-particle and changes to element B. Element B emits a β-particle and is converted to element C. What are the atomic masses and atomic numbers of elements B and C?

> The eighth element, starting from a given one, is a kind of repetition of the first, like the eighth note of an octave in music.
>
> J. A. R. NEWLANDS (1838–1898)

CHAPTER 8 Atomic Structure and the Periodic Table

Таблица II.

Вторая попытка Менделѣева найти естественную систему химическихъ элементовъ. Перепечатана безъ измѣненій изъ „Журнала Русскаго Химическаго Общества", т. III, стр. 31 (1871 г.).

		Группа I.	Группа II.	Группа III.	Группа IV.	Группа V.	Группа VI.	Группа VII.	Группа VIII, переходъ къ группѣ I.
		H=1							
Типическіе элементы.		Li=7	Be=9,4	B=11	C=12	N=14	O=16	F=19	
1-й періодъ.	Рядъ 1-й.	Na=23	Mg=24	Al=27,3	Si=28	P=31	S=32	Cl=35,5	
	— 2-й.	K=39	Ca=40	?=44	Ti=50?	V=51	Cr=52	Mn=55	Fe=56, Co=59 Ni=59, Cu=63
2-й періодъ.	— 3-й.	(Cu=63)	Zn=65	?=68	?=72	As=75	Se=78	Br=80	
	— 4-й.	Rb=85	Sr=87	Yt?=88?	Zr=90	Nb=94	Mo=96	—=100	Ru=104, Rh=104 Pd=104, Ag=108
3-й періодъ.	— 5-й.	(Ag=108)	Cd=112	In=113	Sn=118	Sb=122	Te=128?	J=127	
	— 6-й.	Cs=133	Ba=137	—=137	Ce=138?	—	—	—	—
4-й періодъ.	— 7-й.	—	—	—	—	—	—	—	
	— 8-й.	—	—	—	—	Ta=182	W=184	—	Os=199?, Ir=198? Pt=197, Au=197
5-й періодъ.	— 9-й.	(Au=197)	Hg=200	Tl=204	Pb=207	Bi=208	—	—	
	—10-й.	—	—	—	Th=232	—	Ur=240	—	
Высшая соляная окись		R₂O	R₂O₂ или RO	R₂O₃	R₂O₄ или RO₂	R₂O₃	R₂O₆ или RO₃	R₂O₇	R₂O₈ или RO₄
Высшее водородное соединеніе . . .					(RH₃)	RH₄	RH₃	RH₂	RH	

The periodic table as originally published by Mendeleev.

With more than 100 elements with different characteristics confronting us, the situation is rapidly getting out of hand. Organization is desperately needed. Perhaps we can sort the elements into groups with similar properties so we won't have to remember so much detail.

The property which we choose to measure is the amount of work necessary to remove electrons, one at a time, from the atoms of each kind of element. An unexpected dividend is that we discover that electrons are apparently arranged in levels outside the nucleus of each atom. By grouping elements according to ease of electron removal, we come up with just the device we were looking for—the periodic table. With it, we can approximately predict both the physical properties and chemical reactions of any element because of the family to which it belongs. What a relief!

Materials making up the world have strikingly different properties; they range from the glittering diamond to delights such as a thick juicy steak. One of the prime jobs of the scientist is to organize information about different materials. In this process, proper recognition must be given to both similarities and differences, even when such similarities and differences are not obvious to the casual observer. For example, little similarity is apparent between a chocolate cake and a high-energy rocket fuel, yet both are fuels; they differ only in the type of engine needed to properly utilize them. The energy content of each might be important in planning a space walk.

How do we proceed to organize information on chemical and physical properties of materials? In organizing information on gases, liquids, and solids, the molecular model was very useful. Growth of this model to include atoms and finally atomic structure permitted us to visualize processes ranging from the passage of an electric current through a salt solution to the radioactive decay of a sample of radium. What additional growth is needed in the atomic model to make it useful for organizing chemical information? Chapter 7 indicated that chemical properties are closely related to the arrangement of electrons outside the nucleus of the atom. Let us be like the new tenant hearing about the garbage collector for the first time —we shall accept this proposition *tentatively* while we make additional observations.

8-1 THE LOSS OF ELECTRONS FROM AN ATOM— THE IONIZATION ENERGY OF ATOMS

Perhaps clues to electron arrangement can be found by studying the loss of electrons from atoms. We already know that if an electron is to be pulled away from an atom, energy must be supplied (Chapter 7). We are interested in the process

$$\text{gaseous atom} + \text{energy} \longrightarrow \text{gaseous ion} + \text{gaseous electron}$$
$$\text{Na}(g) + \text{energy} \longrightarrow \text{Na}^+(g) + e^- \qquad (1)$$

The question of interest to us now is: How much energy is needed to remove an electron? This quantity of energy is worthy of a name and is worthy of measurement. The energy associated with the removal of an electron from a neutral *gaseous atom* to give a positively charged *gaseous ion* and an *electron* is known as the **ionization energy of the atom.** How are these ionization energies measured?

8-1.1 Determination of Ionization Energy

The ionization energy for certain atoms may be determined by bombarding the atomic vapor with electrons whose kinetic energy is accurately known. How do we obtain a beam of electrons of known kinetic energy? Section 7-2 stated that a cathode ray tube gives a beam of electrons. The kinetic energy of the electrons can be increased by increasing the voltage used on the tube. The adjustment is not too different from the adjustment for picture brightness on your television tube. To get a brighter picture (more energetic electrons) you turn up the voltage on the picture tube. When the kinetic energy of the bombarding electrons used in the measurement is increased to a certain critical value, singly charged positive ions can be determined electrically.* These ions result from collisions between the atoms being studied and the bombarding electrons. When the bombarding electrons have the same kinetic energy as the energy needed to separate the most loosely bound electron in the atom, a collision will knock the electron from the atom. The energy required to break an electron away from a given target atom is characteristic of that atom—it is a measure of the atom's ionization energy. We can read the voltage on the tube generating the electron beam as soon as positive ions are obtained. The value can be expressed in terms of electron volts or in other energy terms such as calories.

Ionization energies can also be determined even more precisely by using methods of atomic spectroscopy described in Chapter 15. The method used is determined by the system studied.

8-1.2 Stable Electronic Patterns— Trends in Ionization Energies

Ionization energy measurements have been made for each of the elements. The results, in order of increasing atomic number for the first 20 elements, are shown in Table 8-1.

*In actual operation such measurements are made in a mass spectrometer (Appendix 4), where somewhat more elaborate velocity selectors may be used.

TABLE 8-1 FIRST IONIZATION ENERGIES OF ELEMENTS 1 TO 20

Atomic Number	Element	First Ionization Energy (cal/mole)
1	Hydrogen (H)	3.14×10^5
2	Helium (He)	5.68×10^5
3	Lithium (Li)	1.25×10^5
4	Beryllium (Be)	2.14×10^5
5	Boron (B)	1.92×10^5
6	Carbon (C)	2.61×10^5
7	Nitrogen (N)	3.35×10^5
8	Oxygen (O)	3.14×10^5
9	Fluorine (F)	4.02×10^5
10	Neon (Ne)	4.99×10^5
11	Sodium (Na)	1.18×10^5
12	Magnesium (Mg)	1.76×10^5
13	Aluminum (Al)	1.39×10^5
14	Silicon (Si)	1.87×10^5
15	Phosphorus (P)	2.43×10^5
16	Sulfur (S)	2.40×10^5
17	Chlorine (Cl)	3.00×10^5
18	Argon (Ar)	3.65×10^5
19	Potassium (K)	1.00×10^5
20	Calcium (Ca)	1.41×10^5

EXERCISE 8-1

Plot ionization energy against atomic number for the first 20 elements. At what atomic numbers do maxima appear? minima?

STABLE ELECTRONIC PATTERNS

max. - 2, 4, 7, 10, 12, 15, 18.

min. - 3, 5, 8, 11, 13, 16, 19.

If you did Exercise 8-1 carefully, a number of regularities seem to jump at you from the paper! High ionization energies are seen for He (atomic number 2), Ne (no. 10), and Ar (no. 18). Low ionization energies are seen for Li (no. 3), Na (no. 11), and K (no. 19). Apparently, a large amount of energy is needed to pull an electron from the configurations found in helium, neon, and argon. It seems reasonable to describe such electron patterns as stable. *If a large amount of energy is required to pull off an electron, the original configuration is said to be very stable.*

The element Li (no. 3) follows He (no. 2); the element Na (no. 11) follows Ne (no. 10); and the element K (no. 19) follows Ar (no. 18). *Adding one more electron and proton to each of the stable atoms (He, Ne, and Ar) gives new atoms with electron arrangements of relatively low stability (Li, Na, and K).* A relatively small amount of energy is required to pull an electron from Li, Na, or K. The *ion* formed in each case has the same number of electrons as the stable element that precedes it. That is, the lithium *ion* (no. 3) has 2 electrons just like the helium *atom* (no. 2); the sodium *ion* (no. 11) has 10 electrons just like the neon *atom* (no. 10); the potassium *ion* (no. 19) has 18 electrons just like the argon *atom* (no. 18). We now wonder: Does the lithium ion have a stable electron configuration like the helium atom? Does the sodium ion have a stable electron configuration like the neon atom? Does the potassium ion have a stable electron configuration like the argon atom?

These questions can only be answered by experiment. Let us look at the energies required to remove an electron from Li^+, Na^+, and K^+. These values are known as the **second ionization energies** of the atoms and are available from laboratory data. The equations are

$$Li^+(g) + 1.748 \times 10^6 \text{ cal/mole} \longrightarrow Li^{2+}(g) + e^- \qquad (2)$$

$$Na^+(g) + 1.09 \times 10^6 \text{ cal/mole} \longrightarrow Na^{2+}(g) + e^- \qquad (3)$$

$$K^+(g) + 7.3 \times 10^5 \text{ cal/mole} \longrightarrow K^{2+}(g) + e^- \qquad (4)$$

The numbers given answer the earlier questions with a resounding yes! A very large amount of energy is required to pull an electron from each of the ions Li^+, Na^+, and K^+. The value for Li^+ is 15 times the value for Li; the value for Na^+ is more than 9 times the value for Na. These numbers suggest that the ions Li^+, Na^+, and K^+ have the same stable electron configurations that were found with the atoms He, Ne, and Ar. So far, stable electron patterns seem to appear if we have 2, 10, or 18 electrons outside the nucleus.

EXERCISE 8-2

Define the third ionization energy for sodium. Its value is 1.6×10^6 cal. Why is it larger than the value for the second ionization energy?

energy required to pull electron from atom with 2 electrons gone already.

If the proposal just made is true, it should be easy to add an electron to an atom such as fluorine (atomic number 9) to obtain a pattern with 10 extranuclear electrons—the neon arrangement. It should be easy to add an electron to chlorine (no. 17) to obtain a pattern with 18 extranuclear electrons—the argon arrangement. For the processes involving fluorine (no. 9) and chlorine (no. 17), the equations are

$$F(g) + e^- \longrightarrow F^-(g) + 7.9 \times 10^4 \text{ cal/mole} \quad (5)$$
fluorine *atom* + electron $\longrightarrow$ fluoride *ion* + energy

$$Cl(g) + e^- \longrightarrow Cl^-(g) + 8.5 \times 10^4 \text{ cal/mole} \quad (6)$$
chlorine *atom* + electron $\longrightarrow$ chloride *ion* + energy

Energy is not required to add an electron to fluorine or chlorine; instead, energy is *released*. The process in each case goes by itself! The atoms move *spontaneously* toward the neon and argon configurations. This fact adds additional support to the proposal that arrangements of 10 and 18 electrons outside the nucleus represent very stable patterns.

The suggesion can be put to a further test. The next element after lithium is beryllium. Beryllium has two more electrons than the stable element helium. Perhaps beryllium will lose *two* electrons relatively easily to give the helium configuration (two electrons outside the nucleus). On the other hand, removing a third electron from beryllium and thus breaking the stable helium pattern should be very difficult if the helium pattern really has high stability. The appropriate equations and numbers are given below:

$$Be(g) + 2.14 \times 10^5 \text{ cal/mole} \longrightarrow Be^+(g) + e^- \quad (7)$$

$$Be^+(g) + 4.18 \times 10^5 \text{ cal/mole} \longrightarrow Be^{2+} + e^- \quad (8)$$

$$Be^{2+}(g) + 35.3 \times 10^5 \text{ cal/mole} \longrightarrow Be^{3+}(g) + e^- \quad (9)$$

It takes almost twice as much energy to remove the second electron from beryllium as it does to remove the first, but it takes almost 8.5 times as much energy to remove the third electron as it does the second. Again, *the pattern with two electrons outside the nucleus appears to be very stable*.

EXERCISE 8-3

How many electrons can be removed from a calcium atom (no. 20) before a big jump in ionization energy can be observed? an Al atom?

8-1.3 Other Stable Electron Arrangements

All of the foregoing facts support the postulate that 2, 10, or 18 extranuclear electrons represent stable numbers of electrons—these seem to be particularly stable groupings. We are now in a position to try to extend our generalizations. Atoms with atomic numbers 2, 10, and 18 have very high ionization energies. We see that $10 - 2 =$

8, and $18 - 10 = 8$. Do high ionization energies repeat after every eight elements? Will element 26 ($26 - 18 = 8$) have a very high ionization energy? Looking at the list of elements, we see that element number 26 is iron, with the relatively low ionization energy of 1.82×10^5 cal/mole. On the other hand, krypton, element number 36, has an ionization energy of 3.2×10^5 cal/mole—a value twice as high as the value given for iron. It is clear to even the most casual observer that our number 8 has lost its magic.

Our failure to predict the ionization energy for iron has additional meaning in the interpretation of data. Scientists have learned through experience that interpolation between points in a set of data on a curve is usually a fairly reliable procedure. For example, if we know that the ionization energy of aluminum (no. 13) is 1.4×10^5 cal/mole and that phosphorus (no. 15) is 2.4×10^5 cal/mole, the value for silicon should be about 1.9×10^5 cal/mole. Fortunately, the experimental value is also 1.9×10^5 cal/mole. On the other hand, extrapolation of the data—extending it well beyond the available information—is frequently hazardous. The value for iron was lower than predicted.

If extrapolation from limited data is hazardous, we need more data. Values for additional ionization energies are shown in Figure 8-1. Maxima in the curve appear for elements 2, 10, 18, 36, 54, and 86. As we might expect, rubidium (no. 37) and cesium (no. 55) both have low ionization energies.

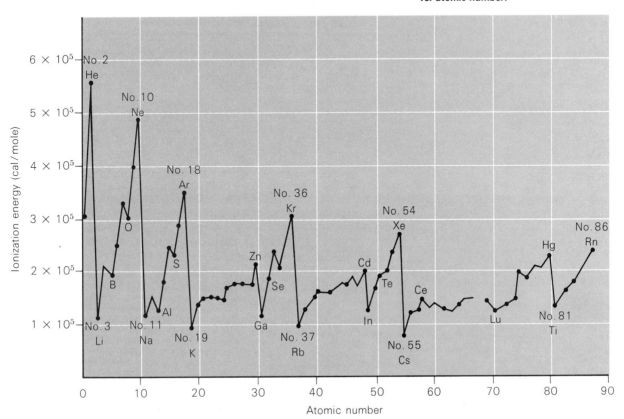

Fig. 8-1 **Plot of ionization energy vs. atomic number.**

EXERCISE 8-4

(1) *Estimate* the ionization energy for element no. 85. 1.8

(2) Would you expect the addition of an electron to an atom of element no. 35 to give off or to absorb energy? Explain the basis for your choice.

Maxima in the curve of Figure 8-1 are summarized in Table 8-2. There is much food for thought in Table 8-2. First, each of the stable electron patterns involves an even number of electrons. Next, we see that there seems to be some regularity in the differences between the number of electrons possessed by a given stable pattern and the number of electrons possessed by its stable predecessor. The first two differences are 8 and the second two differences are 18. The third difference is 32. Would the next stable configuration be $86 + 32 = 118$ electrons? Extrapolation is still dangerous, but this extrapolation is founded on more data than our first. No one will really *know* the answer until element 118 is available for study.

Other even more interesting questions intrigue us. What is the special significance of the numbers 8, 18, and 32? Why do the values of ionization energy decrease in Table 8-2 as we go from He to Rn? What kind of a model for electron arrangement can be developed to "explain" the numbers in Table 8-2? The significance of the numbers 8, 18, and 32 will be explained in Chapter 15. The other questions can be considered here.

TABLE 8-2 STABLE ELECTRON ARRANGEMENTS—ATOMS WITH MAXIMA IN THE IONIZATION ENERGY CURVE

Element	Ionization Energy (cal/mole)	Total Number of Electrons	Change in Number of Electrons
Helium	5.68×10^5	2	
Neon	4.99×10^5	10	$10 - 2 = 8$
Argon	3.63×10^5	18	$18 - 10 = 8$
Krypton	3.22×10^5	36	$36 - 18 = 18$
Xenon	2.78×10^5	54	$54 - 36 = 18$
Radon	2.46×10^5	86	$86 - 54 = 32$

8-2 A MODEL FOR ELECTRON ARRANGEMENT

In seeking an answer to the question of a model for electron arrangement, we return to a consideration of those forces that hold an electron in the atom. The electrons are held by the attraction between the positively charged nucleus and the negatively charged electron. In Section 6-3.3 we found that this force *decreases* as the distance between the charged objects *increases*. In quantitative terms, the relationship obeyed is

$$\left\{\begin{array}{l}\text{force of}\\\text{attraction}\\\text{between}\\\text{nucleus and}\\\text{electron}\end{array}\right\} = \frac{\left\{\begin{array}{l}\text{total positive}\\\text{charge on}\\\text{nucleus}\end{array}\right\}\left\{\begin{array}{l}\text{charge on}\\\text{electron}\end{array}\right\}}{\left\{\begin{array}{l}\text{distance from}\\\text{electron to nucleus}\end{array}\right\}^2} \qquad (10)$$

If we double the distance between nucleus and electron, the force of attraction drops to one-fourth its original value! We note again that *as we progress down in Table 8-2 from helium to radon, the ionization energy drops from 5.68×10^5 cal to 2.46×10^5 cal.* Perhaps the electron lost from radon is farther from the nucleus than the electron lost from helium. If this were true, radon would have a lower ionization energy. The expected trend is observed. This suggestion concerning distance between electron and nucleus was first made by the brilliant Danish physicist Niels Bohr in his original model for the atom. This particular postulate is still accepted in our present-day model for the atom. The electron lost from radon is farther from the nucleus than the electron lost from helium.

Another proposal is worth considering. Suppose electrons are arranged in levels going from the nucleus as in Table 8-3. Very near the nucleus there is a limited amount of room for electrons. Perhaps two electrons would be the maximum number we could place very near the nucleus. After that, a new level would be needed at a larger average distance from the nucleus. The first electron added to this new level would be lost rather easily because it would be at a greater distance from the central positive charge. As extra protons are added to the nucleus to increase the atomic number of the elements (lithium ⟶ neon), the attraction between nucleus and electron would increase if electrons were added to this *same* level at nearly the same average distance from the nucleus. *As the nuclear charge increases, the ionization energy should increase if electrons are being added to the same level.* At neon this second level would be filled

TABLE 8-3 A PROPOSED ELECTRON ARRANGEMENT FOR THE FIRST 20 ELEMENTS

Atomic Number	Element	First Ionization Energy (cal/mole)	Electrons in First Level	Electrons in Second Level	Electrons in Third Level	Electrons in Fourth Level
1	Hydrogen (H)	3.14×10^5	1			
2	Helium (He)	5.68×10^5	2 *(major level full)*			
3	Lithium (Li)	1.25×10^5	2	1 *(lost easily)*		
4	Beryllium (Be)	2.14×10^5	2	2		
5	Boron (B)	1.92×10^5	2	3		
6	Carbon (C)	2.61×10^5	2	4		
7	Nitrogen (N)	3.35×10^5	2	5		
8	Oxygen (O)	3.14×10^5	2	6		
9	Fluorine (F)	4.02×10^5	2	7		
10	Neon (Ne)	4.99×10^5	2	8 *(major level full)*		
11	Sodium (Na)	1.18×10^5	2	8	1 *(lost easily)*	
12	Magnesium (Mg)	1.76×10^5	2	8	2	
13	Aluminum (Al)	1.39×10^5	2	8	3	
14	Silicon (Si)	1.87×10^5	2	8	4	
15	Phosphorus (P)	2.43×10^5	2	8	5	
16	Sulfur (S)	2.40×10^5	2	8	6	
17	Chlorine (Cl)	3.00×10^5	2	8	7	
18	Argon (Ar)	3.65×10^5	2	8	8 *(major level full)*	
19	Potassium (K)	1.00×10^5	2	8	8	1 *(lost easily)*
20	Calcium (Ca)	1.41×10^5	2	8	8	2

and a third level would be started. A drop in ionization energy would be expected when a new level is started because the next added electron would be at a greater distance from the nucleus. Indeed this drop appears at sodium; the second level has been filled at neon. Our tentative model for the electronic arrangement of the first 20 elements can then be summarized as shown in Table 8-3.

This model suggested is consistent with all of the observations we have made. He, Ne, and Ar have electron levels that are full. Such completed levels are described as stable since it is difficult to remove electrons from them. The elements Li, Na, and K have an electron in the next level above He, Ne, and Ar, respectively. This electron can be pulled off relatively easily in each case to give *ions* having electron configurations which are like He, Ne, and Ar, respectively. If Be, Mg, and Ca each lose two electrons, the resulting *ions* have configurations like He, Ne, and Ar, respectively. In every case it is difficult to break up the 2, 10, or 18 pattern.

This crude model for atomic structure makes no effort to tell us precisely where the electrons are or what they are doing. Indeed, we will learn in Chapter 15 that such a question cannot really be answered. Rest assured, however, that the information given here is consistent with the general ideas of the modern quantum theory and is supported by a tremendous amount of chemical and physical evidence. With this thought in mind, let us explore the chemical implications of the model summarized in Table 8-3.

8-3 THE CHEMICAL SIGNIFICANCE OF IONIZATION ENERGIES

The above pattern for electron arrangement offers some justification for the arrangement of atoms in the periodic table (pages 184–185). What are the chemical implications of this arrangement? First let us examine the elements that occupy the peak positions in the ionization energy curve (Figure 8-1). These are helium (no. 2), neon (no. 10), argon (no. 18), krypton (no. 36), xenon (no. 54), and radon (no. 86). They are called the **noble gases.**

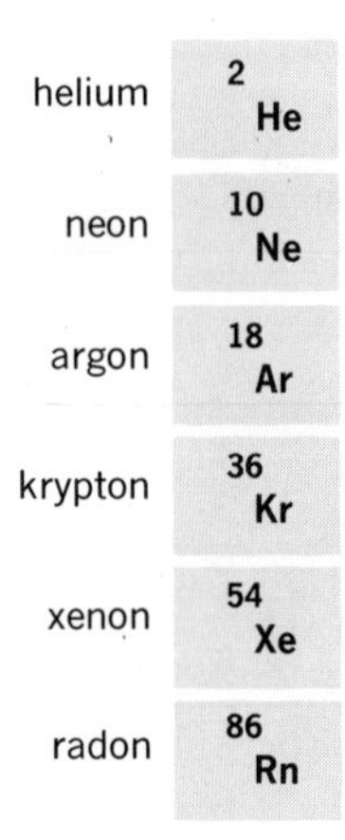

8-3.1 The Noble Gas Family

All six of these elements (He, Ne, Ar, Kr, Xe, and Rn) are gases. In fact, they account for 6 of the 11 elements that are gases at STP. These gases are so strikingly similar that they are conveniently considered as a group called the noble gas family. If we look at the properties of other elements, we find that this recurrence of properties is usual. It is convenient to group all the elements according to their chemistry into families or groups. The elements in a particular group have similar properties (see Figure 8-11). Knowledge about one element in a group then aids in understanding the chemistry of other elements in that group. In the periodic table (see Figure 8-11), each group appears in a vertical column.

As we have noted, all six members of the group with high ionization energies are gases. Each of the other five elements that

Fig. 8-2 **The noble gas family.**

helium	2 He
neon	10 Ne
argon	18 Ar
krypton	36 Kr
xenon	54 Xe
radon	86 Rn

are gases at STP have *two* atoms per molecule. The formulas are H_2, N_2, O_2, F_2, and Cl_2. In sharp contrast, *all members of the noble gas family have molecules that contain only a single atom.* Helium gas is made up of monatomic helium molecules— He, *not* He_2! Neon gas is made up of monatomic neon molecules—Ne, *not* Ne_2. Similarly, argon molecules have the formula Ar; krypton molecules, Kr; xenon molecules, Xe; and radon molecules, Rn.

EXERCISE 8-5

What is the weight of a 22.4-liter sample of xenon, measured at STP? of krypton? 131, 83.8

The atoms of this family are remarkably self-sufficient. They show absolutely no measurable tendency to combine with each other under any conditions. *No real evidence for diatomic molecules in this family has ever been found.* Helium atoms also have little attraction for other atoms. *No stable compounds of helium have yet been found.* This fact, together with the fact that helium has the highest ionization potential of all of the atoms, suggests that the inert character of helium is related to its unusually stable electron configuration. Because helium atoms have such unusually stable electron configurations and tend to have very little interaction with each other, the forces between the monatomic molecules are very low. We would therefore expect helium to have a low boiling point since little energy is needed to disrupt the liquid state. Such is the case. Helium boils at $4.2°K$ ($-268.9°C$) and cannot be solidified at any temperature unless pressure is applied. It becomes solid at $1.1°K$ ($-272.0°C$) under a pressure of 26 atm. It has the lowest boiling point and freezing point of any substance known.

TABLE 8-4 SOME PROPERTIES OF THE NOBLE GASES

Element	Molecular Formula	Atomic Number	Atomic Weight	Boiling Point (°K)	Melting Point (°K)	Ionization Energy (cal/mole)
Helium	He	2	4.00	4.2	—	5.63×10^5
Neon	Ne	10	20.18	27.2	24.6	4.97×10^5
Argon	Ar	18	39.9	87.3	83.9	3.63×10^5
Krypton	Kr	36	83.80	120	116	3.22×10^5
Xenon	Xe	54	131.30	165	161	2.78×10^5
Radon	Rn	86	222	211	202	2.46×10^5

The properties of the other noble gases are summarized in Table 8-4. The gases are all monatomic as suggested by the relatively high ionization potentials, but two points are worthy of comment. As the atomic number goes up and the ionization energy goes down, the boiling points and the freezing points of the noble gases go up (see Figure 8-3, page 170). Apparently, interactions between noble gas atoms, though never strong, become stronger as the atomic number increases and the stability of the electron arrangement decreases. Interactions become stronger as the ionization energy decreases.

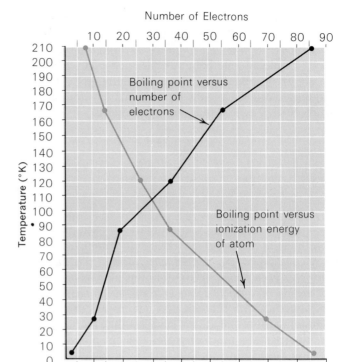

NEIL BARTLETT (1932–)

A native of England, Neil Bartlett received his Ph.D. in chemistry from King's College, Newcastle-upon-Tyne, in 1957. Following a year as senior chemistry master at the Duke's School, Alnwick, he joined the faculty of the University of British Columbia. Since 1966, Bartlett has been a professor of chemistry at Princeton University.

Thus far, Bartlett's outstanding contribution to chemistry is the preparation of the first compound of a noble gas, achieved in 1962. He observed an unknown compound which was produced when platinum was treated with fluorine in glass apparatus, rather than in the less reactive apparatus usually used for fluorine reactions. The development of special techniques of study led to the identification of the compound as an ionic salt, $O_2^+[PtF_6]^-$. The ability of PtF_6 to oxidize molecular oxygen to O_2^+ suggested further experiments with other species of high ionization energy. When PtF_6 was mixed with xenon gas at room temperature, xenon hexafluoroplatinate was formed, as described in Section 8-3.2.

Fig. 8-3 **Boiling points of noble gases plotted against number of electrons and against ionization energy.**

8-3.2 The Non-Inert Noble Gases

Before 1962 all members of the noble gas family were considered inert and the family was called the "inert" gases. The inert character of the group was attributed to the presence of a very stable arrangement of electrons—eight electrons in the outer level. As we noted earlier, the belief that these gases are all inert—i.e., formed no compounds at all—was destroyed in 1962 when Neil Bartlett first succeeded in preparing the yellow solid xenon hexafluoroplatinate ($XePtF_6$). Barlett described the experiment as follows:

> The predicted interactions of xenon and platinum hexafluoride was confirmed in a simple and visually dramatic experiment. The deep red platinum hexafluoride vapor, of known pressure, was mixed by breaking a glass diaphragm, with the same volume of xenon, the pressure of which was greater than that of the hexafluoride. Combination to produce a yellow solid was immediate at room temperature and the quantity of xenon which remained was commensurate with a combining ratio of 1 : 1. (Pictures of the experiment are shown in Figure 8-4.) (See on the next page.)

Following this observation, "inert gas" compounds were prepared in a number of laboratories.* Many xenon compounds are now known,

---

* Research in this area was fast and dramatic. For example, three research teams reported the new and exciting compound XeF_6 almost simultaneously. The teams were headed by Dr. John Malm of Argonne National Laboratories, Dr. George Cady of the University of Washington, and Dr. Bernard Weinstock at the Research Laboratories of the Ford Motor Co.

including XeF_2, XeF_4, XeF_6, XeO_3, $XeOF_2$, $XeOF_3$, $XeOF_4$, $NaXeO_4 \cdot 6H_2O$, $K_4XeO_6 \cdot 2XeO_3$, $Ba_2XeO_6 \cdot xH_2O$, and $XeSbF_6$.

It is now fairly easy to rationalize the existence of these unstable compounds by looking at the ionization potentials. The value for xenon is fairly high, but not exceptionally so in comparison to other elements around it (see Figure 8-1). Rationalization after the fact is easy. (Remember Mr. Jones in the Sherlock Holmes tale.) On the other hand, a number of men had predicted the probable existence of xenon compounds long before Bartlett's brilliant experiments. Among those making such a prediction were Dr. Linus Pauling of the California Institute of Technology and Dr. George Pimentel of the University of California, the editor of the first CHEM Study text.

EXERCISE 8-6

The compound KrF_2 is known. On the basis of the ionization energy data, would you expect KrF_2 to be more or less easily decomposed than XeF_2? Explain. (See Table 8-4.) KrF_2 more easily than XeF_2.

8-4 THE ALKALI METALS

The alkali metals are the six elements following the noble gases. They are lithium (Li), sodium (Na), potassium (K), rubidium (Rb), cesium (Cs), and francium (Fr). Their physical properties contrast sharply with the physical properties of the noble gases (see Table 8-5). All, when pure, are bright, shiny metals and conduct electricity very well. The one extra electron that differentiates the electronic structure of the alkali metals from the noble gases has had a terrific effect upon the properties of the alkali metals. The structure of these metals and the role of this one electron will be considered in a later section of this chapter.

8-4.1 Reactions of the Alkali Metals with Chlorine

When an alkali metal such as sodium is brought into contact with chlorine gas, an alkali metal chloride is formed. For sodium the equation is

$$Na(s) + \tfrac{1}{2}Cl_2(g) \longrightarrow NaCl(s) + \text{energy} \qquad (11)$$

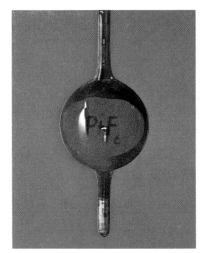

Fig. 8-4 **The reaction of xenon with platinum hexafluoride.** (Above: before reaction; below: after reaction.)

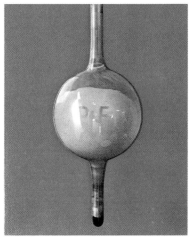

Fig. 8-5 **The alkali metals.**

TABLE 8-5 PHYSICAL PROPERTIES OF ALKALI METALS

Property	Lithium	Sodium	Potassium	Rubidium	Cesium
Atomic number	3	11	19	37	55
Atomic weight	6.94	23.00	39.10	85.47	132.90
Boiling point					
(°K)	1599	1162	1030	952	963
(°C)	1326	889	757	679	690
Melting point					
(°K)	453	371	336.5	311.9	301.8
(°C)	180	98	63.4	38.8	28.7
Atomic volume, solid (cm³/mole of atoms)	13.0	23.7	45.4	55.8	70.0
Density of solid at 20°C	0.535	0.971	0.862	1.53	1.87

In Section 8-1 we saw that relatively little energy is required to remove an electron from a sodium *atom* to give a sodium *ion* (neon configuration). The alkali metals were characterized by the fact that they had lower first-ionization energies than any other group of atoms. We also noted that a chlorine atom with one electron less than the argon configuration will pick up an electron and release energy. Further, we know that when NaCl dissolves in water, $Na^+(aq)$ ions and $Cl^-(aq)$ ions are formed. We also know that molten NaCl conducts electricity. These facts suggest that NaCl(s) may well be built up from Na^+ and Cl^- units. This suggestion has been verified using X-ray diffraction methods (see Chapter 16).

It then appears that one of the steps to be considered in the formation of NaCl involves the transfer of an electron from the outer level of a sodium atom to the incomplete level of a chlorine atom, as in Figure 8-6. Because of the low energy of ionization of sodium and the *release* of energy in the formation of Cl^-, the above process can be carried out at a low net energy cost. These two resulting ions of opposite charge, Na^+ and Cl^-, move toward each other with a large reduction in total energy. *The stability of the sodium chloride crystal can be said to depend upon the electrical attraction of the oppositely charged ions.* The crystal is said to be held together by **ionic bonds.**

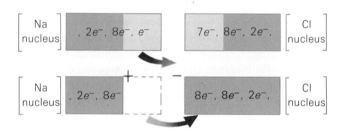

Fig. 8-6 **A representation of the formation of Na^+ and Cl^- ions.**

This chemistry is characteristic of all of the alkali metals. Each of them reacts with chlorine gas in a similar way:

$$Li(s) + \tfrac{1}{2}Cl_2(g) \longrightarrow LiCl(s) + energy \qquad (12)$$

$$Na(s) + \tfrac{1}{2}Cl_2(g) \longrightarrow NaCl(s) + energy \qquad (11)$$

$$K(s) + \tfrac{1}{2}Cl_2(g) \longrightarrow KCl(s) + energy \qquad (13)$$

$$Rb(s) + \tfrac{1}{2}Cl_2(g) \longrightarrow RbCl(s) + energy \qquad (14)$$

$$Cs(s) + \tfrac{1}{2}Cl_2(g) \longrightarrow CsCl(s) + energy \qquad (15)$$

In every case the alkali metal reacts to form a stable ionic solid in which the alkali is present as a noble-gas type positive ion. The product in each case is a crystalline substance with a high solubility in water.

8-4.2 The Nature of Solid Sodium Chloride

What is the nature of solid NaCl? The formation of this solid from gaseous Na^+ and Cl^- ions accounts for the energy released when sodium metal burns in chlorine.

Studies of the crystal using X rays indicate that sodium ions and chloride ions are arranged in a regular geometric array or lattice. The packing arrangement can be seen in Figure 8-7. Each chloride ion is surrounded by six sodium ions, and each sodium ion is surrounded by six chloride ions. *Distinct ions of each element are clearly present.*

The positive and negative ions attract each other very strongly to give a solid with forces of attraction *extending through the entire solid. There are no separate NaCl molecules in a sodium chloride crystal.* The entire mass is a giant molecule; hence the formula NaCl is really just the simplest or empirical formula for the substance. Because the oppositely charged ions are strongly attracted to each other, bonds between ions are strong and the solid is hard with high boiling and melting points. These properties are characteristic of an ionic solid. Because ions cannot move in solid NaCl, it is a very poor conductor of electricity. On the other hand, if it is melted so that ions can move around, it becomes a very good conductor of electricity.

The energy added to the crystal to melt the NaCl tears apart the regular lattice arrangement. When water dissolves NaCl, water molecules surround the Na^+ and Cl^- ions to supply the energy needed to tear apart the lattice.* In summary, ions with the configuration of neon and argon appear to be stable when they are in water solution. They are also stable in a solid ionic lattice of the type formed by NaCl.

8-4.3 Reactions of the Alkali Metals with Water

Sodium metal reacts vigorously with water to form hydrogen gas and an aqueous solution of sodium hydroxide (NaOH), ordinary household lye:

$$2Na(s) + 2H_2O(l) \longrightarrow 2Na^+(aq) + 2OH^-(aq) + H_2(g) + \text{energy} \quad (16)$$

Energy is liberated, often so fast that the temperature rises sharply and the hydrogen, mixing with air, explodes. Thus, sodium metal is dangerous and must be handled with caution. Other alkali metals behave in a similar manner.

In the reaction of an alkali metal with water, electron transfer from the metal to a water molecule or a hydrogen ion is important at some stage in the process.

We see that in this process as well as in the reaction of sodium and chlorine, the ease with which an electron can be removed from sodium contributes strongly to the reactivity characteristic of the system.

---

* Electrical conductivity of an aqueous solution does *not* necessarily imply that the original material dissolved in the water was ionic in form. For example, solid HCl is made up of (distinct) HCl molecules, yet it is pulled apart when dissolved in water to give $H^+(aq)$ and $Cl^-(aq)$. Thus, we cannot safely interpret the conductivity of an aqueous solution to mean that the solid dissolved was an ionic solid. We can, however, state the converse: When an ionic solid dissolves in water, a conducting solution is obtained.

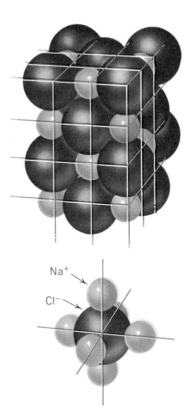

Fig. 8-7.1 **The packing of ions in an ionic crystal.**

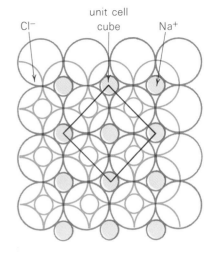

Fig. 8-7.2 **Another view of the NaCl structure.**

8-4.4 Summary of the Chemistry of the Alkali Elements

The alkali metals are extremely reactive. Thus, there is a dramatic change in chemistry as we pass from the noble gases to the alkali metal column in the periodic table. The chemistry of the alkali metals is interesting and often spectacular. These metals react with chlorine and water to form a stable 1+ ion. The 1+ ions, on the other hand, do not have a spectacular chemistry, but their importance and unique characteristics are displayed in more subtle ways. For example, potassium ions are essential to proper plant growth; hence, potassium salts are common constituents of most fertilizers. Similarly, sodium salts are essential to proper animal growth. Salt licks or blocks of salt are frequently supplied to range animals. (A successful man must be "worth his salt.") We have much to learn about the intricacies of alkali metal ion chemistry in biology.

8-5 THE HALOGENS

Fig. 8-8 **The halogens.**

		2 He	3 Li
fluorine	9 F	10 Ne	11 Na
chlorine	17 Cl	18 Ar	19 K
bromine	35 Br	36 Kr	37 Rb
iodine	53 I	54 Xe	55 Cs
astatine	85 At	86 Rn	87 Fr

The lightest halogen, fluorine, occurs just before neon in the periodic table. Chlorine occurs just before argon; bromine appears just before krypton; iodine, just before xenon; and astatine, just before radon. Since astatine is very rare and little is published on its chemistry, our discussion will focus on fluorine, chlorine, bromine, and iodine (see Figure 8-8). We noted earlier that each of these elements has one less electron than the comparable noble gas atom; its outer electron level lacks one electron of being full. Much of the chemistry of this group of elements is concerned with ways of altering the outer electron configuration to obtain a more stable overall electron pattern.

8-5.1 Halogen Molecules and the Covalent Bond

Previously we saw that fluorine, chlorine, bromine, and iodine form diatomic molecules. A sizable amount of energy is needed to disrupt these molecules to give atoms. For chlorine the process is

$$Cl_2(g) + 57{,}800 \text{ cal} \longrightarrow 2Cl(g) \qquad (17)$$

Extremely high temperatures are required to break large numbers of chlorine molecules into atoms. For example, near the surface of the Sun at a temperature of 6000°C, chlorine is present as the monatomic gas Cl; but at room temperature and at most of our common laboratory temperatures, chlorine atoms combine with each other to give Cl_2 molecules.

How can we rationalize the formation of a bond between identical chlorine atoms? It is found that *a pair of electrons, shared between two positively charged nuclei, will be attracted by both nuclei* to give a chemical bond. Thus, in the case of two chlorine atoms each atom can contribute its odd electron (electron number 7) to make a pair that then moves with high probability between the nuclei to make a chemical bond. We might represent our picture schematically as in Figure 8-9, where the two e^-'s represent two electrons that spend a significant part of their time between the nuclei. We do not have positive chlorine ions in the molecule, but unbalanced positive charges on the nucleus of each chlorine atom attract the electron pair between the nuclei to give a bond. We shall have more to say about the details of this bond in a subsequent chapter. A bond in which two atoms are held together by a shared pair of electrons is known as a **covalent bond**.

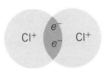

Fig. 8-9 A covalent bond: the bonding of chlorine atoms to give a chlorine molecule.

EXERCISE 8-7

Two argon atoms will *not* form a covalent bond to form Ar_2. Why?   *Outer shell full.*

The consequences of molecule formation are reflected in the physical properties of the halogens shown in Table 8-6.

TABLE 8-6 PROPERTIES OF HALOGENS

Property	Fluorine	Chlorine	Bromine	Iodine	Astatine
Atomic number	9	17	35	53	85
Atomic weight	19.0	35.5	79.9	127	
Molecular formula	F_2	Cl_2	Br_2	I_2	
Boiling point					
(°K)	85	238.9	331.8	457	
(°C)	−188	−34.1	58.8	184	
Melting point					
(°K)	55	172	265.7	387	
(°C)	−218	−101	−7.3	114	
Atomic volume, solid (cm³/mole of atoms)	14.6	18.7	23.5	25.7	

8-5.2 The Chemistry of the Elemental Halogens

The reactions of the alkali metals with chlorine were used to display the similarities of the alkali metals. In a similar way the reactions of the halogens with one of the alkali metals, say sodium, show similarity within this group. The reactions that occur are as follows

$$Na(s) + \tfrac{1}{2}F_2(g) \longrightarrow NaF(s) + energy \quad (18)$$

$$Na(s) + \tfrac{1}{2}Cl_2(g) \longrightarrow NaCl(s) + energy \quad (11)$$

$$Na(s) + \tfrac{1}{2}Br_2(g) \longrightarrow NaBr(s) + energy \quad (19)$$

$$Na(s) + \tfrac{1}{2}I_2(g) \longrightarrow NaI(s) + energy \quad (20)$$

These reactions all proceed readily and they produce ionic solids with the general empirical formula NaX. Each of these solids has a crystal structure made up of positively charged sodium ions and negatively charged halogen ions. These negative ions—F^-, Cl^-, Br^-, and I^-—are called **halide ions.** The stabilities of these ions can be related to the same factors which give rise to the stabilities of the corresponding noble gas atoms.

The halogens also react with hydrogen gas to form the hydrogen halides:

$$H_2(g) + F_2(g) \longrightarrow 2HF(g) \quad \text{HYDROGEN FLUORIDE} \quad (21)$$

$$H_2(g) + Cl_2(g) \longrightarrow 2HCl(g) \quad \text{HYDROGEN CHLORIDE} \quad (22)$$

$$H_2(g) + Br_2(g) \longrightarrow 2HBr(g) \quad \text{HYDROGEN BROMIDE} \quad (23)$$

$$H_2(g) + I_2(g) \longrightarrow 2HI(g) \quad \text{HYDROGEN IODIDE} \quad (24)$$

The reaction of F_2 and H_2 will usually proceed explosively when the gases are just mixed at room temperature. Hydrogen and chlorine can be mixed at room temperature in the dark without reaction, but bright light will initiate an explosion. Why? Some thought tells you that the bonds holding atoms together in the hydrogen and chlorine molecules must be broken if new bonds are to form between hydrogen atoms and chlorine atoms. The first step in starting the reaction is the breaking of the bonds between like atoms. The breaking of bonds in the Cl_2 molecule is achieved by a beam of light or by high temperatures. This starts the reaction. Once started, the combination of hydrogen and chlorine tends to proceed rapidly or explosively without further addition of external heat. The reactions involving Br_2 and I_2 are similar but *much* less energetic than the chlorine reaction.

8-5.3 The Hydrogen Halides

Hydrogen chloride (HCl) and hydrogen fluoride (HF) are frequently prepared by the action of sulfuric acid on a metal salt of the halide:

$$NaCl(s) + H_2SO_4(conc) \longrightarrow HCl(g) + NaHSO_4(s) \quad (25)$$

The HCl leaves the reaction mixture as a gas.

EXERCISE 8-8

$CaF_2 + H_2 \rightarrow 2HF + CaO$

Write an equation for the preparation of HF from crude CaF_2.

All the hydrogen halides are gaseous at room temperature, but hydrogen fluoride liquefies at 19.9°C and 1 atm pressure. All the hydrogen halides dissolve in water to give solutions that conduct electric current. The reactions may be written

$$HF(g) + \text{water} \longrightarrow H^+(aq) + F^-(aq) \quad (26)$$

$$HCl(g) + \text{water} \longrightarrow H^+(aq) + Cl^-(aq) \quad (27)$$

$$HBr(g) + \text{water} \longrightarrow H^+(aq) + Br^-(aq) \quad (28)$$

$$HI(g) + \text{water} \longrightarrow H^+(aq) + I^-(aq) \quad (29)$$

The solutions have similar properties and are called **acid solutions.** The common species in each case is the aqueous hydrogen ion,

$H^+(aq)$, and the properties of aqueous acid solutions are attributed to this ion.

8-5.4 Some Reactions of the Halide Ions

The halogens, because of their ability to form either stable ions or covalent bonds, combine with most of the elements of the table. Thus, we have NaCl, $MgCl_2$, HCl, and CCl_4. When magnesium reacts with chlorine, it gives up two electrons to form a Mg^{2+} ion and two Cl^- ions. Note that the Mg^{2+} ion has the neon electron structure since magnesium metal has two more electrons than neon and it appears in the second group of the periodic table. Calcium will form halides—CaF_2, $CaCl_2$, $CaBr_2$, and CaI_2. Strontium and barium will also form MX_2 compounds where M is the metal ion and X is a halide ion. Aluminum gives us a compound whose simplest formula is $AlCl_3$.

EXERCISE 8-9

Write the formula for the product expected when radium (same family as calcium) combines with bromine. Repeat for the combination of radium with fluorine. $RaBr_2$ or RaF_2

Most metal halides are water soluble. For example, NaCl (ordinary salt), $CaCl_2$ (put on roads to lay dust), and KBr are easily dissolved in water. A few halides such as AgCl, CaF_2, AgBr, and AgI are different. You remember from Experiment 8 that silver chloride has a very low solubility in water. Silver bromide and silver iodide are even less soluble. These low solubilities provide a sensitive test for the presence of chloride ions, bromide ions, and iodide ions in aqueous solutions. When silver nitrate ($AgNO_3$) is dissolved in water, the salt ionizes to give silver ions and nitrate ions in solution:

$$AgNO_3(s) + \text{water} \longrightarrow Ag^+(aq) + NO_3^-(aq) \qquad (30)$$

If this solution is now added to a solution containing chloride ions, a white precipitate of AgCl(s) forms:

$$Ag^+(aq) + Cl^-(aq) \longrightarrow \underset{\text{WHITE SOLID}}{AgCl(s)} \qquad (31)$$

Comparable reactions occur with bromide and iodide ions, except that AgBr is tan and AgI is yellow. Silver fluoride is soluble; therefore, no precipitate forms when a solution containing $Ag^+(aq)$ ions is added to a solution containing $F^-(aq)$ ions.

EXERCISE 8-10

Write the *ionic* equation for the process occurring when a solution of $AgNO_3$ is poured into a solution of potassium iodide (KI). $Ag^+(aq) + NO_3^-(aq) + K^+(aq) + I^-(aq) \rightarrow AgI(s) + NO_3^- + K^+$

EXERCISE 8-11

Devise a procedure for separating $F^-(aq)$ ions and $Br^-(aq)$ ions present in a water solution. Write appropriate equations. $F^-(aq) + Br^-(aq) + Ag^+(aq) + NO_3^-(aq) \rightarrow AgBr(s) + F^- + NO_3^-$

8-5.5 A Summary of Halogen Chemistry

Halogens have two atoms per molecule. Each atom shares its odd electron with the other atom to give a structure in which there is a high probability of finding the electron pair between the nuclei. Electrons between the nuclei give a bond between atoms, since both nuclei are attracted to the electron pair. This bond is called a **covalent bond.** Fluorine is the most reactive of the nonmetals. Halogens frequently produce ionic structures containing the X^- ion or are a part of a covalent structure involving an electron-pair bond. Ionic solids dissolve in water to give solutions that conduct electric current. AgCl, AgBr, and AgI are insoluble in water; addition of silver nitrate to an acid solution can be used as a test for the presence of these halide ions in aqueous solution. Halogen atoms can also form covalent bonds with many elements.

8-6 HYDROGEN—A FAMILY BY ITSELF

Perhaps you are wondering why the element hydrogen was not included among the halogens. It is, after all, an element with one less electron than its neighboring inert gas, helium. We did not consider earlier the process $H + e^- \longrightarrow H^-$. On the other hand, the hydrogen atom has but one electron and, in a sense, it is like an alkali metal. The removal of one electron from an alkali metal atom leaves a very stable electron configuration—that of a noble gas. The removal of one electron from a hydrogen atom leaves it with *no* electrons; this also interacts to give stable components. We shall see both of these influences in the chemistry of hydrogen. This element forms a family by itself; it has some similarities to the halogens and some similarities to the alkalies.

8-6.1 The Properties of Hydrogen Gas

Hydrogen is a diatomic gas like the halogens rather than a metal like the alkalies. Its melting point is 15.9°K and its boiling point is 20.4°K, the lowest boiling point for any element except helium.

EXERCISE 8-12

Describe the bond that holds two hydrogen atoms together in the molecule H_2. *covalent bond*

8-6.2 Some Chemistry of Hydrogen

One of the most distinctive reactions characterizing both the alkalies *and* the halogens is their reaction with each other. The example we have discussed most is the reaction between sodium and chlorine to give sodium chloride. Sodium chloride is an ionic solid that dissolves in water to give positively charged sodium ions, $Na^+(aq)$, and negatively charged chloride ions, $Cl^-(aq)$:

$$Na(s) + \tfrac{1}{2}Cl_2(g) \longrightarrow NaCl(s) + \text{energy} \qquad (11)$$

$$NaCl(s) + \text{water} \longrightarrow Na^+(aq) + Cl^-(aq) \qquad (32)$$

Does hydrogen react like sodium or chlorine? Experiments show that hydrogen can take *either* position, as does chlorine:

$$\tfrac{1}{2}H_2(g) + \tfrac{1}{2}Cl_2(g) \longrightarrow HCl(g) \qquad (33)$$

or as does sodium:

$$Na(s) + \tfrac{1}{2}H_2(g) \longrightarrow NaH(s) \qquad (34)$$

$$HCl(g) + \text{water} \longrightarrow H^+(aq) + Cl^-(aq) \qquad (27)$$

The compound sodium hydride, formed in reaction with sodium, is a crystalline compound with physical properties similar to those of sodium chloride. The chemical properties are very different, however. Whereas sodium burns readily in chlorine, it reacts with hydrogen only on heating to about 300°C. While sodium chloride is a stable substance that dissolves in water to form $Na^+(aq)$ and $Cl^-(aq)$, the alkali hydrides burn in air. In contact with water, a vigorous reaction occurs, releasing hydrogen:

$$NaH(s) + H_2O(l) \longrightarrow H_2(g) + Na^+(aq) + OH^-(aq) \qquad (35)$$

Thus, hydrogen reacts with sodium as a halogen does, but the product, sodium hydride, is very different in its chemistry from sodium chloride.

Hydrogen also reacts as does an alkali metal. Though the product, hydrogen chloride, is not an ionic solid like sodium chloride, it does dissolve in water to give aqueous ions. The formation of $H^+(aq)$ and $Cl^-(aq)$ in water is strikingly like the formation of $Na^+(aq)$ and $Cl^-(aq)$ as salt dissolves in water. In fact, the tendency of hydrogen to form a positively charged ion in water, $H^+(aq)$, and the absence of any evidence for a negatively charged ion in water, $H^-(aq)$, are two of the most significant differences between hydrogen and the halogens.

An overall view of the chemistry of hydrogen requires that it be classified alone—as a separate chemical family. There are some important similarities to halogens. For example, it is a stable diatomic gas. On the other hand, its chemistry is more like that of the alkalies. Hydrogen, therefore, is usually shown on the left side of the periodic table with the alkalies but separated from them to indicate its distinctive character.

8-7 THE ELEMENTS FROM THREE THROUGH TEN— THE SECOND ROW OF THE PERIODIC TABLE

When the first ionization energy for each element is plotted against its atomic number, a cyclic curve is obtained (Figure 8-1). By selecting the element with the maximum ionization energy for each cycle, it was possible to identify a group of relatively nonreactive gaseous elements called the noble gases. All members of this group or family had rather closely related properties. They were identified as a group of the periodic table. A second group was the reactive alkali metal family found at the *minimum* in each cycle of the ionization energy curve. Members of this group appear immediately after the noble gases (Figure 8-5) and exhibit strong similarities within the group. The third group studied was the halogens—the group of elements appearing just before the noble gases (Figure 8-8). Ionization energy values provided strong hints as to the type of chemistry that should be expected for each group, and the hints were fairly useful in considering the chemistry of alkalies and halogens.

Let us now examine the elements in a given cycle—for example, the elements from lithium (no. 3) through neon (no. 10) which make up the second cycle or the second row of the periodic table (Figure 8-10). We pass from Li with a minimum ionization energy to Ne with a maximum ionization energy. What happens in between? Let us examine the facts.

Fig. 8-10 **The placement of the second row of the periodic table.**

1 H									2 He
3 Li	4 Be			5 B	6 C	7 N	8 O	9 F	10 Ne
11 Na	12 Mg			13 Al	14 Si	15 P	16 S	17 Cl	18 Ar
19 K								35 Br	36 Kr
37 Rb								53 I	54 Xe
55 Cs								85 At	86 Rn
87 Fr									

8-7.1 Trends in Physical Properties Across a Row of the Table—The Second Row

Lithium is a typical metal. It is bright and shiny in appearance (metallic luster); it can be rolled into sheets or formed into wire (malleable and ductile); it is a very good conductor of heat and electricity (one-sixth as good as copper); it has a low melting point and is soft enough to be cut with a knife (see Table 8-7). All these properties, except softness and low melting point, are characteristic properties by which we identify a metal. Lithium is truly metallic, yet because of its great chemical reactivity it is not used as a metal in commerce. We never see objects made of metallic lithium.

Beryllium looks like a metal. It is steel gray in color and hard enough to scratch glass and is very brittle. The brittle character is definitely not a metallic characteristic. It is a fair conductor of electricity—about half as good as lithium. This property indicates some metallic character. Further, it combines with other metals to give metallic alloys. It is still quite metallic, although definitely less so than lithium.

With elemental or crystalline boron, most of the characteristics of a metal have disappeared. About the only metallic property remaining at boron is the metallic luster found on the red-to-black crystals. (Many forms of boron are known. One called α-T has red crystals;

the other, called β-R, has black crystals.) Boron is a semiconductor. This means that boron has potential value in electronics. On the other hand, it is a nonconductor compared to lithium. Its conductivity rises with temperature. Pure boron is very hard, being only slightly softer than diamond. It *cannot* be rolled into sheets or drawn into wires, but rods of boron are amazingly strong and flexible and may find uses in commerce.

Carbon exists in two forms, diamond and graphite. You are all familar with both of these. The diamond is strictly nonmetallic. It is the world's standard for hardness; it is brittle and a nonconductor of electricity. It has one of the highest melting and boiling points known to man (see Table 8-7). It is bright, sparkling, and romantic. Graphite, on the other hand, is drab. It is gray to black in color with a very weak metallic luster. It is used to make pencil "leads." It breaks into layers easily and is slippery. For this reason it is used as a dry lubricant and is added to oils. Graphite, in contrast to diamond, is a good conductor of electricity *along the layers,* but a

TRENDS in PHYSICAL PROPERTIES

TABLE 8-7 PHYSICAL PROPERTIES OF THE SECOND ROW OF THE PERIODIC TABLE

Description	Lithium	Beryllium	Boron	Carbon		Nitrogen	Oxygen	Fluorine
				Graphite	Diamond			
Form and appearance	silvery metal	steel-gray metal	black and red shiny crystals	gray-black plates	shiny diamond	colorless diatomic gas	colorless diatomic gas	yellow diatomic gas
Hardness (mohs scale)	very soft 0.6	scratches glass 6–7	very hard 9.3	soft 1.0	extremely hard 10 (std.)	gas —	gas —	gas —
Melting point (°C)	179	1285	2370	3600	3730 (est.)	−210	−218	−223
Boiling point (°C)	1340	2970	—	—	4830 (est.)	−196	−182	−188
*Electrical conductance (mhos)	excellent 10^5	good 5.4×10^4	poor † 2×10^{-5}	good 10^4	no	no	no	no
*Ductile and malleable‡	yes	no— brittle	no	no	no	no	no	no
*Metallic luster	yes	yes	yes	faint	no	no	no	no
Density (g/cm³)	0.53	1.86	2.30	2.1	3.5	0.96 (solid)	1.43 (solid)	1.1 (liquid)
Interesting properties	reactive metal	makes copper hard; alloys valuable	semi- conductor; useful in electronics; rods very strong	soft and slippery	hardest material known; nonmetal; low reactivity	nonmetal of low reactivity	moderately reactive nonmetal	most reactive non- metal known

* Characteristic of metal.
† Semiconductor of *p*-type.
‡ Refer to property that permits substance to be drawn into wires or rolled into sheets.

poor conductor *at right angles to the layers.* You are probably familiar with other forms of carbon, such as lampblack, soot, and charcoal. These are composed, in large part, of very small crystals of graphite randomly arranged. Graphite has a few metallic characteristics, but it really cannot be classed as a metal by any stretch of the imagination.

The elements nitrogen, oxygen, and fluorine are all low-boiling, diatomic gases. Fluorine is probably the prime example of a nonmetal. The transition from metallic lithium to nonmetallic fluorine has been made in distinct steps.

8-7.2 Trends in Chemical Properties Across a Row of the Table—The Second Row

What does the *chemistry* of this row look like? Lithium is a very *reactive metal*. It burns vigorously in air, in chlorine gas, in fluorine gas, and in bromine vapor. It is this very reactivity that hides its metallic character. Typical compounds formed are listed in Table 8-8. At elevated temperatures lithium, like sodium, combines with H_2 to give LiH.

In general, beryllium has a much lower chemical reactivity than does lithium, but it does react with fluorine to give BeF_2, with Cl_2 to give $BeCl_2$, and with oxygen to give BeO. It does not combine directly with hydrogen, but a compound BeH_2 can be made by other means.

TABLE 8-8 FORMULAS OF SOME COMPOUNDS OF THE SECOND-ROW ELEMENTS

Group	I Li	II Be	III B	IV C	V N	VI O	VII F
Fluorides: Ratio F/element	LiF 1	BeF_2 2	BF_3 3	CF_4 4	NF_3 3	OF_2 2	F_2 1
Chlorides: Ratio Cl/element	LiCl 1	$BeCl_2$ 2	BCl_3 3	CCl_4 4	NCl_3 3	OCl_2 2	ClF 1
Oxides: Ratio O/element	Li_2O 0.5	BeO 1.0	B_2O_3 1.5	CO_2 2	N_2O_5 2.5	O_2 1	OF_2 0.5
Hydrides: Ratio H/element	LiH 1	BeH_2 2	B_2H_6 3	CH_4 4	NH_3 3	H_2O 2	HF 1
Calcium: Salts Ca/element	— —	— —	— —	CaC_2 0.5	Ca_3N_2 1.5	CaO 1.0	CaF_2 0.5

EXERCISE 8-13

How many electrons does oxygen need to achieve the neon structure? Justify the formula MgO. *2. Mg lose 2 e⁻ to get stable structure*

Elemental boron is difficult to involve in a reaction. Does elemental boron represent a stable electron pattern like helium or neon? In considering this question, several facts are pertinent. First, boron, particularly if it is impure, will react at higher temperatures with fluorine, chlorine, and oxygen to give BF_3, BCl_3, and B_2O_3, respectively. Second, helium exists as isolated single atoms, not as helium clusters. Boron exists as a hard solid. *Boron atoms combine with themselves to give a stable, interlocked arrangement that is difficult to break up.* It appears that boron atoms, in contrast to helium atoms, are very reactive; however, boron atoms can react with other boron atoms to give a solid form of elemental boron that has low reactivity. Hence, the element *appears* to be inactive. We repeat: The atoms of boron are very reactive and the compounds of boron are common.

Pure crystalline diamond, like pure crystalline boron, is nonreactive; but it will burn in oxygen to give CO and CO_2 if heated to 800°C. With Cl_2 and F_2, CCl_4 and CF_4 can be obtained, although not easily. Impure carbon or soot will react much more readily (compare with boron). Under appropriate conditions, carbon *atoms* are very reactive. They combine with hydrogen to give CH_4 and many other compounds. Chapter 18 is devoted to carbon chemistry. Carbon chemistry is one of the most important phases of modern science.

Nitrogen is a nonreactive gas at room temperature, but it becomes more reactive as the temperature rises. Its inert character is very important to man since it dilutes the oxygen of the air. Pure oxygen can have rather severe physiological effects if breathed for long periods of time. Since nitrogen is a gas, not a solid as is boron, we may properly ask: Is the nitrogen electron configuration inert like helium and neon? Again, the answer is a definite and resounding "No"! Helium *atoms* are inert. They will not combine even with other helium atoms. On the other hand, nitrogen *atoms* are very reactive, but each combines so vigorously with another nitrogen atom that the resulting molecule, N_2, represents a very *stable* structure of relatively low reactivity. The large energy change associated with the formation of N_2 is shown by the equation:

$$2N(g) \longrightarrow N_2(g) + 225{,}800 \text{ cal} \qquad (36)$$

Nitrogen will form a number of compounds with oxygen, including N_2O_3 and N_2O_5. The compounds NCl_3 and NF_3 can be produced by indirect methods. NCl_3 is a dangerously explosive material. NF_3 is more stable and has been considered as an oxidizer for rocket fuels; however, it has undesirable physical properties.

With hydrogen, under appropriate conditions, nitrogen forms the gaseous compound ammonia (NH_3):

$$N_2(g) + 3H_2(g) \longrightarrow 2NH_3(g) \qquad (37)$$

This is the equation for the Haber process, an extremely important commercial process that provides ammonia for everything from fertilizers to explosives, dyes, and nylon. With sodium, nitrogen forms the compound Na_3N.

Oxygen gas is much more reactive than nitrogen; we have already mentioned its combining with H, Li, B, Be, C, and N. It will also form compounds with F—F_2O for example. Its relatively high reactivity, in comparison to nitrogen, can be attributed to the fact

TRENDS in CHEMICAL PROPERTIES

DMITRI MENDELEEV (1834–1907)

Element 101 is named Mendelevium in honor of the great Russian chemist, Dmitri Mendeleev. Born in Tobolska, Siberia, Dmitri distinguished himself in science and mathematics at the University of St. Petersburg (now Leningrad), where he earned his doctorate. Later studies enabled him to visit France, Germany, and the United States.

A professor of chemistry at St. Petersburg when only 32, Mendeleev arranged the elements by their properties. The results led him to propose the periodic table and use it to predict the existence and properties of a number of unknown elements. When they were discovered later, Mendeleev was hailed as a prophet.

Mendeleev, an inspiring teacher and tireless experimenter, was so deeply concerned over social issues that he resigned his professorship rather than obey an order to cease interfering with affairs of government.

The rapid increase of known elements, during and after Mendeleev's lifetime, was made possible by the most important generalization of chemistry, the periodic table.

Chapter 8 / ATOMIC STRUCTURE and PERIODIC TABLE

that two oxygen atoms combine with less vigor than do two nitrogen atoms. Let us compare the energies for the processes:

$$2N(g) \longrightarrow N_2(g) + 225{,}800 \text{ cal} \qquad (36)$$

$$2O(g) \longrightarrow O_2(g) + 117{,}000 \text{ cal} \qquad (38)$$

The fact that oxygen is more reactive than nitrogen reflects differences in the stabilities of the oxygen and nitrogen *molecules*—not necessarily differences in the stabilities of the electron patterns of the oxygen and nitrogen *atoms*. Oxygen gas is essential to most forms of life.

Fluorine is the most reactive element known. It combines directly at room temperature or above with all the elements except oxygen, nitrogen, and some gases of the helium family. As we have already noted, even the compounds OF_2 and NF_3 can be made indirectly, and a number of fluorides of the so-called "inert gas" or noble gas family are known (XeF_2, XeF_4, XeF_6). The energy for the reaction of two fluorine atoms is relatively low, as we might expect:

$$2F(g) \longrightarrow F_2(g) + 37{,}700 \text{ cal} \qquad (39)$$

Fig. 8-11 **The periodic table.**

KEY: Atomic Number → 26, Atomic Weight → 55.85, Symbol of Element → Fe, Name of Element → Iron

1 H 1.0080 Hydrogen									
3 Li 6.940 Lithium	4 Be 9.013 Beryllium								
11 Na 22.991 Sodium	12 Mg 24.32 Magnesium								
19 K 39.100 Potassium	20 Ca 40.08 Calcium	21 Sc 44.96 Scandium	22 Ti 47.90 Titanium	23 V 50.95 Vanadium	24 Cr 52.01 Chromium	25 Mn 54.94 Manganese	26 Fe 55.85 Iron	27 Co 58.94 Cobalt	
37 Rb 85.48 Rubidium	38 Sr 87.63 Strontium	39 Y 88.92 Yttrium	40 Zr 91.22 Zirconium	41 Nb 92.91 Niobium	42 Mo 95.95 Molybdenum	43 Tc (99) Technetium	44 Ru 101.1 Ruthenium	45 Rh 102.91 Rhodium	
55 Cs 132.91 Cesium	56 Ba 137.36 Barium	57–71 * below	72 Hf 178.50 Hafnium	73 Ta 180.95 Tantalum	74 W 183.86 Tungsten	75 Re 186.22 Rhenium	76 Os 190.2 Osmium	77 Ir 192.2 Iridium	
87 Fr (223) Francium	88 Ra (226) Radium	89– ** below							

*LANTHANIDE SERIES

57 La 138.92 Lanthanum	58 Ce 140.13 Cerium	59 Pr 140.92 Praseodymium	60 Nd 144.27 Neodymium	61 Pm (145) Promethium	62 Sm 150.35 Samarium	63 Eu 152. Europium

**ACTINIDE SERIES

89 Ac (227) Actinium	90 Th 232.05 Thorium	91 Pa (231) Protactinium	92 U 238.07 Uranium	93 Np (237) Neptunium	94 Pu (242) Plutonium	95 Am (243) Americium

In summary, the metals from the left side of the table combine with the nonmetals from the right side of the table to give ionic solids. Ions such as M^+ and X^- suggest the stability and importance of the noble gas electron arrangements. This same stability is suggested by the next two elements, one from each side: Be forms Be^{2+} and oxygen forms O^{2-}. Be^{2+} has the helium configuration and O^{2-} has the neon configuration. An ionic lattice is formed. In CF_4, carbon *shares* its four electrons with four fluorine atoms to give four covalent bonds. Intermediate bond types lie between these extremes. The summary of formulas in Table 8-8 indicates the use of the table and of these concepts of bonding in writing formulas for compounds. The few cases such as CaC_2, which seem to be out of line, guarantee that we shall not run out of "wondering why" questions in the weeks ahead.

8-8 THE PERIODIC TABLE

The power of the **periodic table** is evident in the chemistry we have viewed. By arranging the elements as in Figure 8-11, we simplify the

								2 4.003 **He** Helium
		5 10.82 **B** Boron	6 12.011 **C** Carbon	7 14.008 **N** Nitrogen	8 16.0000 **O** Oxygen	9 19.00 **F** Fluorine	10 20.183 **Ne** Neon
		13 26.98 **Al** Aluminum	14 28.09 **Si** Silicon	15 30.975 **P** Phosphorus	16 32.066 **S** Sulfur	17 35.457 **Cl** Chlorine	18 39.944 **Ar** Argon
28 58.71 **Ni** Nickel	29 63.54 **Cu** Copper	30 65.38 **Zn** Zinc	31 69.72 **Ga** Gallium	32 72.60 **Ge** Germanium	33 74.91 **As** Arsenic	34 78.96 **Se** Selenium	35 79.916 **Br** Bromine	36 83.80 **Kr** Krypton
46 106.4 **Pd** Palladium	47 107.880 **Ag** Silver	48 112.41 **Cd** Cadmium	49 114.82 **In** Indium	50 118.70 **Sn** Tin	51 121.76 **Sb** Antimony	52 127.61 **Te** Tellurium	53 126.91 **I** Iodine	54 131.30 **Xe** Xenon
78 195.09 **Pt** Platinum	79 197.0 **Au** Gold	80 200.61 **Hg** Mercury	81 204.39 **Tl** Thallium	82 207.21 **Pb** Lead	83 209.00 **Bi** Bismuth	84 (210)■ **Po** Polonium	85 (210)■ **At** Astatine	86 (222)■ **Rn** Radon

64 157.26 **Gd** Gadolinium	65 158.93 **Tb** Terbium	66 162.51 **Dy** Dysprosium	67 164.94 **Ho** Holmium	68 167.27 **Er** Erbium	69 168.94 **Tm** Thulium	70 173.04 **Yb** Ytterbium	71 174.99 **Lu** Lutetium
96 (248)■ **Cm** Curium	97 (247)■ **Bk** Berkelium	98 (249)■ **Cf** Californium	99 (254)■ **Es** Einsteinium	100 (253)■ **Fm** Fermium	101 (256)■ **Md** Mendelevium	102 (253)■ **No** Nobelium	103 (259)■ **Lr** Lawrencium

problem of understanding the variety of chemical characteristics found in nature. The elements grouped in a vertical column have pronounced similarities. General statements can be made about their chemistries and the compounds they form. Furthermore, the formulas of these compounds and the nature of the bonds that hold them together can be understood in terms of the especially stable electron arrangements of the noble gases.

The periodicity of chemical properties was discovered about 100 years ago. In 1828 J. W. Döbereiner, a German chemist, recognized similarities among certain elements (chlorine, bromine, and iodine; lithium, sodium, and potassium; and so on). He grouped them as "triads." (Remember: "Cylindrical Objects Burn.") J. A. R. Newlands, an English chemist, in 1864 was ridiculed for proposing a "Law of Octaves," which foresaw the differences of eight that we noted in Table 8-2. Simultaneously, Lothar Meyer, a German chemist and physicist, proposed a periodic table similar to Newlands'. Independently and in the same year (the time was ripe for the next step, "Wooden Objects Burn"), D. I. Mendeleev, a Russian, framed the periodic table in more complete form. He even predicted both the existence and properties of elements not then known. The subsequent discovery of these elements and confirmation of their predicted properties solidified the acceptance of the periodic table. It remains, 100 years later, as the most important single correlation of chemistry. It permits us to deal with the great variety of materials we find in nature.

8-9 MODELS AND PHYSICAL PROPERTIES—BONDING PATTERNS ACROSS THE SECOND ROW

It is relatively easy to understand the vaporization of the common gases O_2, N_2, and F_2. Separate, distinct, and self-sufficient molecules with small forces of attraction acting between them make vaporization easy; boiling points and melting points are very low (Table 8-7). But what about the solid elements? How does their structure reflect the properties we summarized in Section 8-7.1? Let us begin with a view of the metals.

8-9.1 Metals

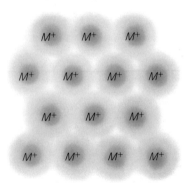

Fig. 8-12 **A segment of the structure of a metal.**

Metals, like ionic solids, have high boiling points. Lithium boils at 1340°C, sodium at 880°C, beryllium at 2970°C, magnesium at 1110°C, and aluminum at 2270°C. These numbers indicate that forces between atoms extend over many atoms throughout the structure. Further, as we go across a row, boiling points for metals go up. Beryllium has a higher boiling point than lithium; magnesium (no. 12) is higher (1110°C) than sodium (no. 11, 880°C); and aluminum (no.13) is higher (2270°C) than magnesium. [As more electrons become available in the bonding level, the boiling points go up. The forces between atoms get stronger.] Further, the volume occupied by 1 mole of the metal gets smaller as we go from sodium to aluminum. The atoms are pulled more tightly together as more electrons are available in the bonding level. These facts, as well as the other physical properties of a metal, are all consistent with a model in which we have metal ions arranged in a regular lattice with a "sea" of electrons moving among many atoms to hold them together in a giant structure (see Figure 8-12). Forces extend over many atoms. This sea of mobile electrons accounts for the electrical conductivity of the metal and for the ease with which it is deformed under pressure (easily deformed metals are said to be malleable and ductile). The more electrons available in the "sea," the stronger the bonds.

8-9.2 Network Solids

What about the diamond? The diamond represents a rigid network solid. Each atom is firmly bound by covalent bonds to four other atoms located at the corners of a regular tetrahedron (see Figure 8-13). This arrangement generates a giant, three-dimensional network; hence, the diamond is called a **network solid.** Because strong covalent bonds hold each atom in a rigid position, it is extremely difficult to break atoms away to obtain molecules or isolated atoms. Further, the structure is extremely hard since atoms are held firmly in a rigid position. Crystalline silicon is also a network solid with the same tetrahedral structure. But it does not have the hardness characteristic of diamond.

Beryllium and boron represent structures between these two extremes. Boron is closer to the network solid, and beryllium is closer to the metallic solid. We shall have more to say about such structures in Chapter 17.

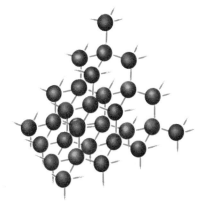

Fig. 8-13 **A segment of diamond structure.**

8-10 HIGHLIGHTS

The energy needed to separate one electron from a gaseous atom to give a singly charged gaseous ion and a free electron is the **first ionization energy** of the atom. The energy required to separate a second electron to get a doubly charged ion and a free electron is the **second ionization energy.** From values for ionization energies we identified stable electron patterns at elements no. 2, no. 10, and no. 18. This gave rise to a suggested model in which electrons for the first 20 elements are arranged in four general levels outside the nucleus. The first level contains two electrons, the second eight and the third eight, and the fourth has two, at which point we stopped development of the model. Using ionization energies as a guide, we could identify three families or groups of elements: the **noble gases,** the **alkali metals,** and the **halogens.** Elements in any group show remarkable similarities. These groups are the families of the **periodic table.**

In studying trends going across a row of the periodic table, we go from metals to nonmetals. The periodic table is very useful in correlating many facts on the chemical and physical properties of the elements. Proper use of the table and our electron model provides us with a basis for understanding the properties of gaseous elements such as N_2, O_2, and F_2; of metals such as Li, Be, Na, Mg, and Al; and of the network solids formed by such elements as C (diamond) and Si.

Chapter 8 / ATOMIC STRUCTURE
and PERIODIC TABLE

QUESTIONS AND PROBLEMS

1 For every atom the first ionization energy is lower than the second ionization energy. Explain.

2 What is the significance of the trends in the boiling points and melting points of the noble gases in terms of attractions among the atoms?

3 The molar heats of vaporization of the noble gases (in kcal/mole) are: He, 0.020; Ne, 0.405; Ar, 1.549; Kr, 2.16; Xe, 3.02; Rn, 3.92. Using the data in Table 8-4, plot the boiling points (on the vertical axis) against the heats of vaporization (on the horizontal axis). Suggest a generalization based upon a simple curve passing near the plotted points. Write an equation for the straight line passing through the origin (that is, through zero) and through the point for radon.

4 An alkali element produces ions having the same electron population as atoms of the preceding noble gas. In what ways do these ions differ from the inert gases? In what ways are they alike?

5 There is a large difference between the energy needed to remove an electron from a neutral, gaseous sodium atom and from a neutral, gaseous neon atom. For 1 mole of each element the equations are:

$$Na(g) + 118.4 \text{ kcal} \longrightarrow Na^+(g) + e^-$$
$$Ne(g) + 497.0 \text{ kcal} \longrightarrow Ne^+(g) + e^-$$

Explain how these energies are consistent with the proposal that the electron arrangements of the noble gases are especially stable.

6 Explain why chemists say that boron has three electrons involved in chemical reactions, and that chlorine has seven.

7 How many electrons of fluorine are involved in chemical reactions? of oxygen? of nitrogen?

8 Refer to the halogen column in the periodic table. How many electrons must each halogen atom gain to have an electron population equal that of an atom of the adjacent noble gas? What property does this population impart to each ion?

9 How do the trends in physical properties for the halogens compare with those for the noble gases? Compare boiling points and melting points with atomic number.

10 Use your knowledge of the usefulness of the periodic table to fill in the blank space in Table 8-6 under astatine. List some chemical reactions expected for astatine.

11 Chlorine is commonly used as a germicide in swimming pools. When chlorine dissolves in water it reacts to form hypochlorous acid (HOCl), as follows:

$$Cl_2(g) + H_2O(l) \rightleftharpoons HOCl(aq) + H^+(aq) + Cl^-(aq)$$

Predict what happens when bromine (Br_2) dissolves in water. Write the equation for the reaction.

12 Zinc metal reacts in a solution of elemental chlorine in water as follows:

$$Zn(s) + Cl_2(aq) \longrightarrow Zn^{2+}(aq) + 2Cl^-(aq)$$

Zinc does not react with a solution of elemental hydrogen in water but it *does* react with an aqueous solution of hydrogen chloride:

$$Zn(s) + 2H^+(aq) + 2Cl^-(aq) \longrightarrow Zn^{2+}(aq) + H_2(g) + 2Cl^-(aq)$$

Recognizing that zinc metal must release electrons to form $Zn^{2+}(aq)$, explain how these reactions demonstrate that hydrogen molecules do not behave like a halogen.

13 Write the molecular formulas of the hydrogen compounds of the third-row elements—Na, Mg, Al, Si, P, S, Cl, and Ar. Indicate, for each compound, the ratio of hydrogen to element.

14 Indicate the electron gain or loss in each kind of atom assuming it attains noble-gas-like electron structure in the following reactions:
 (a) $2Rb + Br_2 \longrightarrow 2RbBr$
 (b) $2Cs + I_2 \longrightarrow 2CsI$
 (c) $Mg + S \longrightarrow MgS$
 (d) $2Ba + O_2 \longrightarrow 2BaO$

15 Which of the following is *not* a correct formula for a substance at normal laboratory conditions?
 (a) $H_2S(g)$ (b) $CaCl_2(s)$ (c) $He(g)$ (d) $NaNe(s)$
 (e) $Al_2O_3(s)$

16 Magnesium metal burns in air, emitting enough light to be useful as a flare and forming clouds of white smoke. Write the equation for the reaction. What is the composition of the white smoke?

17 Use the formulas for magnesium oxide (MgO) and magnesium chloride ($MgCl_2$), together with the periodic table, to decide that magnesium ions have the same number of electrons as each of the following, except (a) neon atoms (Ne), (b) sodium ions (Na^+), (c) fluoride ions (F^-), (d) oxide ions (O^{2-}), (e) calcium ions (Ca^{2+}).

18 Sodium metal reacts with water to form sodium ions (Na^+), hydroxide ions (OH^-), and hydrogen gas (H_2) as follows:

$$2Na(s) + 2H_2O(l) \longrightarrow 2Na^+(aq) + 2OH^-(aq) + H_2(g)$$

Assuming calcium metal reacts in a similar way, write the equation for the reaction between calcium and water. Remember that calcium is in the second column of the periodic table and sodium is in the first.

QUESTIONS and PROBLEMS

19 Use Table 8-8 and the periodic table to write possible formulas for the following compounds: (a) a hydride of barium (no. 56) (b) a chloride of germanium (no. 32) (c) an oxide of indium (no. 49) (d) an oxide of cesium (no. 55) (e) a fluoride of tin (no. 50)

20 All of the isotopes of the element with atomic number 87 are radioactive. Hence, it is not found in nature. Yet, prior to its preparation by nuclear bombardment, chemists were confident they knew the chemical reactions this element would show. Explain. What predictions about the chemistry of this element would you make?

21 How are the boiling points of liquids related to the atomic and bond arrangements in the corresponding solids?

22 Why is the term "a mole of sodium chloride" really an indefinite expression?

23 Why does silicon (no. 14) have a very high boiling point? (Compare with carbon.)

24 Account for the fact that Cs is more reactive than Li; that F_2 is more reactive than I_2.

25 Which ion should have the larger radius, Na^+ or Cs^+? Explain in terms of a model.

26 Use the periodic table to predict the following: (a) the formula of the hydride of Se (no. 34) (b) the element with the highest melting point—Na, Mg, Al, Si, S (c) the formula of a chloride of Ra (no. 88)

27 Lithium forms the following compounds: lithium oxide (Li_2O), lithium hydroxide (LiOH), lithium sulfide (Li_2S). Name and write the formulas of the corresponding sodium and potassium compounds.

28 Fill in the blanks of the following table:

Element	Atomic Number	Particles Per Atom			Mass Number
		Protons	Electrons	Neutrons	
Aluminum (Al)	13	—	—	—	27
Beryllium (Be)	—	4	—	—	9
Bismuth (Bi)	83	—	—	—	209
Calcium (Ca)	—	—	20	20	—
Carbon (C)	—	6	—	6	—
Fluorine (F)	—	—	9	—	19
Phosphorus (P)	15	—	—	16	—
Iodine (I)	—	—	53	—	127

Although a typical chemical reaction . . . may appear far removed from the working of an engine, the same fundamental principles of heat and work apply to both.

F. T. WALL (1912-   )

CHAPTER 9 Energy Changes in Chemical and Nuclear Reactions

The energy of athletics is as surely chemical energy as the energy measured in the laboratory.

Probably you have not given a great deal of thought to the fact that logs are more commonly burned in fireplaces than dynamite or $^{235}_{92}U$. Almost certainly you have never related these facts to the old saying "You get out of something what you put into it," which is squaresville indeed. Yet there *is* a relationship.

The most dramatic—and sometimes the most useful—chemical reactions are those which produce large amounts of energy. We use these reactions to heat our houses, alter our landscape, and run our factories and cars. The amount of energy available from a chemical reaction depends on how much energy is stored in the reactants and products of that reaction. This chapter will explore such energy relationships.

Every chemical reaction has energy changes associated with it. In many cases the energy changes are the most important feature of a chemical process. We have only to recall the fire that warmed "Martin the Martian" in our fable and the gasoline that powers our sports cars. The chemical process in each case is important only for the energy released by the reaction. In the automobile, a *mixture* of gasoline vapor and air leaves the carburetor and is ignited in the cylinders, releasing a predictable amount of energy. How far the energy drives the car depends on the efficiency of the automobile, but the car will not move at all unless *gasoline and air* react. Both gasoline and oxygen are essential for combustion in the cylinders. Even a piece of chocolate cake is important ultimately because of its ability to provide energy to the "human machine." It is "burned" with oxygen in the cells of the body.

Energy changes in chemical processes are always important and frequently even critical. What can we learn about them? The questions come thick and fast. How much energy is associated with a given process? How is the energy change measured? Where does the energy come from? We have touched on some of these questions in earlier chapters. In this chapter we shall examine them in more detail.

ENERGY CHANGES and CHEMICAL REACTIONS

9-1 ENERGY CHANGES AND CHEMICAL REACTIONS

We have studied in the laboratory the heat* released when a given weight of our candle burned. Knowledge of the energy change associated with a laboratory process is helpful in understanding the process. Knowledge of the energy change associated with an industrial process is essential for economic survival. Let us look in on the operation of the fuel-gas industry.

A jet of steam at 600°C is forced through a bed of red-hot coke,† producing a mixture of carbon monoxide and hydrogen known as "water gas." This is an important industrial fuel. The gases coming out of the reactor first contain a high percentage of CO and H_2 and very little unreacted water vapor; but as the steam flow continues, the red-hot coke cools and the yield of water gas drops. Less and less CO and H_2 appear in the product gases, and more unreacted water vapor is found. At this point the operator turns off the steam and blows air through the coke bed. As the coke begins to burn in the air stream, it again becomes red hot. The operator now forces steam into the hot bed of coke and the cycle is repeated. Two qualitative observations seem to be clear.

(1) Energy (heat) is *absorbed* when steam or water vapor combines with hot coke to give water gas. (The coke cools off; therefore, heat is absorbed in water-gas production.)
(2) Energy (heat) is *liberated* when coke is burned in oxygen. (The coke heats up; therefore, heat is released.)

---

* The common term for thermal energy released in a process is heat; however, we shall usually use the more precise terms thermal energy, electrical energy, and so on.
† Coke is almost pure carbon produced by heating coal.

9-1.1 Some Quantitative Energy Relationships

A chemical engineer responsible for the manufacture of water gas cannot be as casual about the energy changes in each process as we have been. He must know how much energy (heat) is involved in each case if his employer is to remain in business. Let us check his lab notebook. His measurements show that in producing water gas 31.4 kcal* of heat are absorbed for every mole of carbon converted to CO. All his observations can be summarized neatly by the equation

$$H_2O(g) + C(s) + 31.4 \text{ kcal} \longrightarrow CO(g) + H_2(g) \qquad (1)$$

We can represent his measurements on the combustion of coke by the equation

$$C(s) + O_2(g) \longrightarrow CO_2(g) + 94.0 \text{ kcal} \qquad (2)$$

A good chemical engineer will try to balance the two processes so that the heat given off in burning coke is just enough to supply the energy needed for the water-gas synthesis—his cycles must be properly timed. If excess coke is burned, valuable energy goes up the chimney or into the room. If the system is not hot enough, the yield of CO and H_2 drops.

The consumer's measurements on the burning of the water-gas fuel can best be summarized by two equations, one describing the burning of CO and the other describing the burning of H_2:

$$CO(g) + \tfrac{1}{2}O_2(g) \longrightarrow CO_2(g) + 67.6 \text{ kcal} \qquad (3)$$

$$H_2(g) + \tfrac{1}{2}O_2(g) \longrightarrow H_2O(g) + 57.8 \text{ kcal} \qquad (4)$$

If the water gas made from 1 mole of carbon is burned, the total heat released is 67.6 kcal + 57.8 kcal = 125.4 kcal. This is much more heat than can be obtained by burning 1 mole of carbon directly:

$$C(s) + O_2(g) \longrightarrow CO_2(g) + 94.0 \text{ kcal} \qquad (2)$$

The consumer seems to be doing better than the chemical engineer in obtaining energy from 1 mole of coke. He obtains 125.4 kcal − 94.0 kcal = 31.4 kcal *more energy* from 1 mole of carbon as water gas than the chemical engineer did from 1 mole of carbon as coke. In comparing the skills of our engineer and consumer, the number 31.4 kcal suddenly seems familiar. This is the very number that appeared in the equation given earlier for the manufacture of water gas:

$$H_2O(g) + C(s) + 31.4 \text{ kcal} \longrightarrow H_2(g) + CO(g) \qquad (1)$$

Refer to Table 9-1. The table shows that extra energy *stored* in the CO and H_2 became available to the consumer when he burned the gas. The chemical engineer obtained the extra energy from burning *extra* coke directly. (Remember: He had a two-cycle process—one

---

* A kilocalorie is 1,000 cal or 10^3 cal. This is a more convenient unit for large energy changes.

cycle absorbed heat and one cycle gave off heat.) *The water gas released more energy per mole of carbon than did coke because extra energy was stored in the water gas when it was made.*

TABLE 9-1 SOME ENERGY CHANGES IN THE MANUFACTURE AND USE OF WATER GAS

Reaction	Debit	Credit
Energy absorbed: $H_2O(g) + C(s) + 31.4 \text{ kcal} \longrightarrow CO(g) + H_2(g)$	31.4 kcal	
Energy released: $CO(g) + \tfrac{1}{2}O_2(g) \longrightarrow CO_2(g) + 67.6 \text{ kcal}$		67.6 kcal
Energy released: $H_2(g) + \tfrac{1}{2}O_2(g) \longrightarrow H_2O(g) + 57.8 \text{ kcal}$		57.8 kcal
Overall reaction (total of above): $C(s) + O_2(g) \longrightarrow CO_2(g)$	31.4 kcal absorbed	125.4 kcal released
Net		125.4 −31.4 94.0
Experimental value for $C(s) + O_2(g) \longrightarrow CO_2(g)$		94.0 kcal released

9-1.2 Enthalpy or Heat Content of a Substance

The example just given shows that energy amounting to 31.4 kcal was stored in the water gas. We see also that the amount of energy stored per mole of carbon is constant. *We added a fixed amount of energy to coke and steam to make a specified amount of carbon monoxide and hydrogen.* The heat is retained by both the CO and H_2. We cannot decide how much of the heat is in the CO and how much is in the H_2 without doing additional laboratory work. However, we can say that the "heat content" of the gas mixture, 1 mole CO + 1 mole H_2, derived from 1 mole of coke and 1 mole of steam, is higher by 31.4 kcal than the "heat content" of 1 mole of coke and 1 mole of steam. With a little more work, we can find out how much of the energy is stored in the CO and how much is stored in the H_2. One mole of carbon monoxide has a characteristic "heat content" just as it has a characteristic mass. Similarly, 1 mole of H_2 has a characteristic heat content. Chemists call the energy stored in a substance the **heat content** or **enthalpy** of the substance. Thus 1 mole of any substance has a characteristic heat content or enthalpy.

This idea provides a good explanation of the energy changes associated with chemical reactions. If the reactants have more energy than the products, energy will be *released* during the reaction. Conversely, if the products have more energy than the reactants, energy will be *absorbed* during the reaction. These two statements can be put into one equation:

$$\left\{\begin{array}{l}\text{change in heat}\\ \text{content or enthalpy}\\ \text{of the system}\end{array}\right\} = \left\{\begin{array}{l}\text{heat content}\\ \text{or enthalpy}\\ \text{of products}\end{array}\right\} - \left\{\begin{array}{l}\text{heat content}\\ \text{or enthalpy}\\ \text{of reactants}\end{array}\right\} \quad (5)$$

The enthalpy of a substance is represented by the letter H. The enthalpy of 1 mole of CO is then indicated as H_{CO}, that of H_2 as H_{H_2} and that of steam as H_{H_2O}. The change in enthalpy in a reaction is represented by ΔH where Δ signifies "difference" or "change." ΔH is often called the **heat of reaction.**

9-1.3 Additivity of Heats of Reaction

Let us apply our newly defined symbolism to the processes involved in the production and use of water gas. In terms of ΔH the energy changes can be summarized as follows: The manufacture of water gas absorbs energy; hence, the enthalpy (heat content) of the products is higher than the enthalpy of the reactants.

$$\Delta H = \text{enthalpy of products} - \text{enthalpy of reactants} \quad (5)$$

we see that ΔH for the process will be positive in this case. We can express this by writing

$$H_2O(g) + C(s) \longrightarrow CO(g) + H_2(g)$$
$$\Delta H = +31.4 \text{ kcal} \quad (1a)$$

The reaction as written is exactly equivalent to the earlier representation

$$H_2O(g) + C(s) + 31.4 \text{ kcal} \longrightarrow CO(g) + H_2(g) \quad (1)$$

If the value of ΔH is positive for a reaction, the process is **endothermic**—energy is absorbed during the process.

For an **exothermic** reaction—one evolving energy—ΔH is negative. In the burning of CO, energy in the form of heat is released. We write

$$CO(g) + \tfrac{1}{2}O_2(g) \longrightarrow CO_2(g) + 67.6 \text{ kcal} \quad (3)$$

This has exactly the same meaning as

$$CO(g) + \tfrac{1}{2}O_2(g) \longrightarrow CO_2(g)$$
$$\Delta H = -67.6 \text{ kcal} \quad (3a)$$

Fig. 9-1 **Enthalpy changes during reactions.**

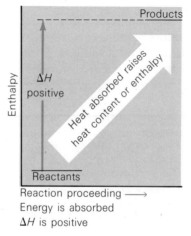
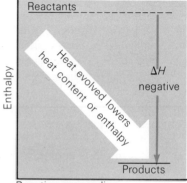

We see that the sign of ΔH is sensible. It is positive when the heat content or enthalpy of the system is rising (heat absorption), and it is negative when enthalpy is dropping (heat given off). This is shown diagrammatically in Figure 9-1.

Let us return to the debit and credit balance we found in our water-gas fuel problem. In terms of ΔH, the energy changes are as follows:

$$H_2O(g) + C(s) \longrightarrow CO(g) + H_2(g)$$
$$\Delta H_{1a} = +31.4 \text{ kcal} \quad (1a)$$

$$CO(g) + \tfrac{1}{2}O_2(g) \longrightarrow CO_2(g)$$
$$\Delta H_{3a} = -67.6 \text{ kcal} \quad (3a)$$

$$H_2(g) + \tfrac{1}{2}O_2(g) \longrightarrow H_2O(g)$$
$$\Delta H_{4a} = -57.8 \text{ kcal} \quad (4a)$$

The overall reaction, $(1a) + (3a) + (4a) = (2a)$, is

$$C(s) + O_2(g) \longrightarrow CO_2(g)$$
$$\Delta H_{2a} = -94.0 \text{ kcal} \quad (2a)$$

We discover that not only is reaction $(2a)$ equal to the sum of reactions $(1a) + (3a) + (4a)$ in terms of atoms, but also that

$$\Delta H_{2a} = \Delta H_{1a} + \Delta H_{3a} + \Delta H_{4a} \quad (6)$$
$$= [31.4 + (-67.6) + (-57.8)] \text{ kcal}$$
$$= (31.4 - 67.6 - 57.8) \text{ kcal}$$
$$= -94.0 \text{ kcal}$$

We see that *when a reaction can be expressed as the algebraic sum of a sequence of two or more other reactions, then the heat of the reaction is the algebraic sum of the heats of these reactions.* This generalization has been found to be applicable to every reaction that has been tested. Because the generalization has been so widely tested, it is called a law—the **Law of Additivity of Heats of Reaction.***

9-1.4 The Measurement of Heats of Reaction

The measurement of the energy changes associated with a given process is achieved by the process of **calorimetry**—a name obviously related to the unit of heat or thermal energy, the calorie. You already have some experience in calorimetry. In Experiment 9 you measured the heat of combustion of a candle and then the heat of solidification of paraffin. Then in Experiment 17 you measured the energy evolved when NaOH reacted with HCl. The device you used was a simple calorimeter. Calorimeters vary in details and are adapted to the particular reaction being studied.

EXERCISE 9-1

If a 1.000-g sample of carbon is burned to CO_2 and liberates 7.833 kcal of energy, what is the heat of combustion of 1 mole of carbon atoms?

---

* This generalization was first proposed in 1840 by G. H. Hess on the basis of his experimental measurements of reaction heats. It is sometimes called **Hess's Law of Constant Heat Summation.**

9-1.5 Predicting the Heat of a Reaction

The heats of many reactions have been measured. (A few are listed in Table 9-2.) With these measured values and the additivity principle, many unmeasured heats of reaction can be predicted.

TABLE 9-2 HEATS OF REACTION BETWEEN ELEMENTS ($T = 25°C$, $P = 1$ atm)

Elements	Compound (Product) Formula	Compound (Product) Name	Heat of Reaction (kcal/mole of product)
$H_2(g) + \frac{1}{2}O_2(g) \longrightarrow$	$H_2O(g)$	water vapor	-57.8
$H_2(g) + \frac{1}{2}O_2(g) \longrightarrow$	$H_2O(l)$	water	-68.3
$S(s) + O_2(g) \longrightarrow$	$SO_2(g)$	sulfur dioxide	-71.0
$H_2(g) + S(s) + 2O_2(g) \longrightarrow$	$H_2SO_4(l)$	sulfuric acid	-194.0
$\frac{1}{2}N_2(g) + \frac{1}{2}O_2(g) \longrightarrow$	$NO(g)$	nitric oxide	$+21.6$
$\frac{1}{2}N_2(g) + O_2(g) \longrightarrow$	$NO_2(g)$	nitrogen dioxide	$+8.1$
$\frac{1}{2}N_2(g) + \frac{3}{2}H_2(g) \longrightarrow$	$NH_3(g)$	ammonia	-11.0
$C(s) + \frac{1}{2}O_2(g) \longrightarrow$	$CO(g)$	carbon monoxide	-26.4
$C(s) + O_2(g) \longrightarrow$	$CO_2(g)$	carbon dioxide	-94.0
$2C(s) + 3H_2(g) \longrightarrow$	$C_2H_6(g)$	ethane	-20.2
$3C(s) + 4H_2(g) \longrightarrow$	$C_3H_8(g)$	propane	-24.8
$\frac{1}{2}H_2(g) + \frac{1}{2}I_2(g) \longrightarrow$	$HI(g)$	hydrogen iodide	$+6.2$

Suppose we are interested in the heat of combustion of nitric oxide (NO). The appropriate equation is

$$NO(g) + \tfrac{1}{2}O_2(g) \longrightarrow NO_2(g) \qquad \Delta H_7 = ? \qquad (7)$$

Since the above equation can be obtained by combining two equations in Table 9-2, we can predict ΔH_7. In Table 9-2 we find the two reactions that involve the compounds $NO(g)$, $NO_2(g)$, and the element O_2;

$$\tfrac{1}{2}N_2(g) + \tfrac{1}{2}O_2(g) \longrightarrow NO(g)$$
$$\Delta H_8 = +21.6 \text{ kcal/mole NO} \qquad (8)$$

$$\tfrac{1}{2}N_2(g) + O_2(g) \longrightarrow NO_2(g)$$
$$\Delta H_9 = +8.1 \text{ kcal/mole NO}_2 \qquad (9)$$

Obviously, the addition of the above two equations will not give $NO(g) + \frac{1}{2}O_2(g) \longrightarrow NO_2(g)$. Since NO is a reactant in this process, we need the reverse of the reaction producing NO. We noted in our work on phase changes (see Section 5-1.1) that the 9.7 kcal of energy put into 1 mole of water to vaporize it was released as the vapor condensed back to a liquid. Thus, if 21.6 kcal of heat are absorbed when 1 mole of NO is formed, then 21.6 kcal of heat will be released when 1 mole of NO is decomposed:

$$NO(g) \longrightarrow \tfrac{1}{2}N_2(g) + \tfrac{1}{2}O_2(g)$$
$$\Delta H_{10} = -21.6 \text{ kcal/mole NO} \qquad (10)$$

Adding the appropriate reactions does give the equation for oxidation of NO to NO_2.

$$\tfrac{1}{2}N_2(g) + O_2(g) \longrightarrow NO_2(g)$$
$$\Delta H_9 = +8.1 \text{ kcal/mole } NO_2 \quad (9)$$

$$NO(g) \longrightarrow \tfrac{1}{2}N_2(g) + \tfrac{1}{2}O_2(g)$$
$$\Delta H_{10} = -21.6 \text{ kcal/mole } NO \quad (10)$$

The overall reaction is

$$NO(g) + O_2(g) + \tfrac{1}{2}N_2(g) \longrightarrow NO_2(g) + \tfrac{1}{2}O_2(g) + \tfrac{1}{2}N_2(g)$$
$$\Delta H_7 = 8.1 + (-21.6) \text{ kcal} \quad (7a)$$

or

$$NO(g) + \tfrac{1}{2}O_2(g) \longrightarrow NO_2(g)$$
$$\Delta H_7 = -13.5 \text{ kcal/mole } NO \quad (7)$$

EXERCISE 9-2

Predict the heat of the reaction

$$CO(g) + \tfrac{1}{2}O_2(g) \longrightarrow CO_2(g)$$

from two appropriate reactions listed in Table 9-2. Compare your results with ΔH_{3a} given in Section 9-1.3.
−67.3

$CO \rightarrow C + \tfrac{1}{2}O_2$
$\Delta H = 26.4$

$C + O_2 \rightarrow CO_2 \quad \Delta H = -94.0$

$\Delta H = -67.6$

We can predict the heat of any reaction that can be obtained by adding two or more of the reactions in the Table 9-2. *A given reaction can be obtained by adding reactions or appropriate multiples of reactions from Table 9-2, provided every compound in the reaction is included in the table.* If appropriate multiples are used, the elements involved in the reactions will appear in proper amounts.

Consider a more complicated example—the oxidation of ammonia (NH_3):

$$NH_3(g) + \tfrac{7}{4}O_2(g) \longrightarrow NO_2(g) + \tfrac{3}{2}H_2O(g) \quad (11)$$

We find three compounds—$NH_3(g)$, $NO_2(g)$, and $H_2O(g)$. These are all found in Table 9-2. Consequently, we are able to calculate ΔH for the oxidation of ammonia.

EXERCISE 9-3

Show that the equation for ammonia oxidation is the result of adding the following equations, and that its $\Delta H = -67.6$ kcal.

$$NH_3(g) \longrightarrow \tfrac{1}{2}N_2(g) + \tfrac{3}{2}H_2(g) \qquad \Delta H = +11.0 \text{ kcal}$$
$$\tfrac{1}{2}N_2(g) + O_2(g) \longrightarrow NO_2(g) \qquad \Delta H = +8.1 \text{ kcal}$$
$$\tfrac{3}{2} \times [H_2(g) + \tfrac{1}{2}O_2(g) \longrightarrow H_2O(g)] \qquad \tfrac{3}{2} \times [\Delta H = -57.8 \text{ kcal}]$$

Thus, when we wish to predict the heat of some reaction, we refer to Table 9-2. If every *compound* in the reaction of interest is in the "compound" column of Table 9-2, then the prediction can be made. Of course, the list in Table 9-2 includes only a small fraction of the known values. Many more heats of reaction are tabulated in

**GILBERT NEWTON LEWIS
(1875–1946)**

Gilbert Newton Lewis was born near Boston, and reared in Nebraska. He was graduated from Harvard University, and taught high school chemistry before returning to Harvard for his Ph.D., received in 1899. Most of his professional life was spent at the University of California, where he built one of the most creative chemistry departments in the world.

Lewis devoted much of his career to the study of molecular structure and of thermodynamics —the energy relations in chemical changes. His understanding of chemical bonding has profoundly influenced modern thinking. One of the first to recognize that energy effects provide a basis for predicting chemical reactions, Lewis awakened chemists to the crucial importance of thermodynamics.

G. N. Lewis enjoyed chemistry, and his enthusiasm and burning interest in it were contagious— many of his students became great scientists. He died in 1946 in the laboratory he loved, surrounded by the beakers and books that were the tools of his trade.

handbooks under "Heat of Formation." The Heat of Formation reported is the ΔH value for the formation of 1 mole of the compound from its elements taken at 25°C and 1 atm.

9-2 THE CONSERVATION OF ENERGY

We have seen how it is possible to measure the heat of a reaction. Using a list of measured values, we can predict the energy changes of many reactions that have not been measured. Thus, the Law of Additivity of Heats of Reaction is a very useful and reliable generalization. But we wonder: Why should this be so? The explanation, as usual, is found by examining a model system—a system of billiard balls.

9-2.1 Conservation of Energy in a Billiard-Ball Collision

In a billiard game the amount of kinetic energy received by the ball is fixed by the amount of work done on it. If the cue stick strikes the ball softly, the ball moves slowly; if the cue stick strikes it hard, the ball moves rapidly. The amount of work done on the ball (W) determines and equals the amount of kinetic energy (KE) which the ball will possess:

$$W = KE \qquad (12)$$

If this ball strikes another ball, the energy that it *loses* in the collision is exactly equal to the energy *gained* by the second ball. Even in a billiard parlor energy is conserved.

9-2.2 Conservation of Energy in the Storage of Billiard Balls—Potential Energy

When billiard balls are stored, energy (work) is used to lift them from the billiard table to the shelf above. If the balls roll off the shelf, the **kinetic energy** (energy of motion) they possess just before hitting the table is equal to the energy used in lifting them to the shelf. While they are stored, this same amount of energy is called **potential energy** (energy of position). When a ball falls from the shelf and strikes the table, the potential energy is converted to kinetic energy, which is in turn transformed into heat. The billiard ball and the table are both warmer than they were before the collision. The amount of heat given to the ball and table is determined by the change in the kinetic energy of the ball. Again, energy is conserved throughout every operation.

EXERCISE 9-4

A rubber band is stretched onto a toy gun. Work, W, is expended in the stretching. Is energy conserved? Explain.

9-2.3 Conservation of Energy in a Chemical Reaction

Figure 9-2 shows an apparatus in which an electric current can be passed through water containing a few drops of sulfuric acid per

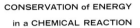

liter. Under the conditions shown, the passage of the current causes the decomposition of the water. As electrical energy is supplied, hydrogen gas and oxygen gas are produced. Measurements of the electric current and voltage show that 68.3 kcal of electrical energy, E_1, must be expended to decompose 1 mole of water. The equation for the reaction is

$$\left\{\begin{array}{l}68.3 \text{ kcal} \\ \text{electrical} \\ \text{energy}\end{array}\right\} + H_2O(l) \longrightarrow H_2(g) + \tfrac{1}{2}O_2(g) \quad (13)$$

Now suppose we measure the heat of the reaction of hydrogen and oxygen in a calorimeter. The results of many experiments show that 68.3 kcal of energy will be liberated for every mole of water formed. The equation for the reaction can then be written as

$$H_2(g) + \tfrac{1}{2}O_2(g) \longrightarrow H_2O(l) + 68.3 \text{ kcal}$$
$$\Delta H = -68.3 \text{ kcal} \quad (14)$$

The energy recovered is exactly equal to that stored in the system when hydrogen and oxygen were generated from the water by the electric current. Symbolically, energy added by the electric current E_1 is equal to the energy $-\Delta H$ released in the combustion:

$$E_1 = -\Delta H = 68.3 \text{ kcal} \quad (15)$$

Hydrogen and oxygen have more potential energy than water, just as the billiard ball on the shelf has more potential energy than the one on the table. As we shall see later, the energy stored in H_2 and O_2 is truly energy of position—energy due to the position of atoms in each substance. Energy is released when the atoms take up new positions to form water. The relationship between atom position and energy content of the system is illustrated in Figure 9-3.

CONSERVATION of ENERGY in a CHEMICAL REACTION

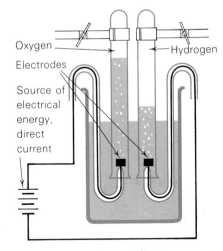

Fig. 9-2 **Apparatus for the electrolytic decomposition of water.**

Fig. 9-3 **Conservation of energy in a chemical reaction.** The energy absorbed in reaction (a),

$$2H_2O(l) + E \longrightarrow 2H_2(g) + O(g)$$

equals the energy evolved in reaction (b),

$$2H_2(g) + O_2(g) \longrightarrow 2H_2O(l) + \Delta H$$

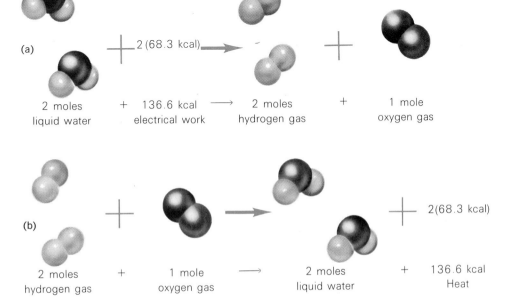

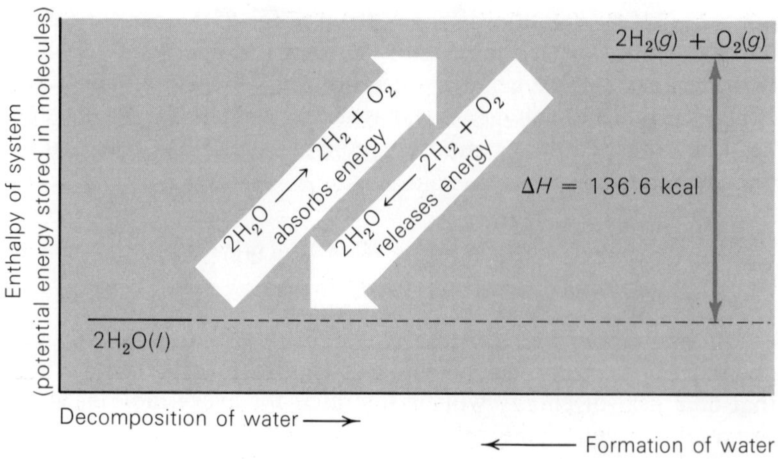

Fig. 9-4 **Enthalpy changes in the decomposition and formation of water.**

[Diagram: Enthalpy of system (potential energy stored in molecules) vs. Decomposition of water → / ← Formation of water. Shows $2H_2O(l)$ at lower enthalpy and $2H_2(g) + O_2(g)$ at higher enthalpy, with $\Delta H = 136.6$ kcal. Arrows: $2H_2O \rightarrow 2H_2 + O_2$ absorbs energy; $2H_2O \leftarrow 2H_2 + O_2$ releases energy.]

The relationships of Figure 9-3 can be represented schematically by another type of diagram shown in Figure 9-4. If the enthalpy of 2 moles of water is represented by a line on this diagram, then the energy of 2 moles of hydrogen plus 1 mole of oxygen should be represented by a line $2 \times 68.3 = 136.6$ kcal higher. The diagram shows that when water is decomposed, energy must be supplied to raise the enthalpy of the system to that of $2H_2$ plus O_2. When hydrogen burns in oxygen, the enthalpy of the system drops down to that for water. *Energy is released.*

Energy is conserved in chemical processes as well as in systems involving billiard balls (and rubber bands, Exercise 9-4). An important idea arises from this discussion. Each molecule has kinetic energy due to its movement through space (indicated by its temperature), and each molecule has potential energy due to the arrangement of atoms in the molecule. In short, each molecule has a certain capacity to store energy; it has a definite heat content or enthalpy at a given temperature. As the temperature is raised, more energy is stored.

9-2.4 The Law of Conservation of Energy

After we recognized both *kinetic* and *potential* energy, it was easy to see that energy was conserved in each of the systems we studied. Our specific observations can be generalized. For all systems studied* by scientists we find that *energy is always conserved!* This is the Law of Conservation of Energy. It is based on thousands and thousands of experiments and can be used to make very accurate predictions. It requires only that we recognize both kinetic and potential energy and keep a careful balance on both.

The Law of Additivity of Heats of Reaction considered earlier for chemical systems is simply a special case of the more general Law of Conservation of Energy. Predictions based on the Law of Additivity of Heats of Reaction have always agreed with experiment. Conservation of energy has always been in agreement with experiment whenever a careful energy balance has been obtained.

---

* For nuclear systems this statement is true when we recognize the equivalence of mass and energy. See Section 9-4.

9-3 THE ENERGY STORED IN A MOLECULE

In Sections 9-2.1 to 9-2.4 we considered energy changes associated with quantities of matter that we could handle in the laboratory. We found it useful to consider kinetic energy, the energy of motion, and potential energy, the energy of position. We also found it useful to talk about such additional forms of energy as heat or thermal energy, the chemical energy stored in molecules such as H_2 and O_2, electrical energy, and mechanical energy associated with particle movement.

In every case, special techniques were required to measure the energy changes in the system. For example, heat energy was measured in a calorimeter; electrical energy would be measured by using a voltmeter, an ammeter, and a watch.

Because different techniques were used in measuring each form of energy, our classification scheme was useful. On the other hand, when we wish to discuss matter and energy on the molecular level—by considering a few molecules rather than a few grams—we find that it is not necessary to use all these different forms of energy to describe molecular behavior. Only the first two forms of energy are needed—kinetic energy and potential energy. We can "explain" all the forms of energy by a molecular model that uses only the energy of motion (kinetic energy) and the energy of position (potential energy) of the molecules and their parts. Let us examine in more detail the ways in which a molecule can hold energy.

9-3.1 The Energy of a Molecule

We shall represent a molecule such as carbon dioxide by three balls of comparable mass held together by springs (see Figure 9-5). The springs represent the bonds between the atoms. Let us start the springs vibrating, then toss the whole assembly through space in an end-over-end motion. There are now three kinds of motion associated with our molecule:

(1) There is the **energy of translation** associated with the motion of the entire molecular assembly through space [see Figure 9-5(a)]. (This is our original *gas-law kinetic energy*.)
(2) There is the **energy of rotation** associated with the end-over-end motion of the molecule as it twirls through space. This is the so-called *rotational kinetic energy* [Figure 9-5(b)].
(3) There is the **energy of vibration** associated with the vibration of the balls on the springs, or the so-called *vibrational kinetic energy* [Figure 9-5(c)].

The above model applies reasonably well to a molecule in the gaseous state, but in the liquid state and (even more so) in the solid state, all the molecular motions are restricted. In these condensed phases the translational molecular motion is restricted almost completely to a back-and-forth motion of the balls about a given point. Rotation and vibration are also frequently restricted.

The ENERGY of a MOLECULE

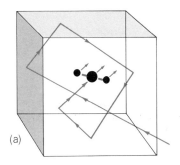

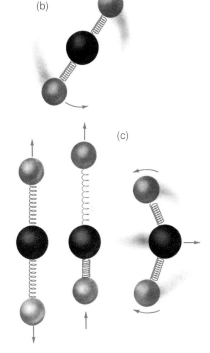

Fig. 9-5 Types of motion of a molecule of carbon dioxide, CO_2. (a) Translational motion; the molecule moves from place to place. (b) Rotational motion; the molecule rotates about its center of mass. (c) Vibrational motion; the atoms move alternately toward and away from the center of mass.

In addition to these three kinds of kinetic energy, there is the potential energy related to the attractive forces acting between molecules. In the gaseous state (molecules far apart) these forces are small. In the liquid state (molecules closer together) the forces are relatively large. In the solid state the forces are even greater.

Next, there is present, within the molecule, chemical energy related to the forces holding the atoms together in the molecule. For example, there is the chemical bond energy—the energy of the covalent bond mentioned in Chapter 8 for H_2, Cl_2, and HCl; or that of the ionic bond mentioned for NaCl. <u>When forces acting between atoms or ions are large, the potential energy of the system is usually low and the system tends to be stable.</u> As we noted earlier, the energy associated with the chemical bond is the direct result of the attraction between nuclei and moving electrons. As you recall, the ionization potential of an atom (Chapter 8) is a measure of the energy required to pull an electron away from an atom. If we assume that ion and electron have zero energy after separation, the ionization energy represents the sum of the kinetic and potential energies of the electron in the atom. Finally, there is present within the nucleus of each atom a store of energy. This energy is related to the forces holding the nuclear particles together (Section 9-4). Since each nucleus remains intact and is apparently unaffected by chemical reactions, this nuclear energy does not change. Hence, the amount of stored energy that the nucleus contributes to the molecular enthalpy does not usually concern us if no nuclear processes are being considered.

The sum of all these forms of molecular energy makes up the *molecular* enthalpy. If we add together the molecular enthalpy of 6.02×10^{23} molecules of a given kind, we obtain the *molar* enthalpy of that substance.

9-3.2 Energy Changes on Warming

In Chapter 5 we considered the general kinetic theory model for liquids and solids. We also used the model to interpret both the vaporization process and the freezing process for a liquid. Let us examine in more detail the heat effects associated with these processes and with the general heating of a solid.

When a solid is warmed, the atoms or molecules making up the solid bounce back and forth more vigorously around their regular crystal positions. The solid expands somewhat. As the temperature rises, these motions disturb the regularity of the crystal more and more. Too much of this random movement destroys the lattice completely. At a definite temperature the kinetic energy of the particles causes so much random movement that the lattice is no longer stable—the crystalline solid melts.

In the liquid each molecule has considerably more freedom of movement, particularly for translation and rotation. Warming the liquid increases the amount of molecular movement. As molecular movement increases, more of the molecules have enough energy to overcome the forces of attraction holding them in the liquid and they move into relatively empty space above the liquid surface. The molecules become more randomly distributed. In the vapor phase the *potential* energy of a molecule is higher than it is in the liquid phase. As the potential energy required for the change from liquid to vapor is supplied (heat of vaporization), the liquid vaporizes. It is interesting to note that the system is going from a state of lower potential energy to a state of higher potential energy. The factor most responsible for the change seems to be the tendency of the molecules to seek a more random arrangement—their tendency to occupy the space above the liquid.

If we continue to warm the substance, we shall reach a point at which the kinetic energies of vibration, rotation, and translation become as large as the chemical bond energies. At this temperature complex molecules fall to pieces. For example, at the temperature of the sun's surface (6000°K), only monatomic and some diatomic molecules remain. All the more complicated structures have fallen to pieces.

The foregoing general discussion indicates that temperature and energy changes associated with chemical processes are usually significantly larger than temperature and energy changes associated with phase changes. In fact, they are usually larger by factors ranging from 10 to over 100. You saw this in the laboratory when you compared the heat released in burning 1 g of wax with that released in the freezing of 1 g of molten wax.

9-4 THE ENERGY STORED IN A NUCLEUS

Our discussion in Chapter 7 indicated that a nucleus is built up of protons and neutrons, and that the number of neutrons per proton is important in determining nuclear stability. We noted earlier in this chapter that nuclei remain intact during even the most vigorous of chemical reactions. The energy involved in a nuclear process is usually at least one million times larger than the energy involved in any chemical process.

9-4.1 The Nuclear Fission Process

One of the first nuclear processes to be used for the production of energy was the nuclear **fission** of uranium-235. Nuclear fission is a process in which a heavy nucleus splits to give two lighter nuclei of roughly comparable mass. If we utilize the symbolism* outlined in Chapter 7, a typical† fission process can be written as

$$^{235}_{92}U + ^{1}_{0}n \longrightarrow ^{141}_{56}Ba + ^{92}_{36}Kr + 3\,^{1}_{0}n + \text{(about)}\ 4.5 \times 10^9\ \text{kcal} \quad (16)$$

We have conserved charge (92 = 56 + 36) and the number of nucleons (235 + 1 = 141 + 92 + 3) in the process. For the process we are considering, one uranium-235 nucleus combines with one neutron, then *immediately* splits (undergoes fission) to give a barium nucleus, a krypton nucleus, and three neutrons. The released neutrons can also bring about the fission of more uranium nuclei. A **chain reaction** is set up. Each

---

* Remember that a given isotope is represented by: (1) the symbol of the element, (2) the *atomic* number at the lower left-hand corner of the symbol, and (3) the mass number at the upper left-hand corner. For uranium the atomic number is 92 and the mass number is 235.

† This is a typical reaction. Many other pairs of products can be obtained.

Chapter 9 / ENERGY CHANGES in CHEMICAL and NUCLEAR REACTIONS

fission process provides neutrons to initiate the splitting of other $^{235}_{92}U$ nuclei. If neutron-loss to the outside is restricted, the process keeps itself going with a continual release of energy. The process can be controlled by regulating the number of neutrons lost or absorbed. Careful control is essential. If on the average each fission process provides neutrons that initiate only one new fission process, then the chain reaction will continue at a steady rate until the supply of fissionable nuclei becomes too small to keep the process going. This is what happens in a **nuclear reactor** (Figure 9-6). But if the neutrons from each fission process initiate more than one new fission process, the chain reaction proceeds at an ever-increasing rate until the liberated heat blows the system to pieces—as in an atomic bomb.

Such a simple item as the size of a piece of uranium is important in determining the number of neutrons that can initiate a new fission process. If we have a large piece of uranium, a large number of uranium nuclei are confined in a unit with relatively *small* surface area per gram. Under these circumstances most of the neutrons released do not escape from the surface, but encounter other uranium nuclei and bring about new fission reactions. The process is self-sustaining. If, on the contrary, we have only a small piece of uranium with a relatively large surface area per gram of uranium, there is a high probability that the neutrons will escape from the uranium without encountering other uranium nuclei. These neutrons are wasted as far as the process is concerned, and the fission reaction is no longer self-sustaining. We say that a piece of fissionable material that is too small to maintain the fission process is below the **critical mass.**

The foregoing discussion on critical mass suggests that nuclear fission reactions can be controlled by controlling the number of neutrons available to initiate new fission processes. If we want to slow down or control nuclear fission, the number of neutrons available to initiate new fission reactions must be reduced. In a nuclear reactor, cadmium rods are used to control the number of neutrons. Cadmium nuclei are effective in absorbing neutrons without undergoing fission or without releasing other neutrons. One might say that they soak up neutrons and provide a means of reaction control.

It is frequently difficult to visualize the amount of energy represented by the amount 4.5×10^9 *kcal* in the fission equation. This analogy may help. The energy released in the fission of 235 g of uranium (about ½ lb) is approximately equal to the energy released in the combustion of some

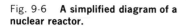

Fig. 9-6 **A simplified diagram of a nuclear reactor.**

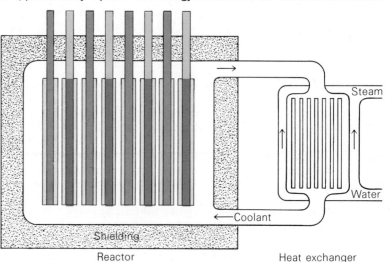

140,000 gal of gasoline. If your car is a gas-guzzling monster and uses as much as 28 gal per week, the 140,000 gal would solve your fuel problems for a little less than 100 years! In short, if we could properly harness the energy of this fission process and use uranium as an automobile fuel, ½ lb of $^{235}_{92}U$ would keep an automobile going for about 100 years!

Where does all this energy come from? The answer is contained in the special theory of relativity, formulated first by Albert Einstein in 1905. One of the consequences of the theory of relativity was a suggestion that mass and energy are different forms of the same fundamental physical entity or "stuff." (We really do not have a good name for it.) The equation relating mass and energy

$$E = mc^2 \qquad (17)$$

is now familiar to most of us. In this equation E is the energy, m is the mass change equivalent to the energy E, and c is the velocity of light or 3.0×10^{10} cm/sec. Because c^2 is so large (9.0×10^{20} cm²/sec²), a very small change in mass corresponds to a staggering change in energy. Conversely, the *mass change* associated with the energy change of a chemical reaction is too small to measure. Consider the mass change that should result from the burning of 1 mole of carbon (12 g) and 1 mole of oxygen (32 g):

$$C(s) + O_2(g) \longrightarrow CO_2(g) + 94 \text{ kcal}$$
$$\Delta H_{2a} = -94 \text{ kcal} \qquad (2a)$$

The 94 kcal of energy corresponds to about 5×10^{-9} (0.000 000 005) g. Although the theory clearly predicts that the CO_2 should be 0.000 000 005 g *lighter* than the 12 g of carbon and 32 g of oxygen with which we started, the difference is just too small to measure at the present time.

Nuclear reactions, in contrast to chemical reactions, do involve the conversion to energy of weighable quantities of matter. Unfortunately, fission reactions involve the formation of so many different products and are so complex that our data do not really permit a good check of the Einstein equation. It seems quite clear, however, that the products are lighter than the reactants. Some mass is converted to energy.

9-4.2 The Nuclear Fusion Process

We have just observed that energy is released when a heavy nucleus such as uranium-235 undergoes fission to give two lighter nuclei. Strangely enough, energy is also released when neutrons and protons combine to give the nuclei of somewhat heavier atoms. Let us examine mass relationships in the hypothetical process

$$2\,^1_1H + 2\,^1_0n \longrightarrow\,^4_2He \qquad (18)$$

Until now we have been content to write the mass of the proton or hydrogen atom as 1, the mass of the neutron as 1, and the mass of the electron as essentially 0. This was acceptable in calculations involving low precision, but if we are to accurately calculate mass-energy relationships, much more precise mass values must be used. Careful measurements made in the mass spectrometer indicate that the hydrogen atom actually has a mass of 1.00782522 mass units. (How many significant figures are there in this value?) The neutron has a mass of 1.00807134 mass units, and the helium atom has a mass of 4.00260361. Let us check the mass balance. Two hydrogen atoms have a mass of

$$2 \times 1.00782522 = 2.01565044$$

Two neutrons have a mass of

$$2 \times 1.00807134 = 2.01614268$$

The total precision mass resulting from a combination of two hydrogen atoms and two neutrons should be

	2.01565044
	+2.01614268
	4.03179312
The measured mass of the helium atom is	−4.00260361*
The difference is	0.02918951

In the hypothetical process cited, 0.02918951 units of mass are converted to energy. This is equivalent to 0.02918951 g per *mole* of helium formed. The arithmetic converting this quantity to a more easily recognized energy form is not hard when we apply $E = mc^2$ and convert to recognizable energy units. If we use 3.0×10^{10} cm/sec as the velocity of light, and round off our mass difference to two significant figures, we can write

$$E = \frac{0.029 \text{ g}}{\text{mole}} \times \left(\frac{3.0 \times 10^{10} \text{ cm}}{\text{sec}}\right)^2 = \frac{2.6 \times 10^{19} \text{ g} \times \text{cm}^2}{\text{mole} \times \text{sec}^2} \qquad (19)$$

The unit of energy given above may be rather strange to you. It is the **erg**.† It can be converted to calories by a factor that changes the units. We have

$$E = \frac{2.6 \times 10^{19} \text{ g} \times \text{cm}^2}{\text{mole} \times \text{sec}^2} \times \frac{2.4 \times 10^{-8} \text{ cal}}{\text{g} \times \text{cm}^2/\text{sec}^2} \qquad (20)$$

$$E = \frac{6.2 \times 10^{11} \text{ cal}}{\text{mole}} \text{ or 620 billion cal}$$

Thus, if 1 mole of helium were to be formed from 2 moles of protons and 2 moles of neutrons, 6.2×10^{11} cal would be released. Conversely, the same amount of energy is required to break up 1 mole of helium nuclei into protons and neutrons. This quantity is called the **binding energy** of a mole of helium nuclei. Similar calculations can be made for other nuclei.

EXERCISE 9-5

If a nitrogen atom weighs 14.00307438 on the atomic weight scale, how much mass is converted to energy in the formation of 1 mole of nitrogen from hydrogen atoms and neutrons? If 1 g of mass is equivalent to 2.16×10^{13} cal, what is the binding energy of the nitrogen nuclei per mole?

A significant comparison between nuclear binding energies can be made if we divide the total binding energy of each nucleus by the number of nucleons in the nucleus. This quantity, the binding energy per particle in the nucleus, varies systematically as the mass number of the nucleus increases. The variation is shown in Figure 9-7.

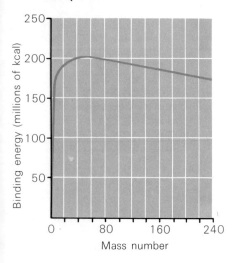

Fig. 9-7 **The binding energy per nuclear particle.**

---

* This is the mass of a neutral helium atom—nucleus plus two electrons.
† The erg is a unit of energy. It has the dimensions of g × cm²/sec². You will recognize this as units indicating mass × velocity². Remember that kinetic energy is $mv^2/2$. One erg is 2.4×10^{-8} cal.

EXERCISE 9-6

The nuclear binding energy per nucleon does not vary in a periodic fashion like the ionization energies of the atoms. Why?

The NUCLEAR FUSION PROCESS

The nuclei with mass numbers around 60 have the highest binding energy per heavy nuclear particle, and are therefore the most stable nuclei. The graph in Figure 9-7 will help us understand the processes of nuclear fission and nuclear **fusion**. Nuclear fusion is a process in which lighter nuclei fuse to give a nucleus of *greater* mass. If the nuclei of the heavier elements, such as uranium and plutonium, are split into two smaller fragments, the binding energy per nucleon is greater in the lighter nuclei. As in every other reaction in which the products are more stable than the reactants, energy is evolved by this process of nuclear fission. Similarly, when light nuclei such as ^1_1H or ^2_1H are brought together, the binding energy per nucleon again increases and energy is released. We see from the graph that the energy released per nucleon is much greater in the fusion process than it is in the fission process (*i.e.*, the curve is much steeper on the left-hand than it is on the right-hand side).

Unfortunately, it has not yet been possible to bring about the direct experimental combination of two protons and two neutrons to give a helium nucleus. It has been possible, however, to bring about the combination of heavy hydrogen or hydrogen-2 (^2_1H) to give heavier nuclei. A typical reaction is

$$^2_1H + {}^2_1H \longrightarrow {}^3_1H + {}^1_1H + 7.5 \times 10^{10} \text{ cal} \qquad (21)$$

The hydrogen-3 nucleus can then combine with another hydrogen-2 nucleus to give helium:

$$^3_1H + {}^2_1H \longrightarrow {}^4_2He + {}^1_0n + 4.0 \times 10^{11} \text{ cal} \qquad (22)$$

The energy evolved is tremendous. It is very difficult to start these reactions; temperatures of 10 million to 100 million degrees are required. If a *fission* bomb reaction is used as a trigger, the fusion process can be initiated to give an uncontrolled *hydrogen* bomb explosion. Such weapons may produce unlimited explosive power, but they are of no value for the production of controlled power.

To achieve the much more difficult objective of a controlled release of energy from fusion demands alternative ways to heat, contain, and compress the hydrogen reactants. Very strong magnetic fields offer some hope. There is a long and difficult road ahead before we can use fusion power in an effectively controlled manner.

EXERCISE 9-7

In the radioactive decomposition of radium-226, the equation for the nuclear process is

$$^{226}_{88}Ra \longrightarrow {}^{222}_{86}Rn + {}^4_2He$$

If the mass of the $^{226}_{88}Ra$ atom is 226.025360 mass units, the mass of $^{222}_{86}Rn$ 222.017530, and the mass of ^4_2He 4.00260361, how much mass is converted to energy in the radioactive decay process? If 1 g of mass is equivalent to 2.16×10^{13} cal, how many calories of energy are released for every mole of ^4_2He liberated?

Let us review the difference between nuclear and chemical reactions. Referring to the fission of uranium-235, the symbol $^{235}_{92}U$ represents not an atom, but a nucleus. The equation for the fission process is written in terms of the nuclei and particles associated with them. A nuclear equation tells us nothing about what compound of radioactive element was disintegrated or what new compounds were formed. We are summarizing only the nuclear changes. During the nuclear change there is much disruption of other atoms because of the tremendous amounts of energy liberated. We do not know in detail what happens, but eventually we return to electrically neutral substances (chemical compounds) and the neutrons are absorbed by other nuclei. Thus, in nuclear reactions, changes in the nuclei take place. In chemical reactions, the nuclei remain intact and the changes are explainable in terms of the electrons outside the nucleus.

9-5 HIGHLIGHTS

Every substance has a definite and characteristic **heat content** or **enthalpy** that varies with temperature. The change in enthalpy of the system is given by

$$\Delta H = \text{(enthalpy of products)} - \text{(enthalpy of reactants)}$$

A positive value of ΔH means that during the process, energy from the surroundings is stored in the products. The reaction is **endothermic**. A negative value of ΔH means that during the process energy is liberated to the surroundings so that the products have less energy than the original reactants. The process is **exothermic**. By proper addition of ΔH values for known reactions, ΔH values for new reactions can be predicted with a high degree of accuracy.

The energy for a system can be classified as kinetic and potential. The total energy for both mechanical and chemical systems is always conserved. Energy is stored in a molecule as **kinetic energy** (energy of motion) and as **potential energy** (energy of position). Potential energy is associated with attractions and repulsions of molecules, atoms, or charged particles making up the substance. Finally, in each nucleus there is a tremendous amount of stored energy.

The very large energy changes associated with nuclear processes have verified the Einstein relationship, $E = mc^2$.

QUESTIONS AND PROBLEMS

1 Given:

$$3C(s) + 2Fe_2O_3(s) + 110.8 \text{ kcal} \longrightarrow 4Fe(s) + 3CO_2(g)$$

Rewrite the equation using 1 mole of carbon and the ΔH notation.

2 Given:

$$\tfrac{1}{2}H_2(g) + \tfrac{1}{2}Br_2(l) \longrightarrow HBr(g)$$
$$\Delta H = -8.60 \text{ kcal}$$

Rewrite the equation for 1 mole of hydrogen gas and include the heat term as part of the equation.

3 Which of the following are endothermic?
(a) $H_2(g) + \tfrac{1}{2}O_2(g) \longrightarrow H_2O(g)$
  $\Delta H = -57.8 \text{ kcal}$
(b) $\tfrac{1}{2}N_2(g) + \tfrac{1}{2}O_2(g) \longrightarrow NO(g)$
  $\Delta H = +21.6 \text{ kcal}$
(c) $\tfrac{1}{2}N_2(g) + O_2(g) + 8.1 \text{ kcal} \longrightarrow NO_2(g)$
(d) $\tfrac{1}{2}N_2(g) + \tfrac{3}{2}H_2(g) \longrightarrow NH_3(g) + 11.0 \text{ kcal}$
(e) $NH_3(g) \longrightarrow \tfrac{1}{2}N_2(g) + \tfrac{3}{2}H_2(g)$
  $\Delta H = +11.0 \text{ kcal}$

4 What is the minimum energy required to synthesize 1 mole of nitric oxide (NO) from the elements? (See Table 9-2.)

5 How much energy is liberated when 0.100 mole of H_2 (at 25°C and 1 atm) is combined with enough $O_2(g)$ to make liquid water at 25°C and 1 atm? (See Table 9-2.)

6 How much energy is consumed in the decomposition of 5.0 g of $H_2O(l)$ at 25°C and 1 atm into its gaseous elements at 25°C and 1 atm?

7 Using Table 9-2, calculate the heat released in burning ethane in oxygen to give CO_2 and water vapor. (*Answer:* $\Delta H = -341$ kcal/mole C_2H_6.)

8 Given:

$$C(diamond) + O_2(g) \longrightarrow CO_2(g)$$
$$\Delta H = -94.50 \text{ kcal}$$

$$C(graphite) + O_2(g) \longrightarrow CO_2(g)$$
$$\Delta H = -94.05 \text{ kcal}$$

Find ΔH for the manufacture of diamond from graphite.

$$C(graphite) \longrightarrow C(diamond)$$

Is heat absorbed or evolved as graphite is converted to diamond?

9 To change the temperature of a particular calorimeter and the water it contains by 1°C requires 1,550 cal. The complete combustion of 1.40 g of ethylene gas [$C_2H_4(g)$] in the calorimeter causes a temperature rise of 10.7°C. Find the heat of combustion per mole of ethylene.

10 The "thermite reaction" is spectacular and highly exothermic. It involves the reaction between ferric oxide (Fe_2O_3) and metallic aluminum. The reaction produces white-hot, molten iron in a few seconds. Given:

$$2Al + \tfrac{3}{2}O_2 \longrightarrow Al_2O_3 \quad \Delta H_1 = -400 \text{ kcal}$$
$$2Fe + \tfrac{3}{2}O_2 \longrightarrow Fe_2O_3 \quad \Delta H_2 = -200 \text{ kcal}$$

Determine the amount of heat liberated in the reaction of 1 mole of Fe_2O_3 with Al. (*Answer:* $\Delta H = -200$ kcal.)

11 How much energy is released in the manufacture of 1.00 kg of iron by the "thermite reaction" mentioned in question 10?

12 How many grams of water could be heated from 0°C to 100°C using the heat liberated in forming 1 mole of aluminum oxide by the "thermite reaction"? (See question 10.)

13 Which would be the better fuel on the basis of the heat released per mole burned, nitric oxide (NO) or ammonia (NH_3)? Assume the products are $NO_2(g)$ for the first equation and $NO_2(g)$ plus $H_2O(g)$ for the second.

14 What is the minimum energy required to synthesize 1 mole of sulfur dioxide from sulfuric acid in the reaction

$$H_2SO_4(l) \longrightarrow SO_2(g) + H_2O(g) + \tfrac{1}{2}O_2(g)$$

(*Answer:* $\Delta H = +65$ kcal.)

15 Why is the Law of Conservation of Energy considered valid?

16 What do you think would happen in scientific circles if a clear-cut, well-verified exception were found to the Law of Conservation of Energy?

17 Is energy conserved when a ball of mud is dropped from your hand to the ground? Explain your answer.

18 What becomes of the energy supplied to water molecules as they are heated in a closed container from 25°C to 35°C?

19 Outline the events and associated energy changes that occur on the molecular level when steam at 150°C and 1 atm pressure loses energy continually until it finally becomes ice at $-10°C$.

20 The heat of combustion of methane (CH_4) is -210 kcal/mole:

$$CH_4(g) + 2O_2 \longrightarrow CO_2 + 2H_2O$$
$$\Delta H = -210 \text{ kcal}$$

Discuss why this fuel is better than water gas if the comparison is based on 1 mole of carbon atoms.

21 In a nuclear reaction of the type called nuclear fusion, two nuclei come together to form a larger nucleus. For example, hydrogen-2 nuclei (^2_1H) and hydrogen-3 nuclei (^3_1H) can "fuse" to form helium nuclei (^4_2He) and a neutron

$$^2_1H + ^3_1H \longrightarrow ^4_2He + ^1_0n$$
$$\Delta H = -4.05 \times 10^8 \text{ kcal}$$

How many grams of hydrogen would have to be burned (to gaseous water) to liberate the same amount of heat as liberated by fusion of 1 mole of ^2_1H nuclei and 1 mole of ^3_1H nuclei? Express the answer in tons (1 ton = 9.07×10^5 g).

22 Which of the following reactions is most likely to have a heat effect of -505 kcal? Which would be -1.7×10^6 kcal? Which would be $+7.2$ kcal?

(a) $UF_6(l) \longrightarrow UF_6(g)$ $\Delta H = ?$ 7.2
(b) $U(s) + 3F_2(g) \longrightarrow UF_6$ $\Delta H = ?$ -505
(c) $^{238}_{92}U + ^1_0n \longrightarrow ^{239}_{92}U$ $\Delta H = ?$ -1.7×10^6

23 Fission of uranium gives a variety of fission products, including praseodymium, Pr. If the process by which praseodymium is formed gives $^{147}_{59}Pr$ and three neutrons, what is the other nuclear product?

$$^{235}_{92}U + ^1_0n \longrightarrow ^{147}_{59}Pr + ? + 3\,^1_0n$$

$^{88}_{33}As$

24 Differentiate fission and fusion processes. What purpose do cadmium control rods serve in a nuclear reactor?

25 A positron has the same mass as an electron. Is the process by which one proton is converted to one positron and one neutron exothermic or endothermic? How much energy in ergs is involved in the conversion? Mass of proton = 1.00727, mass of neutron = 1.00807, and mass of electron = 0.00055.

> ... a molecular system ... [passes] ... from one state of equilibrium to another ... by means of all possible intermediate paths, but the path most economical of energy will be more often travelled.
>
> HENRY EYRING (1901–)

CHAPTER 10 The Rates of Reactions

The candle at the left is burning in air, the candle at the right in pure oxygen. Concentration can affect reaction rates.

Why doesn't a candle burn until lit with a match? Why does it continue to burn after the match is taken away? Why, indeed, does a match not burn until you strike it on something? What is the difference between a candle, which burns fat or wax only at a high temperature and you, who "burn" fat at body temperature?

Why are some chemical reactions extremely rapid and others agonizingly slow? Why do we use a Bunsen burner so frequently in the laboratory and an ice bath so seldom? How can we change the rate of reactions to make them more useful to us?

There are lots of questions—and, theoretically at least—some amazingly simple answers.

A candle remains in contact with air indefinitely without observable reaction, but it burns when lighted by a match. A mixture of household gas and air in a closed room remains indefinitely without reacting, but it may explode violently if a glowing cigarette is brought into the room. Iron reacts quite slowly with air (it rusts), but white phosphorus bursts into flame when exposed to air. All are reactions with oxygen from the air, but they require very different amounts of time. *Reactions proceed at different rates.* We care about reaction rate because we must understand how rapidly a reaction proceeds and what factors determine its rate in order to bring the reaction under control.*

Let us see what the expression "the rate of a reaction" means in terms of an example—the reaction between carbon monoxide (CO) and nitrogen dioxide (NO_2) that produces carbon dioxide (CO_2) and nitric oxide (NO). The equation for the reaction is

$$CO + NO_2 \longrightarrow CO_2 + NO \qquad (1)$$

If we heat a mixture of CO and NO_2 to 200°C, we observe a gradual disappearance of the reddish-brown color of NO_2. Reaction is taking place. We can find the rate of the process by measuring the change in color during successive one-minute periods. Since the other gases are colorless, the color *change* is proportional to the *change* in the number of moles of NO_2 present in a liter. If we divide the change in the quantity of NO_2 per liter by the time interval (one minute), the quotient is the rate of the reaction. We can write symbolically

$$\text{rate} = \frac{\text{quantity } NO_2 \text{ per liter consumed}}{\text{time interval}}$$

$$= \frac{\text{quantity } NO_2 \text{ used/volume container}}{\text{time interval}}$$

Remembering that concentration is quantity of material per unit volume, we can write

$$\text{rate of reaction} = \frac{\text{change in concentration } NO_2}{\text{time interval}}$$

We can express the rate of reaction in terms of the rate of consumption of either CO or NO_2, or in terms of the rate of production of either CO_2 or NO. Which we use depends upon convenience of measurement. If the experimenter prefers to measure the production of CO_2, he will express the rate in the form

$$\text{rate} = \frac{\text{increase in concentration } CO_2}{\text{time interval}}$$

rate CO_2 production = increase in concentration CO_2 per unit of time

The change in concentration can be expressed in partial pressure units if the substance is a gas or in molar concentration units if the

---

* The study of reaction rates is called **chemical kinetics**.

The RATES of REACTIONS

substance is in solution. The time measurement is expressed in whatever units fit the reaction—microseconds for the explosion of household gas and oxygen, seconds or minutes for the burning of a candle, days for the rusting of iron.

10-1 FACTORS AFFECTING REACTION RATES

In the laboratory you observed the reaction of ferrous ion, $Fe^{2+}(aq)$, with permanganate ion, $MnO_4^-(aq)$, and also the reaction of oxalate ion, $C_2O_4^{2-}(aq)$, with permanganate ion, $MnO_4^-(aq)$. These studies show that *the rate of a reaction depends upon the nature of the reacting substances.* In Experiment 18, the reaction between IO_3^- and HSO_3^- shows that *the rate of a reaction depends upon concentrations of reactants and on the temperature.* Let us examine these factors one at a time.

10-1.1 The Nature of the Reactants

Compare two reactions, both of which occur in water solutions:

$$5C_2O_4^{2-}(aq) + 2MnO_4^-(aq) + 16H^+(aq) \longrightarrow$$
$$10CO_2(g) + 2Mn^{2+}(aq) + 8H_2O \quad \text{SLOW} \quad (2)$$

$$5Fe^{2+}(aq) + MnO_4^-(aq) + 8H^+(aq) \longrightarrow$$
$$5Fe^{3+}(aq) + Mn^{2+}(aq) + 4H_2O \quad \text{VERY FAST} \quad (3)$$

Both ferrous ion, $Fe^{2+}(aq)$, and oxalate ion, $C_2O_4^{2-}(aq)$, are able to decolorize a solution containing permanganate ion at room temperature. Yet, there is a great contrast in the time required for the decoloration. The difference lies in specific characteristics of $Fe^{2+}(aq)$ and $C_2O_4^{2-}(aq)$.

Here are two reactions that take place in the gas phase:

$$2NO + O_2 \longrightarrow 2NO_2 \quad \text{MODERATE AT } 20°C \quad (4)$$
$$CH_4 + 2O_2 \longrightarrow CO_2 + 2H_2O \quad \text{EXTREMELY SLOW AT } 20°C \quad (5)$$

The oxidation of nitric oxide (NO), a reaction involved in smog production, is moderately fast at temperatures of (10–100)°C. The oxidation of methane (CH_4), household gas, however, occurs so slowly at room temperature that for all practical purposes there is no reaction. Again, the difference in the reaction rates must depend upon specific characteristics of the reactants, NO and CH_4.

The determination of the molecular characteristics that affect rate behavior is an interesting frontier of chemistry. Chemical reactions which involve the breaking of several chemical bonds and the formation of new chemical bonds tend to proceed slowly at room temperature. The reaction of $C_2O_4^{2-}(aq)$ and $MnO_4^-(aq)$ is of this type—many bonds must be broken in the five $C_2O_4^{2-}$ ions and the two MnO_4^- ions to form the $10CO_2$ and $2Mn^{2+}$. This reaction proceeds slowly, as we might expect. The reaction of CH_4 and O_2 also involves breaking bonds and forming new bonds, and it is slow at room temperature. In contrast, the reaction of $Fe^{2+}(aq)$ and

$MnO_4^-(aq)$ is very rapid, although it involves breaking chemical bonds and thus might be expected to be slow. We see that we cannot be certain of any prediction that a reaction might be slow.

You might properly ask: Do not all reactions involve bond breaking? No, a few appear to proceed without bond breaking. For example, the reaction

$$Fe^{2+}(aq) + Ce^{4+}(aq) \longrightarrow Ce^{3+}(aq) + Fe^{3+}(aq) \quad \text{Very Rapid} \quad (6)$$

appears to involve only the transfer of an electron between ions. We expect it to be very rapid—a prediction that is easily verified in solution. A prediction of this type is usually reliable. On the other hand, the reaction of NO with O_2 requires breaking but one bond and forming two. It has a moderate reaction rate, rapid at high pressures and slow at low pressures.

These and other examples lead to the following rules:

(1) Reactions that do not involve bond rearrangements are usually rapid at room temperature.
(2) Reactions in which bonds are broken tend to be slow at room temperature.

We can say little more about how the nature of the reactants determines the reaction rate until we consider in detail how some reactions take place. For the time being, we shall just say that this is an active field of study and much remains to be learned.

EXERCISE 10-1

Are any of the following three reactions likely to be extremely rapid at room temperature? Are any likely to be extremely slow at room temperature? Explain.

(1) $Cr^{2+}(aq) + Fe^{3+}(aq) \longrightarrow Cr^{3+}(aq) + Fe^{2+}(aq)$ rapid
(2) $3Fe^{2+}(aq) + NO_3^-(aq) + 4H^+(aq) \longrightarrow 3Fe^{3+}(aq) + NO(g) + 2H_2O$ moderate
(3) $C_8H_{18}(l) + 12\tfrac{1}{2}O_2(g) \longrightarrow 8CO_2(g) + 9H_2O(g)$ slow
    GASOLINE

10-1.2 Effect of Concentration—Collision Theory

Henceforth we shall look at one reaction at a time. The nature of the reactants will be held constant while the other factors that affect rates are considered. The first of these factors is *concentration*. A great number of experiments have shown that, for many reactions, raising the concentration of a reactant increases the reaction rate. Often, however, there will be no effect. In this section we shall consider how a rate increase with rising reactant concentration is explained. In Section 10-1.3 we shall explore why some reactions proceed at a rate independent of the concentrations of the reactants. Both explanations are based upon a model of chemical reactions on the molecular scale.

In the molecular model, we assume that two molecules must come close together to react. Therefore, we postulate that chemical reactions depend upon collisions between the reacting particles—

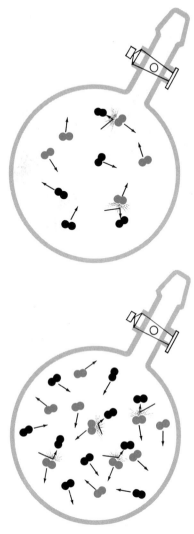

Fig. 10-1 **The number of collisions per second depends upon the concentration of each reactant.**

atoms, molecules, or ions. This model of reaction rate behavior is called the **collision theory** (see Figure 10-1). It provides a successful framework for understanding the effect of concentration. Just as increasing the number of cars in motion on a highway leads to a higher rate of denting fenders, increasing the number of particles in a given volume gives more frequent molecular collisions. The higher frequency of collisions results in a higher rate of reaction.

Consider a homogeneous system, one in which all components are in the same phase. According to the collision theory, increasing the concentration of one or more reactants will result in an increase in the rate of the reaction. Lowering the concentration has the opposite effect. This is exactly the behavior found in the reaction between $HSO_3^-(aq)$ and $IO_3^-(aq)$ seen in Experiment 18. With gases, also homogeneous systems, the concentration of an individual reactant can be raised by admitting more of that substance into the mixture. The concentrations of *all* gaseous components can be raised simultaneously by decreasing the volume occupied by the mixture. Decreasing the volume by compressing the gas raises the concentration of all reactants, hence increases the rates of reactions taking place. Increasing the volume by expanding the gas has the opposite effect on concentrations, hence decreases reaction rates.

In a heterogeneous reaction system, the components are in two or more different phases. As an example, consider

$$\text{wood }(s) + \text{oxygen }(g) \xrightarrow{\text{burning}} \text{carbon dioxide }(g) + \text{water }(g)$$

Here the rate of the reaction depends upon the area of contact between phases. For example, a log burns in air at a relatively slow rate. If the amount of exposed surface of the wood is increased by reducing the log to splinters, the burning is much more rapid. Further, if the wood is reduced to fine sawdust and the sawdust is suspended in a current of air, the combustion takes place explosively. Where one of the reactants is a gas, then, the concentration of the gas is also a factor. A piece of wood burns much more rapidly in pure oxygen than it does in ordinary air, in which the oxygen makes up only about 20 percent of the mixture.

We see that the collision theory provides a good explanation of reaction rate behavior. It is quite reasonable that the reaction rate should depend upon collisions among the reactant molecules. In fact, it is so reasonable that we are left wondering why the concentrations of some reactants in some reactions do not affect the rate. The explanation is found in the detailed steps by which the reaction takes place, the reaction mechanism.

10-1.3 Reaction Mechanism

As has been proposed, *for a chemical reaction to occur, particles must collide.* The particles may be atoms, molecules, or ions. As a result of collisions, there can be rearrangements of atoms, electrons, and chemical bonds, and subsequently production of new species. As an example, let us take another look at the reaction between Fe^{2+} and MnO_4^- in acid solution:

$$5Fe^{2+}(aq) + MnO_4^-(aq) + 8H^+(aq) \longrightarrow$$
$$5Fe^{3+}(aq) + Mn^{2+}(aq) + 4H_2O \qquad (3)$$

The equation indicates that one MnO_4^- ion, five Fe^{2+} ions, and eight H^+ ions (a total of fourteen ions) must react with each other. If this reaction were to take place in a single step, these fourteen ions would have to collide with each other *simultaneously*. The probability of such an event occurring is extremely small, so small that a reaction which depended upon such a collision would proceed at a rate which would be immeasurably slow. Since the reaction does occur at an easily measured rate, it must proceed by some *sequence* of *steps*, rather than by a single step so improbable as the simultaneous collision of fourteen ions. As a matter of fact, the collision of even four molecules is an extremely improbable event if the molecules are at low concentration or if they are in the gas phase. We conclude that a complex chemical reaction which proceeds at a measurable rate probably takes place in a series of simpler steps. The series of reaction steps is called the **reaction mechanism.**

Consider the oxidation of gaseous hydrogen bromide (HBr), a reaction that is reasonably rapid in the temperature range from 400°C to 600°C:

$$4HBr(g) + O_2(g) \longrightarrow 2H_2O(g) + 2Br_2(g) \qquad (7)$$

By the collision theory, we expect that increasing the partial pressure (thus the concentration) of either the HBr or O_2 will speed up the reaction. Experiments show this is the case. Quantitative studies of the rate of the reaction of HBr and O_2 at various pressures and in varying proportions indicate that the reactants O_2 and HBr are equally effective in changing the reaction rate. However, this result raises a question. Since four molecules of HBr are required for every one molecule of O_2, why does a change in the HBr pressure have just the same effect as an equal change in the O_2 pressure?

Considering the details of the process by which the reaction of HBr and O_2 occurs gives us the explanation. The overall reaction brings together five molecules, four of HBr and one of O_2. However, the chance that five gaseous molecules will collide simultaneously is almost zero. The reaction must occur in a series of simpler steps.

All the studies on this process are explained by the following series of reactions:

$$\begin{array}{rcl}
HBr + O_2 & \longrightarrow & \cancel{HOOBr} \qquad\qquad \text{SLOW} \quad (8)\\
\cancel{HOOBr} + HBr & \longrightarrow & \cancel{2HOBr} \qquad\qquad \text{FAST} \quad (9)\\
\cancel{2HOBr} + 2HBr & \longrightarrow & 2H_2O + 2Br_2 \quad\text{FAST} \quad (10)\\
\hline
4HBr + O_2 & \longrightarrow & 2H_2O + 2Br_2 \qquad\qquad\quad (7)
\end{array}$$

Next, note that each step in the sequence requires only two molecules to collide.* Finally, the proposal that the first step in the sequence is slow whereas the other steps are fast explains why HBr and O_2 have the same effect on the reaction rate.

The reaction giving HOOBr is a "bottleneck" in the oxidation of hydrogen bromide. As fast as HOOBr is formed by this slow reaction, it is consumed in the rapid reaction with more HBr. But no

---

* Reaction *(10)* is the reaction $HOBr + HBr \longrightarrow H_2O + Br_2$ taking place twice.

matter how rapid the latter reactions are, they can produce H_2O and Br_2 only as fast as the slowest reaction in the sequence. Hence, the factors that determine the rate of HOOBr formation determine the rate of the overall process.

The sequence of reactions given above is called the reaction mechanism of the overall reaction. The formation of HOOBr is the step that fixes the rate because it is the slowest reaction in the mechanism. *The slowest reaction in a mechanism is called the* **rate-determining step.**

There are two features of this example that are rather common. First, none of the steps in the reaction mechanism requires the collision of more than two particles. *Most chemical reactions proceed by sequences of steps, each involving only two-particle collisions.* Second, the overall or net reaction does *not* show the mechanism. In general, *the mechanism of a reaction cannot be deduced from the net equation for the reaction;* the various steps by which atoms are rearranged and recombined and which make up the mechanism must be determined through experiment.

EXERCISE 10-2

Imagine that a telegram is to be sent on a winter night from New York City to a rancher who lives near the top of a mountain in Nevada. The steps in sending the telegram are the following: (a) The New Yorker calls the telegraph office and dictates his ten-word message. (b) The New York telegraph operator transmits the message to the Reno office. (c) The Reno office types out the message and hands it to a delivery man. (d) The delivery man gets into his car and fights his way 20 miles up the snow-covered road to the rancher's house. How great an increase in the rate of transmitting a message would be observed if:

(1) The secretary receiving the message in New York could type twice as fast? *little*
(2) Electricity (by some undefined and miraculous process) could move twice as fast in the wire? *little*
(3) The Reno office could telephone the message to the rancher rather than send the delivery man? *much faster*

What is the rate-determining step in the original process? *d*

10-1.4 The Quantitative Effect of Concentration

Determining the effect of reactant concentration on the rate of a reaction is a relatively straightforward experimental procedure, but to establish the precise mechanism from this information is a much less certain process. Consider the reaction between gaseous hydrogen (H_2) and gaseous iodine (I_2):

$$H_2(g) + I_2(g) \longrightarrow 2HI(g) \qquad (11)$$

Since $I_2(g)$ is violet in color, we can follow the rate of the reaction under different conditions by measuring the change in color that takes place in a given time interval. (The color of the gas mixture is proportional to its iodine concentration.) Rate measurements show that doubling the concentration of I_2 doubles the reaction rate if the H_2 concentration is kept constant; tripling the concentration of I_2 increases the rate by a factor of three; and so on. The observations can be summarized by the statement: The rate at which I_2 disappears is proportional to I_2 concentration. This statement can be made in a somewhat more formal, mathematical way:

$$\text{rate} = k'[I_2] \quad \text{if } [H_2] \text{ is constant} \tag{12}$$

where k' is a proportionality or rate constant (the rate if $[I_2]$ is 1) and $[I_2]$ is the concentration of I_2 vapor expressed as moles per liter. We can also determine how the reaction rate changes if the H_2 concentration is changed while the I_2 concentration is held constant. It is found that the rate is proportional to the concentration of H_2. We can write

$$\text{rate} = k''[H_2] \quad \text{if } I_2 \text{ is constant} \tag{13}$$

where k'' is a rate constant and $[H_2]$ the concentration of H_2 in moles per liter. If both H_2 and I_2 were to vary, the rate would be proportional to the product of their concentrations:

$$\text{rate} = k'[I_2] \cdot k''[H_2] = k'k''[I_2][H_2] \tag{14}$$

or

$$\text{rate} = k[H_2][I_2] \tag{15}$$

The constant k is a proportionality constant. It is equal to the measured rate when both H_2 and I_2 have a concentration of 1 mole per liter.

10-1.5 The Problems of Reaction Mechanism

The results just cited are those which would be expected for a reaction proceeding by collision between an H_2 molecule and an I_2 molecule:

$$H_2(g) + I_2(g) \longrightarrow 2HI(g) \tag{11}$$

For years chemists throughout the world have accepted and taught this simple mechanism for the reaction of H_2 and I_2. Most chemists were convinced that the reaction was simple; but a few rebels like Dr. Henry Eyring, whose contributions are reviewed in the margin, felt that certain details of the variation of the reaction rate with temperature implied a more complex process. Reports of work published in 1967 now show that the process is more complex than was originally thought. The best current conceptualization of the reaction mechanism suggests that I_2 molecules may undergo dissociation to iodine atoms *before* reacting with H_2. The mechanism would then be

$$I_2 \longrightarrow 2I \tag{16}$$

$$H_2 + 2I \longrightarrow 2HI \tag{17}$$

This mechanism is consistent with all the rate measurements.

The abandonment of time-honored beliefs and concepts in the face of new data is the mark of a vigorous science. Remember the generalization: All Cylindrical Objects Burn.

HENRY EYRING (1901–   )

One of eighteen children, Henry Eyring was born in Chihuahua, Mexico, where his father was a cattle rancher and farmer. The Eyring family was forced to abandon this home in 1912 under threat of the revolutionist Salazar, moving first to Texas, then to Arizona.

Eyring first studied mining, then metallurgy, at the University of Arizona. In 1927 the University of California awarded him the Ph.D. in chemistry. Before becoming Dean of the graduate school and professor of chemistry at the University of Utah, Eyring served on the faculties of the University of Wisconsin, the University of California, and Princeton University.

Eyring's research has been original and frequently unorthodox. He was one of the first chemists to apply quantum mechanics to chemistry. He unleashed a revolution in the treatment of reaction rates by use of detailed thermodynamic reasoning. Having formulated the idea of the activated complex, Eyring proceeded to find a myriad of fruitful applications, from formation of artificial diamonds to conductance, adsorption, catalysis, and diffusion of liquids.

10-1.6 The Effect of Temperature on Reaction Rate— An Application of Collision Theory

In Experiment 18 you discovered that temperature has a marked effect upon the rate of chemical reactions. Thus, raising the temperature speeded up the reaction between IO_3^- and HSO_3^-. This is the same effect, qualitatively, that is observed in the reaction of a candle with air. The match "lighted" the wick by raising its temperature. Once started, the reaction of combustion released enough heat to keep the temperature high, thus keeping the reaction going at a reasonable rate. Raising the temperature speeded up the reaction.

In all these reactions (and in almost all others), increasing the temperature has a very pronounced effect, always speeding up the reaction. Two questions come to mind: Why does a rise in temperature speed up a reaction? And: Why does a rise in temperature have such a large effect? Finding an answer leads us back to the collision theory.

From what we know about molecular sizes, we can calculate that a particular methane molecule (CH_4) collides with an oxygen molecule approximately once every one thousandth of a microsecond (10^{-9} sec). This means that in one second a methane molecule encounters 10^9 oxygen molecules! Yet the reaction does not proceed noticeably. We can conclude either that most of the collisions are ineffective or that the collision theory is not a good explanation. We shall see that the former is the case—we can understand why most collisions might be ineffective in terms of ideas that are consistent with the collision theory.

It is important to note that chemical reactions occur when collisions occur, but *only when the collision involves more than a certain amount of energy.* We can understand this by returning to our analogy of cars bumping each other on a highway. In a line of heavy traffic one may receive gentle bumps from the car in front or the car behind. No damage is done to the cars, only to people's tempers. But occasionally a high-speed collision occurs. If this occurs with enough energy, a bumper may be knocked off a car and a fender may be collapsed. The geometry of the collision is important; a low-velocity collision with a bumper is much less effective than with a fender or door. For a really effective collision, a combination of high velocity and proper geometry is unbeatable.

Just as high-energy collisions at the proper site cause auto damage, high-energy molecular collisions occurring at a vulnerable site on the reacting molecules cause the molecular rearrangement we call a chemical reaction. Just as a certain amount of energy is required to break loose a bumper, a certain amount of energy is required to cause a chemical reaction. If there is more than this threshold energy, the reaction can occur. If there is less, it cannot occur.

This discussion of threshold energy causes us to wonder what energies are possessed by molecules at a given temperature. We have already compared the molecules of a gas with "super-rubber" balls rebounding in a room. When "super-rubber" balls bounce around and collide with each other, some move rapidly and some move slowly. Do molecules behave this way? Experiment provides the answer.

The EFFECT of TEMPERATURE on REACTION RATE

Velocity is commonly measured as the length of time it takes an object to travel a measured distance. In drag racing, for example, the competing cars leave the starting gate at the same moment, and are sorted out at the finish line on the basis of their speed. The faster cars arrive at the finish line first; the slower ones straggle in behind them. The average velocity of each car can be calculated using the relationship: **velocity = distance/time.**

Similarly, the distribution of velocities of molecules or atoms can be measured using the device shown in Figure 10-2. It consists of two disks, D_1 and D_2, rotating rapidly on a common axle, thus at the same speed. They rotate in a vacuum chamber in front of an oven containing molten tin held at a controlled temperature. Vapor streams out of the small opening in the oven and strikes the rotating disk D_1, which lets molecules pass through in small bursts each time the slot passes the opening of the oven. This disk serves as the starting gate for repeated "molecule races." When the disk has

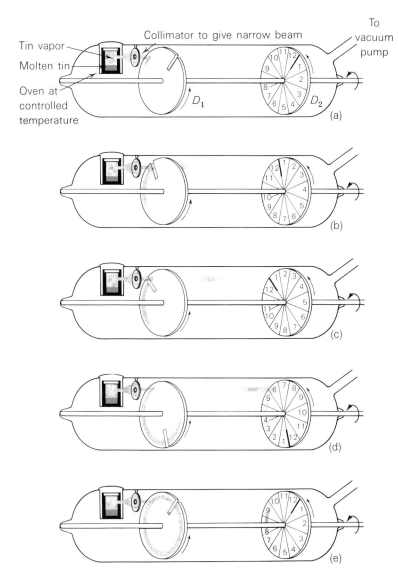

Fig. 10-2 **Rotating disk used to measure atomic or molecular velocities.**

rotated to the position shown in Figure 10-2(b), a small amount of gas has passed the slot in disk D_1. No more gas can get through the slot until it has completed one full rotation and the slot is again opposite the opening of the oven.

A short time later [Figure 10-2(c)], the atoms of tin have traveled part of the way toward the second rotating disk, which is their target, and have spread out from their compact arrangement in Figure 10-2(b). The faster-moving atoms have traveled farther than the others, and are leading the way. The slower-moving atoms are beginning to lag behind. Still later, in Figure 10-2(d), the atoms are spread out in space even more, and the faster atoms have already reached the second rotating disk, and have condensed on it. This disk sorts out the atoms on the basis of their velocity just as the timer in a drag race does. Since disk D_2 is rotating, the atoms condense at positions scattered along the edge of the disk, as shown in Figure 10-2(e). The position at which a given atom condenses on disk D_2 depends upon how long that atom took to travel from D_1 to D_2 and how fast D_2 is rotating.

As the slotted disk, D_1, lets through burst after burst of tin atoms, a layer of tin builds up on the surface of disk D_2. The pattern of this layer is determined by the distribution of velocities of the atoms escaping from the oven at the temperature being maintained. Figure 10-3 shows the disk D_2 divided into sections, like slices of a pie. The fastest-moving atoms have condensed on "pie slices" 3, 4, and 5. The slowest-moving atoms have condensed on pie slices 10 and 11. If the disk is cut up and each slice weighed, the amount of tin can be determined. A plot of the weight of tin against the number of its pie slice indicates the distribution of atomic velocities.

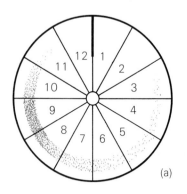

Fig. 10-3 **The relative distribution of atomic or molecular velocities obtained from the rotating disk.**

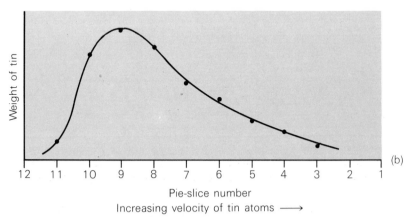

Disk D_2 after many revolutions

EXERCISE 10-3

Why is there no tin deposited on pie slices 1 and 2? on slice 12? What would be the effect of speeding up the rotation of the disks on the amount of tin deposited on each slice? Which slices would have more tin deposited on them? which less? which the same?

We saw in Chapter 4 that the relationship between the velocity of a gas molecule, its mass, and its kinetic energy is expressed by the equation

$$KE = \tfrac{1}{2}mv^2 \qquad (18)$$

The plot of Figure 10-3 contains some information about the distribution of kinetic energies among the molecules which have condensed. From the rate of rotation of the disks and the distance between them, we can calculate the velocity which an atom must have to condense on any one pie slice.* From the atomic mass and our calculated velocity we learn the kinetic energy of the atom. Figure 10-4 shows the result. At temperature T_1 a few atoms have very low kinetic energies and some have very high kinetic energies. Most of them have intermediate kinetic energies, as shown by the black curve. At a higher temperature, T_2, the energy distribution is altered to that shown by the white curve. As can be seen, *increasing the temperature causes a general shift of the molecular distribution toward higher kinetic energies.* Moreover, in going from T_1 to T_2 there is a large increase in the number of molecules having kinetic energies above a certain value, E.

We can apply these curves to our reaction rate problem. Suppose a reaction can proceed only if two molecules collide with a total kinetic energy that exceeds a certain threshold energy, E. Figure 10-4 shows a typical situation. At T_1, the lower temperature, the darkly shaded area is proportional to the number of molecules possessing at least the threshold energy. These molecules have enough energy to stage effective collisions even with molecules of very low kinetic energy. But since only a small number of molecules have as much energy as this, few collisions are effective, and the reaction is slow. Some molecules with energies below E may collide with each other to give a total energy above the threshold, but the probability is rather small. However, if we raise the temperature to T_2, the number of molecules with energy E or greater is raised in proportion to the lightly shaded area. *Only a small temperature change is needed to make a large change in the area out on the tail of the energy distribution curve.* Consequently, the reaction rate is very sensitive to change in temperature.

This argument is based on the "typical" situation in which E is well out on the tail of the curve. Suppose it is not, but is near the maximum of the curve at T_1 or is even to the left of it. Then a large number of molecules have the necessary energy to react, even at the lower temperature T_1. Since collisions occur so rapidly (remember,

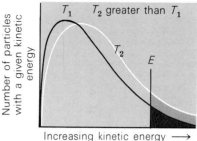

Fig. 10-4 **Effect of temperature on relative distribution of atomic or molecular kinetic energies.**

---

* Here is an example. Suppose the disks are 1.0 cm apart and the disks are rotating at a rate of 1,200 revolutions per minute or 1,200/60 = 20 revolutions per second. One-twentieth of a second is required for a complete revolution. There are 12 sections on the disk, hence each section will be in the target area for $1/12 \times 20 = 1/240$th second. The atoms collecting on slice 1 require 1/240th second to travel the distance of 1.0 cm between disks. Their average velocity will be

$$\frac{1.0 \text{ cm}}{1/240 \text{ sec}} = \frac{240 \text{ cm}}{\text{sec}}$$

Those on 2 require 2/240th of a second; they are going 120 cm/sec. And so on.

one every 10^{-9} seconds or so), the reaction is over in less than the blink of an eye. The circumstances shown in Figure 10-4 are most important in considering a slow reaction.

It should be noted that raising the temperature also increases the reaction rate by increasing the frequency of the collisions. This is, however, a very small effect compared with that caused by the increase in the number of molecules with sufficient energy to cause reaction.

10-2 THE ROLE OF ENERGY IN REACTION RATES

Now that we see the effect of temperature in increasing the fraction of molecular collisions having a total energy greater than the threshold energy, our understanding of the role of energy in fixing reaction rates can be expanded.

10-2.1 Pole-Vaulting and Potential Energy Barriers— The Activation Energy

Suppose it is your ambition to be successful in learning to pole-vault. To achieve this goal, you must take a running start and then raise your body to an altitude of approximately eight feet, go over the bar, and drop into the sandpit, as illustrated in Figure 10-5. Of course, it is in getting over the bar that the difficulties develop. Almost anyone can make a running start and achieve some altitude, but not everyone can get over the bar. Some individuals will clear the bar with room to spare, some will barely squeak over, and many will have to go back and try the running start again. Happiness is getting over the bar. If the bar is high, you will have to increase your distance from the earth; hence, your potential energy will increase. If the bar is high, few individuals will get over. If the bar is low, everyone will meet with success. It is the bar that defines the barrier; if

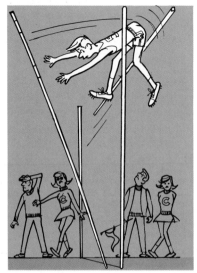
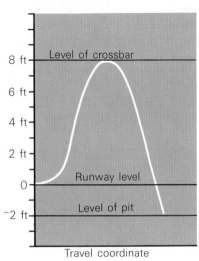

Fig. 10-5 **A strong pole-vaulter with adequate height (potential energy) can jump over the crossbar (the activation energy). A weak pole-vaulter without adequate height cannot jump over the crossbar. Height (potential energy) of successful pole-vaulter vs. distance traveled.**

you do get over it, the remainder of the process, falling from the bar to the sandpit, is sure to follow.

Chemical reactions are similar. As molecules collide and a reaction takes place, the atoms must momentarily take up bonding arrangements that are less stable than either reactants or products. Atoms are separated (as were the earth and the jumper) and the potential energy of the system goes up. These high-energy molecular arrangements are like the bar in Figure 10-5—they place an energy "barrier" between reactants and products. Only if the colliding molecules have enough energy to overcome the barrier imposed by the unstable arrangements can reaction take place. This barrier determines the "threshold energy" or minimum energy necessary to permit a reaction to occur. It is called the **activation energy.** If the activation energy is high, few molecules possess enough energy to react, and the reaction proceeds slowly. If the activation energy is low, more molecules have enough energy to get over the barrier, and the reaction tends to proceed rapidly.

Here again, the geometry of the colliding molecules is involved. Just as the pole-vaulter can get over the bar with a lower jump if he learns the proper position for just slipping over, the activation energy is *the lowest energy necessary to cause a reaction between two molecules colliding in the best geometric arrangement for a reaction to occur* (see Figure 10-6).

POLE-VAULTING and
POTENTIAL ENERGY BARRIERS

Fig. 10-6 A pole-vaulter needs at least eight feet of height to clear the bar. Anything less than eight feet is not enough.

This barrier can be shown graphically by amplifying Figure 10-6, which shows the relative energies of reactants and products. In Figure 10-7 we see that the diagram becomes the equivalent of the route the pole-vaulter takes in our analogy. This diagram applies to the reaction between carbon monoxide (CO) and nitrogen dioxide (NO_2) to produce carbon dioxide (CO_2) and nitric oxide (NO). The horizontal axis of the diagram, called the **reaction coordinate,** shows the progress of the reaction. Proceeding from left to right along this reaction coordinate, we see the CO and NO_2 molecules approaching each other, colliding, and going through an intermediate process that results in the formation of CO_2 and NO. The vertical axis represents the total potential energy of the system. Thus, the curve provides a history of the potential energy change during a collision which results in a reaction. The energy required to overcome the potential energy barrier is usually provided by the kinetic energies of the colliding particles. The kinetic energies of the particles are determined by the temperature of the two reactants, CO and NO_2, in the system.

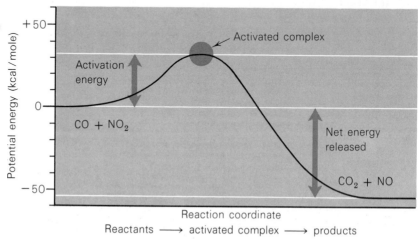

Fig. 10-7 **Potential energy diagram for the reaction**

$$CO + NO_2 \longrightarrow CO_2 + NO$$

Let us move from left to right along this curve and describe the events which occur. Along the flat region at the left, CO and NO_2 are approaching each other. In this region, they possess kinetic energy and their total potential energy shows no change. The beginning of the rise in the curve signifies that the two molecules have come sufficiently close to have an effect on each other. During this approach, the molecules slow down as their kinetic energies furnish the potential energy to climb the curve. If they have sufficient kinetic energy, they can ascend the left side of the barrier all the way up to the summit. Attaining this point is interpreted as follows: CO and NO_2 had sufficient kinetic energy to overcome the mutually repulsive forces between nuclei and between negative electron clouds. Here at the summit the molecular cluster, like the suspended pole-vaulter, is unstable with respect to either the forward reaction (giving CO_2 and NO) or the reverse reaction (restoring CO and NO_2). This transitory arrangement is of prime importance (its potential energy fixes the activation energy). It is called the **activated complex.**

Now there are two possibilities:

(1) the activated complex may separate into the two original CO and NO_2 molecules, which would then retrace their former path on the curve, or
(2) the activated complex may separate into CO_2 and NO molecules.

The latter possibility is represented by moving down the right side of the barrier. In the flat region at the right, CO_2 and NO have separated beyond the point of having any effect on each other and the potential energy of the activated complex has become kinetic energy again.

In the event that the CO and NO_2 molecules do not have sufficient energy to attain the summit, they reach a point only part of the way up the left side of the barrier. Then, repelling one another, they separate again, going downhill to the left.

We have labeled the difference between the high potential energy at the activated complex and the lower energy of the reactants as the activation energy. *The activation energy is the energy necessary to transform the reactants into the activated complex.* This may involve weakening or breaking bonds, forcing reactants close together in opposition to repulsive forces, or storing energy in a vibrating molecule so that it reacts on collision. *Increasing the temperature affects reaction rate by increasing the number of molecular collisions that involve sufficient energy to form this activated complex.* The magnitude of the activation energy for a reaction can be determined by measuring experimentally the change in reaction rate associated with a known change in temperature.

10-2.2 Heat of Reaction

We can deduce the heat of reaction from Figure 10-7. In our example, the reactants are at a higher total energy than are the products. This means that in the course of the reaction there will be a net *release* of energy. This reaction is **exothermic.** Figure 10-7 shows that the reaction releases 54 kcal of heat per mole of carbon monoxide consumed. Notice that the height of the energy barrier between reactants and products has no effect on the net heat release. We must put in an amount of energy equal to the activation energy to get to the top of the barrier, but we get it all back on the way down the other side.

Now let us consider the reverse reaction. We need not draw another reaction diagram, since Figure 10-7 will suffice. Now we are interested in the reaction between CO_2 and NO to produce CO and NO_2:

$$CO + NO_2 \longleftarrow CO_2 + NO \qquad (1a)$$

This reaction begins at the lower energy appropriate to the chemical stability of CO_2 + NO (at the right side of Figure 10-7) and ends at the higher energy appropriate to the chemical stability of CO + NO_2 (at the left side of Figure 10-7). The difference in energy—the heat of this reaction—is just equal to that of the reverse reaction but is opposite in sign. This reaction *absorbs* 54 kcal of heat per mole of carbon monoxide produced. It is **endothermic.**

Figure 10-7 contains one other very interesting piece of information concerning the rate of the reverse reaction between CO_2 and NO. This reaction rate is controlled by the energy barrier confronting the colliding molecules of CO_2 and NO. We see from the diagram that the activation energy for this reaction is higher than that for the reaction we studied earlier. Further, it is higher by exactly the heat of reaction. We conclude that the reaction between CO_2 and NO will be slower, at any given temperature, than the reverse reaction between CO and NO_2, if the rates are compared at the same partial pressures.

The relationship between activation energies for the forward and reverse reactions can be expressed mathematically. The activation energy is denoted by the symbol $\Delta H^{\ddagger}$ (read "delta-H-cross") and the heat of reaction by ΔH. Hence, we may write

$$\Delta H^{\ddagger}_{forward} = \Delta H^{\ddagger}_{reverse} + \Delta H \qquad (19)$$

where

(1) $\Delta H^{\ddagger}_{forward}$ = activation energy for reaction proceeding to the right (energy absorbed),
(2) $\Delta H^{\ddagger}_{reverse}$ = activation energy for reaction proceeding to the left (energy absorbed), and
(3) ΔH = heat absorbed during reaction proceeding left to right (ΔH is positive if endothermic, negative if exothermic).

The heat of reaction, ΔH, is positive if heat is absorbed as the reaction proceeds, left to right. It is negative if heat is evolved. In our example

$$CO + NO_2 \longrightarrow CO_2 + NO \qquad (1)$$

$\Delta H^{\ddagger}_{forward} = +32$ kcal/mole, $\Delta H^{\ddagger}_{reverse} = +86$ kcal/mole, and $\Delta H = -54$ kcal/mole. We see that

$$32 = 86 + (-54)$$

which is in accordance with equation (19). This relationship is important because it implies that we need only *two* of the three quantities $\Delta H^{\ddagger}_{forward}$, $\Delta H^{\ddagger}_{reverse}$, and ΔH, to calculate the third.

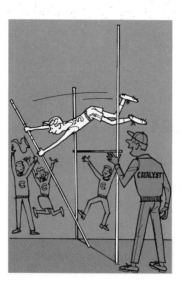

Fig. 10-8 **With the coach's help (catalysis) the weak pole-vaulter can clear the crossbar.**

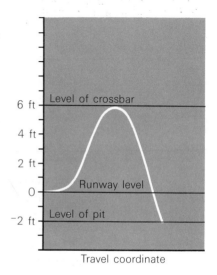

10-2.3 Action of Catalysts

Many reactions proceed quite slowly when the reactants are mixed alone, but can be made to take place much more rapidly by the introduction of other substances. These substances, called **catalysts,** are not used up in the reaction. The process of increasing the rate of a reaction through the use of a catalyst is referred to as **catalysis.** You have seen at least one example of catalytic action, the effect of $Mn^{2+}(aq)$ in speeding up the reaction between $C_2O_4^{2-}(aq)$ and $MnO_4^-(aq)$. The decomposition of hydrogen peroxide to give H_2O and $½O_2$ is also strongly catalyzed by MnO_2.

The action of a catalyst can be explained in terms of our pole-vaulting analogy. In Figure 10-5 we saw a formidable obstacle between our hopeful pole-vaulter and a successful landing in the sandpit. Many people simply cannot get over a bar this high, no matter how many running starts they make. But the coach can help even these people to a successful landing simply by lowering the bar. The net result will be that more people are able to get over the bar and into the sandpit (Figure 10-8). The return trip (from sandpit, over the bar, and back to the starting line) would also be much easier if the bar were lowered.

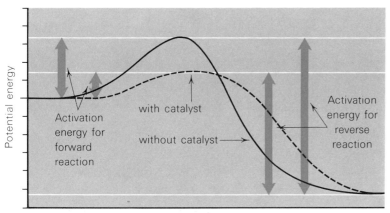

Fig. 10-9 **Effect of a catalyst on a reaction and its reverse.**

Figure 10-9 shows the same situation for a chemical reaction. The potential energy of the system is plotted against the reaction coordinate. The solid curve shows the activation energy barrier that must be overcome for a reaction to take place. When a catalyst is added, there is a new reaction path, hence a different activation energy barrier. The dashed curve shows this situation. This new reaction path corresponds to a new reaction mechanism that permits the reaction to occur via a different activated complex. Hence, more particles can get over the new, lower-energy barrier and the rate of the reaction is increased. The activation energy for the reverse reaction is lowered by exactly the same amount as the forward reaction. This accounts for the experimental fact that a catalyst for a reaction has an equal effect on the reverse reaction. If a catalyst doubles the rate in one direction, it also doubles the rate in the other direction.

10-2.4 Examples of Catalysts

In all cases of catalysis, the catalyst acts by inserting intermediate steps in a reaction, steps that would not occur without the catalyst. *The catalyst itself must be regenerated in a subsequent step.* (An added substance that is continuously used by a reaction is a *reactant,* not a catalyst.) An example is the catalytic action of acid on the decomposition of formic acid (HCOOH). Figure 10-10 shows a model of formic acid. To the carbon atom is attached a hydrogen atom, an oxygen atom, and an OH group.

Figure 10-11 shows how this molecule might decompose. If the hydrogen atom attached to carbon migrates over to the OH group, the carbon-oxygen bond can break to give a molecule of water and a molecule of carbon monoxide. This migration, shown in the center of the drawing, requires a large amount of energy. There is a high activation energy. Hence, the reaction below occurs very slowly.

$$\text{HCOOH} \longrightarrow \text{H}_2\text{O} + \text{CO} \quad (20)$$

If sulfuric acid (H_2SO_4) is added to an aqueous solution of formic acid, carbon monoxide bubbles out rapidly. This also occurs if phos-

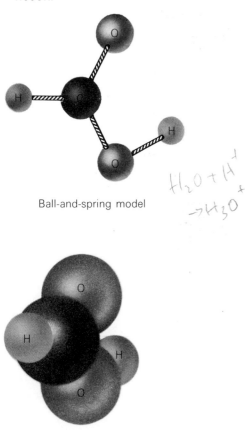

Fig. 10-10 **A model of formic acid, HCOOH.**

Ball-and-spring model

Space-filling model

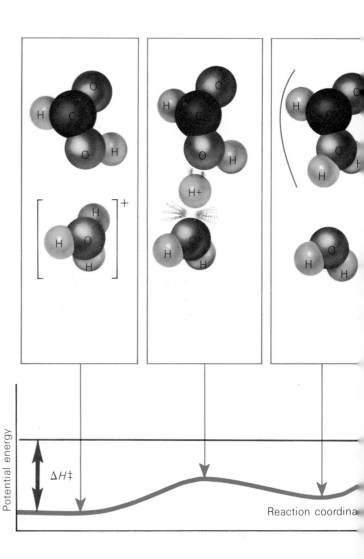

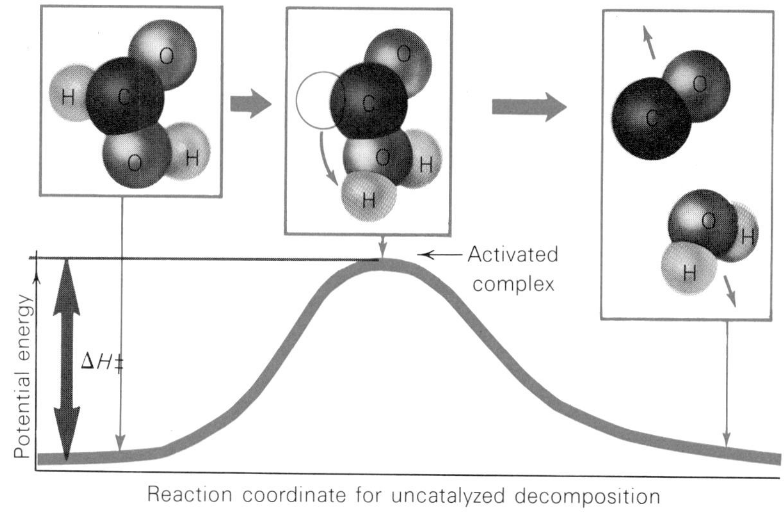

EXAMPLES of CATALYSTS

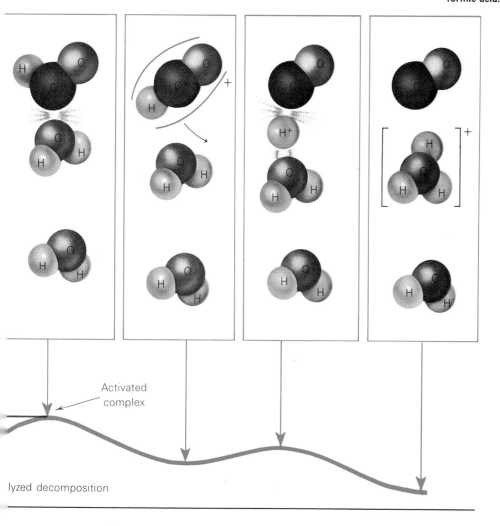

Fig. 10-11 **Potential energy diagram for the uncatalyzed decomposition of formic acid.**

Fig. 10-12 **Potential energy diagram for the catalyzed decomposition of formic acid.**

phoric acid (H_3PO_4) is added instead. The common factor in both cases is that both H_2SO_4 and H_3PO_4 give the ion $H^+(aq)$. On the other hand, careful analysis shows that the concentration of $H^+(aq)$ is constant during the rapid decomposition of formic acid; $H^+(aq)$ is not used up in the reaction. Evidently, hydrogen ion acts as a catalyst in this decomposition. In the presence of $H^+(aq)$, a new reaction path is available. The new reaction mechanism begins with the addition of a hydrogen ion to formic acid, as shown in Figure 10-12. Thus, the catalyst is consumed at first, forming a new species, $(HCOOH_2)^+$. In this new species, one of the carbon-oxygen bonds is weakened. With only a small expenditure of energy, the next reaction shown in Figure 10-12 can occur, producing $(HCO)^+$ and H_2O. Finally, $(HCO)^+$ decomposes to produce CO and H^+. This last reaction of the sequence regenerates the catalyst, $H^+(aq)$.

Each of the steps in this new reaction mechanism is governed by the same principles that govern a simple reaction. Each reaction has an activation energy. The overall reaction has a potential energy diagram that is merely a composite of the simple energy curves of each of the succeeding steps.

The highest energy required in this new reaction path is only 18 kcal, much lower than the activation energy shown in Figure 10-11 for the uncatalyzed reaction. The rate of decomposition, then, is much faster when acid is present.

Notice that the catalyst does not *cause* the reaction. A catalyst can *speed up* a reaction which might be very slow in its absence.

In some cases, the catalyst is a solid substance on whose surface a reactant molecule can be held (adsorbed) in a position favorable for reaction. When a molecule of another reactant reaches the same point on the solid, reaction occurs. Metals such as iron, nickel, platinum, and palladium seem to act in this way in reactions involving gases. There is evidence that in some cases bonds of reactant particles are weakened or actually broken. Such a particle will then react more easily with another reactant particle.

A very large number of catalysts, called enzymes, are found in living tissues. Among the best known examples of these are the digestive enzymes, such as the ptyalin in saliva and the pepsin in gastric juice. A common function of these two enzymes is to hasten the breakdown of large molecules, such as starch and protein, into simpler molecules that can be utilized by body cells. In addition to the relatively small number of digestive enzymes, many other enzymes are involved in biochemical processes.

The specific methods by which catalysts work are not clearly understood in most cases. Finding a catalyst suitable for a given reaction usually requires a long period of laboratory experimentation. Yet, we can look forward to the time when a catalyst can be tailor-made to fit a particular need. This exciting prospect accounts for the great activity on this chemical frontier.

10-3 THE RATE OF NUCLEAR PROCESSES

The rate of a nuclear process, like the rate of a chemical process, is described as the number of particles that undergo change per unit of time. Let us consider the decay of radium-226. The equation for this process is

$$^{226}_{88}Ra \longrightarrow ^{222}_{86}Rn + ^{4}_{2}He \qquad (21)$$

The amount of $^4_2$He produced per hour is proportional to the number of radium-226 atoms present. We can write the equation

rate of helium-4 production = $k \cdot$ number of radium-226 nuclei

or

$$\text{rate of } ^4_2\text{He formation} = k \cdot ^{226}_{88}\text{Ra} \qquad (22)$$

where k is the proportionality constant and $^{226}_{88}$Ra is the number of radium-226 nuclei present. The equation simply states that in a given time period, let us say one hour, a given fraction of the radium-226 nuclei will undergo decomposition. If we double the number of nuclei present, the rate is doubled. As we noted earlier, the time required for one half of the nuclei of a given radioactive isotope to decompose will be characteristic of that isotope. It is known as the half-life of the isotope (see page 88).

Scientists who first studied the rate of nuclear processes were surprised to find that the rate of a nuclear process is not altered by a change in temperature. We may well wonder: If the decay process follows the same concentration-dependence as does a typical chemical reaction, why does it not follow the same temperature-dependence? Why does an increase in temperature not accelerate a nuclear process? The answer lies in the energetics of nuclear processes. We noted in Chapter 9 that energy terms associated with a nuclear reaction are usually millions of times larger than energy terms associated with a chemical process. As you will recall, this is the very reason why the world has such a strong interest in "atomic energy." It is not unreasonable to guess that nuclear processes would also have activation energies millions of times larger than chemical activation energies. In fact, we have evidence to prove that this is so. You recall that fusion reactions can only be initiated at temperatures of 10 million degrees to 100 million degrees. This means that a very high activation energy is involved. In view of this fact, it is not surprising that temperature differences of 100°C or even 500°C have no observable effect on a nuclear process.

10-4 HIGHLIGHTS

Factors determining reaction rate are the kinds of reactants, temperature, and the concentration of the reactants. The rate expression *cannot* be obtained from the equation for the process, only from experiment. The rate expression can be used to suggest a **reaction mechanism.**

Reactions occur when molecules collide. Not all collisions result in reaction; only those collisions of molecules with total energy above the **activation energy** and with proper orientation give rise to reaction.

Catalysts increase the reaction rate by making possible a new mechanism with a lower activation energy. Biological enzymes are catalysts for life processes.

QUESTIONS AND PROBLEMS

1. The rate of movement of an automobile can be expressed in the unit miles per hour. In what units would you discuss the rate of (a) movement of movie film through a projector, (b) rotation of a motor shaft, (c) gain of altitude, (d) consumption of milk by a family, (e) production of automobiles by an auto-assembly plant.

2. Describe two homogeneous and two heterogeneous systems that are not described in the text.

3. Explain why an explosion is possible whenever a large amount of dry, powdered, combustible material is distributed as dust in the air.

4. Explain (at the molecular level) why an increase in concentration of a reactant may cause an increase in rate of reaction.

5. Consider two gases, A and B, in a container at room temperature. These gases are known to react by a mechanism in which one molecule of A and one molecule of B collide in the slow step. What effect will the following changes have on the rate of the reaction between these gases: (a) the pressure is doubled by cutting the volume in half, (b) the number of molecules of gas A is doubled at constant volume, and (c) the temperature is decreased at constant volume?

6. In an important industrial process for producing ammonia (the Haber Process) the overall reaction is

$$N_2(g) + 3H_2(g) \longrightarrow 2NH_3(g) + 24{,}000 \text{ cal}$$

A yield of about 98 percent can be obtained at 200°C and 1,000 atm. The process uses a catalyst which is usually finely divided, mixed iron oxides combined chemically with small amounts of potassium oxide (K_2O) and aluminum oxide (Al_2O_3). (a) Is this reaction exothermic or endothermic? (b) Suggest a reason for the fact that this reaction is generally carried out at a temperature of 500°C and at 350 atm, in spite of the fact that the yield under these circumstances is only about 30 percent. (c) What is the ΔH for the reaction in kilocalories per mole of $NH_3(g)$? (d) How many grams of hydrogen must react to form 1.60 moles of ammonia?

7. Give several ways by which the rate of combustion in a candle flame might be increased. State why the rate would be increased.

8. State three methods by which the pressure of a gaseous system can be increased. How would each method influence the rate of reaction of the gas in the system? Assume that the system is a mixture of A and B and that the rate-determining step is

$$A(g) + B(g) \longrightarrow AB(g)$$

9. Do you expect the reaction

$$C_2H_4(g) + 3O_2(g) \longrightarrow 2CO_2(g) + 2H_2O(g)$$

to represent the mechanism by which ethylene (C_2H_4) burns? Why?

10. A group of students is preparing a ten-page directory. The pages have been printed and are stacked in ten piles, page by page. The pages must be: (1) assembled in order, (2) straightened, and (3) stapled in sets. If three students work together, each performing a different operation, which might be the rate-controlling step? What would be the effect on the overall rate if the first step were changed by ten helpers joining the individual assembling the sheets? What if these ten helpers joined the student working on the second step? the third step?

11. An increase in temperature of 10°C rarely doubles the kinetic energy of particles and hence the number of collisions is not doubled. Yet, this temperature increase may be enough to double the rate of a slow reaction. How can this be explained?

12. In a collision of particles, what is the primary factor that determines whether a reaction will occur?

13. In Figure 10-7, why is kinetic energy decreasing as NO_2 and CO go up the left side of the barrier and why is kinetic energy increasing as CO_2 and NO go down the right side? Explain in terms of conservation of energy and also in terms of what is occurring to the various particles in relation to each other.

14. Phosphorus (P_4) exposed to air burns spontaneously to give P_4O_{10}. The ΔH of this reaction is -712 kcal/mole of P_4. (a) Draw an energy diagram for the net reaction, explaining the critical parts of the curve. (b) How much heat is produced when 12.4 g of phosphorus burns?

15. Considering that so little energy is required to convert graphite to diamond (recall question 8 in Chapter 9), how do you account for the great difficulty found in the industrial process for accomplishing this?

16. Why does a burning match light a candle?

17. Draw an energy diagram for the reaction

$$C(s) + O_2(g) \longrightarrow CO_2(g)$$

(a) when the C is in large chunks of coal. (b) Is the curve changed if very fine carbon powder is used?

18. Sketch a potential energy diagram that might represent an endothermic reaction. (Label parts

of the curve representing activated complex, activation energy, and net energy absorbed.)

19 Why is it difficult to "hardboil" an egg at the top of Pike's Peak? Is it also difficult to cook scrambled eggs there? Explain. (Remember the effect of pressure on boiling temperature.)

20 Give two factors that would increase the rate of a reaction and explain why these do increase the rate.

21 The mechanism proposed for a reaction is:

(1) $A + B \longrightarrow C$ SLOW
(2) $C + 2A \longrightarrow D + E$ FAST
(3) $E \longrightarrow F$ FAST

(a) What is the expression for the rate of the reaction? (b) What is the overall equation for the reaction?

22 Man needs pressures of 350 atm and temperatures of 200°C to make NH_3, yet bacteria in a cold field on a rainy day can operate at (5–25)°C and 0.8 atm of N_2 to synthesize amino acids, which breakdown easily to give NH_3. How do bacteria do it? (If you have the answer, let the Nobel Prize Committee know. They might be interested in you!)

23 Consider the reaction between H_2 and Cl_2. The rate is proportional to the concentration of H_2 and the concentration of Cl_2. What would happen to the rate of reaction of H_2 and Cl_2 if the volume were doubled at constant temperature? Assume total amount of each gas constant right before and right after change in volume.

24 Highly concentrated hydrogen peroxide (96%) can be prepared if the sample is kept extremely pure. Why is it necessary to exclude impurities?

... by ... equilibrium, we mean a state in which the properties of a system, as experimentally measured, would suffer no further observable change even after the lapse of an indefinite period of time. It is not intimated that the individual particles are unchanging.

G. N. LEWIS (1875–1946) and MERLE RANDALL (1888–1950)

CHAPTER 11 Equilibrium in Phase Changes and in Chemical Reactions

The traffic on the road at the top is at equilibrium; the traffic on the road at the bottom is not at equilibrium.

Everyone has had the feeling of being pulled in two directions at once and having to settle for some sort of compromise between alternatives. Strangely enough, chemical reactions sometimes have the same problem. They do not always march neatly down the path of the equation, converting reactants to products in a tidy manner.

Fortunately, the driving forces in chemistry are better understood than those in life situations. One is the tendency to exist in the lowest energy state possible, which is easy to understand by anyone who has frequently fallen *down* but never *up*. The second is the universal tendency toward maximum disorder. Disorder is a much more natural state than order. This may give you some much-needed ammunition the next time you are told to clean up your room, but don't count on it.

In chemistry, the proper combination of these two forces gives rise to the condition known as chemical EQUILIBRIUM.

In Chapter 10 we discussed the rate of the reaction between CO and NO_2,

$$CO(g) + NO_2(g) \longrightarrow CO_2(g) + NO(g) \qquad (1)$$

and the rate of the reverse reaction between CO_2 and NO,

$$CO_2(g) + NO(g) \longrightarrow CO(g) + NO_2(g) \qquad (2)$$

It is only fair to ask: Which reaction *really* goes? In this chapter we shall investigate this and related questions. We shall try to set up procedures for deciding what happens when various reagents are mixed.

If we start by mixing NO_2 and CO, the red-brown color caused by NO_2 fades as the reddish NO_2 is converted into colorless NO. Clearly, reaction (*1*) giving colorless NO is taking place. The equation suggests that the color will continue to fade until the system is colorless and all NO_2 has been used up, but this does *not* happen. Instead, the system reaches a light, red-brown color and the color changes no further. Reaction seems to have stopped; relative amounts of NO_2 and NO are no longer changing. In Chapter 5 we said that a liquid and its vapor are at *equilibrium* when the vapor pressure of the liquid stops changing. If the same criterion—*i.e.*, that the system is not undergoing observable change—is applied here, we would say that our chemical system has reached a state of equilibrium.

Many questions cross our minds.

(1) How do we recognize equilibrium?
(2) What are the molecules doing at equilibrium?
(3) How can we change the equilibrium condition?
(4) How do we define the composition of the gas mixture at equilibrium?
(5) Can equilibrium be treated in a quantitative manner?

Our work is cut out for us as we seek answers to these questions.

11-1 EQUILIBRIUM IN PHASE CHANGES

Let us begin our study by reviewing the use of the word *equilibrium* as we applied it to phase changes.

11-1.1 Vapor-Liquid Equilibrium—Vapor Pressure

In Chapter 5 we constructed a model to explain the vapor pressure of a liquid. We visualized a system in which some molecules leave the liquid and enter the vapor phase while other molecules in the vapor phase return to the liquid surface. We defined equilibrium as the condition existing in this system when the *rate* at which molecules leave the surface is just equal to the *rate* at which they return. By this definition vigorous molecular activity goes on *even after all external signs of change have disappeared.* But we cannot *see* molecules; nor can we measure their rate of leaving or entering the liquid. How do we recognize the equilibrium condition?

Fig. 11-1 Exchange of molecules between liquid and gas. (a) At equilibrium—rate of evaporation is equal to rate of condensation. (b) When the partial pressure of vapor is *below* the vapor pressure—rate of evaporation above rate of condensation. (c) When the partial pressure of vapor is above the vapor pressure—rate of condensation above rate of evaporation.

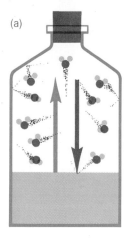

(a)

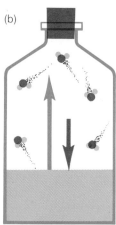

(b)

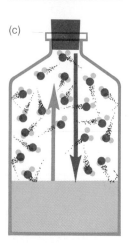

(c)

In our earlier study, we agreed that when the vapor pressure of the liquid becomes constant over a long period of time, when there is no observable change in the system, the system is at equilibrium. In short, equilibrium is characterized by *constancy of macroscopic properties.* For systems at equilibrium, those properties that are observable for weighable amounts of matter do *not* change.

The statement is true, but constancy of properties can be difficult to recognize, particularly when a system is changing very slowly. For this reason we shall return frequently to the question of how to recognize the equilibrium condition. It is a question worthy of our continued attention.

Now let us review the molecular model a little more closely. In Figure 11-1(a), a perfect balance between rates of evaporation and condensation is observed. The system is at equilibrium and no external changes can be seen. The partial pressure of the vapor is *equal to* the equilibrium value. What happens if the partial pressure of the vapor is *less than* the equilibrium value? As Figure 11-1(b) shows, the rate of evaporation exceeds the rate of condensation. As the concentration of molecules in the vapor phase builds up, the rate of condensation builds up; finally, the rate of condensation becomes equal to the rate of evaporation. At that point equilibrium, represented by Figure 11-1(a), is reached. Suppose the equilibrium is disturbed in another way. If we inject an excess of vapor into the bottle or compress the vapor above the liquid, the concentration of molecules in the vapor phase suddenly rises and the rate of condensation increases [Figure 11-1(c)]. As the excess of vapor molecules returns to the liquid phase, the rate of condensation falls off until finally *the rate of condensation equals the rate of evaporation*—the system is at equilibrium and the vapor pressure is the same as in Figure 11-1(a). We repeat: At equilibrium, *microscopic processes continue* but in a balance that yields *no macroscopic* (or large-scale) *changes.*

If we now look closely at the method we used to recognize equilibrium in these cases, we see that we have observed a constancy of properties with time [Figure 11-1(a)]. This is our standard procedure for defining the equilibrium condition. In addition, we have disturbed the equilibrium system and approached it from two dif-

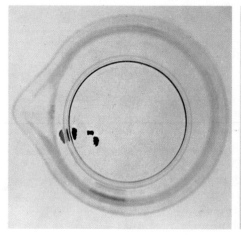

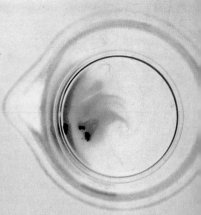

ferent sides. In Figure 11-1(b), the vapor pressure will build *up* to the equilibrium value from a pressure which is *below* the vapor pressure of the system. In Figure 11-1(c) the vapor pressure will drop to the equilibrium value from a pressure which is *above* the vapor pressure of the system. In all three cases the same equilibrium value will be obtained. <u>*If the same pressure is reached when the system approaches equilibrium from a pressure either above or below the equilibrium vapor pressure, the vapor pressure obtained is the equilibrium value.*</u> (Temperature is assumed constant throughout.)

The above discussion answers several of our original questions about equilibrium, inasmuch as the questions apply to a vapor-liquid system. We know what the molecules are doing and we know how to recognize equilibrium. Other questions remain. For example: How can we change the equilibrium condition? We learned in Chapter 5 that we can change the vapor pressure above a given liquid by changing the temperature. As we raise the temperature, the vapor pressure rises; as we lower the temperature, the vapor pressure falls. *Temperature* is an important variable in controlling the equilibrium condition. The vapor pressure is determined by the nature of the liquid and the temperature. The *nature of the liquid* (or *of the system*) is obviously an important factor in determining the equilibrium condition.

11-1.2 Solubility—Solid-Solution Equilibrium

In Figure 11-2 we see the sequence of events that occur if several crystals of iodine are dissolved in a mixture of water and alcohol. At first the liquid is colorless, but very quickly a reddish color appears near the solid. Stirring the liquid causes swirls of the reddish color to move out; solid iodine is dissolving to become part of the liquid. Changes are evident: the liquid takes on an increasing color and the pieces of solid iodine diminish in size as time passes. Finally, however, the color stops changing. Solid remains, but the pieces no longer diminish in size. By our earlier operational definition of equilibrium—that state at which all outward signs of change have disappeared—the solution would appear to be in equilibrium with

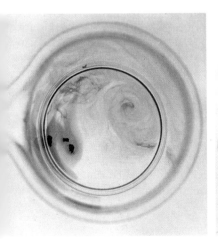
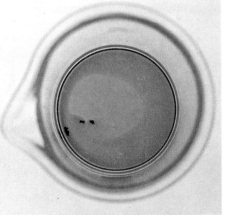

Fig. 11-2 Iodine dissolving in an alcohol-water mixture. Equilibrium is recognized by constant color of the solution.

the solid iodine crystals. Again, in this system equilibrium is characterized by *a constancy of macroscopic* (or large-scale) *properties.*

What does this process look like on the molecular level? Iodine molecules leave the surface of the crystal at a given rate as the crystals are first placed in the water-alcohol solution. The rate at which molecules return to the surface is low at first, since there are very few molecules of I_2 in solution (no color) and there is very low probability that those just leaving will return immediately. As time passes, however, more and more I_2 molecules enter the solution phase and there is a greater probability that a molecule in the solution will encounter the surface of an iodine crystal and become bonded to the crystal. As a result, the *rate of crystallization** increases until finally it is just *equal to* the *rate of solution.* The system is then at equilibrium. No net change is apparent even though action on the molecular level is very vigorous.

In Chapter 5 we found that solubilities ranged from very low (silver chloride in water) to very high (sugar in water). Clearly, the *type of solid and liquid* will be important in determining the concentration at which solubility equilibrium is established. Further, we found that solubilities vary with temperature. *Temperature* is also important in determining the equilibrium concentration.

Our analysis of physical systems at equilibrium has given us a start toward answering our questions about chemical equilibrium. Let us review these answers.

(1) *How do we recognize equilibrium?* Equilibrium is recognized by the constancy of macroscopic properties in a closed system at a uniform temperature. It is important that we specify a *closed* system. The evaporation of water from an open dish does *not* represent equilibrium even though the *rate* of evaporation is *constant.* Further, we found that the equilibrium condition is the same regardless of the side from which it is approached. This knowledge helps greatly in recognizing equilibrium.

(2) *What are the molecules doing at equilibrium?* The molecules participate in both forward and reverse reactions. *At equilibrium the rate of the forward reaction is equal to the rate of the reverse reaction.*

(3) *How can we change the equilibrium condition?* So far, we have identified temperature and the nature of the reactants as two of the important variables determining the equilibrium condition.

The remaining questions, (4) and (5), will take on more significance as we proceed with a discussion of chemical equilibrium.

11-2   EQUILIBRIUM IN CHEMICAL SYSTEMS

11-2.1   The N_2O_4-NO_2 Equilibrium

Let us examine some chemical systems and see if our observations on physical equilibrium can be carried over to chemical systems. Suppose we fill two identical bulbs with equal amounts of red-brown

---

* **Rate of crystallization** is the rate at which molecules or ions bond to crystals.

Fig. 11-3 Nitrogen dioxide gas at different temperatures. Bulb A: N_2O_4 at 0°C. (almost colorless). Bulb B: NO_2 at 100°C. (reddish-brown).

nitrogen dioxide (NO_2). Now let us immerse the first bulb (bulb A) in an ice bath and the second bulb (bulb B) in boiling water (see Figure 11-3). The bulb at 0°C (ice bath) becomes nearly colorless while the bulb at 100°C (boiling water) turns deep red-brown. The deep red-brown color suggests a high concentration of NO_2 molecules, but why is the bulb at 0°C nearly colorless? If we determine the molecular weight of the colorless vapor or nearly colorless liquid present at low temperature, we find that it has *twice* the molecular weight of NO_2; the colorless compound has the formula $(NO_2)_2$ or N_2O_4. The N_2O_4 molecules absorb no visible light and are therefore colorless. The NO_2 molecules absorb some visible light and therefore have a red-brown color.

Now let us transfer these two bulbs to a bath at room temperature as shown in Figure 11-4. Immediately bulb A (cold bulb) becomes a darker brown while bulb B (hot bulb) becomes lighter in color. Our observations can be summarized by two equations:

BULB A (COLD BULB WARMING UP)
$$N_2O_4(g) \longrightarrow 2NO_2(g) \qquad (3)$$

BULB B (WARM BULB COOLING OFF)
$$2NO_2(g) \longrightarrow N_2O_4(g) \qquad (4)$$

Finally, as the two bulbs approach the same temperature, the colors stop changing. The bulbs are now *identical* in color; both have the same shade of brown—the system is at chemical equilibrium. If we now heat the bulb that was cold and cool the bulb that was hot, we can repeat the cycle.

We have watched the two bulbs approach the equilibrium condition from opposite directions. As bulb A was warmed, equilibrium was approached by the dissociation of N_2O_4 to give NO_2; as bulb B was cooled, equilibrium was approached by NO_2 molecules combining to give N_2O_4. If we join the two equations, we can write

$$N_2O_4(g) \longrightarrow 2NO_2(g) \qquad (3)$$

$$N_2O_4(g) \longleftarrow 2NO_2(g) \qquad (4)$$

$$\overline{N_2O_4(g) \rightleftarrows 2NO_2(g)} \qquad (5)$$

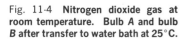
COLORLESS   RED-BROWN

The double arrows indicate that both *dissociation* and *formation* of N_2O_4 are taking place and, hence, chemical equilibrium prevails.* Just as with phase changes at equilibrium, the *microscopic* processes of chemical reactions at equilibrium continue at the same rates, but no *macroscopic* changes are observed. As we noted earlier, equilibrium is approached very rapidly in some systems, while in others it may be approached so slowly that years are required to reach the equilibrium condition. In the latter case, it is difficult to be sure that a system is really at equilibrium because the change is so very slow. How can we be sure, then, that a chemical system is really at equilibrium and is not just changing very slowly? One of the best ways to be sure that we have a true equilibrium condition is to show that we can get the same equilibrium point from either the reactants

* In some cases an equal sign ($=$) replaces the double arrow to indicate equilibrium.

Fig. 11-4 Nitrogen dioxide gas at room temperature. Bulb A and bulb B after transfer to water bath at 25°C.

or the products—that is, we can arrive at the same equilibrium value by beginning from either side. This is what we have done for the N_2O_4-NO_2 system.

11-2.2 The $CaCO_3$-CO_2 Equilibrium

This point can be illustrated further by considering the formation of quicklime or calcium oxide (CaO). Millions of tons of CaO are made in the United States each year for plaster, mortar, and agriculture. If limestone, known chemically as calcium carbonate ($CaCO_3$), is heated in a blast of hot air (800°C), CaO and CO_2 are formed:

$$CaCO_3(s) \xrightarrow{\text{heat}} CaO(s) + CO_2(g) \quad (6)$$

Calcium carbonate breaks down into solid CaO and gaseous CO_2. Since the $CO_2(g)$ is carried away from CaO(s), their recombination to give $CaCO_3$ is prevented. In this dynamic system there is little opportunity for the reverse reaction:

$$CaO(s) + CO_2(g) \longrightarrow CaCO_3(s) \quad (7)$$

The system *cannot* reach equilibrium (see Figure 11-5.1).

The commercial production of CaO is based on the prompt removal of CO_2 from the system. Since fresh $CaCO_3$ is fed in at the bottom of the furnace and newly generated CaO is removed at the top, while CO_2 is swept away rapidly, we might think that the contents of the reactor do not change with time. Is this equilibrium? *No!* We say that the system has reached a **steady state,** but this is *not* equilibrium. It is clear that our steady-state system depends upon the addition of fresh $CaCO_3$ at a constant rate (by a belt on the bottom) and the removal of the resulting CaO at the same constant rate (from the top). If we look beyond the reactor and watch the shrinking pile of $CaCO_3$ and the growing pile of CaO, it is now clear that macroscopic or visible changes are taking place. Even the apparent constancy of conditions in the reactor is a result of adding $CaCO_3$ and removing CaO at identical rates (measured in moles per minute).

How does equilibrium differ from a steady state for a dynamic process? If this system is to reach equilibrium, the reactor must be *closed* so that CO_2 can no longer escape. A suitable vessel is shown in Figure 11-5.2. $CaCO_3$ is placed in the vessel, which has an attached pressure gauge. The air in the vessel is pumped out; then the gas exit valve is closed. If the $CaCO_3$ and container are now heated to 800°C, the pressure of CO_2 inside the vessel builds up until the pressure gauge reads 190 mm. As long as the temperature is held constant at 800°C, the pressure will remain steady at 190 mm. The system appears to be at equilibrium since no changes in macroscopic properties can be detected and the system is closed.

If we wanted to be sure that the system is at equilibrium when the pressure is at 190 mm, we could carry out the reverse reaction in the closed vessel. CaO would be placed in the reactor; air would be removed; then CO_2 would be pumped into the system until the pressure read well above 190 mm. Let us, for example, add CO_2 until

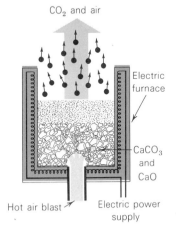

Fig. 11-5.1 *Open system not at equilibrium.* Calcium carbonate heated in stream of hot air. CO_2 generated is swept out of vessel. Recombination of CaO and CO_2 prevented.

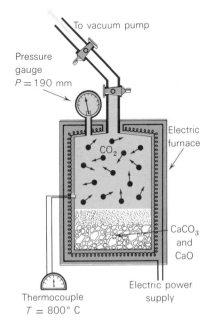

Fig. 11-5.2 *Closed system at equilibrium.* Calcium carbonate heated in evacuated and closed vessel. Pressure reaches a constant value at a given temperature.

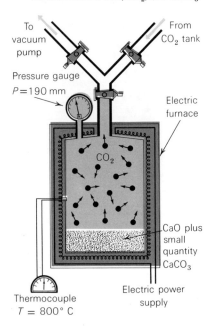

Fig. 11-5.3 CO_2 added to CaO in closed system. Equilibrium established between CaO, CO_2, and $CaCO_3$.

the pressure is 400 mm. (The exact value is not important.) The valve is now closed to seal the system and the temperature is raised. Pressure begins to fall and finally at 800°C, the pressure gauge again becomes steady at 190 mm. *The same equilibrium pressure of CO_2 has been reached starting with either $CaCO_3$ or with CaO plus CO_2* (Figure 11-5.3).

Both the decomposition and formation of $CaCO_3$ are taking place in the container:

$$CaCO_3(s) \longrightarrow CaO(s) + CO_2(g) \qquad (6)$$

$$CaCO_3(s) \longleftarrow CaO(s) + CO_2(g) \qquad (7)$$

$$CaCO_3(s) \rightleftharpoons CaO(s) + CO_2(g) \qquad (8)$$

The system is at equilibrium at 800°C when the pressure of CO_2 above the $CaCO_3$ is 190 mm. This is truly an equilibrium, *not* a steady-state, system: the system is closed; no macroscopic changes, such as growing piles of CaO, are observed; and the constancy of the system is *not* dependent upon the addition or subtraction of reagents at a constant rate. Rates of both forward and reverse reactions are controlled by the environmental and physical conditions such as temperature and nature of the system.

EXERCISE 11-1

Which of the following systems constitute steady-state situations, and which are at equilibrium? For each, a constant property is indicated.

(1) An open pan of water is boiling on a stove. The temperature of the water is constant.
(2) A balloon contains air and a few drops of water. The pressure in the balloon is constant.
(3) A colony of ants is going about its daily routine in an anthill. The population of the anthill is constant.
(4) A Bunsen burner burns in the laboratory to give a well-defined flame. Supplies of gas and air are constant.

11-2.3 The Equilibrium Condition and the Meaning of an Equation

Much of our work in chemical equilibrium utilizes chemical equations. Let us review what is meant by an equation such as the one we discussed in Section 11-2.1:

$$N_2O_4(g) \rightleftharpoons 2NO_2(g) \qquad (5)$$

The equation tells us that *one* molecule of N_2O_4 can decompose to give two molecules of NO_2. It also tells us that two molecules of NO_2 can combine to give one molecule of N_2O_4. The equation does *not* tell us anything about the *relative amounts* of NO_2 and N_2O_4 present in a vessel at equilibrium! It definitely does *not* say that at equilibrium there will be 2 moles of NO_2 for each mole of N_2O_4.

The foregoing statements can be illustrated by a commonplace example. Consider the "reaction" that takes place at a "mixer" at the beginning of any school year. One of the equations for this system might be

$$\text{boy}(s) + \text{girl}(s) \rightleftarrows \text{dancing couple}(s) \qquad (9)$$

What does this equation tell us? It tells us that a boy (if he knows what is good for him) asks one and only one girl to dance at any one time. It tells us that a girl (if she is a real lady) accepts the invitation of one and only one boy at any one time. And it tells us that a dancing couple consists of one boy and one girl. It does *not* say that at any dance there will be equal numbers of boys and girls. It does *not* say that the number of dancing couples will always be equal to the number of nondancing boys in the room. In short, our equation tells us nothing about the **equilibrium condition**—the number of dancing couples and the number of nondancing boys and girls. Experimental conditions such as the quality of the music, the size of the dance hall, the temperature in the room, the age of the group, and even the weather outside must be carefully defined before we can speak with any confidence about the equilibrium condition in the system.

Let us consider a chemical system comparable to the boy-plus-girl reaction—namely, phosphorus trichloride (PCl_3) plus chlorine (Cl_2) to give phosphorus pentachloride (PCl_5). The equation is

$$\text{boy}(s) + \text{girl}(s) \longrightarrow \text{dancing couple}(s) \qquad (9)$$

$$PCl_3(g) + Cl_2(g) \longrightarrow PCl_5(g) \qquad (10)$$

We still have no idea how completely PCl_3 and Cl_2 combine to give PCl_5 at equilibrium. All we know is that 1 mole of phosphorus trichloride will combine with 1 mole of chlorine to give 1 mole of phosphorus pentachloride. All reagents and products are gases at 250°C. However, if we start with 1 mole of PCl_3 and 1 mole of Cl_2 in a vessel under a pressure of 1 atm and a temperature of 250°C, we may bring the system to equilibrium; then we can measure the moles of free PCl_3, the moles of free Cl_2, and the moles of free PCl_5 present *at equilibrium*. At 250°C and 1 atm pressure, such measurements will show 0.71 mole of PCl_3, 0.71 mole of Cl_2, and 0.29 mole of PCl_5. If this is a true equilibrium, we shall obtain exactly the same mixture by starting with 1 mole of PCl_5 instead of 1 mole of PCl_3 and 1 mole of Cl_2. We do, in fact, observe this. All these observations may be summarized by a table:

	$PCl_3(g)$ +	$Cl_2(g)$ $\rightleftarrows$	$PCl_5(g)$
↓ Initial moles	1.0	1.0	0.0
Moles at equilibrium	0.71	0.71	0.29
↑ Initial moles	0.0	0.0	1.0

The composition of the equilibrium mixture is *not* dependent upon the direction from which equilibrium is approached. Further, the equation does *not* indicate relative concentrations at equilibrium.

11-3 ALTERING THE EQUILIBRIUM CONDITION OR THE STATE OF EQUILIBRIUM

We have seen that qualitatively the state of equilibrium for a system is characterized by the relative amounts of products and reactants present. With reference to the decomposition of PCl_5, any change in conditions causing a larger percentage of the PCl_5 to dissociate at equilibrium would change the state of equilibrium for the reaction in favor of the formation of more PCl_3 and Cl_2.

$$PCl_5(g) \rightleftharpoons PCl_3(g) + Cl_2(g) \qquad (10a)$$

What conditions might alter the equilibrium state? *Concentration* and *temperature*, factors that affect the rate of reaction! Equilibrium is attained when the rate of opposing reactions become equal. Any condition that changes the rate of one of the reactions involved in the equilibrium may affect the conditions at equilibrium.

11-3.1 Concentration

Consider the reaction between ferric ion (Fe^{3+}) and thiocyanate ion (SCN^-) which was demonstrated in the laboratory:

$$Fe^{3+}(aq) + SCN^-(aq) \rightleftharpoons FeSCN^{2+}(aq) \qquad (11)$$

We have visual evidence of concentration at equilibrium, since the intensity of the color is fixed by the concentration of the $FeSCN^{2+}$ ion. Increasing either the ferric ion—by adding a soluble salt such as ferric nitrate, $Fe(NO_3)_3$—or thiocyanate ion—by adding, say, sodium thiocyanate, NaSCN—changes the concentration of one of the reactants (see Figure 11-6). Immediately, the color of the solution darkens, indicating an increase in the amount of the colored ion $FeSCN^{2+}$. *The equilibrium concentrations are affected if the concentrations of reactants (or products) are altered.*

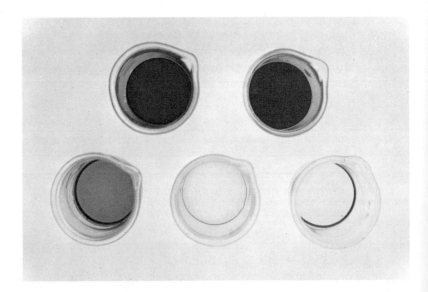

Fig. 11-6 Equilibrium concentrations are affected by the reactant concentrations. As the concentration of $FeSCN^{2+}$ decreases the color of the solution lightens. Top row, left, 0.0010 M KSCN, 0.10 M $Fe(NO_3)_3$; right, 0.0010 M KSCN, 0.040 M $Fe(NO_3)_3$. Bottom row, left, 0.0010 M KSCN, 0.016 M $Fe(NO_3)_3$; center, 0.0010 M KSCN, 0.0064 M $Fe(NO_3)_3$; right, 0.0010 M KSCN, 0.0026 M $Fe(NO_3)_3$.

EXERCISE 11-2

Does the "dancing-couple equilibrium" respond to concentration changes in the same way? What happens if a large group of boys enters the dance hall?

11-3.2 Temperature

We have already considered an example of a change in an equilibrium system as temperature is altered. The relative amounts of NO_2 and N_2O_4 are readily and obviously affected by a change in temperature. *The equilibrium concentrations are affected if the temperature is altered.*

11-3.3 Catalysts Do *Not* Alter Equilibrium

Although catalysts increase the rate of reactions, it is found experimentally that *adding a catalyst to a system at equilibrium does not alter the equilibrium state.* Hence, it must be true that any catalyst has the same effect on the rate of the forward and the rate of the reverse reactions. You will recall that the effect of a catalyst on reaction rates can be discussed in terms of lowering the activation energy. This lowering is effective in increasing the rate in both forward and reverse directions. Thus, a catalyst produces no *net* change in the equilibrium concentrations, even though the system may reach equilibrium much more rapidly.

11-4 ATTAINMENT OF EQUILIBRIUM

We have already noted that some reactions reach equilibrium slowly and that this fact may make it difficult for us to recognize equilibrium. It is extremely important for us to recognize the distinction between a system *at* equilibrium and a system *approaching* equilibrium extremely slowly. In both cases, we may see little in the way of visible change. Consider, for example, the reaction between H_2 and O_2:

$$2H_2(g) + O_2(g) \rightleftharpoons 2H_2O(g) + 115.6 \text{ kcal}$$
$$\Delta H = -115.6 \text{ kcal} \quad (12)$$

There is a large quantity of heat given off. Experiments such as those described for PCl_5 show that as temperature is raised to a very high level (3000°C), a small amount (0.6 percent) of water decomposes to give H_2 and O_2; as temperature is *lowered*, essentially all the H_2 and O_2 combine to give H_2O. Yet a mixture of pure hydrogen and oxygen can remain at room temperature for a long period without apparent reaction or change. *Equilibrium is not attained in this system because hydrogen and oxygen react very slowly at room temperature.* This statement can be checked easily. Let us pass a spark through the H_2 and O_2 mixture. The resulting explosion is dramatic proof that something has happened. The system went rapidly to equilibrium once the process was properly initiated by a

LARS ONSAGER (1903–)

A native of Norway, Lars Onsager received his chemical engineering degree from the Norwegian Technical Institute in 1925. He came to the United States in 1928, and studied at Johns Hopkins and Brown Universities before going to Yale University in 1933. Onsager received the Ph.D. degree in 1935 and has been a member of the faculty of Yale since. In 1931 he published the work for which he received the 1968 Nobel Prize in Chemistry.

Onsager's interest has centered on the thermodynamics or energy relationships of chemical reactions. He has developed a set of equations expressing the heat relationships between various forms of activity (such as voltage and temperature) in reactions *before* equilibrium is reached, or when equilibrium is disturbed. His theory was so advanced that science did not comprehend its significance until after World War II. Onsager has made many other contributions to chemistry, including the development of the gaseous diffusion process which is used to separate different isotopes of the same element.

spark. Apparently, nothing was observed earlier because the initial process was slow.

This distinction between the conditions in a chemical system at equilibrium and the rate at which equilibrium is attained is very important in chemistry. By arguments that we shall consider, a chemist can decide with confidence whether equilibrium favors reactants or products or neither. He cannot predict, however, how rapidly a system will approach equilibrium. This is a matter of reaction rates, and the chemist must perform separate experiments to learn whether a given reaction is rapid or slow.

11-5 PREDICTING NEW EQUILIBRIUM CONCENTRATIONS—LE CHATELIER'S PRINCIPLE

We are not satisfied with the conclusion that certain changes affect the equilibrium concentrations. We would also like to predict the *direction* of the effect (does it favor products or reactants?) and the *magnitude* of the effect (how much does it favor products or reactants?). The first desire, to know the qualitative effects, is satisfied by a generalization first proposed by a French chemist, Henry Louis Le Chatelier (1850–1936) and now called **Le Chatelier's principle**.

Le Chatelier looked for regularities in a large amount of experimental data concerning equilibria. Summarizing the regularities he found, he made this generalization: *If an equilibrium system is subjected to a change, processes occur that tend to counteract partially the imposed change.* To make the meaning of this generalization or principle crystal-clear, we shall restate it as it applies to changes in temperature, concentration, and pressure.

11-5.1 Temperature and Le Chatelier's Principle

Le Chatelier's principle, as it applies to temperature is: *If the temperature of a system at equilibrium is raised by adding thermal energy, that net reaction will take place which will absorb energy. If the temperature of a system at equilibrium is lowered by removing energy, that net reaction will take place which will liberate energy.*

How does this generalization apply to a specific system? Consider the reaction of N_2O_4 to give NO_2:

$$14.1 \text{ kcal} + N_2O_4(g) \rightleftharpoons 2NO_2(g)$$
$$\Delta H = +14.1 \text{ kcal} \quad (13)$$

Our experimental observations indicated that warming a bulb containing NO_2 and N_2O_4 causes a shift of the equilibrium state in favor of the formation of NO_2 (the reddish-brown color deepened). It is easy to see that this is in accord with Le Chatelier's principle. A rise in temperature gives more NO_2. The formation of NO_2 from N_2O_4 *absorbs* heat. An increase in temperature of the equilibrium system results in the formation of more NO_2 by the reaction that absorbs heat. The reverse process, the formation of more N_2O_4 as the temperature is lowered, liberates heat. This is also in accord with the principle of Le Chatelier.

Our arguments are also in accord with the more general statement of Le Chatelier's principle. A rise in temperature is caused by an input of energy. At the higher temperature the equilibrium is changed to form more NO_2. The formation of NO_2 *absorbs* a portion of the energy that caused the temperature rise, hence counteracting partially the imposed change.

Raising the temperature of liquid water raises its vapor pressure. This is also in accord with Le Chatelier's principle, since heat is absorbed as the liquid vaporizes. This absorption of heat, which accompanies the change to the new equilibrium conditions, partially counteracts the temperature rise causing the change. This is seen in the equation

$$H_2O(l) + 9.7 \text{ kcal} \longrightarrow H_2O(g) \tag{14}$$

As heat is added to a closed system, more $H_2O(g)$ is formed *and* the temperature goes up!

11-5.2 Concentration Changes and Le Chatelier's Principle

Le Chatelier's principle can be stated specifically for changes in concentration. *If a reactant or product is added to a system at equilibrium, that reaction will occur which tends to use up the added reactant or product. If a reactant or product is removed, that reaction will occur which tends to generate more of the removed reactant or product.*

We have considered this principle in laboratory experiments. If a soluble thiocyanate salt is added to an equilibrium solution containing both $Fe^{3+}(aq)$ and $SCN^-(aq)$ ions, the color of the solution deepens:

$$Fe^{3+}(aq) + SCN^-(aq) \rightleftharpoons FeSCN^{2+}(aq) \tag{11}$$

COLORLESS   RED

A new state of equilibrium is then attained in which more $FeSCN^{2+}$ is present than before the addition of SCN^-. Increasing the concentration of SCN^- has increased the concentration of the $FeSCN^{2+}$ ion. This agrees with Le Chatelier's principle. The change imposed on the system was an increase in the concentration of SCN^-. This change can be counteracted in part by some Fe^{3+} and SCN^- ions reacting to form more $FeSCN^{2+}$. Some of the added SCN^- was used up! The same argument applies to an addition of ferric ion from a soluble ferric salt. In each case, the formation of $FeSCN^{2+}$ *uses up* a portion of the added reactant, partially counteracting the change.

Le Chatelier's principle can also be applied to gaseous systems. Consider the equilibrium system

$$PCl_3(g) + Cl_2(g) \longrightarrow PCl_5(g) \tag{10}$$

If we were to add more Cl_2 while the volume and temperature of the system were held constant, the pressure would go up, the concentration of Cl_2 would go up, and more $PCl_5(g)$ would be produced. This observation would be in complete accord with Le Chatelier's principle: When a reactant is added, that reaction occurs which uses up the reactant. Similarly, if more PCl_5 were added at fixed volume and temperature, additional PCl$_3$ and Cl$_2$ would be produced. Although pressure would go up in both these systems as more reagents were added, the direction of the equilibrium shift would be determined by the changes in *concentration* resulting from added reagents.

11-5.3 Effect of Volume Changes

Is adding or subtracting reagent the only way to change concentration? If we remember that concentration is number of moles *per unit volume,* we see that changes in the volume can also change the concentrations in a gaseous system. Consider again the PCl$_3$-Cl$_2$-PCl$_5$ equilibrium. Suppose the volume of the equilibrium system were to be reduced suddenly to *half* its original volume. The pressure on this system would go up. (It would be doubled if the system behaved ideally without changing its total number of moles of gas.) The reduction in volume would double the initial concentration of the PCl$_3(g)$ and the PCl$_5(g)$. These concentration changes would then just compensate because they appear on opposite sides of the equation. We have however, an additional concentration change which is not compensated. The change in volume would also double the concentration of Cl$_2$. This additional increase in concentration would have the effect of shifting the equilibrium toward the production of more PCl$_5$. The expected formation of PCl$_5$ does indeed occur. Changes in *volume* of gaseous systems bring about changes in concentrations that then shift the equilibrium.

11-5.4 Pressure As a Variable and Its Relation to Concentration Changes

Changes in volume invariably bring about changes in *pressure* in gaseous systems; hence, people sometimes prefer to consider that pressure is the independent variable. They then consider that pressure changes will shift an equilibrium. *Pressure effects are invariably those which result from changes in concentration and chemical activity.*

It is possible, however, to analyze the foregoing set of changes in the gaseous PCl$_3$-Cl$_2$-PCl$_5$ system in a somewhat different (and perhaps easier) manner if we apply Le Chatelier's principle while considering for convenience that pressure is the independent variable. The answer must be the same as that given for concentration changes, since pressure is really a variable which is proportional to gas concentrations. When we increased the pressure on the system by reducing the volume, the number of moles per unit volume was greater than it had been under the original equilibrium condition. This increase in moles per unit volume could be counteracted in part if some PCl$_3$ and Cl$_2$ were to combine to give PCl$_5$ (if 2 moles were to combine to give 1 mole). Hence, since Le Chatelier's principle tells us that the process which will occur is the one that tends to counteract the imposed change, we expect that PCl$_3$ and Cl$_2$ will combine to give PCl$_5$. The overall conversion of two molecules to one molecule would tend to reduce again the number of molecules per unit volume (for gases this decreases the pressure), thus counteracting in part the effects of increased pressure. Our observations on pressure can be summarized in another special statement of Le Chatelier's principle. If pressure is *increased* on a system at equilibrium, that reaction takes place that tends to give a *decrease* in total volume. In the case just discussed an increase in pressure resulted in the reaction occurring that converts 1 mole of PCl_3 (22.4 liters at STP) and 1 mole of Cl_2 (22.4 liters at STP) to 1 mole of PCl_5 (22.4 liters at STP). The obvious converse of this statement is also true. If pressure is *reduced* on a system at equilibrium, that reaction takes place which tends to give an *increase* in the total volume of the system. A decrease in pressure should favor the conversion of $PCl_5(g)$ to $PCl_3(g)$ and $Cl_2(g)$.

What happens if the reaction does not involve an increase or decrease in the total volume of the system? Our statement would seem to indicate that no change is to be expected. Experiment confirms such a prediction. Consider the first reactions mentioned in this chapter:

$$CO(g) + NO_2(g) \rightleftharpoons CO_2(g) + NO(g) \quad (1, 2)$$

If we increase the pressure on this system by reducing the volume, the concentrations of all four gases would go up, but the composition of the gaseous mixture would not change. The total number of moles of gas is not altered by this chemical reaction; hence, nothing is observed as pressure is increased.

11-6 APPLICATION OF EQUILIBRIUM PRINCIPLES— THE HABER PROCESS

Knowledge of chemical principles pays off! The control of equilibrium by the application of Le Chatelier's principle is literally feeding and clothing a large segment of the population of the world. The large-scale production of ammonia (NH_3) is possible today because we can apply the principles of Le Chatelier. Liquid ammonia for fertilizer, as well as ammonia for the synthesis of nylon, and hundreds of other chemicals can be obtained from the air. Before man learned to control equilibrium, he was dependent upon the erratic whims of nature and international politics for his essential nitrogen compounds.

We observed in Chapter 7 that the nitrogen of the earth's atmosphere is relatively inert at low temperatures. The nitrogen atoms combine so strongly with each other that the N_2 molecule has low reactivity at room temperature. Under proper conditions we can, however, make nitrogen molecules combine with hydrogen molecules to give ammonia:

$$N_2(g) + 3H_2(g) \rightleftharpoons 2NH_3(g) + 22 \text{ kcal}$$
$$\Delta H = -22 \text{ kcal} \quad (15)$$

Can our knowledge of equilibrium and kinetics (Chapter 10) be used to decide what conditions would give the ammonia? First, what about pressure? In the formation of ammonia *one* volume of nitrogen plus *three* volumes of hydrogen give *two* volumes of ammonia (four volumes go to two volumes). Le Chatelier's principle tells us that high pressure (small volume, thus high concentration) should favor the production of ammonia. What about temperature? We note that heat is *given off* when ammonia is formed from N_2 and H_2. Le Chatelier's principle tells us that low temperature will favor a reaction liberating heat; therefore, we want low temperature. It begins to look as though our preferred reactor should be a *refrigerated* tank containing N_2 and H_2 under thousands of pounds of *pressure*. If such a tank is prepared and the N_2-H_2 mixture is allowed to stand for several days, *no NH_3 is found.* If the mixture is allowed to stand for several weeks, *no NH_3 is found.* Even after several years the yield of NH_3 cannot be detected. Apparently N_2 reacts *very* slowly with H_2 at the low temperature which favors production of NH_3. The situation reminds us of the reaction of H_2 and O_2 mentioned earlier. In the H_2-O_2 case, a spark made the reaction proceed explosively. We have no such good luck when we pass a spark through the N_2-H_2 mixture. Other ways must be found to accelerate the process.

We remember from Chapter 10 that reactions can be made to go faster by raising the temperature. We are now face to face with a compromise. Low temperature is required for a desirable equilibrium state and high temperature is necessary for a satisfactory rate. The compromise used industrially involves an intermediate temperature, about 500°C, and even then the success of the process depends upon the presence of a suitable catalyst to achieve a reasonable reaction rate. Very high pressures are desirable, but equipment that will stand up under both high pressure and high temperature is expensive to build. A pressure of about 350 atm is actually used. Under these conditions, 350 atm and 500°C, only about 30 percent of the reactants are converted to NH_3. The NH_3 is removed from the mixture by liquefying it under conditions at which N_2 and H_2 remain as gases. The unreacted N_2 and H_2 are then recycled until the total percentage conversion of expensive hydrogen to ammonia is very high.

Prior to World War I the principal sources of nitrogen compounds were some nitrate deposits in Chile. Fritz Haber, a German chemist, successfully developed the process we have just described, thus allowing the world to use its almost unlimited supply of nitrogen in air. The world eats better and is clothed better because of Fritz Haber.

EXERCISE 11-3

An American corporation is now making nitrogen compounds directly from the reaction of nitrogen with oxygen. The essential equation is

$$N_2(g) + O_2(g) + 21.5 \text{ kcal} \longrightarrow 2NO(g)$$

Using Le Chatelier's principle, select conditions that would give the best yields of NO(g).

11-7 QUANTITATIVE ASPECTS OF EQUILIBRIUM

Le Chatelier's principle permits us to make qualitative predictions about equilibrium states. For example we know that raising the pressure will favor the production of NH_3 from N_2 and H_2. The next question is: By *how much* will an increase in pressure favor

NH₃ formation? Will the yield change by a factor of 10 or by only 0.01 percent? To control a reaction, we need *quantitative* information about equilibrium. Experiments show that quantitative predictions are possible. Furthermore, we are delighted to learn that they can be explained in terms of our models for reacting species on the molecular level.

11-7.1 The Equilibrium Constant

By means of laboratory colorimetric observations (based on estimation of color) you measured the concentration of $FeSCN^{2+}$ in solutions containing ferric and thiocyanate ions, Fe^{3+} and SCN^-. The reaction is

$$Fe^{3+}(aq) + SCN^-(aq) \rightleftharpoons FeSCN^{2+}(aq) \qquad (11)$$

From $[FeSCN^{2+}]$* and the initial values of $[Fe^{3+}]$ and $[SCN^-]$ you calculated the values of $[Fe^{3+}]$ and $[SCN^-]$ at equilibrium. You then made calculations for various combinations of these values. Many experiments just like these show that the ratio

$$\frac{[FeSCN^{2+}]}{[Fe^{3+}][SCN^-]} \qquad (16)$$

comes closest to being a fixed value. Notice that this ratio is the quotient of the equilibrium concentration of the single species produced in the reaction divided by the product of the equilibrium concentrations of the reactants.

Colorimetric analysis is not very exact. Some more accurate data on the H_2-I_2-HI system at equilibrium are shown in Table 11-1. The reaction is

$$2HI(g) \rightleftharpoons H_2(g) + I_2(g) \qquad (17)$$

The data have been expressed in concentrations, although pressure units would be equally good for a reaction involving gases.

* Hereafter we shall regularly use the square bracket notation $[M^+]$ to indicate concentration of the ion $M^+(aq)$ in solution. We read "$[FeSCN^{2+}]$" as *concentration of $FeSCN^{2+}(aq)$ ions in solution.*

TABLE 11-1 EQUILIBRIUM CONCENTRATION AT 698.6°K OF HYDROGEN, IODINE, AND HYDROGEN IODIDE*

Expt. No.	$[H_2]$ (moles/liter)	$[I_2]$ (moles/liter)	$[HI]$ (moles/liter)
1	1.8313×10^{-3}	3.1292×10^{-3}	17.671×10^{-3}
2	2.9070×10^{-3}	1.7069×10^{-3}	16.482×10^{-3}
3	4.5647×10^{-3}	0.7378×10^{-3}	13.544×10^{-3}
4	0.4789×10^{-3}	0.4789×10^{-3}	3.531×10^{-3}
5	1.1409×10^{-3}	1.1409×10^{-3}	8.410×10^{-3}

* Values above the line were obtained by heating hydrogen and iodine together; values below the line, by heating pure hydrogen iodide.

Chapter 11 / EQUILIBRIUM: PHASE CHANGES and CHEMICAL REACTIONS

EXERCISE 11-4

For experiments 4 and 5 in Table 11-1, why is $[H_2]$ equal to $[I_2]$? For experiment 1 in Table 11-1, what were the concentrations of H_2 and I_2 before the reaction occurred to form HI?

TABLE 11-2 VALUES OF $[H_2][I_2]/[HI]$ FOR DATA OF TABLE 11-1

Expt. No.	$\dfrac{[H_2][I_2]}{[HI]}$ (moles/liter)
1	32.429×10^{-5}
2	30.105×10^{-5}
3	24.866×10^{-5}
4	6.495×10^{-5}
5	15.477×10^{-5}

Let us work with these data to compute the value of the ratio

$$\frac{[H_2][I_2]}{[HI]} \qquad (18)$$

We obtain the numbers in Table 11-2. In view of the precision of the data from which these ratios are derived, the ratios are far from constant. Now let us try the ratio

$$\frac{[H_2][I_2]}{[HI]^2} \qquad (19)$$

These calculations are summarized in Table 11-3. The results are most encouraging and imply that with a fair degree of accuracy we can write

$$\frac{[H_2][I_2]}{[HI]^2} = \text{a constant} = 1.835 \times 10^{-2} \text{ at } 698.6°K \qquad (20)$$

Look at this ratio in terms of the reaction

$$2HI(g) \rightleftharpoons H_2(g) + I_2(g) \qquad (17)$$

The ratio is the product of the equilibrium concentrations of the substances produced in the reaction, $[H_2] \times [I_2]$, divided by the *square* of the concentration of the reacting substance, $[HI]^2$. In this ratio, the power to which we raise the concentration of each substance is equal to its coefficient in the reaction.

TABLE 11-3 VALUES OF $[H_2][I_2]/[HI]^2$ FOR DATA OF TABLE 11-1

Expt. No.	$\dfrac{[H_2][I_2]}{[HI]^2}$
1	1.8351×10^{-2}
2	1.8265×10^{-2}
3	1.8359×10^{-2}
4	1.8390×10^{-2}
5	1.8403×10^{-2}
Aver.	1.835×10^{-2}

11-7.2 The Law of Chemical Equilibrium

Let us summarize what we have learned. For the reaction

$$Fe^{3+}(aq) + SCN^-(aq) \rightleftharpoons FeSCN^{2+}(aq) \qquad (11)$$

we found that the concentrations of the molecules involved have a simple relationship:

$$\frac{[FeSCN^{2+}]}{[Fe^{3+}][SCN^-]} = \text{a constant} \qquad (21)$$

Then we considered precise equilibrium data for the reaction

$$2HI(g) \rightleftharpoons H_2(g) + I_2(g) \qquad (17)$$

The concentrations of the molecules in the reaction of 2HI to give H_2 and I_2 (*17*) were found to have a simple relationship:

$$\frac{[H_2][I_2]}{[HI]^2} = \text{a constant} \qquad (20)$$

In each of our simple relationships, (*20*) and (*21*), the concentrations of the products appear in the numerator. In each relationship the concentrations of reactants appear in the denominator. In the reaction for the decomposition of HI (*17*), 2 moles of hydrogen iodide react. This influences the ratio (*20*) because it is necessary to square the concentration of hydrogen iodide, [HI], in order to obtain a constant ratio.

These observations and many others like them lead to the generalization known as the **Law of Chemical Equilibrium**. For a reaction

$$aA + bB \rightleftharpoons eE + fF \quad (22)$$

when equilibrium exists there will be a simple relation between the concentrations of products, $[E]$ and $[F]$, and the concentrations of reactants, $[A]$ and $[B]$:

$$\frac{[E]^e[F]^f}{[A]^a[B]^b} = K = \text{a constant} \quad (\text{temperature constant}) \quad (23)$$

In this generalized equation (*23*), we see again that the numerator is the product of the equilibrium concentrations of the species formed, each raised to the power equal to the number of moles of that species in the chemical equation. The denominator is again the product of the equilibrium concentrations of the reacting species, each raised to a power equal to the number of moles of the species in the chemical equation. The quotient of these two remains constant. The constant K is called the **equilibrium constant**. This generalization is one of the most useful in all chemistry. From the equation for any chemical reaction we can immediately write an expression, in terms of the concentrations of reactants and products, that will be constant at any given temperature. If this constant is known (from measuring concentrations in a particular equilibrium system), then it can be used in calculations for any other equilibrium state of that system at that same temperature.

In Table 11-4 some reactions along with the equilibrium-law relation of concentrations and the numerical values of the equilibrium constants are listed. First, let us verify the forms of the equilibrium-law relation for the concentrations. The very first has an unexpected form. For this reaction,

$$\text{Cu}(s) + 2\text{Ag}^+(aq) \rightleftharpoons \text{Cu}^{2+}(aq) + 2\text{Ag}(s) \quad (24)$$

you do not find

$$\frac{[\text{Cu}^{2+}][\text{Ag}]^2}{[\text{Ag}^+]^2[\text{Cu}]} = K \quad (25)$$

but rather, you find

$$\frac{[\text{Cu}^{2+}]}{[\text{Ag}^+]^2} = K \quad (26)$$

This occurs because the concentrations of solid copper and solid silver are incorporated into the equilibrium constant. The concentration of solid copper (moles of copper per liter) is fixed by the density of the metal—it cannot be altered either by the chemist or by the progress of the reaction. The same is true of the concentration of solid silver. Since neither of these concentrations varies no matter how much solid is added, there is no need to write them each time an equilibrium calculation is made. Equation (26) will suffice.

TABLE 11-4 SOME EQUILIBRIUM CONSTANTS

Reaction	Equilibrium-Law Relation	K at Stated Temperature
$Cu(s) + 2Ag^+(aq) \rightleftharpoons Cu^{2+}(aq) + 2Ag(s)$	$K = \dfrac{[Cu^{2+}]}{[Ag^+]^2}$	2×10^{15} at 25°C
$Ag^+(aq) + 2NH_3(aq) \rightleftharpoons Ag(NH_3)_2^+(aq)$	$K = \dfrac{[Ag(NH_3)_2^+]}{[Ag^+][NH_3]^2}$	1.7×10^7 at 25°C
$N_2O_4(g) \rightleftharpoons 2NO_2(g)$	$K = \dfrac{[NO_2]^2}{[N_2O_4]}$	0.87 at 55°C
$2HI(g) \rightleftharpoons H_2(g) + I_2(g)$	$K = \dfrac{[H_2][I_2]}{[HI]^2}$	0.018 at 423°C
$HSO_4^-(aq) \rightleftharpoons H^+(aq) + SO_4^{2-}(aq)$	$K = \dfrac{[H^+][SO_4^{2-}]}{[HSO_4^-]}$	0.013 at 25°C
$CH_3COOH(aq) \rightleftharpoons H^+(aq) + CH_3COO^-(aq)$	$K = \dfrac{[H^+][CH_3COO^-]}{[CH_3COOH]}$	1.8×10^{-5} at 25°C
$AgCl(s) \rightleftharpoons Ag^+(aq) + Cl^-(aq)$	$K = [Ag^+][Cl^-]$	1.7×10^{-10} at 25°C
$H_2O \rightleftharpoons H^+(aq) + OH^-(aq)$	$K = [H^+][OH^-]$	10^{-14} at 25°C
$AgI(s) \rightleftharpoons Ag^+(aq) + I^-(aq)$	$K = [Ag^+][I^-]$	10^{-16} at 25°C

EXERCISE 11-5

If we assign the equilibrium constant K' to expression (25) and K to expression (26),

$$K' = \frac{[Cu^{2+}][Ag]^2}{[Ag^+]^2[Cu]} \qquad K = \frac{[Cu^{2+}]}{[Ag^+]^2}$$

show that

$$K = K' \times \frac{[Cu]}{[Ag]^2}$$

Another equilibrium constant of unexpected form applies to the reaction

$$H_2O \rightleftharpoons H^+(aq) + OH^-(aq) \tag{27}$$

For this reaction we might have written

$$\frac{[H^+][OH^-]}{[H_2O]} = K \qquad (28)$$

Instead, Table 11-4 lists the equation as

$$[H^+][OH^-] = K \qquad (29)$$

The concentration of water does not appear in the denominator of expression (29). Water is frequently omitted in treating aqueous reactions that consume or produce water. It is justified because the variation in the concentration of water during reaction is very small in dilute aqueous solutions. We can treat $[H_2O]$ as a concentration that does not change. Hence, it can be incorporated in the equilibrium constant.

EXERCISE 11-6

Water has a density of 1 g/ml. Calculate the concentration of water (expressed in moles/liter) in pure water. Now calculate the concentration of water in 0.10 M aqueous solution of acetic acid (CH_3COOH), assuming each molecule of CH_3COOH occupies the same volume as one molecule of H_2O.

In summary, the concentrations of solids and the concentrations of solvent (usually water) can be and usually are incorporated in the equilibrium constant, so they do not appear in the equilibrium-law relation.

Now look at the numerical values of the equilibrium constants in Table 11-4. The K's listed range from 10^{+15} to 10^{-16}, so we see there is a wide variation. We want to acquire a sense of the relation between the size of the equilibrium constant and the state of equilibrium. *A large value of K must mean that in the system at equilibrium there are much larger concentrations of products than of reactants.* Remember that the numerator of our equilibrium expression contains the concentrations of the products of the reaction. The value of 2×10^{15} for the K for the reaction $Cu(s) + Ag^+(aq)$ certainly indicates that if a reaction is initiated by placing metallic copper in a solution containing Ag^+—for example, in silver nitrate solution—silver metal will plate out. When equilibrium is finally reached, the concentration of copper ion, $[Cu^{2+}]$, will be very much greater than the square of the concentration of silver ion, $[Ag^+]^2$.

A small value of K for a given reaction implies that very little of the products have to be formed from the reactants before equilibrium is attained. The value of $K = 10^{-16}$ for the reaction

$$AgI(s) \rightleftharpoons Ag^+(aq) + I^-(aq) \qquad (30)$$

indicates that very little solid AgI can dissolve before equilibrium concentrations are attained. Silver iodide has extremely low solubility. Conversely, if 0.1 M solutions of KI and $AgNO_3$ are mixed, the values of $[Ag^+]$ and $[I^-]$ are large; an equilibrium state cannot be reached until the [Ag$^+$] and [I$^-$] have been greatly reduced by the precipitation of AgI.

11-7.3 The Law of Chemical Equilibrium Derived from Rates of Opposing Reactions

In discussing the equilibrium between a liquid and its vapor, we described the equilibrium state as a dynamic balance between the rate of evaporation and the rate of condensation. An understanding of the law of chemical equilibrium can also be built on this basis if the reaction-rate balance is applied to *each step* in the mechanism for the process. The statement is illustrated in Appendix 10. We cannot, however, use the rate data for only the slow step in a process and arrive at a proper expression for the equilibrium constant.

11-8 THE FACTORS DETERMINING EQUILIBRIUM

At the beginning of this chapter, we raised a number of questions about equilibrium. Some of those questions have been answered. We are now able to recognize equilibrium by the fact that the system does not seem to change. As a double check to differentiate slow change from true equilibrium (remember H$_2$ and N$_2$), we found that the concentrations in the system in true equilibrium are identical (regardless of the side from which equilibrium is approached). We have a picture of equilibrium activity on the molecular level: at equilibrium the rate of the forward reaction is equal to the rate of the reverse reaction. We know how to shift an equilibrium through control of such variables as temperature and concentration. Finally, we know how to express equilibrium relationships in a quantitative fashion through the use of the equilibrium constant.

Although many questions have been answered, the key "why" questions remain. What determines the equilibrium constant? Why does one reaction favor products while another favors reactants? Why does iron react with oxygen to give Fe$_2$O$_3$ while gold does not? These are hard questions, but a great deal of help can be obtained by considering some "common sense" analogies.

11-8.1 Energy and the Equilibrium Condition

We are not surprised, for example, when water or a skier goes downhill. Both move spontaneously from a condition of *higher* potential energy to a condition of *lower* potential energy (the energy of position is less at the bottom of the hill). Similarly, we know in advance that a skateboard will spontaneously carry a rider down a long, smooth hill if the rider can only stay on the board. Our everyday experience tells us that systems tend to move spontaneously from a condition of higher potential energy to a condition of lower potential energy. The potential energy is changed to kinetic energy and ultimately to heat in the process. "Burned" elbows and knees offer grim proof of the conversion of kinetic energy to heat in skateboarding. We even pick up "floor burns" from a fall in the gymnasium.

Fig. 11-7 **Both the skier and the water move spontaneously from a condition of high potential energy to a condition of lower potential energy.**

Chemical systems are not too different from the skateboard or the skis. We agreed in Chapter 9 that each chemical system has a given energy or heat content; furthermore, a portion of this chemical energy is potential energy resulting from the *position* of atoms relative to each other. Chemical systems, like skateboards, tend to move toward a condition of lower energy; further, the chemical energy of chemical systems can be converted to molecular kinetic energy, *i.e.*, heat. These arguments seem to lead from an analogy to a generalization: <u>Chemical systems</u>—like skis and skateboards—<u>*proceed spontaneously from a state of higher energy to a state of lower energy*</u>. The energy involved in the change appears as kinetic energy and ultimately as heat. In short, <u>a reaction will proceed spontaneously if the products will have lower energy than the reactants</u>. This generalization is in accord with much of our experience; it is particularly true for reactions that release a large amount of energy.

Fig. 11-8 **Potential energy is changed to kinetic energy and ultimately to heat as evidenced by burned elbows and knees.**

Unfortunately, this logical generalization based on analogy soon runs into trouble. Like our earlier generalization—All Cylindrical Objects Burn—the generalization about energy conflicts with experiment. Our generalization predicts that only exothermic reactions (reactions liberating heat) will be spontaneous, *yet many endothermic reactions are spontaneous.* We have only to recall the evaporation of a liquid such as water. This is clearly an endothermic process: the system moves spontaneously from a condition of lower potential energy to one of higher potential energy.

Our generalization encounters still another difficulty: *spontaneous chemical reactions do not always go to completion but proceed only until equilibrium is reached.* The system does not always go over completely to the lower energy state but seems to stop at an intermediate equilibrium position. Why?

11-8.2 Randomness and the Equilibrium Condition

Perhaps a factor *in addition to* the energy of the system is important in determining the equilibrium position. What is this additional factor? It is more subtle than energy content as a driving force for a process and is harder to recognize, but it is still very familiar to you from your everyday experience. The new factor is simply the general tendency of objects or particles to get mixed up, to become more *randomly* arranged. One of the most fundamental laws of the universe is that a system tends to become disorganized, randomized, or mixed up, unless work is done on it. Look at your room; unless you, your mother, or someone else does work to keep it neat, it will spontaneously approach a state of maximum disorder as you live in it. If you drop a handful of marbles from the top of a ladder, they will scatter all over the floor; they become randomly distributed. A small child, turned loose in a room, tends to "randomize" the contents.

This *tendency toward randomness* is a characteristic of chemical systems as well as of real-life situations. If a tank of nitrogen is connected to a tank of oxygen of equal pressure and the valve between

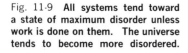

Fig. 11-9 **All systems tend toward a state of maximum disorder unless work is done on them. The universe tends to become more disordered.**

the two is opened, the two gases will mix randomly until a uniform mixture of nitrogen and oxygen is obtained. The mixing process goes on spontaneously because of the random motion of the molecules. Similarly, if we place a few drops of methyl violet solution in a beaker of water, the color will spread gradually throughout the beaker. Because molecular motion is increased by an increase in temperature, the system approaches the uniform color more rapidly as the temperature is raised. Further, we note that there is no tendency for the reverse process to occur spontaneously unless work is done on the system. We never see the nitrogen molecules going back into one tank and the oxygen molecules going back into the other tank spontaneously. Machines and work must be used to separate the components of the nitrogen-oxygen mixture; the process is *not* spontaneous. Similarly, methyl violet molecules, once distributed in the water, will have no tendency to reform crystals. Crystals will form only if external conditions are changed and work is done.

Fig. 11-10 A few drops of methyl violet in a beaker of water tend to become randomly distributed.

It appears that the natural tendency of the universe to approach a state of maximum disorder can be important in determining the equilibrium condition in a system. The degree of disorder is so important in science that it has been given a special name. A property known as **entropy** measures the randomness of the system on the atomic-molecular level. <u>*As the degree of atomic-molecular disorder increases, the entropy of the system increases*</u>. Indeed, the world and even the universe are proceeding toward a condition of complete disorder. When the heat of the universe becomes randomly distributed throughout the universe, all life will cease. This has led some astronomers to refer to our "dying universe." Fortunately, the universe is dying so slowly that mankind has little to worry about. Political and social problems are far more threatening.

How does entropy or disorder apply to an equilibrium system? Let us take our simplest example, the vaporization of a liquid. We notice that the vaporization of a liquid is both *spontaneous* and *endothermic*. Our "super-rubber" ball model for liquids indicated that the balls (or molecules) attract one another when close together in the liquid phase. Separating these balls and moving them into the vapor phase required work; each molecule had to be pushed out of the liquid phase and into the vapor phase, where molecules had higher potential energy. The energy to vaporize the liquid came from the kinetic energy of the molecules; the faster-moving molecules either flew off into the vapor phase or knocked other molecules into the vapor by collision. We noticed that if heat were not supplied from the outside, the liquid became colder; kinetic energy of the faster-moving liquid molecules was converted to the higher potential energy of gaseous molecules. Heat of vaporization had to be supplied to keep the temperature constant.

What is the real driving force in this vaporization process? It is clearly the tendency of the molecules to seek a more random distribution throughout all possible space—*i.e.*, a tendency of molecules to go into the space above the liquid and gain more space in which to move. In our vaporization example, the tendency to seek a random distribution is *opposed* by the energy effect; the process is endothermic. In other cases, however, both energy effects and entropy effects (randomness) combine to drive a process. Since the entropy effects are dependent upon the motion of molecules, we would expect entropy to become a more important factor in the equilibrium as the temperature is raised; more molecules will have the energy to move into the randomized vapor state as the temperature is raised. *Entropy effects become more important as temperature is raised.*

Perhaps a real-life example will be useful in summarizing our equilibrium discussion. Suppose water is poured into a blender. If

Fig. 11-11 Water goes from a condition of higher potential energy (before pouring) to one of lower potential (at the bottom of the blender) after pouring. The kinetic energy supplied to the water by the blender motor at a given speed corresponds to the kinetic energy molecules possess at a given temperature. Speeding up the blender motor corresponds to an increase of temperature and potential energy. A greater degree of randomness is achieved as the water moves into the space above the original liquid surface. Note that at higher speeds water splashes out onto the table. The system is much more randomized.

the blender is not running—corresponding to a very low temperature in nature—the water falls to the bottom. It simply goes from a condition of higher potential energy (from pouring) to lower potential energy. Seeking an energy minimum is the only factor tending to determine the position of the water in the blender: *the water moves to an equilibrium condition of low potential energy.*

If we now start the motor on the blender, some of the liquid splashes about as drops in the space above the liquid. It would be harder to depict the exact distribution of the water in the blender now. The water is more randomly distributed in a larger volume. Because of the kinetic energy supplied by the blender, the liquid is able to assume a more randomized orientation. Furthermore, since some of the liquid is higher in the air, the random orientation corresponds to a higher potential energy for the system. The kinetic energy supplied by the blender motor corresponds to kinetic energy obtained by molecules from an increase in temperature. In liquid-to-vapor phase changes, molecules are vaporized to achieve randomness.

As the motor is speeded up (higher temperatures for the liquid-vapor system), the liquid is distributed in even more space. It splashes higher in the blender and may ultimately splash out of the blender onto the curtains if a lid is not put on quickly. The blender motor provides kinetic energy comparable to that measured by the temperature of a system. When the blender motor is on (higher temperatures), the system achieves greater randomness and a higher potential energy. When the blender motor is off (absolute zero), randomness, or the possibility of randomness, is no longer important. The water simply moves toward a position of minimum potential energy.

Let us summarize the generalizations our model suggests in relation to chemical equilibrium:

(1) All systems tend to approach the equilibrium state.
(2) The first factor important in determining the equilibrium state of a system is its *energy content.* As far as the energy criterion alone is concerned, systems tend to move toward a state of minimum potential energy. Water, for example, runs downhill and strongly exothermic reactions are spontaneous.
(3) The second factor important in determining the equilibrium state is *the degree of randomness or the entropy* of the system. In general, systems tend to move spontaneously toward a state of maximum randomness or disorder—toward maximum entropy.
(4) *The equilibrium state is a compromise between these two factors— minimum energy and maximum randomness, or, in the more sophisticated language of the laboratory, minimum enthalpy and maximum entropy.* At very low temperatures, energy tends to be the more important factor (blender motor off); then equilibrium favors the substance with lowest heat content or lowest energy. At very high temperatures, randomness becomes more important (blender motor on high); then equilibrium favors a random distribution of reactants and products without regard to energy differences.

11-9 HIGHLIGHTS

The physical processes of liquid vaporization and solid solubility reach a state of equilibrium when the rate of the forward reaction is equal to the rate of the reverse reaction in each case. *At equilibrium, no macroscopic changes are visible.* Chemical systems also reach a condition of equilibrium *when the rate of the forward reaction is equal to the rate of the reverse reaction.* Again no macroscopic changes are visible.

Equilibrium states can be altered as predicted by **Le Chatelier's principle:** *If an equilibrium system is subjected to a change, processes occur that tend to counteract partially the imposed change.* Temperature and concentration are the two most important variables involved in determining the equilibrium condition.

Quantitative equilibrium relationships are obtained by use of the **equilibrium constant.** At any temperature, a fixed ratio relates concentrations of products and reactants. This ratio is known as the equilibrium constant, K. The numerator of the ratio is the product of the concentrations of all products, each product concentration being raised to the power corresponding to the number of the product molecules in the balanced equation. Similarly, the denominator is the product of the concentrations of all reactants, each reactant concentration being raised to the power corresponding to the number of the reactant molecules in the balanced equation.

Finally, the equilibrium state was shown to result from a balance of two factors. The first is the tendency of the system to reach minimum energy; the second is the tendency of the system to reach maximum disorder. In more sophisticated terms we might say that the system tends toward minimum **enthalpy** and maximum **entropy**. At low temperatures the tendency of the system to reach minimum energy predominates. At high temperatures the tendency of the system to reach maximum disorder predominates. At very high temperatures the tendency toward maximum disorder is so large that even the order and organization inherent in molecular formation is eliminated. At the temperature of the sun, molecules break up into atoms.

QUESTIONS AND PROBLEMS

1 Which of the following are equilibrium situations and which are steady-state situations? (a) A playing basketball team and the bench of reserves. The number of players *in the game* and the number *on the bench* are constant. (b) The mercury vapor in a thermometer (the temperature is constant). (c) Grand Coulee Dam and the lake behind it. The water level is constant, though a river flows into the lake. (d) A well-fed lion in a cage. The lion's weight is constant. Justify your choices. [*Answer:* (a) and (b) are equilibrium situations.]

2 Which of the following are equilibrium situations and which are steady-state situations? Justify your choices. (a) A block of wood floating on water. (b) During the noon hour, the water fountain constantly has a line of ten persons. (c) When a capillary tube is dipped in water, water rises in the capillary (because of surface tension) to a height h and remains constant there. (d) The capillary system of (c) considered over such a long period that evaporation of water from the end of the capillary cannot be neglected. (e) At a particular point in the reaction chamber of a jet engine, the gas composition (fuel, air, and products) is constant.

3 One drop of water may or may not establish a state of vapor-pressure equilibrium when placed in a closed bottle. Explain.

4 What do the following experiments (done at 25°C) show about the state of equilibrium? (a) One liter of water is added, a few milliliters at a time, to a kilogram of salt, which only partly dissolves. (b) A large salt shaker containing 1 kg of salt is gradually emptied into 1 liter of water. The same amount of solid dissolves as in (a).

5 What *specifically* is "equal" in a chemical reaction that has attained a state of equilibrium?

6 Why are chemical equilibria referred to as "dynamic"?

7 The following chemical equation represents the reaction between hydrogen and chlorine to form hydrogen chloride:

$$H_2(g) + Cl_2(g) \rightleftharpoons 2HCl(g) + 44.0 \text{ kcal}$$

(a) List four important pieces of information conveyed by this equation. (b) What are three important areas of interest concerning this reaction for which no information is indicated?

8 How does a catalyst affect the equilibrium conditions of a chemical system?

9 In any discussion of chemical equilibrium, why are concentrations always expressed in moles rather than grams per unit volume?

10 The phase change represented by

$$\text{energy} + H_2O(l) \rightleftharpoons H_2O(g)$$

has reached equilibrium in a closed system. (a) What will be the effect of a reduction of volume, thus increasing the pressure? (b) What will be the effect of an increase in temperature? (c) What will be the effect of injecting some steam into the closed system, thus raising the pressure?

11 Methanol (methyl alcohol) is made according to the following net equation:

$$CO(g) + 2H_2(g) \rightleftharpoons CH_3OH(g) + \text{energy}$$

Predict the effect on equilibrium concentrations of an increase in (a) temperature, (b) pressure. [*Answers:* (a) decreases CH_3OH, (b) increases CH_3OH.]

12 Consider the reaction

$$4HCl(g) + O_2(g) \rightleftharpoons 2H_2O(g) + 2Cl_2(g) + 27 \text{ kcal}$$

What effect would the following changes have on the equilibrium concentration of Cl_2? Give your reasons for each answer. (a) increasing the temperature of the reaction vessel (b) decreasing the total pressure (c) increasing the concentration of O_2 (d) increasing the volume of the reaction chamber (e) adding a catalyst.

13 (a) Write the equation for the dissociation of $HI(g)$ into its elements. (b) Will HI dissociate to a greater or a lesser extent as the temperature is increased? [$\Delta H = -6.2$ kcal/mole $HI(g)$.] (c) How many grams of iodine will result if at equilibrium 0.050 mole of HI has dissociated?

14 Consider two separate, closed systems, each at equilibrium: (1) HI and the elements from which it is formed, and (2) H₂S and the elements from which it is formed. What would happen in each if the total pressure were increased? Assume conditions are such that all reactants and products are gases.

15 Each of the following systems has come to equilibrium. What would be the effect on the equilibrium concentration (increase, decrease, no change) of each substance in the system when the listed reagent is added?

Reaction	Added Reagent
(a) $C_2H_6(g) \rightleftharpoons H_2(g) + C_2H_4(g)$	$H_2(g)$
(b) $Cu^{2+}(aq) + 4NH_3(g) \rightleftharpoons Cu(NH_3)_4^{2+}(aq)$	$CuSO_4(s)$
(c) $Ag^+(aq) + Cl^-(aq) \rightleftharpoons AgCl(s)$	$AgCl(s)$
(d) $PbSO_4(s) + H^+(aq) \rightleftharpoons Pb^{2+}(aq) + HSO_4^-(aq)$	$Pb(NO_3)_2$ satd/soln
(e) $CO(g) + \frac{1}{2}O_2(g) \rightleftharpoons CO_2(g) + \text{energy}$	heat

[Answer: (a) C_2H_6 (increase), H_2 (increase), C_2H_4 (decrease).]

16 Nitric oxide (NO) releases 13.5 kcal/mole when it reacts with oxygen to give nitrogen dioxide. Write the equation for this reaction and predict the effect of (1) raising the temperature and (2) increasing the concentration of NO (at a fixed temperature) on (a) the equilibrium concentrations, (b) the numerical value of the equilibrium constant, and (c) the speed of formation of NO₂.

17 Given:

$$SO_2(g) + \tfrac{1}{2}O_2(g) \rightleftharpoons SO_3(g) + 23 \text{ kcal}$$

(a) For this reaction discuss the conditions that favor a high equilibrium concentration of SO₃.
(b) How many grams of oxygen gas are needed to form 1.00 g of SO₃? [Answer to (b): 0.200 g of O₂.]

18 Given:

$$CaCO_3(s) \rightleftharpoons CaO(s) + CO_2(g)$$
CLOSED SYSTEM

At a fixed temperature, what effect would adding more CaCO₃ have on the concentration of CO₂ in the region above the solid phase? Explain.

19 Given:

$$H_2(g) + I_2(g) \rightleftharpoons 2HI(g)$$
CLOSED SYSTEM

At 450°C, $K = 50.0$ for the above reaction. What is the equilibrium constant of the reverse reaction at 450°C?

20 Write the expression indicating the equilibrium-law relations for the following reactions:
(a) $N_2(g) + 3H_2(g) \rightleftharpoons 2NH_3(g)$
(b) $CO(g) + NO_2(g) \rightleftharpoons CO_2(g) + NO(g)$
(c) $Zn(s) + 2Ag^+(aq) \rightleftharpoons Zn^{2+}(aq) + 2Ag(s)$
(d) $PbI_2(s) \rightleftharpoons Pb^{2+}(aq) + 2I^-(aq)$
(e) $CN^-(aq) + H_2O(l) \rightleftharpoons HCN(aq) + OH^-(aq)$

21 Equilibrium constants are given for several systems below. In which case does the reaction as written occur to the greatest extent?

Reaction	K
(a) $CH_3COOH(aq) \rightleftharpoons H^+(aq) + CH_3COO^-(aq)$	1.8×10^{-5}
(b) $CdS(s) \rightleftharpoons Cd^{2+}(aq) + S^{2-}(aq)$	7.1×10^{-28}
(c) $H^+(aq) + HS^-(aq) \rightleftharpoons H_2S(aq)$	1.0×10^7

22 In the reaction

$$2HI(g) \rightleftharpoons H_2(g) + I_2(g)$$

at 448°C the partial pressures of the gas at equilibrium are as follows: [HI] = 4×10^{-3} atm, [H₂] = 7.5×10^{-3} atm, and [I₂] = 4.3×10^{-5} atm. What is the equilibrium constant for this reaction?

23 Reactants A and B are mixed, each at a concentration of 0.80 mole/liter. They react slowly, producing C and D:

$$A + B \rightleftharpoons C + D$$

When equilibrium is reached, the concentration of C is measured and found to be 0.60 mole/liter. Calculate the value of the equilibrium constant. (Answer: $K = 9.0$.)

24 At a given temperature HI is 20 percent dissociated in accordance with the equation

$$2HI \rightleftharpoons H_2 + I_2$$

Determine the number of moles of each component present in the mixture if a 0.5-mole sample of HI is placed in a vessel at this temperature.

25 At a specified concentration the substance AO₂ is 10 percent dissociated at a given temperature as follows:

$$4AO_2 \rightleftharpoons 2A_2O_3 + O_2$$

For these conditions, how many moles of each component will be present in a mixture at equilibrium if 2 moles of A₂O₃ and 1 mole of O₂ are present initially?

26 The water-gas reaction

$$CO_2(g) + H_2(g) \rightleftharpoons CO(g) + H_2O(g)$$

was carried out at 900°C with the following results.

Trial No.	Partial Pressure at Equilibrium (Atm)			
	CO	H$_2$O	CO$_2$	H$_2$
1	0.352	0.352	0.648	0.148
2	0.266	0.266	0.234	0.234
3	0.186	0.686	0.314	0.314

(a) Write the equilibrium constant expression.
(b) Verify that the expression in (a) is a constant, using the data given.

27 Select from each of the following pairs the more random system: (a) a brand new deck of cards arranged according to suit and number and (a′) the same deck of cards after shuffling, (b) a box full of sugar cubes and (b′) sugar cubes thrown on the floor, (c) a hay stack and (c′) stacked firewood.

QUESTIONS and PROBLEMS

28 For each of the following reactions, state: (1) whether tendency toward minimum energy favors reactants or products, and (2) whether tendency toward maximum randomness favors reactants or products.

(a) $H_2O(l) \rightleftharpoons H_2O(s)$ $\Delta H = -1.4$ kcal

(b) $H_2O(l) \rightleftharpoons H_2O(g)$ $\Delta H = +10$ kcal
(c) $CaCO_3(s) + 43$ kcal $\rightleftharpoons CaO(s) + CO_2(g)$
(d) $I_2(s) + 1.6$ kcal $\rightleftharpoons I_2$(in alcohol)
(e) $4Fe(s) + 3O_2(g) \rightleftharpoons 2Fe_2O_3(s) + 400$ kcal

. . . solubility . . . depends fundamentally upon the ease with which . . . two molecular species are able to mix, and if the two species display a certain hostility toward mixing . . . saturation [is] attained at smaller concentration . . . .

JOEL H. HILDEBRAND (1881–)

CHAPTER 12 Solubility Equilibrium

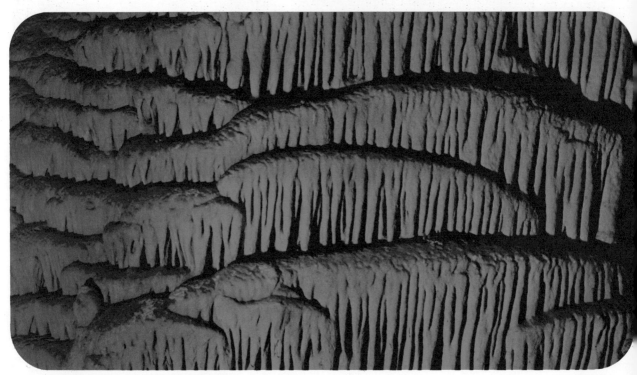

Limestone formations in caves result from the action of equilibrium processes.

Sugar dissolves in water; rock doesn't. In the laboratory you have dissolved lead nitrate and sodium iodide in water to form colorless solutions, yet when you mixed these two solutions the yellow precipitate of lead iodide resulted.

Why do some substances dissolve in water readily, others hardly at all, and still others to a limited extent? Can the factors governing equilibrium be applied to solutions? The answer is an emphatic yes. Differences in solubility offer the chemist tools for identifying chemical substances and separating them from each other. Since chemicals seldom occur in nature in the pure form that we need for our stockroom shelves and industries, application of equilibrium principles to solubility is big business.

In Chapter 11 we saw that a proper application of equilibrium was important in the synthesis of ammonia from the air, a reaction which helps feed the world. Equilibrium would be an important topic even if its usefulness did not extend beyond the ammonia plant; but in actual fact, the principles of equilibrium are important throughout all nature. The solubility of a gas in a liquid, for example, is determined by equilibrium principles, and the solubility of oxygen in water is essential to all fish life. Depletion of oxygen in water is a major problem in lake and stream pollution. The formation of limestone caves, the softening of hard water, the formation of Mammoth Hot Springs in Yellowstone Park, the dating of ancient objects by means of carbon-14 decay, indeed life itself, are all intimately affected by equilibrium processes involving solubility. In this chapter we shall concern ourselves with various **solubility equilibria**—the solution of a gas in a liquid, of a molecular solid in a liquid, and of an ionic solid in a liquid. The general concepts of equilibrium developed in the last chapter can be applied in detail to each of these specific systems.

12-1 SOLUBILITY: A CASE OF EQUILIBRIUM

In the last chapter we developed the idea of an equilibrium constant for a reaction which reaches equilibrium at a constant temperature. For the generalized reaction

$$aA + bB + \cdots \rightleftharpoons eE + fF + \cdots \quad (1)$$

the equilibrium constant can be written

$$K = \text{a constant} = \frac{[E]^e[F]^f \cdots}{[A]^a[B]^b \cdots} \quad (2)$$

Let us see how these generalized concepts can be applied to the dissolving of solid iodine in liquid ethyl alcohol.

12-1.1 The Dissolving of Iodine in Alcohol

You have been told on several occasions that iodine crystals dissolve in liquid ethyl alcohol. The equation representing the process is

$$I_2(solid) \rightleftharpoons I_2(solution) \quad (3)$$

In a saturated solution, a fixed concentration of I_2 molecules will be present in the solution at constant temperature. An application of the general equilibrium law to this system gives

$$K' = \frac{[I_2(solution)]}{[I_2(solid)]} \quad (4)$$

The concentration of iodine molecules in solid iodine (number of moles of iodine per unit volume of solid) *is constant* and is determined by the way iodine molecules are packed in the crystal. It can be calculated from the density of iodine crystals.

The DISSOLVING of
IODINE in ALCOHOL

Fig. 12-1 The growth of formations in limestone caves is a direct product of equilibrium processes.

EXERCISE 12-1

The density of solid iodine is 4.93 g/cm³. How many grams of iodine are present in a liter of actual crystals? How many grams of iodine are present in a mole of I_2 molecules? How many moles of I_2 molecules are present in a liter of iodine crystals?

This value is characteristic of solid iodine and is a constant. If we use our algebra, we can write

$$K'[I_2(solid)] = [I_2(solution)] \quad (5)$$

Since $[I_2(solid)]$ is a constant, we note that $K'[I_2(solid)]$ is also a constant. Let us call it K and write for the iodine solution

$$K = [I_2(solution)] \quad (6)$$

This formalized statement tells us that the concentration of I_2 in a saturated alcoholic solution is *constant* at constant temperature. The statement agrees with our past experience; a given quantity of alcohol *saturated* with *iodine* at a *given temperature* contains a known and reproducible number of moles of iodine. It is also instructive to analyze the solution process in more detail.

12-1.2 The Dynamic Nature of Solubility Equilibrium

The dynamic model for equilibrium indicates that iodine molecules leave the crystal and return to the crystal at the same rate. To understand this dynamic balance, we must consider the factors that determine these two rates.

RATE OF SOLUTION

One of the factors influencing the rate at which a solid dissolves in a liquid is the area of contact, A, between solid and liquid. If many crystals with a large area are dissolving, the rate will be higher than if only a few crystals with small total area are dissolving. *The rate of solution of a solid is proportional to the area of contact between solid and liquid solvent.*

Energy factors are also important. A molecule of I_2 in a crystal, surrounded by other molecules of I_2 in the crystal, is more stable and has lower energy than a molecule of I_2 in the alcohol solution. *The potential energy of an I_2 molecule must rise as it leaves the crystal and enters the solution.* This is true of I_2 molecules in crystals because the interaction of I_2 molecules with each other in the crystal is stronger than the interaction of I_2 molecules with the solvent molecules. We are reminded to some degree of molecules leaving the liquid phase and entering the vapor phase in evaporation. *The rate at which the molecules leave the crystal is proportional to the number of molecules with enough energy to overcome the forces of attraction in the* *solid.* We call this rate k_d. Increasing the temperature increases the *number of molecules* with enough energy to escape. The size of k_d is determined by temperature.

These two factors—area of contact and energy—determine completely the rate of solution:

$$\begin{Bmatrix} \text{rate of} \\ \text{dissolving} \end{Bmatrix} = \begin{Bmatrix} \text{surface} \\ \text{area} \end{Bmatrix} \times \begin{Bmatrix} \text{rate molecules leave} \\ \text{1 cm}^3 \text{ of crystal surface} \end{Bmatrix}$$
$$= A \quad \times \quad k_d \quad (7)$$

RATE OF PRECIPITATION

The rate of precipitation is the rate at which molecules return to the surface and fit into the crystal lattice. To do this, the molecules in solution must strike the crystal surface. Again, the more surface, the more frequently will dissolved molecules encounter the surface. The rate of precipitation is proportional to A, the total surface area of the crystal.

In addition, the rate at which molecules strike the surface depends upon how many molecules there are per unit volume of solution. As the concentration rises, more and more molecules strike the surface per unit time. The rate of precipitation, then, is proportional to the concentration of dissolved iodine, $[I_2]$.

Finally, the remaining factor is the rate at which molecules can pass over the energy barrier and enter the crystal. Thus, we have a rate constant, k_p, that is determined by temperature and the "height" of the energy barrier to precipitation.

Three factors, then, determine the rate of precipitation:

$$\begin{Bmatrix} \text{rate of} \\ \text{precipitation} \end{Bmatrix} = \begin{Bmatrix} \text{surface} \\ \text{area} \end{Bmatrix} \times \begin{Bmatrix} \text{concentration} \\ \text{of dissolved I}_2 \end{Bmatrix} \times \begin{Bmatrix} \text{rate dissolved} \\ \text{molecules pass over} \\ \text{energy barrier and} \\ \text{enter crystal} \end{Bmatrix}$$
$$= A \quad \times \quad [I_2] \quad \times \quad k_p \quad (8)$$

THE DYNAMIC NATURE OF EQUILIBRIUM

At equilibrium, we can equate the rate equations for dissolving and precipitation (see Figure 12-2):

$$\text{rate of dissolving} = \text{rate of precipitation}$$
$$A \times k_d = A \times k_p \times [I_2] \quad (9)$$

The area of contact, A, appears both on the left and on the right sides of the expression. Dividing both sides of (9) by k_p and by A, we obtain

$$\frac{k_d}{k_p} = [I_2] \frac{A}{A} \quad (10)$$

Since $A/A = 1$, we have

$$\frac{k_d}{k_p} = [I_2] \quad (11)$$

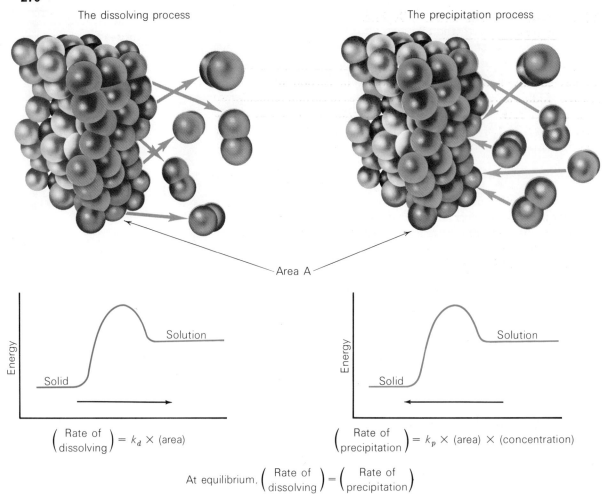

Fig. 12-2 **Solubility equilibrium is dynamic.**

Since k_d and k_p each depend upon temperature, their ratio depends upon temperature. Otherwise, each is constant and their ratio is constant. We can write

$$K = [I_2] \qquad (6)$$

where

$$K = \frac{k_d}{k_p} \qquad (11a)$$

Thus, by taking as our starting point the dynamic balance between the rates of solution and precipitation, we obtain an expression which is identical to that obtained from our more generalized approach to equilibrium. Again, we see that the concentration of I_2 in a saturated alcoholic solution of I_2 will always remain constant at constant temperature.

12-2 THE FACTORS THAT FIX THE SOLUBILITY OF A SOLID

All the foregoing discussion of iodine dissolving in ethyl alcohol applies equally well to iodine dissolving in carbon tetrachloride (CCl_4). Iodine at room temperature dissolves in carbon tetrachloride at a certain rate. At equilibrium, precipitation occurs at a rate exactly equal to the rate of solution. Again, we reach the simple equilibrium expression

$$K = \frac{k_d}{k_p} = [I_2] \qquad (11b)$$

Despite this qualitative similarity in the equation, the solubility of I_2 in CCl_4 is very different from its solubility in alcohol. One liter of alcohol dissolves 0.84 mole of I_2, whereas 1 liter of CCl_4 dissolves only 0.12 mole of I_2 at 20°C:

$$K_{\text{alcohol}} = 0.84 \text{ mole/liter} \qquad (12)$$

$$K_{CCl_4} = 0.12 \text{ mole/liter} \qquad (13)$$

Why are these constants so different? To answer this question, we must turn again to the two factors that control every equilibrium —tendency toward minimum energy and tendency toward maximum randomness.

12-2.1 The Effect of Randomness

In a crystal of iodine, molecules are arranged in a highly regular and ordered pattern. The lattice is so regular that if we know the arrangement of one small segment, known as the unit cell, we can predict the arrangement of atoms in the whole crystal (see Figure 12-3). If we have a perfect crystal at 0°K, where no atomic or molecular vibrations occur (translational kinetic energy of the atoms is zero), the system will have zero entropy—*i.e.*, it will have no randomness, it will be perfectly ordered. As we begin to warm up this crystal, the molecules vibrate back and forth and become disordered. The disorder (or entropy), then, increases as the temperature rises. If the crystal is now put into a liquid, where it dissolves, the beautiful, ordered arrangement of the crystal is destroyed; molecules wander away from the ordered crystal surface; the system becomes *highly randomized* as molecules wander aimlessly through the solution. It is also true that this disorder is greater for dilute than for concentrated solutions. Even the *liquid* arrangement is made more disordered by random injections of foreign molecules into it. <u>The dissolving process *increases* the randomness of the system; hence, it *increases* the entropy of the system.</u> This increase in randomness tends to favor the solution process.

12-2.2 The Effect of Energy

Experiment shows that heat is absorbed as iodine dissolves in alcohol or in carbon tetrachloride. The regular, ideally packed iodine crystal

The EFFECT of ENERGY

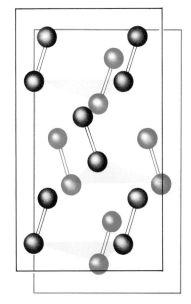

Fig. 12-3 **A unit cell of an I_2 crystal.**

gives an iodine molecule a *lower* potential energy than does the loosely packed solvent environment. Thus, the tendency of a system to seek a minimum energy content (or minimum enthalpy) would favor the growth of iodine crystals.

Our equilibrium condition must now be determined by a balance between the tendency toward maximum randomness (favors solution) and the tendency toward minimum energy (favors crystalline solid). Equilibrium is reached when the concentration is such that these two opposing factors just balance each other.

How much the energy factor favors the crystal depends upon the change in enthalpy or energy content as a mole of solid dissolves (see Figure 12-4). This change is called the **heat of solution.** The heats of solution of iodine in these two solvents have been measured; they are as follows:

$$I_2(s) + 1.6 \text{ kcal} \rightleftharpoons I_2(in\ alcohol) \tag{14}$$

$$I_2(s) + 5.8 \text{ kcal} \rightleftharpoons I_2(in\ CCl_4) \tag{15}$$

Fig. 12-4 **A large energy difference between solid and solution lowers the solubility.**

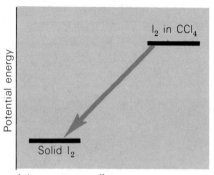
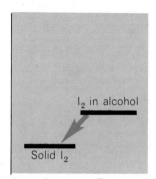

We see that more energy is required to dissolve I_2 in CCl_4 (5.8 kcal) than to dissolve it in alcohol (1.6 kcal). Alcohol molecules interact more vigorously with I_2 molecules than do CCl_4 molecules. We might simply say that I_2 molecules "like" alcohol better than CCl_4 molecules. Yet neither alcohol nor CCl_4 react as vigorously with I_2 as do other I_2 molecules. Thus, the energy factor favoring the crystal is much larger when the solvent is CCl_4 than when it is alcohol. As a consequence, the solubility of I_2 in CCl_4 is not as high as it is in alcohol.

12-2.3 The Effect of Temperature

In the processes considered, it was found that as the temperature increases, more and more of the *molecules* have enough energy to overcome the energy barrier and pass into the more random state (recall vaporization of a liquid). In short, the energy criterion becomes less restrictive and the randomization criterion becomes relatively more important at higher temperatures. This means that higher solubility (increased randomness) is to be expected as the temperature is raised in both alcohol and CCl_4 solutions.

EXERCISE 12-2

The heat of solution of iodine in benzene is +4.2 kcal/mole (heat is absorbed). Assuming that the increase in randomness is the same when equal amounts of I_2 dissolve in equal volumes of either liquid benzene, ethyl alcohol, or CCl_4, justify the prediction that the solubility of I_2 in benzene is higher than in CCl_4 but lower than in alcohol.

The EFFECT of TEMPERATURE

$$I_{2(s)} + 1.6 \text{ kcal} \rightleftharpoons I_2 \text{ (in alcohol)}$$
$$I_{2(s)} + 4.2 \text{ kcal} \rightleftharpoons I_2 \text{ (in benzene)}$$
$$I_{2(s)} + 5.8 \text{ kcal} \rightleftharpoons I_2 \text{ (in } CCl_4\text{)}$$

12-3 THE FACTORS THAT FIX THE SOLUBILITY OF A GAS IN A LIQUID

Gases, too, dissolve in liquids. Let us apply our understanding of equilibrium to this type of system.

12-3.1 The Effect of Randomness

The gaseous state is more random than the liquid state since the molecules move freely through a much larger space as a gas. Hence, randomness *decreases* as a gas dissolves in a liquid. In this case, unlike solids, *the tendency toward maximum randomness favors the gas phase and opposes the dissolving process.*

12-3.2 The Effect of Energy

In a gas the molecules are far apart and they interact very weakly. As a gas molecule enters the liquid, it is attracted to the solvent molecules and the potential energy of the gas molecule is lowered. Again we find a contrast to the behavior of solids. *When a gas dissolves in a liquid, heat is evolved.* The tendency toward minimum energy favors the dissolving process.

Thus, we see that the equilibrium solubility of a gas involves a balance between randomness and energy as it does for a solid, but the effects are opposite. For a gas, the tendency toward maximum randomness favors the gas phase, opposing dissolving. The tendency toward minimum energy favors the liquid state, favoring dissolving.

As an example, consider the solubilities of the two gases oxygen (O_2) and nitrous oxide (N_2O) in water. The heats of solution have been measured and are as follows:

$$O_2(g) \rightleftharpoons O_2(aq) + 3.0 \text{ kcal} \qquad (16)$$
$$N_2O(g) \rightleftharpoons N_2O(aq) + 4.8 \text{ kcal} \qquad (17)$$

Assuming the randomness factor is about the same, the gas with the larger heat effect (favoring dissolving) should have the higher solubility. The measured solubilities at 1 atm pressure and 20°C of O_2 and N_2O in water are, respectively, 1.4×10^{-3} mole/liter and 27×10^{-3} mole/liter. This is consistent with our prediction.

12-3.3 The Effect of Temperature

Raising the temperature always tends to favor the more random state. This means that less gas will dissolve, since the gas is more random than the liquid. The solubility of a gas *decreases* as temperature is raised (see Figure 12-5).

EXERCISE 12-3

From the heat of solution of chlorine in water ($\Delta H = -6.0$ kcal/mole, heat evolved), how would you expect the solubility of chlorine at 1 atm pressure and 20°C to compare with that of oxygen (O_2) and nitrous oxide (N_2O)?

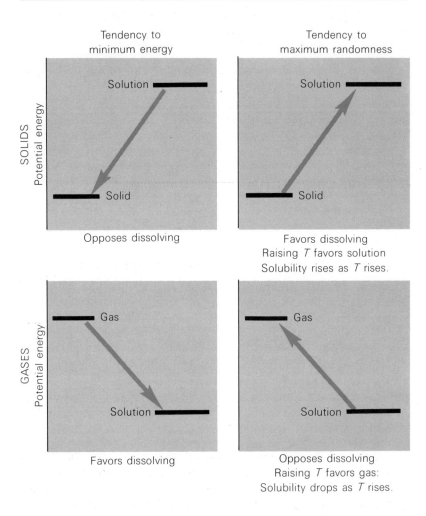

Fig. 12-5 **Maximum randomness versus minimum energy; the solubility of solids and gases.**

12-4 AQUEOUS SOLUTIONS

The expression "$K =$ concentration of solute in solution" is applicable to the solubility equilibria of some substances in water, but not to all. Contrast, for example, water solutions of sugar, salt, and hydrochloric acid. Sugar forms a molecular solid and, as it dissolves in water, the sugar molecules remain intact. These molecules leave the crystal and become a part of the liquid. This is exactly the situation we described for iodine in alcohol, and the expression "$K =$ concentration of sugar in solution" is applicable to aqueous sugar solutions. But sodium chloride (NaCl) behaves quite differently. As we noted in Chapter 6, salt dissolves in water to form positively charged sodium ions and negatively charged chloride ions, each surrounded by water molecules:

$$NaCl(s) \longrightarrow Na^+(aq) + Cl^-(aq) \qquad (18)$$

The existence of these ions made the solution a conductor of electricity. We noticed further that both Na^+ ions and Cl^- ions have the electron configuration of a noble gas. We also noticed in Chapter 6 that hydrogen chloride gas (HCl) dissolves in water to form $H_3O^+(aq)$ and $Cl^-(aq)$ ions. The presence of ions makes the solution a conductor of electricity:

$$HCl(g) + H_2O(l) \longrightarrow H_3O^+(aq) + Cl^-(aq) \qquad (19)$$

You will remember that substances which dissolve in water to give a conductor of the electric current are called *electrolytes*. The equation for the dissolving of an electrolyte is more complex than the equation for the dissolving of a non-electrolyte such as I_2 or sugar. Nevertheless, a careful application of the principles of equilibrium to this more complex process will provide meaningful information about electrolytic solutions.

12-4.1 A Review of Electrolytic Solutions

In Chapters 6, 7, and 8, we discussed the removal of an electron from a metal atom (Na) to form a positive ion (Na^+); we also considered the acceptance of this electron by a nonmetal atom (Cl) to form a negative ion (Cl^-). We noted that these ions may combine to produce an *ionic solid*. It is only natural to expect that such ionic solids might well dissolve to form separate hydrated ions in solution. The equation for the solution of NaCl to form ions, then, is not completely unexpected. Further, we noted that such ionic solids would be expected from compounds formed from metals on the left-hand side of the periodic table (see page 184) and nonmetals on the right-hand side (*e.g.*, NaCl, CsBr, KI, $CaCl_2$). When these solids go into solution, ions are freed.

Conducting solutions are also formed when some substances which are themselves molecular, non-ionic solids, liquids, or gases *react* with water to give ions in solution. The mineral acids are particularly good examples. In addition to hydrogen chloride (HCl), which we know dissolves to give aqueous H_3O^+ and Cl^- ions, other acids such as sulfuric acid (H_2SO_4), nitric acid (HNO_3), and perchloric acid ($HClO_4$) follow this pattern:

$$H_2SO_4(l) + H_2O \longrightarrow H_3O^+(aq) + HSO_4^-(aq) \qquad (20)$$

$$HSO_4^-(aq) + H_2O \longrightarrow H_3O^+(aq) + SO_4^{2-}(aq) \qquad (21)$$

$$HNO_3(l) + H_2O \longrightarrow H_3O^+(aq) + NO_3^-(aq) \qquad (22)$$

$$HClO_4(l) + H_2O \longrightarrow H_3O^+(aq) + ClO_4^-(aq) \qquad (23)$$

Particularly important in our treatment of ionic solutions is a recognition of the fact that a given kind of ion, once it is formed in solution, is the same as other ions of that kind. This is true regardless of the compound from which each ion originated. For example, if a solution is made by dissolving both NaCl and HCl in a single liter of water, the chloride ions are the same *whether they came from NaCl or HCl.* No ion "remembers" the compound from which it originated.

In considering solubility equilibria, it is convenient for us to use special terminology for positive and negative ions. First, we recall that positively charged ions in an electrolysis cell move toward the negative electrode, the cathode, and therefore are known as *cations;* the negatively charged ions move toward the positive electrode, the anode, and are called *anions.* Further, we recall that the number of charges carried by anions in a solution just balances the number of charges carried by cations, so that our solution must be *electrically neutral.* With our memories briefly refreshed about ions in solution, we are ready to consider some qualitative and quantitative relationships involving equilibrium in ionic solutions.

12-4.2 A Qualitative View of Aqueous Solubilities

First, let us consider substances with high solubility. As was stated earlier, chemists consider a substance to be soluble if it dissolves giving a concentration in excess of one tenth of a mole per liter (0.1 M) at room temperature. Using this definition of *solubility,* we can say that a few cations (positive ions) form compounds soluble in water with nearly all anions (negative ions). These cations are the hydrogen ion, $H^+(aq)$, the ammonium ion, NH_4^+, and the alkali ions, Li^+, Na^+, K^+, Rb^+, Cs^+, and Fr^+. Figure 12-6 shows the placement of these ions in the periodic table.

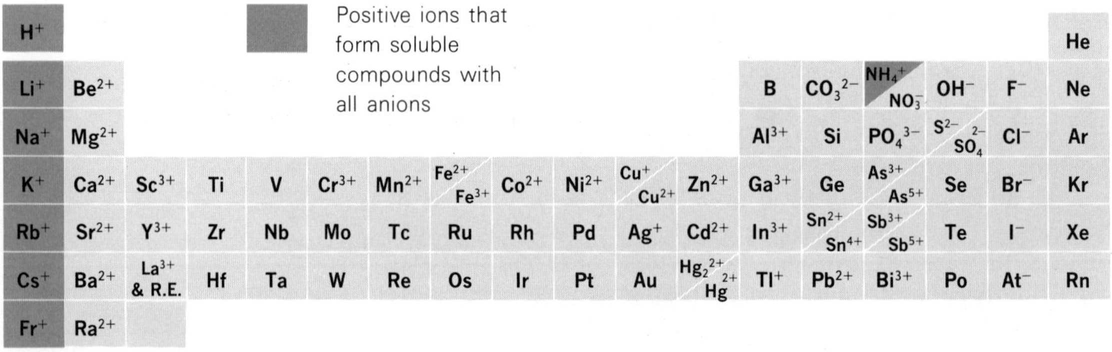

Fig. 12-6 **Positive ions forming soluble compounds with almost all anions.**

The same sort of remark can be made about two anions (negative ions). Almost all compounds involving nitrate ion, NO_3^-, and acetate ion, CH_3COO^-, are soluble in water.*

Other anions (negative ions) form compounds of high solubility in water with some metal cations (positive ions) and compounds of

---

* There are a few compounds of alkalies, nitrate, and acetate that have low solubilities, but most of them are quite complex in composition. For example, sodium uranyl acetate [$NaUO(CH_3COO)$] has low solubility. Silver acetate [$Ag(CH_3COO)$] and chromous acetate [$Cr(CH_3COO)_2$] have low solubilities.

low solubilities with others. Figure 12-7 indicates for five anions the metal ions that form compounds of *low* solubilities. Figure 12-7(a) refers to chlorides, Cl^-, bromides, Br^-, and iodides, I^-. Figure 12-7(b) refers to sulfates, SO_4^{2-}, and Figure 12-7(c), to sulfides, S^{2-}. Notice the difference between Figures 12-6 and 12-7. The dark green shading in Figure 12-6 identifies metal ions that form *soluble* compounds. Figure 12-7 identifies those positive ions that form compounds of *low* solubility.

A QUALITATIVE VIEW of AQUEOUS SOLUBILITIES

Fig. 12-7 **Positive ions forming compounds of low solubilities with anions.**

(a)

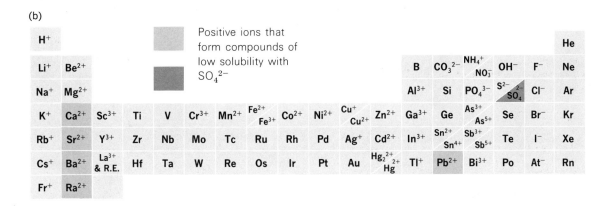

(b)

(c)

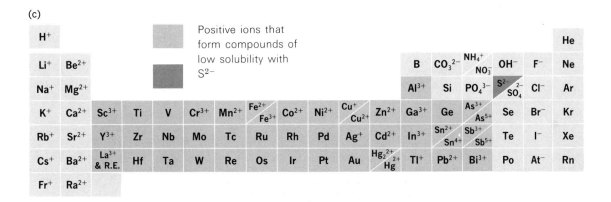

Figure 12-8 continues this pictorial presentation of solubilities. Figure 12-8(a) shows the positive ions that form hydroxides of low solubility. Figure 12-8(b) shows the positive ions that have low solubility when combined with phosphate ion, PO_4^{3-}, carbonate ion, CO_3^{2-}, and sulfite ion, SO_3^{2-}.

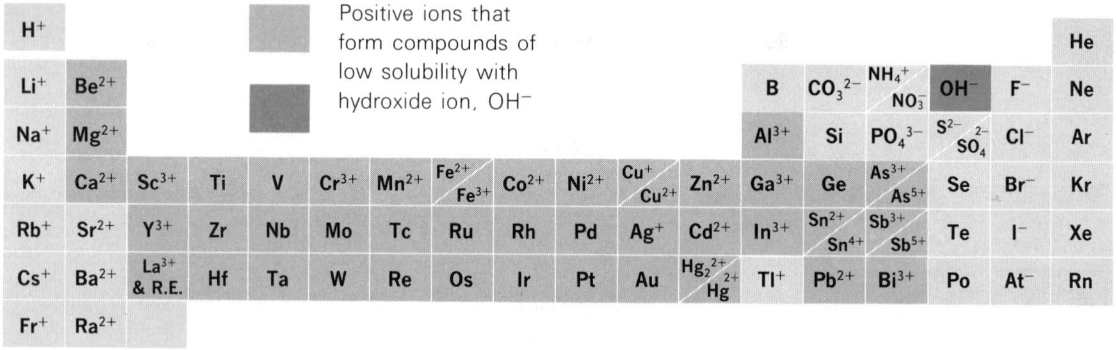

These figures neatly summarize solubility behavior. We see in Figure 12-7(a) that few chlorides have low solubilities. Of the few that do, their cations are metals appearing on the right side of the periodic table (silver ion, Ag^+, cuprous ion, Cu^+, mercurous ion, Hg_2^{2+}, and lead ion, Pb^{2+}), but not in a single column. This irregularity is not unusual in solubility behavior and is seen again in Figures 12-7(b) and 12-8(a). In these figures, the elements in the second column (the alkaline earths) show a trend in behavior. In Figure 12-7(b), we see that beryllium and magnesium ions, Be^{2+} and Mg^{2+}, form soluble sulfates. The others—calcium, strontium, barium, and radium ions (Ca^{2+}, Sr^{2+}, Ba^{2+}, and Ra^{2+})—form sulfates with low solubilities. Except for Ca^{2+} the same ions form soluble compounds with the hydroxide ion, OH^-, as seen in Figure 12-8(a). As for the elements in the middle of the periodic table, they form compounds of low solubilities with the ions sulfide, S^{2-}, hydroxide, OH^-, phosphate, PO_4^{3-}, carbonate, CO_3^{2-}, and sulfite, SO_3^{2-}.

The information in Figures 12-6, 12-7, and 12-8 is summarized in Table 12-1. The generalizations made there, like any generalizations, will have exceptions, but they can be used when quantitative information is not available.

Fig. 12-8 **More positive ions forming compounds of low solubilities with various anions.**

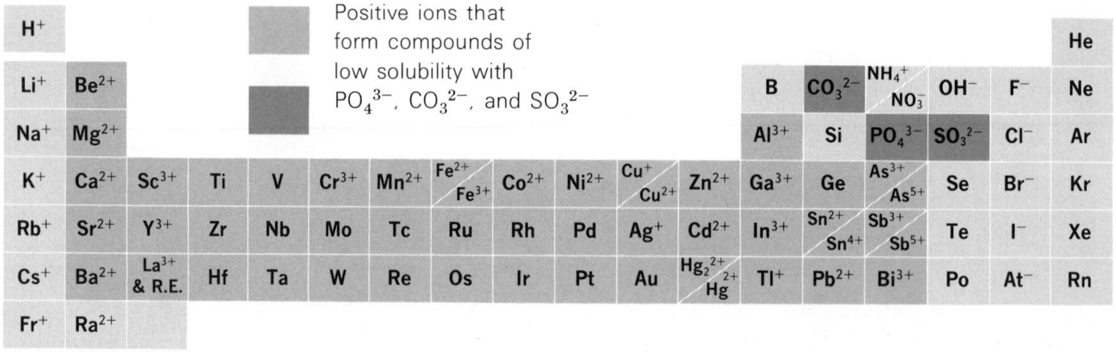

TABLE 12-1 SOLUBILITY OF COMMON COMPOUNDS IN WATER

A QUALITATIVE VIEW of AQUEOUS SOLUBILITIES

Negative Ions (Anions)	+	Positive Ions (Cations)	→	Compounds with the Solubility:
All		alkali ions, Li^+, Na^+, K^+, Rb^+, Cs^+, Fr^+		soluble
All		hydrogen ion, $H^+(aq)$		soluble
All		ammonium ion, NH_4^+		soluble
Nitrate, NO_3^-		all		soluble
Acetate, CH_3COO^-		all		soluble
Chloride, Cl^- Bromide, Br^- Iodide, I^-		Ag^+, Pb^{2+}, Hg_2^{2+}, Cu^+, Tl^+		low solubility
		all others		soluble
Sulfate, SO_4^{2-}		Ca^{2+}, Sr^{2+}, Ba^{2+}, Pb^{2+}, Ra^{2+}		low solubility
		all others		soluble
Sulfide, S^{2-}		alkali ions, $H^+(aq)$, NH_4^+, Be^{2+}, Mg^{2+}, Ca^{2+}, Sr^{2+}, Ba^{2+}, Ra^{2+}		soluble
		all others		low solubility
Hydroxide, OH^-		alkali ions, $H^+(aq)$, NH_4^+, Sr^{2+}, Ba^{2+}, Ra^{2+}, Tl^+		soluble
		all others		low solubility
Phosphate, PO_4^{3-} Carbonate, CO_3^{2-} Sulfite, SO_3^{2-}		alkali ions, $H^+(aq)$, NH_4^+		soluble
		all others		low solubility

EXERCISE 12-4

Use the information in Table 12-1 to determine whether each of the following compounds has high or low solubility in water solution. Write "sol" if the compound is soluble and "low" if it has low solubility.

$Mg(NO_3)_2$ sol $MgCl_2$ sol $MgSO_4$ sol $Mg(OH)_2$ low $MgCO_3$ low
$Ca(NO_3)_2$ sol $CaCl_2$ sol $CaSO_4$ low $Ca(OH)_2$ low $CaCO_3$ low
$Sr(NO_3)_2$ sol $SrCl_2$ sol $SrSO_4$ low $Sr(OH)_2$ sol $SrCO_3$ low

EXERCISE 12-5

Write formulas for each of the following compounds and decide whether the compound will be of high or low solubility in water.

silver carbonate   low
aluminum hydroxide   low
cuprous (Cu^+) chloride   low
lead sulfate   sol
ammonium nitrate   sol
silver sulfide   low
calcium iodide   sol
magnesium sulfate   sol
rubidium iodide   sol
cesium nitrate   sol
barium sulfate   low
silver iodide   low
lead sulfite   low
mercurous bromide   low

Chapter 12 / SOLUBILITY EQUILIBRIUM

$Na^+_{(aq)} + SO_4^{2-}(aq) + Sn^{2+}(aq) + NO_3^-{}_2(aq)$

$\rightarrow 2Na^+_{(aq)} + 2NO_3^-(aq) + SnSO_4(s)$

$NH_4^+(aq) + Cl^-(aq) + Na^+(aq) + NO_3^-(aq)$

$\rightarrow NH_4^+(aq) + NO_3^-(aq) + Na^+(aq) + Cl^-(aq)$

EXERCISE 12-6

Write an equation for the process expected when a solution of sodium sulfate (0.5 M) is poured into a solution of strontium nitrate (0.5 M). Indicate precipitate formation of MX by writing MX(s). Repeat for the pouring of a solution of ammonium chloride into a solution of sodium nitrate. If no precipitate forms, show ions in solution as products.

12-5 AN APPLICATION OF EQUILIBRIUM CONCEPTS TO IONIC SOLUTIONS

So far, we have organized our information in a more or less qualitative fashion so that we can make intelligent estimates of compound solubility. But *qualitative* answers are frequently not enough; we want to know *how much* of an ionic solid dissolves in a liter of water! *Quantitative* models of equilibrium are thus desired. Let us start with the equation for a solubility process.

The low solubility of silver chloride (AgCl) has been considered in the laboratory and makes a convenient place to start. If we write

$$AgCl(s) \rightleftharpoons Ag^+(aq) + Cl^-(aq) \qquad (24)$$

a general equilibrium constant for the system can be written as before:

$$K' = \frac{[Ag^+(aq)][Cl^-(aq)]}{[AgCl(s)]} \qquad (25)$$

Once again, we see the concentration of a solid—AgCl in this case—in the equilibrium constant. Our experience with solid iodine showed that *the concentration of a solid is constant*; therefore, if we consider the concentration of AgCl(s) as constant, we can write

$$K'[AgCl(s)] = \text{a constant} = [Ag^+(aq)][Cl^-(aq)]$$

We can now write

$$K = K'[AgCl(s)] = [Ag^+(aq)][Cl^-(aq)] \qquad (26)$$

This constant, K, which is defined by the relationship

$$K_{sp} = [Ag^+(aq)][Cl^-(aq)] \qquad (26)$$

is known as the **solubility product constant** for silver chloride (in a saturated aqueous solution). A low value of K_{sp} means that the concentration of ions is low at equilibrium; hence, the solubility of the compound must be low. Table 12-2 lists solubility product constants for a few common compounds.

TABLE 12-2
SOME SOLUBILITY PRODUCTS AT ROOM TEMPERATURE

Compound	K_{sp}
TlCl	1.9×10^{-4}
CuCl	3.2×10^{-7}
AgCl	1.7×10^{-10}
TlBr	3.6×10^{-6}
CuBr	5.9×10^{-9}
AgBr	5.0×10^{-13}
TlI	8.9×10^{-8}
CuI	1.1×10^{-12}
AgI	8.5×10^{-17}
SrCrO$_4$	3.6×10^{-5}
BaCrO$_4$	8.5×10^{-11}
PbCrO$_4$	2×10^{-16}
CaSO$_4$	2.6×10^{-4}
SrSO$_4$	7.6×10^{-7}
PbSO$_4$	1.3×10^{-8}
BaSO$_4$	1.5×10^{-9}
RaSO$_4$	4×10^{-11}
AgBrO$_3$	5.4×10^{-5}
AgIO$_3$	2.1×10^{-8}

EXERCISE 12-7

Show from the equation for the solution of lead chloride that K_{sp} for PbCl$_2$ is $K_{sp} = [Pb^{2+}(aq)][Cl^-(aq)]^2$. Write the equation for the dissolving of Ag$_2$CrO$_4$, silver chromate, and write the expression for K_{sp}.

12-5.1 Calculation of the Solubility of Cuprous Chloride in Water

The solubility product is obtained from measurements of solubility. In turn, it can be used to make a quantitative estimate of solubility. In short, a solubility product provides *a method of summarizing very concisely a large amount of experimental information on the solubility of a given ionic substance.* Suppose we wish to know how much cuprous chloride (CuCl) will dissolve in 1 liter of water. How can this information be computed from the value for K_{sp} carried in handbooks and reference sources? We begin by writing the balanced equation for the reaction:

$$CuCl(s) \rightleftharpoons Cu^+(aq) + Cl^-(aq) \qquad (27)$$

After a quick look at this equation, we can write the equilibrium expression:

$$K_{sp} = [Cu^+][Cl^-] \qquad (28)$$

Now the numerical value of K_{sp} for cuprous chloride is found in Table 12-2:

$$K_{sp} = 3.2 \times 10^{-7} = [Cu^+][Cl^-] \qquad (29)$$

The equation for the dissolving process (27) indicates that cuprous chloride (CuCl) continues to dissolve until the product of the molar concentrations of cuprous ion, Cu^+, and chloride ion, Cl^-, is equal to 3.2×10^{-7}.

Now is the time to put our algebra to work. Suppose we designate the solubility of cuprous chloride in water by a symbol, s. This symbol s equals the number of moles of solid cuprous chloride that dissolve to produce 1 liter of saturated cuprous chloride solution. Remembering the equation for the solution process, we see that s moles of solid cuprous chloride will produce s moles of cuprous ion, Cu^+, and s moles of chloride ion, Cl^-. Hence, these concentrations must be equal:

$$[Cu^+] = [Cl^-] = s \text{ moles/liter} = s \text{ moles/liter CuCl dissolved} \qquad (30)$$

Substituting this equality into the K_{sp} expression and then solving for s, we have

$$K_{sp} = 3.2 \times 10^{-7} \text{ moles}^2/\text{liter}^2 = (s) \times (s) = s^2$$
$$s^2 = 3.2 \times 10^{-7} \text{ moles}^2/\text{liter}^2 = 32 \times 10^{-8} \text{ moles}^2/\text{liter}^2$$
$$s = \sqrt{32 \times 10^{-8} \text{ moles}^2/\text{liter}^2} = 5.7 \times 10^{-4} \text{ mole/liter}$$
$$= 0.00057 \, M \qquad (31)$$

EXERCISE 12-8

Calculate the solubility in moles per liter of thallium iodide in water using the solubility product given in Table 12-2.

0.000298 M

12-5.2 Will a Precipitate Form?

In Exercise 12-6, we asked for a qualitative answer to the question: What happens when a solution (0.5 M) of sodium sulfate is poured into a solution (0.5 M) of strontium nitrate? For rather concentrated solutions the answer is clear. A precipitate of insoluble strontium sulfate appears. But is this always true? What happens if the solutions are more dilute? How insoluble does a solid have to be before the precipitate appears? These questions require a more quantitative application of the equilibrium concept. Let us consider quantitative relationships for two examples of mixing solutions of thallium nitrate (TlNO$_3$) and sodium chloride (NaCl).

Example (1) If equal volumes of 0.02 M TlNO$_3$ and 0.004 M NaCl are mixed, will a precipitate form?

The balanced equation for the solution or precipitation of thallium chloride in water is

$$\text{TlCl}(s) \rightleftharpoons \text{Tl}^+(aq) + \text{Cl}^-(aq) \tag{32}$$

The K_{sp} value for TlCl in Table 12-2 is 1.9×10^{-4}. We can write

$$K_{sp} = 1.9 \times 10^{-4} \text{ moles}^2/\text{liter}^2 = [\text{Tl}^+(aq)][\text{Cl}^-(aq)] \tag{33}$$

We must now calculate the concentrations of Tl$^+$(aq) and Cl$^-$(aq) immediately after mixing and before any reaction has had a chance to occur. Since we are mixing *equal volumes* of the two solutions, each ion will be present in twice as much volume; hence, its concentration will be only *half* as great as it was before the solutions were mixed. This being true, we can write

Ion	Concentration Before Mixing	Concentration After Mixing
Tl$^+$(aq)	0.02 M	$\dfrac{0.02\ M}{2}$ = 0.01 M
NO$_3^-$(aq)	0.02 M	$\dfrac{0.02\ M}{2}$ = 0.01 M
Na$^+$(aq)	0.004 M	$\dfrac{0.004\ M}{2}$ = 0.002 M
Cl$^-$(aq)	0.004 M	$\dfrac{0.004\ M}{2}$ = 0.002 M

We must ask next: Is the product [Tl$^+$(aq)][Cl$^-$(aq)] in the solution, after mixing, *larger* or *smaller* than the value of K_{sp} for TlCl? If the product is *larger* than K_{sp}, a precipitate will form: ions will be removed through precipitation until the ion product becomes equal to K_{sp}. If the product [Tl$^+$(aq)][Cl$^-$(aq)] is *smaller* than K_{sp}, no precipitate will form. In our case, [Tl$^+$(aq)] = 0.01 M and [Cl$^-$(aq)] = 0.002 M; we can write

$$[\text{Tl}^+(aq)][\text{Cl}^-(aq)] = 0.01 \times 0.002 = 2 \times 10^{-5} \text{ moles}^2/\text{liter}^2* \tag{34}$$

---

* This value, the product of ion concentrations, is sometimes called the **trial ion product.**

Since 2×10^{-5} is *less than* 1.9×10^{-4}, *no precipitate will form.* This is true since we also know that $NaNO_3$ is water soluble.

Example (2) Equal volumes of $0.08\ M$ $TlNO_3$ and $0.2\ M$ $NaCl$ are mixed. Will a precipitate form?

Again, we may write

Ion	Concentration Before Mixing	Concentration After Mixing
$Tl^+(aq)$	$0.08\ M$	$\dfrac{0.08\ M}{2} = 0.04\ M$
$NO_3^-(aq)$	$0.08\ M$	$\dfrac{0.08\ M}{2} = 0.04\ M$
$Na^+(aq)$	$0.2\ M$	$\dfrac{0.2\ M}{2} = 0.1\ M$
$Cl^-(aq)$	$0.2\ M$	$\dfrac{0.2\ M}{2} = 0.1\ M$

The trial ion product relationship gives

$$[Tl^+(aq)][Cl^-(aq)] = 0.04 \times 0.1 = 4 \times 10^{-3} \text{ moles}^2/\text{liter}^2 \qquad (35)$$

This value is obviously larger than K_{sp}; hence, a precipitate of $TlCl$ will form. Precipitation will occur until the concentrations of $Tl^+(aq)$ and $Cl^-(aq)$ are reduced to a point at which their product is no longer larger than the value of K_{sp}.

EXERCISE 12-9

A 50-ml volume of $0.04\ M$ $Ca(NO_3)_2$ solution is added to 150 ml of $0.008\ M$ $(NH_4)_2SO_4$ solution. Show that a trial value of the calcium sulfate ion product is 6×10^{-5} moles^2/liter^2. Will a precipitate form?

EXERCISE 12-10

Would a precipitate form if 100 ml of $0.04\ M$ $TlNO_3$ were mixed with 300 ml of $0.008\ M$ $NaCl$? Watch the dilution ratios.

The formation of precipitates from solution is of tremendous importance in nature and industry. The oyster in growing his shell must adjust conditions so that the concentration of CO_3^{2-} is large enough to precipitate calcium carbonate ($CaCO_3$) from the surrounding seawater. Coral reefs grow in the same way. Limestone caves are carved out of limestone rock by water in which the concentrations of $Ca^{2+}(aq)$ and $CO_3^{2-}(aq)$ have a product below the K_{sp} for $CaCO_3$. Many beautiful pigments such as yellow lead chromate and "cadmium reds" (cadmium selenide-cadmium sulfide) are prepared by a careful application of the principles of precipitation equilibrium. Failure to pay strict attention to data summarized in values of a K_{sp} table can result in loss of large amounts of expensive raw materials and the contamination of rivers and streams. Chemical engineers cannot afford the luxury of ignoring K_{sp} values.

12-5.3 Precipitations Used for Separations

A chemist is often interested in separating substances in a mixture of solutions. Such a problem is solved by applying equilibrium considerations. Suppose we have a solution known to contain both lead nitrate [$Pb(NO_3)_2$] and magnesium nitrate [$Mg(NO_3)_2$]. The lead and magnesium can be separated by removing from the solution almost all the lead ion (Pb^{2+}) as a solid lead compound. We must avoid precipitation of any magnesium compound. Consulting Figure 12-7(b) or Table 12-1, we see that lead ion and sulfate ion form a compound with low solubility. If enough sodium sulfate (Na_2SO_4) is added, lead sulfate ($PbSO_4$) will precipitate. Since Figure 12-7(b) and Table 12-1 indicate that magnesium sulfate ($MgSO_4$) is soluble, there will be no precipitation of $MgSO_4$. The solid can be removed from the liquid by filtration producing the desired separation.

EXERCISE 12-11

Use Figures 12-7 and 12-8 or Table 12-1 to decide which of the following soluble salts would permit a separation of magnesium and lead through a precipitation reaction: sodium iodide, NaI; sodium sulfide, Na_2S; sodium carbonate, Na_2CO_3.

Let us consider a somewhat more complicated separation. Suppose a solution contains silver nitrate ($AgNO_3$), copper(II) nitrate [$Cu(NO_3)_2$], and magnesium nitrate [$Mg(NO_3)_2$]. How can a separation of the metal ions be achieved? In such a case, the reagents used and the order in which they are added are both important. Referring to Figure 12-8, we see that Ag^+, Cu^{2+}, and Mg^{2+} all form insoluble carbonates (CO_3^{2-}), phosphates (PO_4^{3-}), and hydroxides (OH^-). Clearly, salts with these anions would *not* be helpful to us in accomplishing a separation. On the other hand, we see that Ag^+ forms an insoluble chloride while Cu^{2+} and Mg^{2+} ions form soluble chlorides. This information is useful. Addition of a solution of ammonium or sodium chloride will precipitate AgCl, which can then be filtered off.

We now have only Cu^{2+} and Mg^{2+} ions to separate. Figure 12-7(b) shows that $MgSO_4$ and $CuSO_4$ are both water soluble: addition of Na_2SO_4 will not help us. Figure 12-7(c) provides the answer. MgS is soluble while CuS is insoluble. Adding Na_2S solution to the mixture containing Mg^{2+} and Cu^{2+} will precipitate CuS and leave $Mg^{2+}(aq)$ in solution. The $Mg^{2+}(aq)$ can then be removed as $MgCO_3$ by adding Na_2CO_3 solution.

The order in which the reagents are added is important. If we had added Na_2S solution before we added NaCl solution, both Ag_2S and CuS would have precipitated immediately. If we had added Na_2CO_3 solution at the beginning, $MgCO_3$, $CuCO_3$, and Ag_2CO_3 would have precipitated. Thus, both *the selection of reagents* and *the order in which they are added* is of concern. The segment of chemistry which deals with separations of this type is called **qualitative analysis.** More detailed qualitative-analysis schemes for more difficult separations could be worked out with a knowledge of acids and bases.

EXERCISE 12-12

A liquid contains the following salts in solution: potassium nitrate (KNO_3), lead nitrate [$Pb(NO_3)_2$], copper nitrate [$Cu(NO_3)_2$], and barium nitrate [$Ba(NO_3)_2$]. Show that the addition of the following salts in order would leave only KNO_3 in solution. Identify the precipitate coming out after the addition of each reagent. The order in which they are added is (1) KCl solution, (2) K_2SO_4 solution, and (3) K_2CO_3 solution. Would other orders work as well?

[handwritten: $PbCl_2$, $CuCl_2$. $BaSO_4$]

12-6 HIGHLIGHTS

The extent to which a substance will dissolve depends upon two factors:

(1) the tendency of the system to achieve maximum randomness—a process which favors the solution process for solids and opposes the solution process for gases, and
(2) the tendency of the system to achieve minimum energy content.

Energy changes may favor either the solid or the solution. Energy changes always favor gas solubility. When randomness and energy are balanced, equilibrium is achieved.

General concepts of equilibrium can be applied to problems of solubility. Since concentration of a pure solid phase is constant, this value is conventionally included in the equilibrium constant so that the equilibrium expression takes a rather unexpectedly simple form. For the dissolving of non-electrolytes such as I_2 in a solvent such as alcohol at a constant temperature, the equilibrium constant takes the very simple form

$$K = [I_2(solution)]$$

For ionic materials such as AgCl dissolving in water at a constant temperature, the equilibrium constant takes the form

$$K_{sp} = [Ag^+(aq)][Cl^-(aq)]$$

where K_{sp} is known as the **solubility product constant.**

QUESTIONS AND PROBLEMS

1 Sugar is added to a cup of coffee until no more sugar will dissolve. Does addition of another spoonful of sugar increase the rate at which the sugar molecules leave the crystal phase and enter the liquid phase? Will the sweetness of the liquid be increased by this addition? Explain.

2 In view of the discussion of the factors that determine the rate of solution (Section 12-1.2), propose two methods for increasing the rate at which sugar dissolves in water.

3 When a solid evaporates directly (without melting), the process is called **sublimation.** (Evaporation of Dry Ice, solid CO_2, is a familiar example.) Two other substances that sublime are FCN and ICN:

$$FCN(s) \rightleftarrows FCN(g)$$
$$\Delta H = +5.7 \text{ kcal}$$
$$ICN(s) \rightleftarrows ICN(g)$$
$$\Delta H = +14.2 \text{ kcal}$$

(a) In sublimation, does the tendency toward *maximum randomness* favor solid or gas? (b) In sublimation, does the tendency toward *minimum energy* favor solid or gas? (c) The vapor pressure of solid FCN is 760 mm at 201°K. In view of (b), would you expect solid ICN to have a lower or higher vapor pressure than solid FCN at this same temperature, 201°K?

4 Liquid chloroform, $CHCl_3$, and liquid acetone, CH_3COCH_3, dissolve in each other in all proportions. (They are said to be **miscible**.)

(a) When pure $CHCl_3$ is mixed with pure CH_3COCH_3, is randomness increased or decreased?

(b) Does the tendency toward maximum randomness favor reactants or product in the reaction

$$CHCl_3(l) + CH_3COCH_3(l) \longrightarrow \text{1:1 solution}$$
$$\Delta H = -495 \text{ cal}$$

(c) Considering the sign of ΔH shown in (b), does the tendency toward minimum energy favor reactants or product?

(d) In view of your answers to (b) and (c), discuss the experimental fact that these two liquids are miscible.

5 Assume the following compounds dissolve in water to form separate, mobile ions in solution. Write the formulas and names for the ions that can be expected. (a) HI (b) $CaCl_2$ (c) Na_2CO_3 (d) $Ba(OH)_2$ (e) KNO_3 (f) NH_4Cl

6 Write the equation for the reaction that occurs when each of the following electrolytes is dissolved in water: (a) lithium hydroxide(*solid*) (b) nitric acid(*liquid*) (c) potassium sulfate(*solid*) (d) sodium nitrate(*solid*) (e) ammonium iodide (*solid*) (f) potassium carbonate(*solid*). [Answer to (a): $LiOH(s) \longrightarrow Li^+(aq) + OH^-(aq)$.]

7 Which of the following substances can be expected to dissolve in the indicated solvent to form, primarily, ions? Which would form molecules? (a) sugar in water (b) RbBr in water (c) $CHCl_3$ in water (d) $CsNO_3$ in water (e) HNO_3 in water (f) S_8 in carbon disulfide, CS_2 (g) ICl in ethyl alcohol.

8 Which of the substances listed in question 7 would be called electrolytes?

9 Write the empirical formulas for each of the following compounds and indicate which have low solubilities in water: (a) silver sulfide (b) potassium sulfide (c) ammonium sulfide (d) nickel sulfide (e) ferrous sulfide (Fe^{2+}).

10 Use the solubility charts as a basis for making *qualitative* predictions about the results of the following experiments: (a) A concentrated solution of $AgNO_3$ in water is poured into a saturated solution of NaCl in water. (b) Write the equation for the process indicated in (a). (c) A concentrated solution of barium nitrate, $Ba(NO_3)_2$, is poured into a saturated solution of sodium sulfate. Write the equation. (d) A concentrated solution of KNO_3 is poured into a fairly concentrated solution of $Cu(NO_3)_2$. (e) A concentrated solution of $MgSO_4$ is poured into a concentrated solution of Na_2CO_3. (f) A solution of $AgNO_3$ is poured into a solution of Na_2S. (g) A solution of $MgSO_4$ is poured into a solution of $ZnCl_2$ (concentrations about 1 M). If possible, verify each of your predictions by laboratory observation.

11 Which of the following statements are *false*? Modify such statements so they will be true (see Figures 12-6, 12-7, and 12-8). (a) All nitrates are soluble except those of Pb^{2+}, Cu^{2+}, Sr^{2+}, and Ba^{2+}. (b) All chlorides are *in*soluble except those of Ag^+, Hg_2^{2+}, and Pb^{2+}. (c) All sulfates are soluble. (d) All acetates are soluble.

12 What ions could be present in a solution if samples of it gave: (a) a precipitate when either $Cl^-(aq)$ or $SO_4^{2-}(aq)$ is added? (b) a precipitate when $Cl^-(aq)$ is added but none when $SO_4^{2-}(aq)$ is added? (c) a precipitate when $SO_4^{2-}(aq)$ is added but none when $Cl^-(aq)$ is added?

13 Write net ionic equations for any reactions that will occur upon mixing equal volumes of solutions of the following pairs of compounds: (a) silver nitrate and ammonium bromide (b) $SrBr_2$ and $NaNO_3$ (c) sodium hydroxide and aluminum chloride (d) NaI and $Pb(NO_3)_2$ (e) barium chloride and sodium sulfate. [Answer to (a): $Ag^+(aq) + Br^-(aq) \longrightarrow AgBr(s)$.]

14 Write the solubility product expression for each of the following reactions.

(a) $BaSO_4(s) \rightleftarrows Ba^{2+}(aq) + SO_4^{2-}(aq)$
(b) $Zn(OH)_2(s) \rightleftarrows Zn^{2+}(aq) + 2OH^-(aq)$
(c) $Ca_3(PO_4)_2(s) \rightleftarrows 3Ca^{2+}(aq) + 2PO_4^{3-}(aq)$

15 Write the solubility product expression applicable to the solubility of each of the following substances in water: (a) calcium carbonate (b) silver sulfide (c) aluminum hydroxide

16 The solubility of nickel sulfide, NiS, is 1×10^{-12} mole/liter. (a) Write the equation representing the solubility equilibrium between NiS(s) and its saturated solution. (b) What is the concentration (moles/liter) of $Ni^{2+}(aq)$ in the solution? (c) What is the concentration of $S^{2-}(aq)$ in the solution? (d) What do you compute to be the value of K_{sp} for NiS?

17 The solubility of $Ca(OH)_2$ is 0.02 mole/liter. Calculate the solubility product constant for this substance.

18 The solubility product for silver bromide, AgBr, is 6.4×10^{-13} at a given temperature. (a) Write the equation for the solution process and the equation for K_{sp}. (b) Express the value of K_{sp} as a number $\times 10^{-14}$. (c) Using the value in

(b) and the equation in (a), calculate the solubility of AgBr in moles/liter. (d) What is the solubility of AgBr in g/liter? (e) How many mg of AgBr dissolve in 20 liters of water? [Answer to (d): 1.5×10^{-4} g/liter.]

19 The solubility product of AgCl is 1.4×10^{-4} at 100°C. Calculate the solubility of AgCl in boiling water. Give the units.

20 Experiments show that 0.0059 g of $SrCO_3$ will dissolve in 1.0 liter of water at 25°C. What is K_{sp} for $SrCO_3$? (Answer: 1.6×10^{-9} moles^2/liter^2.)

21 The solubility of AgCl is so low that all but a negligible amount of it is precipitated when excess NaCl solution is added to $AgNO_3$ solution. What would be the weight of the precipitate formed when 100.0 ml of 0.500 M NaCl is added to 50.0 ml of 0.100 M $AgNO_3$? (Answer: 0.717 g.)

22 To 1 liter of 0.002 M H_2SO_4 solution is added 1 liter of 0.002 M $Pb(NO_3)_2$ solution. (a) What would be the concentration of $SO_4^{2-}(aq)$ in the final solution? (Assume no precipitate has formed.) (b) What would be the concentration of $Pb^{2+}(aq)$? (Assume no precipitate has formed.) (c) What would be the value of the trial product for $[Pb^{2+}][SO_4^{2-}]$? (d) Using the solubility product constant for lead sulfate from Table 12-2, predict whether a precipitate will form.

23 Suppose 10 ml of 1.0 M $AgNO_3$ is diluted to 1 liter with tap water. If the chloride concentration in the tap water is about 10^{-6} M, will a precipitate form?

24 The test described in question 23 does not give a precipitate if the laboratory distilled water is used. What is the maximum chloride concentration that could be present?

25 Will a precipitate exist at equilibrium if 0.5 liter of a 2×10^{-3} M $AlCl_3$ solution and 0.5 liter of a 4×10^{-2} M solution of sodium hydroxide are mixed and diluted to 10^3 liters with water at room temperature ($K_{sp} = 5 \times 10^{-33}$)?

QUESTIONS and PROBLEMS

26 Use Figures 12-7 and 12-8 or Table 12-1 to decide which of the following soluble substances would permit a separation of aqueous magnesium and barium ions. For those that are effective, write the equation for the reaction that occurs. (a) ammonium carbonate (b) sodium bromide (c) potassium sulfate (d) sodium hydroxide.

27 To a solution containing 0.1 M of each of the ions Ag^+, Cu^+, Fe^{2+} and Ca^{2+} is added 2 M NaBr solution, giving precipitate A. After filtration, a sulfide solution is added to the solution and a black precipitate forms, precipitate B. This precipitate is removed by filtration and 2 M sodium carbonate solution is added, giving precipitate C. What is the composition of each precipitate, A, B, and C?

28 What cations from the fourth row of the periodic table could be present in a solution with the following behavior? (a) No precipitate is formed with hydroxide ion. (b) A precipitate forms with hydroxide ion and with sulfate ion. (c) A precipitate forms with hydroxide ion and with sulfide ion. (d) A precipitate forms with carbonate ion but none with sulfide ion.

29 Magnesium sulfite hexahydrate, $MgSO_3 \cdot 6H_2O$, dissolves in water to the extent of 1.2 g/100 ml at 0°C. Will a precipitate form if equal volumes of 0.2 M Na_2SO_3 and 0.2 M $MgSO_4$ are mixed?

30 (a) Calculate the solubility products of $BaCrO_4$ and of $Pb(OH)_2$ if their solubilities (at 18°C) are 1.5×10^{-5} moles/liter and 4×10^{-4} moles/liter, respectively. (b) What is the largest concentration of $Ba^{2+}(aq)$ which can exist in a 0.1 M Na_2CrO_4 solution without the formation of a precipitate? (c) What is the largest concentration of $Pb^{2+}(aq)$ ions which can exist in tap water where the $OH^-(aq)$ ion concentration is 10^{-7} mole/liter?

Out of the simplest things ye shall know the truth.

GREGOR JOHANN MENDEL (1822–1884)

CHAPTER 13 Ionic Equilibrium: Acids and Bases

Most common fruits that we eat are acidic.

Acids have quite a reputation, but their bark is worse than their bite. If you're naturally curious, you may already have asked your teacher which acid is strongest. It's a common question. Everyone knows that acids must be handled with great care, yet oranges contain citric acid, and boric acid is so mild that it is used as an eyewash! What gives? What makes some acids dangerous and others tame? And what is a base? You guessed it—it's all a matter of equilibrium. Read on!

We have investigated equilibrium as it applies to gaseous systems and have learned that the commercial preparation of ammonia (NH_3) is based on an application of equilibrium principles. We were able to use equilibrium concepts in discussing solubility relationships. For the solubility of an ionic solid we wrote

$$AgCl(s) \rightleftharpoons Ag^+(aq) + Cl^-(aq) \qquad (1)$$

Since the concentration of $AgCl(s)$ is constant in the above system, the equilibrium constant has a particularly simple form at constant temperature. It is known as the **solubility product constant**, K_{sp}, and is written as

$$\underbrace{K'[AgCl(s)]}_{\text{CONSTANT}} = K_{sp} = [Ag^+(aq)][Cl^-(aq)] \qquad (2)$$

You may now wonder: What about an ionic equilibrium which does not involve a solid phase? In Experiment 19 we considered such an equilibrium involving the red $FeSCN^{2+}(aq)$ ion. We wrote

$$Fe^{3+}(aq) + SCN^-(aq) \rightleftharpoons Fe(SCN)^{2+}(aq) \qquad (3)$$

and

$$K = \frac{[Fe(SCN)^{2+}(aq)]}{[Fe^{3+}(aq)][SCN^-(aq)]} \qquad (4)$$

In this chapter we shall consider additional ionic equilibria. The discussion will lead us naturally into a study of acids and bases, a topic of great importance to chemists.

13-1 ELECTROLYTES—STRONG AND WEAK

Our model describing the electrical conductivity of aqueous solutions pictured positive and negative ions moving through the liquid to carry the electric current (Chapter 6). Such a model suggested that electrical conductivity would rise as the concentration of ions increased and would drop to zero if no ions were present. On the basis of this model we can infer that those solutions having very good electrical conductivity furnish a large supply of ions for each mole of material dissolved in water. For example, NaCl and HCl appear to dissolve to give *only* ions in dilute solution. These solutions are very good conductors. A solution of sodium hydroxide (NaOH) is also a good conductor. Its dissociation in dilute solution to give $Na^+(aq)$ and $OH^-(aq)$ appears to be essentially complete:

$$NaOH(s) \longrightarrow Na^+(aq) + OH^-(aq) \qquad (5)$$

There is no clear experimental evidence for the presence of a significant number of NaOH *molecules* in a dilute solution of NaOH in water or for a significant number of HCl molecules in a dilute solution of HCl in water. *A substance which dissociates almost completely*

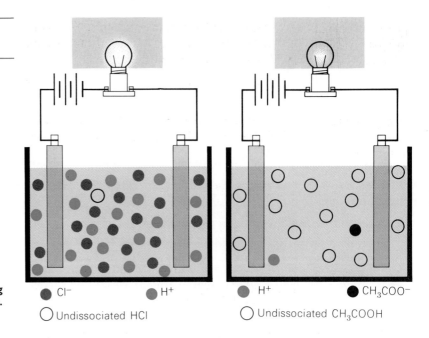

Fig. 13-1 **Conductivity of a strong electrolyte and of a weak electrolyte.**

to give ions in solution is known as a **strong electrolyte.** (See Figure 13-1.1.)

The very name "strong electrolyte" suggests that we may have weak electrolytes too—substances which dissociate to only a limited extent when they are dissolved in water. Acetic acid, found in vinegar, is such a material. A 0.1 M acetic acid solution is a *much poorer* conductor of electricity than a 0.1 M hydrochloric acid solution. This can be understood most easily by assuming that acetic acid does not dissociate completely when dissolved in water. Our proposal can be summarized by an equation:

$$CH_3COOH(aq) \underset{\longleftarrow}{\overset{H_2O(l)}{\rightleftharpoons}} H^+(aq) + CH_3COO^-(aq) \qquad (6)$$

The long arrow pointing to the left tells us that very little of the acetic acid ionizes to give hydrogen ions, $H^+(aq)$, and acetate ions, $CH_3COO^-(aq)$. In contrast, our equation for HCl in water would be written as

$$HCl(g) \xrightarrow{H_2O(l)} H^+(aq) + Cl^-(aq) \qquad (7)$$

The reverse process to give HCl is too small to indicate with an arrow. *A substance like acetic acid which dissolves, yet which undergoes only partial dissociation to ions, is called a* **weak electrolyte.** (See Figure 13-1.2.)

13-1.1 The Dissociation of Water, a Very Weak Electrolyte

Relatively crude measurements show that pure water does *not* conduct electric current. If, however, the conductivity of pure water is measured with an extremely sensitive meter, a tiny electrical conductivity is found in even the purest water.* Such an observation suggests that water itself forms a few ions by the process

$$H_2O(l) \rightleftharpoons H^+(aq) + OH^-(aq) \qquad (8)$$

Clearly, the concentration of ions is *very* small and the equilibrium lies well toward undissociated water. The equilibrium constant can be written

$$K' = \frac{[H^+(aq)][OH^-(aq)]}{[H_2O(l)]} \qquad (9)$$

Since only the *tiniest trace* of water is ionized, essentially all the water is present as $H_2O(l)$. In 1 liter of water we have 1,000 g. Since 1 mole of water weighs 18.0 g (16 + 1 + 1), we have

$$\frac{1,000 \text{ g } H_2O}{18.0 \text{ g } H_2O/\text{mole}} = 55.6 \text{ moles } H_2O$$

In most water solutions the concentration of water is virtually constant at 55.6 M; consequently, we can incorporate this number into the constant, just as we did when we considered the solubility of a solid. We can write

$$\underset{\text{CONSTANT}}{K'[H_2O]} = [H^+(aq)][OH^-(aq)] \qquad (10)$$

The value $K'[H_2O]$ is a constant since $[H_2O]$ is constant. We thus write

$$K'[H_2O] = K_w = [H^+(aq)][OH^-(aq)] = 10^{-14} \text{ mole}^2/\text{liter}^2 \qquad (11)$$

13-1.2 The Nature of $Li^+(aq)$ and $H^+(aq)$ and a Comment on the Symbols Used

When we consider the ionization of water and the existence of H+ in solution, questions about the actual nature of H+(aq) arise. What is really known about the nature of the proton in aqueous solution?

Let us approach this problem by considering the behavior of a related species in aqueous solution. Lithium chloride provides a good example. As we noted in Chapter 8, lithium chloride is composed of lithium cations, Li+, and chloride anions, Cl−, held together in a rigid lattice by electrostatic forces of attraction. If we heat this solid to a sufficiently high temperature, 613°C, the kinetic energy of the ions is large enough to overcome the attraction of the positively and negatively charged particles. Under these circumstances, the tendency of a system to achieve maximum randomness takes over. The ions scatter in helter-skelter fashion; in short, the solid LiCl melts and Li+ and Cl− ions are freed. These ions can then carry current through the molten mass.

---

*The fact that pure water dissociates to a small degree can be shown qualitatively by the use of a 2-watt neon glow lamp.

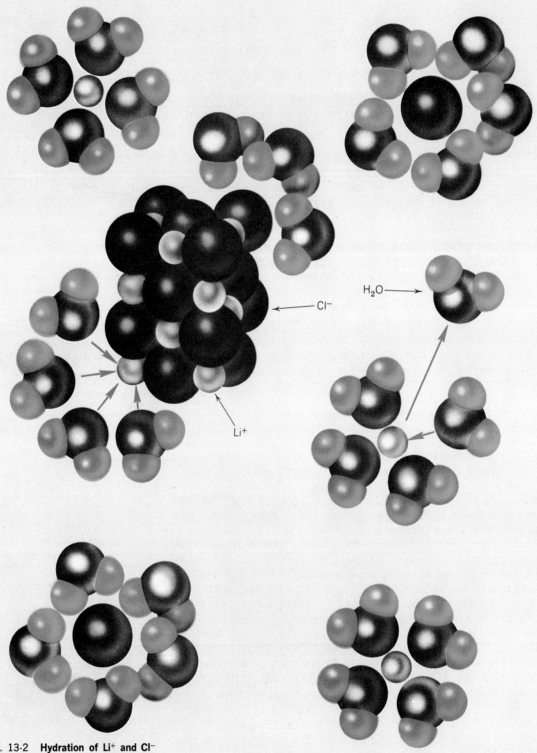

Fig. 13-2 **Hydration of Li$^+$ and Cl$^-$ ions from LiCl.**

In contrast to the high temperature required to produce ions in a molten salt, lithium chloride dissolves in water at 25°C to produce Li$^+$(aq) and Cl$^-$(aq) ions in aqueous solution (see Figure 13-2, page 292). Where does the energy to separate the ions come from? What takes the place of the large amounts of energy required in the melting process (system heated to 613°C)? The most reasonable assumption is that Li$^+$(g) and Cl$^-$(g) interact very strongly with water. It is known that each water molecule has a negative end and a positive end (we use the symbols "−" and "+" for these poles) and therefore is known as a **polar molecule.** The negative end of this polar molecule will be attracted toward the positively charged Li$^+$ ion, while the positive end will be attracted toward the negative Cl$^-$ ion. The interaction between the positive Li$^+$ and the negative end of the water molecule releases energy. Similarly, the interaction between the negative Cl$^-$ and the positive end of the water molecule releases energy. This energy, known as the **energy of hydration,** provides the energy required to break up the lattice of Li$^+$ ions and Cl$^-$ ions, even at room temperature.

When we write Li$^+$(aq) in solution, we indicate that many water molecules are grouped around the Li$^+$ ion and that their negative end is pointing inward in a manner that releases energy. At present, there is no completely reliable way of counting the number of water molecules around any single Li$^+$ ion in solution. Water molecules seem to move in and out around the Li$^+$ with considerable ease, and different methods of estimating numbers provide different answers. One of the best estimates, based on the size of Li$^+$, the size of water molecules, and the formulas for salts such as LiCl · 4H$_2$O, places four water molecules around each Li$^+$ (as in Figure 13-3). We could write Li(H$_2$O)$_4^+$, but we really are not sure 4H$_2$O is right! It is easier to hide our ignorance by using a less definite symbolism,

$$\text{LiCl}(s) \xrightarrow{\text{H}_2\text{O}(l)} \text{Li}^+(aq) + \text{Cl}^-(aq) \qquad (12)$$

where Li$^+$(aq) indicates lithium ions surrounded by properly oriented water molecules and Cl$^-$(aq) indicates chloride ions surrounded by properly oriented water molecules. Water molecules interact strongly with Li$^+$ and Cl$^-$ to liberate enough energy to break up the crystal.

The solution of HCl in water is only slightly different. Liquid HCl is not a good conductor of electricity; it contains no sizable concentration of ions in the liquid. Similarly, if HCl is dissolved in benzene, the system is a nonconductor—no ions are formed. The HCl in both cases is present as HCl molecules. As in the lithium chloride case, a very large amount of energy is required to pull a proton away from the Cl$^-$ ion to give H$^+$ and Cl$^-$. In the gaseous state, temperatures far in excess of 1000°C are required. Since this process proceeds readily in water at 25°C, it is logical to conclude that the protons and the chloride ions interact strongly with water molecules to provide the required energy. A real chemical reaction to produce ions from HCl and water takes place.

As in the case of Li$^+$, we cannot really determine how many water molecules surround a proton. We might write H(H$_2$O)$_4^+$ by analogy to Li(H$_2$O)$_4^+$ (Figure 13-3). But there is a big difference: the H$^+$ ion is a bare proton—it has no external electrons and differs markedly from the much larger Li$^+$ ion. This fact has led many chemists to suggest that the proton attaches itself rather strongly to a *single* water molecule to give H$_3$O$^+$, and that this positive ion then interacts with additional water molecules to give H$_3$O$^+$(aq). As seen in Figure 13-4, all three protons of H$_3$O$^+$ would be equivalent, and the positive ion would resemble the well-known NH$_4^+$ ion (Figure 13-5). It is further argued that some chemists have found evidence for the ion H$_3$O$^+$ in crystalline HClO$_4$ · H$_2$O.

The NATURE of Li$^+$(aq) and H$^+$(aq)

Fig. 13-3 Possible tetrahedral arrangements of water molecules around Li$^+$ and H$^+$ ions.

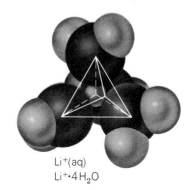

Li$^+$(aq)
Li$^+$·4H$_2$O

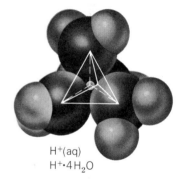

H$^+$(aq)
H$^+$·4H$_2$O

Fig. 13-4 A model for the hydronium ion, H$_3$O$^+$.

Fig. 13-5 **A model for ammonium ion, NH_4^+.**

Despite these arguments, we still do not *know* how many water molecules surround a proton. Many models, each with different numbers of water molecules, are possible; hence, we again find it convenient to hide our ignorance by using a noncommittal symbol, $H^+(aq)$. As an additional convenience in notation, we shall indicate concentration of $H^+(aq)$ in solution as $[H^+]$. The concentration of $OH^-(aq)$ will be written as $[OH^-]$. The fact that ions are hydrated is implied by the symbols.

A little later we shall want to emphasize the fact that water *has* actually served to break off a proton from a molecule. In these cases we shall find it more convenient to specify one molecule of water as one of the reactants and write

$$HX(aq) + H_2O(l) \longrightarrow H_3O^+(aq) + X^-(aq) \qquad (13)$$

Remember that this is a matter of convenience only; it does not really say anything about the true condition of the proton in water solution.

One other point of notation is important. In the introduction we indicated by relative arrow lengths the predominating species in a solution. While this scheme is helpful in a general sense, it is not quantitative and may be inconvenient in repeated use. For this reason we shall simply use *double arrows of equal length* to indicate a system at equilibrium. As we shall see later, *the size of the equilibrium constant will indicate which species predominates in the container.* With these conventions, the ionization of water is now indicated as

$$H_2O(l) \rightleftharpoons H^+(aq) + OH^-(aq) \qquad K_w = 10^{-14} \text{ at } 25°C \qquad (14)$$

13-1.3 The Size and Significance of K_w— Its Variation with Temperature

When *pure water* dissociates, one $H^+(aq)$ is produced for every $OH^-(aq)$. Hence, in *pure* water, where *the only source of ions is the dissociation of water*, the concentrations of $H^+(aq)$ and $OH^-(aq)$ must be equal.* For *pure water* we can write

$$[OH^-] = [H^+] \qquad (15)$$

If we combine this with our expression for K_w, we can write

$$K_w = [H^+][OH^-] = [H^+][H^+] = [H^+]^2 \qquad (11a)$$

or

$$[H^+] = \sqrt{K_w} = \sqrt{1.00 \times 10^{-14}} \quad \text{at } 25°C$$
$$[H^+] = 10^{-7} M$$
$$[OH^-] = 10^{-7} M$$

At equilibrium and 25°C, the concentration $[H^+]$ in pure water is 10^{-7} mole per liter, and the concentration $[OH^-]$ is 10^{-7} mole per liter. *Solutions in which $[H^+] = [OH^-]$ are called* **neutral**.† In neutral solutions at 25°C, $[H^+] = [OH^-] = 10^{-7} M$. At any given time only approximately two per

---

*This is the same as saying that at a formal party where only *couples* attend, the number of nondancing girls will be equal to the number of nondancing boys. This is not generally true of a dance for singles.

†This use of the word *neutral* for a solution with equal amounts of H^+ and OH^- has its disadvantages because the same word is used in reference to electrical neutrality. Aqueous solutions are *always* electrically neutral, whether there is an excess of either H^+ or OH^-. Other positive or negative ions always restore neutrality.

billion molecules are ionized. This should not surprise us because it explains the very low conductivity of pure water.

So far we have considered the process

$$H_2O(l) \rightleftharpoons H^+(aq) + OH^-(aq) \qquad K_w = 10^{-14} \quad \text{at } 25°C \qquad (14)$$

What happens to the value of K_w if the temperature of the system is raised to 30°C? What happens if the temperature is lowered to 20°C? Experiments show that the ionization reaction for water absorbs energy:

$$H_2O(l) + 13.7 \text{ kcal} \rightleftharpoons H^+(aq) + OH^-(aq)* \qquad (14a)$$

The measured energy term can now be used to predict how K_w changes with temperature. According to Le Chatelier's principle, an increase in temperature shifts the equilibrium in the direction of the process which absorbs heat. In this case, an increase in temperature should give a larger concentration of ions; hence, K_w should *increase* as the temperature rises. Experimental values given in Table 13-1 show that K_w does in fact increase with an increase in temperature. Conversely, decreasing the temperature should reduce the number of ions. This is also observed.

TABLE 13-1 VALUES OF K_w AT VARIOUS TEMPERATURES

Temperature (°C)	K_w
0	0.114×10^{-14}
10	0.295×10^{-14}
20	0.676×10^{-14}
25	1.00×10^{-14}
60	9.55×10^{-14}

EXERCISE 13-1

Using the data in Table 13-1, show that the concentration of $H^+(aq)$ in pure water at 60°C is 3.1×10^{-7} M. What is the concentration of $OH^-(aq)$ in this pure water at 60°C?

13-1.4 Concentrations of $H^+(aq)$ and $OH^-(aq)$ in Solutions of HCl and NaOH

In the earlier discussion we focused on the concentrations of $H^+(aq)$ and $OH^-(aq)$ in pure water. Pure water is neutral; hence, we were able to write

$$[H^+] = [OH^-] \qquad (15)$$

Does the concentration of $H^+(aq)$ always have to equal the concentration of $OH^-(aq)$ in water solution? Let us investigate this question. Suppose we add a solution of HCl containing $H^+(aq)$ and $Cl^-(aq)$ to pure water.† The original equilibrium for the water alone was

$$H_2O(l) \rightleftharpoons H^+(aq) + OH^-(aq) \qquad (14)$$

If we add $H^+(aq)$, some of the $OH^-(aq)$ will be used up until the equilibrium relationship is re-established:

$$K_w = [H^+][OH^-] \qquad (11)$$

Similarly, if we add NaOH, containing $Na^+(aq)$ and $OH^-(aq)$, $H^+(aq)$ will be used up until the equilibrium relationship is again

---

* It is usually most convenient to measure the value 13.7 kcal by measuring the energy released in the reverse process when 1 mole of water is formed from 1 mole of $H^+(aq)$ and 1 mole of $OH^-(aq)$. (Refer to Experiment 17.)

† This would be comparable to some boys coming without dates to the formal, "couple" party described earlier. The concentration of nondancing boys would be higher than the concentration of nondancing girls.

attained. *In both these cases the concentration of H^+ (aq) is no longer equal to the concentration of OH^-(aq). $[H^+]$ is equal to $[OH^-]$ only in neutral solutions.*

Let us consider some actual examples of systems containing HCl or NaOH. Suppose 0.1 mole of HCl is dissolved in enough water to give 1.0 liter of solution at 25°C. Since HCl is completely dissociated to give $H^+(aq)$ ions and $Cl^-(aq)$ ions, 0.1 mole of HCl in 1 liter will give 0.1 mole of $H^+(aq)$ and 0.1 mole of $Cl^-(aq)$ per liter. The concentration of $OH^-(aq)$ at equilibrium with the $H^+(aq)$ can be calculated easily from the equilibrium law

$$[H^+][OH^-] = 10^{-14} \text{ mole}^2/\text{liter}^2$$

or

$$[OH^-] = \frac{10^{-14} \text{ mole}^2/\text{liter}^2}{[H^+]} = \frac{10^{-14} \text{ mole}^2/\text{liter}^2}{10^{-1} \text{ mole/liter}} = 10^{-13} \text{ mole/liter}$$

Addition of 0.1 mole of HCl lowered the $[OH^-]$ from 10^{-7} to 10^{-13} mole/liter, a decrease by a factor of one million. Strangely enough, a solution of HCl in water still contains a measurable concentration of $OH^-(aq)$ ions, as demanded by the water ionization equilibrium.

Let us look at another experiment. Suppose we add 0.1 mole of NaOH to enough water to give 1 liter of solution at 25°C. Since sodium hydroxide is a strong electrolyte—almost 100 percent dissociates into $Na^+(aq)$ and $OH^-(aq)$—the concentration of both $Na^+(aq)$ and $OH^-(aq)$ will be 0.1 mole per liter. Since water is the solvent, the water equilibrium must be established:

$$H_2O(l) \rightleftharpoons H^+(aq) + OH^-(aq) \tag{14}$$

$$K_w = [H^+][OH^-] = 10^{-14} \text{ mole}^2/\text{liter}^2 \tag{11}$$

$$[OH^-] = 0.1 \text{ mole/liter}$$

Hence,

$$K_w = [H^+] \times 0.1 \text{ mole/liter} = 10^{-14} \text{ mole}^2/\text{liter}^2$$

$$[H^+] = \frac{10^{-14} \text{ mole}^2/\text{liter}^2}{0.1 \text{ mole/liter}} = 10^{-13} \text{ mole/liter}$$

In these two ordinary laboratory solutions, the concentration of $H^+(aq)$ has varied from 0.1 mole per liter to 10^{-13} mole per liter. The range is almost too large to comprehend. This would just be a chemical curiosity but for one fact: *the ions H^+(aq) and OH^-(aq) take part in many important reactions that occur in aqueous solution.* Many biological processes in solution are extremely sensitive to the concentration of $H^+(aq)$. For example, the human blood must be kept very close to the $H^+(aq)$ concentration of 6.0×10^{-8} mole per liter or severe damage and even death may result. In many solution reactions either the $H^+(aq)$ or the $OH^-(aq)$ ions take part as reactants. It is not hard to see that a million-fold or a trillion-fold change in concentration of one of the reactants could have a serious effect on the equilibrium and the course of the reaction. Furthermore, there are many reactions for which either the hydrogen ion or the hydroxide ion is a catalyst. For example, catalysis of the decomposition of formic acid by $H^+(aq)$ was discussed in Chapter 10. Formic acid is reasonably stable until the $H^+(aq)$ concentration is raised; then the rate of decomposition becomes very rapid. The concentration of $H^+(aq)$ is one of the most important variables in the study of aqueous solutions.

EXERCISE 13-2

Show that the addition of 0.010 mole of NaOH(s) to enough water to give 1 liter of solution reduces the $[H^+]$ to 1.0×10^{-12} M.

$$[H^+] = \frac{10^{-14}}{[OH^-]} = \frac{10^{-14}}{0.01} = 10^{-12}$$

EXERCISE 13-3

Suppose that 3.65 g of HCl are dissolved in enough water to give 10.0 liters of solution. What is the value of $[H^+]$? Use the expression 0.01M

$$K_w = [H^+][OH^-] = 10^{-14}$$

to show that $[OH^-] = 1.00 \times 10^{-12}$ M.

EXERCISE 13-4

In the laboratory you saw that the color of a solution of potassium chromate (K_2CrO_4) changes to the color of a solution of potassium dichromate ($K_2Cr_2O_7$) when we add a few drops of HCl solution. Write the balanced equation for the reaction between $CrO_4^{2-}(aq)$ and $H^+(aq)$ to produce $Cr_2O_7^{2-}(aq)$ and H_2O. Then explain the color change on the basis of Le Chatelier's principle.

$$2CrO_4^{2-}(aq) + 2H^+ \rightarrow Cr_2O_7^{2-}(aq) + H_2O$$

13-2 EXPERIMENTAL INTRODUCTION TO ACID AND BASE SYSTEMS

13-2.1 Experimental Identification of Acids

In trying to classify materials they found around them, early experimenters identified a group of substances having some common properties. These materials dissolved in water to give solutions which

(1) were electrical conductors;
(2) reacted with an active metal such as Zn to liberate hydrogen gas;
(3) turned blue litmus solution red and altered the color of many organic dyes, including such common liquids as the juice of red cabbage; and
(4) had a sour taste.*

Substances having these experimental properties are called **acids**.

---

* Many chemicals, however, are poisonous. Hydrogen cyanide, one of the most poisonous materials, is an acid. Obviously, taste should *not* be used by chemists to identify a substance.

青化鉀

13-2.2 A Model for Explaining the Properties of Acids

What makes an acid behave as it does? What common characteristic or structural feature can account for the properties of an acid? Let us investigate this question more carefully.

You are familiar with hydrochloric acid (HCl), nitric acid (HNO_3), acetic acid (CH_3COOH), sulfuric acid (H_2SO_4), and phosphoric acid (H_3PO_4). These acids have in common hydrogen atoms in combined form. Is hydrogen important to the behavior of acids? It might be, since Zn gives hydrogen gas in reacting with acids. Since acids conduct electric current, it is reasonable to propose that acids produce ions when dissolved in water. Since all acids have a similar group of properties, we might suggest that all contain a common ion. Based on the evidence before us, the ion $H^+(aq)$ is a likely choice. We postulate: _A substance has the properties of an acid if it can release hydrogen ions in water solution._

13-2.3 Experimental Identification of Bases

Another group of materials identified by early experimenters resemble acids in a few ways, but most of their properties contrast sharply with the properties of acids. For example, many compounds of this group resemble acids in that they contain combined hydrogen and dissolve in water to give solutions that conduct electricity. On the other hand, these materials change red litmus to blue and cause many dyes to assume a color quite different from that found in acidic solution. The substances have a bitter rather than a sour taste* and feel slippery. (Like acids, these substances are corrosive to the skin.) Finally, when we add one of these compounds to an acid, the identifying properties of both acid and the compound disappear; only electrical conductivity remains as a characteristic property of the mixture.

Materials in this second class are called **bases.** Typical bases include sodium hydroxide (NaOH), potassium hydroxide (KOH), calcium hydroxide [$Ca(OH)_2$], magnesium hydroxide [$Mg(OH)_2$], sodium carbonate (Na_2CO_3), and ammonia (NH_3).

13-2.4 A Model for Explaining the Properties of Bases

Using the line of reasoning we applied to acids, we seek a common factor that accounts for the similarities of bases. Because of their electrical conductivity, we might look for an ion. Because of their ability to counteract the properties of acids, we might look for an ion that can remove the acidic hydrogen ion, $H^+(aq)$.

Sodium hydroxide (NaOH), when dissolved in water, gives a solution having basic properties. The hydroxides of many elements from the left side of the periodic table behave in the same way. Perhaps they dissolve to form ions of the sort

$$NaOH(s) \rightleftharpoons Na^+(aq) + OH^-(aq) \qquad (16)$$

---

*If you have forgotten the danger of tasting chemicals, reread the preceding footnote.

$$KOH(s) \rightleftharpoons K^+(aq) + OH^-(aq) \quad (17)$$

$$Mg(OH)_2(s) \rightleftharpoons Mg^{2+}(aq) + 2OH^-(aq) \quad (18)$$

$$Ca(OH)_2(s) \rightleftharpoons Ca^{2+}(aq) + 2OH^-(aq) \quad (19)$$

The hydroxide ion, $OH^-(aq)$, could react with hydrogen ion to account for the second property of bases, the removal of acidic properties:

$$OH^-(aq) + H^+(aq) \rightleftharpoons H_2O(l) \quad (14)$$

The similarities among the hydroxides are obvious. Let us compare sodium carbonate (Na_2CO_3) and ammonia (NH_3). Na_2CO_3 dissolves in water to give a solution having the properties of a base. Quantitative studies of the solubilities of carbonates show that the carbonate ion, CO_3^{2-}, reacts with water. The reactions are

$$Na_2CO_3(s) \rightleftharpoons 2Na^+(aq) + CO_3^{2-}(aq) \quad (20)$$

$$CO_3^{2-}(aq) + H_2O(l) \rightleftharpoons HCO_3^-(aq) + OH^-(aq) \quad (21)$$

Equation (21) indicates that the presence of the carbonate ion, $CO_3^{2-}(aq)$, in water increases the hydroxide ion concentration, $[OH^-]$. The hydroxide ion is present in the solutions of NaOH, KOH, $Mg(OH)_2$, and $Ca(OH)_2$. The reaction of $CO_3^{2-}(aq)$ with H_2O to give $OH^-(aq)$ also provides a basis for understanding the removal of acidic properties. If $CO_3^{2-}(aq)$ readily forms bicarbonate ion, $HCO_3^-(aq)$,* then the probable reaction is

$$CO_3^{2-}(aq) + H_2O(l) \rightleftharpoons HCO_3^-(aq) + OH^-(aq) \quad (21)$$

We see that the existence of the stable bicarbonate ion, $HCO_3^-(aq)$, produces the chemical species $OH^-(aq)$, the same ion found in solutions of the hydroxides. We can postulate that $OH^-(aq)$ accounts for the slippery feel and bitter taste of the basic solutions. The stability of the bicarbonate ion also explains the removal of acidic properties through direct reaction between the carbonate ion and a hydrated proton to give the bicarbonate ion:

$$CO_3^{2-}(aq) + H^+(aq) \rightleftharpoons HCO_3^-(aq) \quad (22)$$

We have listed ammonia (NH_3) as a base. Ammonia readily forms ammonium ion, NH_4^+. Ammonia reacts with water,

$$NH_3(g) + H_2O(l) \rightleftharpoons NH_4^+(aq) + OH^-(aq) \quad (23)$$

and with hydrogen ion,

$$NH_3(g) + H^+(aq) \rightleftharpoons NH_4^+(aq) \quad (24)$$

The formation of $NH_4^+(aq)$ explains the fact that ammonia has basic properties. The reaction of ammonia with water produces hydroxide ion, which, by our postulate, accounts for the taste and feel

---

* Bicarbonate ion, HCO_3^-, is also termed "hydrogen carbonate" or "monohydrogen carbonate."

of basic solutions. The direct reaction of an ammonia molecule and a proton to give the ammonium ion shows how ammonia can act to destroy the acidic properties of any solution which contains hydrogen ions.

Investigation of the reactions of other compounds having basic properties shows that each compound can produce hydroxide ions in water. The $OH^-(aq)$ ions may be produced directly [as when $NaOH(s)$ dissolves in water] or by adding a substance which reacts chemically with water [as when $Na_2CO_3(s)$ and $NH_3(g)$ dissolve in water]:

$$NaOH(s) \rightleftharpoons Na^+(aq) + OH^-(aq) \qquad (16)$$

$$CO_3^{2-}(aq) + H_2O(l) \rightleftharpoons HCO_3^-(aq) + OH^-(aq) \qquad (21)$$

$$NH_3(g) + H_2O(l) \rightleftharpoons NH_4^+(aq) + OH^-(aq) \qquad (23)$$

Furthermore, *any substance that can produce hydroxide ions in water can also combine with hydrogen ions*:

$$OH^-(aq) + H^+(aq) \rightleftharpoons H_2O(l) \qquad (14)$$

$$CO_3^{2-}(aq) + H^+(aq) \rightleftharpoons HCO_3^-(aq) \qquad (22)$$

$$NH_3(g) + H^+(aq) \rightleftharpoons NH_4^+(aq) \qquad (24)$$

Since production of $OH^-(aq)$ and reaction with $H^+(aq)$ go hand-in-hand in aqueous solutions, we can describe a base *either* as a substance that produces $OH^-(aq)$ *or* as a substance that can react with $H^+(aq)$. In solvents other than water, the latter description is generally more useful. Therefore, the more general postulate is: *A substance has the properties of a base if it can combine with hydrogen ions.*

13-2.5 Acids and Bases: A Summary

Let us repeat our two definitions and explanations.

DEFINITIONS

An *acid* is a hydrogen-containing substance that has the following properties when dissolved in water:

(1) It is an electrical conductor.
(2) It reacts with Zn to give $H_2(g)$.
(3) It turns blue litmus red.
(4) It tastes sour.

A *base* is a substance that has the following properties when dissolved in water:

(1) It is an electrical conductor.
(2) It reacts with an acid, removing the acidic properties.
(3) It turns red litmus blue.
(4) It tastes bitter.
(5) It feels slippery.

EXPLANATIONS

(1) *A substance is an acid if it can release hydrogen ions, H^+ (aq).*
(2) *A substance is a base if it can react with hydrogen ions, H^+ (aq).*

EXERCISE 13-5

Sodium acetate (CH_3COONa) serves as a base in water solution.

(1) Write the equation by which sodium acetate (or its active component) removes hydrogen ions from water solution.
(2) Why is sodium acetate classed as a base?

13-3 CONCEPTUAL AND OPERATIONAL DEFINITIONS

Looking back on the earlier sections reveals that we have used two different definitions of an acid and two different definitions of a base. The first definition of each came from the laboratory; it told us what an acid or base will do and how to recognize each. We can summarize such descriptions by listing the properties. Let us compare acids and bases in terms of these properties:

Acid	*Base*
(1) electrical conductor	(1) electrical conductor
(2) reacts with Zn to give $H_2(g)$	(2) destroys the properties of acids
(3) makes blue litmus red	(3) makes red litmus blue
(4) tastes sour	(4) tastes bitter
(5) frequently corrosive to skin	(5) frequently corrosive to skin
	(6) feels slippery

These are called **operational definitions.** To understand this term, consider the meaning of the word *definition*. According to one dictionary, *definition* means "a statement of what a thing is." By using a definition, we can sort the universe into two piles, one containing those objects that fit the definition and another containing those that do not. Our *operational* definition gives the criteria by which this sorting process can be carried out. An operational definition is, then, one that lists the measurements or observations (the *operations*) by which we decide if an object belongs in a given group.

The second type of definition we have used is a **conceptual definition.** It is more concerned with the question "Why?" It seeks to define the group in terms of an explanation of *why the class has its properties.* When we state that an acid is a substance that releases hydrogen ions to aqueous solution, we are using a *conceptual* definition. We are on less secure experimental ground than with the operational definition, but the conceptual definition is far more useful in the construction of chemical models and in *expansion of the definitions to solvents other than water.* Conceptual definitions lead to new research and to the development of hidden likenesses, or regularities.

Each type of definition has its merits; neither is *the* definition.

We shall see that, as more and more complicated systems are considered, the operational and conceptual definitions of acids and bases must be expanded. The concept of an acid is a device used by chemists to correlate different kinds of observations. The definition of an acid has been expanded accordingly. We shall confine our attention to water solutions and to the operational and conceptual definitions given thus far. Remember, however, that there is no *absolute* definition of an acid; we use this classification scheme because it is appropriate to the system we are studying. Every musician knows that there is no *one* key in which a symphony must be written; the key is chosen to evoke the desired mood. Similarly, definitions of acids are selected by the chemist on the basis of their ability to explain or simplify the system he is studying.

13-4 ACID-BASE TITRATIONS

One of the observations we used in defining a base was that a base can neutralize or destroy certain properties characteristic of an acid. The conceptual definition we used explains this ability in terms of a simple reaction:

$$OH^-(aq) + H^+(aq) \longrightarrow H_2O(l) \qquad (14a)$$
$$\text{BASE} \qquad \text{ACID} \qquad \text{WATER}$$

Let us see how this concept applies to mixing HCl and NaOH in the same solution.

13-4.1 HCl and NaOH in the Same Solution: Excess HCl

Suppose that to 0.100 liter of 1.00 M HCl solution we add 0.090 mole of NaOH(s). By adding solid NaOH we keep the volume of the solution essentially constant. Now we have both $H^+(aq)$ and $OH^-(aq)$ in relatively high concentrations in the *same* solution. What will happen? Immediately after the sodium hydroxide dissolves, the concentrations of $H^+(aq)$ and $OH^-(aq)$ far exceed the equilibrium values. The trial product $[H^+][OH^-]$ far exceeds 1.00×10^{-14} mole²/liter²:

$$\text{initial } [H^+] = 1.00 \ M$$

$$\text{initial } [OH^-] = \frac{0.090 \text{ mole}}{0.100 \text{ liter}} = 0.90 \ M$$

$$\text{trial product} = [H^+] \times [OH^-] = 9.00 \times 10^{-1} \text{ mole}^2/\text{liter}^2$$

The equilibrium relationship, $[H^+][OH^-] = 10^{-14}$, can be achieved most easily by removing both $H^+(aq)$ and $OH^-(aq)$. The reaction of these two ions to form water is clearly indicated.

$$OH^-(aq) + H^+(aq) \longrightarrow H_2O(l) \qquad (14a)$$

Since K_w is so small, almost all the $OH^-(aq)$ is consumed when the HCl is present in excess. In our example $[H^+]$ initially exceeds $[OH^-]$ by 0.1 mole per liter:

$$\text{initial } [H^+] - \text{initial } [OH^-] = \text{excess } [H^+]$$
$$1.00 \ M \quad - \quad 0.90 \ M \quad = \quad 0.10 \ M$$

If [H$^+$] is 0.1 mole per liter, then

$$[H^+][OH^-] = 10^{-14} \text{ mole}^2/\text{liter}^2$$

$$[OH^-] = \frac{10^{-14} \text{ mole}^2/\text{liter}^2}{10^{-1} \text{ mole/liter}} = 10^{-13} \text{ mole/liter}$$

The concentration of OH$^-$(aq), 10^{-13} mole per liter, is a million times smaller than the OH$^-$(aq) concentration of pure water. Nearly all the OH$^-$(aq) has been *neutralized* or removed by the HCl present.

EXERCISE 13-6

Suppose that 0.099 mole of NaOH(s) is added to 0.100 liter of 1.00 *M* HCl.

(1) How many moles of ionized HCl are present in the final solution? ~~0.001~~
(2) From the moles of HCl and the volume, calculate the concentration of excess H$^+$(aq).   0.01 M
(3) Calculate the concentration of OH$^-$(aq) at equilibrium (see your calculations for Exercise 13-3). 10^{-12} mole/litre

13-4.2 HCl and NaOH in the Same Solution: Excess NaOH

Returning to our original 0.100 liter of 1.00 *M* HCl, let us now consider the addition of 0.101 mole of NaOH(s). Again, we have added both H$^+$(aq) and OH$^-$(aq) to the *same* solution, and the concentrations immediately after mixing do not satisfy the equilibrium expression:

$$\text{initial } [H^+] = 1.00 \ M$$

$$\text{initial } [OH^-] = \frac{0.101 \text{ mole}}{0.100 \text{ liter}} = 1.01 \ M$$

$$\text{initial product} = [H^+] \times [OH^-] = 1.01 \text{ moles}^2/\text{liter}^2$$

far exceeding 1.00×10^{-14} mole^2/liter^2.

This solution contains excess hydroxide ion, OH$^-$(aq); therefore, essentially all the H$^+$(aq) will be consumed, forming water:

$$\text{initial } [OH^-] - \text{initial } [H^+] = \text{excess } [OH^-]$$
$$1.01 \ M \quad - \quad 1.00 \ M \quad = \quad 0.01 \ M$$

In Exercise 13-2 we calculated the equilibrium concentration of H$^+$ in a solution containing [OH$^-$] = 0.010 *M*. You will recall that

$$[H^+] = \frac{10^{-14} \text{ mole}^2/\text{liter}^2}{10^{-2} \text{ mole/liter}} = 10^{-12} \text{ mole/liter}$$

13-4.3 HCl and NaOH in the Same Solution: No Excess of Either

In each example used in this section, a number of moles of NaOH was added to 0.100 liter of 1.00 *M* HCl. Either HCl or NaOH was in excess. Reaction between H$^+$(aq) and OH$^-$(aq) consumes essentially all the constituent not in excess. Let us now consider the case in which there is an excess of *neither* HCl nor NaOH.

Suppose we add 0.100 mole of NaOH to 0.100 liter of 1.00 M HCl. The initial values of [H$^+$] and [OH$^-$] are equal and their product far exceeds 1.00×10^{-14}:

$$\text{initial [H}^+\text{]} = 1.00 \ M$$

$$\text{initial [OH}^-\text{]} = \frac{0.100 \text{ mole}}{0.100 \text{ liter}} = 1.00 \ M$$

$$\text{initial product} = \text{[H}^+\text{]} \times \text{[OH}^-\text{]} = 1.00$$

Reaction between H$^+(aq)$ and OH$^-(aq)$ must again occur, forming water:

$$\text{OH}^-(aq) + \text{H}^+(aq) \rightleftharpoons \text{H}_2\text{O}(l) \qquad (14)$$

Since 1 mole of OH$^-(aq)$ consumes 1 mole of H$^+(aq)$, the concentrations [H$^+$] and [OH$^-$] remain equal as the neutralization reaction between H$^+(aq)$ and OH$^-(aq)$ proceeds. When equilibrium is reached, they will still be equal. This is exactly the situation in pure water. As we saw in Section 13-1.3,

$$[\text{H}^+] = [\text{OH}^-] = \sqrt{K_w} = 1.00 \times 10^{-7} \ M \qquad (15a)$$

A solution containing exactly equivalent amounts of acid and base is neither acidic nor basic; it is neutral. *If equal numbers of moles of HCl and NaOH are added to water, the final solution is neutral.*

13-4.4 Progressive Addition of NaOH to HCl: A Titration

Now we have considered the progressive addition of NaOH to a fixed amount of HCl solution. The results are compiled in Table 13-2. We see that [H$^+$] changes by a factor of 10^{10} as the initial [OH$^-$] is changed from 0.99 M to 1.01 M. Since [H$^+$] can be easily followed experimentally, it provides a sensitive means of observing the mixing of an acid and a base. The process is called a **titration,** *the progressive addition of a base to an acid.**

In an acid-base titration, carefully measured amounts of a basic solution are added to a known volume of an acidic solution. The acidic solution contains some substance that provides visual evidence of the magnitude of [H$^+$]. The dye litmus is one such substance. As mentioned in Sections 13-2.1 and 13-2.3, litmus is red in solutions with excess H$^+(aq)$ and blue in solutions with excess OH$^-(aq)$. Fortunately, the dye color is extremely sensitive to very small amounts of H$^+(aq)$. In solutions containing [H$^+$] of $10^{-6} \ M$, the dye is red. In solutions containing [H$^+$] of $10^{-8} \ M$, the dye is blue. As we see in Table 13-2, the *slightest* addition of excess OH$^-$ causes a great change in [H$^+$] and a marked change in color.

There are many dyes whose color is very sensitive to the concentration of H$^+(aq)$. Litmus is one of the poorer examples. Such dyes are called **acid-base indicators.** All do not change color where [H$^+$] = [OH$^-$]. Some change on the acidic side, some on the basic

---

* The addition of an acid to a base is also a titration.

side. Litmus and bromthymol blue change at the neutral point. In Figure 13-6, (on next page) the colors of other acid-base indicators in solutions of different [H⁺] are shown.

13-5 THE pH SCALE

Chemists and biologists have found it convenient to use a quantity known as the *pH* for indicating in shorthand form the hydrogen-ion concentration in any solution. The *pH* is officially defined by the expression

$$pH = -\log_{10} [H^+] \qquad (25)$$

The *pH* values for the solutions we have been dealing with are very easily obtained from values for [H⁺]. Consider the following example: If the solution contains 0.01 *M* HCl (100 percent ionized), we write [H⁺] = 0.01 *M* = 10^{-2} *M*. Once the hydrogen-ion concentration has been expressed in this exponential form (10^{-2}), the *pH* is easily obtained by simply taking the exponent of 10 (-2, in this case) and changing its sign: *pH* = 2. Values of *pH* for solutions in Table 13-2 are shown in that table. Note that:

(1) A *pH* value of 7 is neutral.
(2) *pH* values less than 7 are acidic.
(3) *pH* values greater than 7 are basic or alkaline.

TABLE 13-2 CONCENTRATIONS OF H⁺ AND OH⁻ IN SOLUTIONS CONTAINING HCl AND NaOH

Initial [H⁺] (M)	Initial [OH⁻] (M)	Excess [H⁺] or [OH⁻] (M)	[H⁺][OH⁻] = 10^{-14} Calc [H⁺] (M)	Calc [OH⁻] (M)	Acidic or Basic	log H⁺	pH	Color of Solution (litmus)
1.00	none	1.00 H⁺	10^0 or 1.00	10^{-14}	acidic	0	0	red
1.00	0.90	0.10 H⁺	10^{-1}	10^{-13}	acidic	-1	1	red
1.00	0.99	0.01 H⁺	10^{-2}	10^{-12}	acidic	-2	2	red
1.00	1.00	none	10^{-7}	10^{-7}	neutral	-7	7	purple
1.00	1.01	0.01 OH⁻	10^{-12}	10^{-2}	basic	-12	12	blue

13-6 STRENGTHS OF ACIDS

Earlier in this chapter we noted that HCl ionizes almost completely in water solution:

$$HCl(aq) \longrightarrow H^+(aq) + Cl^-(aq) \qquad (7)$$

For all practical purposes, the reaction goes completely to H⁺(*aq*) and Cl⁻(*aq*). The concentration of HCl(*aq*) remaining in the solution is *extremely small*. Because the concentration of HCl(*aq*) is so small, the equilibrium constant for HCl ionization is *very large*.

$$K_{HCl} = \frac{[H^+(aq)][Cl^-(aq)]}{[HCl(aq)]} \Bigg\} = \text{VERY LARGE} \qquad (26)$$

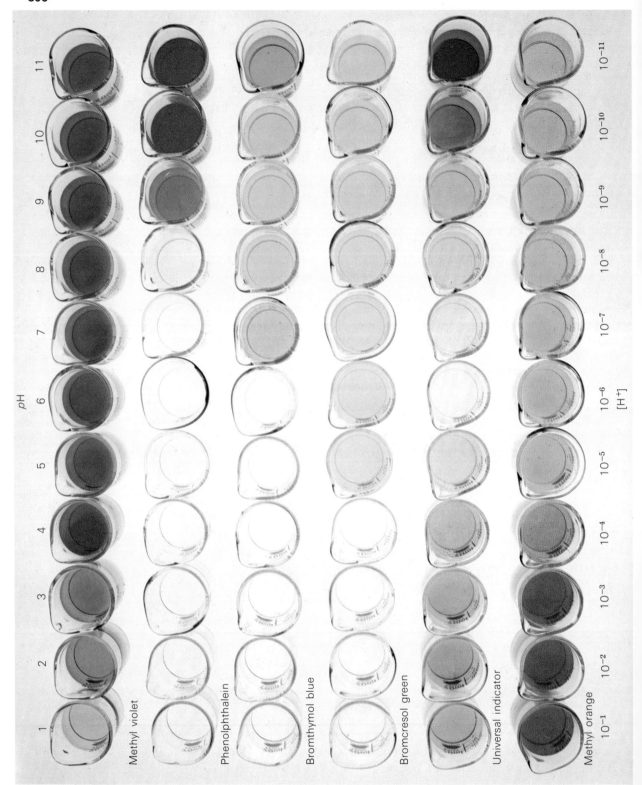

Fig. 13-6 Some acid-base indicators in solutions with different H⁺ concentrations.

An acid which gives a large equilibrium constant is known as a strong acid.

We also considered the ionization of acetic acid, which ionizes to a much smaller degree. We wrote

$$CH_3COOH(aq) \rightleftharpoons H^+(aq) + CH_3COO^-(aq) \qquad (6)$$

Conductivity studies showed that very little acetic acid was ionized. The equilibrium constant for this process takes the form

$$K_{CH_3COOH} = \frac{[H^+][CH_3COO^-]}{[CH_3COOH]} = 1.8 \times 10^{-5} \qquad (27)$$

The *small value* of the equilibrium constant means that very little acetic acid is ionized. As we noted earlier, this value, sometimes called K_A or the **acid equilibrium constant,** is a precise way of indicating acid strength. It is a most useful number for characterizing an acid.

If we now compare quantitatively the electrical conductivity of 0.1 M solutions of CH_3COOH and HF, we find that HF has about ten times as many ions present as acetic acid (ten times the conductivity). HF is a stronger acid than CH_3COOH. The equilibrium constant reveals this fact:

$$K_{HF} = \frac{[H^+][F^-]}{[HF]} = 6.7 \times 10^{-4} \qquad (28)$$

The value 6.7×10^{-4} is significantly larger than 1.8×10^{-5}.

We can express these ideas in terms of a general acid, HB:

$$HB(aq) \rightleftharpoons H^+(aq) + B^-(aq) \qquad (29)$$

where

$$K_A \text{ for acid } HB = K_{HB} = \frac{[H^+(aq)][B^-(aq)]}{[HB(aq)]} \qquad (30)$$

Values of K_A for a number of acids are listed in Table 13-3. The strong acids have large values of K_A; the weak acids have small values of K_A.

EXERCISE 13-7

Which of the following acids is the strongest acid and which the weakest?

Nitrous acid (HNO_2) $K_{HNO_2} = 5.1 \times 10^{-4}$ *weakest*
Sulfurous acid (H_2SO_3) $K_{H_2SO_3} = 1.7 \times 10^{-2}$ *strongest*
Phosphoric acid (H_3PO_4) $K_{H_3PO_4} = 7.1 \times 10^{-3}$

13-6.1 Determination of K_A

How do we determine values of this useful number, K_A, from laboratory data? The experiments must provide a measurement of hydrogen-ion concentration. A number of procedures are available.

TABLE 13-3 RELATIVE STRENGTHS OF ACIDS IN AQUEOUS SOLUTION AT ROOM TEMPERATURE, $K_A = [H^+][B^-]/[HB]$

Acid	Strength	Reaction	K_A
HCl	very strong	$HCl(l) \longrightarrow H^+(aq) + Cl^-(aq)$	very large
HNO_3	↓	$HNO_3(l) \longrightarrow H^+(aq) + NO_3^-(aq)$	very large
H_2SO_4	very strong	$H_2SO_4 \longrightarrow H^+(aq) + HSO_4^-(aq)$	very large
HSO_4^-	strong	$HSO_4^-(aq) \longrightarrow H^+(aq) + SO_4^{2-}(aq)$	1.3×10^{-2}
HF	weak	$HF(aq) \longrightarrow H^+(aq) + F^-(aq)$	6.7×10^{-4}
CH_3COOH	↓	$CH_3COOH(aq) \longrightarrow H^+(aq) + CH_3COO^-(aq)$	1.8×10^{-5}
H_2CO_3 $(CO_2 + H_2O)$	↓	$H_2CO_3(aq) \longrightarrow H^+(aq) + HCO_3^-(aq)$	4.4×10^{-7}
H_2S	weak	$H_2S(aq) \longrightarrow H^+(aq) + HS^-(aq)$	1.0×10^{-7}
NH_4^+	↓	$NH_4^+(aq) \longrightarrow H^+(aq) + NH_3(aq)$	5.7×10^{-10}
HCO_3^-	↓	$HCO_3^-(aq) \longrightarrow H^+(aq) + CO_3^{2-}(aq)$	4.7×10^{-11}
H_2O	very weak	$H_2O(l) \longrightarrow H^+(aq) + OH^-(aq)$	1.8×10^{-16}*

* The equilibrium constant, K_A, for water equals

$$\frac{K_w}{[H_2O]} = \frac{1.00 \times 10^{-14}}{55.6} = \frac{[H^+][OH^-]}{[H_2O]}$$

(See Section 13-1.1.)

One of the most precise methods involves a careful measurement of cell voltage. The method will be discussed in the next chapter. Acid-sensitive dyes (see Figure 13-6) offer the easiest estimate of $H^+(aq)$ when electronic meters are not available. Let us consider a specific example.

Benzoic acid (C_6H_5COOH) is a white solid with only moderate solubility in water. The aqueous solution has all the characteristic properties of acids listed in Section 13-2.1. The ionization equation and acid equilibrium constant, $K_{C_6H_5COOH}$, are

$$C_6H_5COOH(aq) \rightleftharpoons H^+(aq) + C_6H_5COO^-(aq) \qquad (31)$$

$$K_{C_6H_5COOH} = \frac{[H^+][C_6H_5COO^-]}{[C_6H_5COOH]} \qquad (32)$$

To determine $K_{C_6H_5COOH}$, we need to obtain experimental values for the concentrations of $H^+(aq)$, $C_6H_5COO^-(aq)$, and $C_6H_4COOH(aq)$ in a given water solution. Once these values are obtained, the calculation of the equilibrium constant is simple. The following experiment gives us all the information needed.

A 1.22-g sample of solid benzoic acid is dissolved in enough water to give 1 liter of solution at 25°C. Several drops of methyl orange acid-base indicator are dropped into this solution, and the color of the resulting liquid is compared to those of solutions containing methyl orange and known concentrations of $H^+(aq)$ (see Figure 13-6). The best color match is obtained with a known solution containing 8×10^{-4} mole $H^+(aq)$/liter. We can therefore conclude that the $H^+(aq)$ concentration of the benzoic acid is about 8×10^{-4} M. We write

$$[H^+] = 8 \times 10^{-4} M$$

Equation (31) tells us that when pure benzoic acid is dissolved in water, the concentration of $H^+(aq)$ is the same as that of the benzoate ion, $C_6H_5COO^-(aq)$,* since each mole of benzoic acid produces 1 mole of $H^+(aq)$ and 1 mole of $C_6H_5COO^-(aq)$ when it dissociates. We now have

$$[H^+] = 8 \times 10^{-4}\ M$$

and

$$[C_6H_5COO^-] = 8 \times 10^{-4}\ M$$

The only information still needed is the concentration of undissociated $C_6H_5COOH(aq)$. We dissolved 1.22 g of benzoic acid to make 1 liter of solution. Since the molar weight of benzoic acid is 122 g/mole, the number of moles of benzoic acid is

$$\frac{1.22\ \text{g benzoic acid}}{122\ \text{g}\ \dfrac{\text{benzoic acid}}{\text{mole}}} = 0.010\ \text{mole}$$

We have, then, 0.010 mole of benzoic acid in a liter of solution. The original benzoic acid concentration is 0.0100 M. To generate an $H^+(aq)$ concentration of 8×10^{-4} mole per liter, we had to have 8×10^{-4} mole per liter of the original benzoic acid ionize. [See equation (31) for ionization process.] The final concentration of undissociated benzoic acid in the solution will be

$$\begin{Bmatrix} \text{original concn} \\ \text{benzoic acid} \\ \text{in solution} \end{Bmatrix} - \begin{Bmatrix} \text{concn} \\ \text{benzoic acid} \\ \text{lost by} \\ \text{ionization} \end{Bmatrix} = \begin{Bmatrix} \text{final} \\ \text{equilibrium concn} \\ \text{benzoic acid} \end{Bmatrix} \quad (33)$$

or

$$0.0100\ M\ -\ 0.0008\ M\ =\ 0.0092\ M$$

We can now write

$$[H^+] = 8 \times 10^{-4}\ M$$
$$[C_6H_5COO^-] = 8 \times 10^{-4}\ M$$
$$[C_6H_5COOH] = 0.0092 = 9 \times 10^{-3}\ M$$

Then $K_{C_6H_5COOH}$ is

$$K_{C_6H_5COOH} = \frac{[8 \times 10^{-4}][8 \times 10^{-4}]}{9 \times 10^{-3}} = 7 \times 10^{-5}$$

Since the concentration of $H^+(aq)$ was measured to only one significant figure, the answer can have no more than one significant figure.

---

*A solution containing only pure benzoic acid may be compared to the formal dance involving *only couples,* where the concentration of nondancing boys equals the concentration of nondancing girls.

13-6.2 The Use of K_A Values—The Calculation of $[H^+]$ in an Acid Solution

Let us use the value of K_A we just measured to determine the $[H^+]$ in a solution that contains 0.41 g per liter of benzoic acid. In this solution (pure benzoic acid in water), the concentration of $H^+(aq)$ is again equal to the concentration of benzoate ion, $C_6H_5COO^-(aq)$.

$$[H^+] = [C_6H_5COO^-] \qquad (34)$$

The concentration of benzoic acid before ionization took place was 0.0033 mole/liter (0.41 g/122 g/liter). If we remember the relationship

$$\begin{Bmatrix} \text{original concn} \\ \text{benzoic acid} \\ \text{in solution} \end{Bmatrix} - \begin{Bmatrix} \text{concn} \\ \text{benzoic acid} \\ \text{lost by} \\ \text{ionization} \end{Bmatrix} = \begin{Bmatrix} \text{final} \\ \text{equilibrium concn} \\ \text{benzoic acid} \end{Bmatrix} \qquad (33)$$

and the relationship from equation (34),

$$\begin{Bmatrix} \text{concn} \\ \text{benzoic acid} \\ \text{lost by} \\ \text{ionization} \end{Bmatrix} = \begin{Bmatrix} \text{concn } H^+(aq) \\ \text{formed in} \\ \text{solution} \end{Bmatrix} \qquad (34a)$$

we can write

$$\begin{Bmatrix} \text{original concn} \\ \text{benzoic acid} \\ \text{in solution} \end{Bmatrix} - \begin{Bmatrix} \text{concn} \\ \text{benzoic acid} \\ \text{lost by} \\ \text{ionization} \end{Bmatrix} = \begin{Bmatrix} \text{final} \\ \text{equilibrium concn} \\ \text{benzoic acid} \end{Bmatrix} \qquad (33)$$

$$0.0033\ M \qquad\qquad [H^+(aq)] \qquad\qquad [C_6H_5COOH(aq)]$$

We can then write

$$\frac{[H^+][C_6H_5COO^-]}{[C_6H_5COOH]} = 7 \times 10^{-5} = \frac{[H^+][H^+]}{0.0033 - [H^+]}$$

The equation

$$\frac{[H^+]^2}{0.0033 - [H^+]} = 7 \times 10^{-5}$$

can be solved, but the algebra can be greatly simplified by making an assumption that permits us to obtain an approximate answer easily. The quantity $0.0033 - [H^+]$ is probably not too different from 0.0033, since a relatively small percentage of benzoic acid is lost by ionization. As you will remember, in our calculation of K_A we found that $0.0100 - 0.0008 = 0.0092\ M$ benzoic acid at equilibrium. The final value of 0.0092 differs from the original value 0.0100 by less than 10 percent! Let us then assume as a reasonable approximation that $0.0033 - [H^+]$ is very close to 0.0033.

We can then write

$$\frac{[H^+]^2}{0.0033} = 7 \times 10^{-5}$$

$$[H^+]^2 = 23 \times 10^{-8}$$

$$[H^+] \approx 5 \times 10^{-4} \, M$$

The concentration of H^+ is about 5×10^{-4} mole per liter. To complete the calculation we must check our assumption. We see that 0.0005 is small compared to 0.0033. The assumption is a reasonably good one.

13-6.3 Another Example of the Calculation of [H⁺]

Other calculations using K_A are frequently made. As an example, suppose a chemist needs to know the hydrogen-ion concentration in a solution containing both 0.010 M benzoic acid (C_6H_5COOH) and 0.030 M sodium benzoate (C_6H_5COONa). Of course, he could go to the laboratory and investigate the colors of indicator dyes placed in the solution. However, it is easier to calculate the value of $[H^+]$ using the accurate value of K_A listed in Appendix 6.

Sodium benzoate is a strong electrolyte; its aqueous solutions contain sodium ions, $Na^+(aq)$, and benzoate ions, $C_6H_5COO^-(aq)$. Hence, the equilibrium involved is the same as before:

$$C_6H_5COOH(aq) \rightleftharpoons H^+(aq) + C_6H_5COO^-(aq) \qquad (31)$$

At equilibrium, the concentrations must be in accord with the equilibrium expression. That is,

$$K_A = \frac{[H^+][C_6H_5COO^-]}{[C_6H_5COOH]} = 6.6 \times 10^{-5} \qquad (32)$$

In Section 13-6.2, we calculated the $[H^+]$ in a solution of benzoic acid. We noted that the concentration of undissociated benzoic acid was

$$\begin{Bmatrix} \text{original concn} \\ \text{benzoic acid} \\ \text{in solution} \end{Bmatrix} - \begin{Bmatrix} \text{concn} \\ \text{benzoic acid} \\ \text{lost by} \\ \text{ionization} \end{Bmatrix} = \begin{Bmatrix} \text{final} \\ \text{equilibrium concn} \\ \text{benzoic acid} \end{Bmatrix} \qquad (33)$$

Again, we shall neglect the concentration of benzoic acid lost through ionization and write

$$[C_6H_5COOH] \approx \begin{Bmatrix} \text{original concn} \\ \text{benzoic acid} \\ \text{in solution} \end{Bmatrix} = 0.010 \, M$$

In the same way we can write for C_6H_5COONa

$$\begin{Bmatrix} \text{final} \\ \text{equilibrium concn} \\ C_6H_5COO^- \end{Bmatrix} = \begin{Bmatrix} \text{original concn} \\ C_6H_5COONa \\ \text{in solution} \end{Bmatrix} + \begin{Bmatrix} \text{concn } C_6H_5COO^- \\ \text{gained from} \\ \text{ionization} \\ \text{benzoic acid} \end{Bmatrix} \qquad (35)$$

ANOTHER EXAMPLE of the CALCULATION of [H⁺]

SVANTE AUGUST ARRHENIUS (1859–1927)

The early career of this great Swedish chemist was filled with a battle for acceptance. In his doctoral research, Arrhenius collected voluminous data on the passage of electricity through aqueous solutions. He then formulated a carefully considered hypothesis that aqueous solutions contain charged species, ions. This was so revolutionary that his degree was only grudgingly granted.

Years were spent in travel, writing, and research, seeking acceptance for his theory, and much controversy surrounded him. In 1893, however, he was appointed a professor in Stockholm, and within two years was elected President of the University and had received the Nobel Prize. In 1905, Arrhenius was named Director of the Nobel Institute for Physical Chemistry and continued as a tireless experimenter and versatile scientist until his death.

Arrhenius' success must be credited not only to his brilliance as a scientist, but also to his conviction. His understanding of the electrical properties of aqueous solutions was so far ahead of contemporary thought that it would have been ignored but for his confidence in it and his refusal to abandon it.

The concentration of $C_6H_5COO^-(aq)$ from ionization of benzoic acid will be very small because benzoic acid is a weak acid. Thus, we can write

$$[C_6H_5COO^-] \approx \begin{Bmatrix} \text{initial concn} \\ C_6H_5COONa \end{Bmatrix} = 0.030\ M$$

We can then use the expression for the acid ionization constant:

$$K_A = 6.6 \times 10^{-5} = \frac{[H^+][C_6H_5COO^-]}{[C_6H_5COOH]} = \frac{[H^+][0.030]}{0.010} \quad (32)$$

Multiplying both sides of the equation by 0.010 and dividing both sides by 0.030, we obtain

$$[H^+] = \frac{6.6 \times 10^{-5} \times 0.010}{0.030} = 2.2 \times 10^{-5}\ M$$

The calculation is not completed until we validate the assumption. Was it reasonable to assume that the concentrations of benzoate ion and benzoic acid were not changed by the ionization of benzoic acid? Let us put some numbers into equation (33) for the concentration of undissociated benzoic acid:

$$\begin{Bmatrix}\text{final} \\ \text{equilibrium concn} \\ C_6H_5COOH \end{Bmatrix} = \begin{Bmatrix}\text{original concn} \\ C_6H_5COOH \end{Bmatrix} - \begin{Bmatrix}\text{concn }C_6H_5COOH \\ \text{lost by} \\ \text{ionization} \end{Bmatrix} \quad (33)$$

$$\begin{Bmatrix}\text{final} \\ \text{equilibrium concn} \\ C_6H_5COOH \end{Bmatrix} = 0.010\ M - 0.000022\ M \approx 0.010\ M$$

The concentration loss due to ionization is equal to the concentration of $H^+(aq)$, since every $H^+(aq)$ formed by ionization takes away one molecule of C_6H_5COOH. The final concentration of C_6H_5COOH is then 0.010 M within the limits of measurement. The correlation is less than the uncertainty in the measured value of the original $[C_6H_5COOH]$.

EXERCISE 13-8

Show that the concentration of benzoate ion is determined to a good approximation by the sodium benzoate added.

13-7 AN EXPANSION OF ACID-BASE CONCEPTS

13-7.1 The Proton-Transfer Concept of Acids and Bases

We have explained the properties of acids in terms of their abilities to release hydrogen ions, $H^+(aq)$. Thus, acetic acid is a weak acid because the ionization reaction releases $H^+(aq)$ only slightly:

$$CH_3COOH(aq) \rightleftharpoons H^+(aq) + CH_3COO^-(aq) \quad (6)$$

We have explained the properties of bases in terms of their ability to react with hydrogen ion. Ammonia is therefore a base:

$$NH_3(aq) + H^+(aq) \rightleftharpoons NH_4^+(aq) \quad (24)$$

Now consider the result of mixing aqueous solutions of acetic acid and ammonia. The reaction that occurs can be broken down into a sequence of reactions:

$$CH_3COOH(aq) \rightleftharpoons H^+(aq) + CH_3COO^-(aq) \quad (6)$$

$$NH_3(aq) + H^+(aq) \rightleftharpoons NH_4^+(aq) \quad (24)$$

The net reaction is

$$CH_3COOH(aq) + NH_3(aq) \rightleftharpoons CH_3COO^-(aq) + NH_4^+(aq) \quad (36)$$

Practically, the sum of reactions (6) and (24) is reaction (36). *Acetic acid acts as an acid* in giving a proton to ammonia to form the ammonium ion, $NH_4^+(aq)$, just as it gives a proton to water to form the hydronium ion, $H_3O^+(aq)$. In either case, acetic acid releases hydrogen ions: in its ionization in water, acetic acid releases hydrogen ions and forms $H^+(aq)$; and in its reaction with ammonia, acetic acid releases hydrogen ions to NH_3, forming NH_4^+. Similarly, *ammonia acts as a base* by reacting with the hydrogen ion released by acetic acid. The reaction between acetic acid and ammonia is, then, an acid-base reaction, although the net reaction, (36), does not explicitly show $H^+(aq)$.

By going one step further, we can view acid-base reactions more broadly. Suppose we mix aqueous solutions of ammonium chloride (NH_4Cl) and sodium acetate (CH_3COONa). One sniff tells us that ammonia has been formed. The following reaction has occurred:

$$NH_4^+(aq) + CH_3COO^-(aq) \rightleftharpoons CH_3COOH(aq) + NH_3(aq) \quad (37)$$

This is exactly the reverse of reaction (36). We see that it, too, is an acid-base reaction! Once again there is an acid that releases H^+, NH_4^+, and a base that accepts H^+, CH_3COO^-. Once again, the net effect of the reaction is transfer of a hydrogen ion from one species to another.

We see that the acid-base reaction between acetic acid and ammonia gives two products—one an acid, NH_4^+, and the other a base, CH_3COO^-. A little thought will convince you that every acid-base reaction does so. The transfer of a hydrogen ion from an acid to a base necessarily implies that it might be "handed back." The reaction of "handing it back," the reverse reaction, is just as much a **hydrogen-ion transfer,** or an **acid-base reaction,** as is the original transfer.

Notice that we are now referring to reactions in which a hydrogen ion is transferred from an acid to a base without specifically involving the aqueous species $H^+(aq)$. A hydrogen ion, H^+, is nothing more than a proton. Consequently, we can frame a more general view of acid-base reactions in terms of **proton transfer.** The main value of this view is that it is applicable to a wider range of chemical systems, *including nonaqueous systems.*

Let us generalize our view of the acid-base reaction. In our example, the following equation applies:

$$\underset{\text{ACID}}{CH_3COOH} + \underset{\text{BASE}}{NH_3} \rightleftharpoons \underset{\text{ACID}}{NH_4^+} + \underset{\text{BASE}}{CH_3COO^-} \quad (36)$$

The acetic acid acts as an acid, giving up its proton, to form acetate, CH_3COO^-, a substance that can act as a base. We can write the acetic acid-ammonia reaction in a general form:

$$HB_1 + B_2 \rightleftharpoons HB_2 + B_1$$
$$acid_1 + base_2 \rightleftharpoons acid_2 + base_1 \qquad (38)$$

We see that *an acid and a base react, through proton transfer, to form another acid and another base.*

We can use this more general view to discuss the strengths of acids. In our generalized acid-base reaction, the proton transfer implies that the chemical bond in HB_1 must be broken and that the bond in HB_2 must be formed. If the HB_1 bond is easily broken, then HB_1 will be a strong acid. Equilibrium will then tend to favor a proton transfer from HB_1 to some other base, B_2. If, on the other hand, the HB_1 bond is extremely stable, HB_1 will be a weak acid. Equilibrium will tend to favor a proton transfer from some other acid, HB_2 to base B_1, forming the stable HB_1 bond.

13-7.2 Hydronium Ion in the Proton-Transfer Theory of Acids and Bases

In the proton-transfer view of acid-base reactions, an acid and a base react to form another acid and another base. Let us see how this theory encompasses the elementary reaction between $H^+(aq)$ and $OH^-(aq)$ and the reaction in which acetic acid dissociates:

$$H^+(aq) + OH^-(aq) \rightleftharpoons H_2O(l) \qquad (14)$$

$$CH_3COOH(aq) \rightleftharpoons H^+(aq) + CH_3COO^-(aq) \qquad (6)$$

It does so by making a specific assumption about the nature of the species $H^+(aq)$. It assumes that $H^+(aq)$ is more properly written with the molecular formula $H_3O^+(aq)$. Thus, when HCl dissolves in water, the reaction is written

$$HCl(g) + H_2O(l) \rightleftharpoons H_3O^+(aq) + Cl^-(aq) \qquad (39)$$

instead of

$$HCl(g) \rightleftharpoons H^+(aq) + Cl^-(aq) \qquad (7)$$

Whenever $H^+(aq)$ might appear in an equation for a reaction, it is replaced by the **hydronium ion**, H_3O^+, and a molecule of water is added to the other side of the equation. We write the ionization equation in the form

$$CH_3COOH(aq) + H_2O \rightleftharpoons H_3O^+(aq) + CH_3COO^-(aq) \qquad (40)$$

Now the dissociation of acetic acid can be regarded as an acid-base reaction. The acid CH_3COOH transfers a proton to the base H_2O

---

*This more general view of acids and bases is named the **Brønsted-Lowry theory** after the two scientists who proposed it independently in 1923, J. N. Brønsted (1879–1947) and T. M. Lowry (1874–1939).

forming the acid H_3O^+ and the base CH_3COO^-. The neutralization of $H^+(aq)$ by $OH^-(aq)$ now takes the form

$$H_3O^+(aq) + OH^-(aq) \rightleftharpoons H_2O + H_2O \qquad (41)$$

The acid H_3O^+ transfers a proton to the base OH^-, forming an acid, H_2O, and a base, H_2O. We see that within the proton-transfer theory, the molecule H_2O must be assigned the properties of an acid as well as a base.

No definite experimental evidence demands that we write $H^+(aq)$ as $H_3O^+(aq)$. Nevertheless, the *convenience* of this assumption amply justifies its use in this representation of acid-base reactions.

13-7.3 Other Extensions of Acid-Base Theory

We noted earlier that by a slight redefinition of terms, we could expand our concept of acids and bases to encompass solutions using liquids other than water as a solvent. This idea has been extensively applied by chemists so that we now have at least four additional definitions of acids and bases.

One of the more useful acid-base definitions is that suggested by G. N. Lewis. A *base* is defined as *an electron-pair donor* and an *acid* as *an electron-pair acceptor*. Using these definitions, the concept of acids and bases can be extended to systems not involving protons. (See Appendix 11 for an example of a Lewis acid-base interaction not involving H^+.)

We shall not pursue these ideas further here, but it is worthwhile to note that the best definition of an acid is dependent upon the system being studied and upon the ideas being considered.

13-8 HIGHLIGHTS

It is reasonable to write equilibrium constants (also called **ionization constants**) for ionization processes. **Strong electrolytes** have *very large* equilibrium constants, whereas **weak electrolytes** have *small* equilibrium constants. About two water molecules in every billion ionize to give $H^+(aq)$ and $OH^-(aq)$. In every water solution the concentration of non-ionized water is almost constant so we can write

$$[H^+][OH^-] = K_w \qquad (11)$$

At 25°C, $K_w = 10^{-14}$. Le Chatelier's principle can be used to show that K_w should increase as the temperature rises. The value of K_w can be used to calculate concentrations of $OH^-(aq)$ in acidic solutions and of $H^+(aq)$ in basic solutions.

Acids and **bases** can be defined using both operational and conceptual definitions. **Operational definitions** list the observations defining the group. **Conceptual definitions** explain why the group has its properties. *There is no one best definition of an acid or a base.*

The **hydrogen-ion concentration** of aqueous solutions is an extremely important property. It can be measured by organic dyes whose color is sensitive to the concentration of $H^+(aq)$ (see Figure 13-6). Such dyes are useful in experimentally determining acid-equilibrium constants.

The **pH value** is a convenient shorthand representation of the concentration of $H^+(aq)$. It is defined as $pH = -\log_{10}[H^+]$. If $H^+(aq)$ has a concentration of 10^{-7} M, the pH is 7. In the process of **titration,** acids and bases are mixed and the neutralization process occurs. A number of useful quantitative relationships can be based on equilibrium constants.

Other acid-base definitions can be used. A base can be defined as a species which *receives* a proton; an acid can be defined as a species which *donates* a proton. Other definitions of acids and bases can also be used, depending upon the system under study.

QUESTIONS AND PROBLEMS

1 You are given two solutions. Describe briefly experiments which would be useful in classifying each of these solutions as (1) a strong electrolyte, (2) a weak electrolyte, or (3) a non-electrolyte.

2 Why is the notation $Li^+(aq)$ preferred to $Li(H_2O)_4^+$ to describe a lithium ion in water solution?

3 What is $[H^+]$ in an aqueous solution in which $[OH^-] = 1.0 \times 10^{-3}$ M?

4 A 100-ml sample of the HCl solution described in Exercise 13-3 is diluted with water to 1.00 liter. What is $[H^+]$? What is $[OH^-]$ in this solution?

5 Vinegar, lemon juice, and curdled milk all taste sour. What other properties would you expect them to have in common?

6 Give the name and formula of two hydrogen-containing compounds that are *not* classified as acids. State for each compound one or more properties common to acids that it does not possess.

7 As a solution of barium hydroxide is mixed with a solution of sulfuric acid, a white precipitate forms and the electrical conductivity decreases markedly. Write equations for the reactions that occur and account for the conductivity change. [Review Figure 12-7(b).]

8 Devise an operational and a conceptual definition of a gas.

9 An eyedropper is calibrated by counting the number of drops required to deliver 1.0 ml. Twenty drops are required. (a) What is the volume of one drop? (b) How many moles of HCl will be contained in one drop of 0.20 M HCl? [Remember: Number moles = volume (liters) × molarity.] (c) If one drop of 0.20 M HCl is added to enough water to give 100.0 ml of solution, what is $[H^+]$ in this new solution? (d) What will $[H^+]$ be if two drops of solution are used? if three drops are used? (e) By what *factor* did $[H^+]$ change when one drop of HCl was added to the neutral water in (c)? [Answer to (e): 1,000 × concn in H_2O.]

10 Suppose drops (from the same eyedropper) of 0.10 M NaOH are added, one at a time, to the 100 ml of HCl in question 9(c). (a) What will $[H^+]$ be after one drop is added? (*Hint:* Calculate the number of moles of NaOH added [see the hint for question 9(b)], then assume that the NaOH removes an equal number of moles of HCl. Use the number of moles of excess HCl or NaOH and the volume to calculate concentration.) (b) What will be $[H^+]$ after two drops are added? (c) What will be $[H^+]$ after three drops are added?

11 Calculate $[H^+]$ and $[OH^-]$ in a solution made by mixing 50.0 ml of 0.200 M HCl and 49.0 ml of 0.200 M NaOH. (*Answer:* $[OH^-] = 5 \times 10^{-12}$ M.)

12 Calculate $[H^+]$ and $[OH^-]$ in a solution made by mixing 50.0 ml of 0.200 M HCl and 49.9 ml of 0.200 M NaOH.

13 How much more 0.200 M NaOH solution must be added to the solution in question 11 to change $[H^+]$ to 10^{-7} M?

14 An acid is a substance that can form $H^+(aq)$ in the equilibrium

$$HB(aq) \rightleftharpoons H^+(aq) + B^-(aq)$$

(a) Does equilibrium favor reactants or products for a strong acid?
(b) Does equilibrium favor reactants or products for a very weak acid?
(c) If acid HB_1 is a stronger acid than acid HB_2, is K_1 a larger or smaller number than K_2?

$$K_1 = \frac{[H^+][B_1^-]}{[HB_1]} \qquad K_2 = \frac{[H^+][B_2^-]}{[HB_2]}$$

15 (a) Which of the following acids is the strongest and which is the weakest: ammonium ion, NH_4^+ (in an NH_4Cl solution); bisulfate ion, HSO_4^- (in a $KHSO_4$ solution); and hydrogen sulfide, H_2S? (b) If 0.1 M solutions are made of NH_4Cl, $KHSO_4$, and H_2S, in which will $[H^+]$ be highest and in which will it be lowest?

16 (a) Nitric acid is a very strong acid. What is $[H^+]$ in a 0.050 M HNO_3 solution? (b) Hydrogen peroxide (H_2O_2) is a very weak acid. What is $[H_2O_2]$ in a 0.050 M H_2O_2 solution?

17 From a study of Appendix 6, what generalization can you make concerning acids which contain more than one ionizable hydrogen in their molecules or ions?

18 A 0.25 M solution of benzoic acid (symbolize it HB) is found to have a hydrogen ion concentration $[H^+] = 4 \times 10^{-3}$ M.
(a) Assuming the simple reaction

$$HB(aq) \rightleftharpoons H^+(aq) + B^-(aq)$$

calculate K_A for benzoic acid.
(b) Compare the values of $[HB]$, $[H^+]$, $[B^-]$, and K_A used in this problem to the corresponding quantities in the benzoic acid calculation presented in Section 13-6.1.

19 If 23 g of formic acid (HCOOH) are dissolved in 10.0 liters of water at 20°C, the $[H^+]$ is found to be 3.0×10^{-3} M. Calculate K_A.

20 A chemist dissolved 25 g of CH_3COOH in enough water to make 1 liter of solution. What is the concentration of this acetic acid solution? What is the concentration of $H^+(aq)$? Assume a negligible change in $[CH_3COOH]$ because of dissociation to $H^+(aq)$. (K_A for $CH_3COOH = 1.8 \times 10^{-5}$.)

21 When sodium acetate (CH_3COONa) is added to an aqueous solution of hydrogen fluoride (HF), a reaction occurs in which the weak acid HF loses H^+.
(a) Write the equation for the reaction. (b) What ion is competing with F^- for H^+?

22 (a) Write the equation for the reaction that shows the acid-base reaction between hydrogen sulfide (H_2S) and the carbonate ion, CO_3^{2-}. (b) What are the two acids competing for H^+? (c) From the values of K_A for these two acids (see Table 13-3), predict whether the equilibrium favors reactants or products.

23 The acid hydrogen fluoride (HF) is about 10 percent dissociated in a 0.1 M solution. Calculate the pH of a 0.1 M solution of HF.

24 What is the pH of a solution that contains 1×10^{-8} mole of $OH^-(aq)$ per liter?

25 Solution A has a pH of 3. Solution B has a pH of 5. On the basis of this information, we know that: (a) the $[H^+]$ for A is three fifths that for B, (b) the $[H^+]$ for A is 100 times that for B, or (c) the concentration of $[H^+]$ for A is five thirds that for B.

26 If the pH of a solution is 5, what is $[H^+]$? Is the solution acidic or basic?

27 What is $[H^+]$ in a solution of pH $= 8$? Is the solution acidic or basic? What is $[OH^-]$ in the same solution?

28 Write the equations for the reaction between each of the following acid-base pairs. For each reaction, predict whether reactants or products are favored. (Use the values of K_A given in Appendix 6.)
(a) $HNO_2(aq) + NH_3(aq) \rightleftharpoons$
(b) $NH_4^+(aq) + F^-(aq) \rightleftharpoons$
(c) $C_6H_5COOH(aq) + CH_3COO^-(aq) \rightleftharpoons$
[Answer to (a): $HNO_2(aq) + NH_3(aq) \rightleftharpoons NO_2^-(aq) + NH_4^+(aq)$. Products, $NO_2^-(aq)$ and $NH_4^+(aq)$, are favored.]

29 Write the equations for the reactions between each of the following acid-base pairs. For each reaction, predict whether reactants or products are favored.
(a) $H_2SO_3(aq) + HCO_3^-(aq) \rightleftharpoons$
(b) $H_2CO_3(aq) + SO_3^{2-}(aq) \rightleftharpoons$
(c) $H_2SO_3(aq) + SO_3^{2-}(aq) \rightleftharpoons$

30 In separation of metal ions, it is known that the concentration of $S^{2-}(aq)$ in a solution can be reduced by adding $H^+(aq)$. Write the equation for the reaction by which $[S^{2-}]$ is reduced when HCl solution is poured into a solution of Na_2S in water. Would this make a sulfide like FeS more or less soluble? (Write the equation for the dissolving of FeS and apply Le Chatelier's principle.)

Chemical thermodynamics enables one to state what may happen when two substances react.

WENDELL M. LATIMER (1893–1955)

CHAPTER 14 **Oxidation-Reduction Reactions**

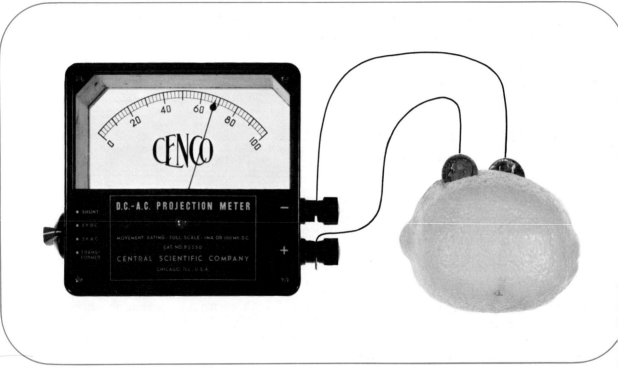

A piece of silver, a piece of copper, and a lemon make a simple electrochemical cell.

By this time the tendencies driving chemical reactions in opposite directions may well have you feeling a bit pushed around yourself. Can anyone predict whether a chemical reaction will or won't "go" in the direction you want it to? Or do you have to try each one out in the laboratory?

A clue is found in the electrical nature of chemical reactions. If we force the electrons transferred in a chemical reaction to detour through a wire, we can measure "electron pressure" easily with a voltmeter. Comparing the voltages produced in different reactions gives us real insight into the "go power" of each. It's really rather similar to listing acids in order of strength, isn't it? Let's try.

We have used the principles of equilibrium in two general types of reactions. First, we took up the question of equilibrium reactions involving a solid and a solution. In Chapter 13 we turned to reactions occurring entirely in solution and involving proton transfer. Now we shall take a more general view of equilibrium in aqueous solutions—a view provided by an investigation of the chemistry of an **electrochemical cell.**

14-1 ELECTROCHEMICAL CELLS

Electrochemical cells are familiar to you: a flashlight operates on current drawn from electrochemical cells called dry cells; automobiles are started with the aid of a battery, a set of electrochemical cells hooked together in series. The last time you changed the dry cells in a flashlight because the old ones were "dead," did you wonder what happened inside those cells? Why does electric current flow from a new dry cell but not from one that has been used many hours? We shall see that this is an important question in chemistry. By studying the chemical reactions taking place in an electrochemical cell, we discover a basis for predicting whether equilibrium in a chemical reaction favors reactants or products. The reactions in an electrochemical cell are of the type called **oxidation-reduction reactions.** This type of reaction is the subject of this chapter.

14-1.1 The Chemistry of an Electrochemical Cell

Let us begin our investigation of an electrochemical cell by assembling one. Fill a beaker (beaker A) with a dilute solution of silver nitrate (0.1 M) and another beaker (beaker B) with dilute copper sulfate (also 0.1 M). Put a *silver rod* in the $AgNO_3$ *solution* and a *copper rod* in the $CuSO_4$ *solution*. With a wire, connect the silver rod to one terminal of an **ammeter.** (Ammeters measure electric current.) Connect the other terminal of the ammeter to the copper rod. (See Figure 14-1.1, page 320.) Finally, connect the two solutions. One way to make this connection is by a salt bridge as shown in Figure 14-1.2 (page 321). The glass tube, plugged at each end with cotton, contains a sodium nitrate ($NaNO_3$) solution, and it completes the electric circuit by furnishing a path for the ions in the solutions to move from beaker A to beaker B.

As soon as the connection is made, the ammeter needle deflects —electric current is moving through the wires. The wires become warm; the cell is doing work as it forces electrons through the wires. In beaker B, the copper rod is dissolving and the copper sulfate solution is becoming a deeper blue. In beaker A, the silver rod is being covered with a loosely adhering deposit of metallic silver crystals.* As time goes by, the ammeter registers less and less current until finally there is none. Our cell is now "dead."

Let us repeat the experiment but be more quantitative. We shall weigh both metal rods in the beginning and at the completion

---

* If the crystals are very small, the deposit may appear black in color; but it is still metallic silver.

of the experiment. We find that during the experiment the copper rod has *lost* 0.635 g and the silver rod has *gained* 2.16 g. How many moles of copper and silver are involved in the reaction? (Review Section 3-1.) We remember that a mole of copper or silver atoms is defined as its atomic weight in grams; so we can write

$$\text{moles Cu dissolved} = \frac{\text{wt Cu dissolved}}{\text{wt Cu per mole}} \quad (1)$$

$$= \frac{0.635 \text{ g}}{63.5 \text{ g/mole}}$$

$$= 0.0100 \text{ mole}$$

$$\text{moles Ag deposited} = \frac{\text{wt Ag deposited}}{\text{wt Ag per mole}} \quad (2)$$

$$= \frac{2.16 \text{ g}}{108 \text{ g/mole}}$$

$$= 0.0200 \text{ mole}$$

We see that there is a simple relationship between the moles of copper dissolved and the moles of silver deposited. One mole of copper dissolves in beaker B for every 2 moles of silver deposited in beaker A. We also noticed that the solution in the beaker containing the copper rod became a deeper blue as the cell operated. The deepening blue color offers strong evidence for an increasing concentration of $Cu^{2+}(aq)$ in the solution.

These facts—the disappearing copper metal, the electric current in the wire, and the deepening blue of the solution around the copper rod—are best summarized by the equation

$$Cu(s) \longrightarrow \underset{\text{BLUE COLOR}}{Cu^{2+}(aq)} + 2e^{-} \quad (3)$$

This represents the process in beaker B. The deposition of silver on the electrode in beaker A suggests that the two electrons from each

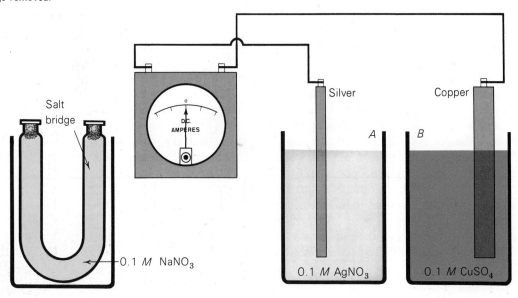

Fig. 14-1.1 **An electrochemical cell with salt bridge removed.**

copper atom are being used to deposit silver. The process in beaker *A* can be summarized by the equation

$$2Ag^+(aq) + 2e^- \longrightarrow 2Ag(s) \qquad (4)$$

The electrons must then flow from copper to silver through the wire between beakers; these electrons are picked up by silver ions at the silver electrode surface to give silver metal. The overall process is the combination of the reactions in the individual beakers:

Beaker *B*	$Cu(s) \longrightarrow Cu^{2+}(aq) + 2e^-$	(3)
Beaker *A*	$2e^- + 2Ag^+(aq) \longrightarrow 2Ag(s)$	(4)
Net reaction	$Cu(s) + 2Ag^+(aq) \longrightarrow 2Ag(s) + Cu^{2+}(aq)$	(5)

We see that half the reaction is taking place in beaker *A* and half in beaker *B*; the overall reaction is the *sum* of the two **half-reactions.** The reaction in each beaker is called a **half-cell reaction** or a **half-reaction.**

There are several interesting features about these half-reactions:

(1) *The two half-reactions are written separately.* In this electrochemical cell, the half-reactions occur in separate beakers. As the name implies, there must be two such reactions.
(2) *Electrons are shown as part of the reaction.* The ammeter shows that electrons are involved. Electrons flow when the reaction starts and flow no longer when the reaction stops. Earlier arguments showed that electrons flow from copper to silver through the external wire.
(3) *New chemical species are produced in each half of the cell.* The copper rod is converted to copper ions (the rod loses weight) and the silver ions are changed to metal (the silver rod gains weight). The new species can be explained in terms of *loss of electrons* (by copper) and *gain of electrons* (by silver ions).

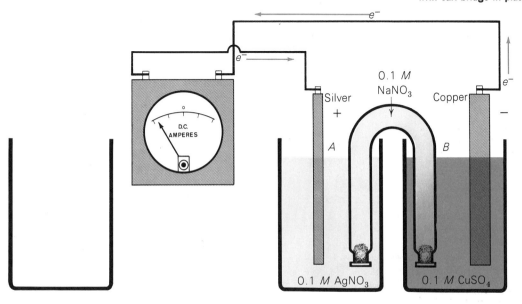

Fig. 14-1.2 **An electrochemical cell with salt bridge in place.**

(4) *The half-reactions, when combined, express the overall reaction.* The equation for the overall reaction, (5), is obtained by combining the equations for the two half-reactions, (3) and (4), so that electrons lost (by copper atoms) equal electrons gained (by silver ions). Since the same number of electrons appearing as reactants also appear as products, we can eliminate them from the overall equation. Electrical measurements confirm that the electrochemical cell operates without accumulating or consuming electric charge, that *the number of electrons lost equals the number of electrons gained.* Two moles of electrons were released for each mole of copper dissolved. One mole of electrons was used up for each mole of silver deposited. Thus, 2 moles of silver were deposited for each mole of copper dissolved. This analysis is in agreement with earlier quantitative observations. Finally, charge build-up at any point *within the solution* is minimized by appropriate movement of positive and negative ions. We shall come back to this point in a moment.

We see that the overall chemical reaction that occurs in an electrochemical cell is conveniently described in terms of two *types* of half-reactions. In one, electrons are lost; in the other, electrons are gained. To distinguish these half-reactions, we need two identifying names.

(1) *The half-reaction in which the reactant accepts electrons is called* **reduction.**

$$2Ag^+(aq) + 2e^- \longrightarrow 2Ag(s) \qquad (4)$$

(2) *The half-reaction in which the reactant releases electrons is called* **oxidation.**

$$Cu(s) \longrightarrow Cu^{2+}(aq) + 2e^- \qquad (3)$$

(3) *The overall reaction is called an* **oxidation-reduction reaction.**

$$Cu(s) + 2Ag^+(aq) \longrightarrow Cu^{2+}(aq) + 2Ag(s) \qquad (5)$$

Such half-cell reactions are not new to us; you will remember that in Chapter 6 we explained the electrical conductivity of a solution of $CuCl_2$ in terms of half-cell reactions occurring at the junction of solution and electrode. We wrote

$$2e^- + Cu^{2+}(aq) \longrightarrow Cu(s) \qquad (6)$$

$$2Cl^-(aq) \longrightarrow Cl_2 + 2e^- \qquad (7)$$

Half-cell reactions are of great importance in the analysis of all electrochemical processes.

EXERCISE 14-1

Consider the electrolysis of $CuCl_2$. At the electrode where copper is being deposited, is the process oxidation or reduction? Explain. Is the process by which chloride is converted to chlorine oxidation or reduction? Explain.

14-1.2 Operation of an Electrochemical Cell—Some Terminology

Let us pause at this point and take a more detailed look at an electrochemical cell. Figure 14-2 shows in cross section a slightly modified cell that also uses the reaction between copper and silver nitrate. The only difference between this cell and that shown originally in Figure 14-1.2 is that the salt bridge of Figure 14-1.2 has been replaced by a porous porcelain cup. The cup prevents bulk mixing of the silver nitrate and copper sulfate solutions, but permits ions to seep through the cup to establish electrical contact. The processes are identical to those of our original cell (Section 14-1.1).

The cell as shown has a number of parts. Some names will help us to identify parts and discuss cell operation more easily. As you will recall, the metal strips of copper and silver dipping into the solution are called the **electrodes.** Electrons accumulate on the copper electrodes as copper atoms enter the solution to become ions:

$$Cu(s) \longrightarrow Cu^{2+}(aq) + 2e^- \qquad (3)$$

Using our new terms (Section 14-1.1), this is an *oxidation* process because the copper metal *releases electrons* as it forms copper ions. *Chemists call <u>the electrode at which oxidation occurs the</u>* **anode.** Because extra electrons are supplied to the anode by the oxidation process, the anode carries a negative charge in this cell. It would be the electrode labeled "−" ("negative") in a commercial cell.

If a wire joins the copper electrode to the silver electrode, electrons flow down the wire to the silver electrode (Figure 14-2). Here *silver ions* make contact with the silver metal surface and pick up the needed electrons to become silver metal:

$$2Ag^+(aq) + 2e^- \longrightarrow Ag(s) \qquad (4)$$

We can consider this a *reduction* reaction because the silver ions *gain electrons* to become silver metal. *Chemists call <u>the electrode at which reduction occurs the</u>* **cathode.** Because electrons are being used up here as fast as they can be supplied, this electrode carries a positive charge relative to the copper electrode. This electrode would be labelled "+" ("plus" or "positive") in a commercial cell. Electrons flow from the copper electrode to the silver electrode down the wire.

Now what happens in solution? Around the copper anode, positive copper ions are being generated. Around the silver cathode, positive silver ions are being used up. Consequently, positive ions show a net drift from around the copper electrode, where they are being formed, to around the silver electrode, where they are being used up. To help maintain electrical neutrality, the negative ions will also move in solution away from the region around the silver electrode where positive ions have been removed and toward the region around the copper electrode where positive ions are being generated. Notice that the movement of both positive and negative ions is a result of the accumulation of positive ions around one electrode and their removal from the other electrode. In this cell, ion movement has nothing to do with the signs on the electrodes. The whole process

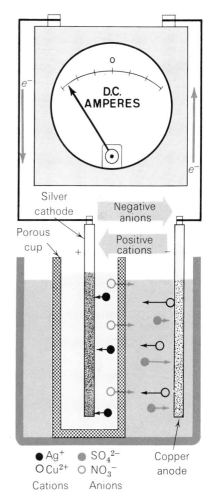

Fig. 14-2 The operation of an electrochemical cell.

Reduction reaction at silver cathode:

$$2Ag^+(aq) + 2e^- \longrightarrow 2Ag(s)$$

Oxidation reaction at copper anode

$$Cu(s) \longrightarrow Cu^{2+}(aq) + 2e^-$$

is driven by the oxidation-reduction reaction. Because the positive ions move toward the silver cathode where they are being used up, they are called **cations**. Because negative ions are attracted by excess copper ions around the copper electrode and move toward the copper anode, they are called **anions**. In short, *cations move toward the cathode; anions move toward the anode*. This movement of ions in the solution completes the circuit; electrons move through the external wires; ions move through the solution; and exchange of electrons with ions occurs at the electrode-solution interface.

These processes and names are summarized in Figure 14-2. Clearly, an electrochemical cell is just an oxidation-reduction reaction in which oxidation occurs in one region and reduction occurs in another, electrons flow from the oxidation region to the reduction region along an external wire, and ions move through solution to complete the circuit.

The names used in the foregoing description can be conveniently summarized for future use:

(1) **Electrode:** The solid conductors at the surface of which reactions involving electron transfer between solid and solution occur.
(2) **Anode:** The electrode at which oxidation occurs.
(3) **Cathode:** The electrode at which reduction occurs.
(4) **Anion:** A negatively charged ion which moves toward the anode.
(5) **Cation:** A positively charged ion which moves toward the cathode.

EXERCISE 14-2

We wish to make an electrochemical cell using the reaction

$$Zn(s) + 2Ag^+(aq) \longrightarrow Zn^{2+}(aq) + 2Ag(s)$$

Construct a diagram similar to Figure 14-1. Indicate the substances in each beaker, the direction of flow of electrons in the wire, the half-reactions at each electrode, and the direction of movement of both positive and negative ions in the solution. Identify the cathode and the anode in each cell. Which electrode is negative?

Now that we see how an electrochemical cell works and what we mean by half-cell reactions, let us examine other oxidation-reduction reactions in terms of the half-cell concept.

14-1.3 Oxidation-Reduction Reactions in a Beaker— Copper and Silver Nitrate Solution

These ideas concerning an electrochemical cell have great importance in chemistry because they also apply to chemical reactions that occur in a single beaker. Without having an obvious electric circuit, we can duplicate the chemical changes that occur in a cell using only a *single* solution (see Figure 14-3). For example, in Experiment 7 we saw that when a copper wire was immersed in $AgNO_3$ solution, silver crystals and a blue color appeared in solution. The blue color suggested the formation of $Cu^{2+}(aq)$ ions. The ratio of Cu dissolved

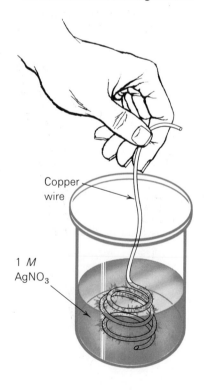

Fig. 14-3 **Oxidation and reduction reactions can occur in a single beaker.**

to Ag deposited was the same as that in the cell; hence, the net reaction was probably the same.

Thus, Experiment 7 involved the same oxidation-reduction reaction but *the electron transfer must have occurred locally between individual copper atoms* (in the metal) *and individual silver ions* (in the solution near the metal surface). This *local transfer* replaces the wire in the cell, which carries electrons from one beaker (where they are released by copper) to the other (where they are accepted by silver ions).

14-1.4 Oxidation-Reduction Reactions in a Beaker—Zinc Oxidized by H⁺(aq)

One of the reactions mentioned in Section 13-2.2 characterizing an acid was the reaction between the acid and a metal, liberating hydrogen. Zinc was a useful metal for the process:

$$Zn(s) + 2H^+(aq) \longrightarrow Zn^{2+}(aq) + H_2(g) \qquad (8)$$

Each zinc atom loses two electrons in changing to a zinc ion; therefore, zinc is oxidized. Each hydrogen ion gains an electron, changing to a hydrogen atom; therefore, hydrogen is reduced. (After reduction, two hydrogen atoms combine to form molecular H_2.) As before, the reaction can be separated into two half-reactions:

Oxidation $\qquad Zn(s) \longrightarrow Zn^{2+}(aq) + 2e^- \qquad (9)$

Reduction $\qquad 2e^- + 2H^+(aq) \longrightarrow H_2(g) \qquad (10)$

Net reaction $\quad Zn(s) + 2H^+(aq) \longrightarrow Zn^{2+}(aq)* + H_2(g) \qquad (8)$

Thus, the reaction by which a metal dissolves in an acid is conveniently discussed in terms of oxidation and reduction.

Not all metals react with aqueous acids. Of the common metals, magnesium, aluminum, iron, and nickel liberate H_2 as zinc does. Other metals, including copper, mercury, silver, and gold, do not produce measurable amounts of hydrogen, even though we make sure that the equilibrium state has been attained. Apparently, some metals release electrons to H^+ (as zinc does) and others do not (see Figure 14-4).

14-1.5 Oxidation-Reduction Reactions in a Beaker—Zinc Oxidized by Cu²⁺(aq)

For a third example of oxidation-reduction, let us place a strip of metallic zinc in a solution of copper nitrate [$Cu(NO_3)_2$]. The strip becomes coated with reddish metallic copper and the bluish color of the solution disappears. The presence of zinc ion, Zn^{2+}, among the products can be shown by the following experiment.

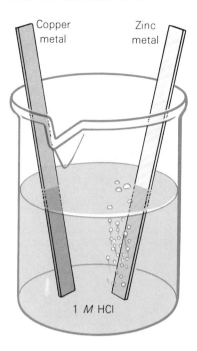

Fig. 14-4 Some metals release electrons to H⁺ and others do not.

---

*For the remainder of this chapter, we shall consider only aqueous solutions and therefore not specify (aq) for each ion.

When the Cu^{2+} color is gone, we pass hydrogen sulfide (H_2S) gas through the mixture. Immediately, white zinc sulfide (ZnS) precipitates if the solution is neutral. The reaction between metallic zinc and the aqueous copper nitrate is

$$Zn(s) + Cu^{2+} \longrightarrow Zn^{2+} + Cu(s) \qquad (11)$$

Zinc loses electrons to form Zn^{2+}:

$$Zn(s) \longrightarrow Zn^{2+} + 2e^- \qquad (9)$$

Zinc is oxidized. If zinc is oxidized, releasing electrons, something must be reduced. Copper ion is reduced:

$$Cu^{2+} + 2e^- \longrightarrow Cu(s) \qquad (6)$$

This time, copper ion *gains* electrons from the zinc, in contrast to reaction (5), where copper metal loses electrons to silver ions.

What about the state of equilibrium for the reaction of Zn and Cu^{2+}? Let us try the reverse process by placing a strip of metallic copper in a zinc sulfate solution. No visible reaction occurs. Attempts to detect the presence of cupric ion by adding H_2S to produce the black color of cupric sulfide (CuS) fail. Cupric sulfide has such low solubility that this is an extremely sensitive test; yet the amount of Cu^{2+} formed by the reverse process, $Cu(s) + Zn^{2+}$, cannot be detected. It appears that the state of equilibrium for reaction (11) greatly favors the products, $Cu(s)$ and Zn^{2+}, over the reactants, $Zn(s)$ and Cu^{2+}.

14-1.6 Competition for Electrons

These reactions can be viewed as a competition between two kinds of atoms or ions for electrons. Equilibrium is then attained when this competition reaches a balance between opposing reactions. In the case of the reaction of copper metal with silver nitrate solution, the $Cu(s)$ releases electrons fairly easily and the Ag^+ accepts them so readily that equilibrium greatly favors the products, Cu^{2+} and $Ag(s)$.

What factors determine the point at which equilibrium is reached? Again, we check randomness and energy. The number of particles is the same on both sides of the equation; hence, differences in randomness are relatively small. However, as we shall see, differences *do exist* that favor the reactants, $Cu(s) + 2Ag^+$, over the products, Cu^{2+} and $2Ag(s)$. We know that ions distributed randomly in solution have a greater degree of disorder than atoms in an ordered metal lattice. For reactants, we have *two* ions in a random solution and *one* atom in the ordered metal. For products, we have only *one* ion in a random solution and *two* atoms in the ordered metal. Therefore, the reactants, $2Ag^+$ and $Cu(s)$, are favored by randomness because they are more disordered. Since the equilibrium lies far to the side of the products, the energy term must favor products. The energy must decrease very significantly as electrons move from copper to silver ions. If we regard the reaction between copper and silver ions as a competition for electrons, we can say that the silver ions accept electrons more readily than do copper ions. This same sort of competition for electrons is involved in reaction (11), in which $Zn(s)$ releases electrons and Cu^{2+} accepts them. This time the competition for electrons is such that equilibrium favors Zn^{2+} and $Cu(s)$. These two situations are shown in Figure 14-5 (page 329).

By way of contrast, compare the foregoing with the reaction of metallic cobalt placed in a nickel sulfate solution:

$$Co(s) + Ni^{2+} \rightleftharpoons Co^{2+} + Ni(s) \qquad (12)$$

At equilibrium, chemical tests show that both Ni^{2+} and Co^{2+} are present in moderate concentrations. In this case, neither the reactants, $Co(s)$ and Ni^{2+}, nor the products, Co^{2+} and $Ni(s)$, are greatly favored.

This competition for electrons is reminiscent of the competition among bases for protons. The similarity suggests that we might develop a table in which metal ions are listed in order of their tendency to pick up electrons. This would resemble a list of bases arranged in order of their ability to pick up protons. We can already make some comparisons. We might begin by listing some of the half-reactions in this chapter. We shall write them to show ions or atoms picking up electrons. We shall arrange them to show their decreasing ability to pick up electrons.

First, we remember that a silver ion picks up electrons from copper. Therefore, we shall write our first two reactions in the following order:

$$Ag^+ + e^- \longrightarrow Ag(s) \qquad (13)$$

$$Cu^{2+} + 2e^- \longrightarrow Cu(s) \qquad (6)$$

Listing the Ag^+–$Ag(s)$ half-reaction first indicates that a silver ion picks up electrons more readily than does a cupric ion.

Now consider the reaction of a zinc and copper ion. Since a copper ion picks up electrons from zinc, we must put the zinc half-reaction below copper:

$$Ag^+ + e^- \longrightarrow Ag(s) \qquad (13)$$

$$Cu^{2+} + 2e^- \longrightarrow Cu(s) \qquad (6)$$

$$Zn^{2+} + 2e^- \longrightarrow Zn(s) \qquad (14)$$

Listing the Zn^{2+}–$Zn(s)$ half-reaction last tells us that Zn^{2+} has less tendency to pick up electrons than does Cu^{2+}. But if this is true, we know that Zn^{2+} will be a poorer competitor for electrons than Ag^+. *The list leads us to expect that zinc metal will release electrons to silver ion, reacting to produce zinc ion and silver metal.* We should test this proposal! We dip a piece of zinc metal in a solution of silver nitrate. The result confirms our expectations: zinc metal dissolves and bright crystals of metallic silver appear.

Our data allow us to make one more addition to the list. Zinc reacts with H^+ to give Zn^{2+} and $H_2(g)$. The H^+ ion is a better competitor for the electron than is the zinc ion; hence, the half-reaction H^+–H_2 must be placed above Zn^{2+}–Zn in our list. How far above? To answer that, remember that copper metal does *not* react with H^+ to give H_2; hence, Cu^{2+} is a better competitor for electrons than is H^+. We shall place the half-reaction H^+–H_2 between Cu^{2+}–Cu and Zn^{2+}–Zn.

TABLE 14-1 DECREASING ABILITY OF IONS TO COMPETE FOR ELECTRONS

Half-Reaction
$Ag^+ + e^- \longrightarrow Ag(s)$
$Cu^{2+} + 2e^- \longrightarrow Cu(s)$
$2H^+ + 2e^- \longrightarrow H_2(g)$
$Ni^{2+} + 2e^- \longrightarrow Ni(s)$
$Zn^{2+} + 2e^- \longrightarrow Zn(s)$

Now we can expand our list, as in Table 14-1. The value of this list is obvious. Any half-reaction can be combined with the reverse of another half-reaction (making sure that electron gain is equal to electron loss) to give us a possible chemical reaction. Table 14-1 permits us to predict which ion is the better competitor. From this information we can decide whether the equilibrium favors reactants or products. We would like to expand the list and make it more quantitative. Let us obtain more data.

EXERCISE 14-3

(1) You were told that nickel metal reacts with H^+ to give H_2. Which is the better competitor for electrons, H^+ or Ni^{2+}? Will Ni^{2+}–Ni go above or below H^+–H_2 in Table 14-1?
(2) You are told that zinc metal reacts with Ni^{2+} to give metallic nickel and Zn^{2+}. Which is the better competitor for electrons, Zn^{2+} or Ni^{2+}? Place the half-reaction Ni^{2+}–Ni in Table 14-1.

14-2 ELECTRON TRANSFER AND PREDICTING REACTIONS

The half-reactions listed in Table 14-1 are arranged in order of decreasing ability of the ions to attract electrons. We can make rather rough, qualitative predictions from this table, but much more precise quantitative predictions are desirable. Perhaps such quantitative predictions can be made if we have a quantitative measure of the ability of an ion to attract electrons.

14-2.1 The Ability of Ions to Attract Electrons

The cells we have constructed in some ways resemble cells that power flashlights and start cars. If we put a small light bulb of proper voltage in place of the ammeter of Figure 14-2, the bulb will glow, giving off heat and light. The cell is doing electrical work in forcing the electric current (stream of electrons) through the bulb. As the electrons leave the bulb, they must have lower potential energy than they had when they entered. This drop in potential energy appears as heat and light.* The potential-energy change associated with the chemical reaction of the cell is directly related to the voltage of the cell. The **voltage** of a chemical cell measures its capacity to do electrical work; as such, it serves as a measure of the tendency of the cell reaction to go in one direction.

Experiment 27 shows us that different cells have different voltages. Consider first the cell based on the reactions

$$Zn(s) \longrightarrow Zn^{2+} + 2e^- \qquad (9)$$

$$2e^- + Cu^{2+} \longrightarrow Cu(s) \qquad (6)$$

$$Zn(s) + Cu^{2+} \longrightarrow Zn^{2+} + Cu(s) \qquad (11)$$

---

* Remember that when billiard balls fell off the high shelf in Chapter 9, the change in their potential energy appeared as heat after impact. The balls were warmer.

The experimental arrangement is shown in Figure 14-5(a). It involves one half-cell consisting of a zinc rod dipping in 1 M ZnSO$_4$, and another half-cell consisting of a copper rod in 1 M CuSO$_4$ solution. As we noted earlier, the electron flow will be *from zinc,* through the meter, *to copper.* Current flowing in this direction causes the **voltmeter** to deflect to the *right* (clockwise), as shown. The voltmeter reads close to 1.1 volts when the zinc half-cell is connected. The value is recorded in Table 14-2 (page 330).

One interesting feature of the cell we are using is that the Cu^{2+}–Cu half-cell is completely contained in the center porous cup, while the Zn^{2+}–Zn half-cell is completely contained in the beaker outside the porous cup. This arrangement makes it possible to unhook the copper electrode from the voltmeter and *to transfer the entire half-cell over to a Ag$^+$–Ag half-cell contained in a similar outer beaker.* The result of this transfer is shown in Figure 14-5(b). Our earlier analysis has shown that the reaction between copper metal and silver nitrate is given by the equations

$$\text{Cu}(s) \longrightarrow \text{Cu}^{2+} + 2e^- \quad (3)$$

$$2e^- + 2\text{Ag}^+ \longrightarrow 2\text{Ag}(s) \quad (4)$$

$$\overline{\text{Cu}(s) + 2\text{Ag}^+ \longrightarrow \text{Cu}^{2+} + 2\text{Ag}(s)} \quad (5)$$

We see now that electrons are given up by copper and the electron flow is *from copper,* through the meter, *to silver.* We observe that the

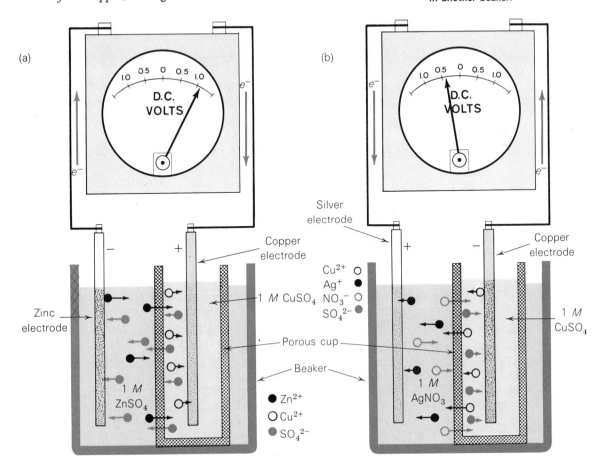

Fig. 14-5 Two electrochemical cells involving copper. In each case the porous cup containing the copper electrode can be removed and placed in another beaker.

meter deflects in the opposite direction, toward the *left* (counterclockwise). The reading on the voltmeter is recorded as left 0.4 volt. This value is placed in Table 14-2. If our standard Cu–Cu^{2+} half-cell is placed in a beaker containing an identical Cu–Cu^{2+} half-cell, the voltmeter reads 0.0 and no deflection is observed. That observation is also recorded in the table. Table 14-2 can be expanded by systematically placing the center Cu–Cu^{2+} half-cell in other half-cells contained in external beakers. With a Ni–Ni^{2+} half-cell, the reading is right 0.60 volt. Other cells can be constructed and the results added to Table 14-2.

TABLE 14-2 HALF-CELL VOLTAGES AGAINST COPPER REFERENCE CELL

Half-Reaction in Outer Beaker	Voltmeter Reading
$Ag^+ + e^- \longrightarrow Ag(s)$	left 0.4 volt
$Cu^{2+} + 2e^- \longrightarrow Cu(s)$	0.00 (no deflection)
$2H^+ + 2e^- \longrightarrow H_2(g)$	right 0.3 volt
$Ni^{2+} + 2e^- \longrightarrow Ni(s)$	right 0.6 volt
$Zn^{2+} + 2e^- \longrightarrow Zn(s)$	right 1.1 volts

A special cell is required if we want to enter the value for the half-reaction

$$2e^- + 2H^+ \longrightarrow H_2(g) \qquad (10)$$

because gaseous hydrogen does not give us a place to attach the wire as does metallic copper or silver. For this reason, an auxiliary platinum electrode is placed in the 1 molar solution of H$^+$. Then, hydrogen gas at 1 atm pressure is bubbled around the platinum metal electrode. The arrangement is shown in Figure 14-6. When this electrode is hooked up to the Cu–Cu^{2+} standard, the reading is right 0.3.

14-2.2 Selection of a Standard Half-Cell for Comparison

Let us study the results summarized in Table 14-2. These numbers tell us, in effect, how each ion competes with Cu^{2+} for electrons. Silver is clearly a better competitor and is placed first. Instead of indicating "right" and "left" readings, perhaps we could designate "left" as "positive" and "right" as "negative." This is a reasonable but *arbitrary* choice. A positive value means that the metal *ion* is a better competitor than Cu^{2+} *ion* for electrons; a negative value, that Cu^{2+} is a better competitor for electrons than the metal ion being considered (for example, Zn^{2+}). The results appear in Table 14-3.

If we review the last experiment, we suddenly realize that copper was chosen as a comparison electrode because copper and copper sulfate solution were on hand. Suppose we had chosen a Ni–Ni^{2+} half-cell for comparison. All half-cells in Table 14-3 would have had the same relative positions, but the numbers would have been different. The Ni–Ni^{2+} half-cell would obviously be 0.0 and the Cu–Cu^{2+} half-cell would read left 0.6 because the electrons would be

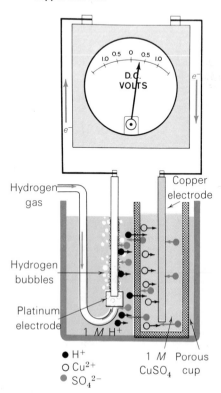

Fig. 14-6 An electrochemical cell made of a hydrogen half-cell and a copper half-cell.

TABLE 14-3 HALF-CELL VOLTAGES AGAINST COPPER REFERENCE CELL

Half-Reaction in Outer Beaker	Voltmeter Reading
$Ag^+ + e^- \longrightarrow Ag(s)$	+0.4 volt (left)
$Cu^{2+} + 2e^- \longrightarrow Cu(s)$	0.0 volt (no deflection)
$2H^+ + 2e^- \longrightarrow H_2(g)$	−0.3 volt (right)
$Ni^{2+} + 2e^- \longrightarrow Ni(s)$	−0.6 volt (right)
$Zn^{2+} + 2e^- \longrightarrow Zn(s)$	−1.1 volts (right)

TABLE 14-4 HALF-CELL VOLTAGES AGAINST NICKEL REFERENCE CELL

Half-Reaction in Outer Beaker	Voltmeter Reading
$Ag^+ + e^- \longrightarrow Ag(s)$	+1.0 volt (left)
$Cu^{2+} + 2e^- \longrightarrow Cu(s)$	+0.6 volt (left)
$2H^+ + 2e^- \longrightarrow H_2(g)$	+0.3 volt (left)
$Ni^{2+} + 2e^- \longrightarrow Ni(s)$	0.0 volt (no deflection)
$Zn^{2+} + 2e^- \longrightarrow Zn(s)$	−0.5 volt (right)

flowing from the central reference half-cell. This is just the reverse of when copper was in the center and nickel outside. Values shown in Table 14-4 indicate that all readings would be just 0.6 volt more positive if the Ni–Ni^{2+} half-cell were our comparison standard. Any half-cell in the list could have been used as the standard for determining the half-cell voltage readings, but *we must use a reference half-cell. Without a standard half-cell, there is no place to fasten one voltmeter wire and the circuit cannot therefore be completed to get a reading!*

Faced with this situation, chemists have arbitrarily selected the H$^+$–H$_2$ half-cell as a standard and have compared all other half-cells to it. This means that a H$^+$–H$_2$ half-cell, measured against an identical H$^+$–H$_2$ half-cell used as a standard, generates 0.0 volt. Other half-cells will give the values shown in Table 14-5 (page 333), *if the measurement is properly made*.

EXERCISE 14-4

Suppose a Ag$^+$–Ag half-cell were chosen as the standard. What would all values in Table 14-4 read?

14-2.3 The Importance of Concentration in Defining Cell Voltage—An Application of Le Chatelier's Principle

The qualifier "if the measurement is properly made" suggests that in defining the reference half-cell and the measured half-cells, we must tell precisely how each half-cell is made! You remember from your laboratory work that the concentration of the solution surrounding the metal electrode is important in determining the cell voltage. You studied a cell based on the reaction

$$Zn(s) + Cu^{2+}(aq) \longrightarrow Zn^{2+}(aq) + Cu(s) \qquad (11)$$

If *more* CuSO$_4$ were dissolved in the solution surrounding the copper electrode, the cell voltage *increased*. If a solution containing *less* ZnSO$_4$ were used around the zinc electrode, the voltage *also increased*. A decrease in the concentration of Cu^{2+} or an increase in the concentration of Zn^{2+} made the voltage drop. Are not these observations in complete agreement with predictions we would make on the basis of Le Chatelier's principle? The tendency of the reaction to go from left to right as written is measured by the *positive* voltage value for the cell. Le Chatelier's principle tells us that if the Cu^{2+} concentration is reduced by precipitation of Cu^{2+} as CuS, there will be less tendency for the reaction to go. (The equilibrium will shift toward Zn + Cu^{2+}.) The voltage should decrease—a prediction in complete agreement with experimental findings.

EXERCISE 14-5

A cell is based on the reaction

$$Ni(s) + 2Ag^+ \longrightarrow Ni^{2+} + 2Ag(s)$$

What would be the effect on cell voltage if

(1) more AgNO$_3$ were dissolved in the solution around the silver electrode?
(2) more NiSO$_4$ were dissolved in the solution around the nickel electrode?
(3) Ag$_2$S were precipitated by adding H$_2$S?

The effects of concentration changes on cell voltage are very familiar to you, although you probably were not aware of the meaning of your observations when you made them. You have all watched a flashlight go dim after continued use or heard the battery of a car gradually run down as someone tried to start it on a cold morning. What happens? Why does the cell or battery stop generating electrical energy?

To answer this question, let us first consider another question. What happens to any cell or battery as it operates? Observation shows that the *voltage decreases* as the cell operates until finally the voltage of the cell reaches zero. We say that the cell is "dead." Equilibrium has been attained, and the reaction that has been producing the energy has the same tendency to proceed as its reverse reaction. Again, the observed voltage measures the *net* tendency for the reaction to occur. At equilibrium there is a balance between forward and reverse reactions; hence, there is *no* tendency for further reaction either way. The voltage of a cell at equilibrium is zero; the cell is "dead."

14-2.4 Standard Half-Cell Potentials—E^0

Since concentrations in a cell do influence its voltage, concentrations of all reagents in a cell should be specified if the measured voltages are going to have maximum value for predicting chemical reactions. Because of this, chemists have introduced the idea of **standard state.** The standard state for gases is taken as a pressure of 1 atm at 25°C; the standard state for ions is taken as a 1 M solution,* and the standard state for other pure substances is taken as the state in which the pure substance exists at 25°C and 1 atm.

STANDARD HALF-CELL POTENTIALS—E^0

TABLE **14-5** SELECTED STANDARD REDUCTION POTENTIALS FOR HALF-REACTIONS*

Half-Reaction		E^0
Oxidized State	Reduced State	(volts)
$MnO_4^- + 8H^+ + 5e^- \longrightarrow$	$Mn^{2+} + 4H_2O$	+1.52
$Cl_2(g) + 2e^- \longrightarrow$	$2Cl^-$	+1.36
$MnO_2(s) + 4H^+ + 2e^- \longrightarrow$	$Mn^{2+} + 2H_2O$	+1.28
$Br_2(l) + 2e^- \longrightarrow$	$2Br^-$	+1.06
$Ag^+ + e^- \longrightarrow$	$Ag(s)$	+0.80
$I_2(s) + 2e^- \longrightarrow$	$2I^-$	+0.53
$Cu^{2+} + 2e^- \longrightarrow$	$Cu(s)$	+0.34
$2H^+ + 2e^- \longrightarrow$	$H_2(g)$	0.00
$Ni^{2+} + 2e^- \longrightarrow$	$Ni(s)$	−0.25
$Co^{2+} + 2e^- \longrightarrow$	$Co(s)$	−0.28
$Zn^{2+} + 2e^- \longrightarrow$	$Zn(s)$	−0.76

* A more complete list is given in Appendix 7.

PETER J. DEBYE
(1884–1966)

We can now specify more precisely how half-cell potentials should be measured. Values in Table 14-5 were obtained by measuring the cell voltages with all reagents in their standard state. Thus, if we set up a cell based on the reaction

$$Zn(s) + 2H^+ (1\ M) \longrightarrow Zn^{2+} (1\ M) + H_2 (1\ \text{atm}) \qquad (8a)$$

a reading of 0.76 (right) will be made on the voltmeter if the measurement is made at 25°C. Results for a number of half-cells are listed in Table 14-5. This value is known as the E^0 value. <u>The E^0 **value** is the half-cell potential associated with a half-reaction taking place between substances in their standard states.</u> (The superscript zero means standard state.) By international agreement,† the standard state H^+–H_2 half-cell is assigned a value of zero and used as a standard of comparison.

A large number of standard half-cell potentials,‡ E^0 values, have been determined. Some are given in Table 14-5. A more complete list is shown in Appendix 7. What do they mean and what can they

During his brilliant scientific career, Peter Debye added richly to our knowledge of physical chemistry. Born in the Netherlands, he received his doctorate from the University of Munich in 1908. He occupied many distinguished professorships in Europe, and was director of the Max Planck Institute in Berlin prior to World War II. In 1940, he became professor of chemistry at Cornell University.

While in Europe, Debye developed the still accepted theories of the specific heat of solids and of the interactions among polar molecules. His research also covered X-ray scattering, interatomic distances, magnetic cooling, the theory of electrolytes, and dipole theory. He received the 1936 Nobel Prize in Chemistry for this work. At Cornell, Debye's research turned toward the structure and particle size of the high polymers.

A professor emeritus at the time of his death, Debye was in great demand as a consultant and lecturer, presenting complex subjects with penetrating clarity.

---

* Some slight refinement is needed to define the standard state for ions precisely, but we shall not be concerned with that refinement here.

† The electrode signs used in this text are those adopted by the International Union of Pure and Applied Chemistry.

‡ The term *potential* refers to the voltages we have been measuring. It emphasizes the relationship between voltage and potential energy. **Voltage** is the energy supplied per elementary charge in driving the charge around the circuit of the cell. Voltage multiplied by charge is electrical energy. One volt driving one electron is 1.6×10^{-12} **ergs.** Remember that the erg is a work term.

be used for? *The standard half-cell potential is a quantitative measure of the electron-attracting ability of an ion in its standard state in the half-cell as compared to the electron-attracting ability of the hydrogen ion in the standard hydrogen reference half-cell.* If the sign of E^0 is *positive*, it means that *the ion of the half-cell equation is better at attracting an electron than is the hydrogen ion.* Thus, an E^0 value of $+0.80$ volt for the half-cell having the half reaction $Ag^+ + e^- \longrightarrow Ag(s)$ means that silver ion in solution has a stronger attraction for electrons than does hydrogen ion in solution. Therefore, silver ions can pull electrons away from hydrogen gas to give hydrogen ions and silver metal. The overall equation then goes spontaneously as written with a voltage of 0.80 volt:

$$2Ag^+ + H_2(g) \longrightarrow 2Ag(s) + 2H^+ \qquad E^0 = +0.80 \text{ volt} \qquad (15)$$

We have not added 1 M for the concentration of each ionic species, nor have we indicated 25°C and 1 atm for H_2. All this is implied by the symbol E^0.

If the sign of a half-cell is *negative, the hydrogen ion has a stronger attraction for electrons than does the ion of the half-cell.* Thus, for the half-cell with the half-reaction

$$Zn^{2+} + 2e^- \longrightarrow Zn(s) \qquad E^0 = -0.76 \text{ volt} \qquad (14)$$

the negative sign means that Zn^{2+} has less attraction for electrons than does H^+. Accordingly, we would expect that H^+ would be able to take electrons away from metallic zinc to give hydrogen gas. The process has been observed in the laboratory:

$$Zn(s) + 2H^+ \longrightarrow H_2(g) + Zn^{2+} \qquad E^0 = +0.76 \text{ volt} \qquad (8)$$

The negative sign for $Zn^{2+} + 2e^- \longrightarrow Zn(s)$ means that the half-reaction does not go spontaneously as written but in the reverse direction, $Zn(s) \longrightarrow Zn^{2+} + 2e^-$. Thus, if the half-reaction is written in reverse, the value of E^0 becomes positive:

$$Zn^{2+} + 2e^- \longrightarrow Zn(s) \qquad E^0 = -0.76 \text{ volt} \qquad (14)$$

$$Zn(s) \longrightarrow Zn^{2+} + 2e^- \qquad E^0 = +0.76 \text{ volt} \qquad (9)$$

Note that when the direction of the reaction is reversed, the sign is reversed.

14-2.5 Using Standard Half-Cell Potentials for Understanding Cells

Table 14-5 lists some standard half-cell potentials. Does it agree with our laboratory experience? Taking four reactions from the table, we find

$$Ag^+ + e^- \longrightarrow Ag(s) \qquad E^0 = +0.80 \text{ volt} \qquad (13)$$

$$Cu^{2+} + 2e^- \longrightarrow Cu(s) \qquad E^0 = +0.34 \text{ volt} \qquad (6)$$

$$2H^+ + 2e^- \longrightarrow H_2(g) \qquad E^0 = 0.00 \text{ volt} \qquad (10)$$

$$Zn^{2+} + 2e^- \longrightarrow Zn(s) \qquad E^0 = -0.76 \text{ volt} \qquad (14)$$

These half-cell reactions, listed in order of decreasing E^0 value, are in the *same* order as those of Table 14-1. (Remember that Table 14-1 was based only on our qualitative laboratory observations!)

Another way to see the usefulness of Table 14-5 is to compare its voltage predictions with voltages we can measure. For example, a rapid laboratory measurement shows that a standard cell based on the reaction

$$Zn(s) + Ni^{2+} \longrightarrow Zn^{2+} + Ni(s) \qquad (16)$$

has a voltage of 0.5 volt with the zinc electrode as the negative electrode or anode in the cell. The two half-reactions are

$$Ni^{2+} + 2e^- \longrightarrow Ni(s) \qquad E^0 = -0.25 \text{ volt} \qquad (17)$$
$$Zn^{2+} + 2e^- \longrightarrow Zn(s) \qquad E^0 = -0.76 \text{ volt} \qquad (14)$$

From the size of the E^0 values, we see that Ni^{2+} has a greater attraction for electrons than does Zn^{2+}; hence, Ni^{2+} should be able to take electrons from metallic Zn to give Zn^{2+} and metallic Ni. This suggests that the half-reaction for zinc should be reversed to get the actual equation for the process. We then write

$$Zn(s) \longrightarrow Zn^{2+} + 2e^- \qquad E^0 = +0.76 \text{ volt} \qquad (9)$$
$$2e^- + Ni^{2+} \longrightarrow Ni(s) \qquad E^0 = -0.25 \text{ volt} \qquad (17)$$
$$\overline{Zn(s) + Ni^{2+} \longrightarrow Zn^{2+} + Ni(s) \qquad E^0 = +0.51 \text{ volt} \qquad (16)}$$

Note that we changed the sign of E^0 for the Zn^{2+}–Zn half-reaction when we reversed it. The half-cell potentials predict that metallic zinc will react with Ni^{2+} to give Zn^{2+} and metallic nickel and that the measured E^0 value will be 0.51 volt. Experiments agree with this.

Let us try another prediction. Another cell we studied is based on the reaction

$$Zn(s) + 2Ag^+ \longrightarrow Zn^{2+} + 2Ag(s) \qquad (18)$$

The two half-reactions involved are:

$$Ag^+ + e^- \longrightarrow Ag(s) \qquad E^0 = +0.80 \text{ volt} \qquad (13)$$
$$Zn^{2+} + 2e^- \longrightarrow Zn(s) \qquad E^0 = -0.76 \text{ volt} \qquad (14)$$

From the half-cell potentials we conclude that Ag^+ has a greater tendency to pick up electrons than does Zn^{2+}; therefore, silver ion should remove electrons from metallic zinc to give zinc ions and metallic silver. We then have

$$2e^- + 2Ag^+ \longrightarrow 2Ag(s) \qquad E^0 = +0.80 \text{ volt} \qquad (4)$$
$$Zn(s) \longrightarrow Zn^{2+} + 2e^- \qquad E^0 = +0.76 \text{ volt} \qquad (9)$$
$$\overline{2Ag^+ + Zn(s) \longrightarrow Zn^{2+} + 2Ag(s) \qquad E^0 = +1.56 \text{ volts} \qquad (18)}$$

Our conclusions again agree with experiment. In the cell zinc metal dissolves and silver metal precipitates. The voltage is about 1.5 volts. Finally, experiment shows that 1 mole of zinc reacts with 2 moles of silver ion, as indicated by reaction (18).

Notice that we did *not* double E^0 for the reaction $Ag^+ + e^- \longrightarrow$ Ag in obtaining the voltage. The voltage of a half-reaction does *not* depend on how many moles we consider, but it does depend upon the concentration. Thus,

$$Ag^+ + e^- \longrightarrow Ag(s) \quad E^0 = +0.80 \text{ volt} \quad (13)$$
$$2Ag^+ + 2e^- \longrightarrow 2Ag(s) \quad E^0 = +0.80 \text{ volt} \quad (4)$$
$$3Ag^+ + 3e^- \longrightarrow 3Ag(s) \quad E^0 = +0.80 \text{ volt} \quad (19)$$

This is simply a statement that the voltage of a cell is not determined by how big the cell is. A tiny cell (battery) for an electric watch can give a voltage as high as that of a massive storage battery based on the same reaction.

You might wonder what we would have learned if we had assumed that either of these two cells operates in the reverse reaction. Suppose we had proposed a cell based on the oxidation of nickel [$Ni(s) \longrightarrow Ni^{2+} + 2e^-$] and the reduction of zinc [$Zn^{2+} + 2e^- \longrightarrow Zn(s)$]:

$$Ni(s) \longrightarrow Ni^{2+} + 2e^- \quad E^0 = +0.25 \text{ volt} \quad (20)$$
$$2e^- + Zn^{2+} \longrightarrow Zn(s) \quad E^0 = -0.76 \text{ volt} \quad (14)$$
$$\overline{Zn^{2+} + Ni(s) \longrightarrow Zn(s) + Ni^{2+} \quad E^0 = -0.51 \text{ volt} \quad (21)}$$

Equation (*21*) is just the reverse of equation (*16*) and the voltage is the same *except for the opposite sign*. As we noted earlier, the negative voltage (−0.51 volt) implies that equilibrium in the reaction favors *reactants, not products*. We obtain the same prediction as we did before since the voltage is negative: The reaction will tend to go *in the reverse direction;* that is, it will dissolve zinc metal and will precipitate nickel metal.

14-2.6 Predicting Reactions from Table 14-5

The ideas we have developed for reactions occurring in electrochemical cells are also applicable to many other chemical reactions. Therefore, *half-cell potentials can be used to predict what chemical reactions can occur spontaneously.*

If someone wishes to know whether zinc can be oxidized when it is placed in contact with a solution of nickel sulfate, the values of E^0 help him decide. The standard half-cell potential, or E^0 value, for Zn^{2+}–Zn is −0.76 volt; that of Ni^{2+}–Ni, −0.25 volt. Apparently, Ni^{2+} has a stronger attraction for electrons than does Zn^{2+}. Therefore, $Zn(s)$ can transfer electrons to Ni^{2+}. We predict from this information that zinc metal will react with nickel ion to give nickel metal and zinc ion. Zinc is oxidized and nickel is reduced. This is exactly what happens.

Successful predictions save lots of hard work. Let us try another example. Will silver be oxidized if it is immersed in copper sulfate? The standard half-cell potential for silver ($Ag^+ + e^- \longrightarrow Ag$) is +0.80 volt and that for copper ($Cu^{2+} + 2e^- \longrightarrow Cu$) is +0.34 volt. We see that silver ion has a stronger attraction for electrons than does copper ion (E^0 for Ag^+–Ag is greater than that for Cu^{2+}–Cu).

Thus, copper ions would *not* be able to remove electrons from silver metal; the reaction will *not* tend to proceed spontaneously; silver will *not* be oxidized to an appreciable extent in copper sulfate solution. Experiments confirm this prediction.

EXERCISE 14-6

Use the values of E^0 to predict whether cobalt metal will tend to dissolve in a 1 M solution of acid, H^+. Now predict whether cobalt metal will tend to dissolve in a 1 M solution of zinc sulfate (reacting with Zn^{2+}).

We can generalize now on the use of Table 14-5. A substance on the left in Table 14-5 reacts by gaining electrons. A substance on the right reacts by losing electrons. We may draw the following conclusions:

(1) *An oxidation-reduction reaction must involve a substance from the left side of the equations in Table 14-5 (something which can be reduced) and a substance from the right side (something which can be oxidized).*
(2) *A substance on the left side of the equations in Table 14-5 tends to react spontaneously with any substance on the right side which is lower in the table.*

Applying these rules, we would predict: Copper metal could be oxidized to Cu^{2+} by $Br_2(l)$ or $MnO_2(s)$, but not by Ni^{2+} or Zn^{2+}. Of course, copper metal cannot be oxidized by either zinc metal or nickel metal because neither zinc metal nor nickel metal can accept electrons to be reduced (as far as we know from Table 14-5).

EXERCISE 14-7

Use Table 14-5 to decide which of the following substances tend to oxidize bromide ion, Br^-, to give the element bromine: $Cl_2(g)$, H^+, Ni^{2+}, MnO_4^-.

EXERCISE 14-8

Use Table 14-5 to decide which of the following substances tend to reduce $Br_2(l)$ to $2Br^-$: Cl^-, $H_2(g)$, $Ni(s)$, Mn^{2+}.

14-2.7 Predictions and the Effect of Concentrations

The foregoing predictions have been based upon the values of E^0 that apply to standard conditions. Yet we often wish to carry out a reaction at conditions other than standard ones. The prediction then must be adjusted in accord with Le Chatelier's principle as we deviate from standard conditions.

For example, by comparing E^0's, we predicted that zinc would react with nickel sulfate solution. These E^0's apply, however, only if zinc ion and nickel ion are both present at a 1 M concentration:

$$Zn(s) + Ni^{2+} \longrightarrow Zn^{2+} + Ni(s) \qquad (16)$$

However, no zinc ion is present in this case. How does this fact affect our prediction? By Le Chatelier's principle, the removal of Zn^{2+} tends to shift equilibrium toward the products. Therefore, removing Zn^{2+} *increases* the tendency for zinc to dissolve in a solution of nickel ion. The prediction is not altered by changes in ion concentration.

This will not always be the case, however. Consider the question: Will silver metal dissolve in $1\ M\ H^+$? According to Table 14-5,

$$2Ag(s) \longrightarrow 2Ag^+ (1\ M) + 2e^- \qquad E^0 = -0.80 \text{ volt} \qquad (22)$$

$$2e^- + 2H^+ (1\ M) \longrightarrow H_2(g) \qquad E^0 = 0.00 \text{ volt} \qquad (10a)$$

$$2Ag(s) + 2H^+ (1\ M) \longrightarrow 2Ag^+ (1\ M) + H_2(g) \quad E^0 = -0.80 \text{ volt} \quad (23)$$

The negative E^0 shows that the state of equilibrium favors the reactants rather than the products for reaction (23). For standard conditions, the reaction will not tend to occur spontaneously. However, if we place $Ag(s)$ in fresh $1\ M\ H^+$, the initial Ag^+ concentration will be zero, not $1\ M$. By Le Chatelier's principle, we know that this will increase the tendency to form products. Will this tendency be enough to counteract the E^0 prediction (no reaction)? Experiments show that some silver will dissolve, though only a *very small amount*. The tendency for silver metal to release electrons is much less than the tendency for H_2 to release electrons. The equilibrium concentration of Ag^+ is so small that no silver chloride precipitate is formed when chloride ions are present in the solution. Very little Ag^+ is present since silver chloride has a very low solubility.

That some silver does dissolve to form Ag^+ can be shown experimentally by adding a little KI to the solution. Silver iodide has an even lower solubility than does silver chloride. The experiment shows that the amount of dissolved silver is sufficient to precipitate some AgI but not AgCl. This places the Ag^+ concentration below $10^{-10}\ M$ but above $10^{-17}\ M$. Either concentration is so small that we can consider our prediction for the standard state to be applicable. Silver metal does not dissolve appreciably in $1\ M$ HCl. In general, whether a prediction based upon the standard state can be applied to other conditions depends upon how large E^0 is. If E^0 for the overall reaction is only 0.1 or 0.2 volt (either positive or negative), then deviations from standard conditions may invalidate predictions that do not consider these concentration or temperature changes.

14-2.8 Reliability of Predictions

There is one more limitation on the reliability of predictions based upon E^0's. To see it, we shall consider the three reactions

$$Cu(s) + 2H^+ \longrightarrow Cu^{2+} + H_2(g) \qquad E^0 = -0.34 \text{ volt} \qquad (24)$$

$$Fe(s) + 2H^+ \longrightarrow Fe^{2+} + H_2(g) \qquad E^0 = +0.44 \text{ volt} \qquad (25)$$

$$3Fe(s) + 2NO_3^- + 8H^+ \longrightarrow 3Fe^{2+} + 2NO(g) + 4H_2O$$
$$E^0 = +1.40 \text{ volts} \qquad (26)$$

The three values of E^0 are easily calculated from half-cell potentials. We can predict with confidence that the reaction of copper metal with dilute H^+ will not occur to any appreciable extent. The negative value of E^0 indicates that equilibrium strongly favors the reactants. Moreover, we can predict that the reaction of iron metal with dilute acid *might* occur, and that the reaction of iron metal with dilute nitric acid to give NO *might* occur. The positive values of E^0 for reactions (25) and (26) indicate that the products are strongly favored in equilibrium. Experiments are warranted. A piece of iron is immersed in dilute HCl. Bubbles of hydrogen appear; the reaction does occur. A piece of iron is immersed in a 1 molar nitric acid solution. Although bubbles of hydrogen may appear, no nitric oxide (NO) gas appears. The reaction between iron and nitrate ion in acid solution to give NO does not occur immediately; the reaction rate is extremely slow. This slow rate could *not* be predicted from the E^0's.

So the equilibrium predictions based on E^0's do not eliminate the need for experiments. They provide no basis, for example, for anticipating reaction rate; only experiment provides this information. The E^0's do, however, provide definite and reliable guidance concerning the equilibrium state, thus eliminating the need for many experiments. A multitude of reactions do not need to be tried. We can predict their failure by considering the values of E^0.

14-2.9 E^0 and the Factors that Determine Equilibrium

We see that E^0 furnishes a basis for predicting the equilibrium state. In Chapter 11 and in subsequent chapters, equilibrium was treated in terms of two opposing tendencies—a tendency toward minimum energy and a tendency toward maximum randomness. What is the connection between E^0 and these two tendencies?

Consider two reactions for which E^0 shows that products are favored, one exothermic and the other endothermic. For the exothermic reaction, reactants, having been mixed, are driven toward equilibrium by their tendency toward minimum energy. Contrast this with the endothermic reaction, for which E^0 shows that equilibrium favors products. When these reactants are mixed, they approach equilibrium *against* the tendency toward minimum energy (since heat is absorbed). This reaction is driven by the tendency toward maximum randomness. Neither the tendency toward minimum energy nor the tendency toward maximum randomness *considered alone* will give us a reliable guide to the direction in which a reaction will go. But E^0 does give us in effect a way of combining the two tendencies.

In summary, E^0 measures quantitatively a proper combination of the tendency toward minimum energy and the tendency toward maximum randomness under the standard-state conditions.

14-3 BALANCING OXIDATION-REDUCTION EQUATIONS

A chemical reaction neither destroys nor produces atoms. Therefore, there must be the same number and types of atoms among the reactants as among the products. A chemical reaction also neither destroys nor produces electric charge. Therefore, the sum of the charges appearing on the reactants must be the same as the sum of the charges appearing on the products. When a reaction is properly written, it expresses both these conservation principles. The process of writing a correct chemical equation and making sure that these two conservation principles are observed is called **balancing** the chemical equation.

Oxidation-reduction equations must be balanced if correct predictions are to be made. As in most matters, there are several ways to reach the desired goal. We shall discuss two ways to balance oxidation-reduction equations—firstly, by using half-reactions and, secondly, by using oxidation numbers which we shall discuss shortly. Which method is better will depend somewhat upon your own preference, as well as upon the type of problem under consideration.

14-3.1 Using Half-Reactions to Balance Oxidation-Reduction Reactions

Suppose we want to describe by an equation what happens when pure lithium metal is added to a 1 M HCl solution. Our first step must be to decide what the products will be. We may know them already. If we do not, we can look them up in a reference book, perform the experiment ourselves, or proceed with some assumed products based on tables of half-cell reactions.

For lithium metal in 1 M HCl, the observed facts are that the metal dissolves spontaneously and hydrogen gas bubbles out of the solution. From Appendix 7 we select the two half-reactions (notice that the half-reactions are already "balanced" in both charge and number of atoms):

$$e^- + Li^+ \longrightarrow Li(s) \tag{27}$$

$$2e^- + 2H^+ \longrightarrow H_2(g) \tag{10}$$

Both these half-reactions show acceptance of electrons. We observed in Section 14-2 that one of the equations must be reversed. Experiment and E^0 values show us that hydrogen gas is a product while lithium metal dissolves to form ions. Therefore, equation (27) must be reversed. Thus,

$$Li(s) \longrightarrow Li^+ + e^- \tag{28}$$

$$2e^- + 2H^+ \longrightarrow H_2(g) \tag{10}$$

These equations now agree with observation.

Next, the equations must be properly combined. The first step is to make them indicate the same number of electrons. We "balance the electrons." By *inspection,* we find that balance is achieved by doubling equation (28):

$$2Li(s) \longrightarrow 2Li^+ + 2e^- \tag{29}$$

$$2e^- + 2H^+ \longrightarrow H_2(g) \tag{10}$$

Now we add the two equations to get the net reaction:

$$2H^+ + 2Li(s) \longrightarrow 2Li^+ + H_2(g) \tag{30}$$

As a final check, let us verify the conservation of charge:

$$2H^+ + 2Li(s) \longrightarrow 2Li^+ + H_2(g) \tag{30}$$
$$(2+) + \quad 0 \quad\quad (2+) + \quad 0$$
$$(2+) = (2+)$$

As a more complex case, suppose we want to write the equation for the reaction that occurs when hydrogen sulfide gas (H_2S) is bubbled into an acidified potassium permanganate solution ($KMnO_4$). When we do this, we observe that the purple color of the MnO_4^- ion disappears and that the mixture turns cloudy (sulfur particles). From Appendix 7 we can write the two half-reactions:

$$H_2S(g) \longrightarrow S(s) + 2H^+ + 2e^- \qquad (31)$$

$$Mn^{2+} + 4H_2O \longrightarrow MnO_4^- + 8H^+ + 5e^- \qquad (32)$$

Since we know that sulfur is formed, we shall use the equation for the oxidation of H_2S as it is written. However, the purple MnO_4^- is being changed to almost colorless Mn^{2+}, so we shall reverse equation (32):

$$H_2S(g) \longrightarrow S(s) + 2H^+ + 2e^- \qquad (31)$$

$$5e^- + 8H^+ + MnO_4^- \longrightarrow Mn^{2+} + 4H_2O \qquad (33)$$

Balancing the electrons is the next step. Using 10 as the least common multiple for the numbers of electrons, we multiply equation (31) by 5 and equation (33) by 2:

$$5H_2S(g) \longrightarrow 5S(s) + 10H^+ + 10e^- \qquad (34)$$

$$10e^- + 16H^+ + 2MnO_4^- \longrightarrow 2Mn^{2+} + 8H_2O \qquad (35)$$

$$\overline{5H_2S(g) + 16H^+ + 2MnO_4^- \longrightarrow 5S(s) + 10H^+ + 2Mn^{2+} + 8H_2O} \qquad (36)$$

When we add equations (34) and (35), the ten electrons on each side add to zero. Both atoms and charges are properly balanced in equation (36), but there are excess H^+ ions included among the reactants as well as the products. Subtracting $10H^+$ from both sides of the equation, we obtain the final balanced reaction

$$5H_2S(g) + 2MnO_4^- + 6H^+ \longrightarrow 5S(s) + 2Mn^{2+} + 8H_2O \qquad (37)$$

Before leaving the equation, let us check the electric-charge balance:

$$5H_2S(g) + 2MnO_4^- + 6H^+ \longrightarrow 5S(s) + 2Mn^{2+} + 8H_2O \qquad (37)$$
$$5(0) \ + \ 2(1-) \ + 6(1+) \qquad\quad 5(0) \ + \ 2(2+) \ + \ 8(0)$$
$$(2-) \ + \ (6+) \qquad\qquad\qquad (4+)$$
$$(4+) = (4+)$$

14-3.2 Balancing Half-Reactions

When potassium chlorate solution ($KClO_3$) is added to hydrochloric acid, chlorine gas is evolved. Although we can find the half-reaction $Cl_2(g) + 2e^- \longrightarrow 2Cl^-$ in Appendix 7 we find no equation involving ClO_3^-. We can guess that ClO_3^- accepts electrons and releases chlorine. Let us write a partial half-reaction in which we indicate an unknown number of electrons and in which we conserve only chlorine atoms:

$$ClO_3^- + xe^- \longrightarrow \tfrac{1}{2}Cl_2(g) + \text{other products} \qquad (38)$$

From experience we know that in acid solution, the oxygen in such oxidizing agents as MnO_4^- and $Cr_2O_7^{2-}$ combines with H^+ to give water. Let us incorporate our knowledge by showing $6H^+$ among the reactants and $3H_2O$ among the products:

$$ClO_3^- + 6H^+ + xe^- \longrightarrow \tfrac{1}{2}Cl_2(g) + 3H_2O \qquad (39)$$

Finally, we have to remember that charge is conserved. Since the products are neutral molecules, x must be 5 if the total charge among the reactants is to be zero. Our desired half-reaction is

$$ClO_3^- + 6H^+ + 5e^- \longrightarrow \tfrac{1}{2}Cl_2(g) + 3H_2O \qquad (40)$$

This balanced half-reaction can now be combined with the half-reaction for chloride oxidation to produce a balanced equation representing the reaction between HCl and $KClO_3$. If whole numbers of moles of chlorine are to be used, equation (40) becomes

$$2ClO_3^- + 12H^+ + 10e^- \longrightarrow Cl_2 + 6H_2O \qquad (41)$$

Combining this equation with the proper equation for chloride oxidation gives

$$(5 \times 2)Cl^- \longrightarrow 5Cl_2 + (5 \times 2)e^- \qquad (42)$$
$$2ClO_3^- + 12H^+ + 10e^- \longrightarrow Cl_2 + 6H_2O \qquad (41)$$
$$\overline{10Cl^- + 2ClO_3^- + 12H^+ \longrightarrow 6Cl_2 + 6H_2O} \qquad (43)$$

Notice that both numbers of atoms and the numbers of charges now balance.

14-3.3 Oxidation Numbers— An Electron Bookkeeping Device

The term oxidation has been used repeatedly. It and the word oxygen have the same root, but does the process have a connection to the element? How did we come to identify oxidation with a loss of electrons?

A look at a simple oxidation process is instructive. When magnesium metal burns in air, it combines with oxygen:

$$2Mg(s) + O_2(g) \longrightarrow 2MgO(s) \qquad (44)$$

Magnesium is **oxidized** when it combines with oxygen. For reasons given in Chapter 8, we can picture this reaction as a transfer of two electrons from the magnesium atom to the oxygen atom. Ions of charge $2+$ and $2-$ are formed which combine in a NaCl type of crystal lattice. X-ray studies reveal an ionic arrangement similar to that of NaCl. We can then say that magnesium loses two electrons to oxygen when it is oxidized. From this fact it is logical to describe oxidation of magnesium as a loss of electrons. Magnesium also loses electrons, or is oxidized, when it combines with chlorine:

$$Mg + Cl_2 \longrightarrow MgCl_2 \qquad (45)$$

Positively charged magnesium ions are again formed. The same type of electron reaction goes on in both cases. It was thus fairly logical for earlier chemists to define *oxidation* as a loss of electrons, and to call oxygen the *oxidizing agent* because it oxidized the magnesium. These same arguments identify chlorine as an oxidizing agent. *An **oxidizing agent** brings about the oxidation of another species.*

The same line of reasoning identifies the reducing agent. O_2 is reduced when it is converted to $2O^{2-}$; each atom picks up two electrons. Cl_2 is reduced when it is converted to $2Cl^-$. In each case, magnesium serves as the reducing agent. *A **reducing agent** brings about the reduction of another species.*

EXERCISE 14-9

An oxidation-reduction reaction is represented by the equation

$$Fe^{3+} + Cu^+ \longrightarrow Fe^{2+} + Cu^{2+}$$

Identify the oxidizing agent in this process.

Arguments given above show that magnesium has an apparent charge of $2+$ in MgO and oxygen has an apparent charge of $2-$. Since these came about in an oxidation-reduction process, it is logical to call these numbers **oxidation numbers.** *The oxidation numbers add to zero* in MgO. *Electrons lost equal electrons gained;* charge is conserved.

EXERCISE 14-10

What is the oxidation number of lithium (atomic number 3) in lithium oxide? (Review Chapter 8.)

Can the idea of oxidation numbers be extended beyond ionic solids? Let us try. If elemental sulfur is burned in air, the equation for the process is

$$S(s) + O_2(g) \longrightarrow SO_2(g) \qquad (46)$$

Remembering what happened with magnesium, it is tempting to suggest that each sulfur donates, or shares, a pair of electrons with each of the two oxygen atoms. Extrapolating from the magnesium oxide case, we can *assign* each oxygen an oxidation number of $2-$ and the sulfur an oxidation number of $4+$. As in MgO, the oxidation numbers add to zero.

$$\begin{Bmatrix} \text{oxidation} \\ \text{number} \\ \text{sulfur} \end{Bmatrix} + 2 \begin{Bmatrix} \text{oxidation} \\ \text{number} \\ \text{oxygen} \end{Bmatrix} = \text{zero} \qquad (47)$$

$$[4+] \quad + \quad [2 \times (2-)] \quad = \quad 0$$

Does this notation mean that each sulfur has given away four electrons and is now a $4+$ ion and that each oxygen is a $2-$ ion? Not at all! SO_2 is a gas, not an ionic solid like MgO. *The oxidation number is just a convenient number for electron bookkeeping.* It suggests here that four of the six outermost electrons of sulfur are involved in the formation of SO_2.

If four out of six electrons are involved, can we involve the other two electrons and use six out of six electrons? Under the influence of a catalyst such as NO_2, gaseous SO_2 will combine with oxygen to give SO_3, a liquid whose boiling point is 45°C:

$$SO_2(g) + \tfrac{1}{2}O_2(g) \xrightarrow{NO_2} SO_3(l) \qquad (48)$$

This is clearly an oxidation of SO_2 in the classical sense. Again, extrapolation from the simpler cases suggests that we can assign an oxidation number of $6+$ to sulfur in SO_3 if each oxygen is assigned its regular number of $2-$. Once again,

$$\left\{\begin{array}{c}\text{oxidation}\\\text{number}\\\text{sulfur}\end{array}\right\} + 3\left\{\begin{array}{c}\text{oxidation}\\\text{number}\\\text{oxygen}\end{array}\right\} = \text{zero} \qquad (49)$$

$$[6+] \;+\; [3 \times (2-)] \;=\; [0]$$

The oxidation numbers add to zero in a neutral molecule. The oxidation number of sulfur has gone up to $6+$ when SO_2 was oxidized to SO_3 using elemental oxygen. *Apparently, oxidation produces an increase in oxidation number.*

If SO_3 is allowed to react with calcium oxide, one molecule of each combines to form a new molecule:

$$CaO + SO_3 \longrightarrow CaSO_4 \qquad (50)$$

Has sulfur changed its oxidation number in this case? We can check. Calcium, like magnesium, should have an oxidation number of $2+$. (Why?) According to the rule carried over from the MgO and SO_2 cases, each oxygen should have an oxidation number of $2-$. What is then left for sulfur?

$$\left\{\begin{array}{c}\text{oxidation}\\\text{number}\\\text{calcium}\end{array}\right\} + \left\{\begin{array}{c}\text{oxidation}\\\text{number}\\\text{sulfur}\end{array}\right\} + 4\left\{\begin{array}{c}\text{oxidation}\\\text{number}\\\text{oxygen}\end{array}\right\} = \text{zero} \qquad (51)$$

$$\begin{array}{rcccccl}[2+] &+& x &+& [4 \times (2-)] &=& [0]\\ [2+] &+& x &+& [8-] &=& [0]\\ & & & & x &=& [6+]\end{array}$$

Sulfur did not change its oxidation number when it combined with CaO. This makes sense, since sulfur had already used all its six outer electrons in SO_3. The oxide ion which combined with sulfur had two electrons from calcium to complete its octet. (Sulfur trioxide plus calcium oxide exhibits Lewis acid-base behavior, *not* that of an oxidation-reduction reaction.) The assignment of oxidation numbers to molecules or ions enables us to keep track of electrons and to identify an oxidation process. *Oxidation causes an increase in oxidation number. Reduction causes a decrease in oxidation number.*

People differ in the extent to which they accept the existence of the charges indicated by oxidation numbers; but nearly all people agree that oxidation numbers provide a reasonable device for electron bookkeeping.

Based on arguments of the type outlined above, a set of rules has been developed for assigning oxidation numbers. We should emphasize that these numbers do not represent ionic charges, since it is quite probable that ions of the type required by oxidation numbers such as $6+$ do *not* exist in compounds—there is no S^{6+} in SO_3, for example. Nevertheless, oxidation numbers provide a convenient basis for keeping track of electrons in oxidation-reduction reactions. Here are the rules. You will realize that many were derived from ion-formation studies such as that discussed for MgO.

(1) *The oxidation number of a monatomic ion is equal to the charge on the ion.* Chloride (Cl^-), for example, has an oxidation number of $1-$; oxide, O^{2-}, an oxidation number of $2-$; phosphide in Na_3P, an oxidation number of $3-$; Fe^{2+}, $2+$; Fe^{3+}, $3+$.

(2) *The oxidation number of any element is zero.* The oxidation number of chlorine (Cl_2) is zero; of oxygen (O_2), zero; of magnesium (Mg), zero.

(3) *The oxidation number of members of the alkali metal family (Li, Na, K, Rb, Cs, and Fr) in compounds is $1+$.* Sodium is $1+$ in NaCl and lithium is $1+$ in Li_2O. Both are zero in the metallic form.

(4) *The oxidation number of Be, Mg, Ca, Sr, Ba, and Ra in compounds is $2+$.* Calcium is $2+$ in CaO; Mg is $2+$ in $MgCl_2$; Ba is $2+$ in $BaCl_2$.

(5) *The oxidation number of oxygen in compounds is taken to be $2-$* (except in peroxides and superoxides, which contain an oxygen-oxygen bond).

(6) *The oxidation number of hydrogen in compounds is taken to be $1+$* (except in a few hydrides such as NaH). Hydrogen is $1+$ in HCl, H_2O, H_2S, and so on.

(7) *The oxidation numbers of any other element or elements in a molecule or ion are selected to make the sum of the oxidation numbers equal to the charge on the molecule or ion.* You will recall that for SO_3 we had

$$\left\{\begin{array}{l}\text{oxidation}\\ \text{number}\\ \text{sulfur}\end{array}\right\} + 3\left\{\begin{array}{l}\text{oxidation}\\ \text{number}\\ \text{oxygen}\end{array}\right\} = \left\{\begin{array}{l}\text{charge on}\\ \text{molecule}\\ \text{or ion}\end{array}\right\} \quad (49a)$$

$$[6+] \quad + \quad [3 \times (2-)] = \quad [0]$$

For the SO_4^{2-} ion, with a charge of $2-$, the rule can be written as

$$\left\{\begin{array}{l}\text{oxidation}\\ \text{number}\\ \text{sulfur}\end{array}\right\} + 4\left\{\begin{array}{l}\text{oxidation}\\ \text{number}\\ \text{oxygen}\end{array}\right\} = \left\{\begin{array}{l}\text{charge on}\\ SO_4^{2-} \text{ ion}\end{array}\right\} \quad (52)$$

$$[6+] \quad + \quad [4 \times (2-)] = \quad [2-]$$

For the ammonium ion (NH_4^+), the rule takes the form

$$\left\{\begin{array}{l}\text{oxidation}\\ \text{number}\\ \text{nitrogen}\end{array}\right\} + 4\left\{\begin{array}{l}\text{oxidation}\\ \text{number}\\ \text{hydrogen}\end{array}\right\} = \left\{\begin{array}{l}\text{charge on}\\ NH_4^+ \text{ ion}\end{array}\right\} \quad (53)$$

$$[3-] \quad + \quad [4 \times (1+)] = \quad [1+]$$

The next rule is simply a statement of the principle that charge is conserved.

(8) *In any overall reaction the net change in oxidation numbers must be zero.*

$$2Mg(s) + O_2(g) \longrightarrow 2MgO(s) \qquad (44)$$

Mg changes from 0 to $2+ = (2+) \times 2$
O_2 changes from 0 to $2- = (2-) \times 2$
Net change $= \quad 0$

$$SO_2(g) + \tfrac{1}{2}O_2(g) = SO_3(l) \qquad (54)$$

Sulfur changes from $4+$ to $6+ = 2+$
One oxygen changes from 0 to $2- = 2-$
Net change $= 0$

Do not worry about the exceptions in rules (5) and (6). Your attention will be called to them later when substances involving them are considered.

EXERCISE 14-11

Use the above rules to obtain the oxidation numbers of

(a) P in Na_3PO_4 (d) Br in NaBr
(b) Cr in K_2CrO_4 (e) S in S_8
(c) N in HNO_3 (f) Cr in $Cr_2O_7^{2-}$.

[*Answers:* (a) $5+$, (c) $5+$, (e) 0, (f) $6+$.]

Let us use the concept of oxidation number in a more complex case to help us identify an oxidation process. When ferric ion (Fe^{3+}) is treated with hydrogen sulfite ion (HSO_3^-) in water solution, ferrous and sulfate ions form in solution. The equation is

$$2Fe^{3+} + HSO_3^- + H_2O \longrightarrow 2Fe^{2+} + SO_4^{2-} + 3H^+ \qquad (55)$$

Using rule (1), we see that the charge on each iron ion has changed from $3+$ to $2+$. The oxidation number of iron has decreased; the iron has been reduced. What about hydrogen and oxygen? By rule (5), oxygen is $2-$ in all compounds and ions in the equation. By rule (6), hydrogen is $1+$ in all compounds and ions in the equation. To make the net change in oxidation number for the reaction zero—rule (8)—sulfur *must* change because we have already noted a change of $2-$ in the oxidation number for ferric ions.

Let us check this change by calculating the oxidation numbers for sulfur in HSO_3^- and in SO_4^{2-}. Using rule (7), we write for HSO_3^-

$$\left\{\begin{matrix}\text{oxidation}\\\text{number}\\\text{sulfur}\end{matrix}\right\} + \left\{\begin{matrix}\text{oxidation}\\\text{number}\\\text{hydrogen}\end{matrix}\right\} + 3\left\{\begin{matrix}\text{oxidation}\\\text{number}\\\text{oxygen}\end{matrix}\right\} = \left\{\begin{matrix}\text{charge}\\\text{on ion}\end{matrix}\right\} \qquad (56)$$

$$x \quad + \quad [1+] \quad + \quad [3 \times (2-)] \quad = \quad [1-]$$
$$x \quad = \quad [4+]$$

For SO_4^{2-}, we can write

$$\left\{\begin{array}{l}\text{oxidation}\\\text{number}\\\text{sulfur}\end{array}\right\} + 4 \left\{\begin{array}{l}\text{oxidation}\\\text{number}\\\text{oxygen}\end{array}\right\} = \left\{\begin{array}{l}\text{charge}\\\text{on ion}\end{array}\right\} \qquad (52)$$

$$x \quad + \quad [4 \times (2-)] \quad = \quad [2-]$$
$$x = [6+]$$

Sulfur has gone from an oxidation number of $4+$ in HSO_3^- to $6+$ in SO_4^{2-}, a change of $2+$. Two ferric ions have each changed from $3+$ to $2-$ or a net change of $2 \times (1-) = 2-$. Then, by applying rule (8), the net change in oxidation number of sulfur ($2+$) plus the net change in oxidation number of iron ions ($2-$) equals zero. *Sulfite has been oxidized to sulfate by the ferric ion.*

EXERCISE 14-12

The reactions by which SO_2 and SO_3 dissolve in water are not considered to be oxidation-reduction reactions:

$$SO_2(g) + H_2O \longrightarrow HSO_3^-(aq) + H^+(aq)$$
$$SO_3(g) + H_2O \longrightarrow HSO_4^-(aq) + H^+(aq)$$

Show that neither process is an oxidation-reduction reaction by checking oxidation numbers.

14-3.4 Using Oxidation Numbers to Balance Oxidation-Reduction Equations

Oxidation numbers are a useful device for keeping track of electrons and for recognizing oxidation-reduction processes. Oxidation-reduction reactions involve changes of oxidation number. Furthermore, the net change of oxidation numbers in any process must be zero [rule (8)]. Consequently, oxidation numbers provide just as good a basis for balancing equations as do half-reactions. Let us see how this works.

Just as before, the first step in balancing a reaction is to determine the products. Again, experiment provides the answer. Let us reconsider one of the examples we balanced previously by the half-reaction method. Then we shall already know the products. In the second example of Section 14-3.1, we find H_2S gas reacts with MnO_4^- to give solid sulfur, Mn^{2+}, and H_2O:

$$MnO_4^- + H_2S(g) \longrightarrow S(s) + Mn^{2+} \qquad (57)$$

First, we assign oxidation numbers to each element, using rules (1)–(6). We find

$$MnO_4^- + H_2S(g) \longrightarrow S(s) + Mn^{2+} + H_2O \qquad (57)$$
$$7+, 2- \quad 1+, 2- \qquad 0 \qquad 2+ \qquad 1+, 2-$$

Since manganese changes from an oxidation number of $7+$ to a new value of $2+$, it has undergone *a change of* $5-$. Sulfur, on the other hand, has undergone a change from 2− to 0, or *a change of* 2+. If the net change in oxidation number for the overall process is to be zero, the gain in oxidation number by sulfur must be equal to the loss by manganese. We must take two MnO_4^- [net change = $2 \times (5-) = 10-$] and five H_2S [net change = $5 \times (2+) = 10+$] to make the net change zero:

$$2MnO_4^- + 5H_2S(g) \text{ gives } 5S(s) + 2Mn^{2+} \tag{58}$$

Now we proceed to ensure conservation of oxygen atoms. There are eight oxygen atoms on the left side of the equation; hence, we must add eight molecules of H_2O to the right side. (The reaction occurs in aqueous solution, so there is a large quantity of H_2O.)

$$2MnO_4^- + 5H_2S(g) \text{ gives } 5S(s) + 2Mn^{2+} + 8H_2O \tag{59}$$

Next we must ensure conservation of hydrogen atoms. There are ten hydrogen atoms on the left side of the equation (in $5H_2S$) and sixteen on the right (in $8H_2O$). In aqueous solutions (in neutral or acidic solutions), we assume that these six additional hydrogen atoms on the left are provided by H^+:

$$2MnO_4^- + 5H_2S(g) + 6H^+ \longrightarrow 5S(s) + 2Mn^{2+} + 8H_2O \tag{37}$$

The equation is balanced now, but experience advises that a check always be made on the charge balance:

$$2(1-) + 5(0) + 6(1+) \longrightarrow 5(0) + 2(2+) + 8(0) \tag{60}$$
$$(2-) \qquad\qquad (6+) \qquad\qquad\qquad (4+)$$
$$(4+) = (4+)$$

Of course, the oxidation-number method gives the same balanced equation as the half-reaction method.

As mentioned at the beginning of this section, your choice of one of these two methods will usually be based on the type of problem considered. Many chemists have found the following rules of thumb useful in deciding between the two methods:

(1) Oxidation-reduction reactions occurring in aqueous solutions are conveniently treated in terms of half-reactions showing transfer of electrons.
(2) Under more general conditions (gaseous state, other solvents, and so on), it is more convenient to treat oxidation-reduction reactions in terms of oxidation numbers.

14-4 ELECTROLYSIS

Earlier in this chapter we dealt with processes in which the potential energy decrease associated with a chemical reaction appeared as electrical energy. In Chapter 6 we considered the reverse process, where an externally generated electric current (electrical energy) was passed through a solution to give products of higher potential energy. Let us compare and contrast the spontaneous process (**electrochemical** or **voltaic cell**) and the nonspontaneous one (**electrolytic cell**).

Many features of the voltaic and electrolytic cells are the same. Each cell involves *both* oxidation and reduction processes. In both cells the *anode* is the electrode at which *oxidation* occurs and the *cathode* is the electrode at which *reduction* occurs. Electric current moves in the external wires of both cells as a result of net *electron* flow in one direction. In both cells electric current flowing through the solution of the cell results from migration of ions. Because the mechanism for carrying the current through the solution differs from that for carrying the current in a wire, a chemical reaction must take place at *each* junction between metal electrode and solution. Oxidation or reduction processes occur at these junctions. In both cells, *cations* move through the solution toward the *cathode*, and *anions* move through the solution toward the *anode*.

Despite the many common features shared by electrochemical and electrolytic cells some real and significant differences exist between them. These differences can be underscored by a reconsideration of two voltaic cells. The first cell (Figure 14-7.1, page 350) is based on the reaction

$$Ni(s) + Cu^{2+} \longrightarrow Ni^{2+} + Cu(s) \qquad (61)$$

The half-reactions and the E^0 values are

$$Ni(s) \longrightarrow Ni^{2+} + 2e^- \qquad E^0 = +0.25 \text{ volt} \qquad (20)$$
$$2e^- + Cu^{2+} \longrightarrow Cu(s) \qquad E^0 = +0.34 \text{ volt} \qquad (6)$$
$$\overline{Ni(s) + Cu^{2+} \longrightarrow Ni^{2+} + Cu(s)} \qquad E^0 = +0.59 \text{ volt} \qquad (61)$$

Nickel is the negative electrode and copper is the positive electrode. The observed voltage is 0.59 volt. The second cell (Figure 14-7.2, page 350) uses the reaction

$$Ni(s) + 2Ag^+ \longrightarrow Ni^{2+} + 2Ag(s) \qquad (62)$$

Appropriate half-cell reactions and E^0 values are

$$Ni(s) \longrightarrow Ni^{2+} + 2e^- \qquad E^0 = +0.25 \text{ volt} \qquad (20)$$
$$2e^- + 2Ag^+ \longrightarrow 2Ag(s) \qquad E^0 = +0.80 \text{ volt} \qquad (4)$$
$$\overline{Ni(s) + 2Ag^+ \longrightarrow Ni^{2+} + 2Ag(s)} \qquad E^0 = +1.05 \text{ volts} \qquad (62)$$

Nickel is again the negative electrode and silver the positive electrode.

Experience shows that when the cell is operated, the voltage will be 1.05 volts. What happens if the two cells are connected in opposition to each other? To find out, we place a wire between the two nickel electrodes, both of which are negative electrodes of their respective cells. Next, we connect one wire of the voltmeter to the copper electrode of cell one and the other wire to the silver electrode of cell two (refer to Figure 14-7.3 page 351). In each of the original cells operating separately nickel dissolved to release electrons into the external wire. If the same situation existed here, the two nickel electrodes would release electrons into opposite ends of the connecting wire, one cell opposing the other. But we cannot have two flows of electrons in opposite directions at the same time in the same wire.

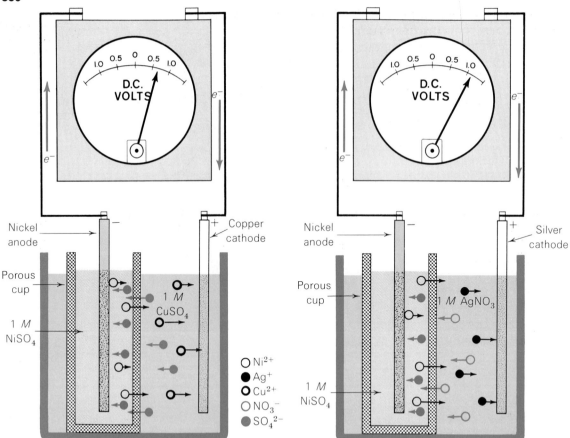

Fig. 14-7.1 **(Left)** A voltaic cell based on the reaction

Ni(s) + Cu^{2+} ⟶ Cu(s) + Ni^{2+}
E^0 = 0.59 volt

Fig. 14-7.2 **(Right)** A voltaic cell based on the reaction

Ni(s) + 2Ag$^+$ ⟶ 2Ag(s) + Ni^{2+}
E^0 = 1.05 volts

Experiment shows that under these conditions, the cell with the larger E^0 value continues to function as a voltaic cell, while the cell with the lower E^0 is forced to operate in the reverse direction—*it operates as an electrolytic cell.* Thus, the net reaction in the cell with the lower E^0 value is

$$2e^- + Ni^{2+} \longrightarrow Ni(s) \qquad (18)$$

$$Cu(s) \longrightarrow Cu^{2+} + 2e^- \qquad (3)$$

$$Cu(s) + Ni^{2+} \xrightarrow[\text{energy}]{\text{electrical}} Cu^{2+} + Ni(s) \qquad (63)$$

The voltage observed in the circuit will be the difference between the voltages of the two cells:

$$E^0 = E^0_{\text{Ni-Ag}} - E^0_{\text{Ni-Cu}} = 1.05 - 0.59 = +0.46 \text{ volt} \qquad (64)$$

This experimental observation can be understood by considering the net reaction taking place in the entire circuit. If the net reaction in the electrochemical cell and the net reaction in the electrolytic cell are added together, the net reaction for the combined process is obtained:

Voltaic   Ni(s) + 2Ag$^+$ ⟶ Ni^{2+} + 2Ag(s) E^0 = +1.05 volts   (62)

Electrolytic   Cu(s) + Ni^{2+} ⟶ Ni(s) + Cu^{2+} E^0 = −0.59 volt   (63)

Net for cell   2Ag$^+$ + Cu(s) ⟶ Cu^{2+} + 2Ag(s) E^0 = +0.46 volt   (5)

The E^0 value for the net reaction can be calculated separately from E^0 values for the silver and copper half-cells:

$$2e^- + 2Ag^+ \longrightarrow 2Ag(s) \qquad E^0 = +0.80 \text{ volt} \qquad (4)$$

$$Cu(s) \longrightarrow Cu^{2+} + 2e^- \qquad E^0 = -0.34 \text{ volt} \qquad (3)$$

$$\overline{2Ag^+ + Cu(s) \longrightarrow Cu^{2+} + 2Ag(s) \qquad E^0 = +0.46 \text{ volt} \qquad (5)}$$

We see that:

(1) The E^0 value for the Ni–Ni^{2+} half-cell cancels out in considering the net reaction for the overall process.
(2) The measured voltage for the overall process can be calculated from the net equation for the overall process.

Let us now focus our attention on the electrolytic cell, the cell containing nickel and copper half-cells. The process is in many ways like the electrolysis process discussed in Chapter 6. Electrons forced through the wire make the nickel electrode negatively charged. At the nickel electrode, Ni^{2+} ions make contact and electrons are transferred to give nickel metal:

Reduction $\qquad 2e^- + Ni^{2+} \longrightarrow Ni(s) \qquad (17)$

At the copper electrode electrons flow out to the positive electrode of the voltaic cell:

Oxidation $\qquad Cu(s) \longrightarrow Cu^{2+} + 2e^- \qquad (3)$

Positive ions that formed around the copper electrode now drift toward the negatively charged nickel electrode, while negative ions left in excess around the nickel electrode by removal of Ni^{2+} drift

Fig. 14-7.3 Two electrochemical cells connected in opposition.

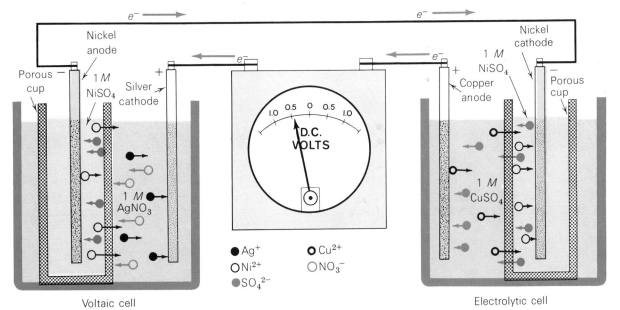

toward the positively charged copper electrode. *We see that an electrolytic cell is simply a voltaic cell forced to operate in the reverse direction by a voltage larger than that which the original cell could produce.*

How does this idea apply to the electrolysis of nickel bromide? Let us review what occurs when an electric current is passed through a solution of nickel bromide (see Figure 14-8). The generator or external battery forces electrons onto the nickel electrode. Nickel ions making contact with the electrode pick up electrons to become nickel metal:

$$2e^- + Ni^{2+} \longrightarrow Ni(s) \tag{6}$$

At the platinum electrode, electrons are removed by the generator or external battery. Bromide ions touching the electrode give up an electron to become bromine atoms:

$$Br^- \longrightarrow Br + e^- \tag{65}$$

Two bromine atoms combine to give a molecule of bromine:

$$2Br \longrightarrow Br_2 \tag{66}$$

Fig. 14-8.1 **(Left) When the generator's output is applied the electrochemical cell works as an electrolytic cell.**

Fig. 14-8.2 **(Right) When the generator's output is removed the electrochemical cell works as a voltaic cell.**

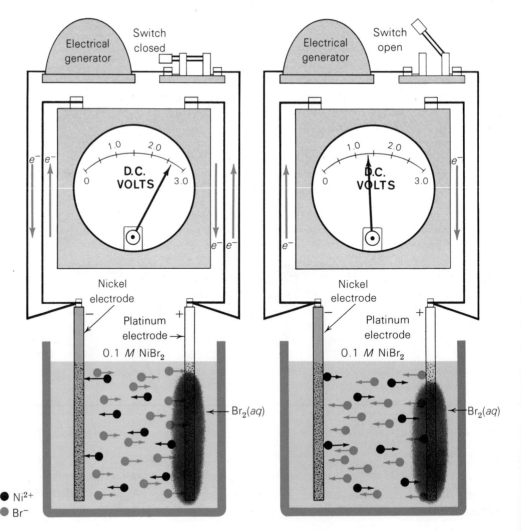

Molecular bromine accumulates around the platinum electrode where it is formed. The positive ions drift toward the negative electrode where they become neutral atoms and are deposited and negative ions drift toward the positive electrode where they become neutral atoms, then molecules of bromine. As electrolysis proceeds, metallic nickel deposits at one electrode and elemental bromine accumulates around the other.

If now the current is interrupted by disconnecting the generator, the voltmeter suddenly swings over to indicate a voltage in the opposite direction. The voltage observed is close to $+1.31$ volts, the voltage expected for a cell based on the reaction

$$Ni(s) \longrightarrow Ni^{2+} + 2e^- \qquad E^0 = +0.25 \text{ volt} \qquad (20)$$

$$2e^- + Br_2 \longrightarrow 2Br^- \qquad E^0 = +1.06 \text{ volts} \qquad (67)$$

$$\overline{Ni(s) + Br_2(aq) \longrightarrow Ni^{2+} + 2Br^-} \qquad E^0 = +1.31 \text{ volts} \qquad (68)$$

When nickel bromide solution is electrolyzed, a nickel-bromine cell is set up rapidly by the processes at the electrodes. The external voltage supplied must be larger than the E^0 value of this cell or electrolysis will stop. The nickel-bromine system is like the nickel-copper cell. It is simply a voltaic cell being forced backward by an opposing voltage larger than the cell voltage. If the current is stopped, the cell will function as a voltaic cell until the Br_2 around the platinum electrode is used up. Using this idea, we can estimate the voltage needed for any given electrolysis operation.

If more than one reaction is possible in a cell, that reaction will take place which requires the lowest voltage. For example, in the electrolysis of a table salt solution, two reduction reactions are possible at the negative electrode:

$$Na^+ + e^- \longrightarrow Na(s) \qquad E^0 = -2.71 \text{ volts} \qquad (69)$$

$$2H^+ + 2e^- \longrightarrow H_2(g) \qquad E^0 = 0.00 \text{ volt} \qquad (10)$$

Hydrogen gas must be liberated much more easily than sodium if both ions are present at a 1 molar concentration. Decreasing the concentration of hydrogen ion should make hydrogen formation more difficult; but because of the very high voltage needed for sodium deposition (-2.71 volts), hydrogen gas is still more easily produced than sodium metal in any water solution. For this reason, the electrolysis of NaCl in water solution gives H_2 and Cl_2, *not* Na and Cl_2.

EXERCISE 14-13

Which ion will be deposited first if a solution containing 1 M concentrations of Cu^{2+} and Ni^{2+} is electrolyzed?

14-5 HIGHLIGHTS

A chemical reaction involving electron transfer from one species to another can be broken down into two **half-reactions**. The half-reaction in which electrons are lost is called **oxidation**. The half-reaction in which electrons are gained is called **reduction**. If oxidation and reduction are carried out in separate parts of a cell, *electrons will flow through a wire* from one part to the other, and *ions will flow through the solution* from one part of the cell to the other.

During its operation, the **electrochemical** or **voltaic cell** passes from a state of higher potential energy to a state of lower potential energy and electrical work is done. The voltage generated by an **electrochemical cell** is a measure of the driving force for the cell reaction. If the voltage for the H_2–$H^+(aq)$ half-cell is taken as zero, the voltage associated with other half-cell reactions can be measured by using carefully defined experimental conditions.

E^0 **values** or standard half-cell potentials can be collected in a table such as that shown in Appendix 7. E^0 values are useful in estimating cell voltages and in deciding whether equilibrium favors a given reaction or its reverse. A positive E^0 value favoring reactivity may not guarantee that the process will go as written, but a proper negative E^0 value indicates with certainty that the process will not go.

In **balancing** oxidation-reduction reactions, both atoms and charge must be conserved. A method for balancing oxidation-reduction reactions can be based on half-cell reactions. A second method can be based on **oxidation numbers.**

Finally, we saw that the process of **electrolysis,** first discussed in Chapter 6, involves a voltaic cell which is forced to run backward by an opposing potential larger than the potential of the cell.

QUESTIONS AND PROBLEMS

1 You want to use the reaction indicated below to make a voltaic cell:

$$Zn(s) + Fe^{2+} \longrightarrow Zn^{2+} + Fe(s)$$

The process is spontaneous as indicated. (a) Draw a diagram of a cell which would use this process. (b) Identify anode and cathode. (c) Indicate the direction of electron flow in the external circuit. (d) Write the equation for the half-reaction taking place at each electrode. (e) Calculate E^0 for this cell.

2 One method of obtaining copper metal is to let a solution containing Cu^{2+} ions trickle over scrap iron. Write the equations for the two half-reactions involved. Assume the iron becomes Fe^{2+}. Indicate in which half-reaction oxidation is taking place.

3 Aluminum metal reacts with aqueous acidic solutions to liberate hydrogen gas. Write the two half-reactions and the net ionic reaction.

4 Can 1 M $Fe_2(SO_4)_3$ solution be stored in a container made of nickel metal? Explain your answer.

5 What happens if an aluminum spoon is used to stir an $Fe(NO_3)_2$ solution? What happens if an iron spoon is used to stir an $AlCl_3$ solution?

6 The following observations are made (equations for the processes are given):

$$Mn(s) + Zn^{2+} \longrightarrow Zn(s) + Mn^{2+}$$
$$Fe(s) + Co^{2+} \longrightarrow Fe^{2+} + Co(s)$$
$$Fe(s) + Zn^{2+} \longrightarrow \text{no reaction}$$

Use these observations to arrange the ions Mn^{2+}, Zn^{2+}, Co^{2+}, and Fe^{2+}, in *decreasing* order of their attraction for electrons.

7 In acid solution the following are true: H_2S will react with oxygen to give H_2O and sulfur. H_2S will not react in the corresponding reaction with selenium or tellurium. H_2Se will react with sulfur giving H_2S and selenium but it will not react with tellurium. Arrange the elements S, O_2, Te, and Se in decreasing order of their tendency to attract electrons to form the compounds H_2O, H_2S, H_2Se, and H_2Te.

8 Suppose chemists had chosen to assign to the $2e^- + I_2 \longrightarrow 2I^-$ half-cell the potential of zero. (a) What would be the value of E^0 for $e^- + Na^+ \longrightarrow Na(s)$? (b) How much would the net potential for the reaction $2Na + I_2 \longrightarrow 2Na^+ + 2I^-$ change? (See Appendix 7.)

9 Complete the following equations. Determine the net potential of such a cell and decide whether a reaction can occur.
(a) $Zn(s) + Ag^+ \longrightarrow$
(b) $Cu(s) + Ag^+ \longrightarrow Cu^{2+}$
(c) $Sn(s) + Fe^{2+} \longrightarrow$
(d) $Hg(l) + H^+ \longrightarrow Hg^{2+}$

10 For each of the following, (1) write the half-reactions, (2) determine the net reaction, and (3) predict whether the reaction can occur, giving the basis for your prediction:
  (a) $Mg(s) + Sn^{2+} \longrightarrow$
  (b) $Mn(s) + Cs^+ \longrightarrow$
  (c) $Cu(s) + Cl_2(g) \longrightarrow Cu^{2+}$
  (d) $Zn(s) + Fe^{2+} \longrightarrow$
  (e) $Fe(s) + Fe^{3+} \longrightarrow$

11 A half-cell consisting of a palladium rod dipping into a 1 M $Pd(NO_3)_2$ solution is connected with a standard hydrogen half-cell. The cell voltage is +0.99 volt and the platinum electrode in the hydrogen half-cell is the anode. Determine E^0 for the reaction
$$2e^- + Pd^{2+} \longrightarrow Pd(s)$$

12 If a piece of copper metal is dipped into a solution containing Cr^{3+} ions, what will happen? Explain using E^0 values.

13 When copper is placed in concentrated nitric acid, vigorous bubbling takes place as a brown gas evolves. The copper disappears and the solution changes from colorless to a greenish-blue. The brown gas is nitrogen dioxide (NO_2) and the color of the solution is caused by the presence of copper(II) ion, Cu^{2+}. Using half-reactions from Appendix 7, write the net ionic equation for this reaction.

14 Suppose water is added to each of the porous cups containing copper sulfate in the two electrochemical cells shown in Figure 14-5. What change will occur in the voltage in each cell? Explain.

15 (a) If a neutral atom becomes positively charged, has it been oxidized or reduced? Write a general equation using M for the neutral atom. (b) If an ion X^- acquires a 2− charge, has it been oxidized or reduced? Write a general equation.

16 A cell is based on the reaction
$$Zn(s) + Cu^{2+} \longrightarrow Zn^{2+} + Cu(s)$$
H_2S is passed into the solution of $ZnSO_4$. Does the E^0 value of the cell increase or decrease over the original E^0?

17 Determine the oxidation numbers of carbon in the compounds carbon monoxide (CO), carbon dioxide (CO_2), and in diamond.

QUESTIONS and PROBLEMS

18 Determine the oxidation number of uranium in each of the known compounds UO_3, U_3O_8, U_2O_5, UO_2, UO, K_2UO_4, and $Mg_2U_2O_7$.

19 Use half-reactions to give a balanced equation for each of the following reactions:
  (a) $H_2O_2 + I^- + H^+$ gives $H_2O + I_2$
  (b) $Cr_2O_7^{2-} + Fe^{2+} + H^+$ gives $Cr^{3+} + Fe^{3+} + H_2O$
  (c) $Cu + NO_3^- + H^+$ gives $Cu^{2+} + NO + H_2O$
  (d) $MnO_4^- + Sn^{2+} + H^+$ gives $Mn^{2+} + Sn^{4+} + H_2O$

20 Use oxidation numbers to give a balanced equation for each of the following reactions:
  (a) $HBr + H_2SO_4$ gives $SO_2 + Br_2 + H_2O$
  (b) $NO_3^- + Cl^- + H^+$ gives $NO + Cl_2 + H_2O$
  (c) $Zn + NO_3^- + H^+$ gives $Zn^{2+} + NO_2 + H_2O$
  (d) BrO^- gives $Br^- + BrO_3^-$

21 Use oxidation numbers to balance the reaction between iron(II) ion, Fe^{2+}, and permanganate ion, MnO_4^-, in acid solution to produce iron(III) ion, Fe^{3+}, and manganous ion, Mn^{2+}.

22 Show the arbitrariness of oxidation numbers by balancing the reaction discussed in Problem 21 with the assumption that the oxidation number of manganese in MnO_4^- is 2+. Compare with the result obtained in Problem 21.

23 In order to make Na(s) and $Cl_2(g)$, an electric current is passed through NaCl(l). What does the energy supplied to this reaction do?

24 If you wish to replate a silver spoon, would you make it the anode or cathode in an electrolytic cell? Use half-reactions in your explanation. How many moles of electrons are needed to plate the spoon with 1.0 g of Ag?

25 In the electrolysis of aqueous cupric bromide ($CuBr_2$), 0.500 g of copper is deposited at one electrode. How many grams of bromine are formed at the other electrode? Write the anode and cathode half-reactions. [*Answer:* 1.26 g of $Br_2(l)$.]

26 Estimate the voltage required to electrolyze a 1 M solution of $CuBr_2$.

Treat electrons like light waves. Though they are bullets, they are also waves.

LOUIS de BROGLIE (1892–)

CHAPTER 15 Electromagnetic Radiation and Atomic Structure

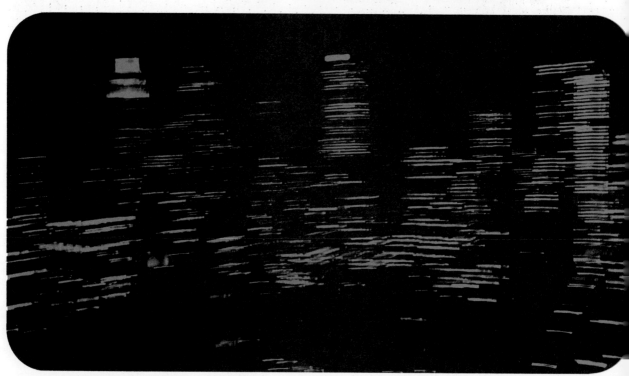

The familiar phenomenon, light, can tell us much about atoms.

One of the concepts you'll have to give up in this chapter is the belief that electrons are particles which whiz rapidly around the nucleus in discrete orbits. It's a nice, neat concept, but it doesn't happen to agree with the observed facts. So it's time to shed your security blanket, see what the facts are, and what new model we can build to agree with them. The clue here is in the similarity of the behavior of electrons to that of light, of all things. Good grief! But light at least is familiar to us; we can see it and measure it, and even though the idea of the electron behaving as a wave is foreign to us, we can come up with a model that leaves us somewhat less in the dark.

So far a relatively simple model of the atom has been adequate for all our discussion of the properties of substances. Building on the original Rutherford model, we visualized a very small, very dense nucleus composed of protons and neutrons; surrounding this nucleus we pictured just enough electrons to neutralize the positive nuclear charge. On the basis of ionization-energy measurements, we were able to achieve a rough subdivision of the electrons of the atom into groups; but little detailed information on electronic arrangement was available.

Our situation is not too different from that facing the scientific community in 1912. Following Rutherford's proposal of a nuclear model for the atom (review Section 7-1.2), three questions arose:

(1) What is the structure of the nucleus?
(2) What keeps the nucleus from exploding as a result of the mutual repulsion of its components? (It contains densely packed protons, all bearing the same positive charge.)
(3) How are the electrons arranged around the nucleus?

In 1969 the answers to questions (1) and (2) still elude us. But a large amount of information is available on question (3). Since the electronic structure of the atom is of major importance in chemistry, we shall examine the evidence and the new ideas associated with the details of atomic structure.

The evidence for the electronic arrangement of atoms comes from a rather unexpected source—*a study of color*. What can be learned about light and color?

EXERCISE 15-1

Review electron groupings shown in Chapter 8.

(1) Complete the following table.

Noble Gas	Number Electrons in Atom	Electrons Needed to Give Next Noble Gas
—	0	2
Helium	2	8
Neon	10	8
Argon	18	8
Krypton	36	18
Xenon	54	18
Radon	86	32

(2) Assume another noble gas were found. What would be a reasonable estimate of its atomic number? Explain.

(3) Can you predict the next number beyond 32 in the third column of the table? Explain how you did it.

15-1 LIGHT AND COLOR

A copper sulfate solution is deep blue; a cobalt sulfate solution is red. A neon sign *glows* with a bright red light. A mercury vapor lamp *gives off* light with a faint violet shade; a sodium vapor lamp, which is used to light dangerous sections of some highways, *gives off* a distinctive yellow light. A color television tube *glows* red, green, and blue as the television set operates.

If we dip a clean platinum wire in a sodium chloride solution and then place it in the flame of a Bunsen burner, the flame becomes bright yellow. It is just the same shade of yellow as the sodium vapor lamp. If we dip the platinum wire in a strontium chloride solution instead of a sodium chloride solution, we see a red color when the wire is placed in the flame of the Bunsen burner. We shall see a deep red color if lithium chloride is used, a brick red color with calcium chloride, a green color with barium chloride, and a blue-green color with copper chloride. Solutions and flames of several different colors are illustrated in Figure 15-1.

The common element in all these strikingly different observations is *color*. In some cases, the system displays color best when it is placed in white light ($CuSO_4$ solution and $CoSO_4$ solution). In other cases, the substance *gives off* colored light if it is "excited" by a source of energy such as the flame of a Bunsen burner or the electric current operating the neon tube, the mercury vapor lamp, the sodium vapor light, and the television tube.

Many "wondering why" questions flood our minds. Why is a copper sulfate solution blue? Why does a sodium vapor lamp glow with a yellow light while a neon sign is bright red? Why does a color television tube glow red, green, and blue? These questions are not easy to answer, and we shall not be able to give a complete and quantitative answer to any of them; but they and related experiments do lead us to a more detailed model for the atom.

Let us begin our intellectual adventure with a study of light. What is the nature of light? Is it made up of little particles, or is it a continuous wavy stream? Is it a "solid ray," as a solid "beam" of sunlight might be? Be prepared for some exciting new thoughts and ideas as we explore this question. Some of our models based on "super-rubber" balls, eggs, and cars may have to be modified as we talk about very small items such as electrons and photons. We are about to meet the modern **quantum theory.** Its concepts revolutionized much of man's view of physics. The first question of interest is: What is light?

15-1.1 The Nature of Light

Men have wondered about light since the dawn of recorded history. To the ancient Egyptians, the sun was a god, Ra, and he blessed the earth with light. To succeeding generations of philosophers, light was a challenge; they all tried to explain it, but none were successful. Light was first considered in a truly scientific sense by Sir Isaac Newton (1642–1727), who suggested that a light beam is made up of very small particles or bundles of light. He proposed that a glowing or luminous body gives off a stream of these "light particles." A surprisingly large number of the properties of light can be explained by such a particle model.

Yet, despite its successes, the early **particle model** of light was unable to explain a number of everyday observations. For example,

The NATURE of LIGHT

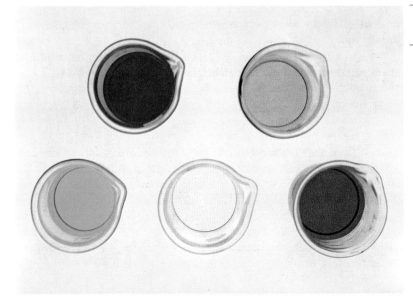

Fig. 15-1.1 **The colored solutions of different compounds in water from left to right: top row [Cu(NH$_3$)$_4$]SO$_4$, [Co(NH$_3$)$_6$]Cl$_3$; bottom row, CuSO$_4$, K[AuCl$_4$], CoSO$_4$.**

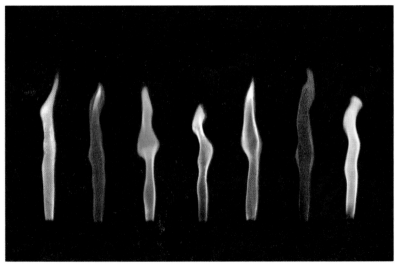

Fig 15-1.2 **Flame tests for different ions; from left to right: Na$^+$ in NaCl, Li$^+$ in LiCl, Cu^{2+} in CuCl$_2$, K$^+$ in KCl, Ca^{2+} in CaCl$_2$, Sr^{2+} in SrCl$_2$, and Ba^{2+} in BaCl$_2$.**

a film of oil on water shows many different colors when light strikes it properly; and a soap bubble shows definite, but changing, colors as one looks at it from different angles. The particle model cannot tell us why. You may have seen another phenomenon that cannot be explained by a particle model. In recent years, so-called diffraction jewelry has appeared on the market. Samples may be available in your classroom. The surface of such diffraction jewelry displays a variety of colors which change as you look at it from different angles. The surface is marked with many fine, closely spaced lines. The color arises from the interaction of white light with these closely spaced lines. (See Figure 15-2.)

The phenomena shown by the oil film, the soap bubble, and the jewelry are examples of what we call **diffraction** or **interference**. There are many other examples, *yet none can be explained satisfactorily by a particle model for light.* On the other hand, all interference and diffraction can be explained in a straightforward fashion

Fig. 15-2 A piece of diffraction jewelry.

if we assume that light is *wave-like*. By making this assumption scientists during the last century developed an impressive wave theory of light. In modified form it is still in use.

If light is to be described in terms of waves, it will be helpful to examine some of the characteristics of waves.

15-1.2 Waves and Their Characteristics

Have you ever watched waves rolling into the shore? If so, you have noticed that some waves are large, others are small. With large waves, the water surface at the top, or crest, is many feet above the water surface at the bottom, or trough. In technical language we say that the large wave has a large **amplitude,** the *vertical distance from equilibrium point to crest;* small waves have a small amplitude. We could also measure the *horizontal distance between crests.* In technical language this is the **wavelength.** (See Figure 15-4.) For ocean waves, meters would be a suitable unit for wavelength.

We are interested in one other property of ocean waves, the **frequency.** It may be determined by counting the *number of crests* that pass a particular point *in a given period of time.* A suitable unit for the frequency of ocean waves would be *wave crests per 15 minutes.* We could describe frequency of other waves such as light in terms of *vibrations per second* or in terms of *cycles per second,* where a cycle means a complete vibration. Note that the unit has time in the denominator in all cases.

The frequency of light waves is given by the Greek letter nu (ν). Sometimes the word *vibration* in the expression for frequency is understood and the frequency, ν, is given in units of $1/\text{sec}$ or sec^{-1}. The wavelength of light is called lambda (λ) and is given in units of length. Centimeters can be used, but shorter units such as the micron (10^{-4} cm), millimicron (10^{-7} cm), or the ångström (10^{-8} cm) are more convenient for the short wavelengths found with light.

What is the relationship between frequency and wavelength? Consider a wave traveling from the sea toward the shore. The number of crests reaching the shore in a given period of time (frequency of wave) is determined by how fast the waves are coming in and by the distance between crests. If the waves are coming in at a fast

Fig. 15-3 **Waves on the ocean.**

rate, the number of crests reaching the shore per unit of time will be large. If, however, each crest is coming in fairly slowly with a long distance between crests, the number of crests reaching the shore in a given period of time will be small. We can express this relationship as

$$\frac{\left\{\begin{array}{c}\text{number of crests}\\ \text{reaching shore}\end{array}\right\}}{\text{unit of time}} = \text{frequency} = \frac{\text{velocity of ocean wave}}{\text{wavelength of ocean wave}} \quad (1)$$

(See Figure 15-4.) For light this relationship is given in a very useful form by the equation

$$\text{frequency of light} = \frac{\text{velocity of light}}{\text{wavelength of light}} \quad (2)$$

or

$$\nu = \frac{c}{\lambda} = \frac{3 \times 10^{10} \text{ cm/sec}}{\lambda} \quad (2a)$$

where c = velocity of light, ν = frequency of light, and λ = wavelength of light.

EXERCISE 15-2

What is the frequency of light with a wavelength of $5,000 \times 10^{-8}$ cm (5,000 Å)? $\frac{3 \times 10^{10} \text{ cm/sec}}{5 \times 10^{-5} \text{ cm}}$ .6 × 10¹⁵ , 6 × 10¹⁴

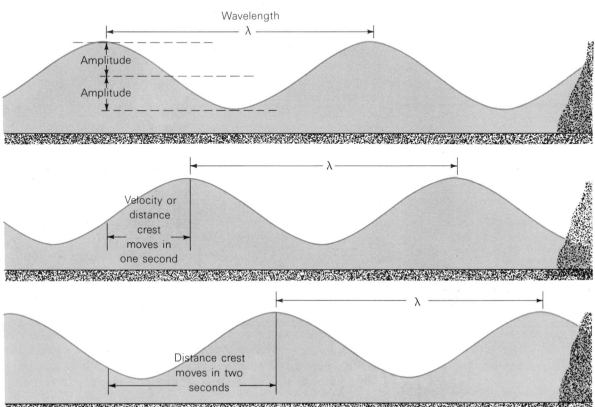

Fig. 15-4 **Amplitude, wavelength, and wave velocity for ocean waves.**

Fig. 15-5 **Deflection of a compass needle by a bar magnet.**

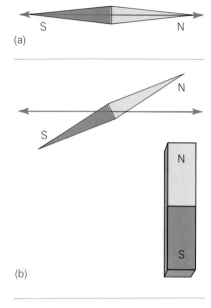
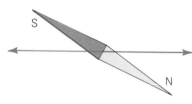
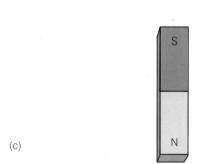

15-1.3 What Waves in Light Waves?

In an ocean, water moves up and down to give *waves;* when a flag *waves* in the wind, the cloth moves back and forth; and so on. With these easily visualized waves, something moves. This seems to be common sense. If, then, light is a wave, what waves? What moves to define the wave of light?

This question plagued early physicists. They hypothesized that all space was filled with a mystical "luminiferous aether" which allowed the transmission of light waves. The "aether" supposedly moved when light waves traveled. Many experiments were done in an attempt to prove the existence of the aether, but all failed. Ultimately, the idea of a mystical aether ran into conflict with Einstein's theory of special relativity. Modern day physicists no longer believe it necessary to postulate the "luminiferous aether." There is no evidence for its existence. We are back now to our original question. If light is a wave, what waves? What moves? The answer to this question is not easily found, but an analogy based on a simple experiment may be helpful. First, we need some new concepts.

15-1.4 Oscillating Magnetic and Electric Fields

Let us take a long compass needle and a very strong bar magnet, as shown in Figure 15-5. If the *north pole* of the *bar magnet* is turned toward the *north end* of the *needle,* the north end of the needle *moves away* from its original rest position [Figure 15–5(b)]. If the magnet is now reversed and the *south pole* of the *magnet* is turned toward the *north end* of the *needle,* the north end of the needle *moves in* toward the magnet [Figure 15-5(c)].

How is the push or pull of the bar magnet carried to the needle? Perhaps air is important. Let us repeat our experiment inside a vacuum chamber from which most of the air has been removed. Our observations are the same inside as they were outside the chamber. Clearly, the push or pull of the magnet is carried just as well through empty space as it is through air. *The attractive or repulsive force of a magnet is carried through empty space.*

Scientists are fond of giving names to phenomena when they cannot construct a simple model to describe what they see. Let us do that here. Although we cannot explain in a simple form *how* the magnet attracts the needle, we can assign a name to describe what we see. We say that there is a **magnetic field** arising and radiating from the bar magnet which pushes or pulls the compass needle. The field is strongest very near the end or pole of the magnet and rapidly grows weaker as we move away from the pole. If the bar magnet in Figure 15-5 is slowly rotated, we shall see that the needle is first pushed out from the bar magnet and then pulled toward it. The magnetic field is changing from a push to a pull and back again.

This is seen if we plot the position of the needle against time as the magnet rotates slowly at constant speed, as in Figure 15-6. The green line showing the position of the tip of the north pole traces out a wave pattern as the needle moves from one side of the rest position to the other. The changing magnetic field, which makes the needle move in a wave-like pattern, might be described as an "oscillating magnetic field" which swings from a push to a pull, or a **magnetic wave.** The magnetic wave is being carried through empty space as an alternating push or pull, or as an **oscillating force field.** We can identify the **source** of the oscillating magnetic field as the rotating magnet. The oscillating needle represents the **receiver** or detector. In many of our studies of light and other forms of radiation, we shall be able to identify a *source* (a light bulb for light, a broadcasting antenna for a radio, or television station) and a *receiver* (the eye for light or your receiving set for radio and television).

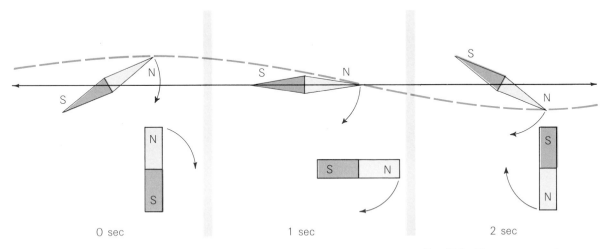

Fig. 15-6 **The orientation of a compass needle near a rotating bar magnet at equal intervals of time.**

A somewhat similar experiment can be carried out using a negatively charged plate as the source instead of the bar magnet and a small, charged, metallized styrofoam ball as the receiver instead of the compass needle. If the negatively charged plate is moved back and forth, the styrofoam ball will also swing back and forth. Again, our observations can be given a name but not a simple explanation. The ball is being pushed back and forth by an oscillating **electric field.** The oscillating nature of the field is generated by the movement of the charged plate. Oscillating electric and magnetic fields have meaning in terms of simple experiments. Such fields can be propagated through empty space.

Finally, what happens when a *charged* body moves? Whenever a charged body moves, it generates a magnetic field in addition to the electric field which surrounds it. Anyone who has watched an electric motor run has seen experimental proof of this statement. Electrons flowing through the wires of the motor (moving charges) generate magnetic fields which push and pull on electromagnets in the armature to make the motor turn. A moving charged particle has both electric and magnetic fields associated with it.

What is the relationship of these experiments to the wave-like nature of light? Again, we shall reason by analogy and consider some closely related experimental results, but the connection is not direct.*

---

* In the experiments just described, field strength varies inversely with the square of the distance between source and receiver. With the electromagnetic wave, field strength varies inversely with the first power of the distance. Furthermore, electromagnetic waves can be directed in space (*e.g.*, a radio beam).

15-1.5 Electromagnetic Radiation

During the period 1864–1873, James Clerk Maxwell,* a gifted Scottish physicist, made a detailed mathematical analysis of electric and magnetic fields. He predicted from his equations that if an electrically charged body is accelerated (speeded up or slowed down), *oscillating electric* and *magnetic fields* will radiate from the charged particle at a velocity of 3×10^{10} cm/sec. This ingenious prediction was verified by Heinrich Hertz (1857–1894) eight years after Maxwell's death.

It is now well known that if a charged particle is suddenly displaced, a "ripple" of electric and magnetic force is generated in much the same manner as a water ripple is generated when a stone is dropped into a pond. However, the **electromagnetic radiation** differs from water waves and similar types of disturbances in several important ways. First, electromagnetic radiation is a wave of electric and magnetic force comparable to those described in Section 15-1.4. It is not a material or tangible wave. The push and pull of the magnetic field is at right angles to the push and pull of the electric field. (See Figure 15-7.) Because it is made up of traveling and oscillating force fields, electromagnetic radiation can travel through a vacuum at the predicted velocity of 3×10^{10} cm/sec. *No material body is needed for the transmission of electromagnetic radiation.* The waves are transmitted through empty space. In fact, we now say that one of the properties of space is its ability to carry energy in electromagnetic form and to support the waves of electric and magnetic fields making up visible light, radio signals, and X rays.

*James Clerk Maxwell (1831–1879) was a Scotsman who served on the faculties of both Scottish and British universities. At the time of his death, he was Professor of Experimental Physics at Cambridge, England. His work on electromagnetic theory must surely stand as one of the most brilliant intellectual achievements in the history of man.

Fig. 15-7 **The electric and magnetic force fields of electromagnetic radiation such as sunlight are at right angles to each other.**

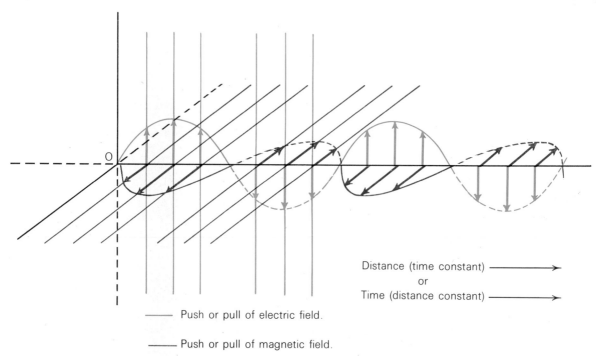

The simple facts we have just discussed stand at the core of our radio and television industries. A radio or television transmitting station pumps charges (electrons) along an antenna. (The charges speed up, slow down, stop, then speed up in the reverse direction.) Power from the station is used to pump the charges back and forth. As these change speed, electromagnetic radio waves break away from the antenna and travel through space in all directions at 3×10^{10} cm/sec. The waves have the *frequency* of the oscillating charges in the antenna. Your radio or television set, when properly tuned, serves as the *detector*. It picks up the signal and converts the weak electromagnetic waves into sound or pictures.

The color of compounds has its origin in a surprisingly similar set of events. The movement of electrons between levels of differing energies generates electromagnetic waves. Electromagnetic radiation of short wavelength results. Light is absorbed or emitted.

We can now summarize the answer to our earlier question, What waves in electromagnetic radiation? There is no reason to say that any substance is waving. Electromagnetic waves are like water waves, but they are different too. We have been able to show that the fundamental properties of the world can frequently be understood in terms of simple laboratory models. On the other hand, as we saw in Chapter 1, there is usually a real difference between model and system. As you recall, molecules have perfectly elastic collisions while "super-rubber" balls do not.* This gives rise to differences in properties between molecular systems and systems of "super-rubber" balls. Similarly, water waves and electromagnetic waves are not identical; there are differences. What is remarkable and wonderful is that they are enough alike that we can learn much about electromagnetic waves from observing water waves!

15-1.6 The Electromagnetic Spectrum

Our description of electromagnetic radiation indicates that radio waves and visible light are different examples of the same fundamental phenomenon. How do these waves differ? Experiments on ordinary sunlight help provide the answer. If a very narrow "beam" of sunlight passes through a glass prism, it spreads out into a bright **spectrum** of colors—violet, indigo, blue, green, yellow, orange, red. The colors are those of the rainbow; they are best seen when displayed on a white screen (see Figure 15-8, following page). (The spectrum can be recorded on a photographic plate if study at a later date is desired.) Detailed studies of prism action indicate that the prism separates light of different wavelengths into different bands. Red light has the longest wavelength (λ) and the lowest frequency (ν) (remember: $\nu = c/\lambda$); a beam of red light is bent the least in going through the prism. Violet light has the shortest wavelength (λ) and the highest frequency (ν); a beam of violet light is bent the most in going through the prism. The wavelengths and frequencies of the visible spectral colors are shown in Table 15-1.

---

*For simple monatomic gases all collisions are perfectly elastic; for complex molecules collisions are perfectly elastic *on the average*. If this were not so, the pressure in a tank of gas would gradually fall to zero.

Fig. 15-8.1 (Top) The spectrum of visible light.

Fig. 15-8.2 (Bottom) The action of a prism on white light.

TABLE 15-1 SPECTRAL COLORS AND THE WAVELENGTHS OF VISIBLE LIGHT

Color	Wavelength (λ) (cm)	Frequency (ν) (vibrations/sec)
Violet	shorter than 4.5×10^{-5}*	more than 6.7×10^{14}
Blue	4.5 to 5.0×10^{-5}	6.7 to 6.0×10^{14}
Green	5.0 to 5.7×10^{-5}	6.0 to 5.2×10^{14}
Yellow	5.7 to 5.9×10^{-5}	5.2 to 5.1×10^{14}
Orange	5.9 to 6.1×10^{-5}	5.1 to 4.9×10^{14}
Red	longer than 6.1×10^{-5}	less than 4.9×10^{14}

* If wavelengths were given in ångström units (1 Å = 10^{-8} cm), that of violet light would be shorter than 4,500 Å.

In order to resolve a beam of light into its spectral colors in the laboratory, a somewhat more elaborate prism arrangement is usually used. Such an instrument is known as a **spectrometer** or **spectrograph** (an instrument which measures spectra). It is shown in Figure 15-9.

Even a casual inspection of Table 15-1 shows that the visible electromagnetic spectrum covers only a very narrow range of frequencies. The numbers run from about 4.5×10^{14} vibrations/sec to about 7.0×10^{14} vibrations/sec. We immediately ask: Are there frequencies higher than 7.0×10^{14} vibrations/sec, the frequency characteristic of violet light? Are there frequencies lower than the 4.5×10^{14}, the value typical of red light? The answer is found by examining a photographic plate upon which the spectrum of sunlight has been registered. The plate is darkened at positions corresponding to frequencies higher than that of violet (the **ultraviolet**

region) and at positions corresponding to frequencies lower than that of red (the **infrared region**). The visible spectrum is only a part of the total electromagnetic spectrum, shown schematically in Figure 15-10 and summarized in Table 15-2 (see next page). X rays have a very short wavelength while heat (infrared) is long-wavelength radiation.

The ELECTROMAGNETIC SPECTRUM

EXERCISE 15-3

The light produced by a sodium vapor lamp shows two lines with wavelengths of 5.89×10^{-5} cm and 5.90×10^{-5} cm. Identify the color of the sodium vapor lamp using the data assembled in Table 15-1. What color is the light from a potassium vapor lamp if the spectral lines for potassium have wavelengths of 4.044×10^{-5} and 4.047×10^{-5} cm? What color is a mercury vapor lamp if the frequencies of the lines are 6.6×10^{14} vibrations/sec and 5.5×10^{14} vibrations/sec? Estimate the wavelength of one of the most intense lines in the *neon* spectrum. Remember the color of neon light.

yellow

violet

blue & green

Fig. 15-9 **A spectrograph operating with the light produced by a hot tungsten ribbon.**

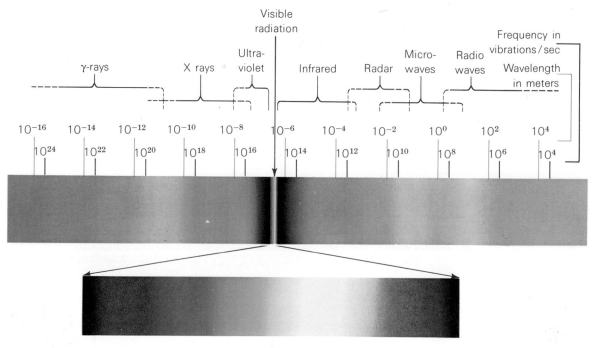

Fig. 15-10 **The whole electromagnetic spectrum.**

TABLE 15-2 SPECTRUM OF ELECTROMAGNETIC RADIATION

Name of Radiation	Approximate Range of Wavelength
Radio waves	a few meters and up
Microwaves	a few millimeters to a few meters
Infrared waves	$7{,}500 \times 10^{-8}$ cm to 0.01 cm
Visible light	4,000 to 7,500 Å*
Ultraviolet light	100 to 4,000 Å
X rays	0.1 to 500 Å
Gamma rays	less than 0.5 Å

*$1 \text{ Å} = 10^{-8}$ cm.

15-1.7 A Summary of the Nature of Light

Simple experiments show that the push or pull of a magnet can pass through empty space and act on another magnet. This push or pull, known as the **magnetic field,** can increase or decrease systematically as the magnet moves. If we plot against time the push or pull exerted by a rotating bar magnet on another magnet, we see a wave-like pattern for this push or pull (the magnetic field is oscillating from push to pull). Similar experiments define an **oscillating electric field** and show that a wave-like pattern exists when an oscillating electric field is generated by a charged rod moving back and forth. A moving electric charge generates a magnetic field, as well. Light is related in a general way to these oscillating fields.

If an electric charge (electron or proton) has its velocity changed suddenly (is accelerated), **electromagnetic waves** radiate from the charge. These waves are a combination of electric and magnetic fields vibrating at right angles to each other. The waves move in a vacuum at the speed of light or 3×10^{10} cm/sec (symbolized by c). Such electromagnetic radiation can be characterized by the radiation **frequency** (ν), which is the number of field oscillations or vibrations per second, and by the **wavelength** (λ) of the radiation, which is the distance between wave crests. These numbers are related by a common-sense formula which can be verified on any good beach:

$$\text{frequency of light} = \nu = \frac{c}{\lambda} = \frac{\text{velocity of light}}{\text{wavelength}} \qquad (2b)$$

Red light has relatively low frequency and long wavelength. Violet light has high frequency and short wavelength. **Ultraviolet** radiation has higher frequency than violet; **infrared** radiation has lower frequency than red.

15-2 ENERGY AND AN INTERPRETATION OF THE HYDROGEN SPECTRUM

15-2.1 Electromagnetic Radiation and Energy

Sunlight is a form of energy. Remember how warm it gets in the sunshine, even on a cool day? The sun feels even warmer if you have on dark-colored clothes, which absorb the sunlight. It is not really hard to believe that sunlight is a form of energy when you feel the heat absorbed by dark clothing. Further, if a large amount of sunlight is focused by means of a hand lens onto a smaller area, very high temperatures result. Many of you have used a hand lens to focus light rays on paper until the paper catches on fire. With larger lenses we can build huge "solar furnaces" capable of melting substances that have very high melting points. Sunlight is indeed a form of energy.

The connection between light and kinetic energy is easy to visualize if we remember the simple experiment in which a fluctuating magnetic field made the compass needle move. The compass needle picked up kinetic energy (energy of motion) from the fluctuating magnetic field. In a similar manner, motions of electrons and of atoms in molecules may be induced by the fluctuating electromagnetic fields present in radiation. Under appropriate conditions radiation can increase the molecular energy very substantially.

Recognizing that light is a form of energy, we might now ask: How much energy is carried by light? Can we obtain a workable *quantitative* relationship between some obvious property of light and the amount of energy it carries? This problem troubled many scientists. The first successful answer was unexpectedly simple in form but revolutionary in concept. It provided the basis for our modern quantum theory of atomic structure.

Let us return to the state of physics near the turn of the century, when Max Planck (1858–1947) was studying the light and heat (electromagnetic radiation) given off by a hot body such as a piece of hot iron. He found that he was unable to obtain a quantitative explanation for what he saw *unless* he made an unusual and unpopular assumption, an assumption that recalled Newton's particle model for light. Planck proposed that light is made up of *bundles of energy* which he called **quanta.** *The energy in any one bundle is dependent upon the color of the light.* Violet light has more energy per bundle than does red light, red light has more energy than does infrared, and infrared has more energy than does a radio wave. His assumption can be given in quantitative terms by a simple relationship between energy and the frequency of the radiation:

$$\left\{\begin{matrix}\text{energy per quantum}\\ \text{of light (or radiation)}\end{matrix}\right\} = \left\{\begin{matrix}\text{proportionality}\\ \text{constant}\end{matrix}\right\} \times \left\{\begin{matrix}\text{frequency}\\ \text{of light}\end{matrix}\right\} \quad (3)$$

Using symbols, we write

$$E = h\nu \quad (3a)$$

The letter h stands for a simple proportionality constant and is known as **Planck's constant of action.** This simple equation, $E = h\nu$, is one of the most important and powerful statements of modern science. It helped to solve Planck's quantitative problem and became the key to atomic structure. You will meet it many times if you continue your study of physical science.

15-2.2 Energy Changes and the Hydrogen Spectrum

You may be wondering what light has to do with atomic structure. To answer such a question let us look at the light given off by a "hydrogen light." A hydrogen light is related to a neon light. You will recall that a tube containing low-pressure neon gas glows with a beautiful red color when electricity is passed through it. If the light from the neon tube is passed through the spectrometer, bright red lines appear. (A color photo of the spectrum of the neon light is shown in Figure 15-11.) If a tube containing low-pressure hydrogen is used instead of low-pressure neon, a light with a purplish color is emitted. When this light is passed through the spectrometer, the series of lines shown in Figure 15-12 appears. Each line in the spectrum corresponds to a given pure color or frequency given off by hydrogen atoms. Each space between two lines on the film corresponds to frequencies or colors *not* emitted by the hydrogen atom.

Let us examine the spectrum of hydrogen more carefully. We find that *every* hydrogen discharge lamp emits the *same small group of frequencies*—each hydrogen spectrum has the same set of lines. Further, we note that the frequencies corresponding to the lines on the film are spaced systematically. There are two groups of lines in the part of the spectrum shown in Figure 15-13—one group is in the visible part and one is in the ultraviolet region. Within each group there is a regular decrease in the spacing between successive lines as the frequency increases.

Perhaps regularities in the data exist. How can we best express these regularities? This question was answered by a physicist named J. R. Rydberg (1854–1919) after an extensive trial-and-error fitting of spectral data. He found that the frequency of the lines in any small group could be given by a simple relationship:

$$\nu = R\left(\frac{1}{n_1^2} - \frac{1}{n_2^2}\right) \quad (4)$$

Fig. 15-11 **The spectrum of neon.**

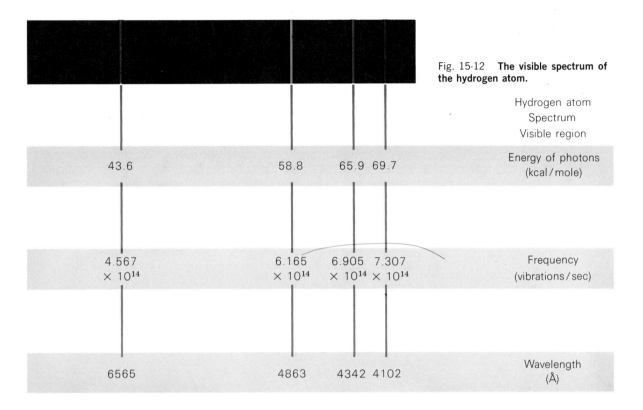

Fig. 15-12 **The visible spectrum of the hydrogen atom.**

Hydrogen atom Spectrum Visible region				
43.6		58.8	65.9 69.7	Energy of photons (kcal/mole)
4.567×10^{14}		6.165×10^{14}	6.905×10^{14} 7.307×10^{14}	Frequency (vibrations/sec)
6565		4863	4342 4102	Wavelength (Å)

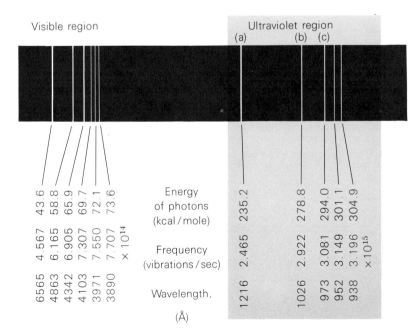

Fig. 15-13 **The visible and ultraviolet spectrum of the hydrogen atom.**

Visible region / Ultraviolet region (a) (b) (c)

Energy of photons (kcal/mole): 43.6 58.8 65.9 69.7 72.1 73.6 235.2 278.8 294.0 301.1 304.9

Frequency (vibrations/sec): 4.567 6.165 6.905 7.307 7.550 7.707×10^{14} 2.465 2.922 3.081 3.149 3.196×10^{15}

Wavelength (Å): 6565 4863 4342 4103 3971 3890 1216 1026 973 952 938

In this formula, v is the frequency of the line, R is a constant known as the **Rydberg constant**,* and n_1 and n_2 are whole numbers.

Let us see how this formula works by considering the lines in the ultraviolet region of the hydrogen spectrum shown in Figure 15-13. As in the visible spectrum, the lines in the ultraviolet spectrum are spaced far apart at first; but as we proceed from left to right, the lines in a given group become closer. Finally, the lines run together and the ultraviolet spectrum terminates in an identifiable limit. The frequency of the line at the limit is 3.287×10^{15} vibrations/sec. If we use this value of 3.287×10^{15} vibrations/sec as R in the Rydberg equation, then

$$v = R\left(\frac{1}{n_1^2} - \frac{1}{n_2^2}\right) = (3.287 \times 10^{15})\left(\frac{1}{n_1^2} - \frac{1}{n_2^2}\right) \quad (4a)$$

and the frequency of every other line in the ultraviolet spectrum can be calculated. To make these calculations, we *assign* the value of 1 to n_1; then we assign successive whole number values to n_2. For example, the frequency of line (a) in Figure 15-13 is computed using the values $n_1 = 1$ and $n_2 = 2$; for line (b), $n_1 = 1$ and $n_2 = 3$; and so on.

In Table 15-3, experimental values for three lines in the ultraviolet spectrum are compared with values calculated from the Rydberg equation. Note that the value n_1 for the ultraviolet region is held constant at 1. We can easily show that the frequency of every line in the ultraviolet spectrum is obtained by assigning to n_2 the successive values 2, 3, 4, 5, 6, ..., to infinity (∞). When n_2 equals infinity, the calculated frequency corresponds to the observed frequency of the series limit. This fact is seen by making the following substitutions:

$$v = R\left(\frac{1}{1^2} - \frac{1}{\infty^2}\right) = R(1 - 0) = R \quad (4b)$$

Remember that $1/\infty^2 = 0$.

TABLE 15-3 FREQUENCIES OF LINES IN THE ULTRAVIOLET REGION OF THE HYDROGEN SPECTRUM —A CHECK ON THE VALIDITY OF THE RYDBERG EQUATION

Value of n_2	Equation	Frequency Calculated (vibrations/sec)	Frequency Observed (vibrations/sec)
2	$v = (3.287 \times 10^{15})\left(\frac{1}{1^2} - \frac{1}{2^2}\right)$	2.465×10^{15}	2.465×10^{15}
3	$v = (3.287 \times 10^{15})\left(\frac{1}{1^2} - \frac{1}{3^2}\right)$	2.922×10^{15}	2.922×10^{15}
4	$v = (3.287 \times 10^{15})\left(\frac{1}{1^2} - \frac{1}{4^2}\right)$	3.081×10^{15}	3.081×10^{15}

The Rydberg equation gives an amazingly accurate representation of much experimental data. Furthermore, it suggests new ex-

---

* R can be given in different units if $1/\lambda$ is desired instead of v. This value, called R' here would be equal to R/c.

periments and new correlations. For example, if n_1 in the equation were assigned the value of 2 rather than 1, a new series of lines would be expected. By allowing n_2 to take the values 3, 4, 5, 6, ..., ∞, the frequencies of lines in the new series can be calculated. These are actually the lines of the visible spectrum. Experiments show that every line appears just as predicted.

EXERCISE 15-4

Calculate the frequencies of two lines in the *visible* spectrum. Check your results against the values in Figure 15-13. (Remember that for all lines in the visible spectrum, $n_1 = 2$.)

The Rydberg equation makes additional predictions. It suggests that another series of lines should be found in which n_1 has the value 3. The frequencies of successive lines in this series could be calculated by letting n_2 run from 4 to infinity. This prediction has been verified. Experiment shows that every line appears as predicted.

The existence of many additional series can be forecast; furthermore, the location of each line in each series can be predicted. It is interesting to note that several series of lines in the hydrogen spectrum were discovered *after* the Rydberg equation had been used to predict their existence.

EXERCISE 15-5

(1) Calculate the frequencies of two more lines in the visible spectrum of hydrogen and compare your calculated values with the observed values.
(2) Suppose we let $n_1 = 3$. Calculate the frequency of the lowest frequency line in this group. Where would this line appear in the electromagnetic spectrum (i.e., ultraviolet, visible, infrared, and so on)? See Table 15-2.

[handwritten annotations:] $\nu_5 = 3.287 \times 10^{15} \left(\frac{1}{4} - \frac{1}{49}\right) = 7.546 \times 10^{14}$ $\nu_6 = 3.287 \times 10^{15} \left(\frac{1}{4} - \frac{1}{64}\right) = 7.668 \times 10^{14}$

infrared

The Rydberg equation passes the test for an excellent regularity. It is the simplest method of recording the observations made on the hydrogen spectrum. It is difficult to see how the equation was derived from the raw numbers. Recall, however, that in Chapter 2 a plot of pressure versus 1/volume was useful in obtaining the equation "$PV = $ a constant." In Rydberg's study, plots of different types were also useful in suggesting the form of the final equation. Although deriving the first equation was difficult, the equation itself was very easy to use once it had been presented. Complicated matters frequently appear simple after someone points out an easy method of approach. (Remember Mr. Wilson in the Sherlock Holmes tale in Chapter 1.)

Before continuing, let us go over the ground we have covered in studying the light given off when electricity is passed through a tube containing hydrogen gas at low pressure. First, the light was passed through a spectrograph, which broke it down into lines having distinct and reproducible frequencies. Then, the frequencies of many lines were recorded. Using a regularity developed by J. R. Rydberg, we correlated data from existing experiments and predicted the results of new experiments.

The next step in a scientific study is "wondering why." Can we build a model of the hydrogen atom that will account for the Rydberg equation? It is worthwhile to try.

Let us start by changing the frequency, ν, of spectral lines into energy terms. Remember: $E = h\nu$. The original hydrogen atom has some definite amount of energy; let us call it $E_{initial}$. When light of frequency ν is emitted, energy, $h\nu$, is carried away. The hydrogen atom now has less energy than it did before; let us call this smaller amount of energy E_{final}. Since energy must be conserved in atomic physics, we say that the energy *given off* in the radiation must be exactly equal to the difference between initial and final energy states of the atom.

$$\begin{Bmatrix} \text{energy } \textit{given off} \\ \text{in radiation} \end{Bmatrix} = \begin{Bmatrix} \text{energy of atom } \textit{before} \\ \text{radiation emitted} \end{Bmatrix} - \begin{Bmatrix} \text{energy of atom } \textit{after} \\ \text{radiation emitted} \end{Bmatrix} \quad (5)$$

or, symbolically:

$$h\nu = E_{initial} - E_{final} \quad (5a)$$

Perhaps the spectral lines for hydrogen can be interpreted if we assume that hydrogen atoms exist in distinct and reproducible energy states. The form of the equation for hydrogen frequencies is similar in many ways to the form of the Rydberg equation. The idea deserves further analysis.

15-2.3 The Quantum Model for the Hydrogen Atom

The light given off by an excited hydrogen atom corresponds to a change in the energy of the hydrogen atom from a condition of higher to a condition of lower energy. Since only certain lines appear in the spectrum, the hydrogen atom seems to be using only *a limited number of energy levels*. A model will help to make this point clearer. Let us assume that the modernistic bookcase shown in Figure 15-14 has been designed by an avante-garde artist. Let us now put a large paperweight on the bottom shelf of this bookcase. The paperweight cannot fall anywhere; the system is in its lowest energy state, or in its "ground state." If, however, we pick up the paperweight and place it on shelf 2 of the bookcase, work must be done to lift the weight from the bottom shelf, 1, to the next shelf, 2. If the paperweight falls from shelf 2 to shelf 1, energy will be released. The energy released will be exactly equal to the amount of energy used to raise the paperweight from shelf 1 to shelf 2. The kinetic energy of the paperweight just before it hits the bottom shelf will be equal to the difference in potential energy of the paperweight on shelves 2 and 1. There are no intermediate levels or shelves between shelves 2 and 1, so the paperweight can have only potential energy 2 or 1, so far. The weight will always gain the *same amount* of kinetic energy as it falls from level 2 to level 1. When the weight hits the floor and stops, its kinetic energy will be converted to heat. The weight and the floor will be a little warmer after collision than they were before.

The QUANTUM MODEL for the HYDROGEN ATOM

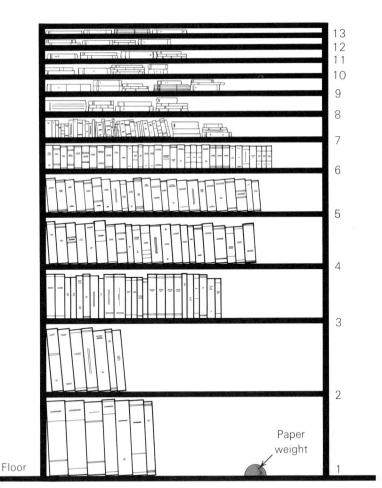

Fig. 15-14 **A paperweight in a bookcase can rest only on the shelves and at no place in between.**

Suppose we now raise the weight to level 3 and let it fall to level 1. The kinetic energy of the weight just before it hits the bottom shelf will be greater than it was when the weight only fell from level 2. The potential energy of the weight is greater in position 3 than in position 2, so more potential energy will be converted into kinetic energy as the weight falls from 3 to 1.

We can now write

$$\begin{Bmatrix} \text{thermal energy} \\ \text{released on} \\ \text{impact} \end{Bmatrix} = \begin{Bmatrix} \text{kinetic energy} \\ \text{of weight} \\ \text{before impact} \end{Bmatrix} = \begin{Bmatrix} \text{potential energy} \\ \text{of weight on} \\ \text{top shelf (2 or 3)} \end{Bmatrix} - \begin{Bmatrix} \text{potential energy} \\ \text{of weight on} \\ \text{bottom shelf (1)} \end{Bmatrix} \quad (6)$$

or

$$\text{thermal energy released on impact} = E_2 - E_1 \quad (6a)$$
$$\text{thermal energy released on impact} = E_3 - E_1 \quad (6b)$$

and so on. If the temperature increase of the weight is proportional to the heat released on impact, we can write

$$\begin{Bmatrix} \text{thermal energy} \\ \text{released on} \\ \text{impact} \end{Bmatrix} = \begin{Bmatrix} \text{proportionality} \\ \text{constant} \end{Bmatrix} \times \begin{Bmatrix} \text{temperature} \\ \text{change} \end{Bmatrix} \quad (6c)$$

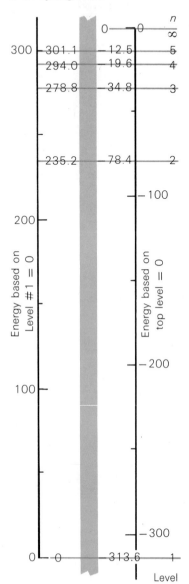

Fig. 15-15 **The energy-level scheme of the hydrogen atom.**

We can see that *the same increase in temperature will always be observed as the weight falls from level 2 to level 1*. A different and higher temperature increase will be observed if the weight falls from level 3 to level 1.

Our picture of the hydrogen atom is much like our picture of the bookcase. The electron of the hydrogen atom can exist at different **energy levels** (see Figure 15-15). The electron can be placed in levels 1, 2, 3, 4, and so on, but *not* at intermediate points where no levels exist (just as with the bookcase shelves). When the electron falls from level 2 to level 1, it releases energy as electromagnetic radiation. The energy released is just equal to the difference in energy between level 2 and level 1:*

energy of electromagnetic radiation = energy of level 2 − energy of level 1

Symbolically, we write

$$E_{released} = E_2 - E_1 = E_{initial} - E_{final} \qquad (6d)$$

Since the energy released is proportional to the frequency of the radiation, ($E_{released} = h\nu$), we can write

$$h\nu = E_2 - E_1 \qquad (6e)$$

Light of the same frequency (or color) will be given off every time an electron falls from level 2 to level 1. Light of a different frequency (or color) will be given off every time an electron falls from level 3 to level 1. Another spectral line will be seen.

The general form of the expression obtained from the "bookcase model" of the hydrogen atom,

$$h\nu = E_2 - E_1 \qquad (6e)$$

is identical to the expression obtained in the Rydberg representation of the data. So far, the bookcase model passes the first test of a good theory: it describes the experimental regularities quantitatively and in detail. If we assume that the energy of the hydrogen atom is quantized—that is, it exists in separate and distinct levels—we can "explain" the hydrogen spectrum.

15-3 MECHANICAL ATOMIC MODELS

So far all we have done is state that the electron in a hydrogen atom can exist in *any one* of many different and distinct *energy levels*. This same conclusion was implied by regularities in ionization energies discussed in Chapter 8. So far the model has not been very specific about how the atom looks and how one energy level differs from

---

* There are some differences between our bookcase model and the hydrogen atom. The energy of the weight on different shelves of the bookcase is all potential energy. In contrast, the energy of the electron in level 1 or 2 of hydrogen is made up of both kinetic and potential energy.

another. Man's attempt to understand the atom—first in mechanical, then in mathematical terms—is a fascinating story of a frequently frustrating process. Let us retrace the last 50 or 60 years of this problem. We have the advantage of hindsight in reviewing the history. Physicists in 1912 had no such advantage.

15-3.1 Mechanical Models Based on Classical Laws

The Rutherford model of the atom which resulted from experiments on alpha-particle scattering (see Chapter 7) pictured the atom as a tiny but very dense, positively charged nucleus surrounded by enough electrons to neutralize the positive nuclear charge. If the nuclear charge was known, the number of electrons required to complete the atom was also known. In 1912 Newton's laws of mechanical motion were universally accepted; they were adequate to explain with wonderful accuracy all observations on the movements of large particles such as golf balls or planets in the solar system. This information was the backbone of classical physics. It seemed logical to believe that such laws should describe the behavior of particles moving in the atom. Furthermore, thanks to Maxwell and his contemporaries, much was known about electricity and electrically charged bodies. We have already noted that if a moving electron is accelerated, electromagnetic waves radiate out from the system. You will remember that radio waves are a result of pumping electric charges back and forth along a broadcasting antenna. <u>When the linear velocity of a charged body is changed, electromagnetic waves are absorbed or released</u>.

With all this information, it would seem that the physicists of 1912 were in a position to describe the movement of electrons and nuclei in great detail. The time was ripe to build a simple model of an atom and to use the model to explain atomic properties. Rutherford had started the model-building process as a result of his alpha-particle scattering experiments. He had information on the size, charge, and mass of the nucleus, but no information on the distribution or position of electrons. Lacking distinct experimental evidence, Rutherford endorsed an earlier, speculative suggestion made by the Japanese physicist Nagaoka in 1904. Nagaoka had envisioned an atom in which a central nucleus was surrounded by electric charges revolving in rings similar to the rings of Saturn. In Rutherford's model electrons moved around a very small nucleus like planets around the sun. The attraction of the proton for the electron was just counteracted by the tendency of the electron to continue moving in a straight line. This is the same tendency a car has to continue moving in a straight line when it travels in a curve. According to the classical picture, it was the revolution of the electron about the nucleus which prevented it from being pulled into the nucleus by the positive charges of the nuclear protons.

Up to this point the successes of the mechanical model of the atom were most encouraging (see Figure 15-16). Tragedy struck, however, when the model ran into conflict with additional, known information. An electron going in a circle has its linear direction of motion changed constantly. If such a directional change did not occur, the electron would go off in a straight line, just as a revolving weight on the end of a string does when the string breaks (see Figure 15-17). *This circular path can be followed only if the electron is accelerated or has its direction of motion changed constantly toward the nucleus.* According to laws of classical electricity, an electron or other charged body which is accelerated or has its linear velocity changed will give off or absorb electromagnetic radiation. According to Rutherford's model, the electron should *give off* electromagnetic radiation. Such radiation would drain off the electron's energy so the electron would gradually spiral into the nucleus (see Figure 15-18). By making the electron move around the nucleus in circles, then, Rutherford had not really prevented it from falling into the nucleus! His model made it fall in a little slower, but it must still fall in if classical laws are to be obeyed. The Rutherford atom can last only about 2×10^{-11} sec. Such atoms are far from eternal! Furthermore, this picture of the hydrogen atom tells us that the atom should emit light of *all* frequencies, not the few selected frequencies we have seen in the hydrogen spectrum. Something was wrong; the Rutherford atom could not be correct. A completely new start was needed.

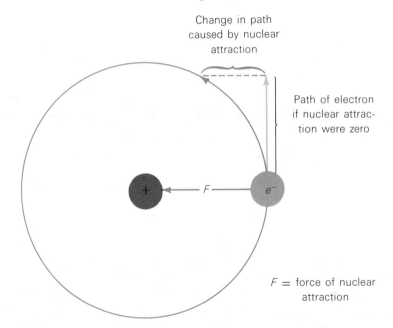

Fig. 15-16 **Diagram of the Rutherford atom.**

Fig. 15-17 **When the string breaks the rock continues moving but in a straight line.**

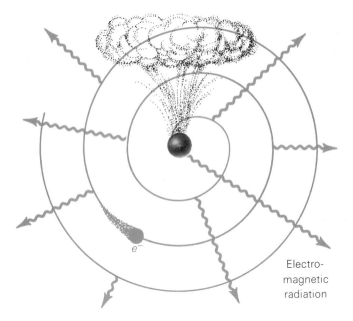

Fig. 15-18 **Tragedy strikes the Rutherford atom.**

15-3.2 Quantized Mechanical Models—The Bohr Atom

At this point a Danish physicist, Niels Bohr (1885–1962), made one of the most daring and significant suggestions of modern science. He questioned the validity of applying to electrons and similar small bodies the electrical and mechanical laws which described so accurately the behavior of baseballs, planets, and radio antennas. With remarkable foresight, *he suggested that a new set of laws might be needed to describe the particles inside an atom.* One of his most original and controversial suggestions recognized the fact that electrons in atoms exist in discrete energy levels, or "stationary states" as he called them. He established this fact from the hydrogen spectrum. Our earlier treatment followed Bohr's presentation. He simply rejected the mechanical model that permitted the electron to have *any* and *all* energy values. He stated quite flatly that a model in which the electron spirals into the nucleus cannot be correct because atoms do exist. Further, those classical laws which predict atomic collapse must be rejected because of the simplest observation—matter exists! Bohr then proceeded to apply Planck's new ideas on the quantization of radiant energy to the quantization of energy in the hydrogen atom. He developed a mechanical model for the hydrogen atom using some classical physical laws while rejecting others. He visualized the electrons as revolving around the nucleus in any one of a group of "allowed" orbits. Each orbit represented a given energy level, and energy was released when the electron jumped from an outer to an inner orbit. Bohr hypothesized, for example, that the lines in the *ultraviolet* region arose when the electron jumped from level 2 to level 1, from level 3 to level 1, from level 4 to level 1, and so on. The lines in the *visible* spectrum arose when an electron jumped from level 3 to level 2, from level 4 to level 2, from level 5 to level 2, and so on. (See Figure 15-19.) Bohr's success was truly

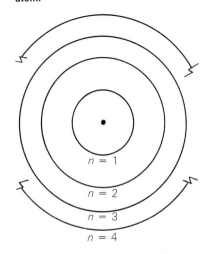

Fig. 15-19 **The orbits of the Bohr atom.**

amazing. He calculated the value of the Rydberg constant from first principles with great accuracy. This interpretation of the hydrogen spectrum seemed to have the problem of electron arrangement completely solved.

Bohr's work was fine as long as he restricted it to hydrogen *atoms*. When he turned his attention to hydrogen *molecules* or to *complicated atoms*, his theory did not apply. Today we cannot accept Bohr's model of the atom because it tries to tell us more about electrons in an atom than nature permits us to discover. We say that the **Bohr model** conflicts with the "uncertainty principle" and does not *properly* recognize the true nature of the electron. This will be discussed further in later sections.

Because the Bohr model conflicts with the uncertainty principle, Bohr's mechanical picture of the hydrogen atom is no longer accepted. We do, however, accept and use Bohr's scheme for quantization of energy in the atom. We recognize that the lowest energy state of the atom is the most stable state. The electron in the ground level characterizes a hydrogen atom in its most stable state. Atoms exist in this state until they are *excited* to higher energy levels by absorption of light or by violent collision with other atoms or electrons. In the excitation process electrons jump to higher energy levels. After a brief time in the higher level the electron falls back to the ground state and electromagnetic radiation (or light) is given off. The energy change as the excited atom falls back to its ground state must be equal to the energy of the light emitted, or, as our earlier equation indicated,

$$E_2 - E_1 = h\nu \qquad (6e)$$

Unfortunately, after all this discussion, we do not have a pat mechanical model for the hydrogen atom, or for *any* atom. *We can only describe the atom in terms of energy levels.* As Bohr so clearly pointed out, our classical laws have failed to explain several phenomena. Models based on bouncing "super-rubber" balls and "orbiting electrons" do not correctly formulate a picture of the inside of an atom.

Bohr's rejection of many laws of classical physics was unheard of. His ideas were so revolutionary that they would not have been accepted except for the fact that he was able to calculate accurately the energy levels of the hydrogen atom. *Within ten years his methods of calculation were completely replaced by better methods. However, his postulate that energy is quantized in the hydrogen atom is recognized today as one of the cornerstones of modern physics.* Bohr was truly a scientific giant.

15-4 WAVE MODELS FOR THE HYDROGEN ATOM

15-4.1 De Broglie's Matter-Waves

The next development in man's attempt to understand atomic structure originated in a suggestion made in the 1920's by the French physicist Louis de Broglie (1892–   ). Planck had adopted the notion that light is not only wave-like but particle-like in character

—that is, it is made up of little bundles of energy. After pondering Planck's suggestion that light has particle-like character, de Broglie asked a significant question: If waves have some particle character, do particles have some *wave* character? Do electrons, protons, baseballs, and freight trains have a wave nature as well as a particle nature? De Broglie concluded that particles do indeed have wavelike character and he derived the equation

$$\begin{Bmatrix} \text{wavelength of} \\ \text{the matter-wave} \\ \text{associated with} \\ \text{a particle} \end{Bmatrix} = \frac{\text{Planck's constant}}{\text{mass of particle} \times \text{velocity of particle}} \quad (7)$$

or

$$\lambda = \frac{h}{mv} \quad * \quad (7a)$$

This relationship was soon verified experimentally for electrons by Davisson and Germer, who showed in 1927 that electrons will give diffraction patterns like light. Perhaps the best description of the atom can be given in terms of waves rather than particles? Indeed, if we use the de Broglie relationship to calculate the wavelength of an electron in an atom, we find that the electron's wavelength is comparable to the size of the atom. *Wave properties are important for electrons in atoms!* To the atom the electron appears as a wave. On the other hand, if we use the de Broglie relationship to calculate the theoretical wavelength of a baseball, we find that its wavelength is very short because the baseball is heavy and m is therefore large. The wavelength of a baseball cannot be measured by any diffraction experiment. *The baseball appears only as a particle, not as a wave;* we cannot do *any* experiments which permit us to measure or see the wave-like character of a baseball. A baseball is a particle, *not* a wave!

Here then in 1927 was a justification for Bohr's bold prediction of 1912. He said that the classical laws governing the motion and properties of massive bodies such as apples, oranges, and baseballs

---

* The mathematical formalism leading to de Broglie's relationship is not difficult. You will recall that Planck had written $E = h\nu$, where E = energy, h = Planck's constant, and ν = frequency radiation. Einstein had offered $E = mc^2$, where m = mass converted to energy and c = the velocity of light. We saw earlier that

$$\nu = \frac{c}{\lambda}$$

We can then write

$$\frac{hc}{\lambda} = mc^2$$

or

$$\lambda = \frac{h}{mc}$$

The quantity mc is the momentum of a **photon**. If we place the momentum of a massive particle in the expression in place of mc, the momentum of a photon, we obtain

$$\lambda = \frac{h}{mv}$$

or

$$\text{wavelength} = \frac{\text{Planck's constant}}{\text{momentum of particle}}$$

DE BROGLIE'S MATTER-WAVES

may not be applicable to electrons in atoms. Indeed, this seems to be so! Apples, oranges, and baseballs have wavelengths which are *far* too short to see or detect. Their mass is large, so the wavelength is short. On the other hand, electrons in atoms are very light and have wavelengths comparable to atomic sizes. Many experiments on electrons in atoms can be explained best if the electron is treated as a wave.

15-5 THE QUANTUM THEORY

Atoms are described today by a new form of mechanics called **wave mechanics** or **quantum mechanics**. The origin of both names should now be apparent. One reflects the wave-like nature of the electron; the other, the quantization of energy in the atom. Unfortunately, quantum mechanics (or wave mechanics), unlike classical mechanics, is fundamentally mathematical, and the physical picture of the atom it suggests is far different from the mechanical models. We shall try to summarize the results of the quantum-mechanical analysis without getting lost in the mathematical jungle.

15-5.1 The Principal Quantum Number, n

One of the very significant results of **quantum theory** is that the distinct energy levels we have mentioned can be identified by numbers called **quantum numbers.** In our bookcase analogy, numbers increased as the shelves went from bottom to top. The shelf numbers so assigned would correspond to the **principal quantum numbers** in the bookcase model. Each principal quantum number corresponds to a particular energy level. It has been deduced (by methods beyond us at this point) that the energy for any principal quantum number, n, can be derived from the following formula:

$$E_n = -\frac{313.6}{n^2} \text{ kcal/mole} \qquad (8)$$

The negative sign in front of 313.6 is significant and deserves comment. Simple algebra shows that as n becomes larger, the energy becomes less negative and as n approaches infinity, the energy approaches zero. Why is the energy of a hydrogen atom negative in its lowest state? To answer this question look again at Figure 15-15. You will notice two energy scales; in the scale on the left, the bottom level is assigned the value zero; in the scale on the right, the top level is assigned the value zero. Choice of *either* assigned value would be understandable, but physical scientists have determined that calling the top level zero offers a number of advantages. Thus, *by definition, the hydrogen atom has zero energy when the electron and the nucleus are separated by an infinite distance.* As the electron approaches the proton, energy is released; the system has *less* energy than it did at the defined zero of energy. In view of this fact, all energies recorded for the hydrogen atom must carry the negative sign.

15-5.2 Orbitals

The number n is the principal quantum number. It defines the energy and tells something about where the electron moves in space. But it does not describe an electron path.

In attempting to translate this mathematical picture of energy levels into physical terms, scientists have identified and named the region of space in which an electron in a given energy level is most likely to be found. This spacial distribution of an electron in a given energy level is called an **orbital.** An orbital, in contrast to an orbit, is not based on classical physics—it is based on probability. Let us be more specific about the quantum numbers and orbitals of the hydrogen atom.

The nature of the information furnished about an atom by quantum mechanics can be illustrated by the following analogy. Picture a bird-feeding station in a large, snow-covered park in the dead of winter. In this park there is a single bird (good chemistry but poor ornithology), whose movements we want to study. We have an excellent camera mounted on a very high platform, so that a picture of the entire park can be taken at any time during the day. Any one picture of the bird will not be very helpful in deciding where the bird spends most of his daylight hours. On the other hand, suppose we arranged to photograph the park and the bird every five minutes throughout the day. After we have done this several days, we can place the photographic negatives on top of each other so that the bird-feeding station in all pictures always appears at the same point. In this way a composite print is obtained. In those areas where the bird has spent the most time, the composite picture will be relatively dark in color. In those areas where the bird has appeared only infrequently, the photograph will be light in color. If the bird has behaved normally, the picture is darkest around the feeding station

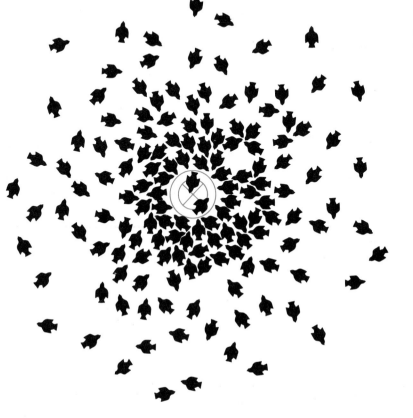

Fig. 15-20.1 **A composite picture of a bird at a bird-feeder.**

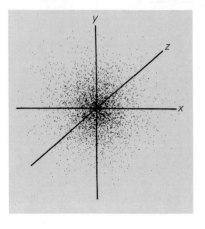

Fig. 15-20.2 A computer generated plot of a 1s orbital. A computer plotted the position of a 1s electron at 4000 consecutive instants of time. This and the following five figures are courtesy of Don T. Cromer.

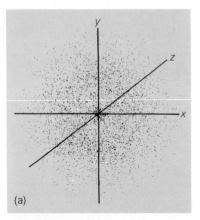
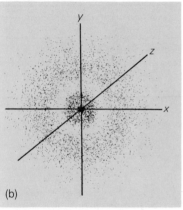

Fig. 15-21 Computer generated plots of (a) the 2s orbital and (b) the 3s orbital.

and fades out toward the edges of the park. The *probability* that the bird will be found in a given spot can be determined by how dark the film is at that spot. The composite picture tells us the *relative probability* that the bird will be present in a given spot at any given instant. (Figure 15-20.1 gives a schematic representation of what this composite photograph would look like.)

Quantum mechanics gives us the same type of information about an electron. *In the case of electron distribution, the properties of the orbital tell us the probability that an experiment designed to locate the electron will find it, in a given unit of volume, at a particular distance from the nucleus.* Our composite picture of the bird in the park does not tell us how the bird flies from spot to spot. Similarly, the quantum mechanical picture of the atom does *not* tell us *how the electron moves* from point to point, its **trajectory.** Though quantum mechanics does not tell us the electron trajectory, it *does* tell us *how the orbital changes as* n *increases.* It also indicates that *for each value of* n *there are* n^2 *different orbitals.* For the hydrogen atom, the n^2 orbitals associated with a given *n* all have the same energy:

$$E_n = -\frac{313.6}{n^2} \text{ kcal/mole} \qquad (8)$$

How then do orbitals with the same value of *n* differ? They differ in the spacial distribution of electrons. Orbital "shapes" are quite different. Let us examine characteristic **orbital shapes** for the hydrogen atom.

s ORBITALS

Consider the lowest energy level of a hydrogen atom with $n = 1$. We have just learned that there are n^2 levels with this energy; then, since $n = 1$, there is only one level. This orbital corresponds to an electron distribution that is spherically symmetrical around the nucleus. A "composite picture" of this orbital is shown in Figure 15-20.2. Notice how closely it resembles the distribution for the bird in the park (Figure 15-20.1). We call this a **1s orbital.** An electron moving in an *s* orbital is called an *s* electron. *All* s *orbitals are spherically symmetrical* (like a ball). The probability of finding the electron in a given unit of volume decreases regularly in a spherical pattern as we go away from the nucleus. This is seen in Figure 15-20.2. The fact has an interesting consequence: *the atom does not have a definite size;* it tends to "fade away" as the distance from the nucleus increases. In short, a 1s electron has the *greatest probability* per unit of volume *at the nucleus*. The probability *decreases* as the distance from the nucleus increases.

The next energy level corresponds to $n = 2$. According to the rule, there are $n^2 = 2^2$ or four different spacial arrangements having the same energy. The energy value is $-313.6/4$ or -78.4 kcal/mole. One of these spacial arrangements or orbitals is again spherically symmetrical and is called the **2s orbital** (Figure 15-21). As we might have reasoned, the higher energy of the 2s electron results from the fact that it spends more time farther from the nucleus.

p ORBITALS

We have described the *2s* orbital, which is just *one* of *four* orbitals having the principal quantum number 2. The other three orbitals are known as **2p orbitals.** Each *p* **orbital** tends to concentrate electron probability in the direction of one of the coordinate axes. As Figure 15-22 shows, a $2p_x$ orbital concentrates electron density along the *x* axis in a dumbbell-shaped pattern. Note that the probability of finding a *p* electron in a unit of volume at the nucleus is zero, whereas the probability of finding an *s* electron in a unit of volume at the nucleus is significant. As you might have guessed, we also have p_y and p_z orbitals (Figure 15-20.2). They differ only in orientation of the dumbbell-shaped electron clouds. The p_y has the axis of the dumbbell along the *y* axis. The p_z has the axis of the dumbbell along the *z* axis (see Figure 15-22). *Every energy level with* n *above 1 has three* p *orbitals.* These are np_x, np_y, and np_z. As *n* increases, electrons in *p* orbitals are, on the average, farther and farther from the nucleus; but the *overall shape* of the electron distribution along the *x*, *y*, and *z* axes does not change as *n* increases.

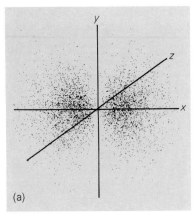

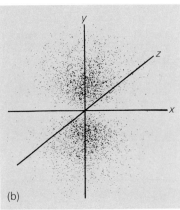

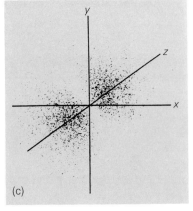

Fig. 15-22 **Plots of the three 2p orbitals: (a) $2p_x$ orbital, (b) $2p_y$ orbital, (c) $2p_z$ orbital.**

d ORBITALS

At this point we might summarize our information on orbitals by constructing the diagram shown in Figure 15-23. Each orbital is

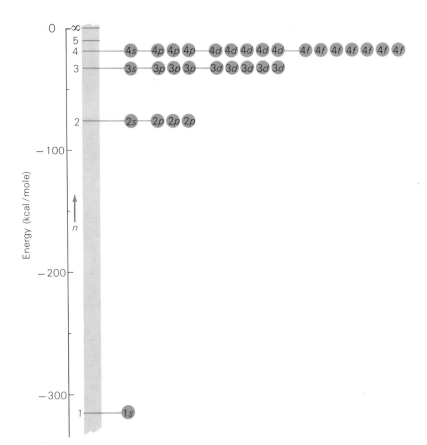

Fig. 15-23 **The energy-level scheme of the hydrogen atom.**

represented as a circle into which electrons may be placed. For a given value of n, there are n^2 total orbitals; for example, for $n = 3$, there are 3^2 or nine orbitals. We can immediately identify one **3s** and three **3p orbitals.** The five remaining orbitals are called **3d orbitals.** They have somewhat more complicated spacial distribution; their shape will not be considered here.

EXERCISE 15-6

How many orbitals would be expected for $n = 4$? How many orbitals are accounted for by the 4s, 4p, and 4d groups? The remaining electrons in the fourth quantum level are f electrons. How many 4f electrons are there?

15-6 ELECTRONIC STRUCTURE AND THE PERIODIC TABLE

If the orbital picture for hydrogen were only useful in explaining the spectrum of hydrogen, we would expect to find the foregoing discussion in a book on spectroscopy but not in a book on general chemistry. The fact that the discussion appears in a book on chemistry should suggest to you that the concept of orbitals can be extended well beyond hydrogen. Indeed, this idea can serve as the basis for constructing the entire periodic table; it can also provide a sound basis for understanding many of the chemical trends first summarized in Chapter 8.

15-6.1 Many-Electron Atoms

All atoms display line spectra. In general these spectra are much more complicated than the atomic hydrogen spectrum shown in Figure 15-13 on page 371, but they can be interpreted just as we interpreted the hydrogen spectrum, if certain reasonable assumptions are made. Such assumptions are logically supported by correlations with the periodic table. Let us list our assumptions:

(1) Atoms of all elements have orbitals and energy levels qualitatively like those of hydrogen.
(2) The lines in the spectra of all atoms can be understood in terms of an electron jumping between two definite energy levels. Such jumps are called **electronic transitions** between levels.
(3) The orbital energy-level diagrams for many-electron atoms resemble the hydrogen energy-level diagram *except that all orbitals with a given value of* n *no longer have the same energy.* For example, in many-electron atoms s orbitals are lower in energy than p orbitals. Figure 15-24 shows a schematic energy-level diagram of a many-electron atom for levels up through $6p$. We sometimes speak of $2s$ and $2p$ sublevels. (Compare diagram 15-24 with 15-23. What differences do you see?)
(4) A single orbital of any atom can accommodate *no more than* two electrons.* Each electron in an orbital may be visualized as

---

*This is just a statement, in its simplest form, of an empirical generalization known as the **Pauli Exclusion principle.**

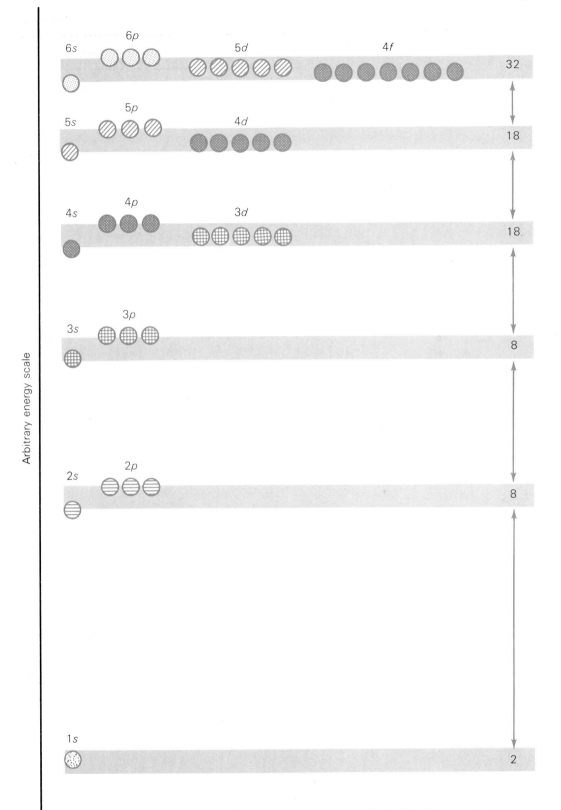

Fig. 15-24 **The energy-level scheme of a many-electron atom.**

spinning around its own axis. The first electron in an orbital spins in one direction; the second must spin in the reverse direction. Such electrons are said to differ in spin or to have "paired spins." There are only two directions of spin; hence, each orbital can accommodate only two electrons.

(5) All orbitals of equal energy will acquire one electron before any orbital accepts two electrons.*

These five assumptions permit a sound interpretation of more complicated atomic spectra.

15-6.2   The First Eighteen Elements

Let us now use these assumptions to suggest the electron arrangement for an atom in the "ground" state or state of lowest energy for that atom. We do this by filling orbitals with electrons until the nuclear charge is just counterbalanced by the extranuclear electron charge. We shall always place each additional electron in the lowest unfilled orbital available. Thus, in hydrogen with only a single electron, one electron is placed in the $1s$ level. The number of electrons in the level is designated by a superscript to the right of the orbital symbol. We write $1s^1$ as the **electron configuration** for hydrogen. In helium, with a nuclear charge of $2+$, we need two electrons. Both can be accommodated in the $1s$ orbital; hence, for helium we write $1s^2$.

For lithium, with a nuclear charge of $3+$, we need three electrons outside the nucleus. Two are placed in the $1s$ orbital. Since two electrons fill an orbital, the third electron must be placed in the $2s$ orbital. The electron configuration of Li is then $1s^2 2s^1$. Despite the nuclear charge of $3+$ in the lithium atom, the last electron is rather weakly bound because the $2s$ electron of lithium spends most of its time farther away from the nucleus than do the $1s$ electrons. The $2s$ electron should be fairly easily lost to give Li^+. Indeed, we have already noticed (Chapter 8) that gaseous alkali metal atoms lose electrons rather easily to give gaseous metal ions. Alkali metal atoms have a low **ionization energy**.

The beryllium atom has one more electron than the lithium atom. The fourth electron for beryllium is put in the $2s$ level; hence, its electron configuration is $1s^2 2s^2$. Removal of the two $2s$ electrons gives the Be^{2+} ion. The Be^{2+} ion has the same configuration as the noble gas helium.

With a boron atom five electrons are needed to balance the nuclear charge; hence, the configuration is $1s^2 2s^2 2p_x^1$.

Continuing this process, we obtain the following configurations:

Carbon atom	$1s^2$	$2s^2 2p_x^1 2p_y^1$
Nitrogen atom	$1s^2$	$2s^2 2p_x^1 2p_y^1 2p_z^1$
Oxygen atom	$1s^2$	$2s^2 2p_x^2 2p_y^1 2p_z^1$
Fluorine atom	$1s^2$	$2s^2 2p_x^2 2p_y^2 2p_z^1$
Neon atom	$1s^2$	$2s^2 2p_x^2 2p_y^2 2p_z^2$

* If two electrons are in the same orbital rather than in each of two equivalent orbitals, the electrons will repel each other to give a higher total energy for the atom.

We note that at the stable electron configuration of helium (high ionization energy), the level of principal quantum number 1 is completed. The two required electrons are present. The next configuration with high ionization energy is found at neon, where the levels of principal quantum number 2 as well as principal quantum number 1 are completed. Eight electrons go in the four orbitals (2^2) of principal quantum number 2; hence, neon has eight more electrons than helium. Argon represents completion of the one $3s$ and the three $3p$ orbitals. Also, argon has eight more electrons than neon. The "magic number" of eight electrons separating neon and helium and argon and neon is clearly tied to the eight electrons required to fill the one s and the three p orbitals.

The construction of the first 11 elements is shown schematically in Figure 15-25 (next page). Electrons spinning in one direction are shown by filling the upper half of the circle; electrons spinning in the opposite direction are shown by filling the lower half. Only two electrons are allowed per orbital. We made this assumption earlier, listing it as assumption (4).

If we proceed stepwise beyond neon, we come to sodium. Again, we are forced to use an orbital of higher quantum number, the $3s$:

Sodium atom $\qquad 1s^2 \quad 2s^2 2p_x^2 2p_y^2 2p_z^2 \quad 3s^1$

The $3s$ electron spends more time away from the nucleus than does the electron in any other orbital; hence, it should be more easily removed to give Na^+. Perhaps the chemistry of sodium is like the chemistry of lithium. A review of Chapter 8 shows that this is so. In many other cases the properties of the first 20 elements are suggested by their electronic patterns. Electronic structure is based on available energy levels. The close agreement between the predicted chemical properties based on spectroscopy and the actual chemical properties found by experiment adds to our faith in these models.

EXERCISE 15-7

Magnesium has one more electron than sodium. Write the electron configuration for magnesium. What ion is suggested for magnesium? Write the electron configuration for argon (atomic number 18). Write the electron configuration for potassium (no. 19).

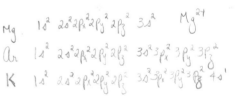

15-6.3 Elements Beyond Argon

For elements beyond argon we follow the same procedure. The only modification is in the appearance of the energy-level diagram used as we advance to higher levels (see Figure 15-24). Levels having the *same* principal quantum number are symbolized by the same type of interior pattern. As we go beyond the $3p$ orbitals, we find, to our surprise, that the $3d$ levels lie *above* the $4s$ levels but *below* the $4p$ levels.* Continuing to add electrons to the lowest

---

* The arrangement of levels shown in Figure 15-24 is an acceptable simplification. In actual fact, the relative position of levels changes as the nuclear charge and number of electrons change. The order given here accounts for the form of the periodic table.

390 Chapter 15 / ELECTROMAGNETIC RADIATION and ATOMIC STRUCTURE

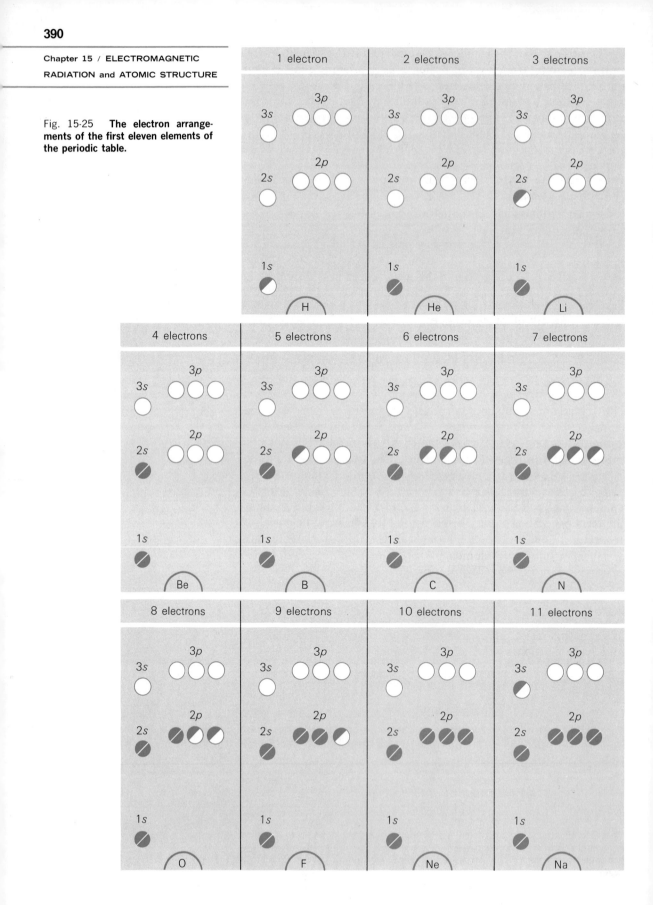

Fig. 15-25 The electron arrangements of the first eleven elements of the periodic table.

available level, then, the first electron beyond the closed shell of argon goes into the $4s$ level. Thus, the potassium atom (atomic number 19) has the configuration

Potassium atom $\qquad \underbrace{1s^2 \quad 2s^2 2p^6 \quad 3s^2 3p^6}_{\text{ARGON CONFIGURATION}} \quad 4s^1$

ELEMENTS BEYOND ARGON

Potassium is the first element of the fourth row of the periodic table. With the next element, calcium (no. 20), the twentieth electron is added to complete the $4s$ level; hence, its configuration is

Calcium atom $\qquad \underbrace{1s^2 \quad 2s^2 2p^6 \quad 3s^2 3p^6}_{\text{ARGON CONFIGURATION}} \quad 4s^2$

When another electron is added to form the next element, scandium, the available orbital of lowest energy is one of the $3d$ levels (remember, the $3d$ levels are a little lower in energy than the $4p$ levels). Scandium is thus represented as

Scandium atom $\qquad \underbrace{1s^2 \quad 2s^2 2p^6 \quad 3s^2 3p^6 3d^1}_{\text{ARGON CONFIGURATION}} \quad 4s^2$

As electrons are added to form other elements, they enter the $3d$ orbitals until the ten available spaces in these orbitals are filled.

The elements that are formed when the $3d$ electrons are added are called the **transition metals** *or the* **transition elements.** Since their chemical properties and electronic structures are unlike those of the lighter elements we have discussed, it is reasonable that these elements should be considered together. The first group of transition elements include the ten elements of the table, from scandium to zinc: scandium (Sc), titanium (Ti), vanadium (V), chromium (Cr), manganese (Mn), iron (Fe), cobalt (Co), nickel (Ni), copper (Cu), and zinc (Zn).* In building up these ten elements additional electrons are added to $3d$ levels until the $3d$ level is filled. Zinc then has the same configuration as calcium except that the $3d$ level is filled in zinc. The electron configuration of zinc is

Zinc atom $\qquad \underbrace{1s^2 \quad 2s^2 2p^6 \quad 3s^2 3p^6 3d^{10}}_{\text{ARGON CONFIGURATION}} \quad 4s^2$

After zinc the $4p$ levels are the lowest orbitals available. Gallium (no. 31), then, has the structure

Gallium atom $\qquad \underbrace{1s^2 \quad 2s^2 2p^6 \quad 3s^2 3p^6 3d^{10}}_{\text{ARGON CONFIGURATION}} \quad 4s^2 4p^1$

* Occasionally people argue over the placement of zinc in the transition group. Its properties are not really those of the transition elements. It is in an intermediate region.

Gallium has an outer electronic pattern similar to that of aluminum, except that outer electrons occupy the 4s and 4p levels in gallium but the 3s and 3p levels in aluminum:

Aluminum atom $\quad 1s^2 \quad 2s^22p^6 \quad 3s^23p^1$

Also, gallium has a completed 3d level.

The similarities suggest that gallium and aluminum might have many similar chemical properties; for example, both form oxides of the form M_2O_3 and halides of the empirical formula MX_3. On the other hand, the differences in electronic structure result in important differences in properties that cannot be overlooked. Such similarities and differences must still be established by experiment. Chemistry is still an experimental science!

When all the 4p levels have been completed, the elements gallium (Ga), germanium (Ge), arsenic (As), selenium (Se), bromine (Br), and krypton (Kr) will have been formed. The configuration of krypton is

Krypton atom $\quad \underline{1s^2 \quad 2s^22p^6 \quad 3s^23p^63d^{10}} \quad 4s^24p^6$

ARGON CONFIGURATION

The eighteen elements running from potassium to krypton make up the fourth row of the periodic table. We see that this fourth row contains eighteen rather than eight elements because the five 3d orbitals have energies that are about the same as the energies of the 4s and 4p levels. Since ten electrons can be placed in the 3d levels, the fourth row has eighteen elements.

We might well guess that the fifth row could have a group of ten transition elements with the 4d level existing in varying stages of completion. Indeed, if we look at the periodic table, we see ten elements running from yttrium (Y) to cadmium (Cd), the second group of transition elements.

In the third group of transition elements, the 5d level is being filled. This group includes lanthanum (La), hafnium (Hf), tantalum (Ta), tungsten (W), rhenium (Re), osmium (Os), iridium (Ir), platinum (Pt), gold (Au), and mercury (Hg).

You will notice that we did not include the elements from lanthanum (no. 57, La) through lutetium (no. 71, Lu) in the transition metals. These fourteen elements result when the 4f levels start to fill. The 4f levels have about the same energy as the 5d and 6s levels; hence, they start to fill at the element having atomic number 57. The fourteen elements formed in the filling of the seven 4f levels are called the **inner transition elements.** Sometimes older names such as the **rare earths** or the **lanthanides** are used for these elements. They all have very similar chemistry and are usually placed at the bottom of the periodic table outside the normal arrangement.

15-6.4 The Periodic Table in Retrospect

We see that the rows of the periodic table arise from filling orbitals of approximately the same energy. When all orbitals of a given

n value are filled (two electrons per orbital), the next electron must be placed in an s orbital of higher principal quantum number and a new row or *period* of the table starts. We can summarize the relation between the number of elements in each row of the table and the number of available orbitals of approximately equal energy by the information shown in Table 15-4.

TABLE 15-4 NUMBER OF ELEMENTS IN EACH ROW OF THE PERIODIC TABLE

Row of Table	Number of Elements	Lowest Energy Orbitals Available to Be Filled			
1	2	1s			
2	8	2s,	2p		
3	8	3s,	3p		
4	18	4s,	3d,	4p	
5	18	5s,	4d,	5p	
6	32	6s,	4f,	5d,	6p
7		7s,	5f,	6d,	7p

15-7 IONIZATION ENERGIES, ENERGY LEVELS, AND THE PERIODIC TABLE

15-7.1 Ionization Energies of Atoms— The First Ionization Energy

On several occasions we have mentioned the ionization energies of the elements. Let us repeat our definition: *The **first ionization energy** of an atom is the amount of energy required to remove the most loosely bound electron from the gaseous atom.* It is the energy required for the process

$$\text{gaseous atom} + \text{energy} \longrightarrow \text{gaseous ion}^+ + \text{electron}^- \qquad (9)$$

Since energy is involved in the definition of the ionization process, we should be able to represent ionization energy on the energy-level diagram. To do this, consider the energy-level diagram for the hydrogen atom (Figure 15-26, next page). As our eye travels up the diagram, the levels become closer and closer together. As the principal quantum numbers grow larger, the differences between levels become smaller, until differences are too small to see on the diagram— the levels run together and the energy of the system approaches zero. Remember, the energy was negative in the original atom. As you will recall, zero was arbitrarily selected as the energy value for the state in which an electron is completely removed from the atom. Thus, if we put enough energy into the system to remove the electron from its "ground-state" orbital and raise it to the zero energy, we shall have ionized the atom. In terms of the energy-level diagram, *the **ionization energy** is the energy necessary to lift an electron from the highest occupied orbital up to the limit corresponding to $n = $ infinity (∞)*.

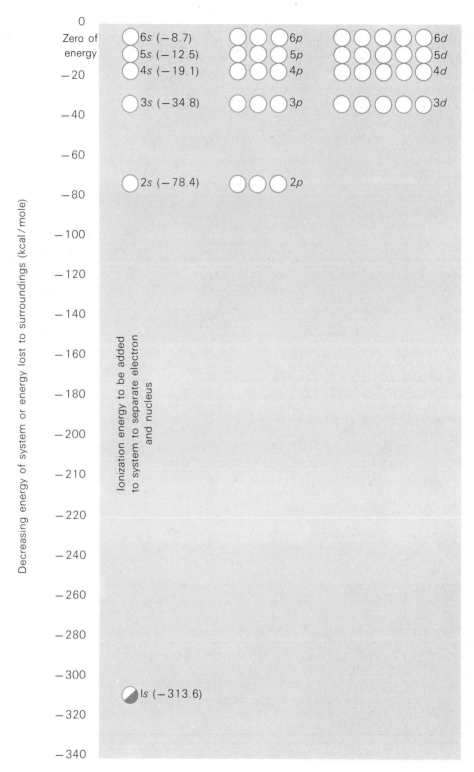

Fig. 15-26 **The ionization energy and the energy-level diagram of the hydrogen atom.**

For the hydrogen atom, we should be able to calculate the difference in energy between these two levels. The hydrogen electron in the ground state (1s orbital) has energy given as

$$E = -\frac{313.6}{n^2} \text{ kcal/mole} = -\frac{313.6}{1^2} \text{ kcal/mole} = -313.6 \text{ kcal/mole} \quad (8a)$$

The free electron with quantum number ∞ has energy given as

$$E = -\frac{313.6}{\infty^2} \text{ kcal/mole} = 0 \quad (8b)$$

The difference between these two states is then $E_{final} - E_{initial}$ or $0 - (-313.6) = 313.6$ kcal/mole. The ionization energy for the hydrogen atom is 313.6 kcal/mole, the quantity we have been using as the constant in our energy equation for hydrogen; a total of 313.6 kcal must be put into a mole of gaseous hydrogen atoms to remove electrons to infinity. The ionization energies of the first nineteen elements are shown in Table 15-5.

EXERCISE 15-8

Sketch in qualitative form the energy-level diagram for lithium in its ground state (indicate individual electrons) and show its ionization energy on the diagram. If 1 kcal is equal to 4.18×10^{10} ergs, what frequency of light ($E = h\nu$) would cause ionization of lithium ($h = 6.6 \times 10^{-27}$ erg $\times$ sec)? See Table 15-5 for the ionization energy. (Answer: $\nu = 1.3 \times 10^{15}$ vibrations/sec.)

15-7.2 Ionization Energies of Ions— The Second Ionization Energy

As you will recall, the term ionization energy can also be applied to ions (atoms that have already lost an electron). For example, the ionization energy of $Mg^+(g)$ is the energy in the process

$$Mg^+(g) + \text{energy} \longrightarrow Mg^{2+}(g) + e^-(g) \quad (9)$$

Since the above process removes the *second* electron from a magnesium atom, the ionization energy of $Mg^+(g)$ is called the **second ionization energy** of magnesium. Because the second electron must be pulled from a *positive ion* (Mg^+) rather than from a *neutral atom* (Mg), the second ionization energy for an atom is *always larger* than its first ionization energy. For example, the first ionization energy for magnesium is 175 kcal/mole:

$$Mg(g) + 175 \text{ kcal} \longrightarrow Mg^+(g) + e^-(g) \quad (10)$$

while its second ionization energy is 345 kcal/mole:

$$Mg^+(g) + 345 \text{ kcal} \longrightarrow Mg^{2+}(g) + e^-(g) \quad (9a)$$

The effect of the positive charge on Mg^+ is clearly apparent.

TABLE 15-5 IONIZATION ENERGIES OF THE ELEMENTS

Atomic Number	Element	Ionization Energy (kcal/mole)
1	H	313.6
2	He	566.7
3	Li	124.3
4	Be	214.9
5	B	191.2
6	C	259.5
7	N	335
8	O	313.8
9	F	401.5
10	Ne	497.0
11	Na	118.4
12	Mg	175.2
13	Al	137.9
14	Si	187.9
15	P	241.7
16	S	238.8
17	Cl	300
18	Ar	363.2
19	K	100.0

15-7.3 Energy-Level Diagrams and Trends in Ionization Energies

We first noted in Chapter 8 that the ionization energies of the elements in a given row will rise from a minimum for an alkali metal (Li = 124.3 kcal/mole) to a maximum for the noble gas which completes that row (Ne = 497.0 kcal/mole) (see Figure 15-27). If we focus attention on the second row—from Li to Ne—we find each atom is losing an electron of principal quantum number 2. As our detailed diagram (Figure 15-28) of energy levels for the first ten elements shows, the 2s and 2p levels display a general decrease in energy as we go across the table. The ionization energies, then, must rise from Li to Ne. Such behavior can only be related to the build up of positive charge in the nucleus. As the nuclear charge increases in the atom, it should become more difficult to remove an electron from a given quantum level. An ionization-energy curve should show a *general rise* from Li to Ne. Indeed, a general rise is seen in the values from Li to Ne. But this simple picture does not explain the irregularities seen at boron and oxygen. We shall consider some of these minor irregularities in the next section.

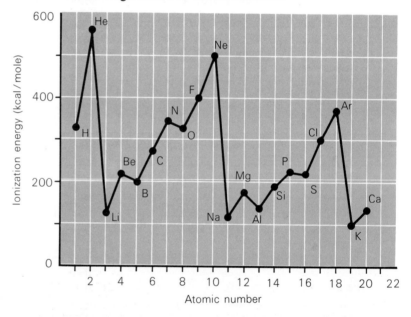

Fig. 15-27 Ionization energy as a function of atomic number.

15-7.4 Ionization Energies and Valence Electrons

In Chapter 8 we noted that the most loosely bound electrons can be lost or shared when one atom combines with another. These relatively loosely bound electrons are called **valence electrons.** A more detailed examination of the ionization energies of valence electrons will help us relate chemical properties to atomic structure.

Consider the elements lithium, beryllium, and boron. For each of these elements we know several ionization energies corresponding to processes such as

$Li(g) \longrightarrow Li^+(g) + e^-(g)$	first ionization energy = 124 kcal	(*11*)
$Li^+(g) \longrightarrow Li^{2+}(g) + e^-(g)$	second ionization energy = 1,739 kcal	(*12*)

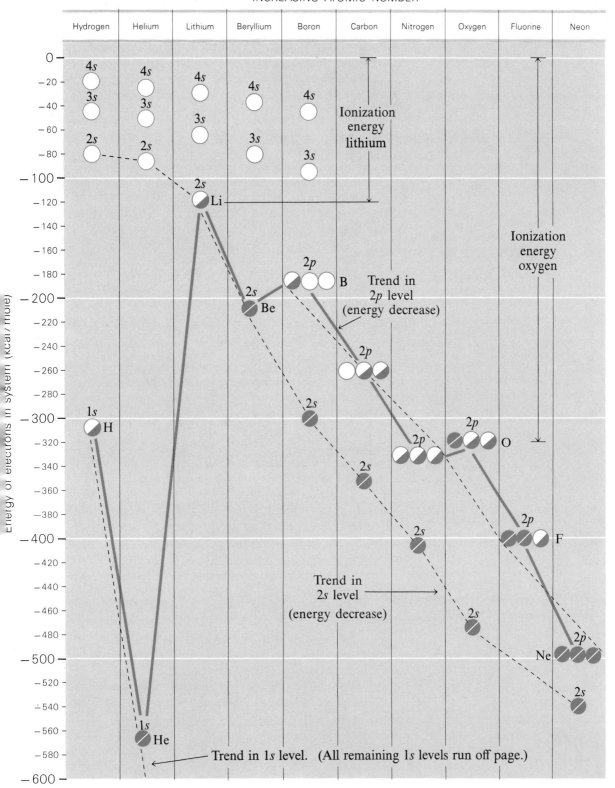

Fig. 15-28 **Energy levels and atomic number for the first ten elements.**

TABLE 15-6 SUCCESSIVE IONIZATION ENERGIES OF LITHIUM, BERYLLIUM, AND BORON

Element	Electron Configuration for Neutral Atom	Ionization Energy (kcal/mole)			
		E_1	E_2	E_3	E_4
Lithium	$1s^2 2s^1$	123	1,739	2,815	—
Beryllium	$1s^2 2s^2$	214	419	3,540	5,004
Boron	$1s^2 2s^2 2p^1$	191	577	872	5,964

The experimental values of these energies are shown in Table 15-6. Let us compare lithium and beryllium. For each, the first electron to be removed is a $2s$ electron. The attraction between the nucleus and the $2s$ outer electron will be higher for beryllium (four nuclear protons) than for lithium (three nuclear protons); hence, the higher first ionization energy of beryllium is understandable:

$$\text{Li}(g) + 124 \text{ kcal} \longrightarrow \text{Li}^+(g) + e^-(g) \qquad (11a)$$
$$\text{Be}(g) + 215 \text{ kcal} \longrightarrow \text{Be}^+(g) + e^-(g) \qquad (12)$$

The second ionization, however, reverses the situation; lithium has the higher second ionization energy:

$$\text{Li}^+(g) + 1,739 \text{ kcal} \longrightarrow \text{Li}^{2+}(g) + e^-(g) \qquad (12a)$$
$$\text{Be}^+(g) + 419 \text{ kcal} \longrightarrow \text{Be}^{2+}(g) + e^-(g) \qquad (13)$$

The energy required to remove the second electron from lithium is four times that of the energy required to remove the second electron from beryllium. Why? Why is the second ionization energy of lithium so much larger than the second ionization energy of beryllium? The answer is found in the energy-level diagram of Figure 15-28. *The second electron removed from beryllium is still a $2s$ electron, but the second electron removed from lithium is a $1s$ electron.* Notice how much lower in energy a $1s$ electron is than a $2s$ electron. (See Figure 15-28. The $1s$ levels run off the page.) The $1s$ electron is generally much closer to the nucleus than is the $2s$ electron.

Continuing to boron, we see that its first ionization energy is below that of beryllium, even though boron has a higher nuclear charge. The answer to this apparent deviation is also found in Figure 15-28. The first electron removed in boron is a $2p$ electron, which lies well above the $2s$ energy level. It is easier to remove a $2p$ electron from boron than it is to remove a $2s$ electron from beryllium, even though boron has the higher nuclear charge.

If we continue to remove electrons from boron, we discover a *very large increase* in ionization energy when the fourth electron is removed. We can understand this large jump in view of the fact that all electrons with quantum number 2 have been removed from boron when the third electron has gone. The fourth electron must come from the very low-energy $1s$ level. The energy jump from $1s$ to infinity is enormous.

By remembering how we placed electrons in the lowest empty orbitals, we can draw up a guideline for determining the number of valence electrons a given atom possesses. The number of electrons placed in the orbitals that form the *highest partially filled* cluster of energy levels is the number of valence electrons. These electrons are most easily removed or shared. Lithium has one valence electron; beryllium has two; boron, three; carbon, four; and nitrogen, five. In short, for this group the valence electrons are the electrons in the second level. These numbers will help us understand the chemistry of these elements.

EXERCISE 15-9

Explain why chemists say that aluminum has three valence electrons and chlorine seven. How many valence electrons does sodium have? magnesium?

15-8 HIGHLIGHTS

Light is a form of **electromagnetic radiation.** As such, it is also a form of energy which can be propagated across empty space as **oscillating electric** and **magnetic fields.** Historically, a study of the light given off by hydrogen gas in a discharge tube led to various *mechanical models* for the hydrogen atom. All conflicted with the laws of classical physics and were finally abandoned.

A new *mathematical model* for the hydrogen atom based on the **wave** nature of the electron emerged. As a result of work in atomic spectroscopy, an energy-level diagram suitable for considering elements in general was obtained. By using this energy-level diagram and a few additional rules, the form of the periodic table can be understood.

Ionization energies, first considered in Chapter 8, can be represented on energy-level diagrams. Many of the details of ionization energies are understandable in terms of the energy-level diagram. The **valence electrons** of an atom are the electrons in an incompletely filled cluster of energy levels of roughly comparable energy.

A number of useful and significant relationships have been given. These are summarized here. For light:

$$\nu = \frac{c}{\lambda}$$

where $c = 3 \times 10^{10}$ cm/sec. For light quantization:

$$E = h\nu$$

For frequency of lines in the hydrogen spectrum:

$$\nu = 3.287 \times 10^{15} \left(\frac{1}{n_1^2} - \frac{1}{n_2^2} \right) \text{ vibrations/sec}$$

The road to our understanding of the atom has been a winding and tortuous one. Today we are sure of only one thing: there are still many curves ahead. You are certain to meet them as you advance in science.

HIGHLIGHTS

**LUIS W. ALVAREZ
(1911–)**

The son of a prominent physician, Luis Alvarez was born in San Francisco and raised in Rochester, Minnesota. Educated at the University of Chicago, he received the Ph.D. in 1936. Most of his professional career has been spent at the University of California, where he is senior physicist of the Lawrence Radiation Laboratory.

Alvarez' scientific contributions are many and varied, ranging from the development of sophisticated atom accelerators to a color television system. Perhaps his most important work has been in the development of the hydrogen bubble chamber, with photographic processes and systems of data analysis, to study new subatomic particles. He has discovered a number of new particles, sometimes called "resonances" or higher energy levels of particles previously known.

For his many contributions to the sciences, Luis Alvarez was awarded the 1968 Nobel Prize in Physics.

Chapter 15 / ELECTROMAGNETIC
RADIATION and ATOMIC STRUCTURE

QUESTIONS AND PROBLEMS

1 In what way is the "heat" given off by an infrared lamp similar to the X rays used in the dentist's office? In what way does the "heat" differ from the X rays?

2 Which of the following statements concerning visible light is *false*? (a) It is a form of energy. (b) All photons possess the same amount of energy. (c) It cannot be bent by a magnet. (d) It includes the part of the spectrum called X rays.

3 A chemist describes the radiation he is using as having a frequency of 6×10^{13} vibrations/second. What is the wavelength of this radiation in centimeters? Use tables 15-1 and 15-2 to identify this radiation as red, blue, infrared, ultraviolet, and so on. Estimate the energy in ergs for one photon of this radiation. Planck's constant is 6.6×10^{-27} erg $\times$ sec. The velocity of light is 3×10^{10} cm/sec. (*Answer*: 4×10^{-13} erg/photon.)

4 Carbon monoxide absorbs light at frequencies near 1.2×10^{11}, near 6.4×10^{13}, and near 1.5×10^{15} vibrations/second. It does not absorb at intermediate frequencies. (a) Name the spectral regions in which it absorbs (see Figure 15-10). (b) Explain why carbon monoxide is colorless.

5 The wavelength and frequency of light are related by the expression $\lambda = c/\nu$, where λ = wavelength in centimeters, ν = frequency in vibrations/second, and c = velocity of light = 3.0×10^{10} cm/sec. Calculate the wavelength corresponding to each of the three frequencies absorbed by CO (see question 4). Express each answer first in centimeters, and then in ångströms (1 Å = 10^{-8} cm). (*Answer*: 1.5×10^{15} cycles/sec: 2.0×10^{-5} cm/cycle = 2.0×10^3 Å/cycle.)

6 The oxygen molecule undergoes molecular vibrations at a frequency of 2.4×10^{13} vibrations/sec. If the pressure is such that an oxygen molecule has about 10^9 collisions per second, how many times does the molecule vibrate between collisions?

7 Electrons in an X-ray tube are accelerated to a velocity of 6×10^9 cm/sec before striking the target of the tube and stopping. (a) Why should radiation be emitted as the electron hits the target? (b) If the mass of the electron is about 10^{-27} g and the kinetic energy of a body is $\frac{1}{2}mv^2$, calculate the kinetic energy of the electron in ergs—that is, g $\times$ cm²/sec². (c) If Planck's constant is 6.6×10^{-27} erg $\times$ sec, what is the highest frequency of the X rays which can be emitted by this tube? (This is known as the Duane-Hunt limit.) (d) What is the wavelength of these X rays? (e) How could X rays of higher frequency be obtained? [*Answer to (b)*: 1.8×10^{-8} erg; *to (d)*: 1.1 Å.]

8 Use the energy-level diagram in Figure 15-15 to calculate the energy required to raise the electron in a hydrogen atom from level 1 to level 2, from level 1 to level 3, and from level 1 to level 4. Compare these energies with the spectral lines shown in Figure 15-13.

9 Complete the following table for all the ultraviolet lines listed in Figure 15-13. Plot the energy spacing against the arbitrary spacing number assigned in the last column to assure yourself that there is regularity in the spacings of these lines. Make the same sort of a table for the lines in the visible group.

Group	Energy per Mole of Photons (kcal)	Energy Spacing (kcal)	Spacing Number
Ultraviolet lines	235.2 kcal 278.8 294.0	43.6 kcal	1 2 3

10 Your plot in question 9 suggested that the energy levels given in Figure 15-15 are systematically related. To explore this relationship further, divide the energy of each level by that of the first level (use the right-hand scale). How are the fractions so obtained related to the numbers of the energy levels?

11 Calculate the frequency of a line in the hydrogen spectrum corresponding to a transition from $n = 5$ to $n = 4$. The ionization energy of hydrogen is 2.18×10^{-11} erg/atom.

12 According to the quantum mechanical description of the 1s orbital of the hydrogen atom, what relation exists between the surface of a sphere centered about the nucleus and the location of an electron?

13 What must be done to a 2s electron to make it a 3s electron? What happens when a 3s electron becomes a 2s electron?

14 If the energy difference between two electron states is 46.12 kcal/mole, what will be the frequency of light emitted when the electron drops from the higher to the lower state? Planck's constant = 9.52×10^{-14} (kcal $\times$ sec)/mole.

15 Determine the value of E_n for $n = 1, 2, 3$, and 4 for a hydrogen atom using the relation $E_n = 313.6$ kcal/mole/n^2. For each E_n, indicate how many orbitals have this energy.

16 The quantum mechanical description of the 1s orbital is similar in many respects to a description of the holes in a much-used dart board. For example, the "density" of dart holes is constant anywhere on a circle centered about the bullseye, and the "density" of dart holes reaches zero only at a distance very far from the bullseye (effectively, at infinity). What are the corresponding properties of a 1s orbital?

In view of your answer, point out erroneous features of the following models of a hydrogen atom (both of which were used before quantum mechanics demonstrated their inadequacies): (a) a ball of uniform density, (b) a "solar system" atom with the electron circling the nucleus at a fixed distance.

17 Name the elements that correspond to each of the following electron configurations:

$1s^2$ He
$1s^2$ $2s^1$ Li
$1s^2$ $2s^2 2p^1$ Boron
$1s^2$ $2s^2 2p^3$ Nitrogen
$1s^2$ $2s^2 2p^6$ $3s^2 3p^6$ $4s^1$ Potassium

18 Make a table listing the principal quantum numbers (through 3), the types of orbitals, and the number of orbitals of each type.

19 What trend is observed in the first ionization energy as you move from lithium down the first-column metals? On this basis, can you suggest a reason why potassium or cesium might be used in preference to sodium or lithium in photoelectric cells?

20 Consider these two electron populations for neutral atoms:

(1) $1s^2$ $2s^2 2p^6$ $3s^1$
(2) $1s^2$ $2s^2 2p^6$ $6s^1$

Which of the following is *false*? (a) Energy is required to change (1) to (2). (b) (1) represents a sodium atom. (c) (1) and (2) represent different elements. (d) Less energy is required to remove one electron from (2) than from (1).

21 How many valence electrons has carbon? silicon? phosphorus? hydrogen? Write the electron configurations for neutral atoms of each element.

22 The first four ionization energies of boron atoms are as follows: $E_1 = 191$ kcal/mole, $E_2 = 577$, $E_3 = 872$, and $E_4 = 5,964$. Explain the magnitudes in terms of the electron configuration of boron and deduce the number of valence electrons of boron.

For the nature of the chemical bond is the problem at the heart of all chemistry.

BRYCE CRAWFORD, JR., (1914–)

CHAPTER 16 **Molecular Architecture: Gaseous Molecules**

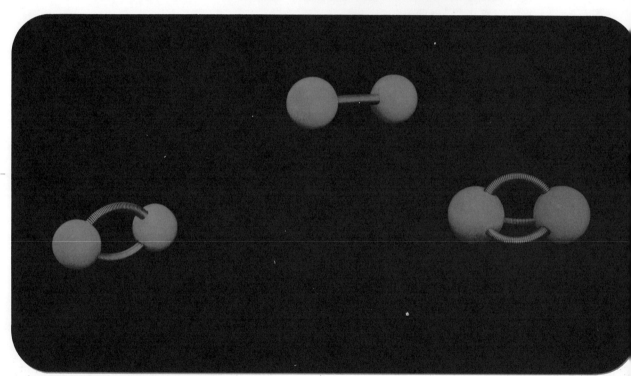

Single, double, and triple bonds are frequently found in gases.

By now, you are well aware of the atom as a positively charged nucleus surrounded by negatively charged electrons. Since electrons repel each other, the prospect of two atoms approaching each other seems exceedingly remote. But take heart. Under that repulsive exterior beats an attractive nucleus! Molecules do exist! The attractive forces between atoms overcome the repulsions to form a chemical bond.

In this chapter we explore the attractive and repulsive forces within molecules and their relationship to the geometry of gaseous molecules.

In Chapter 8 we discussed the bonding in some molecules in a rather simplified manner. Since then we have learned much more about atomic structure and about the fundamental particles making up atoms. We must now apply the newly developed concepts of atomic structure to obtain a more detailed view of the chemical bond.

We shall now consider two important questions:

(1) Why does the cluster of atoms persist?
(2) Why does the cluster have characteristic properties?

In this chapter we shall restrict our attention to molecules as they exist in the gaseous phase. Then, in Chapter 17, we shall consider the additional ideas needed to understand the forces causing the formation of liquids and solids.

16-1 THE COVALENT BOND

In Chapter 8 the connection between the atoms in H_2 or F_2 was identified as an electron pair shared between atoms. This linkage was called a **covalent bond.** Let us examine the covalent bond in gaseous H_2 molecules more carefully.

16-1.1 The Covalent Bond in the Hydrogen Molecule

Under normal conditions of temperature and pressure, hydrogen forms diatomic molecules. At a temperature of several thousand degrees, highly energetic intermolecular collisions knock hydrogen molecules apart:

$$H_2(g) + 103.4 \text{ kcal} \rightleftarrows H(g) + H(g) \qquad (1)$$

Since energy is absorbed in the dissociation process, the molecule H_2 is more stable (has a lower energy) than a system containing two separate hydrogen atoms.* To learn why the energy is lower when the atoms are near each other, we must examine interactions among the electric charges of the atoms.

Quantum mechanics tells us that the 1s orbital of an isolated hydrogen atom has spherical symmetry *before reaction*. On the other hand, if two hydrogen atoms (atoms 1 and 2) are brought together, the electron of atom 1 will be pulled toward the nucleus of atom 2. Similarly, the electron of atom 2 will be pulled toward the nucleus of atom 1. Both electrons 1 and 2 will spend a sizable fraction of their time in the space *between the two nuclei*. In this region each electron is attracted to *both* nuclei. Such *attraction* is the "glue" which holds two atoms together. <u>*The chemical bond in H_2 forms because each of the two electrons is attracted to two protons simultaneously. This arrangement is energetically more stable than the separated atoms in which each electron is attracted to only one proton*</u>.

* The film "Chemical Bonding" shows that the combination of hydrogen atoms liberates enough energy to make the platinum foil red hot.

But it is well to remember that there are also *repulsions* caused by the approach of the two atoms. The two electrons repel each other, as do the two protons. These repulsions tend to push the two atoms apart.

Which are more important, the attraction or the repulsion terms? Experiment shows that the attraction terms are greater—a stable chemical bond is formed. Why is this so? We find an explanation in the mobility of the electrons. The electrons do not occupy fixed positions but move about the molecule. The electrons move away from positions in which they would be near each other, although they still occupy positions between two nuclei. They are said to "correlate" their motion so as to remain apart. In this way electron-electron repulsion is minimized while proton-electron attraction remains high. The stable bond length in the hydrogen molecule is determined by a balance between the forces of attraction and the forces of repulsion.

We can also represent the chemical bond in the hydrogen molecule by using the visual representations of quantum mechanics. In Figure 16-1(a), we picture the electron distribution of an isolated atom in cross section. The electron distribution extends far from the

Fig. 16-1 Schematic representation of an atom and the interaction between two atoms.

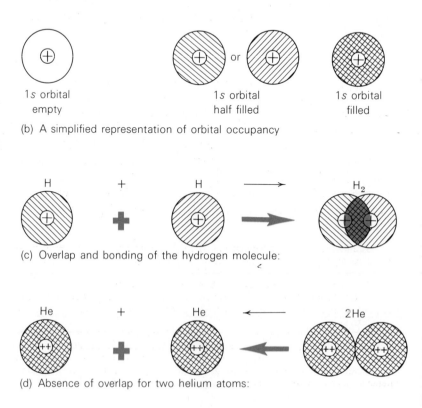

nucleus; it is uniform in all directions but concentrated near the nucleus. Therefore, we should focus attention on the center region of the 1s orbital. We do this by representing the 1s orbital by a circle having a radius large enough to contain most of the electron distribution.* An orbital can accommodate either one or two electrons but no more. Figure 16-1(b) offers a way of showing a 1s orbital empty, with one electron, and with two electrons.

Now, in Figure 16-1(c), consider the interaction of two hydrogen atoms. Each atom has a single electron in a 1s orbital. As the two hydrogen atoms approach, the circles are distorted and tend to overlap each other. *In this region of overlap the two electrons are shared by the two protons* (as shown by the crosshatched and shaded area). If this sharing occurs, the two electrons can be near *both* protons a good part of the time. This causes the chemical bond. When a bond arises from equal sharing, it is called a **covalent bond**.

16-1.2 Interaction Between Helium Atoms— No Bond Formation

A measurement of the density of helium gas shows that it is a monatomic gas. Molecules of He_2 do *not* form. What difference between hydrogen and helium atoms accounts for the absence of bonding in helium? The answer to this question also must lie in the attractive and repulsive electrical interactions between two helium atoms when they approach. Each separate helium atom has two electrons attracted to its nucleus. This gives two attractive interactions for each helium atom. Further, *each* atom has two electrons repelling each other to give one repulsive interaction in *each* helium atom. In two separate He atoms there are four attractive interactions and two repulsive interactions. Now what happens if two atoms are brought together? Taking score, we find eight attractive interactions, four more than in the two separated atoms, and seven repulsive interactions, five more than in the two separated atoms.† Again appealing to experiment, we learn that the four new attractive terms are *not* sufficient to counterbalance the five new repulsive terms. A chemical bond does *not* form.

Thus, the explanation of bonding for H_2 and the absence of it for He_2 lies in the *relative magnitudes* of attractive and repulsive terms. Quantum mechanics can be put to work with the aid of advanced and difficult mathematics to calculate these relative magnitudes. Unfortunately, the mathematics is so difficult that only a

---

* The fraction of the total electron cloud enclosed within the circle increases as the circle is made larger. We arbitrarily plan to enclose 90 percent of the cloud by drawing a circle to scale.

† The details of the interactions can be summarized as follows.

For the attractive interactions, nucleus of atom 1 attracts all four electrons; similarly, nucleus of atom 2 attracts all four electrons; thus, there are a total of eight attractive interactions.

In counting repulsive interactions, we start with the fact that the two nuclei repel each other. This gives one repulsive interaction. The repulsive interactions between electrons can best be schematized by numbering the electrons from 1 to 4. The repulsions are then given as 1-2, 1-3, 1-4, 2-3, 2-4, and 3-4. Notice that there are six electron interactions, giving a total of seven repulsive interactions.

handful of the very simplest molecules have been treated with high accuracy. Nevertheless, for some time chemists have been able to decide whether chemical bonds can form without appealing to a digital computer. They use simple models.

Refer to Figure 16-1(d), which is a simplified diagram of the interaction of two helium atoms. Unlike the diagram for hydrogen [Figure 16-1(c)], each helium atom is crosshatched *before* the two atoms approach. This indicates there are *already* two electrons in the 1s orbital. The rule for orbital occupancy tells us that the 1s orbital can contain *only* two electrons. Consequently, when the two helium atoms approach, their valence orbitals cannot overlap significantly because each orbital is already filled. Filled orbitals *cannot* overlap enough to share electrons. As a result the helium atom forms no chemical bonds.

16-1.3 Representations of Chemical Bonding

We propose, then, that <u>*chemical bonds can form if two atoms can share valence electrons using partially filled orbitals*</u>. We need a shorthand notation which aids in the use of this rule. Such a shorthand notation is called a **representation of the bonding.**

Our rule about covalent bond formation can be shown quite simply for H_2 and He_2 through an **orbital representation:**

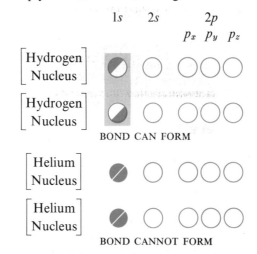

With this representation we need not consider the next higher energy-level clusters—the 2s and 2p orbitals—which for hydrogen and helium are much higher in energy and can only give rise to extremely weak attractions.

You will remember that the sharing of electrons can be shown by representing valence electrons as dots placed between the atoms:

$$H\cdot + H\cdot \longrightarrow H\!:\!H \qquad (2)$$

With this notation the symbol H must represent a bare proton. *The symbol of the atom as it appears in dot formulas represents the atomic core—the atom minus the valence electrons.* Valence electrons are shown by the dots. We shall use both **orbital** and **electron dot representations** to show chemical bonding.

16-1.4 Fluorine Atoms and Fluorine Molecules

A fluorine atom has seven valence electrons; that is, seven electrons occupy the outermost, partially filled cluster of energy levels. Using the electron dot method for representing the atom, we write fluorine as

$$\cdot \ddot{\underset{\cdot\cdot}{\text{F}}} :$$

You will recall from Chapter 8 that a gaseous fluorine atom *releases* energy when it picks up an electron to give a gaseous fluoride ion:

$$\cdot \ddot{\underset{\cdot\cdot}{\text{F}}} : + e^- \longrightarrow : \ddot{\underset{\cdot\cdot}{\text{F}}} :^- + 79 \pm 2 \text{ kcal/mole of F atoms} \qquad (3)$$

The energy change associated with this process is known as the **electron affinity** of the fluorine atom; it is symbolized by the letter E.

$$\Delta H = -E \qquad (4)$$

The fact that energy is released tells us in experimental terms that the *attraction* between the fluorine nucleus and the extra electron *is greater than* the *repulsion* between electrons of the fluorine atom and the added electron. Indeed, this energy release suggests that there is some stability associated with completing the valence shell of fluorine. A fluoride *ion*, with a *completed* 2 level, is lower in energy than a fluorine *atom*, with an *incomplete* 2 level. In our analysis of the bonding of fluorine atoms, we shall be concerned with discovering the different ways in which some or all of the stability resulting from a completed 2 level can be gained in the bonding of two fluorine atoms.

Let us examine the bonding of two fluorine atoms with this thought in mind. A fluorine atom has the orbital configuration

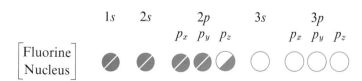

Each atom has a valence electron in a half-filled p_z orbital. Suppose two gaseous fluorine atoms (atoms 1 and 2) collide. We can imagine these two atoms orienting so that the half-filled p_z orbitals overlap in space. The half-filled valence orbital of atom 1 shares one valence electron with atom 2. Thus, *part* of the electron affinity of fluorine atom 1 is satisfied without pulling the electron away from fluorine atom 2. Meanwhile, atom 2 derives the same *energy benefit* from the valence electron of atom 1. Each fluorine atom has acquired *part interest* in another electron. The *most* energy we could expect to be released by such an electron-sharing procedure would be double the electron affinity of fluorine, or $2 \times 79 = 158$ kcal/mole of F_2. But this value does not take into account the amount of work done in bringing the two positive nuclei near each other. Nor can we expect to gain the whole electron affinity under conditions of electron sharing. We would be lucky to gain half since each atom has only a half interest in the electron pair. Experimentally, the energy released when two fluorine atoms form a bond is only 37.7 kcal/mole of F_2 formed—a reasonable fraction of the maximum energy possible.

The orbital representation for the two fluorine atoms in the F_2 molecule is

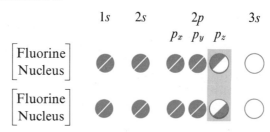

The overlapping half-filled p_z orbitals are enclosed in the shaded area. Now we can see why the chemical bond forms between two fluorine atoms. The electron affinity of a fluorine atom indicates that it is energetically favorable for a fluorine atom to acquire one more electron. Two fluorine atoms can realize a part of this energy decrease by *sharing* electrons. *All chemical bonds form because one or more electrons are placed so that they are attracted to two or more positive nuclei simultaneously.*

The interaction can also be represented easily by using the dot model of the atom. The electron-sharing is seen in Figure 16-2. In this discussion of covalent bonding in F_2, we have properly assumed that the 1s orbital in the fluorine atom is so tightly bound that it plays little role in the chemistry of fluorine. Only electrons in the 2 level are important. In the fluorine *molecule* each separate fluorine atom has two electrons in the 2s and six electrons in the three 2p orbitals. Some of the stability associated with a completed 2 level has been gained in the formation of the fluorine molecule from two fluorine atoms.

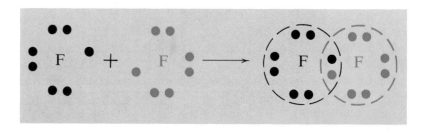

Fig. 16-2 **Sharing of electrons in the fluorine molecule.**

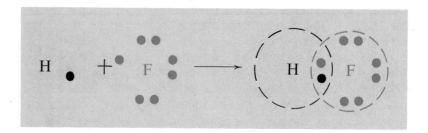

Fig. 16-3 **Sharing of electrons in hydrogen fluoride.**

All methods of representation show that when two atoms of fluorine share electrons, they each have "filled" all their valence orbitals and they have no additional bonding capacity. Hence, F_2 does not add a third or fourth atom to form F_3 or F_4. Fluorine is **univalent.**

Hydrogen fluoride can also be represented by the electron dot and orbital notations (see Figure 16-3). A census of the number of electrons near each atom in the molecule shows that the hydrogen atom has only two electrons nearby, whereas fluorine has eight. This is energetically desirable, however, because hydrogen has only one valence orbital, the $1s$ orbital. Two electrons just fill this orbital. A chemical bond results.

EXERCISE 16-1

Justify, using the orbital representation, the fact that no compound H_3F is formed.

16-2 THE BONDING CAPACITY OF THE SECOND-ROW ELEMENTS

In Chapter 8 a number of compounds of the second-row elements were identified. Formulas such as H_2O, NH_3, CH_4, CCl_4, and BF_3 were written. Now we have a basis for explaining in more detail why these compounds are formed.

16-2.1 The Bonding Capacity of Oxygen Atoms

The neutral oxygen atom has eight electrons. Six of these electrons occupy the $2s$ and $2p$ orbitals. They are much more easily removed than the two electrons in the $1s$ orbital because they are in the outer energy level, which is not yet filled. Oxygen, then, has six valence electrons. The $2s$ and $2p$ orbitals are the valence orbitals. They can accommodate the valence electrons in two ways, as follows:

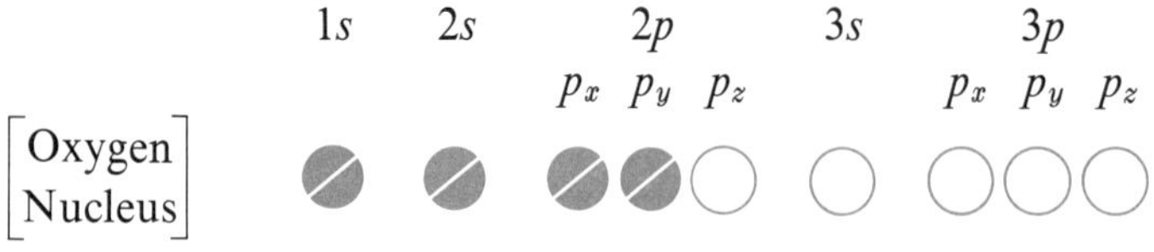

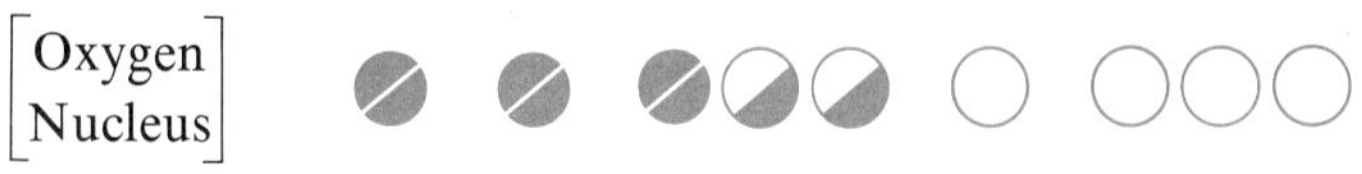

Since electrons repel each other, that electron configuration which keeps the electrons farther apart will be the lower in energy. A configuration with one electron in each of two separate orbitals (different regions in space) keeps the electrons farther apart than a configuration with two electrons in a single orbital (same region in space). We would expect the configuration for the oxygen atom having two unpaired electrons to be lower in energy than the configuration having all electrons paired. Experiment confirms this prediction. Much of the chemistry of oxygen can be interpreted more easily using the orbital model having two unpaired electrons in separate, half-filled orbitals:

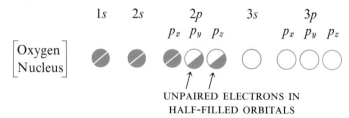

The differences in configuration between the two models are easily represented by the electron dot method. The arrangements below

$$\cdot \ddot{\underset{..}{O}} :$$

has a significantly lower energy than this configuration

$$\ddot{\underset{..}{O}} :$$

Suppose a single hydrogen atom approaches an oxygen atom in the lower energy state. Each atom has partially filled valence orbitals. Electron sharing can occur, placing electrons close to two nuclei simultaneously; hence, a stable bond can occur:

$$\text{H} \cdot + \cdot \ddot{\underset{..}{O}} : \longrightarrow \text{H} : \ddot{\underset{..}{O}} : \quad * \tag{5}$$

— HALF-FILLED ORBITAL

We see that the species HO† has one unpaired electron left. Combination of HO with another hydrogen atom gives water:

$$\text{H} \cdot + \text{H} : \ddot{\underset{..}{O}} : \longrightarrow \text{H} : \ddot{\underset{..}{O}} : \\ \phantom{\text{H} \cdot + \text{H} : \ddot{\underset{..}{O}} : \longrightarrow \text{H} : \ddot{}}\text{H} \tag{6}$$

The electron model suggests that the residual bonding capacity of HO has now been used up. We are not surprised to find that H₂O is very stable; it stands in striking contrast to the very reactive HO molecule. <u>Oxygen is said to be **divalent** in water</u>. Each atom in H₂O has filled its valence orbitals by electron sharing.

What happens if the HO unit combines with another HO unit? The electron dot diagram indicates the formula and the way the atoms are joined in the resulting compound. The equation is

$$\ddot{\underset{..}{O}} \cdot + \cdot \ddot{\underset{..}{O}} : \longrightarrow \left(:\ddot{\underset{..}{O}}\!:\!\ddot{\underset{..}{O}}: \right) \tag{7}$$

(with H atoms attached)

The compound produced by two HO units combining is the well-known bleaching agent hydrogen peroxide (H₂O₂). The model accurately predicts the existence of an O—O bond in H₂O₂.

---

* Electron color is convenient for purposes of electron bookkeeping. Remember all electrons in the molecule are indistinguishable except for their energy.

† A molecular species with a half-filled valence orbital is frequently called a **free radical**. The HO· unit is called the **hydroxyl radical**. It has been identified and studied as an intermediate in many reactions, such as those taking place in high temperature flames.

EXERCISE 16-2

Predict the structure of the compound S_2Cl_2 from the electron dot representation of the atoms.

The BONDING CAPACITY of NITROGEN ATOMS

These analyses of the chemistry of oxygen and fluorine can be used to predict the compound F_2O. The electron dot representation is

$$\overset{..}{:}\overset{..}{O}:\overset{..}{F}:$$
$$:\overset{..}{F}:$$

Again oxygen is divalent.

EXERCISE 16-3

Draw orbital and electron dot representations of each of the following molecules: OF, F_2O_2, HOF, and HFO_2. Which of these would you expect to be the most reactive? H_2O_2 decomposes easily to give H_2O and O_2. Why?

16-2.2 The Bonding Capacity of Nitrogen Atoms

As with the oxygen atom, the nitrogen atom is most stable when it has the maximum number of partially filled valence orbitals because the electrons are then as far apart as possible. The most stable state of the five valence electrons of the nitrogen atom is

[diagram: Nitrogen Nucleus with 1s (filled), 2s (filled), 2p (p_x, p_y, p_z each half-filled)]

It is straightforward to predict that nitrogen will form a stable hydrogen compound with formula NH_3. The similar compound NF_3 will also be formed. The electron dot formula for NH_3 is

$$\begin{array}{c} H \\ :\overset{..}{N}:H \\ H \end{array}$$

and the formula for NF_3 is

$$\begin{array}{c} :\overset{..}{F}: \\ :\overset{..}{N}:\overset{..}{F}: \\ :\overset{..}{F}: \end{array}$$

Nitrogen is **trivalent**.

EXERCISE 16-4

The molecule NH_2 has residual bonding capacity and is extremely reactive. The hydrazine molecule (N_2H_4) is much more stable. Draw an electron dot representation of the bonding of hydrazine. Draw its structural formula, showing which atoms are bonded to each other.

16-2.3 The Bonding Capacity of Carbon Atoms

Like oxygen and nitrogen, the carbon atom also is most stable when it has the maximum number of partially filled p orbitals. It has been experimentally shown that the lowest state for the carbon atom has a configuration of one electron in each p_x and p_y orbital. In this way electrons achieve a maximum average distance of separation.

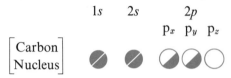

Now we can predict the chemistry of the carbon atom in this state. It should be divalent, forming CH_2 and CF_2. Let us consider CH_2:

$$H\!:\!\overset{..}{C}\!:\!H$$

The orbitals used are

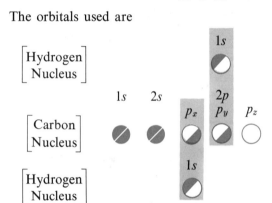

Here is a situation we have not yet met. After using the two available partially filled orbitals to form covalent bonds with hydrogen atoms, one vacant p orbital remains, and there is a filled, nonbonding s orbital.

What are the experimental consequences of this arrangement? It is found that both CH_2 and CF_2 can form as reaction intermediates at high temperatures, *but neither molecule has ever been obtained in a pure state. Both CH_2 and CF_2 are very reactive molecules.* This experimental fact can best be understood if we picture one of the electrons in the carbon 2s orbital as moving up to the $2p_z$ orbital with only a slight increase in the energy of the system. This process is called **promoting the electron.** As a result of the promotion process, the carbon atom has two more half-filled orbitals and thus the capacity to form *two* more covalent bonds.

$$\cdot\overset{\cdot}{\underset{H}{C}}\!:\!H + 2H\cdot \longrightarrow H\!:\!\overset{\overset{H}{\cdot\cdot}}{\underset{\underset{H}{\cdot\cdot}}{C}}\!:\!H \tag{8}$$

Each covalent bond formed increases the stability of the system significantly. The decrease in the energy of the system due to new bond formation more than compensates for the small energy required to promote the 2s electron to the $2p_z$ level. The net result is that carbon is **tetravalent.**

EXERCISE 16-5

Draw the electron dot representation of the reaction between CF_2 and two F atoms. How would the CF_2 molecule be pictured in the electron dot formulation *prior to* and *after* electron promotion?

EXERCISE 16-6

Draw electron dot formulas for the molecules CH_3, CF_3, CHF_3, CH_2F_2, and CH_3F. Which will be extremely reactive?

EXERCISE 16-7

Draw an electron dot and a structural formula for the ethane molecule (C_2H_6), which forms when two CH_3 molecules are brought together. Explain why C_2H_6 is much less reactive than CH_3.

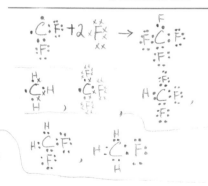

16-2.4 The Bonding Capacity of Boron Atoms

The boron atom presents the same sort of option in orbital occupancy as does carbon:

But the promoted electron configuration is somewhat higher in energy than the lower electron state, in which full use is made of the low-energy 2s orbital. In return for this higher energy the boron atom gains bonding capacity from promotion. Whereas a boron atom can form only one covalent bond in the configuration before promotion, it can form three covalent bonds in the configuration after promotion. Since each bond lowers the energy, the chemistry of boron is determined by the promoted electron configuration.

Now we can expect that boron will be trivalent. We predict that there should be a molecule such as BF_3:

$$:\!\ddot{F}\!: \\ B:\ddot{F}: \\ :\!\ddot{F}\!:$$

Furthermore, the fact that there is one *empty* orbital left in BF_3 suggests possible additional reactivity for this species. What kind of reactivity is expected? In BF_3 there is *an empty orbital but no unused full orbital.* Thus, it is *not* possible for BF_3 to generate half-filled orbitals by a simple electron promotion process. The result is that BF_3 is a stable molecule which does *not* add other fluorine atoms to form additional covalent bonds.

Let us review and contrast the reactions of CF_2 and BF_3. In CF_2 we have a structure with *one unused orbital* and *one unused electron pair* on carbon. When one electron of the unused pair moves into the unused orbital, *two half-filled orbitals are obtained.* These half-filled orbitals make CF_2 very reactive toward any species having one half-filled orbital. For example, it combines with F atoms or other CF_2 groups. In contrast, BF_3 has *one empty orbital* but *no unused electron pair* on boron. Therefore, it is *not* possible to obtain half-filled orbitals by promoting an unused electron. Consequently, BF_3 is *not* readily attacked by F atoms, H atoms, or even other BF_3 molecules. BF_3 can be obtained as a pure compound of relatively high stability. But the compound BF_3 is *not inert;* it combines readily with molecules or ions, such as NH_3 or F^-, which can share a complete electron pair. This point is discussed in Appendix 11.

You may be wondering about the hydrides of boron. The line of reasoning we used in considering compounds of hydrogen with fluorine, oxygen, nitrogen, and carbon suggests that hydrogen and fluorine behave similarly. A compound BH_3 should exist just as does BF_3. We had H_2O, F_2O, HF, F_2, CH_4, CF_4, and so on. Why have we not discussed BH_3? Strangely enough, no stable BH_3 species is known. The simplest boron hydride is B_2H_6, known as diborane. This molecule and other boron hydrides in the series challenged many chemical theories. It is only recently that progress in understanding these molecules has been made. We shall discuss the boron hydrides briefly in the next chapter, after we consider the structure of elemental boron.

EXERCISE 16-8

What reaction, if any, would you expect between an HO molecule and a CH_3 molecule?

16-2.5 The Bonding Capacity of Beryllium Atoms

The beryllium atom in its lowest energy configuration shows two electrons in the $2s$ orbital. All $2p$ orbitals are vacant. Like boron and carbon, beryllium can promote a $2s$ electron to a $2p$ level:

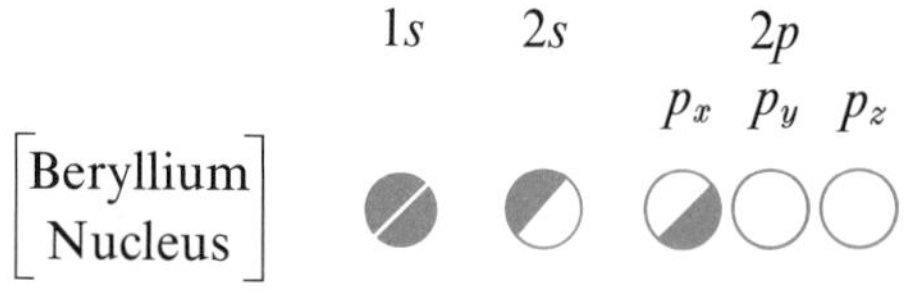

After promotion we would expect the compound BeF_2, which is indeed found at temperatures above $1000°K$. At lower temperatures, experiment shows that BeF_2 molecules tend to combine with each other to give a three-dimensional polymeric solid structure. BeH_2 is also a solid. It can best be explained in subsequent chemistry courses.

16-2.6 The Bonding Capacity of Lithium Atoms

The bonding capacity of a lithium atom is almost predictable from the foregoing arguments. Since the Li atom has just one valence electron, an LiF molecule might be expected to form at very high temperatures. Observation confirms the prediction.

16-3 TREND IN BOND TYPE AMONG THE SECOND-ROW FLUORIDES

All chemical bonds occur because electrons can be placed simultaneously near two nuclei. Yet, it is often true that this electron sharing is not exactly *equal* sharing. Sometimes the electrons tend to distribute somewhat nearer to one of the nuclei. We can understand this by comparing chemical bonding in gaseous fluorine (F_2) and in gaseous lithium fluoride (LiF).

16-3.1 The Bonding in Gaseous Lithium Fluoride

We have already treated the bonding in an F_2 molecule (Section 16-1.4). Since neither fluorine atom can completely pull an electron away from the other, they compromise and share a pair of electrons equally in a covalent bond. How does the chemical bonding in the lithium fluoride molecule compare?

As we have said, the Li atom has one valence electron, hence can share a pair of electrons with one fluorine atom. Thus, we can expect a stable gaseous molecular species, LiF.

$$\text{Li} : \ddot{\underset{..}{\text{F}}} :$$

However, the Li and F atoms attract electrons differently. This is shown by the ionization energies of these two atoms and the electron affinity of F.

$$401.5 \text{ kcal} + F(g) \longrightarrow F^+(g) + e^-(g) \quad (9)$$

$$124.3 \text{ kcal} + Li(g) \longrightarrow Li^+(g) + e^-(g) \quad (10)$$

$$F(g) + e^- \longrightarrow F^-(g) + 79 \text{ kcal} \quad (11)$$

Clearly, the F atom holds electrons much more strongly than does the Li atom. As a result, the electron pair in the gaseous LiF bond is more strongly attracted to the F than to the Li atom. The energy is lowered when the electrons spill toward the F atom. *When the bonding electrons move closer to one of the two atoms, the bond is said to have* **ionic character.**

In the most extreme situation, the bonding electrons move so close to one of the atoms that this atom has virtually the electron distribution of the negative ion. This is the case in gaseous LiF. Using an electron dot representation, we might write

$$Li^+ : \ddot{\underset{..}{\text{F}}} :^-$$

or

$$Li^+ \quad F^-$$

When a formula showing ions provides a useful basis for discussing the properties of a molecule, the bond in that molecule is said to be an **ionic bond.**

Remember that there is but one principle governing the formation of a chemical bond between two atoms: *All chemical bonds form because electrons are placed simultaneously near two positive nuclei.*

The term **covalent bond** indicates that the most stable distribution of the electrons (in terms of *energy*) between the two atoms is symmetrical. When the bonding electrons are somewhat closer to one of the atoms, the bond is said to have **ionic character.** The term **ionic bond** indicates the electrons are displaced so much toward one atom that the bonded atoms must be considered a pair of ions that are near each other. Figure 16-4 shows schematically how the valence electron distributions are pictured in covalent (F—F), partially ionic (F—H), and ionic (F—Li) bonds. Figure 16-4 also shows how the valence electrons might look in an instantaneous snapshot. In each type of bond, the electron-nucleus attractions account for the energy stability of the molecule.

Fig. 16-4 Electron distribution in various bond types.

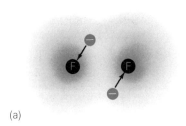

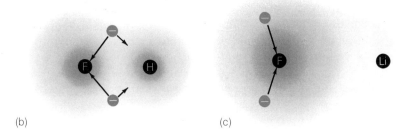

(a)　　　　　　　　　　(b)　　　　　　　　　　(c)

16-3.2 The Electric Dipole of the Ionic Bond

The spilling of negative electric charge toward one of the atoms in the ionic bond causes a charge separation. This can be represented crudely as in the last drawing in Figure 16-5. The LiF molecule is electrically positive at the lithium end and electrically negative at the fluorine end. It is said to possess an **electric dipole** and the molecule is called a **polar molecule.** The forces between polar molecules are much stronger than those between nonpolar molecules. The arrow representation of Figure 16-5 is commonly used and it is the simplest way of showing a bond dipole. The arrow indicates that the negative charge is mainly at one end of the bond. The directional property of the arrow implies that the force this molecule exerts on another molecule depends upon the direction of approach of the second molecule.

Fig. 16-5 Representations of the electric dipole of gaseous lithium fluoride.

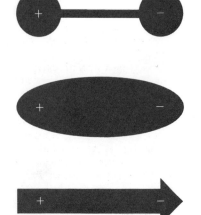

16-3.3 Ionic Character in Bonds to Fluorine

We can expect the effects just discussed in LiF to be at work in other fluorides. Ionization energies give us a rough clue to the electron-nucleus attractions in these bonds. Table 16-1 compares the ionization energies of each element of the second row with that of fluorine. The last column describes the type of chemical bond.

The trend from covalent to ionic-bond type shown in Table 16-1 greatly influences the trend in properties of the fluorine compounds. This trend is caused by the increasing difference between ionization energies of the two bonded atoms.

16-3.4 Ionic Character in Bonds to Hydrogen

In Chapter 8 the element hydrogen was characterized as a family by itself. Often its chemistry distinguishes it from the rest of the periodic table. We find this is the case when we attempt to predict the ionic character of bonds to hydrogen.

The ionization energy of the H atom, 313.6 kcal/mole, is quite close to that of the F atom; so we expect a covalent bond between these two atoms in HF. Actually, the properties of HF show that the molecule has a significant electric dipole, indicating significant ionic character in the bond. The same is true in the O—H bonds of water and, to a lesser extent, in the N—H bonds of ammonia. Examination of the properties of a number of compounds involving hydrogen indicates that the ionic character of bonds to hydrogen are roughly like that of bonds to an element having an ionization energy near 200 kcal/mole. *We cannot predict, then, the ionic character of bonds to hydrogen from its measured ionization energy.* The C—H bond has only a slightly ionic character. At the other end of the periodic table, gaseous lithium hydride is known to have a significant electric dipole, but with the electric dipole turned around. In LiH the electrons are spilled toward the H atom, leaving the Li atom with a partial positive charge. This is in accord with the low ionization energy of Li, 124.3 kcal/mole, which is well below the value of 200 kcal/mole that we have assigned to hydrogen. For our purposes, *it suffices to discuss the bonding of hydrogen in terms of an apparent ionization energy near 200 kcal/mole.*

16-4 MOLECULAR GEOMETRY

Molecular properties result from both the identity of the atoms in the molecule and their geometrical arrangement. Shapes of many molecules have been determined experimentally using methods summarized briefly in the last section of this chapter. On the basis of

**LINUS C. PAULING
(1901–   )**

No living chemist has contributed more to our understanding of chemical bonding than Linus C. Pauling. His ideas pervade every aspect of chemistry. They have won him many high awards, including the 1954 Nobel Prize in Chemistry.

Born in Portland, Oregon, Pauling received his Ph.D. from the California Institute of Technology, where he has pursued his illustrious career. One of the earliest chemists to recognize the importance of quantum mechanics in chemistry, Pauling focused his interest on the chemical bond. Many notions of modern chemistry—the quantitative meaning of electronegativity, ionic and covalent character of bonds, hydrogen bonding, metallic bonding, effective sizes of atoms in molecules and crystals—were developed in detail by Pauling. His later work on protein structure has been important in biochemistry.

Pauling has worked energetically to awaken the conscience of society to its new responsibilities in the nuclear age. His activities have at times attracted ridicule, when fear made his cause an unpopular one. In recognition of them, however, Pauling was awarded the 1962 Nobel Peace Prize, and thus became the second person in history to receive the Nobel Prize twice.

TABLE 16-1 BOND TYPES IN SOME FLUORINE COMPOUNDS

Compound	Bond	Ionization Energies (kcal/mole)		Bond Type
		Element Bonded to F	Fluorine	
F_2	F—F	F	401.5	401.5	COVALENT ↑ increasing ionic character
OF_2	O—F	O	313.8	401.5	
NF_3	N—F	N	335	401.5	
CF_4	C—F	C	259.5	401.5	
BF_3	B—F	B	191.2	401.5	increasing covalent character ↓
BeF_2	Be—F	Be	214.9	401.5	
LiF	Li—F	Li	124.3	401.5	IONIC

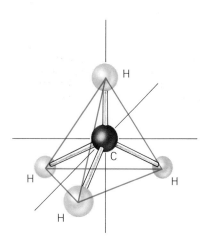

Fig. 16-6.1 **The geometry of the methane molecule, CH_4.**

Fig. 16-6.2 **A space-filling model of the methane molecule.**

this structural information, a number of different electron models have been proposed to help us understand and even predict molecular geometry.*

Chemists disagree rather vigorously over the relative values of these models. Some like one model; some like another. The plain truth of the matter is that we do not have a "correct" model for a system as complicated as a multi-electron atom. In every case a general theoretical picture is constructed *after* a large amount of experimental information about structure has been accumulated. The theory is then used to predict an unknown structure. Experimental information about structure for the particular molecule is obtained and compared with that predicted by the general theory. The value of the theory is judged by its ability to correlate and anticipate the new experimental results. In this book we shall focus our attention on a relatively simple, but very effective, old theoretical model which has been vigorously revived recently.† The effectiveness of this model is matched only by its simplicity.

16-4.1 Arrangement of Electrons in Orbitals

In our discussion of orbital shapes for the gaseous hydrogen *atom*, we recognized one spherical $2s$ and three dumbbell shaped $2p$ orbitals. The $2p$ orbitals were at a 90° angle to each other. This known orbital geometry is restricted to the hydrogen atom because of its makeup—one proton and one electron. Additional electrons and protons create electrical fields which greatly distort the electron or orbital arrangement. For example, in CH_4 we might expect an H—C—H angle of 90° since the $2p$ carbon orbitals forming the C—H bonds presumably make a 90° angle. (Section 16-2.3.) In fact the experimental H—C—H bond angle is 109°28′, not the 90° predicted by p-orbital geometry. Further, we do not find that the three hydrogen atoms bound to p orbitals differ from the hydrogen bound to an s orbital. *Experiment shows that every hydrogen atom in CH_4 is the same. All H—C—H angles in the molecule CH_4 are 109°28′.* The molecule has four hydrogen atoms arranged at the corners of a regular tetrahedron. The carbon atom is in the center of the tetrahedron. (See Figure 16-6.) What kind of model can interpret these and related observations?

The theory we shall consider proposes that *the arrangement of atoms around any given central atom is determined primarily by the repulsive interactions between electron pairs in the valence shell of that central atom.* For example, the geometry of CH_4 would be determined by the repulsion between electron pairs making up each of the four C—H bonds. The electron pairs move as far apart as possible and the tetrahedron results.

---

* The *Journal of Chemical Education* for December, 1968, contains two papers describing different approaches to molecular geometry. See L. S. Bartell, page 754, and H. S. Bent, page 768. Molecular geometry is still an area of active research.
† Two of the most articulate and able spokesmen for this point of view are Professor R. J. Gillespie of McMaster University in Canada and Professor R. Nyholm of University College, London, England. See *J. Chem. Ed.,* **40,** 295 (1963).

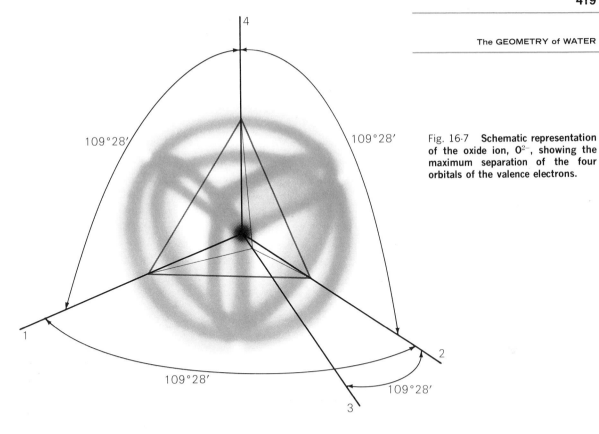

Fig. 16-7 Schematic representation of the oxide ion, O^{2-}, showing the maximum separation of the four orbitals of the valence electrons.

The geometry of CH_4 is fairly obvious. Can the theory work to give us the structure of a less obvious molecule such as H_2O? In visualizing the geometry of water, let us start with an imaginary oxide ion containing eight valence electrons. Since only two electrons can exist in any one orbital, we must have four orbitals. Thus, for an oxide ion the one $2s$ and three $2p$ orbitals would combine or be "hybridized" to give *four identical orbitals,* each containing two electrons. We know that electron clouds repel each other; hence, it would seem logical to assume that the electron clouds defining these four identical orbitals of O^{2-} would assume a position with a maximum distance between the clouds. The line defining the central axis of each orbital would make an angle of 109°28′ with the central axis of every other orbital in that atom. (See Figure 16-7.) We would expect an essentially spherical distribution of electron charge around the oxygen core in O^{2-}. Theory and experiment agree for those structures with eight electrons in the outer level. Oxide ions are spherically symmetrical!

16-4.2 The Introduction of Protons into Orbitals— The Geometry of Water

Now what happens if a proton is made to approach an oxide *ion?* As the proton approaches, two of the electrons in one of the orbitals will be pulled in toward the approaching proton. *The formerly spherical electron cloud is distorted.* The electron cloud which represented this particular "hybrid orbital" will stretch out toward the proton and shrink downward toward the line between O and H nuclei. The electron cloud of the bonding orbital will become longer and thinner as the proton approaches. As the orbital shrinks in diameter, the other electron clouds expand to utilize the newly available space (see Figure 16-8.1).

Let us now bring a second proton toward the OH⁻ unit just formed. Again, an electron pair in one of the three unused orbitals will become involved in bond formation with the proton. Again, the electron cloud in this newly utilized bonding orbital will stretch out toward the proton and become thinner. The free electron pairs will force together the two somewhat shrunken bonding orbitals linked to each of the two protons (Figure 16-8.2).

The introduction of the proton causes a contraction in the angle defined by the H, O, and H atoms. The two electron clouds forming bonds with protons are contracted and forced together. The resulting H—O—H angle in water should be *less than* 109°28'. The experimental value is 104°30'. (See Figure 16-8.2.) Theory and experiment are in qualitative agreement because the theory was constructed from the observed facts.

The foregoing discussion tells us what the process might be when a water molecule is formed from an oxide ion and two protons; yet the final geometry is the same, no matter how the water molecule is formed. We note that the geometry of water can be understood

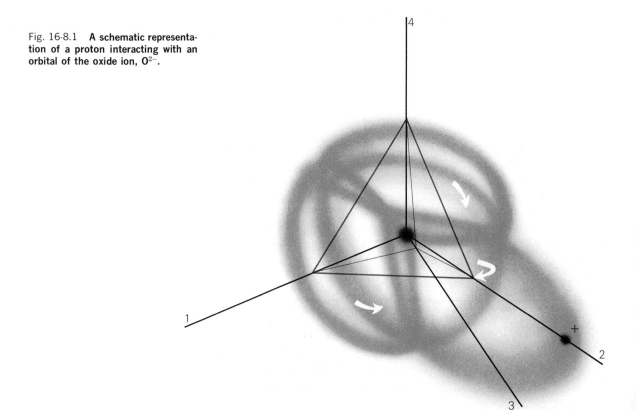

Fig. 16-8.1 **A schematic representation of a proton interacting with an orbital of the oxide ion, O^{2-}.**

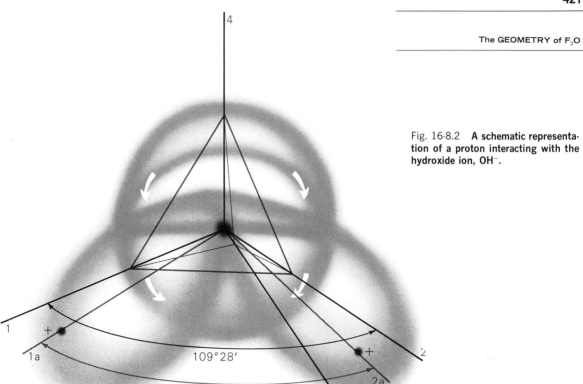

Fig. 16-8.2 **A schematic representation of a proton interacting with the hydroxide ion, OH⁻.**

in terms of the repulsion of the four pairs of valence electrons around the oxygen core (O^{6+}). Two pairs of electrons are used in binding protons and two pairs are free. The free pairs are the largest in size.

16-4.3 The Geometry of F_2O

It seems quite clear that if the approaching atom or ions attract electrons more strongly than does a proton of water, the electron cloud will stretch out and shrink in diameter to an even greater degree than in the case of water. If the electron clouds used in bonding are contracted more during bond formation, they can be pushed closer together by the expanding clouds of the unused electron pairs. This argument leads us to believe that if we visualize interaction of an F^+ with an O^{2-}, the bonding electron cloud will be compressed somewhat more along the bond axis. Electrons will then actually move well out toward F^+. This follows because of the very great attraction of F^+ for electrons. (Remember that the fluorine atom has an ionization energy of 402 kcal/mole. This value is second only to neon in the second row of the periodic table.) The electron-cloud model then suggests that the electron cloud as it moves out toward F^+ will become thinner and longer. It will then be pushed together by the expanding free electron clouds. An angle of even less than 104°30′ is expected. The experimental value is 103°. While a contraction in angle is observed, the change is so small that its significance may be questioned.

The molecules can be conveniently represented using these models:

$\angle$ H—O—H = 104°30'

$\angle$ F—O—F = 103°

A line is drawn between the oxygen atom and each hydrogen atom in water to indicate that a chemical bond holds these two atoms together. No line is drawn between the two hydrogen atoms since we think they are not directly bonded to each other. The same is true of F_2O.

16-4.4 The Geometry of NH_3 and of NF_3

In considering the geometry of NH_3 and NF_3, it is convenient to imagine the interaction of protons or of *positive* fluorine *ions** with a hypothetical nitride ion, N^{3-}. If three protons were to approach the nitride ion one at a time, electron clouds in three orbitals would shrink in diameter around each N—H bond. *The unused electron pair would then expand, pushing the compressed electron clouds together.* The H—N—H angle in ammonia (NH_3) should be less than 109°28'. The experimental value is 107°.

Identical arguments indicate that NF_3, having three fluorine-nitrogen bonds and one free electron pair, should have a bond angle smaller than 107° (Figure 16-9). The experimental value is 102°, almost equal to the F—O—F angle in F_2O. One prediction of the theory—a smaller angle in NF_3 than in NH_3—is verified.

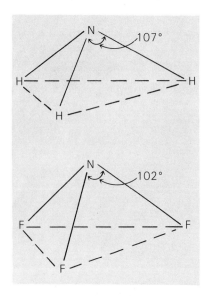

Fig. 16-9 The geometries of NH_3 and NF_3.

16-4.5 The Geometry of CH_4 and CF_4

In both CH_4 and CF_4 all four orbitals are identical. We are not surprised that in CH_4 the angle between *any two* C—H bonds is 109°28' and in CF_4 the angle between *any two* C—F bonds is 109°28'. This structure is called **tetrahedral** because the four hydrogen atoms occupy the positions at the corners of a *regular tetrahedron*. The structure is shown in Figure 16-6.

16-4.6 The Geometry of BF_3 and of BeF_2

In BF_3 we have to consider only three pairs of electrons around boron. Each electron pair is involved in forming a bond with fluorine. The arrangement permitting maximum distance between three electron pairs would be a **planar molecule** with a fluorine at each vertex of an *equilateral triangle*. Experiment confirms this. All atoms of the molecule lie in a plane. (See Figure 16-10.)

Fig. 16-10 The structures of BF_3 and BeF_2.

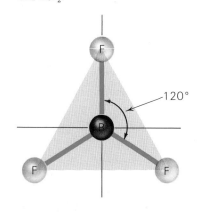

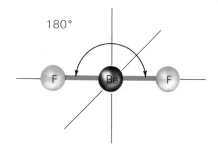

* This does not imply that we have F^+ ions or N^{3-} ions in the final molecule. Electrons move out to form a covalent bond. The cloud ultimately is probably closer to the fluorine than the nitrogen.

In gaseous BeF$_2$ there are only two electron pairs to keep apart and no free electron pairs. Of course, each electron pair is involved in a bond to fluorine. The arrangement giving maximum distance between electron clouds would be that of the linear molecule shown in Figure 16-10. Gaseous BeF$_2$ is a **linear molecule.**

16-4.7 A Summary of Molecular Geometry

In summary, the structure of most molecules can be derived by using the **valence-shell electron-pair repulsion theory.** The following procedure permits application of the theory.

(1) Write an electron dot formula for the molecule showing electron pair bonds and free electron pairs.
(2) Count the number of electron pair bonds and the number of free electron pairs around the central atom.
(3) If the molecule contains only two atoms, it must be linear. If the molecule contains three atoms and the number obtained in (2) is 2, the molecule is linear [Figure 16-11(a)]. If the molecule has three atoms and the number obtained in (2) is 3, the angle is somewhat less than 120° [Figure 16-11(b)].

Shapes corresponding to various numbers of atoms (X) and electron pairs (E) around a central atom (A) are shown in Figure 16-11.

16-4.8 A Comment on Notation

As you will remember, the orbital description of the beryllium atom used in forming BeF$_2$ showed one electron in a 2s orbital and one electron in a 2p orbital. Experiment shows that the BeF$_2$ molecule is linear. Chemists sometimes summarize these facts by saying that the combination of one s and one p orbital gives two **linear orbitals.** The bonding is sometimes called ***sp*** **bonding.** A linear molecule is implied by the *sp* notation.

In considering boron bonding, we noted that the promoted boron atom has one electron in a 2s orbital and two electrons in two 2p orbitals. This combination of one s and two p orbitals gave a molecule with triangular geometry. For this reason the bonding is sometimes called ***sp*^{2}** **bonding.** The *sp*^{2} notation implies **planar triangular geometry.**

In forming CH$_4$ we had to use one s and three p orbitals. This leads to the designation ***sp*^{3}** **bonding** for **tetrahedral geometry.**

16-5 MOLECULAR SHAPE AND ELECTRIC DIPOLES

Consider the fluorides of the second-row elements. There is a continuous increase in ionic character of the bonds formed by fluorine with elements F, O, N, C, B, Be, and Li. (Look up ionization energies of these elements in Table 16-1.) This ionic character results in an electric dipole in the bond. The *molecular dipole* will be determined by the sum of the dipoles of all bonds if this addition takes

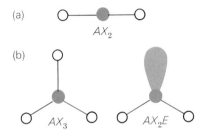

Fig. 16-11 The general shapes of molecules of the nontransition elements.

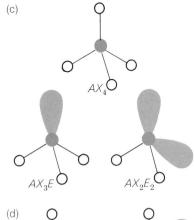

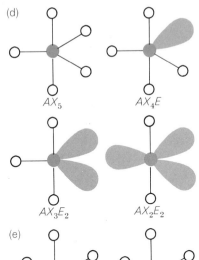

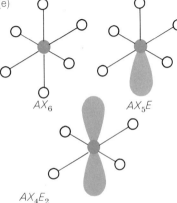

into account the geometry of the molecule. Since the properties of the molecule are strongly affected by the molecular dipole, we shall investigate its relationship to the molecular architecture and the ionic character of the individual bonds. For this study we shall begin at the left side of the periodic table. Information on geometry and bonding is summarized in Table 16-2.

TABLE 16-2 BONDING AND MOLECULAR GEOMETRY FOR SOME FLUORINE COMPOUNDS

Element	Bonding Orbitals	Bonding Capacity	Number of Free Electron Pairs in Valence Shell	Shape of Fluoride Molecule	Formula
He	none	0	0	no fluoride known	He
Li	s	1	0	linear, diatomic	LiF
Be	sp	2	0	linear, triatomic	BeF_2
B	sp^2	3	0	planar, triangular	BF_3
C	sp^3	4	0	tetrahedral	CF_4
N	sp^3*	3	1	pyramidal	NF_3
O	sp^3*	2	2	angular, bent	OF_2
F	p	1	3	linear	F_2
Ne	none	0	4	no fluoride known	Ne

* Nearly sp^3 with one orbital containing a free pair of electrons.

16-5.1 The Molecular Dipole of LiF

The lithium fluoride bond is highly ionic in character because of the large difference in ionization energies of lithium and fluorine. Consequently, gaseous lithium fluoride has an unusually high electric dipole.

16-5.2 The Molecular Dipole of BeF$_2$

The beryllium-fluorine bond is also highly ionic in character. However, there are two such Be—F bonds, and the electrical properties of the entire molecule depend upon how these two bonds are oriented to each other. We must find the "geometrical sum" of these two

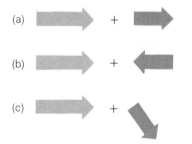

Fig. 16-12 The geometrical sum of dipoles.

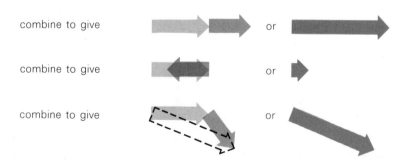

bond dipoles. The geometrical sum of two arrows can be understood with the aid of Figure 16-12. Figure 16-12(a) shows how two arrows that point in the same direction combine to give a longer arrow. Figure 16-12(b) shows how two arrows that point in opposite directions combine to give a shorter arrow. Figure 16-12(c) shows how two arrows that are not parallel add to give an arrow in a new direction. In the linear, symmetrical BeF₂ molecule, the two bond dipoles point in *opposite* directions. Since the two bonds are equivalent, their sum is zero, as shown in Figure 16-13; hence, the molecule has no *net* dipole; the molecular dipole is zero.

Fig. 16-13 **The absence of a molecular dipole in BeF₂.**

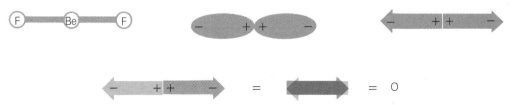

16-5.3 The Molecular Dipoles of BF₃ and CF₄

Both these molecules are thought to have moderate amounts of ionic character in each bond. Yet each molecular dipole is exactly zero. Careful consideration of the molecular geometry shows that there is a complete cancellation of the bond dipoles. This cancellation is shown in Figure 16-14 for BF₃.

Fig. 16-14 **The absence of a molecular dipole in BF₃.**

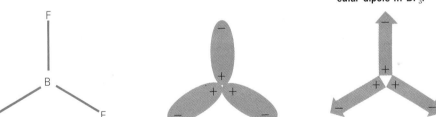

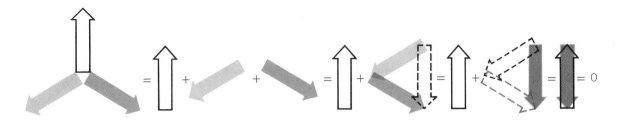

16-5.4 The Molecular Dipole in F₂O

Since F₂O is a bent molecule, the two bond dipoles do not cancel each other as they do in BeF₂ (Figure 16-13). On the other hand, the ionization energies of oxygen and fluorine are not very different, so the electric dipole of each bond is small in magnitude. These add together, according to their geometry or the "direction" of their

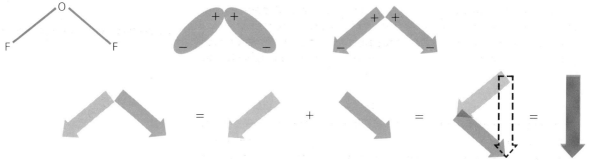

Fig. 16-15 The molecular dipole of F_2O.

"arrows," to give a polar molecule, as shown in Figure 16-15. (The polar molecule has a net molecular dipole.)

EXERCISE 16-9

Water has a high molecular dipole while CO_2 has zero molecular dipole. Can you rationalize these facts?

16-6 MULTIPLE BONDS

In determining the bonding capacity of a given atom from the second row, we counted the number of hydrogen or fluorine atoms with which the atom will combine. Oxygen combines with *two* hydrogen atoms to give water; oxygen is described as **divalent**. Carbon combines with *four* hydrogen atoms to give methane (CH_4); carbon is described as **tetravalent**. In each case, the bond formed between the hydrogen atom and the central element is a single bond—two atoms share an electron pair. Similarly, one can recognize a single-bonded structure, giving C—H and C—C bonds, in the structure of the compound ethane (C_2H_6). Its structure is represented as

$$\begin{array}{c} H\ \ H \\ H:\overset{..}{C}:\overset{..}{C}:H \\ H\ \ H \end{array} \quad *$$

Note that all bonds are single bonds; they are made with a *single* electron pair.

16-6.1 The Carbon-Carbon Double Bond

Another compound containing hydrogen and two carbon atoms is ethylene (C_2H_4). Note that it has *two* hydrogen atoms *less than* does ethane (C_2H_6). Suppose we write an electron dot representation of ethylene following the pattern established for ethane:

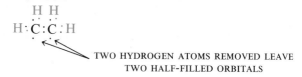

TWO HYDROGEN ATOMS REMOVED LEAVE TWO HALF-FILLED ORBITALS

---
* Remember again that color is only a bookkeeping convenience. All H atoms are identical and all C atoms are identical.

This formula has two unpaired electrons, thus unused bonding capacity. This objectionable situation could easily be rectified by the pairing of the unused electrons, thus forming an additional shared-electron bond. Now the carbon atoms are joined by a double bond:

$$\text{H}\!:\!\overset{\cdot\cdot}{\underset{\cdot\cdot}{C}}\!:\!:\!\overset{\cdot\cdot}{\underset{\cdot\cdot}{C}}\!:\!\text{H}$$

Much evidence supports this proposal. The carbon-carbon bond length in C_2H_4 is 1.34 Å while that in C_2H_6 is 1.54 Å. The bond in C_2H_4 is 0.20 Å shorter as a result of increased electron density between carbons. The vibrational frequency of the $\text{C}=\text{C}$ bond (1,650 cm^{-1}) is higher than that of the $-\text{C}-\text{C}-$ bond (1,200 cm^{-1}). If we remember the relationship $E = h\nu$, we shall see that a higher vibrational frequency corresponds to the utilization of more energy in the vibration of carbon atoms joined by a double bond. *The $C=C$ double bond in C_2H_4 is stronger than the $C-C$ single bond in C_2H_6.* More energy is required to break the $\text{C}=\text{C}$ bond in C_2H_4 (146 kcal/mole) than the $-\text{C}-\text{C}-$ bond in C_2H_6 (83 kcal/mole). On the other hand, we shall find that although the breaking of a $\text{C}=\text{C}$ bond to give $\text{C}\!:$ fragments requires much energy, the $\text{C}=\text{C}$ bond is readily attacked and *opened up by many reagents* to give a more stable molecule. Such a new molecule contains two *new covalent* C—X bonds instead of the one extra C—C bond. Consider, for example, the reaction with hydrogen:

$$\overset{\text{H}}{\underset{\text{H}}{>}}\!\text{C}=\text{C}\!\overset{\text{H}}{\underset{\text{H}}{<}} + \text{H}-\text{H} \longrightarrow \text{H}-\overset{\text{H}}{\underset{\text{H}}{\text{C}}}-\overset{\text{H}}{\underset{\text{H}}{\text{C}}}-\text{H} \qquad (12)$$

The formation of two new C—H bonds more than compensates for the destruction of one of the two C—C bonds in ethylene and of the H—H bond in gaseous hydrogen.

For this reason, *the double bond is a source of reactivity* in a molecule. In ethane (C_2H_6) all bonds are normal single bonds. Experiment shows that ethane is fairly unreactive. It reacts only when treated with quite reactive species (such as free chlorine atoms), or when it is raised to excited energy states by heat (as in combustion). Ethylene (C_2H_4), on the other hand, reacts readily with many chemical reagents. Any reagent which can utilize electrons from one of the double-bond pairs to form two new covalent bonds will attack ethylene very readily. Chlorine or bromine, for example, will be picked up very easily by ethylene to give dichloroethane or dibromoethane:

$$\overset{\text{H}}{\underset{\text{H}}{>}}\!\text{C}=\text{C}\!\overset{\text{H}}{\underset{\text{H}}{<}} + \text{Cl}-\text{Cl} \longrightarrow \text{H}-\overset{\text{H}}{\underset{\text{Cl}}{\text{C}}}-\overset{\text{H}}{\underset{\text{Cl}}{\text{C}}}-\text{H} \qquad (13)$$

428

Chapter 16 / MOLECULAR ARCHITECTURE: GASEOUS MOLECULES

The electrons of the double bond seem to be accessible. We find that the typical reactions of ethylene are with those reagents which *seek* electrons, oxidizing agents. The double bond is readily oxidized by such oxidizing agents as potassium permanganate or potassium dichromate at ordinary temperatures. But at ordinary temperatures ethane, *without* the double bond, is completely unreactive to the same reagents.

16-6.2 The Geometric Consequences of the Carbon-Carbon Double Bond—Structural Isomers

Experiment shows that the ethylene molecule is planar—the four hydrogen atoms and the two carbon atoms all lie in one plane. The twisting of CH_2 groups around the C—C axis is difficult; otherwise, the molecule would not retain its flat form. The ball, spring, and stick model shown in Figure 16-16 shows the rigidity of the double-bonded structure. Compare it with a ball-and-stick model for ethane, in which rotation around the C—C bond is relatively easy.

It is possible to replace hydrogen atoms of ethylene with halogen atoms. One such compound has the formula $C_2H_2Cl_2$. Examination of the model for ethylene suggests three possible ways

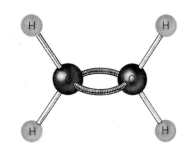

Fig. 16-16.1 A model of an ethylene molecule, C_2H_4.

Fig. 16-16.2 A model of an ethane molecule, C_2H_6.

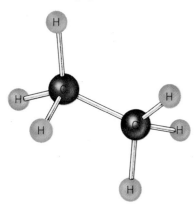

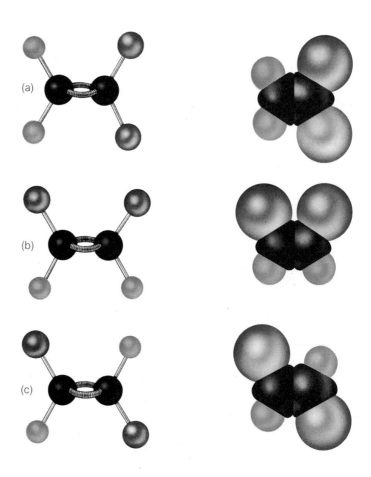

Fig. 16-17 Models of the three isomers of dichloroethylene, $C_2H_2Cl_2$.

in which the two chlorine atoms can be arranged in $C_2H_2Cl_2$. This is shown in Figure 16-17. Experimentally, three different compounds with this formula are found, and structural studies confirm the existence of these three configurations. Note that all three compounds have the same formula, $C_2H_2Cl_2$. All are called dichloroethylene. *Different compounds with the same formula are called* **isomers.**

In one $C_2H_2Cl_2$ molecule two chlorines are attached to the *same* carbon [Figure 16-17(a)]. In the remaining two molecules [Figure 16-17(b) and (c)] chlorines are attached to *different* carbon atoms. Isomers in which the actual bonding arrangements among atoms differ—for example, two chlorines to a single carbon in one case, and one chlorine to each of two carbons in the other—are called **structural isomers.** The structural isomer with chlorines attached to two different carbons has two different forms—the *cis*-form—two chlorines on the *same side* of the double bond [Figure 16-17(b)]—and the *trans*-form—chlorine atoms on *opposite sides* of the double bond [Figure 16-17(c)].

16-6.3 The Triple Bond in Nitrogen

Multiple bonds are not restricted to carbon compounds. Let us investigate this by writing an electron dot formula for nitrogen. Suppose nitrogen had a single bond; the structure would be

$$:N:N:$$

The problem we recognized in C_2H_4 is present here, only it is worse because each nitrogen atom has *two* half-filled orbitals. The solution suggested for ethylene can be applied here to give a **triple bond:**

$$:N:::N:$$

or $N\equiv N$. The triple bond between nitrogen atoms in elemental nitrogen is very strong and accounts for the very high dissociation energy of N_2.

$$N_2(g) + 226 \text{ kcal} \longrightarrow 2N(g) \qquad (14)$$

The energy of dissociation is six times larger than the energy of the single N—N bond in hydrazine:

$$H_2N-NH_2 + 38 \text{ kcal} \longrightarrow 2H_2N \qquad (15)$$

Indeed, we can understand the very inert nature of nitrogen at low temperatures in terms of the very strong triple bond in the elemental nitrogen molecule. Similarly, a good part of the high reactivity of fluorine can be ascribed to the very weak single bond between two fluorine atoms:

$$F_2(g) + 37 \text{ kcal} \longrightarrow 2F(g) \qquad (16)$$

16-7 METHODS OF STRUCTURE DETERMINATION

Much has been said about molecular structure in earlier parts of this book. How do scientists know just where atoms are in a molecule? How sure are we that the models constructed really represent molecular geometry? Many types of experiments give structural information. Most such experimental techniques are so complex that a detailed study of them here is impossible. We shall, however, summarize the general ideas associated with several well-known methods for establishing structure. Some of these techniques will be applied to specific compounds in Chapter 18. Chemical evidence, infrared spectroscopy, and nmr spectroscopy will be used to establish the structure of an organic molecule.

16-7.1 Structural Information from Chemical Reactions

You learned in Section 16-6.2 that three distinct substances of formula $C_2H_2Cl_2$ could be formed. You also saw that the model for ethylene indicates the existence of three and only three distinct isomers. Agreement between chemical fact and structural prediction based on a given model (*i.e.*, number of isomers) provides experimental support for the model. Indeed, most early information on structure in chemistry was obtained using this very method. Today the chemical information is usually supplemented by more direct physical methods which can reveal actual atomic positions and identify individual isomers.

16-7.2 X-Ray Diffraction Methods

As you will remember from Chapter 15, X rays are light waves of frequencies near 10^{18} vibrations/sec and wavelengths near 10^{-8} cm. Such X-ray waves, when reflected from the surface of a crystal, give patterns on a photographic film. The appearance of the pattern is determined by the spacings of the atoms in the crystal and their geometric arrangement. The pattern is obtainable only with X rays because it results from scattering effects that occur only if the wavelength of the light is close to the atomic separations within the crystal. Therefore, a knowledge of the wavelength of the X-ray light permits an interpretation of the pattern in terms of atomic packing.

This method of study, when properly conducted, provides final evidence of molecular structure. After a proper single crystal X-ray study, atomic arrangement, bond angles, and bond distances are known with a high degree of certainty. The method is particularly applicable to crystalline solids. X-ray methods established the structure of NaCl, diamond, graphite, and recently, many biologically important substances.

16-7.3 Electron Diffraction

As you will recall from Chapter 15, moving electrons have wave character and will undergo diffraction just like X rays. Electron-diffraction methods are particularly appropriate to a study of gases and give the same type of structural information for gases that X rays give for solids.

16-7.4 Infrared Spectroscopy

Again, you will recall from Chapter 15 that light of longer wavelength than red light is called infrared radiation. Frequencies of infrared radiation range from about 2×10^{13} to 12×10^{13} vibrations/sec. When photons with this much energy* are absorbed by molecules, the atoms vibrate back

---

* Remember that $E = h\nu$, where E is the energy of a photon, h is Planck's constant, and ν is light frequency.

and forth against each other.* The specific frequency needed to set a system in motion is determined by the type of atomic motion, the mass of the atoms, the shape of the molecule, and the strength of the chemical bonds linking the atoms. Through a study of the precise frequencies at which absorption occurs, it is often possible to obtain structural information. Infrared data are easy to obtain but provide less detailed structural information than do X-ray data.

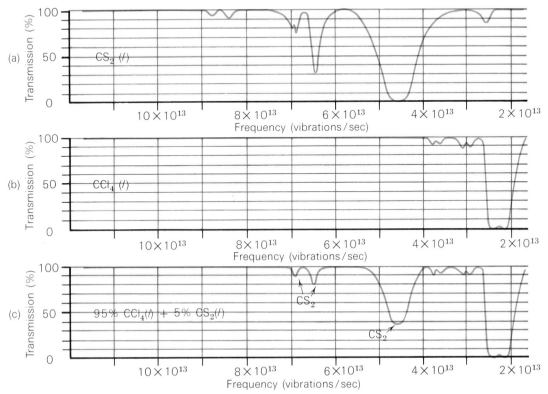

Infrared spectra are very widely used for compound identification. Each molecule has a characteristic spectrum, which can be used to recognize molecular species. In Figure 16-18(a), we see the spectrum of carbon tetrachloride (CCl_4); in Figure 16-18(b), the spectrum of carbon disulfide (CS_2); and in Figure 16-18(c), that of a mixture containing 95 percent CCl_4 and 5 percent CS_2. As you can easily see, the presence of CS_2 in CCl_4 is not difficult to recognize.

Fig. 16-18 Infrared absorption spectra of liquid carbon tetrachloride, CCl_4, carbon disulfide, CS_2, and a mixture of the two.

The value of infrared spectra in identifying substances, verifying purity, and quantitative analysis rivals their usefulness in learning about molecular structure. The infrared spectrum is as important as melting point in characterizing a pure substance. Thus, infrared spectroscopy has become an important addition to the many techniques used by the chemist.

16-7.5 Microwave Spectroscopy

If you check the electromagnetic spectrum in Table 15-2 on page 368, you will find a group of *frequencies* lower than infrared but higher than radio waves. This region of the spectrum, known as the **microwave region,** covers frequencies from about 3×10^9 to 3×10^{11} vibrations/sec. Photons of this frequency carry enough energy to make a gaseous molecule *rotate* around various axes, but not quite enough energy to induce molecular *vibrations*. Precise knowledge of the microwave absorption frequencies permits identification of a geometric pattern which can be associated with

* Remember that a fluctuating magnetic field can make a compass needle vibrate.

the observed energy of rotation (see Figure 16-19). Very precise information on both the structure and electric dipole of a molecule can be obtained from microwave spectroscopy, but the technique is applicable only to a limited number of gaseous molecules.

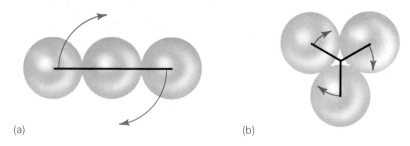

Fig. 16-19 The energy required to rotate a molecule is larger in the extended form (a) than in the compact form (b).

16-7.6 Nuclear Magnetic Resonance

In the past fifteen years another spectroscopic technique has found extensive use in structural studies and compound identification. This is the technique called **nmr**, or **nuclear magnetic resonance** spectroscopy. A nucleus is a charged body. *If a charged body moves, it becomes a small electromagnet.* Since some nuclei appear to spin, they behave as small electromagnets and will interact with an applied magnetic field. Consider that the nucleus is a small bar magnet which can be placed between the poles of a very strong, large magnet. The small bar magnet will orient itself with the north pole pointing toward the south pole of the big magnet and the south pole pointing toward the north pole (see Figure 16-20). If we want to turn the little magnet around, we must grasp it and turn it—energy is expended; work is done on the system.

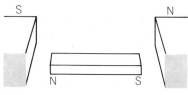

Fig. 16-20 A small bar magnet in a large magnetic field.

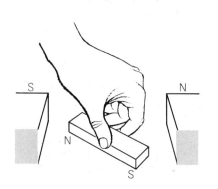

In dealing with a nucleus, our hand is replaced by radiation of appropriate frequency. Hence, one examines the frequencies of the electromagnetic radiation which will flip over the nuclear magnet (make it change energy levels). This is the **resonance absorption frequency** of the nucleus. This frequency is dependent upon the strength of the field of the large magnet, the nature of the small nuclear magnet, and the arrangement of other nuclei and electrons in the region of the nucleus under study. The fact that the frequency is partially dependent upon the nuclei and electrons in the region of the nucleus under study makes the technique of great value in determining molecular structure and in identifying compounds. Nuclear magnetic resonance is now as widely used as infrared spectroscopy in most chemical laboratories. Its application will be illustrated in Chapter 18.

16-8 HIGHLIGHTS

We have explored the nature of that interaction between atoms giving rise to a chemical bond. A **covalent bond** results from a shared electron pair. When an electron is transferred more or less completely from one atom to another, ions are formed. Electrostatic attraction between ions gives an **ionic bond.** All chemical bonds form because one or more electrons are placed so as to "feel" electrostatic attraction of two or more positive nuclei simultaneously. Using either an **orbital representation** or an **electron dot representation** of bonding, the bonding capacity of the second-row elements can be discussed and their chemistry rationalized.

Unequal sharing of electrons between atoms gives rise to charge separation; a **bond dipole** is created. Proper addition of bond dipoles gives the **molecular dipole**. Such addition must take account of both size and orientation of the bond dipoles. Molecular geometry for simple molecules can be understood using a model which maximizes the distance between electron pairs. Both free pairs and pairs involved in bond formation must be considered. A **single bond** results from the sharing of an electron pair between two atoms; a **double bond**, when two electron pairs are shared; and a **triple bond**, when three electron pairs are shared.

Methods of structure determination are summarized.

QUESTIONS and PROBLEMS

QUESTIONS AND PROBLEMS

1 Which one of the following statements is *false* as applied to this equation?

$$H_2(g) \rightleftharpoons H(g) + H(g) \quad \Delta H = 103.4 \text{ kcal}$$

(a) The positive ΔH means the reaction is endothermic. (b) Two grams of $H(g)$ contain more energy than 2 g of $H_2(g)$. (c) Weight for weight, $H(g)$ is a better fuel than $H_2(g)$. (d) The spectrum of $H_2(g)$ is the same as the spectrum of $H(g)$.

2 Determine the number of attractive and repulsive forces in LiH.

3 What energy condition must exist if a chemical bond is to form between two approaching atoms?

4 Give the orbital and also the electron dot representations for the bonding in these molecules: Cl_2, HCl, and Cl_2O.

5 Using the electron dot representation, show a neutral, a negatively charged, and a positively charged HO group.

6 Draw the orbital representation of the hydrazine molecule (N_2H_4).

7 Draw the electron dot representation of hydroxylamine (H_2NOH).

8 Would you expect N_2H_4 to react with BF_3? Explain using an electron dot representation.

9 Draw the electron dot representation of boron tri-iodide. Predict the structure of boron tri-iodide.

10 Suppose that by some mysterious process* you were able to obtain a large number of CH_2 molecules in a given container. What reaction would you anticipate?

11 Predict the formula and structure of the chloride of silicon. What orbitals does silicon use for bonding?

*Such molecules are formed at temperatures near 1500°K.

12 Predict the structure for the product obtained when two NO_2 molecules combine to give a single molecule, N_2O_4.

13 Draw the orbital representations of gaseous (a) sodium fluoride and (b) beryllium fluoride (BeF_2).

14 In general, what conditions cause two atoms to combine to form: (a) a bond that is mainly covalent, (b) a bond that is mainly ionic, and (c) a polar molecule?

15 What type of bonding would you expect to find in gaseous MgO? Explain.

16 Considering comparable oxygen compounds, predict the shape of H_2S and H_2S_2 molecules. What bonding orbitals are used?

17 Predict the formula and molecular shape of a hydride of phosphorus.

18 Draw an electron dot representation for the NH_4^+ ion. What shape do you predict this ion will have?

19 Predict the type of bonding and the shape of the ion BF_4^-.

20 Consider the two compounds CH_3CH_3 (ethane) and CH_3NH_2 (methylamine). Why does CH_3NH_2 have an electric dipole while CH_3CH_3 does not?

21 Consider the following series: CH_4, CH_3Cl, CH_2Cl_2, $CHCl_3$, and CCl_4. In which case(s) will the molecule have an electric dipole? Support your answer by considering the bonding orbitals of carbon, the molecular shape of the molecules, and the resulting symmetry.

22 Predict the structure of the compound N_2F_2 from the electron dot representation of the atoms and the molecule. Should isomers exist?

23 Which of the isomers of dichloroethylene shown in Figure 16-17 will be polar molecules?

24 Draw structural formulas for all the isomers of ethylene (C_2H_4) in which two of the hydrogen atoms have been replaced by deuterium atoms. Label the *cis*- and the *trans*- isomers.

> We are all agreed that the theory is crazy. The question that divides us is whether it is crazy enough to have a chance of being correct.
>
> NIELS BOHR (1885–1962)

CHAPTER 17 Molecular Architecture: Liquids and Solids

The natures of familiar liquids and solids are determined by their molecular architecture.

The architecture of gaseous molecules, as discussed in Chapter 16, leaves us, as usual, with a few answers and a lot of new questions. We now have knowledge of the forces *within* molecules in the gas phase in terms of our model of electron arrangement in orbitals. Can we apply this model to the forces *between* molecules in the liquid and solid phases? You bet! Distribution of the electron population in valence orbitals turns out to be a valuable key to understanding the bonding in metals, molecular solids, network solids, and ionic solids—and why each type of substance has its own characteristic properties.

In Chapter 8 a rather primitive bonding model was used to describe network solids, metals, and ionic solids. How does this model change as more sophisticated ideas of atomic structure are applied to the problem? Let us begin the present discussion of solids and liquids by considering bonding between atoms in more detail.

Two or more atoms remain near each other in a particular arrangement because energy is lowest in that arrangement. This is true whether the cluster of atoms is strongly or weakly bound, whether it contains a few atoms or 6.02×10^{23} atoms, whether the arrangement is regular (as in a crystal) or irregular (as in a liquid). The cluster of atoms is stable if, and only if, the energy is lower when the atoms are together than when they are apart. Remember that randomness always favors the breaking up of clusters.

Furthermore, the magnitude of the attractive forces between atoms varies greatly, depending on how closely the valence electrons are able to approach the positive nuclei. As we noticed in Chapter 16, this approach distance is determined by the electron occupancy of the valence orbitals.

Thus, the occupancy of the valence orbitals is the clue we shall use to predict whether a substance is a high-melting, salt-like crystal; a metal; or a low-melting, molecular crystal like Dry Ice or solid iodine. This is an ambitious program. Let us see how far we can go toward our goal.

17-1 MOLECULAR SOLIDS—VAN DER WAALS FORCES

In Chapter 16 you were told that the diatomic molecule of fluorine (F_2) does not form larger molecules such as F_3 or F_4. We explained this by noting that each fluorine atom has only *one* partially filled valence orbital; when two atoms combine by sharing electrons from the two formerly half-filled orbitals, the valence orbitals in both fluorine atoms are completely filled. A very stable electron configuration has been achieved and there is no additional electron-nucleus interaction.

But at sufficiently low temperatures fluorine molecules condense to give a liquid! And at still lower temperatures a solid may form! The observation is general. *Any pure gas, when cooled sufficiently under appropriate pressure, will condense to give a liquid; at lower temperatures it will give a solid.* Even helium gas will condense to liquid helium.

The same forces that cause molecular fluorine to condense at 85°K cause noble gases such as helium to condense. These forces are named **van der Waals forces,** after the Dutch physicist Johannes D. van der Waals (1837–1923), who studied them. When the outer valence orbitals of all atoms in a molecule are filled, the electrons of another molecule cannot come close to the nuclei in that molecule. When molecules of this kind approach each other, the energy is lowered only a few tenths of a kilocalorie per mole. This weak interaction is typical of van der Waals forces. We shall consider van der Waals forces in more detail, but they are still poorly understood.

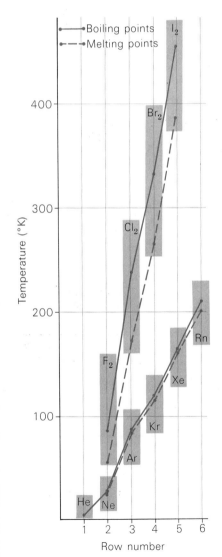

Fig. 17-1 Elements that form molecular crystals bound by van der Waals forces.

Fig. 17-2 Melting and boiling points of noble gases and halogens.

Under the simple attraction of van der Waals forces, a weakly bound molecular liquid or a low-melting molecular crystal will form for a given substance. <u>When the atoms of a molecule have an electron configuration with all valence orbitals filled, then only van der Waals interactions among such molecules remain.</u>

The *elements* forming van der Waals liquids and solids are concentrated in the upper right-hand corner of the periodic table (see Figure 17-1). These elements are capable of forming stable molecules that satisfy completely the bonding capacity of each atom. Distinct molecules, such as CH_4, C_2H_4, and those found in gasoline are held together in the liquid state by van der Waals forces.

17-1.1 Van der Waals Forces and the Number of Electrons

In Chapter 8 we noticed that the boiling points of the noble gases increase as their atomic numbers increase. Boiling points range from $4.2°K$ for helium (atomic number 2) to $211°K$ for radon (no. 86). Furthermore, the boiling points of the halogens increase from $85°K$ for F_2 to $457°K$ for I_2. These data, along with corresponding melting points, are plotted in Figure 17-2. The horizontal axis shows the row number, which is an index of the total number of electrons per element. In Figure 17-3 the same type of information is plotted for compounds having the formula CX_4. The horizontal axis shows the row number of the periodic table for the outermost atoms in the molecule, since these are the atoms that "rub shoulders" with neighboring molecules. As far as van der Waals forces are concerned, it is quite important that CBr_4 has atoms from the fourth row (bromine) on the outer "surface" of the molecule, and it is somewhat less important that the central atom is carbon from the second row. The outermost atoms are the most influential in determining intermolecular forces because these forces act over very short distances. The data indicate that for substances in closely related groups, <u>van der Waals forces increase as the number of electrons in the molecule increases.</u>

A further and closely related generalization can be made. If we compare similar molecules, we find that the larger the molecule, the higher its boiling point. For example, if we compare methane (CH_4) and ethane (C_2H_6), the exterior atoms are the same—hydrogen atoms. Still, the boiling point of C_2H_6, $185°K$, is higher than that of CH_4, $112°K$. This difference can be attributed to the fact that there must be greater contact surface between two C_2H_6 molecules than between two CH_4 molecules. The same effect is found for C_2F_6 (boiling point, 195°K) and CF_4 (145°K).*

EXERCISE 17-1

Gaseous phosphorus is made up of P_4 molecules with four phosphorus atoms arranged at the corners of a regular tetrahedron. With such a geometry, each phosphorus atom is bound to three other phosphorus atoms. Would you expect this gas to condense to a solid with a low, intermediate, or high melting point? After making a prediction on the basis of the valence orbital occupancy, check the melting point of phosphorus in Table 19-2 on page 501. (*Hint:* Compare with I_2.) *high.*

EXERCISE 17-2

Natural gas used for fuel is largely CH_4. Gasoline is a mixture of molecules in the same family, with formulas ranging from about C_5H_{12} to $C_{11}H_{24}$. Explain why dry fuel gas does not freeze even in winter, while gasoline is a liquid in the middle of the summer. *CH_4 has low boiling pt., gasoline has higher boiling pt.*

17-1.2 Van der Waals Forces and Molecular Shape

A substance whose structure has a high degree of symmetry generally has a higher melting point than a closely related compound whose structure lacks this symmetry. For example, consider the two structural isomers of formula C_5H_{12}, called pentane and neopentane. Their molecular shapes differ drastically, as Figure 17-4 (next page) shows. The zigzag shape of normal pentane allows van der Waals forces to act between the external envelope of hydrogen atoms of one molecule and those of adjacent molecules. This large surface contact gives a relatively *high boiling point*. On the other hand, this flexible, snake-like molecule does not pack readily in a regular lattice, so its crystal has a *low melting point*. Compare normal pentane with the highly compact, symmetrical neopentane. This ball-like molecule readily packs in an orderly crystal lattice which, because of its stability, has a rather *high melting point*. Once melted, however, neopentane forms a liquid that boils at a temperature lower than does normal pentane. Neopentane has less surface contact with its neighbors and hence is more volatile; it has a relatively *low boiling point*.

Most carbon compounds condense to molecular liquids and solids. Their melting points are generally low (below about 300°C) and many carbon compounds boil below 100°C. The similar chemistry of the liquid and solid phases suggests that the basic geometry of the organic molecule is not changed as we go from liquid to solid. Van der Waals forces hold both liquid and solid together.

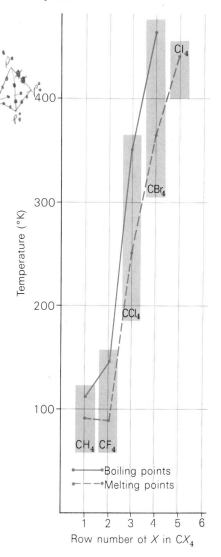

Fig. 17-3 Melting and boiling points of CX_4 molecules.

---
* Notice that these two factors, number of electrons and molecular size, might lead to another generalization—that the boiling point increases in proportion to molecular weight. Molecular weight, molecular size, number of electrons, and boiling point all tend to increase together. This molecular weight-boiling point correlation has some usefulness with molecules of similar composition and general shape, but chemists do not feel that there is a direct causative relation between molecular weight and boiling point.

438

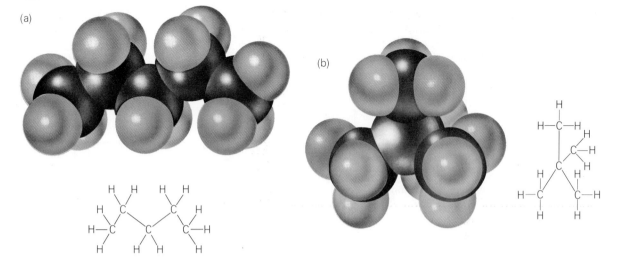

Fig. 17-4 Molecular shape influences melting and boiling points. Normal pentane (a) melts at $-130°C$ and boils at $36°C$; neopentane (b) melts at $-20°C$ and boils at $9°C$.

17-2 COVALENT BONDS AND NETWORK SOLIDS

As we noted in Chapter 8, solid carbon exists in two forms—the diamond and graphite structure.* Diamond is the classic example of the **three-dimensional network solid**; graphite is typical of the **two-dimensional network solid.** Let us review their geometries.

17-2.1 The Three-Dimensional Network Solid

In the diamond each carbon atom is surrounded by four others arranged *tetrahedrally*. Since one $2s$ and three $2p$ orbitals are used to give a tetrahedral arrangement of single bonds around carbon, the bonding can be described as sp^3. The very high melting point, extreme hardness, and low conductivity of a diamond can be interpreted in terms of this structure [Figure 17-5(a)].

* A new form of carbon was reported in *Science,* **161,** 363 (July 26, 1968). It resembles graphite.

Fig. 17-5 **Carbon: (a) diamond and (b) graphite.**

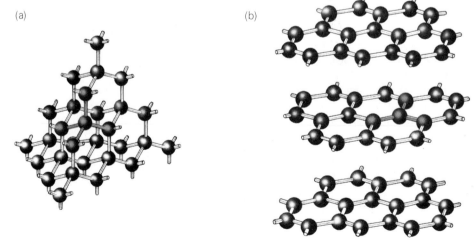

The compound silica (SiO$_2$) also has a covalent network structure. In one form silicon atoms will replace carbon atoms in a diamond lattice, giving rise to a structure with oxygen atoms lying between each pair of silicon atoms. This compound is hard and has a very high melting point.

17-2.2 The Two-Dimensional Network Solid

In contrast to diamond and silica, graphite has a two-dimensional, layered structure [Figure 17-5(b)]. Each carbon is surrounded by three others *in a plane* (instead of four atoms arranged in a tetrahedron, as in diamond). The geometry suggests the *sp*^{2} bonding notation we used for BF$_3$ (Section 16-2.4). Let us draw the electron dot representation for this structure. We see that each carbon is using only three electrons and orbitals. In the following structure, we are short one electron per carbon:

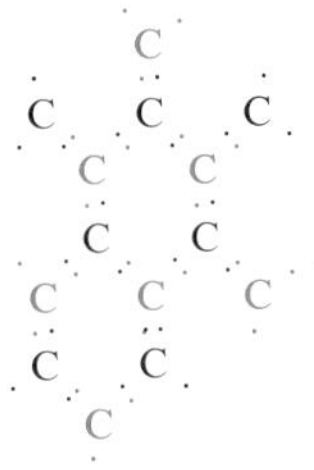

[electron dot structure of carbon network]

What is to be done with the extra electron and orbital for each atom? One solution would be to use the extra electrons to form multiple bonds, as we did with C$_2$H$_4$ and N$_2$. Then there would be ten extra electrons (indicated below by ×'s) for the ten carbons in the two rings:

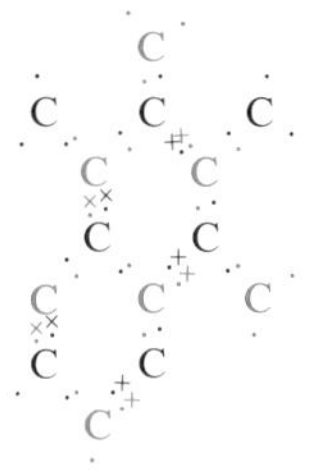

[electron dot structure of carbon network with extra electrons marked by ×'s]

Since there is only one extra orbital and one extra electron per carbon, *only one of the three bonds per carbon can become a double bond at any one time.* A possible graphite structure is then

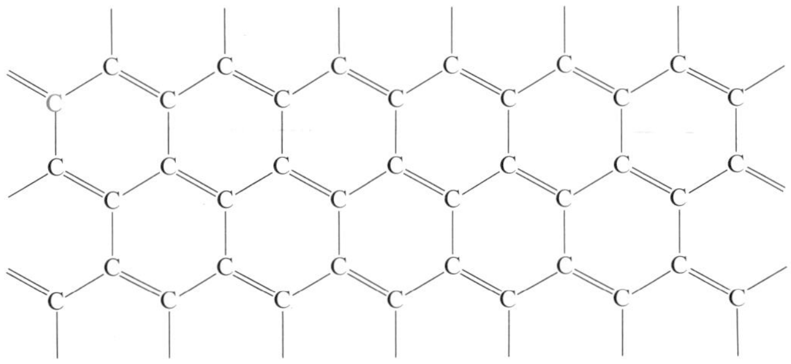

[graphite hexagonal network structure]

By using this arrangement all carbons have filled their 2s and 2p valence levels. But this is not the only arrangement that gives carbon atoms with filled valence orbitals. There is a second arrangement:

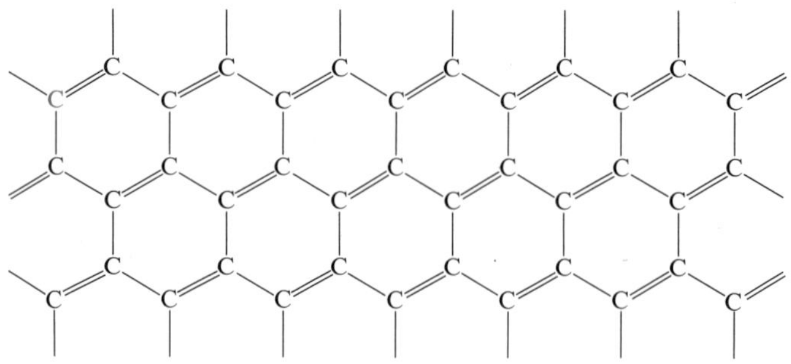

A third representation is equally valid:

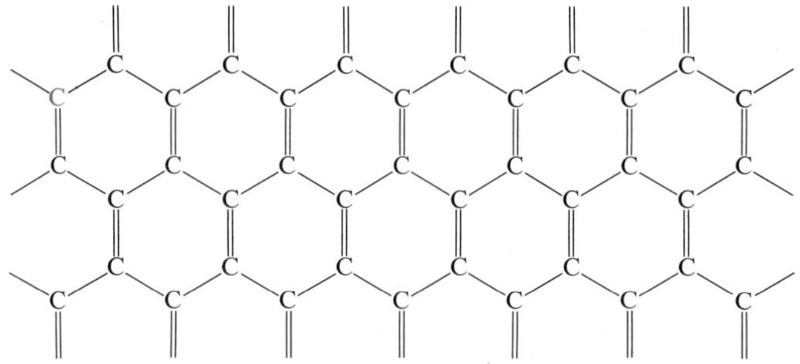

Notice that the double bond can move around; *any two adjacent atoms in the sheet can be joined by a double bond.* This really means that the electrons making up the double bonds can move easily between atoms in the sheet. Such mobile electrons are described as **delocalized electrons;*** they are responsible for the high electrical conductivity of graphite *along the sheet.* On the other hand, there is no easy route for electrons to move between layers. Graphite is a *poor* electrical conductor in a direction *perpendicular to* the layers.

The layers themselves are bonded together by weak van der Waals forces; hence, the graphite crystal is soft and slippery. It breaks easily into layers; on the other hand, bonds between atoms in any *one* layer are strong.

Other elements forming network solids of various types with covalent bonds are shown in Figure 17-6.

17-3 THE METALLIC STATE

How do we recognize a metal? Let us review the properties of a metal and apply our expanding concept of chemical bonding.

---

* Graphite can also be described in another way. Note that the s, p_x, and p_y orbitals are used to make the planar sheets with three triangularly arranged bonds around each carbon. This is an sp^2 structure. The p_z orbitals on each carbon remain perpendicular to the graphite sheet. These p_z orbitals on adjacent carbons can overlap to make giant orbitals extending over the whole sheet. Electrons can then move easily over the sheets using these so-called "π orbitals."

Fig. 17-6 Elements forming network solids with covalent bonding.

17-3.1 Properties of a Metal

First and foremost, metals have luster; they are bright and shiny. With few exceptions (*e.g.*, gold, copper, bismuth, manganese) metals have a silvery-white color because they reflect *all* frequencies of light. We remember from Chapter 15 that light absorption is closely related to energy levels and the movement of an electron between these levels. Reflection, or lack of permanent light absorption, also has implications for energy levels. Apparently all metals share a common electron configuration that accounts for their high reflectivity.

Metals have high electrical and thermal conductivity. In Section 17-2.2 we noted that the electrical conductivity of graphite is related to delocalized electrons in the carbon layers. In Chapter 8 we suggested that metals themselves might be described as positive ions immersed in a "sea" of electrons. This "sea" of electrons would account for the ability of metals to conduct electricity in any direction. Thermal conductivity would also seem to be closely related to these mobile electrons.

Finally, all true metals can be drawn into wires or hammered into sheets without shattering—that is, metals are ductile and malleable; they are workable. Mobile, nonrigid electron clouds would seem to be consistent with the physical characteristics of metals. What bonding patterns are suggested by all these characteristics of metals?

17-3.2 Bonding in Metals

Atoms forming metals are characterized by their appearance on the left-hand side of the periodic table (see Figure 17-7, following page). Elements in this region have vacant valence orbitals and low ionization energies for valence electrons. Vacant valence orbitals and low ionization energies, then, are the two conditions necessary for metallic bonding. We shall see how these characteristics are essential to forming the metallic bond.

Like other bonds, the metallic bond forms because electrons can move near two or more positive nuclei simultaneously. Our problem is to obtain some insight into the special way in which electrons in metals do this. Consider the lithium atom. Because each

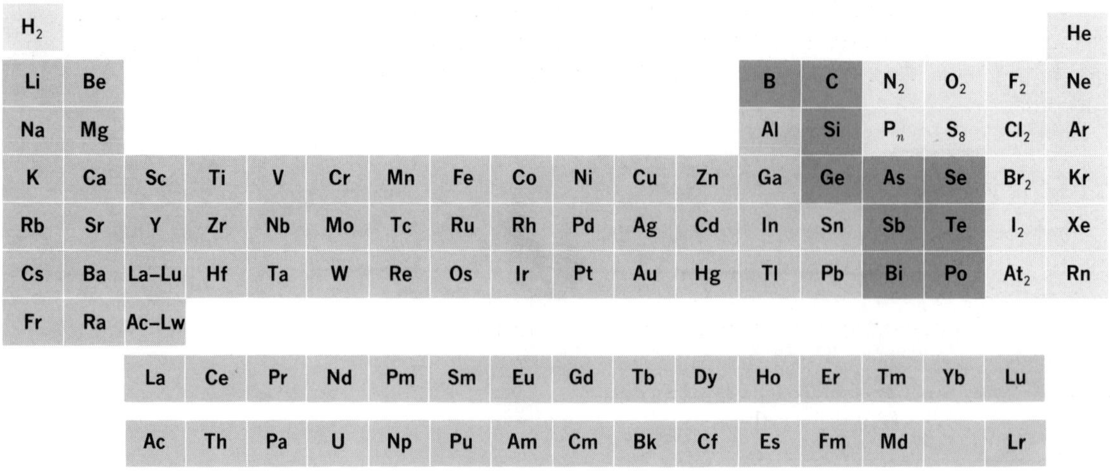

H$_2$																	He
Li	Be											B	C	N$_2$	O$_2$	F$_2$	Ne
Na	Mg											Al	Si	P$_n$	S$_8$	Cl$_2$	Ar
K	Ca	Sc	Ti	V	Cr	Mn	Fe	Co	Ni	Cu	Zn	Ga	Ge	As	Se	Br$_2$	Kr
Rb	Sr	Y	Zr	Nb	Mo	Tc	Ru	Rh	Pd	Ag	Cd	In	Sn	Sb	Te	I$_2$	Xe
Cs	Ba	La–Lu	Hf	Ta	W	Re	Os	Ir	Pt	Au	Hg	Tl	Pb	Bi	Po	At$_2$	Rn
Fr	Ra	Ac–Lw															

| | La | Ce | Pr | Nd | Pm | Sm | Eu | Gd | Tb | Dy | Ho | Er | Tm | Yb | Lu |
| | Ac | Th | Pa | U | Np | Pu | Am | Cm | Bk | Cf | Es | Fm | Md | | Lr |

Fig. 17-7 **The metallic elements.**

lithium atom has a single 2s electron, we would expect one single covalent bond between two lithium atoms to give Li$_2$.

$$\text{Li:Li}$$

Indeed, Li$_2$ is observed at high temperatures.

Notice, however, that the gaseous Li$_2$ still has *three* unused valence orbitals in each atom. The existence of just one unused orbital per atom conferred a special reactivity upon the CH$_2$ molecule. We saw that one electron of an unused pair on carbon was promoted to the empty valence orbital, and two new covalent bonds could then form. CH$_2$ was a very reactive molecule. We might, then, anticipate that Li$_2$ could be a reactive molecule. Three empty valence orbitals are available.

But the molecule Li$_2$ lacks bond-forming electrons; the valence electron for each atom is already involved in a bond between two lithium atoms; furthermore, there are no free pairs. If extra bonds *between Li$_2$ molecules* are to form, the valence electrons in Li$_2$ must do double duty. They must spend some time in the empty valence orbitals of lithium atoms from other Li$_2$ molecules. Simple logic suggests that *chances for this electron sharing would be best if the electrons between lithium atoms were rather weakly held.* Sharing would be achieved most easily if the atoms had a low ionization energy. Under these conditions, the electron pair between nuclei could wander toward other nuclei, all of which have empty orbitals.

Experiment suggests that this is indeed what happens. Everywhere the electron moves it finds itself between two positive nuclei with orbitals available to form a bond. Orbital geometry adjusts to give relatively high electron density between any two atoms. (Remember that in CH$_4$ and C$_2$H$_4$ the orientation of orbitals was determined by the nuclei and electrons surrounding a given atom.) The space around a central atom is a region of almost uniformly low potential energy. Each valence electron is virtually free to make its way throughout the crystal.

This argument leads us to picture a metal as an array of positive ions located at crystal lattice sites and immersed in a "sea" of mobile electrons. A metallic structure of this type will be formed by atoms having many vacant valence orbitals and low ionization energy for valence electrons. These are the very features which identify metals.

17-3.3 Properties of Metals and the Metallic Bond

The above idea emphasizes an important difference between metallic and covalent bonding. In molecular covalent bonds the electrons are concentrated in certain regions of space between two atoms. In contrast, the valence electrons in a metal are spread almost uniformly throughout the crystal. The carbon atoms in diamond are held together by these strongly localized covalent bonds. The diamond is very hard and rigid. Metals, on the other hand, are malleable and ductile because of the mobility of the electrons—metals can be worked. Under stress, one plane of atoms may slip over another (see Figure 17-8); but as it does so, electrons can move easily between planes to maintain bonding. If the easy movement of electrons is blocked, metals become hard and brittle. Metals can be hardened by alloying them with elements which have a strong attraction for electrons or which lack empty bonding orbitals, and thus reduce the easy mobility of metallic electrons. Often, just a trace of carbon, phosphorus, or sulfur will turn a relatively soft and workable metal into a brittle solid, much as a dam is able to control water movement over the whole stream, even though it occupies a very small

(a)

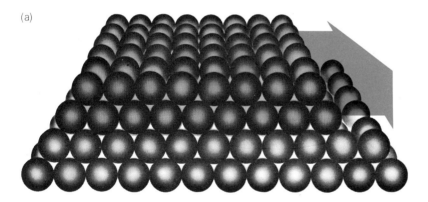

(b)

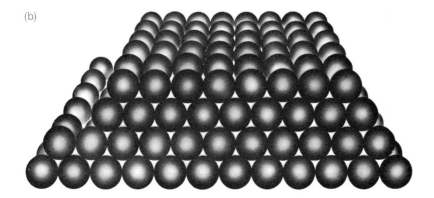

Fig. 17-8 **Slippage of planes of metal atoms.**

fraction of the volume of the stream bed. In a sense, carbon, phosphorus, and sulfur atoms can be regarded as "dams" in the mobile electron sea.

The excellent heat conductivity of metals is also due to mobile electrons. Electrons that are in regions of high temperature can acquire large amounts of kinetic energy. These electrons move through the crystal very rapidly and transfer energy to the atoms in the cooler regions. If electron movement is fairly restricted between atoms, thermal energy can be passed stepwise from atom to atom, but the process is much slower than in metals.

Why do the foregoing characteristics of metals disappear as we go across the periodic table from left to right? First, because as nuclear charge on the atoms increases, the ionization energy for valence electrons rises rapidly. It becomes more and more difficult for such atoms to release electrons for delocalized bonding. Furthermore, as additional orbitals are filled, the ability of an atom in a molecule to accept a share in an electron pair from other atoms decreases. For these reasons, the localized regions between two nuclei become more attractive to electron pairs in elements going from left to right across the periodic table. Consequently, there is less and less possibility for delocalization and metallic-bond formation.

17-3.4 Strength of the Metallic Bond

Is a metallic bond weak because electrons are mobile? We can obtain some idea of the effectiveness of this electron "sea" in binding atoms if we compare the energy necessary to vaporize 1 mole of a metal with the energy required to vaporize 1 mole of a covalently bonded solid. (Free atoms are present in the vapor in both cases.) We find that for the alkali metals the energy is only one fourth to one third that needed for 1 mole of ordinary, covalently bonded solid. This is not surprising. The ionization energy of a free alkali metal atom is small. This means that the valence electron in the free atom does not experience a strong attraction to the nucleus. Since the electron is not strongly attracted by one alkali metal nucleus, it is not strongly attracted by two or three such nuclei in the metallic crystal. Thus, the binding energy between electrons and nuclei in the alkali metal crystals is rather small, and the resulting metallic bonds are rather weak.

We might expect, however, that the metallic bond would become stronger in those elements that have a greater number of valence electrons and a greater nuclear charge. In these cases, there are both more electrons in the "sea" and a stronger attraction to the nucleus because of the increased nuclear charge. This argument is supported by the experimental heats of vaporization in Table 17-1. To discuss a particular case let us compare the heat of vaporization of magnesium with the heat of vaporization of aluminum. The higher value for aluminum shows that the metallic bond is indeed stronger when both the number of valence electrons and the charge on the nucleus increase. Thus, the strength of the metallic bond tends to increase going from left to right along a row in the periodic table. The transition metal elements are harder and melt and boil at higher

TABLE 17-1 HEATS OF VAPORIZATION OF METALS

Row of Periodic Table	Heats of Vaporization (kcal/mole)		
Second row	Li 32.2	Be 53.5	B 129
Third row	Na 23.1	Mg 31.5	Al 67.9
Fourth row	K 18.9	Ca 36.6	Sc 73
Fifth row	Rb 18.1	Sr 33.6	Y 94
Sixth row	Cs 16.3	Ba 35.7	La 96

temperatures than the alkali or alkaline earth metals. This is explained by the fact that transition elements have *d* electrons and *d* orbitals that may be called upon in the process of forming bonds.

17-3.5 The Geometry of Metals

In Chapter 16 it was possible to relate molecular structure and electron orbital configuration for a number of gaseous molecules. What can be said about the electronic structure and detailed geometry of metals?

Metals, like molecules, have a characteristic geometry. For example, the metals lithium, sodium, potassium, rubidium, and cesium all have a structure that is described as "body-centered cubic." A portion of this structure is shown in Figure 17-9. Notice that the structural unit is a simple cube with an atom at each corner and another atom inside the cube. The name **body-centered cubic arrangement** describes what we see in Figure 17-9.1. An atom is centered in the body of each cube. Since the actual structure of a metal is obtained by putting together thousands of cubes, we find that each atom in the structure is the same; there is no difference between central and corner atom. Each atom, regardless of position in the alkali metal structure, has eight neighbors. This can be seen by focusing on the atom in the center of the cube in Figure 17-9.2.

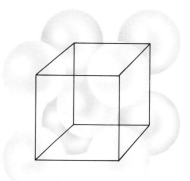

Fig. 17-9.1 **Body-centered cubic structure.**

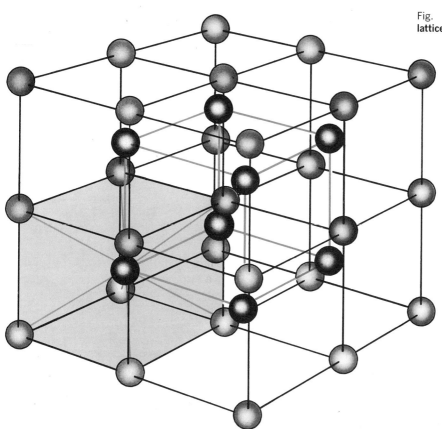

Fig. 17-9.2 **Body-centered cubic lattice.**

We further remember that each alkali metal atom uses a single *s* electron in bond formation. Just as the use of one *s* and one *p* orbital by beryllium gave a *linear sp* molecule, we can say that the use of a single *s* electron in metallic bond formation seems to favor a *body-centered cubic structure*.

What is known about the structure of metals such as beryllium and magnesium? Both beryllium and magnesium have two electrons in the *s* orbital of the valence level. In our earlier treatment of beryllium, we decided that an *s* electron must be promoted to a *p* level before electron-pair bonds could be formed. Magnesium is similar. For Mg we write

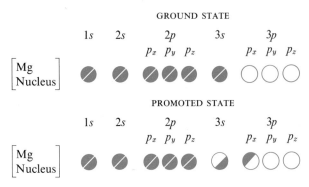

Promoted electrons in the $3s$ and $3p_x$ levels can be used for bonding. Once again it is assumed that the energy gained in promotion will be offset by the energy released in bond formation. When *sp* electrons are used, a structure known as the **hexagonal close-packed arrangement** has the lowest energy and appears to be favored. In a hexagonal close-packed arrangement, each atom is surrounded by *twelve* other atoms. There are six atoms in a plane around any given atom, plus three atoms above and three atoms below the plane (see Figure 17-10.1). The number of atoms per unit of volume is

Fig. 17-10.1 **Hexagonal close-packed structure: normal and exploded views.**

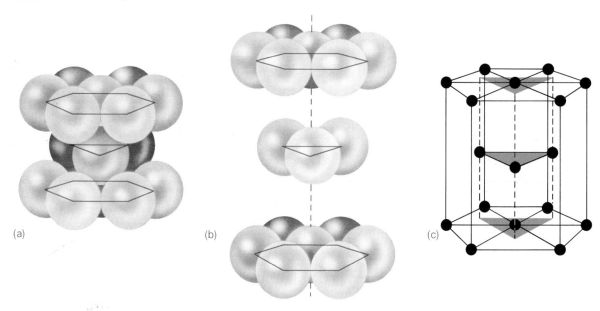

slightly greater than in the body-centered cubic structure, so if atoms of nearly equal weight were used, the density of a hexagonal close-packed structure would be slightly higher than the density of a body-centered cubic structure.

For aluminum the ground state electronic structure is

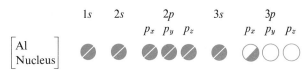

To form metallic bonds, one electron in the $3s$ is promoted to the $3p_y$ orbital. The promoted electronic structure for aluminum is

Aluminum metal, using sp^2 electrons, displays still a third type of geometry, shown in Figure 17-10.2—the **face-centered cubic arrangement**. The name derives from the fact that there is an extra atom in the center of each cubic face. This structure is sometimes called close-packed cubic, but we shall use face-centered cubic because it is more descriptive.

Orbital concepts provided helpful nomenclature for describing the geometry of gaseous molecules. Methane (CH_4), with sp^3 orbitals, was described as tetrahedral. Similarly, orbital concepts

Fig. 17-10.2 **Face-centered cubic structure.**

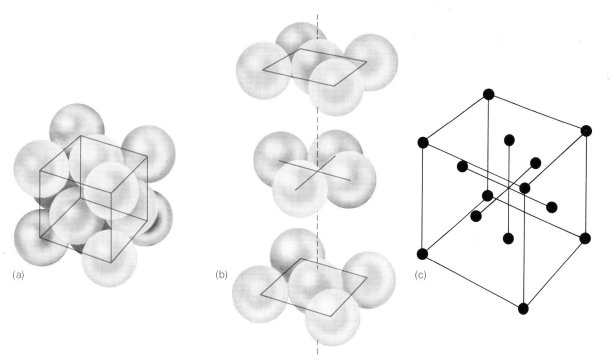

**ALFRED E. STOCK
(1876–1946)**

Born in Danzig, Poland, Alfred E. Stock received the Ph.D. from the University of Berlin in 1899. Soon after, he began what was to be his lifetime work—a study of the chemistry of boron. Although at the time the element boron was thought to be thoroughly investigated, Stock reasoned that this neighbor of the versatile carbon atom could not possibly have the dull and limited chemistry popularly assumed. With his discovery of a succession of hydrides of boron (for example, B_2H_6, B_4H_{10}, B_5H_9, and $B_{10}H_{14}$) whose very existence baffled chemists, Stock reopened the study of the chemistry of boron. At the time of Stock's death in 1946, chemists still had no convincing explanation of the absence of the prototype molecule, BH_3; and discussion of bonding in diborane was based upon an incorrect assumption of its structure. Stock's amazing exploratory study went far beyond the expectations and predictions of other inorganic chemists of his day.

assist in describing the geometry of metals. Let us summarize these ideas and their application:

(1) The number of electron-pair bonds that an atom can form is determined by the number of unpaired electrons that it has.
(2) Electrons may be promoted above the ground state to give more unpaired electrons if the energy of bonding more than compensates for the promotion energy.
(3) Each electron bonding configuration can be associated with a metallic structure of minimum energy that is favored. Three common metallic structures have been considered here.

These structural ideas are particularly useful to metallurgists in considering alloy formation.

17-4 BORON—THE ELEMENT WHERE METALS AND NETWORK SOLIDS MEET

For many years after its discovery, the element boron and many of its compounds were an enigma to chemists. The structure of the element appeared to be exceedingly complex. Furthermore, the hydrides of boron presented a challenge to the bonding rules that worked so well for carbon compounds. Why? What is unusual about boron? First, boron has four valence orbitals but only three valence electrons. *It does not have enough electrons to use all its valence orbitals in normal covalent bond formation.* In this sense it is "electron deficient" and markedly different from carbon. How do elements such as lithium and beryllium, having even fewer electrons, handle this "problem" of "electron deficiency"? They form metallic bonds in which a single pair of electrons moves between many atoms. But boron cannot form metallic bonds with any ease because boron has only one unused valence orbital and a fairly high ionization potential for at least two of its three electrons.

Boron is an *in-between element.* It does not have enough electrons to bond like carbon, nor does it have enough empty orbitals and easily available electrons to bond like metals. What is the result of this dilemma? The answer was first provided by an X-ray structural study done by Professor J. L. Hoard and his colleagues at Cornell University in 1958.

17-4.1 The Structure of Elemental Boron

Boron, like carbon, exists in more than one form, but all these forms have some structural features in common. Structural differences are not as extreme as the differences between diamond and graphite. One form of boron is shown in Figure 17-11.1. Notice the four distinct clusters of boron atoms. Each cluster contains 12 boron atoms arranged in a regular group called an **icosahedron,** which is seen in expanded form in Figure 17-11.2. In this small cluster many of the features of a developing metallic structure appear. Some electron pairs are shared among a number of boron atoms in the

STRUCTURE of ELEMENTAL BORON

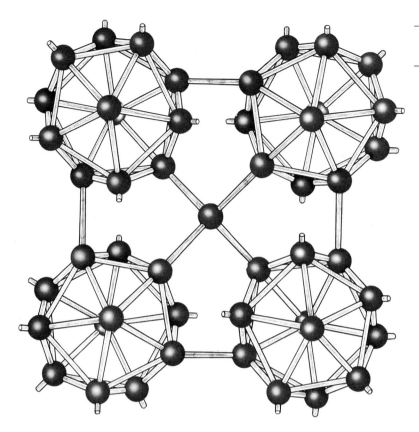

Fig. 17-11.1 **Structure of elemental boron (α-tetragonal form).**

cluster of 12, not between just two atoms as in the normal covalent bond, and not among an almost infinite number of atoms as in a true metal. Electrons *within this cluster of 12 borons* are delocalized to a large extent. The behavior within the cluster is strongly suggestive of many features of metallic bonding. But the boron atoms in the cluster soon run out of empty orbitals and of electrons with low ionization energies. The metallic structure cannot be continued.* At this point each boron in the icosahedron will usually form one normal covalent bond directed outward from the icosahedron. This extra bond is formed with a boron in another icosahedron or with a boron atom positioned between icosahedrons.† The net result is that all forms of elemental boron are made up of icosahedral units linked in various ways. The icosahedron—a remnant of the metallic structure—appears in all known forms of elemental boron!

Boron is neither a metal nor a normal covalent solid such as diamond. It is an in-between structure with semimetallic bonds in the icosahedral clusters and with normal covalent bonds between clusters. In retrospect this is almost what we would have expected

---

* An icosahedral array provides an extremely efficient way to pack 12 spheres around a point. The hole in the middle is about 30 percent smaller than any other standard packing arrangement such as hexagonal close-packing, but this increase in packing efficiency for 12 is achieved at the expense of a serious loss of three-dimensional packing efficiency for the icosahedral units. Such geometry seems to be ideal for an element such as boron where a small compact cluster is important.

† In Figure 17-11.1 the boron in the center makes four regular covalent bonds to four separate icosahedrons. Each center boron must then receive a total of five electrons from the four icosahedrons around it to permit such bond formation.

Fig. 17-11.2 **An icosahedral cluster of boron.**

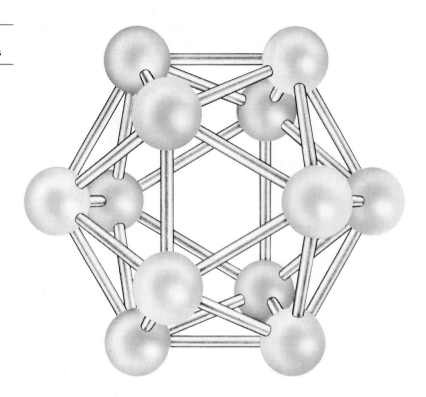

for boron—it is neither metal nor nonmetal. It is in-between and therefore must have characteristics of both a metal and a network nonmetal.

17-4.2 The Boron Hydrides

We noted earlier that the compound BH_3 does *not* exist in isolatable form. Instead, a **dimer,** or molecule containing two empirical formulas of BH_3, *i.e.*, B_2H_6 is the simplest boron hydride ever isolated at room temperature. Chemists can now rationalize this formerly mystifying fact in the following way: Each boron combines with three separate hydrogens to form a BH_3 unit, but when this BH_3 unit approaches the boron of another BH_3 unit, the relatively loosely held electron pair involved in one B—H bond is pulled over to share the unused orbital on the boron of the second BH_3 unit. The electrons now move in a region such that they are attracted by two boron nuclei and one hydrogen nucleus. The electrons are delocalized. (Remember how the Li:Li bonding pair was "delocalized" into a metal structure.)

$$\begin{array}{ccccccc} H & & H & & H & & H \\ & B & & B & & B & & B \\ H & & H & & H & & H \end{array} \longrightarrow \begin{array}{ccc} H & H & H \\ B & & B \\ H & H & H \end{array} \qquad (1)$$

The result is a bond in which one pair of electrons links three atoms —one hydrogen and two boron atoms. It is called a **three-center bond.** As you can see in the electron dot representation above, two so-called three-center bonds fasten together the two BH_3 units. The experimentally determined structure for B_2H_6 appears in Figure 17-12.

Apparently, the in-between bonding characteristics of boron account for the unexpected behavior of its hydrides. Let us support this thought with the following summary:

(1) A structure in which one electron pair joins together many atoms is characteristic of metals.
(2) A structure in which one or more electron pairs link *two* atoms is characteristic of nonmetallic network solids.
(3) A structure in which one electron pair links three, four, or five atoms is characteristic of behavior intermediate between metals and nonmetallic network solids.

Case (3) is characteristic of boron. Indeed, many of the boron compounds testify to the fact that boron is more metallic than carbon and less metallic than beryllium.

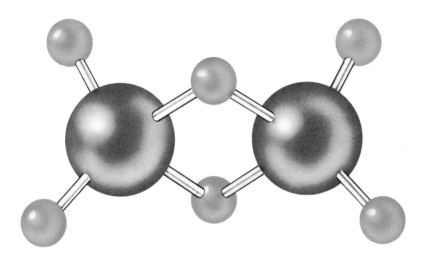

Fig. 17-12 **Structure of diborane.**

Boron forms a whole family of hydrides in addition to B_2H_6. Typical formulas are B_5H_{11}, B_6H_{10}, and $B_{10}H_{14}$. Furthermore, ions such as $B_{12}H_{12}^{2-}$ have been prepared in the last few years. The compounds B_5H_{11}, B_6H_{10}, and $B_{10}H_{14}$ each have a structure in which the boron arrangement is a *fragment* of the icosahedron of elemental boron. The open edges where the fragment was broken from the icosahedron are closed by B—H—B three-center bonds. The $B_{12}H_{12}^{2-}$ ion is an icosahedron with a hydrogen covalently bonded to each boron. The ion is shaped almost like a ball. (See Figure 17-11.2 on page 450.) In retrospect, the chemistry of boron does not seem so irregular if we remember where it lies in the periodic table.

EXERCISE 17-3

By referring to the diagram of an icosahedron in Figure 17-11.2, suggest a reasonable arrangement for boron atoms in a hydride of formula B_4H_{10}.

17-5 IONIC STRUCTURES—THE IONIC BOND AND ITS CONSEQUENCES

We have not yet considered the effects that arise from charge separations. The most extreme case is represented by the formation of ionic solids. Usually, these can be considered arrays of positive and negative ions, neatly stacked so that each positive ion has only negative-ion neighbors and each negative ion has only positive-ion neighbors. Figure 8-7 on page 173 shows such a crystal arrangement, that of sodium chloride. Why does such a solid form, and what are its properties? These are the questions we shall try to answer here.

17-5.1 The Stability of Ionic Crystals

In discussing the bonding in the gaseous LiF molecule in Section 16-3.1, the electric dipole of the molecule was explained in terms of the different ionization energies of lithium and fluorine atoms. Although the lithium fluoride molecule (LiF) holds together because the bonding electrons are near both nuclei, the energy favors an electron distribution concentrated toward the F atom. We may write $Li^+\cdots F^-$ for gaseous LiF. When molecules of gaseous $Li^+\cdots F^-$ approach each other the electrostatic attraction between positive Li^+ and negative F^- ions generates three-dimensional ionic clusters in which each Li^+ is surrounded by six F^- and each F^- by six Li^+. The structure is like that of NaCl in which there are no distinct molecules. Just as the atoms in metals are more stable when surrounded by other atoms in a metallic crystal, the ions in ionic solids are more stable when each has several neighboring ions of appropriate charge around it. Completely delocalized electrons are characteristic of solid metals. Solid LiF is, however, significantly different from metals. Half the atoms have high ionization energies because F atoms hold their electrons tightly. Therefore, the characteristic electron mobility of metals is not present in the ionic solids. This absence of mobile electrons implies that LiF has none of the metallic properties. Let us review the properties of ionic solids.

17-5.2 Properties of Ionic Crystals

Ionic solids, such as lithium fluoride and sodium chloride (NaCl), form regularly shaped crystals with well-defined crystal faces. Pure samples of these solids are usually transparent and colorless, but color may be caused by traces of impurities or crystal defects. Most ionic crystals have high melting points and their liquids have high boiling points. NaCl melts at 801°C and boils at 1413°C; LiF melts at 842°C and boils at 1676°C.

Molten LiF and NaCl have electrical conductivities which are lower than metallic conductivities by several factors of 10. Molten NaCl at 801°C has a conductivity about 10^{-5} times that of copper metal at room temperature. Perhaps the electric charge does not move by the same mechanism in molten NaCl as in metallic copper. Experiments show that the charge is carried in molten NaCl by slow-moving Na^+ and Cl^- ions. Some electrical conductivity in the liquid state is a property characteristic of molten ionic substances. In contrast, molecular crystals generally melt to form molecular liquids that do not conduct electricity. Metals have higher conductivity because electrons move more rapidly and easily than Na^+ and Cl^- ions.

We have considered covalent molecules and ionic lattices. But most molecules fall between these two extremes, being held together by bonds that are largely covalent but still having enough charge separation to affect the properties of the substance. These are the molecules we have called **polar molecules.**

Chloroform ($CHCl_3$) is an example of a polar molecule. It has four tetrahedrally oriented bonds (as in Figure 16-6.1 on page 418), three C—Cl bonds and one C—H bond. The C—Cl bonds do not have the same dipole as does the C—H bond. When the dipoles of the C—Cl and C—H bonds are added, there is a molecular dipole remaining (see Section 16-5). Such electric dipoles are important to chemists because they affect chemical properties such as solvent action. Let us see how a strongly polar solvent such as water interacts to dissolve an ionic crystal such as NaCl.

17-5.3 Solubility of Ionic Solids in Water

The dissolving of ionic solids in water is one of the most extreme and most important solvent effects that can be attributed to electric dipoles. Crystalline NaCl is quite stable, as indicated by its high melting point, yet it dissolves readily in water. To break up the stable crystal arrangement, there must be a strong interaction between water molecules and the ions that are released in the solution. This interaction can be explained in terms of the properties of the dipolar water molecule.

The energy is lower if an electric dipole, when brought near an ion, is oriented so that unlike charges are close together. Hence, water molecules tend to orient around ions with the positive end of the water dipole pointing inward toward a negative ion and with the negative end pointing toward a positive ion. This process is called **hydration** (refer to Section 13-1.2). Figure 17-13 (following page) shows it schematically.

Orientation of water dipoles around the ions has two effects. First, the energy is lowered because the orientation brings unlike charges near each other. This lowering of ionic energy by hydration provides the energy needed to break up the lattice. The hydration energies for the ions of NaCl are very close to the lattice energy for the solid. As a result, ΔH for the solution process for NaCl is only $+0.9$ kcal/mole. Notice that the positive sign for ΔH indicates that the energy term has a very small tendency to oppose the solution process. The hydration energy is not quite so large as lattice energy.

Why then does NaCl dissolve in water? The most important consequence must be a marked *increase in randomness* of the system when the regular NaCl lattice is broken up and the Na^+ and Cl^- ions are distributed randomly throughout the liquid phase.* *The*

---

* The randomness change, when reduced to energy terms at 25°C, amounts to about $-8,400$ cal—a term that favors solution.

water molecules surrounding each ion reduce the force of attraction between ions in solution and stabilize this more randomized arrangement. There are smaller randomization decreases due to ordering of water molecules around the ions, but these are well counterbalanced by the large change in randomness associated with the breakup of the NaCl lattice.

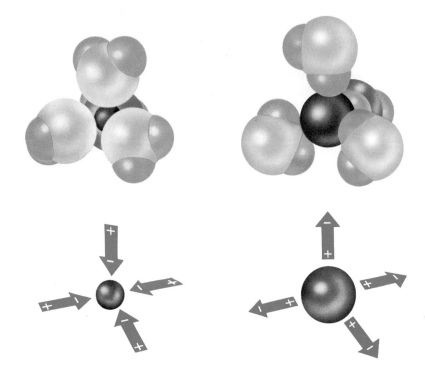

Fig. 17-13 **Hydration of ions, showing orientation of water dipoles.**

These two effects of ion hydration—compensation for the lattice energy of the crystalline solid and stabilization of ions in a more randomized arrangement—give water distinctive properties as a solvent for ionic solids and for other electrolytes such as HCl and H_2SO_4. These two effects help explain, for example, why some salts (such as NH_4Cl) absorb heat as they dissolve in water while some (such as NaOH) release heat as they dissolve. <u>The tendency toward maximum randomness is the dominant factor promoting solubility of all solutes in liquid solvents.</u>

17-6 HYDROGEN BONDS

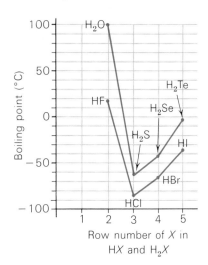

Fig. 17-14 **The boiling points of some hydrides.**

In Figure 17-3 on page 437 we saw that the boiling points of symmetrical molecules increase regularly as we move down the periodic table. Figure 17-14 shows the corresponding plot for some molecules possessing electric dipoles. Consider first the boiling points of HI, HBr, HCl, and HF. Hydrogen fluoride (HF) is far out of line, boiling at 19.9°C instead of below −95°C, as we would predict from an extrapolation of the other three boiling points. There is an even larger discrepancy between the boiling point of H_2O and the value we would predict from the trend suggested by the boiling points of H₂Te, H₂Se, and H₂S.

Could the extremely high boiling points of HF and H₂O be due to the fact that these are the smallest molecules of their respective series? No, this is not the explanation because there are no corresponding discrepancies in the data plotted in Figure 17-3. In fact, smaller molecules usually have lower boiling points. There must be some other explanation for these high boiling points. There must be additional new forces between the molecules of H₂O and HF that tend to keep molecules in the liquid phase.

These forces are evident with solid compounds also. The most familiar example is solid H₂O, or ice. Ice has a structure in which the oxygen and hydrogen atoms are distributed in a regular hexagonal crystal lattice. Each O atom, like each C atom in diamond, is surrounded by four other O atoms in a tetrahedral arrangement. The H atoms are found on the lines extending between the oxygen atoms:

$$\diagup\text{O}-\text{H}\cdots\text{O}\diagdown$$

The attractive force between O—H and O must be the bond that joins the water molecules together in the crystal lattice of ice. This bond is a **hydrogen bond.**

17-6.1 Energy of Hydrogen Bonds

The hydrogen bond is usually represented by O—H····O, in which the solid line represents the original O—H bond in the parent compound, as in water (HOH) or methyl alcohol (CH₃OH), and the dotted line represents the second bond formed by hydrogen, or the bond called the hydrogen bond. It is dotted to indicate that it is much weaker than a normal covalent bond. Consideration of the boiling points in Figure 17-14, on the other hand, shows that the interaction must be much stronger than van der Waals forces. Experiments show that most hydrogen bonds release between 3 kcal/mole and 10 kcal/mole upon formation: $\Delta H = -3$ to -10 kcal/mole. The energy of this bond places it between van der Waals attractions and covalent bonds. Roughly speaking, the energies are in the ratio

$$\begin{Bmatrix}\text{van der Waals}\\ \text{attractions}\end{Bmatrix} : \begin{Bmatrix}\text{hydrogen}\\ \text{bonds}\end{Bmatrix} : \begin{Bmatrix}\text{covalent}\\ \text{bonds}\end{Bmatrix}$$
$$\quad\ 1 \quad\quad\quad : \quad 10 \quad : \quad 100$$

17-6.2 Where Hydrogen Bonds Are Found

Hydrogen bonds are found between only a few elements of the periodic table. The most common are those in which H connects two atoms from the group F, O, N, and, less commonly, Cl.

The hydrogen bond to fluorine is clearly evident in most of the properties of hydrogen fluoride. The high boiling point of HF, compared with boiling points of the other hydrogen halides, is one of several data indicating that HF does not exist in the liquid phase as separate HF molecules but as *aggregates of molecules*, which we describe in general terms as

$(HF)_x$. Gaseous HF contains the "floppy ring" species H_6F_6, as well as some single HF molecules and perhaps a few percent of molecules of intermediate type such as H_2F_2.*

The H_6F_6 species can be represented in a descriptive formula such as the following:

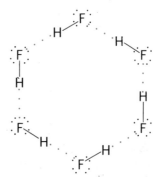

An extreme example of the fluorine-hydrogen bond is found in the hydrogen difluoride ion, HF_2^-. This ion exists in acidic solutions of fluorides,

$$H^+(aq) + F^-(aq) \rightleftharpoons HF(aq) \qquad (2)$$

$$HF(aq) + F^-(aq) \rightleftharpoons HF_2^-(aq) \qquad (3)$$

and in the ionic crystal lattice of salts such as KHF_2. The HF_2^- ion may be regarded as consisting of two negatively charged fluoride ions held together by a proton:

This arrangement is not typical, however, because few hydrogen bonds form with the proton equidistant from the two atoms to which it is bonded.

17-6.3 The Nature of the Hydrogen Bond

In the hydrogen bond the hydrogen atom is attached to two other atoms. Yet the bonding rules tell us that the hydrogen atom, with only the 1s orbital for bond formation, cannot form two covalent bonds. We must seek an explanation of this second bond.

The simplest explanation for the hydrogen bond is based upon the polar nature of F—H, O—H, and N—H bonds. In a molecule such as H_2O, the electron pair in the O—H bond is displaced toward the oxygen nucleus and away from the hydrogen nucleus. This partial ionic character of the O—H bond gives the hydrogen atom some positive character, permitting electrons from another oxygen atom to approach closely to the proton, even though the proton is already bonded. A second, weaker link is formed.

17-6.4 The Significance of the Hydrogen Bond

Hydrogen bonds are important in fixing properties such as solubilities, melting points, and boiling points, and in determining the form and

---

* Based on a 1968 electron diffraction study by Jay Janzen and L. S. Bartell, The University of Michigan.

stability of crystalline structures. They play a crucial role in biological systems. For example, water is so common in living matter that it must influence the chemical behavior of many biological molecules, most of which can also form hydrogen bonds. Water can attach itself by hydrogen bonding either by donating the proton, as in

$$\text{H}\diagup \text{O} - \text{H} \cdots \text{O} = \text{C} \diagdown$$

or by accepting the proton, as in

$$\text{N} - \text{H} \cdots \text{O} \diagup^{\text{H}}_{\text{H}}$$

Furthermore, hydrogen bonding within a single molecule is one of the chief factors in determining the structure of important biological substances such as DNA, as will be discussed in Chapter 22.

17-7 HIGHLIGHTS

Molecules in which all atoms have completed valence shells will liquefy and freeze at relatively low temperatures because of comparatively weak **van der Waals forces** acting between molecules. Van der Waals forces increase as the number of electrons in an atom increases and as the area of contact between two molecules in the liquid increases. The existence of strong covalent bonds between atoms in a solid gives rise to solids with very high melting temperatures, such as diamond and graphite.

Solids of metallic structure are formed by atoms having empty orbitals and electrons with low ionization energies. The properties of **metals** can be treated in a qualitative manner using this model. Boron is in between metallic and covalent network structures. This accounts for the fact that its chemistry has long been considered unusual.

Ionic solids have strong inter-ionic forces. **Polar solvents** such as water interact with ions to dissolve the solid. The **hydration** energy almost compensates for the lattice energy, and the solution process is driven by the tendency of the system to achieve maximum randomness.

Hydrogen bonds are able to form between two molecules if each contains an atom of F, O, N, or sometimes Cl and if there is a hydrogen atom between the two atoms. Many unexpected physical and chemical properties can be interpreted in terms of hydrogen bonds.

QUESTIONS AND PROBLEMS

1 Construct a table that contrasts the melting and boiling points of LiF, Li, and F_2, expressing the temperatures on the Celsius scale. (See Table 8-7 and Section 17-5.)

2 Without looking in your textbook, do the following:
(a) Draw an outline of the periodic table showing the rows but not the individual elements.
(b) Place a number at the left of each row indicating the number of elements in that row.
(c) Fill in the symbols for as many of the first 18 elements as you can. (Leave blank any that

you forget.) (d) Draw two diagonal lines across the table to divide it into three regions. In each region write either "metals," "nonmetals," or "covalent solids." Now compare your diagram with Figure 17-7 on page 442.

3 Sulfur exists in a number of forms depending upon the temperature and, sometimes, upon the past history of the sample. Let us call three of these forms A, B, and C. A exists at room temperature; it changes to B above the melting point of A, 113°C. B changes to C above 160°C.

Form A
Crystalline solid
Yellow color, no metallic luster
Melting point = 113°C
Dissolves in CS_2, not in water
Electrical insulator

Form A changes above 113°C to Form B:

Form B
Liquid
Clear, straw color
Viscosity (fluidity) about the same as water
Electrical insulator

Form B changes above 160°C to Form C:

Form C
Liquid
Dark color
Very viscous (syrupy)
Electrical insulator

Which of the following structures would be most likely to account for the observed properties of each of the three forms described above? (a) a metallic crystal of sulfur atoms, (b) a network solid of sulfur atoms, (c) an ionic solid of S^+ and S^- ions, (d) a molecular crystal of S_8 molecules, (e) a metallic liquid like mercury, (f) a molecular liquid of S_8 molecules, (g) a molecular liquid of S_n chains, where n is a very large number, (h) an ionic liquid of S^+ and S^- ions.

4 Contrast the bonds between atoms in metals, in molecular solids, and in network solids in regard to (a) bond strength, (b) orientation in space, and (c) number of orbitals available for bonding.

5 Aluminum, silicon, and sulfur are near each other in the third row of the periodic table, yet their electrical conductivities are widely different. Aluminum is a metal; silicon has much lower conductivity and is called a semiconductor; sulfur has such low conductivity it is called an insulator. Explain these differences in terms of valence-orbital occupancy.

6 Sulfur is made up of S_8 molecules; each molecule has a circular structure. Phosphorus contains P_4 molecules; each molecule has a tetrahedral structure. On the basis of molecular size and shape, which do you expect to have the higher melting point?

7 Discuss the conductivity of heat by copper (a metal) and by glass (a network solid) in terms of the valence-orbital occupancy and electron mobility.

8 The elements carbon and silicon form oxides with similar empirical formulas, CO_2 and SiO_2. CO_2 vaporizes at $-78.5°C$; SiO_2 melts at about 1700°C and boils at about 2200°C. Given this large difference, propose the types of solids these are. Draw an electron dot or orbital representation of the bonding in CO_2 consistent with your answer.

9 How do you account for the following properties in terms of the structures of the solids? (a) Both graphite and diamond are made of carbon (or C) atoms. Both are high melting, yet the diamond is very hard while graphite is a soft, greasy solid. (b) When sodium chloride crystals are shattered, plane surfaces appear on some of the fragments. (c) Silicon carbide, or carborundum, (SiC) is a very high-melting, hard substance used as an abrasive.

10 If you were given a sample of a white solid, describe some simple experiments that you would perform to help you decide whether the solid was held together by ionic bonds, covalent bonds, or van der Waals forces.

11 Would you expect elemental boron to be a good conductor of heat? Explain.

12 In what way can you justify the statement that boron is in between a metallic solid and a network solid?

13 A hydride of boron can be prepared of formula B_5H_{11}. Look at Figure 17-11.2 on page 450 and suggest a possible arrangement of the five boron atoms. Number them on your drawing.

14 What kind of structure would you predict for metallic K? for metallic Be? Explain.

15 If elements A, D, E, and J have atomic numbers of 6, 9, 10, and 11 respectively, write the formula for a substance you would expect to form between the following: (a) D and J, (b) A and D, (c) D and D, (d) E and E, and (e) J and J. In each case describe the forces involved between the building blocks in the solid state.

16 Consider each of the following in the solid state: sodium, germanium, methane, neon, potassium chloride, and water. Which would be an example of (a) a solid held together by van der Waals forces that melts far below room temperature; (b) a solid with a high degree of electrical conductivity that melts near 200°C; (c) a high-melting, network solid involving covalently bonded atoms; (d) a nonconducting solid that becomes a good conductor upon melting; (e) a substance in which hydrogen bonding is pronounced?

17 Predict the order of increasing melting point of these substances containing chlorine: HCl, Cl$_2$, NaCl, CCl$_4$. Explain the basis of your prediction.

18 Identify all the types of bonds you would expect to find in each of the following crystals: (a) argon, (b) water, (c) methane, (d) carbon monoxide, (e) Si, (f) Al, (g) CaCl$_2$, (h) KClO$_3$, (i) NaCl, (j) HCN.

19 Each of three bottles on the chemical supply shelf contains a colorless liquid. The labels have fallen off the bottles. They read as follows:

Label No. 1
n-butanol
CH$_3$CH$_2$CH$_2$CH$_2$OH
mol wt = 74.12

Label No. 2
n-pentane
CH$_3$CH$_2$CH$_2$CH$_2$CH$_3$
mol wt = 72.15

Label No. 3
diethyl ether
CH$_3$CH$_2$OCH$_2$CH$_3$
mol wt = 74.12

The three bottles are marked A, B, and C, and the measurements in the column on the right are made on the three liquids to permit identification.

(Consider hydrogen-bonding possibilities for determining solubility in water.) Which liquid should be given Label No. 1, Label No. 2, and Label No. 3? Explain how each type of measurement influenced your choices.

QUESTIONS and PROBLEMS

Property	Liquid A	Liquid B	Liquid C
Melting point (°C)	−131.5	−116	−89.2
Boiling point (°C)	36.2	34.6	117.7
Density (g/cm^3)	0.63	0.71	0.81
ΔH of vaporization (cal/g)	85	89.3	141
Solubility in water (g/100 ml)	0.036	7.5	7.9

20 Maleic and fumaric acids are cis- and trans-isomers having two carboxyl groups, HOOC—CH=CH—COOH. Maleic acid gives up its first proton more readily than does fumaric acid. However, the opposite is the case for the second proton. Account for this in terms of structure.

The property of carbon which distinguishes it most conspicuously is its tendency to form non-ionizing links both with other carbon atoms and with atoms of different elements.

R. C. FUCSON (1895–) and H. R. SNYDER (1924–)

CHAPTER 18

The Chemistry of Carbon Compounds

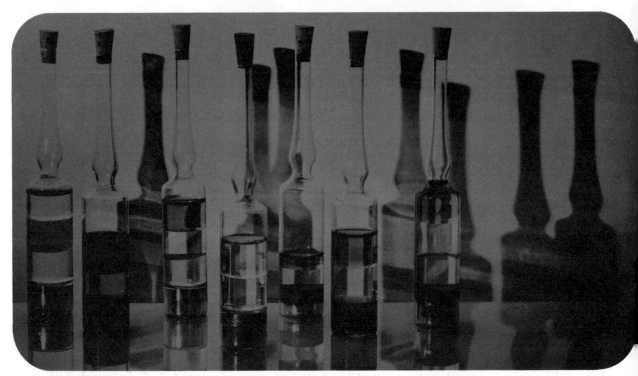

Carbon combines to form a nearly endless variety of liquids (shown here), solids, and gases.

What is so special about the chemistry of carbon compounds that it merits a chapter all its own—indeed, a chemistry all its own? One answer is, of course, that carbon compounds are the basis of the chemistry of living things. Another is that there are far more carbon compounds than all of the non-carbon compounds put together. Carbon is unique in its ability to form long-chain compounds.

In this chapter we will explore methods of learning about the molecular structure of carbon compounds and will learn about the kinds of molecules that comprise hydrocarbons, alcohols, organic acids, aspirin, nylon, and other "organic" substances.

The compounds of carbon furnish one of the most intriguing aspects of chemistry. They especially interest us because they play a dominant role in the chemistry of living things, both plant and animal. Also, there are many carbon compounds useful to man as dyes, drugs, detergents, plastics, perfumes, fibers, fabrics, flavors, and fuels. The manufacturing of these compounds has given rise to a huge chemical industry requiring millions of tons of raw materials every year. Where do we find the enormous quantities of carbon and carbon compounds needed to feed this giant industry? Let us begin our study of carbon chemistry by looking at the chief sources of carbon and carbon compounds.

18-1 SOURCES OF CARBON COMPOUNDS

18-1.1 Coal

Coal, a black mineral of vegetable origin, is believed to have come from the accumulation of decaying plant material in swamps during prehistoric eras when a warm, wet climate permitted rapid growth of plants. The cycles of decay, new growth, and decay caused successive layers of plant material to gradually build up vast deposits. The top layers of this material and of sedimentary rocks kept air from the lower material and subjected it to enormous pressures. In time the layers were compressed into hard beds composed chiefly of the carbon that was present in the original plants, and also containing appreciable amounts of oxygen, hydrogen, nitrogen, and some sulfur. Thus, coal is not pure carbon. The "hardest" coal, anthracite, contains from 85 to 95 percent carbon by weight; the "softest" coal, peat, is not really coal but one of the early stages in the geological history of coal. Peat still contains unchanged plant remains and may contain no more than 50 to 60 percent carbon.

When coal is heated to a high temperature *in the absence of air,* it decomposes; volatile products (coal gas and coal tar) distill away and a residue called **coke** remains. Coke is a valuable industrial material used chiefly in the reduction of iron ore (iron oxide) to iron for the manufacture of steel. Coke is essentially carbon that still contains the mineral substances present in all coals. About eight gallons of **coal tar** are obtained from a ton of coal. Coal tars are very complex mixtures; over 200 different carbon compounds have been isolated from them, many of which are used as industrial raw materials.

18-1.2 Petroleum and Natural Gas

Petroleum is a complex mixture that may range from a light, volatile liquid to a heavy, tarry substance. Like coal, petroleum also has its origin in living matter that has undergone chemical changes over a period of time. It is found in porous rock formations called oil pools, which lie between impervious rock formations that seal off the pools. **Natural gas** is a mixture of low-molecular weight compounds of hydrogen and carbon **(hydrocarbons)** found in underground "fields" of sandstone or other porous rock.

18-1.3 Certain Plant and Animal Products

Plants and animals are themselves highly effective chemical factories, and they synthesize many carbon compounds useful to man. These include sugars, starches, plant oils and waxes, fats, gelatin, dyes, drugs, and fibers.

Because these carbon compounds have their origin in living matter, plant or animal, *the chemistry of carbon is called* **organic chemistry.** *Compounds containing carbon are called* **organic compounds.** This term includes all compounds of carbon except CO_2, CO, and a handful of ionic substances such as sodium carbonate (Na_2CO_3) and sodium cyanide (NaCN). You may wonder how many organic substances are known. The number is actually so large it is difficult to provide a reliable estimate. Carbon compounds outnumber the compounds of all other elements with the possible exception of hydrogen (which is present in most carbon compounds). There are undoubtedly over one million different carbon compounds known. The number of *new* organic compounds synthesized in any one year (about 100,000 compounds) exceeds the total number of compounds known that contain no carbon!

18-2 MOLECULAR STRUCTURE OF CARBON COMPOUNDS

How can there be so many compounds containing this one element? The answer lies in molecular structure. In our discussion of bonding, we saw that carbon atoms have a strong tendency to form covalent bonds to other carbon atoms, resulting in long chains, branched chains and rings of atoms. (Remember the structures of diamond and graphite.) Each different atomic arrangement gives a molecule having distinct properties. To understand why a particular substance has certain characteristic properties, its structure must be known. Thus, the determination of the molecular structure of carbon compounds is one of the central problems of organic chemistry. This problem is usually solved by applying ideas and techniques we have already considered. Let us follow the process in detail for two compounds, ethane and ethanol.

18-2.1 The Composition and Structure of Carbon Compounds

Ethane and ethanol* are two common carbon compounds. Ethane is a gas that usually makes up about 10 percent of the household gas used for heating and cooking. Its useful chemistry is almost wholly restricted to the combustion reaction. Ethanol is a liquid that participates in a variety of useful chemical reactions. It has great value in the manufacture of chemicals, and it bears little chemical resemblance to ethane. Yet, the similarity of the two names suggests that these

---

* Ethanol is also known as ethyl alcohol.

compounds are related. This is so. To understand how they are related and why their chemistries are nevertheless so different, we must learn their molecular structures. We must find out what atoms are present in each substance, how many atoms there are per molecule, and what their bonding arrangement is.

Usually many experiments must be performed before the molecular structure of a compound is known with certainty. This fascinating problem of carbon chemistry involves three basic experimental steps—to determine first the *empirical formula,* then the *molecular formula,* and finally the *structural formula.* We shall begin by reviewing the information conveyed by each of these formulas.

As you will recall, the **empirical formula** tells only *the relative number of atoms* of each element in a molecule. The empirical formula of ethane is CH_3; that of ethanol is C_2H_6O. Ways in which the empirical formula can be determined will be considered in the next section.

The **molecular formula** tells *the total number of atoms* of each element in a molecule. The molecular formula is derived from the empirical formula by multiplying it by a whole number. This relationship can be represented as (empirical formula)$_n$, where n is a whole number. For example, the empirical formula of ethane is CH_3. One CH_3 unit has a formula weight of 15, but ethane has a molecular weight of 30; hence, there must be *two* empirical formulas in one molecular formula. Then, the molecular formula of ethane is $(CH_3)_2$ or C_2H_6. To convert the empirical formula to a molecular formula, the molecular weight must be known.

EXERCISE 18-1

A compound of carbon and hydrogen contains two hydrogen atoms for every carbon atom. The molecular weight is found to be 84. Write the empirical formula and the molecular formula.

The **structural formula** tells *which atoms are connected* in the molecule. Consider the compound having the molecular formula $C_2H_4F_2$. There are two ways in which atoms can be connected in this molecule without violating the bonding rules developed in Chapter 16. Either two fluorine atoms can be attached to one carbon,

$$\ddot{\underset{..}{F}}:H$$
$$:\ddot{\underset{..}{F}}:\underset{..}{\overset{..}{C}}:\underset{..}{\overset{..}{C}}:H$$
$$H\ H$$

or one fluorine atom can be attached to each carbon.

$$H\ H$$
$$:\ddot{\underset{..}{F}}:\underset{..}{\overset{..}{C}}:\underset{..}{\overset{..}{C}}:\ddot{\underset{..}{F}}:$$
$$H\ H$$

As you will recall from the discussion in Chapter 16 concerning $C_2H_2Cl_2$, compounds having the same molecular formula but different structural formulas are called **structural isomers.** Structural isomers differ in properties. The structural formula permits us to represent such differences, even though the two compounds have identical molecular formulas.

EXERCISE 18-2

The molecular formula of ethane is C_2H_6. How many structural formulas can be drawn for ethane without violating the bonding rules of Chapter 16?

The structural formula provides the most explicit representation for the molecule. How is the structural formula established? In Section 16-7 we listed some of the ways in which a structural formula can be determined. Let us examine the application of some of these methods to a specific problem—the determination of the structural formula of ethanol.

EXERCISE 18-3

What kind of forces would you expect to be acting between molecules of $C_2H_4F_2$? Make an estimate of the general boiling-point range that you would expect for $C_2H_4F_2$.

EXERCISE 18-4

The boiling point of the compound HF_2CCH_3 is $-25°C$. The boiling point of FH_2CCH_2F is $10°C$. Make a model of each of these two molecules. Which one is most nearly like a ball? Are the differences in boiling point consistent with differences in van der Waals forces predicted on the basis of molecular symmetry? (Refer to Section 17-1.)

18-2.2 The Experimental Determination of the Molecular Formula

The compound ethanol is isolated in the laboratory as a colorless liquid that boils at the constant and reproducible temperature of $78.5°C$ under 1 atm pressure. It freezes at $-117.4°C$. The constancy of its properties during different phase changes indicates that we are dealing with a pure substance. The first step in establishing a formula for this substance is analysis. We must know what atoms and how many of each are present.

Most organic compounds can be analyzed by burning the substance in pure oxygen. If the compound contains only carbon and hydrogen, only carbon dioxide and water will be produced. If the compound contains some nitrogen as well, nitrogen gas or one of the nitrogen oxides will be produced. Another way of finding out which elements are in a compound is to allow the compound to react with hot, liquid sodium metal. If the compound contains nitrogen, sodium cyanide (NaCN) will be formed; if it contains sulfur, sodium sulfide (Na_2S) will be produced.

Once such reactions show which elements are in the compound, relative numbers of atoms of each element (the empirical formula) can be determined. For example, suppose we burn 46 g of ethanol in oxygen and obtain 88 g of CO_2 and 54 g of water. How many *moles* of carbon *atoms* and how many *moles* of hydrogen *atoms* were present in the original sample? Since all carbon was added in the ethanol, and since each mole of CO_2 contains 1 mole of carbon atoms, we see that the number of moles of carbon atoms in the ethanol is equal to the number of moles of CO_2 obtained:*

$$\text{number moles } CO_2 = \frac{\text{wt } CO_2}{\text{molar wt } CO_2} = \frac{88 \text{ g } CO_2}{44 \frac{\text{g } CO_2}{\text{mole } CO_2}} = 2.0 \text{ moles } CO_2 \quad (1)$$

$$\text{number moles } CO_2 = 2.0 \text{ moles } CO_2$$

$$\text{number moles } CO_2 = \text{number moles C atoms} = 2.0$$

Since water contains 2 moles of hydrogen atoms per mole of water, we determine the number of moles of water and multiply the result by 2.0 to determine the number of moles of hydrogen atoms:

$$\text{number moles } H_2O = \frac{\text{wt } H_2O}{\text{molar wt } H_2O} = \frac{54 \text{ g } H_2O}{18 \frac{\text{g } H_2O}{\text{mole } H_2O}} = 3.0 \text{ moles } H_2O \quad (2)$$

$$\text{number moles H atoms} = 3.0 \text{ moles } H_2O \times \frac{2.0 \text{ moles H}}{\text{mole } H_2O}$$

$$\text{number moles H atoms} = 6.0$$

The total sample weight accounted for so far is

$$\left(2 \text{ moles C atoms} \times \frac{12 \text{ g}}{\text{mole C atoms}}\right) + \left(6 \text{ moles H atoms} \times \frac{1 \text{ g}}{\text{mole H atoms}}\right)$$

$$= 24 \text{ g} + 6 \text{ g} = 30 \text{ g} \quad (3)$$

The above numbers account for only 30 g of the original 46 g of sample. Since only CO_2 and H_2O were obtained as products and pure oxygen was the other reactant, the sample itself must have contained 16 g of oxygen (46 − 30). No other element can account for the difference of 16 g between weight of original sample and weight of carbon plus hydrogen in the products. Using this information, we write

$$\text{number moles oxygen atoms} = \frac{16 \text{ g O}}{16 \frac{\text{g O}}{\text{mole O}}} = 1.0 \text{ mole O} \quad (4)$$

$$\text{number moles oxygen atoms} = 1.0$$

From this information the *empirical* formula must be C_2H_6O.

This example has been much simplified by our selection of 46 g of sample. In actual practice less than 1 g of sample would be used and whole numbers of moles would not be obtained. A typical set of experimental data is given in Exercise 18-5 on the next page. In most cases the sample is analyzed in a special **analytical laboratory.**

---

* Symbolically we can represent this by the equation

$$C_nH_yO_z + aO_2 \longrightarrow nCO_2 + \frac{y}{2}(H_2O)$$

Note that the number of carbon atoms, n, is equal to the number of moles of CO_2.

Chapter 18 / The CHEMISTRY of CARBON COMPOUNDS

Data are usually reported by chemists in this laboratory as *percentage* carbon by weight, *percentage* hydrogen by weight, and *percentage* oxygen, or other elements in the sample, by weight.

EXERCISE 18-5

Automobile antifreeze often contains a compound called ethylene glycol. Analysis of pure ethylene glycol shows that it contains only carbon, hydrogen, and oxygen. A sample of ethylene glycol weighing 15.5 mg is burned and the weights of CO_2 and H_2O resulting are 22.0 mg and 13.5 mg respectively.

(1) What is the empirical formula of ethylene glycol?
(2) Calculate the percentage by weight of carbon, hydrogen, and oxygen in the sample.

The molecular formula for ethanol must be a whole number multiple of the empirical formula. We can write $n \times (C_2H_6O) =$ molecular formula. What is n? Each unit of C_2H_6O weighs 46 g. If n is 1, the molecular weight of ethanol will be 46; if n is 2, the molecular weight of ethanol will be 92; and if n is 3, the molecular weight of ethanol will be 138. We need an experimental value for the molecular weight of ethanol!

Since ethanol can be vaporized fairly easily, a vapor-density method for molecular weight is applicable. The procedure is very much like that which you used in Experiment 5. To apply the method to a liquid such as ethanol, a temperature above the boiling point is needed. A weighed amount of liquid is placed in a gas-collecting device held at an easily regulated temperature. (A steam condenser around the device provides a convenient way of holding the temperature at 100°C.) When the substance has vaporized completely, its pressure and volume are measured (see Experiment 5). This provides a measurement of the weight per unit volume of gaseous ethanol at a known temperature and pressure. Again, this weight is compared with the weight of the same volume of a reference gas (usually O_2) at the same temperature and pressure.

Suppose such a vapor-density measurement shows that a given volume of ethanol at 100°C and 1 atm weighs 1.5 times as much as the same volume of oxygen gas at 100°C and 1 atm. Since equal volumes contain equal numbers of molecules at the same temperature and pressure (Avogadro's hypothesis), one molecule of the unknown gas must weigh 1.5 times the weight of one molecule of oxygen gas. Therefore,

$$\text{mol wt of unknown gas} = (1.5) \times (\text{mol wt } O_2)$$
$$= 1.5 \times 32 \text{ g/mole} = 48 \text{ g/mole} \quad (5)$$

Even though this number is not very accurate, it will suffice for the purpose of deciding that the molecular formula is C_2H_6O. Clearly, formulas requiring molecular weights of 92 ($n = 2$) or 138 ($n = 3$) conflict seriously with the measured molecular weight of about 48. Only a value of $n = 1$ is consistent.

Freezing and boiling points of a solution can also be used to determine the molecular weight of a solute. Use of the mass spectrograph represents still another procedure for molecular weights. These methods will be described in subsequent courses in chemistry.

EXERCISE 18-6

Ethylene glycol, the example treated in Exercise 18-5, has an empirical formula of CH_3O. (Is this what you obtained?) A sample that weighs 0.49 g is vaporized completely at 200°C and 1 atm. The volume measured under these conditions is 291 ml. The same volume, 291 ml, of oxygen gas at 200°C and 1 atm weighs 0.240 g. What is the molecular formula for ethylene glycol, CH_3O, $C_2H_6O_2$, $C_3H_9O_3$, $C_4H_{12}O_4$, or some higher multiple of CH_3O?

18-2.3 The Experimental Determination of the Structural Formula—Chemical Evidence

You were told in Chapter 16 that much structural information can be obtained from an intelligent interpretation of the reactions which a compound undergoes. How can this assertion be applied to ethanol? We know that its empirical formula is C_2H_6O and its molecular formula is also C_2H_6O. It remains to discover the structural formula, the arrangement and connections of the atoms.

To begin, let us eliminate some structures we are sure are incorrect. Ethanol is not simply ethane with an oxygen atom somehow attached to a carbon atom. In ethane all four bonds of each carbon are satisfied, so there is no way in which an additional bond can form. We say that ethane is a **saturated compound.*** Nor can the oxygen atom just be attached somehow to a hydrogen atom. Each hydrogen atom in ethane has its bonding capacity already satisfied.

We have rejected a structural formula that pictures ethanol as ethane with an oxygen atom tacked onto it. By applying our bonding theory, let us start with an oxygen atom and try to build a molecule around it having two carbon atoms and six hydrogen atoms. We already know that the oxygen atom is commonly divalent, and that it makes bonds to hydrogen atoms, as in water. Let us start our molecular construction with a bond between one hydrogen atom and the oxygen atom:

$$O-H$$

The other bond the oxygen atom can make must be to a carbon atom, since if it were to another hydrogen atom we would simply have a water molecule. Therefore, we write

$$\begin{array}{c} C \\ \diagdown \\ O-H \end{array}$$

---

* This use of the word *saturated* shows that chemists, like other people, sometimes use the same word with two entirely different meanings. In Section 12-1.1 this word was used to describe a solution that contains the equilibrium concentration of a dissolved substance. In reference to organic compounds, saturated means that all bonds between carbon atoms are single bonds.

The carbon atom we have added must form three additional bonds to satisfy its tetravalent bonding capacity. If all these bonds were to hydrogen atoms, we would have the completed molecule CH_3OH and two hydrogen atoms and a carbon atom left over. Therefore, one of the bonds our first carbon atom forms must be to the other carbon atom, and the two other bonds must be to hydrogen atoms. We then have

$$C-C\begin{matrix}H\\H\\O-H\end{matrix}$$

We can easily complete the structure by adding three bonds from our last carbon atom to the three hydrogen atoms we have left. The result is

$$H-\underset{H}{\overset{H}{C}}-\underset{O-H}{\overset{H}{C}}-H$$

We have now used all six hydrogen atoms, the two carbon atoms, and the oxygen atom required by the molecular formula of ethanol.

Since all bonding rules are satisfied, this structure is a *possible* structural formula for ethanol. Now we must decide whether this structural formula is the *only* possible one for a molecule having the molecular formula C_2H_6O. A little reflection shows it is not. Instead of beginning with one oxygen-carbon and one oxygen-hydrogen bond, why not start with two oxygen-carbon bonds?

$$C\overset{O}{\diagup\diagdown}C$$

Since we have six hydrogen atoms at our disposal, and each carbon atom must form three more bonds, we complete the structure by writing

$$\underset{H}{\overset{H}{\diagdown}}\underset{H}{\overset{}{C}}\overset{O}{\diagup\diagdown}\underset{H}{\overset{H}{C}}\underset{}{\overset{}{\diagup}}H$$

Check to make certain that this structure violates no bonding rules and conforms to the empirical and molecular formula of ethanol.

We now have found all possible structural formulas for the ethanol molecule that agree with our bonding theory. While structures in conflict with current bonding theory might be considered in some situations, evidence to support them would have to be very strong. Bonding theory summarizes a large amount of experience.

According to bonding theory, the oxygen atom is either directly bonded to one carbon atom or to two carbon atoms. *Once a choice between these two possibilities is made, the structure of the rest of the molecule can be determined from the molecular formula and the bonding rules.* The two possible structures are shown in Figure 18-1. Once again we have a choice between two structural isomers.* The

---

* Review Section 16-6.2 if you have forgotten the meaning of structural isomers.

STRUCTURAL FORMULA—
CHEMICAL EVIDENCE

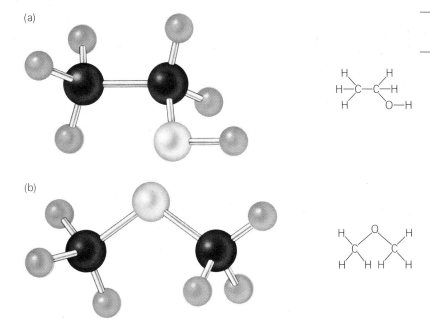

Fig. 18-1 **The structural formulas and models of the C_2H_6O isomers.**

existence of two compounds having the formula C_2H_6O perplexed chemists for decades. Now we recognize the crucial importance of determining the structural as well as molecular formula of a substance.

Our problem, then, is to decide whether ethanol has structure (a) or (b). How can we tell which is correct? Let us see what preliminary ideas we can get from an examination of the structural formulas.

In structure (b) all hydrogen atoms are the same—each is bonded to a carbon atom, which is bonded to the oxygen atom. In structure (a), however, one of the hydrogen atoms is quite different from the others—it is bonded to oxygen and not to carbon. Of the remaining five, two are bonded to the carbon bonded to the oxygen and three are bonded to the other carbon. Structures (a) and (b) should have very different chemistries. Which one should correspond to the chemistry of ethanol?

We can offer several kinds of evidence. Some come from the behavior of ethanol in chemical reactions and some from the determination of certain physical properties. Let us consider the reactions first.

Clean sodium metal reacts vigorously with ethanol, giving hydrogen gas and an ionic compound, sodium ethoxide, which has the empirical formula C_2H_5ONa. The reaction is quite similar to the behavior of sodium and water to give hydrogen and the ionic compound sodium hydroxide (NaOH). This suggests, but certainly does not prove, that ethanol shows some structural similarity to water. In water two hydrogen atoms are bonded to one oxygen atom, and in structure (a) one hydrogen atom is bonded to the oxygen atom. This chemical evidence supports structure (a) for ethanol.

More quantitative evidence can be obtained by using an excess of sodium and a weighed amount of ethanol and measuring the amount of hydrogen gas evolved. It is found that 46 g of ethanol

(1 mole) will produce only 0.5 mole of hydrogen gas. We can therefore write a balanced chemical equation for the reaction of sodium with ethanol:

$$Na(s) + C_2H_6O(l) \longrightarrow \tfrac{1}{2}H_2(g) + C_2H_5ONa(s) \qquad (6)$$

According to equation (6), 1 mole of ethanol produces 0.5 mole of hydrogen gas. Hence, 1 mole of ethanol must contain 1 mole of hydrogen atoms that are uniquely capable of undergoing reaction with sodium. Apparently, one molecule of ethanol contains one hydrogen atom that is capable of reacting with sodium and five that are not. Let us now consider structures (a) and (b) in the light of this information. In structure (b) all six hydrogen atoms are structurally equivalent, whereas in structure (a) there is one hydrogen atom in the molecule that is structurally unique because it is bonded to the oxygen atom. Structure (a) is therefore consistent with the experimental fact that only one hydrogen atom per molecule of ethanol will react with sodium, but structure (b) is not.

We can find further evidence that structure (a), CH_3CH_2OH, is the correct structural formula for ethanol. It is known that compounds that contain only carbon and hydrogen, such as ethane (C_2H_6), do not react readily with metallic sodium to produce hydrogen gas. In these compounds the hydrogen atoms are all bonded to carbon atoms, as in the structural formula for ethane in Section 16-6; so we can deduce that hydrogen atoms bonded to carbon atoms generally do not react with sodium to produce hydrogen gas. In structure (b), CH_3OCH_3, all hydrogen atoms are bonded to carbon atoms, so we do not expect it to react with sodium. But ethanol reacts with sodium, so it is unlikely that ethanol has structure (b). Our confidence in this reasoning is strengthened by noting that the other known structural isomer of formula C_2H_6O does *not* react with sodium at all. Its chemistry is consistent with structure (b).

Let us consider one other reaction of ethanol. If ethanol is heated with aqueous HBr, a volatile compound is formed. This compound is only slightly soluble in water and it contains bromine; by analysis and molecular weight, its molecular formula is found to be C_2H_5Br (ethyl bromide, or bromoethane). Using the bonding rules, we can see that there is only one possible structure for this compound. This result is verified by the fact that only one isomer of C_2H_5Br has ever been discovered.

Fig. 18-2 The structural formula and model of ethyl bromide (bromoethane).

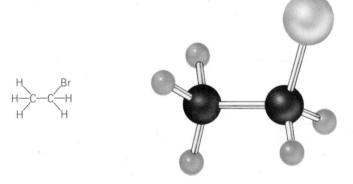

Now we can ask how this chemical reaction furnishes a clue to the structure of ethanol. Structure (a) could give the structure shown in Figure 18-2 by breaking the carbon-oxygen bond. In contrast, bromoethane can be obtained from structure (b) only by a complicated rearrangement of atoms. Many years of laboratory experience show that such complicated reshufflings of atoms rarely occur. Therefore, the reaction between ethanol and HBr to form bromoethane provides more evidence that ethanol has structure (a).

18-2.4 The Experimental Determination of the Structural Formula—Physical Evidence

The evidence cited so far has been associated with the chemistry of ethanol. Its boiling point provides a different sort of evidence, which also supports a choice of structure (a). Ethanol is a liquid with a boiling point of 78°C. This can be compared with the boiling points of ethane (C_2H_6), which is -172°C, and water, 100°C. Ethanol is, then, more like water than like ethane in terms of boiling point. Once again, this can be explained better by structure (a), which, like H_2O, has oxygen linked to hydrogen. The high boiling point of water is explained in terms of an abnormally large intermolecular attraction of such an OH group to surrounding water molecules. The interaction is called hydrogen bonding (see Section 17-6). If ethanol also has the OH group, as in structure (a), then it too can exert the same abnormally large attraction to neighboring ethanol molecules. Thus, structure (a) provides an explanation of the fact that the boiling point of ethanol is so high.

This possibility of forming hydrogen bonds should cause a strong attraction between water and a compound of structure (a). If there is strong attraction, then ethanol should have high solubility in water. Experiment shows that they are **miscible,** that they dissolve in all proportions. Again, the evidence tends to support structure (a).

Various spectroscopic techniques provide more direct evidence. The **infrared spectrum** of the material tentatively identified as ethanol, structure (a), can be obtained in the gas phase. It shows absorption frequencies associated with hydrogen bonded to carbon, and with hydrogen bonded to oxygen. The other structural isomer of C_2H_6O shows only the absorption frequency of hydrogen bonded to carbon.

Nuclear magnetic resonance (nmr) spectroscopy provides even clearer evidence for the structure of ethanol. As noted in Section 16-7.6, the frequency of the electromagnetic radiation needed to flip over the small electromagnet created by a spinning nucleus is determined by the environment in which the nucleus finds itself. This means that for the compound

$$3 \longrightarrow \begin{array}{c} \text{H} \\ | \\ \text{H—C—C—O—H} \\ | \\ \text{H} \end{array} \begin{array}{c} \text{H} \longleftarrow 2 \\ | \\ \longleftarrow 1 \\ | \\ \text{H} \longleftarrow 2 \end{array}$$

three different nmr frequencies corresponding to hydrogens in positions *1, 2,* and *3* should be found. The intensity of the signal is proportional to the number of hydrogen atoms in that position. For a compound of structure (b), CH_3OCH_3, only a single nmr frequency should be observed since all six hydrogens are identical. In Figure 18-3 the nmr spectra for CH_3CH_2OH and for CH_3OCH_3 are displayed. Structure (a) shows the expected three absorptions and structure (b) shows only one.

Notice that many types of evidence lead to the same conclusion, that ethanol has the structure

$$\begin{array}{c} \text{H} \quad \text{H} \\ | \quad \ \ | \\ \text{H}-\text{C}-\text{C}-\text{O}-\text{H} \\ | \quad \ \ | \\ \text{H} \quad \text{H} \end{array}$$

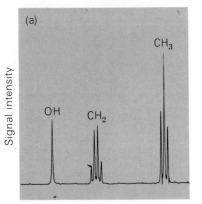

Fig. 18-3 **The nmr spectra for (a) CH_3CH_2OH and (b) CH_3OCH_3.**

No fact by itself gives absolute proof of the structure, but all facts taken together establish the structure with a high degree of certainty. An X-ray diffraction or electron diffraction study would give even more detailed structural information, such as distances between nuclei in the molecule.

The second isomer with structure (b) is known as dimethyl ether.

EXERCISE 18-7

Ethylene glycol has the empirical formula CH_3O and the molecular formula $C_2H_6O_2$. Using the usual bonding rules (e.g., carbon is tetravalent, oxygen is divalent, hydrogen is monovalent), draw some of the structural formulas possible for this compound.

EXERCISE 18-8

Decide which of your structures in Exercise 18-7 best fits the following properties observed for pure ethylene glycol: (a) It is a viscous (syrupy) liquid that boils at 197°C. (b) It is miscible with water—that is, it dissolves, forming solutions, in all proportions. (c) It is miscible with ethanol. (d) It reacts with sodium metal, producing hydrogen gas. (e) A 6.2-g sample of ethylene glycol reacts with an excess of sodium metal to produce 2.4 liters of hydrogen gas at 1 atm and 25°C. (f) The low resolution nmr signal for ethylene glycol shows two absorptions—one twice as intense as the other.

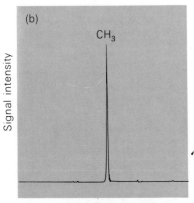

18-2.5 Structural Components—The Ethyl Group

All the reactions and the physical properties of ethanol have been explained on the basis of the behavior of the OH group in structure (a), CH_3CH_2OH. This is true of most of the reactions of ethanol—the reaction centers at the OH group or **hydroxyl group,** and the remainder of the molecule, CH_3CH_2—, remains intact. These reactions suggest that there are two parts to the ethanol molecule, the

$$\begin{array}{c} \text{H} \quad \text{H} \\ | \quad \ \ | \\ \text{H}-\text{C}-\text{C}- \\ | \quad \ \ | \\ \text{H} \quad \text{H} \end{array}$$

group, *which is unchanged during reactions,* and the —OH group, *which can change.* This concept of *the structural integrity of the hydrocarbon group* is important in organic chemistry. It focuses attention on the groups that *do* change, the so-called **functional groups.**

If we understand the chemistry of a particular functional group for one compound, we can assume that its chemistry is true of other compounds containing this same functional group. Thus, compounds with the OH group are given a *family name*, **alcohols.** The rest of the molecule, the carbon skeleton, has relatively little effect, and remains intact during the reactions of the functional group.

We mentioned earlier that when ethanol reacts with hydrogen bromide, ethyl bromide is formed. Similar treatment of ethanol with hydrogen chloride or hydrogen iodide gives us the ethyl halides:

$$CH_3CH_2OH + HBr \longrightarrow CH_3CH_2Br + H_2O \qquad (7)$$

$$CH_3CH_2OH + HCl \longrightarrow CH_3CH_2Cl + H_2O \qquad (8)$$

$$CH_3CH_2OH + HI \longrightarrow CH_3CH_2I + H_2O \qquad (9)$$

We say that the hydroxyl group has been *displaced* and that the halogen atom has been substituted for it. You can see that the group CH_3CH_2- has remained intact in all these reactions. Indeed, this group has appeared in most of our discussion so far, sometimes attached to oxygen, as in ethanol and sodium ethoxide, sometimes attached to other atoms, as in the ethyl halides. You will recall that earlier we became acquainted with ethane (C_2H_6). Looking at the structural formula of ethane, you see it is simply the CH_3CH_2- group attached to hydrogen:

$$\begin{array}{c} H\ \ H \\ |\ \ \ | \\ H-C-C-H \\ |\ \ \ | \\ H\ \ H \end{array}$$

or CH_3CH_2-H. This group, CH_3CH_2- (also written C_2H_5-), is called the **ethyl group.**

Because ethyl bromide and ethyl alcohol (ethanol) can be considered as derived from ethane (by the substitution of $-Br$ and $-OH$ for one of its hydrogens), we speak of these as *derivatives* of ethane, and we say that ethane is the *parent hydrocarbon* for a series of related compounds. The name "ethyl" is derived from the name of the parent hydrocarbon, ethane. In the same way the name of the **methyl group** (CH_3-) is derived from that of methane (CH_4), and the name of the **propyl group** ($CH_3CH_2CH_2-$) is derived from propane ($CH_3CH_2CH_3$).

It is important to realize that these groups are not substances that can be isolated and bottled. They are simply parts of molecules that remain intact in composition and structure during reactions. We find this way of classifying organic groups useful and convenient; but we must keep in mind that in the reactions we have described, the ethyl group is not actually formed as a distinct substance. Table 18-1 (page 484) gives more examples of group names.

18-3 SOME CHEMISTRY OF ORGANIC COMPOUNDS

18-3.1 Some Chemistry of Ethyl and Methyl Bromide

We can use ethyl and methyl bromide to illustrate one kind of organic reaction. Ethyl bromide is not particularly reactive, but it does react with bases such as NaOH or NH_3. If we mix ethyl bromide and aqueous sodium hydroxide solution and then heat the mixture for about an hour, we find that sodium bromide and ethanol are formed:

$$C_2H_5Br + OH^-(aq) \longrightarrow C_2H_5OH + Br^-(aq) \qquad (10)$$

This reaction may seem similar to the reaction between aqueous HBr and NaOH, but there are two important differences. The ethyl bromide reaction is very slow (about one hour is needed for the reaction) and the reactants are a covalent molecule (C_2H_5Br) and an ion (OH^-). In contrast, the reaction between HBr and NaOH in water occurs in a fraction of a second and it involves ions only:

$$H^+(aq) + OH^-(aq) \longrightarrow H_2O(l) \qquad (11)$$

Let us describe the course of the reaction of methyl bromide and OH^- in terms of a model. We shall use *methyl* bromide to simplify the description, but the reaction of ethyl bromide is of the same type. The equation for the process we shall study is

$$CH_3Br + OH^- \longrightarrow CH_3OH + Br^- \qquad (12)$$

First, let us recount a few of the experimental facts:

(1) Methyl bromide is a compound in which the chemical bonds are predominantly covalent. An aqueous solution of methyl bromide does not conduct electricity; hence, it does not form significant quantities of ions (such as CH_3^+ and Br^- ions) in aqueous solutions.
(2) The reaction takes a measurable time for completion.
(3) Experiments show that the rate of the reaction is increased by increasing the concentration of OH^- and also by raising the temperature.

These observations remind us of Chapter 10, in which we considered the factors that determine the rate of a chemical reaction. Of course, these same ideas apply here. We can obtain qualitative information about the mechanism of the reaction by applying the collision theory. A quantitative study of the effects of temperature and concentration on the rate should enable us to construct potential energy diagrams like those shown in Figure 10-11.

Figure 18-14 shows the mechanism chemists have deduced for reaction (*12*). The steps of this mechanism are:

(1) the approach of the hydroxide ion [Figure 18-4(a)],
(2) the formation of the atomic arrangement thought to be the activated complex [Figure 18-4(b)], and
(3) formation of the final products from the activated complex [Figure 18-4(c)].

In the activated complex the O—C bond is beginning to form and the C—Br bond is beginning to break. The potential energy curves

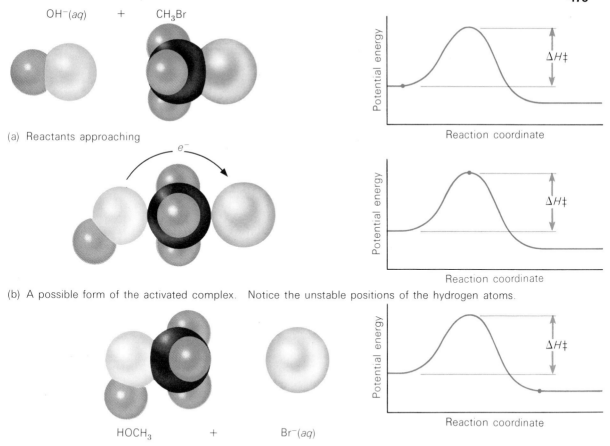

Fig. 18-4 The mechanism and potential energy diagrams for the reaction $CH_3Br + OH^-(aq) \longrightarrow CH_3OH + Br^-(aq)$.

for the reaction are shown alongside the molecular models in Figure 18-4. The slow rate suggests that activation energy is needed. One reason why activation energy must be supplied is that in the activated complex the bond angles have been distorted from their (stable) configurations and forced into an unstable arrangement.

18-3.2 Oxidation of Organic Compounds— Formation of Aldehydes

By far the majority of the million known compounds of carbon also contain hydrogen and oxygen. There are several important types of oxygen-containing organic compounds; they can be studied as an oxidation series. For example, the compound, methanol (CH_3OH), is very closely related to methane, as their structural formulas show (Figures 18-5.1 and 18-5.2). Methanol can be regarded as the first step in the complete oxidation of methane to carbon dioxide and water.

Methanol (and other alcohols) react with common inorganic oxidizing agents such as potassium dichromate ($K_2Cr_2O_7$). <u>When an acidic, aqueous solution of potassium dichromate reacts with methanol, the solution turns from bright orange to muddy green because of the production of the green chromic ion Cr^{3+}</u>. The solution then has a strong odor easily identified as that of formaldehyde.

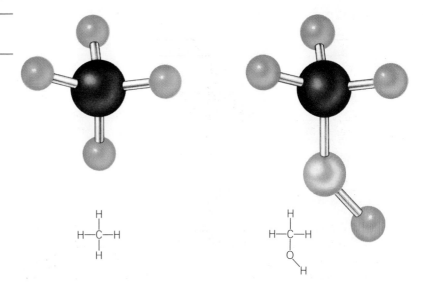

Fig. 18-5.1 (Left) The structural formula and model of methane.

Fig. 18-5.2 (Right) The structural formula and model of methanol.

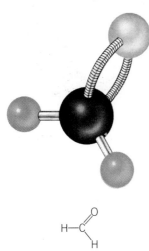

Fig. 18-6.1 The structural formula and model of formaldehyde.

Its formula, CH_2O, represents the structure in Figure 18-6.1. Notice that the bond between the carbon and oxygen is a double bond (see Section 16-6), and that the two C, the H, and the O atoms lie in the same plane.

The balanced net reaction for the formation of formaldehyde is

$$3CH_3OH + Cr_2O_7^{2-}(aq) + 8H^+(aq) \longrightarrow \\ 3CH_2O + 2Cr^{3+}(aq) + 7H_2O \qquad (13)$$

Since the dichromate ion on the left side of the equation has been reduced to chromic ion, Cr^{3+}, on the right side, the conversion of methanol to formaldehyde must involve oxidation.

To show more clearly that methanol has been oxidized, let us balance this reaction by the half-reaction method. We have encountered the half-reaction involving dichromate and chromic ions before (question 19 in Chapter 14):

$$Cr_2O_7^{2-}(aq) + 14H^+(aq) + 6e^- \longrightarrow 2Cr^{3+}(aq) + 7H_2O \qquad (14)$$

To balance the methanol-formaldehyde half-reaction, we begin with

$$CH_3OH \text{ gives } CH_2O$$

This statement does not yet show the fact that hydrogen atoms are conserved in the reaction, since there is a deficiency of two hydrogen atoms on the right. This can be remedied by adding two hydrogen ions:

$$CH_3OH \text{ gives } CH_2O + 2H^+(aq)$$

Now the equation is chemically but not electrically balanced. The addition of two electrons to the right-hand side completes the balancing procedure, and the completed half-reaction is

$$CH_3OH \longrightarrow CH_2O + 2H^+(aq) + 2e^- \qquad (15)$$

This equation shows that the methanol molecule has lost electrons and has thus been oxidized. Formaldehyde is the second member in the oxidation series of methane.

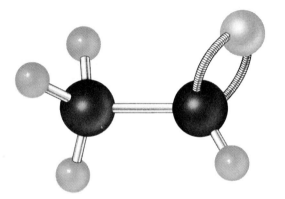

Fig. 18-6.2 The structural formula and model of acetaldehyde.

In a similar manner, ethanol can be oxidized by the dichromate ion to form a compound called acetaldehyde (CH_3CHO). The molecular structure of acetaldehyde, which is similar to that of formaldehyde, is shown in Figure 18-6.2. We see that the molecule is structurally similar to formaldehyde. The methyl group, CH_3—, replaces one of the hydrogens of formaldehyde. The balanced equation for the formation of acetaldehyde from ethanol is

$$3CH_3CH_2OH + Cr_2O_7^{2-}(aq) + 8H^+(aq) \longrightarrow \\ 3CH_3CHO + 2Cr^{3+}(aq) + 7H_2O \quad (16)$$

18-3.3 Oxidation of Organic Compounds—
The Formation of Carboxylic Acids

Another oxidation product can be obtained from the reaction of an acidic aqueous solution of potassium permanganate with methanol. The product has the formula HCOOH and is called formic acid. The structural formula of formic acid is shown in Figure 18-7.1. Formic acid is also related to formaldehyde structurally. If one of the hydrogen atoms of formaldehyde is replaced by an OH group, the resulting molecule is formic acid. The balanced equation for the formation of formic acid from methanol is

$$5CH_3OH + 4MnO_4^-(aq) + 12H^+(aq) \longrightarrow \\ 5HCOOH + 4Mn^{2+}(aq) + 11H_2O \quad (17)$$

The half-reaction involving methanol and formic acid can be obtained by using the three steps outlined in the previous example of the methanol-formaldehyde half-reaction. Again, we begin with

$$CH_3OH \text{ gives } HCOOH$$

Chemically balanced this is

$$CH_3OH + H_2O \text{ gives } HCOOH + 4H^+(aq)$$

and with charge balanced,

$$CH_3OH + H_2O \longrightarrow HCOOH + 4H^+(aq) + 4e^- \quad (18)$$

From this completed half-reaction we see that the conversion of methanol to formic acid involves the loss of four electrons. Since

Fig. 18-7.1 The structural formula and model of formic acid.

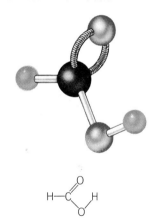

Fig. 18-7.2 The structural formula and model of acetic acid.

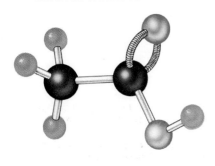

the oxidation of methanol to formaldehyde was only a two-electron change, it is clear that formic acid is a more highly oxidized compound of carbon than either formaldehyde or methanol.

EXERCISE 18-9

Balance the half-reaction for the conversion of formaldehyde (CH_2O) to formic acid ($HCOOH$).

Just as methanol can be oxidized to formic acid, ethanol can be oxidized to acetic acid (CH_3COOH). The molecular structure of acetic acid is shown in Figure 18-7.2. The atomic grouping —COOH is called the **carboxyl group,** and acids containing this group are called **carboxylic acids.**

The balanced equation for production of acetic acid from ethanol is

$$5CH_3CH_2OH + 4MnO_4^-(aq) + 12H^+(aq) \longrightarrow$$
$$5CH_3COOH + 4Mn^{2+}(aq) + 11H_2O \quad (19)$$

Acetic acid can also be obtained by the oxidation of acetaldehyde (CH_3CHO):

$$5CH_3CHO + 2MnO_4^-(aq) + 6H^+(aq) \longrightarrow$$
$$5CH_3COOH + 2Mn^{2+}(aq) + 3H_2O \quad (20)$$

The further oxidation of acetic acid is difficult to accomplish. It does not react in solutions of $K_2Cr_2O_7$ or $KMnO_4$. Vigorous treatment, such as burning, causes its complete oxidation to carbon dioxide and water. Formic acid also can be oxidized to carbon dioxide and water by combustion with oxygen.

EXERCISE 18-10

There is a compound called propanol having the structural formula $CH_3CH_2CH_2OH$. If it is oxidized carefully, an aldehyde called propionaldehyde is obtained. Vigorous oxidation gives an acid called propionic acid. Draw structural formulas like those shown in Figures 18-6 and 18-7 for propionaldehyde and propionic acid.

EXERCISE 18-11

Balance the half-reaction involved in the oxidation of ethanol to acetic acid. Compare the number of electrons released per mole of ethanol with the number per mole of methanol in the equivalent reaction. How many electrons would be released per mole of propanol in the oxidation to propionic acid?

18-3.4 The Oxidation of Organic Compounds— The Formation of Ketones

The bonding rules permit us to draw two acceptable structural formulas for an alcohol containing three carbon atoms, $CH_3CH_2CH_2OH$ and $CH_3CHOHCH_3$. In the first isomer (considered in Exercises 18-10

OXIDATION—
The FORMATION of KETONES

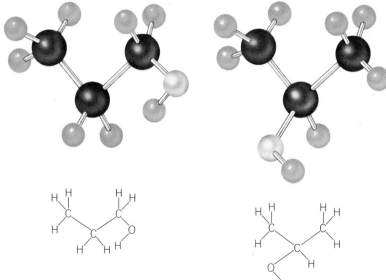

Fig. 18-8.1 (Left) The structural formula and model of 1-propanol.

Fig. 18-8.2 (Right) The structural formula and model of 2-propanol.

and 18-11), the OH group is attached to the end carbon atom. In the second isomer, the OH group is attached to the second carbon atom. They are both called propanol because they are both derived from propane ($CH_3CH_2CH_3$). They are distinguished by numbering the carbon atom to which the functional group, the OH, is attached. Thus, $CH_3CH_2CH_2OH$ is called 1-propanol because the OH is attached to the end carbon atom in the chain. The other alcohol, $CH_3CHOHCH_3$, is called 2-propanol because the OH is attached to the second carbon atom. The structures of these two alcohols are shown in Figure 18-8. We have already considered the oxidation of 1-propanol in Exercise 18-10. The second isomer, 2-propanol, can also be oxidized, and the product is called acetone:

$$3CH_3CHOHCH_3 + Cr_2O_7^{2-}(aq) + 8H^+(aq) \longrightarrow$$
$$3CH_3COCH_3 + 2Cr^{3+}(aq) + 7H_2O \quad (21)$$

Fig. 18-9 The structural formula and model of acetone, the simplest ketone.

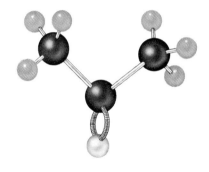

Acetone is the simplest member of a class of compounds called **ketones**. They are quite similar in structure to the aldehydes, since each contains a carbon atom doubly bonded to an oxygen atom* (see Figure 18-9). They differ in that the aldehyde has a hydrogen atom attached to this same carbon atom, whereas the ketone does not. (Compare Figures 18-6 and 18-9.) Since this hydrogen atom is not present, a ketone cannot be oxidized further to an acid.

Figure 18-10 summarizes the successive oxidation products that can be obtained from alcohols. When the hydroxyl or OH group is attached to an end carbon atom, oxidation will give an aldehyde or a carboxylic acid. When the hydroxyl group is on a carbon atom attached to other carbon atoms, oxidation will give a ketone. Huge

* The group

$$>\!\!C\!\!=\!\!O$$

is called the **carbonyl group.**

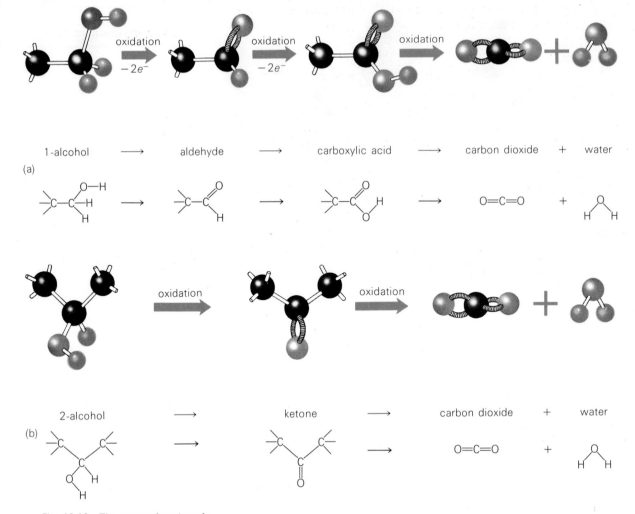

Fig. 18-10 The successive steps in the oxidation of alcohol to carbon dioxide and water.

amounts of aldehydes and ketones are used industrially in a variety of chemical processes. Furthermore, these functional groups are important in chemical syntheses of medicines, dyes, plastics, and fabrics.

18-3.5 The Functional Group

The reactive groups we have encountered thus far, such as —Br, —OH, —CHO, and —COOH, are called **functional groups**. They are the parts from which the molecules get their characteristic chemical behavior. For example, the ability to undergo a reaction in which the —OH group is displaced by a halogen atom, such as in the reaction of CH_3OH and HBr or C_2H_5OH and HCl, is common to all alcohols. Characteristic of aldehydes is the ability to be oxidized to acids, as in the oxidation of formaldehyde to formic acid:

$$CH_2O + \text{source of oxygen atoms gives HCOOH}$$

These common types of behavior are shown by using the general symbol R— for the part of the molecule that does not change and by writing a reaction in such a way as to focus attention on the functional group. For example,

$$RCH_2OH + HBr \longrightarrow RCH_2Br + H_2O \tag{22}$$

$$RCH_2Br + OH^-(aq) \longrightarrow RCH_2OH + Br^-(aq) \tag{23}$$

$$3RCHO + Cr_2O_7^{2-}(aq) + 8H^+(aq) \longrightarrow$$
$$3RCOOH + 2Cr^{3+}(aq) + 4H_2O \tag{24}$$

The symbol $R-$ in these formulas represents any **alkyl group,** such as CH_3- or C_2H_5-.

18-3.6 Amines

Alcohols can be related to water by imagining that an alkyl group (such as CH_3-) has been substituted for one of the two hydrogen atoms of water. In the same way, **amines** are related to ammonia:

$$\begin{array}{ccc} \text{H} & \text{H} & \text{H} \\ \text{H}\!\!>\!\!\text{N}\!-\!\text{H} & \text{H}\!\!>\!\!\text{N}\!-\!R & \text{H}\!\!>\!\!\text{N}\!-\!CH_2CH_3 \\ \text{AMMONIA} & \text{AMINE} & \text{ETHYLAMINE} \end{array}$$

Amines can be prepared by direct reaction of ammonia with an alkyl halide, such as CH_3Br or CH_3CH_2I. Iodides react fastest, and an excess of ammonia is often used to help control formation of undesired alternate products:

$$\underset{\text{ETHYL IODIDE}}{CH_3CH_2I} + 2NH_3 \longrightarrow \underset{\text{ETHYLAMINE}}{CH_3CH_2NH_2} + NH_4I \tag{25}$$

$$R\!-\!I + 2NH_3 \longrightarrow RNH_2 + NH_4I$$

The above equations represent a net change that occurs when an excess of ammonia reacts with an alkyl iodide. The actual reaction proceeds in two steps. The first step is analogous to the attack of the hydroxide ion on an alkyl halide (see Figure 18-4 on page 475):

$$NH_3 + RI \longrightarrow RNH_3^+(aq) + I^-(aq) \tag{26}$$

The second step is a proton-transfer reaction (see Section 13-7):

$$NH_3 + RNH_3^+(aq) \longrightarrow RNH_2 + NH_4^+(aq) \tag{27}$$

18-3.7 Acid Derivatives—Esters

We see from the equation for the oxidation of CH_2O to $HCOOH$ that oxidation of an aldehyde will give an organic acid. All these organic acids contain the functional group $-COOH$, the **carboxyl group.** The bonding in this group is as follows:

$$-C\!\!\!\begin{array}{c}\nearrow\!\!O\\ \searrow\!\!O\!-\!H\end{array}$$

The carboxyl group readily releases a proton, so it is an acid. Acetic acid, for example, dissolved in water gives a conducting solution; it turns blue litmus red; it is sour; and it exhibits the other properties of an acid. The reaction

$$CH_3COOH + H_2O \longrightarrow CH_3COO^-(aq) + H_3O^+(aq) \qquad (28)$$

has an equilibrium constant of 1.8×10^{-5}.

In addition to this acidic behavior, an important characteristic of carboxylic acids is that the entire OH group can be replaced by other groups. The resulting compounds are called **acid derivatives**. We shall consider only two types of acid derivatives, **esters** and **amides**.

Compounds in which the —OH of an acid is transformed into —OR (such as —OCH$_3$) are called esters. They can be prepared by the direct reaction between an alcohol and the acid:

$$CH_3OH + CH_3C\overset{O}{\underset{OH}{\diagdown}} \longrightarrow CH_3C\overset{O}{\underset{O-CH_3}{\diagdown}} + H_2O \qquad (29)$$

$$\{\text{methyl alcohol}\} + \{\text{acetic acid}\} \longrightarrow \{\text{methyl acetate}\} + \{\text{water}\}$$

The method of naming the product is indicated by the bold-faced parts of the names for the reactants and products.

EXERCISE 18-12

Write equations for the reaction of ethanol and formic acid, propanol and propionic acid, and methanol and formic acid. Name the esters produced in each case.

When equilibrium is reached in the formation of methyl acetate, appreciable concentrations of all reactants may be present. If methyl acetate alone (the product on the right) is dissolved in water, it will react with water slowly to give acetic acid and methyl alcohol until equilibrium is attained:

$$CH_3C\overset{O}{\underset{O-CH_3}{\diagdown}} + H_2O \longrightarrow CH_3OH + CH_3C\overset{O}{\underset{OH}{\diagdown}} \qquad (30)$$

Of course, the usual equilibrium considerations apply. For example, if we add more methyl alcohol, equilibrium conditions will shift, consuming the added reagent and acetic acid to produce more methyl acetate and water, in accordance with Le Chatelier's principle. Thus, an excess of methyl alcohol causes most of the acetic acid to be converted to methyl acetate.

EXERCISE 18-13

Write the equilibrium expression relating the concentrations of reactants and products in the reaction of methyl acetate with water. Notice that the concentration of water must be included because it is not necessarily large enough to be considered constant in these nonaqueous solutions.

The reaction between methanol and acetic acid is slow, but it can be greatly accelerated if a catalyst is added. For example, adding a strong acid such as hydrochloric acid or sulfuric acid will speed up the reaction by catalysis. As mentioned in Section 11-3.3, the catalyst does not alter the equilibrium state (that is, the concentration of the reactants at equilibrium); it only enables equilibrium to be attained more rapidly.

EXERCISE 18-14

A strong acid such as hydrochloric acid or sulfuric acid will catalyze the formation of an ester from alcohol and acid. Explain why this implies that these acids will catalyze the reaction between an ester and water as well. (Refer to Section 10-2.3.)

Esters are important substances. The esters of the low-molecular weight acids and alcohols have fragrant, fruit-like odors, and are used in perfumes and artificial flavorings. Esters are useful solvents; this is the reason they are commonly found in "model airplane dope" and fingernail polish remover.

18-3.8 Acid Derivatives—Amides

A compound in which the —OH group of an acid is replaced by —NH$_2$ is called an **amide**. When the —OH is replaced by —NHR, the product is called a nitrogen-substituted amide or, abbreviated, an N-substituted amide. Amides can be produced from the reaction of ammonia (or an amine) with an ester:

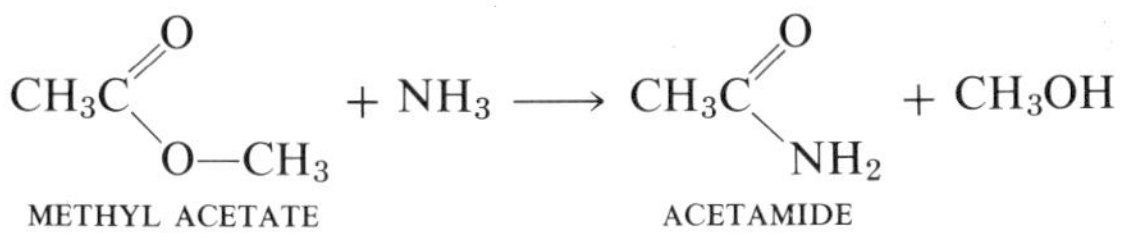

$$CH_3C(=O)O-CH_3 + NH_3 \longrightarrow CH_3C(=O)NH_2 + CH_3OH \quad (31)$$
METHYL ACETATE → ACETAMIDE

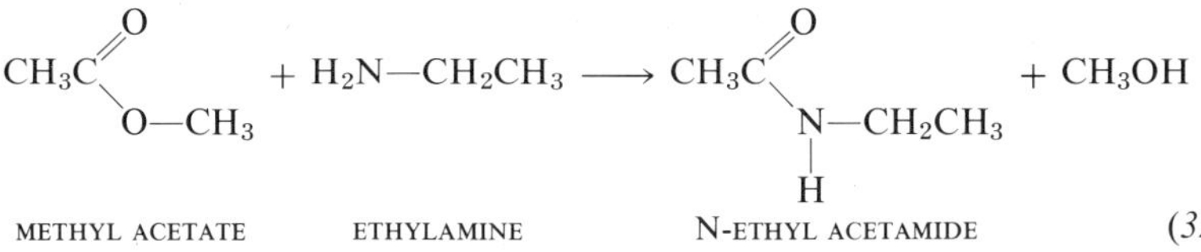

$$CH_3C(=O)O-CH_3 + H_2N-CH_2CH_3 \longrightarrow CH_3C(=O)N(H)-CH_2CH_3 + CH_3OH \quad (32)$$
METHYL ACETATE ETHYLAMINE N-ETHYL ACETAMIDE

Note the similarity of the two reactions. Amides are of special importance because the amide grouping

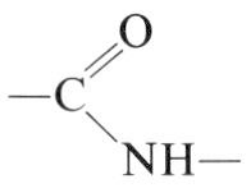

$$-C(=O)NH-$$

is the basic structural element in the long-chain molecules that make up proteins and enzymes in living matter. Hydrogen bonding between two amide groups helps determine the protein structure. This will be discussed in Chapter 22.

18-4 NOMENCLATURE

The names of organic compounds have some system. Each functional group defines a family (for example, alcohols and amines), and a specific modifier is added to identify a particular member in the family (for example, *ethyl* alcohol and *ethyl* amine). As an alternate naming system, the family may be identified by a specific ending (for example, alcohol names end in *-ol*), and a particular member of the family may be indicated by an appropriate stem (ethyl alcohol would be *ethan*ol).

TABLE 18-1 REGULARITIES IN NAMES OF ALKANES, ALCOHOLS, AND AMINES

Number of Carbon Atoms	Alkanes	Alcohols	Amines
1	CH_4 methane	CH_3OH methyl alcohol methanol	CH_3NH_2 methylamine
2	CH_3CH_3 ethane	CH_3CH_2OH ethyl alcohol ethanol	$CH_3CH_2NH_2$ ethylamine
3	$CH_3CH_2CH_3$ propane	$CH_3CH_2CH_2OH$ propyl alcohol 1-propanol	$CH_3CH_2CH_2NH_2$ 1-propylamine
4	$CH_3CH_2CH_2CH_3$ butane	$CH_3CH_2CH_2CH_2OH$ butyl alcohol 1-butanol	$CH_3CH_2CH_2CH_2NH_2$ 1-butylamine
8	$CH_3(CH_2)_6CH_3$ octane	$CH_3(CH_2)_6CH_2OH$ octyl alcohol 1-octanol	$CH_3(CH_2)_6CH_2NH_2$ 1-octylamine

These naming systems are illustrated in Tables 18-1 and 18-2. An examination of these tables will reveal that the key to developing nomenclature is the name of the alkane, which is systematically modified to the names of its acid derivatives. From pentane (C_5H_{12}) and hexane (C_6H_{14}), the alkane names themselves are quite regular, deriving their prefixes from the Greek words for their number of carbon atoms. Compounds with more complicated shapes and more than one functional group are described by a straightforward naming system based on numbering of carbon atoms.

TABLE 18-2 REGULARITIES IN NAMES OF ACIDS, AMIDES, AND ESTERS

Number of Carbon Atoms	Acids	Amides	Esters—Acids with Methanol
1	HCOOH formic acid	$HCONH_2$ formamide	$HCOOCH_3$ methyl formate
2	CH_3COOH acetic acid	CH_3CONH_2 acetamide	CH_3COOCH_3 methyl acetate
3	CH_3CH_2COOH propionic acid	$CH_3CH_2CONH_2$ propionamide	$CH_3CH_2COOCH_3$ methyl propionate
4	$CH_3CH_2CH_2COOH$ butyric acid	$CH_3CH_2CH_2CONH_2$ butyramide	$CH_3CH_2CH_2COOCH_3$ methyl butyrate
8	$CH_3(CH_2)_6COOH$ octanoic acid caprylic acid	$CH_3(CH_2)_6CONH_2$ octanamide caprylamide	$CH_3(CH_2)_6COOCH_3$ methyl octanoate methyl caprylate

18-5 HYDROCARBONS

SATURATED HYDROCARBONS

Compounds that contain only hydrogen and carbon are called **hydrocarbons** (a contraction of hydrogen-carbon). The hydrocarbons that have only single bonds all have a similar chemistry and they are called, as a family, the **saturated hydrocarbons**. With carbon-carbon double bonds, the reactivity is much enhanced (Section 16-6); hence, hydrocarbons containing one or more double bonds are named, as a distinct family, **unsaturated hydrocarbons**. Both saturated and unsaturated hydrocarbons can occur in chain-like structures or in cyclic structures. Each of these families will be considered.

18-5.1 Saturated Hydrocarbons

Ethane is a saturated hydrocarbon, which, as we stated above, can occur in chains, branched chains, and cyclic structures. The chain and branched-chain saturated hydrocarbons make up a family called the **alkanes**. Some saturated hydrocarbons with five carbon atoms are shown in Figure 18-11. Figure 18-11(a), containing no branches, is called normal pentane or *n*-pentane. Figure 18-11(b) has a single branch at the end of the chain. Such a structural type is commonly identified by the prefix *iso-*. Hence, this isomer is called isopentane. Figure 18-11(c) also contains a five-carbon molecule, but it has the

Fig. 18-11 Structural formulas and models for some five-carbon saturated hydrocarbons; (a) *n*-pentane, (b) isopentane, (c) cyclopentane.

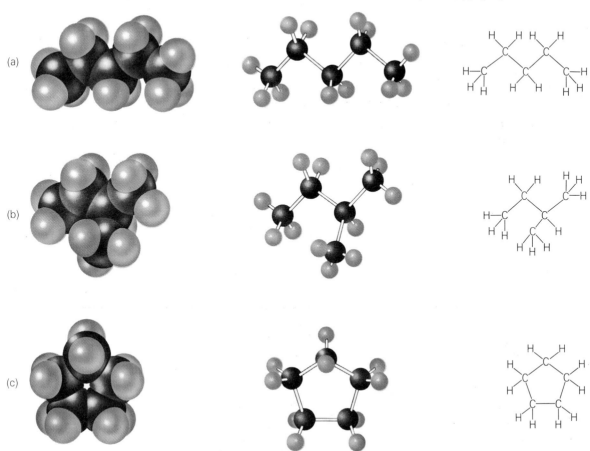

distinctive feature of a cyclic carbon structure. Such a compound is identified by the prefix *cyclo-* in its name—in the case shown, cyclopentane.

EXERCISE 18-15

What are the empirical formulas of the three compounds shown in Figure 18-11? the molecular formulas? Which are structural isomers?

EXERCISE 18-16

There is one more alkane having the molecular formula C_5H_{12} called neopentane. Draw its structural formula.

The alkanes are the principal compounds present in natural gas and in petroleum. The low-molecular weight compounds are gases under normal conditions (their boiling points are shown in Table 18-3). The composition of gasoline is mainly complex, branched alkanes with from 6 to 10 carbon atoms. Paraffin waxes are usually alkanes with from 20 to 35 carbon atoms.

TABLE 18-3 SOME PROPERTIES OF SATURATED HYDROCARBONS

Saturated Hydrocarbon	Molecular Formula	Melting Point (°C)	Boiling Point (°C)	Heat of Combustion of Gas (kcal/mole)
methane	CH_4	−182.5	−161.5	−212.8
ethane	CH_3CH_3	−183.3	−88.6	−372.8
propane	$CH_3CH_2CH_3$	−187.7	−42.1	−530.6
n-butane	$CH_3CH_2CH_2CH_3$	−138.4	−0.5	−687.7
isobutane	$CH_3-\underset{\underset{CH_3}{\mid}}{CH}-CH_3$	−159.6	−11.7	−685.7
n-hexane	C_6H_{14}	−95.3	68.7	−1,002.6
cyclohexane	C_6H_{12}	+6.6	80.7	−944.8
n-octane	C_8H_{18}	−56.8	125.7	−1,317.5
n-octadecane	$C_{18}H_{38}$	+28.2	316.1	−2,891.9

The saturated hydrocarbons are relatively inert except at high temperatures. For this reason sodium metal is usually stored immersed in an alkane such as kerosene (8 to 14 carbon atoms) to protect it from reaction with water or oxygen. Combustion is almost the only important chemical reaction of the alkanes, but that reaction makes the hydrocarbons one of the most important energy sources of our modern technology.

EXERCISE 18-17

Using the data given in the last column of Table 18-3, plot the heat released during combustion per mole of carbon atoms against the number of carbon atoms per molecule for the normal alkanes. Consider the significance of this plot in terms of the molecular structures of these compounds.

In a sense, the absence of reactivity of saturated hydrocarbons, whether cyclic or not, is a crucial aspect of their chemistry. This inertness accounts for the fact that the chemistry of organic compounds is mainly concerned with the functional groups. The functional groups are usually so much more reactive than the carbon "skeleton" that it can be assumed that the skeleton will remain intact and unchanged by the reaction.

18-5.2 Unsaturated Hydrocarbons

Unsaturated compounds are the organic compounds in which fewer than four atoms are attached to one or more of the carbon atoms. Ethylene (C_2H_4) is an unsaturated compound; and because it has only carbon and hydrogen atoms, it is an unsaturated hydrocarbon. Propylene, the next most complicated unsaturated hydrocarbon, has the molecular formula C_3H_6. The structural formulas of ethylene and propylene are shown in Figure 18-12. Cyclic hydrocarbons also can involve double bonds. The structural formula of a cyclic unsaturated hydrocarbon is also shown in Figure 18-12.

Unsaturated hydrocarbons are quite reactive in contrast to the relatively inert saturated hydrocarbons. As we learned earlier, this

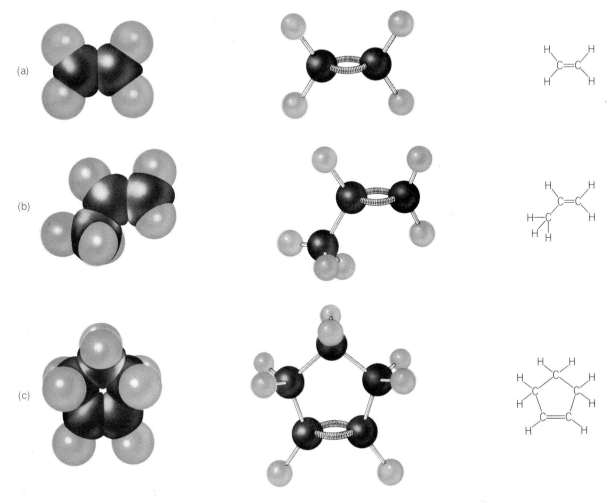

Fig. 18-12 Structural formulas and models for some unsaturated hydrocarbons; (a) ethylene, (b) propylene, (c) cyclopentene.

reactivity is associated with the double bond. In the most characteristic reaction, called "addition," one of the bonds of the double bond opens and a new atom becomes bonded to each of the carbon atoms (see Section 16-6.1). Some of the reagents that will add to the double bond are H_2, Br_2, HCl, and H_2O. The following examples use ethylene:

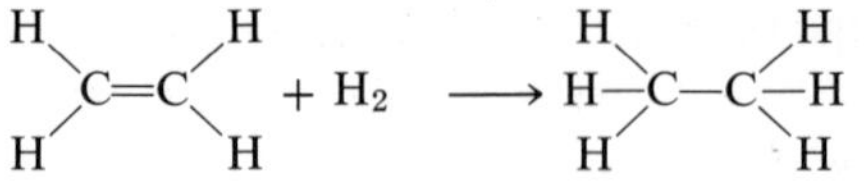

$$\begin{array}{c}H\\H\end{array}\!\!>\!\!C\!\!=\!\!C\!\!<\!\!\begin{array}{c}H\\H\end{array} + H_2 \longrightarrow H\!\!-\!\!\overset{H}{\underset{H}{C}}\!\!-\!\!\overset{H}{\underset{H}{C}}\!\!-\!\!H \tag{33}$$

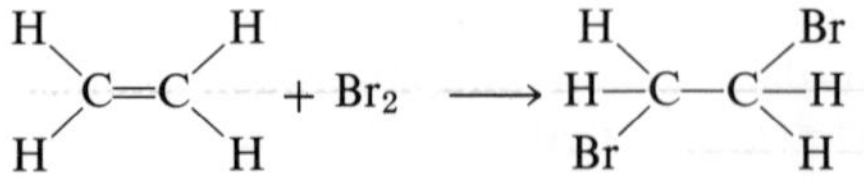

$$\begin{array}{c}H\\H\end{array}\!\!>\!\!C\!\!=\!\!C\!\!<\!\!\begin{array}{c}H\\H\end{array} + Br_2 \longrightarrow H\!\!-\!\!\overset{H}{\underset{Br}{C}}\!\!-\!\!\overset{Br}{\underset{H}{C}}\!\!-\!\!H \tag{34}$$

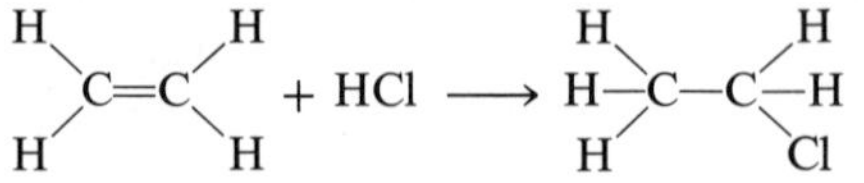

$$\begin{array}{c}H\\H\end{array}\!\!>\!\!C\!\!=\!\!C\!\!<\!\!\begin{array}{c}H\\H\end{array} + HCl \longrightarrow H\!\!-\!\!\overset{H}{\underset{H}{C}}\!\!-\!\!\overset{H}{\underset{Cl}{C}}\!\!-\!\!H \tag{35}$$

$$\begin{array}{c}H\\H\end{array}\!\!>\!\!C\!\!=\!\!C\!\!<\!\!\begin{array}{c}H\\H\end{array} + H_2O \longrightarrow H\!\!-\!\!\overset{H}{\underset{H}{C}}\!\!-\!\!\overset{H}{\underset{OH}{C}}\!\!-\!\!H \tag{36}$$

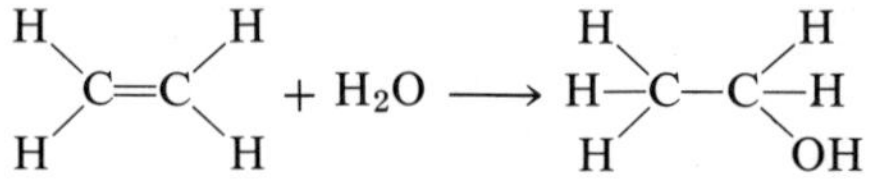

Oxidizing agents also attack the double bond. When a reaction between an unsaturated compound and the permanganate ion occurs, for example, the violet color of permanganate fades. This color change, as well as the reaction with bromine in which a change from orange to colorless occurs (bromine is orange), is used as a qualitative test for the presence of double bonds in compounds of unknown structure.

18-5.3 Benzene—Its Structure and Representation

There is another important class of cyclic compounds that is different from the two classes just described. The simplest example is the compound benzene, a cyclic compound having six carbon atoms in the ring and the formula C_6H_6. The **benzene ring** is planar with 120° angles between each pair of carbon-carbon bonds. Thus, experiment tells us the molecule is a regular hexagon with the following atomic arrangement:

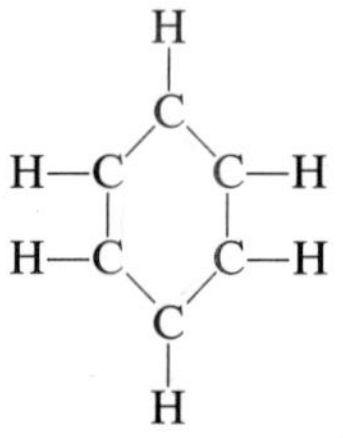

A count of electrons and atoms shows us that there must be three double bonds in a benzene ring, but a difficulty arises when we attempt to represent the bonding in the usual way. We might use either of the notations

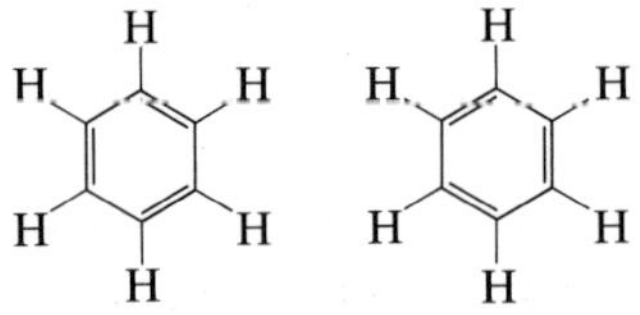

Both structures satisfy the formal valence rules for carbon, but each has a serious fault. Each structure shows three of the carbon-carbon bonds as double bonds and three as single bonds. A wealth of experimental evidence indicates that this is not the case. Any one of the six carbon-carbon bonds in benzene is the same as any other. Apparently the fourth bond of each carbon atom is shared equally with each adjacent carbon. This makes it difficult to represent the bonding in benzene by the usual diagrams. Benzene seems to be best represented as the "superposition" or "average" of the two structures. For simplicity, chemists use either structure. Expressed in a shorthand form omitting carbon and hydrogen atoms, the notation is either of the following:

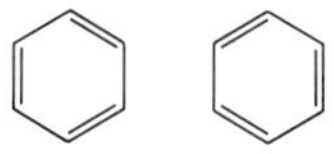

Another shorthand symbol sometimes used is

Whichever symbol he uses, the chemist always remembers that the carbon-carbon bonds are actually all the same and that they have properties unlike either simple double or single bonds. The situation is very much like that which we faced with graphite, where more than one arrangement of the double bonds provided equally good representations of the structure. In the case of graphite, we said that the electrons were "delocalized," that they could wander *across the surface of the sheet* of carbon atoms. We supported this hypothesis by the fact that graphite is a good conductor of electricity along the sheets, but a poor conductor perpendicular to the sheets. Similarly, with benzene, a small fragment of the graphite lattice, electrons should move easily *around the ring*—electrical effects should be transmitted easily through the delocalized electronic cloud. (This is sometimes called the **π cloud**; it is above and below the ring surface.) Indeed, much of the chemistry of ring compounds reflects this mobility of electrons in the ring. All compounds, such as benzene, containing rings with delocalized electrons, are called **aromatic compounds.**

18-5.4 Substitution Reactions of Benzene

Benzene shows neither the typical reactivity nor the usual addition reaction of ethylene. Benzene does react with bromine (Br_2), but in a different type of reaction:

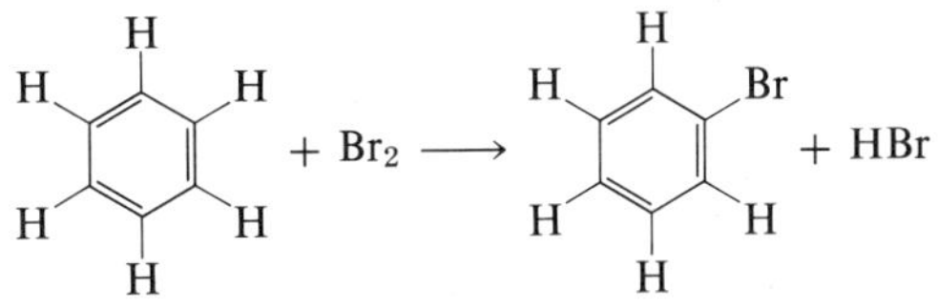

$$\text{BENZENE} + Br_2 \longrightarrow \text{BROMOBENZENE} + HBr \qquad (37)$$

In this reaction, called bromination, one of the hydrogen atoms has been replaced by a bromine atom. Notice that the double-bond structure is not affected—this is not an addition reaction. Nitric acid causes a similar reaction, called nitration:

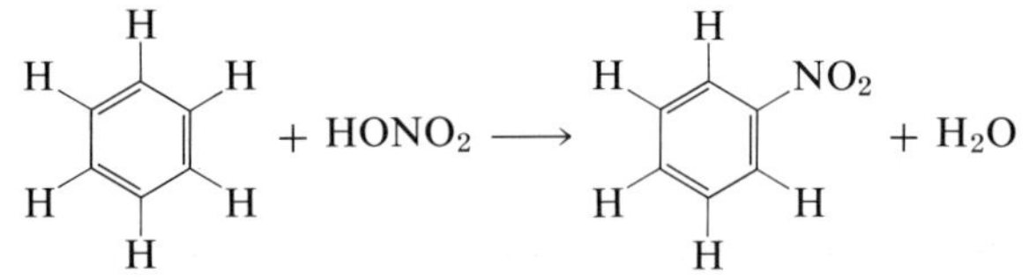

$$\text{BENZENE} + HONO_2 \longrightarrow \text{NITROBENZENE} + H_2O \qquad (38)$$

Reactions of the type shown above are called **substitution reactions.** *The substitution reaction is the characteristic reaction of benzene and its derivatives* and is the way in which a multitude of compounds are prepared by the organic chemist. By substitution he is able to introduce functional groups, which can then be modified in various ways.

18-5.5 Modification of Functional Groups on the Benzene Ring

One of the most important derivatives of benzene is nitrobenzene. The nitro- group is —NO_2. Nitrobenzene is important chiefly because it is readily converted by reduction into an aromatic amine, aniline. One procedure uses zinc as the reducing agent:

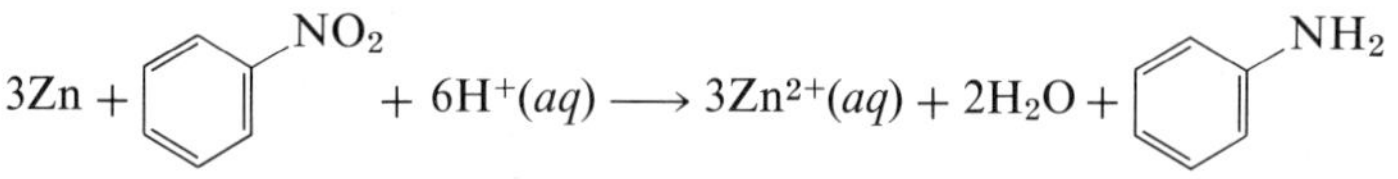

$$3Zn + \text{NITROBENZENE} + 6H^+(aq) \longrightarrow 3Zn^{2+}(aq) + 2H_2O + \text{ANILINE} \qquad (39)$$

Aniline and other aromatic amines are valuable industrial raw materials. They form an important starting point from which many of our dyestuffs, medicinals, and other valuable products are prepared. For example, you have used the indicator methyl orange in your laboratory experiments. Methyl orange is an example of an aniline-derived dye, although it is used more as an acid-base indicator than for dyeing fabrics. The structure of methyl orange is as follows:

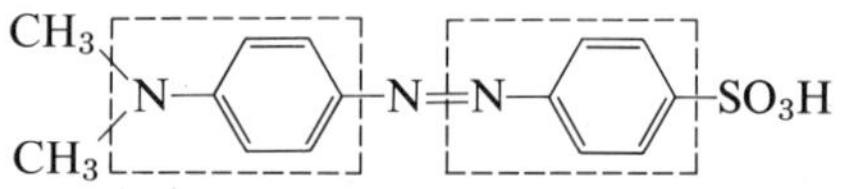

METHYL ORANGE

The portions of the methyl orange molecule set off by the broken lines come from aromatic amines like aniline. Aniline is the starting material from which methyl orange and related dyes ("azo dyes") are made.

Another useful aniline derivative is acetanilide, which is simply the amide formed from aniline and acetic acid:

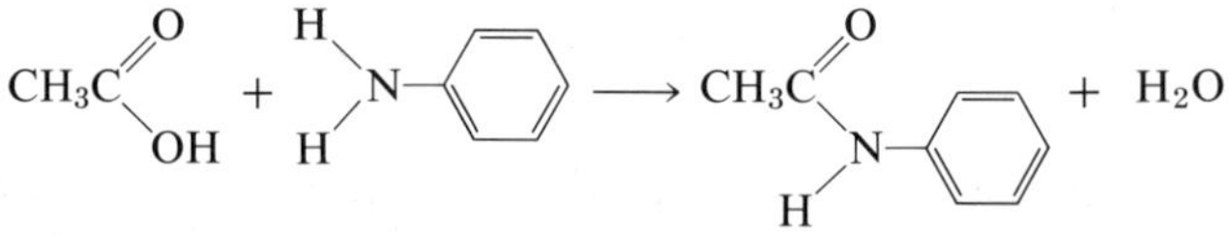

$$\{\text{acetic acid}\} + \{\text{aniline}\} \longrightarrow \{\text{acetanilide}\} + \{\text{water}\} \qquad (40)$$

Acetanilide has been used medicinally as a pain-killing remedy.

Hydroxybenzene, or phenol, is another important derivative:

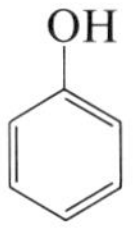

$$\text{C}_6\text{H}_5\text{OH}$$

Most phenol is now made industrially from benzene, which is chlorinated as a first step:

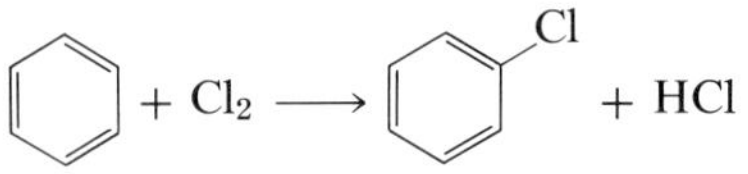

$$C_6H_6 + Cl_2 \longrightarrow C_6H_5Cl + HCl \qquad (41)$$

CHLOROBENZENE

The reaction of chlorobenzene with a base gives phenol:

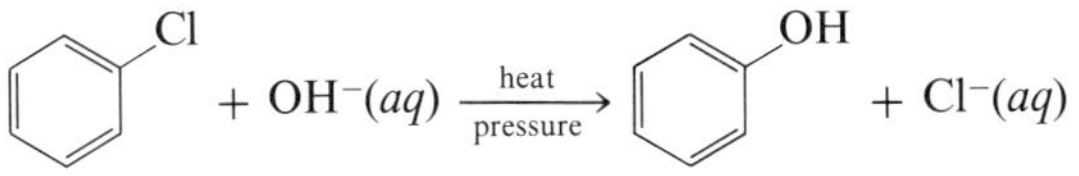

$$C_6H_5Cl + OH^-(aq) \xrightarrow[\text{pressure}]{\text{heat}} C_6H_5OH + Cl^-(aq) \qquad (42)$$

Phenol, a germicide and disinfectant, was first used by the English surgeon Joseph Lister (1827–1912) in 1867 as an antiseptic in medicine. More effective and less toxic antiseptics have since been discovered.

Perhaps the most widely known compound prepared from phenol is aspirin. If phenol, sodium hydroxide, and carbon dioxide are heated together under pressure, the sodium salt of salicylic acid is formed:

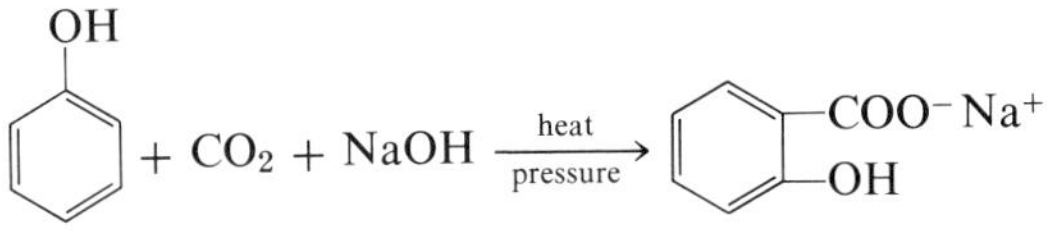

$$C_6H_5OH + CO_2 + NaOH \xrightarrow[\text{pressure}]{\text{heat}} C_6H_4(COO^-Na^+)(OH) \qquad (43)$$

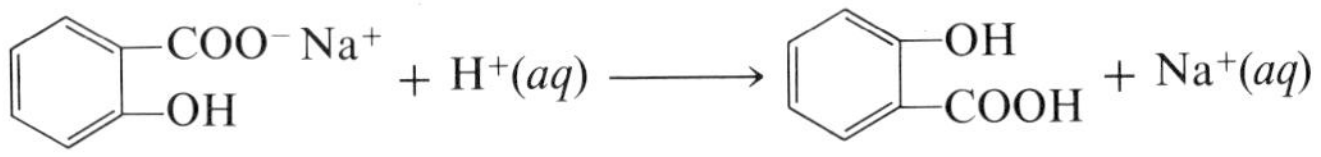

$$C_6H_4(COO^-Na^+)(OH) + H^+(aq) \longrightarrow C_6H_4(OH)(COOH) + Na^+(aq) \qquad (44)$$

SALICYLIC ACID

Salicylic acid is quite useful. Its methyl ester has a sharp, characteristic odor and is called "oil of wintergreen." (See Table 5-5 on page 101 for the vapor pressure of methyl salicylate.) The acid itself (or the sodium salt) is a valuable drug in the treatment of arthritis. But the most widely known derivative of salicylic acid is aspirin, which has the following structure:

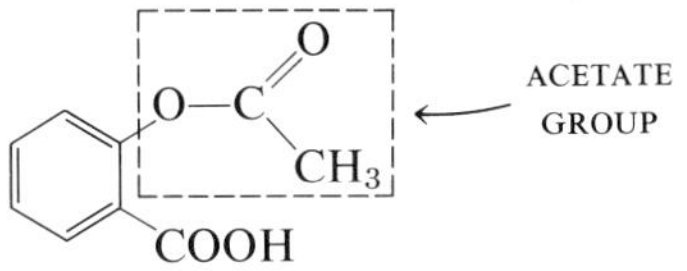

$$C_6H_4(O\text{-}CO\text{-}CH_3)(COOH) \quad \longleftarrow \text{ACETATE GROUP}$$

By examining this structure, you will see that aspirin is an ester of acetic acid. Aspirin is the most widely used drug. Over 20 million pounds of aspirin, or about 150 five-grain tablets for every person, are manufactured each year in the United States alone!

Table 18-4 shows the structures of a few simple benzene derivatives that are important commercially. Study these structures so that you can see their relationship with the simple compounds from which they are derived.

TABLE 18-4 STRUCTURES AND USES OF SOME BENZENE DERIVATIVES

Structure	Name	Use
(benzene ring with OH, OCH₃, and CHO groups)	vanillin ("vanilla")	flavoring material
(benzene ring with OCH₂CH₃ and NHC(O)CH₃ groups)	phenacetin	pain-reliever (in headache remedies)
(benzene ring with two OH groups, para)	hydroquinone	photographic developer
(benzene ring with NH₂ and C(O)OCH₂CH₂N(C₂H₅)₂ groups)	procaine ("Novocain")	local anaesthetic
(benzene ring with CH=CH₂ group)	styrene	monomer for preparation of polystyrene plastics

18-6 POLYMERS

Table 18-3 on page 486 shows that the melting points of the normal alkanes tend to increase as the number of carbon atoms in the chain increases. Ethane (C_2H_6) is a gas under normal conditions; octane (C_8H_{18}) is a liquid; octadecane ($C_{18}H_{38}$) is a solid. We noted in Chapter 17 that van der Waal's forces increase as the area of molecular surface in contact and the size of the molecules increase. We see that desired physical properties in a solid can be obtained by controlling the length of the chain. Functional groups attached to the chain provide additional variability, including chemical reactivity. In fact, by adjusting the chain length and the percentage of high-molecular weight compounds, chemists have produced a multitude of organic solid substances called **plastics.** These have been tailored for a wide variety of uses, giving rise to an enormous chemical industry.

The key to this chemical treasure chest is the process by which extended chains of atoms are formed. Inevitably, it is necessary to begin with relatively small chemical molecules and with carbon chains of only a few atoms. These small units, called **monomers,** must be bonded together, time after time, until the desired range of molecular weight is reached. Often the desired properties are obtained only with giant molecules, each containing hundreds or even thousands of monomers. These giant molecules are called **polymers,** and the process by which they are formed is called **polymerization.**

18-6.1 Types of Polymerization

Polymerization involves the chemical combination of a number of identical or similar molecules to form a complex molecule of high molecular weight. The small units may be combined by **addition polymerization** or **condensation polymerization.**

Addition polymers are formed by the reaction of the monomeric units without the elimination of atoms. The monomer is usually an unsaturated organic compound such as ethylene ($H_2C=CH_2$), which in the presence of a suitable catalyst will undergo an addition reaction to form a long chain molecule such as polyethylene. A general equation for the first stage of such a process is

$$\underset{H}{\overset{H}{>}}C=C\underset{H}{\overset{H}{<}} + \underset{H}{\overset{H}{>}}C=C\underset{H}{\overset{H}{<}} \longrightarrow H-\underset{H}{\overset{H}{C}}-\underset{H}{\overset{H}{C}}-C=C\underset{H}{\overset{H}{<}} \quad (45)$$

The same addition process continues, and the final product is the very familiar polymer polyethylene:

$$H-\underset{H}{\overset{H}{C}}-\left(\underset{H}{\overset{H}{C}}\right)_n-\underset{}{\overset{H}{C}}=C\underset{H}{\overset{H}{<}}$$

where n is a very large number.

When one or more of the hydrogens are replaced by groups such as fluorine (F), chlorine (Cl), methyl (CH_3), or methyl ester ($COOCH_3$), polymers such as "Teflon," "Saran," "Lucite," and "Plexiglas" result. It is thus possible to create molecules with custom-built properties for various uses as plastics or fibers.

Condensation polymers are produced by reactions in which some simple molecule (such as water) is eliminated between functional groups (such as alcoholic OH or acidic COOH groups). In order to form long chain molecules, two or more functional groups must be present in each of the reacting units. For example, when ethylene glycol ($HOCH_2CH_2OH$) reacts with *para*phthalic acid,

$$HOOC-\underset{}{\bigcirc}-COOH$$

TYPES of POLYMERIZATION

ROBERT BURNS WOODWARD (1917–)

Robert Woodward is one of the outstanding American synthetic chemists of all time. Entering the Massachusetts Institute of Technology at the age of 16, he earned his bachelor's degree in three years and the Ph.D. in one additional year.

Woodward's first major success was the synthesis of quinine, a highly complex molecule with two benzene rings, a tricyclic structure, a double bond, an OH group, and a methyl ether linkage, all in a particular geometrical configuration. Other important syntheses achieved by Woodward and his students have been cholesterol, cortisone, chlorophyll, and a number of antibiotics. He has added to our knowledge of the polymerization of amino acids into proteins, and some of his synthetic protein-like polymers have properties similar to those of silk and wool fibers. For his many contributions to chemistry, Woodward was awarded the 1965 Nobel Prize.

All of these accomplishments illustrate a dedication to chemistry and a capacity for intense work. To Robert Woodward, chemistry is an exciting adventure for which there isn't enough time.

a polyester of high molecular weight called "Dacron" is produced. The diagram below shows the first stages of this process:

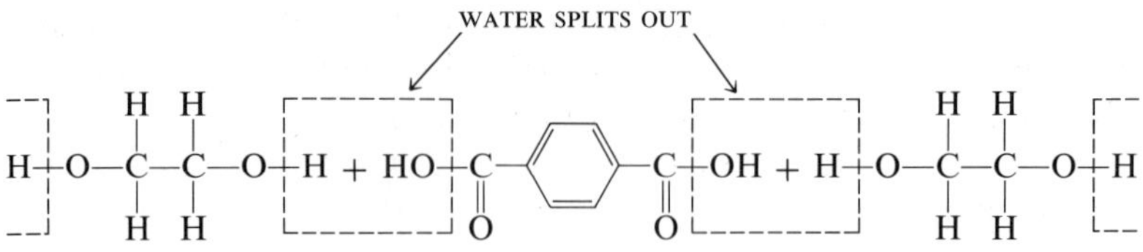

WATER SPLITS OUT

H–[O–C(H)(H)–C(H)(H)–O]–H + HO–[C(=O)–⌬–C(=O)]–OH + H–[O–C(H)(H)–C(H)(H)–O]–H

18-6.2 "Nylon," a Polymeric Amide

"Nylon," the material widely used in plastics and fabrics, is a condensation polymer. It consists of molecules of extremely high molecular weight, and it is made up of small units joined in a long chain of atoms. The reaction by which the units become bonded together is a conventional amide formation. However, the reaction must take place again and again to form an extended, repeating chain. To accomplish this, we select reactants with two functional groups. Thus, polyamides can be made from one compound with two acid groups,

$$\text{HOOC—CH}_2\text{—CH}_2\text{—CH}_2\text{—CH}_2\text{—COOH}$$

ADIPIC ACID

and another with two amine groups,

$$\text{H}_2\text{N—CH}_2\text{—CH}_2\text{—CH}_2\text{—CH}_2\text{—CH}_2\text{—CH}_2\text{—NH}_2$$

1,6-DIAMINOHEXANE

These molecules can react repeatedly, removing water each time, and forming amide linkages at both ends:

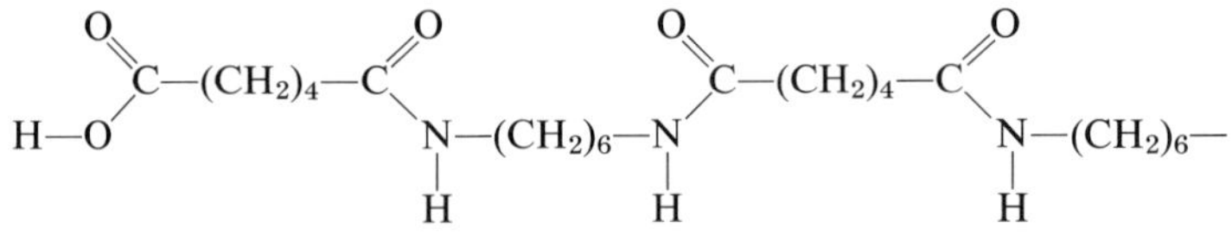

H—O—C(=O)—(CH$_2$)$_4$—C(=O)—N(H)—(CH$_2$)$_6$—N(H)—C(=O)—(CH$_2$)$_4$—C(=O)—N(H)—(CH$_2$)$_6$—

Other polyamides can be made from different acids and other amines, giving a variety of properties suited to a variety of uses.

18-6.3 Protein, Another Polymeric Amide

A most important class of polyamides is **proteins,** the essential structures of all living matter. They are necessary to the human diet as a source of the "monomeric" units, the amino acids, from which living protein materials are made.

Proteins are polyamides formed by the polymerization, through amide linkages, of α-amino acids. Three of the 25 to 30 important natural α-amino acids are shown in Figure 18-13. Each acid has an amine group, —NH₂, attached to the α-carbon, the carbon atom immediately adjacent to the carboxylic acid group.

The protein molecule may involve hundreds of such amino acid molecules connected through the amide linkages. A portion of this chain might be represented as in Figure 18-14.

An amide is decomposed by aqueous acids to an acid and the ammonium salt of an amine:

$$\underset{\text{N-METHYL ACETAMIDE}}{CH_3C(=O)NHCH_3} + H_3O^+(aq) \longrightarrow \underset{\text{ACETIC ACID}}{CH_3C(=O)OH} + \underset{\text{METHYL AMMONIUM ION}}{CH_3NH_3^+(aq)} \quad (46)$$

In this same type of reaction, a protein can be broken down into its constituent amino acids. In this way most of what we call the "natural" amino acids have been discovered. Studies have been made of proteins from many sources—egg yolk, milk, animal tissues, plant seeds, gelatin, and so on—to learn of what amino acids they are composed. In this way about 30 of the natural amino acids have been identified. When you consider how many different ways up to 30 amino acids can be combined in long chains of 100 or more amino acid units, you will see why there are so many known proteins, and why different living species of plants and animals can have in their tissues a great many different proteins. **Enzymes,** the biological catalysts, are also proteins. Each enzyme, of course, has its own particular structure, determined by the order and spacial arrangement of the amino acids from which it is formed.

Perhaps the most marvelous part of the chemistry of living organisms is their ability to synthesize just the right protein structures from the myriad of structures possible and to reproduce structures like themselves. Determining the double helical structure of deoxyribonucleic acid (DNA)—a complex molecule that transmits

PROTEIN, ANOTHER POLYMERIC AMIDE

Fig. 18-13 Molecular structures of α-amino acids.

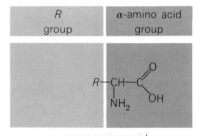

an α-amino acid (general formula)

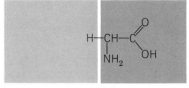

glycine

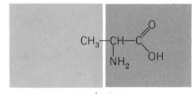

alanine

glutamic acid

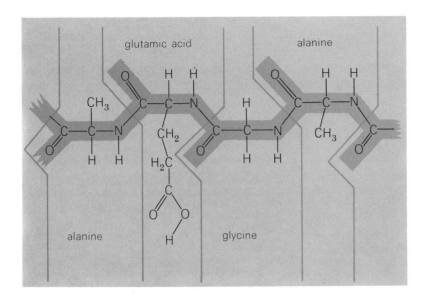

Fig. 18-14 The structure of protein showing the amide chain.

biological information from one generation to another—earned F. H. C. Crick, J. D. Watson, and M. H. F. Wilkins the 1962 Nobel Prize for Medicine and Physiology.

EXERCISE 18-18

Take the letters *A, B, C,* and see how many different three-unit combinations you can make; for example, *ABC, BAC, AAC, CBC.* This will convince you that a chain made of hundreds of groups with up to 30 different kinds of units in each group can have an almost unlimited number of combinations.

18-7 HIGHLIGHTS

The **structural formula** for a compound is a summary of a great deal of explicit information about a compound. From the structural formula we can deduce the **empirical formula,** the **molecular formula,** and many chemical and physical properties of the molecule. The structural formula is obtained from many laboratory observations.

Hydrocarbons are compounds containing carbon and hydrogen. Straight chain, **saturated hydrocarbons** have only single bonds in the molecule and each carbon atom is bonded to four atoms such as carbon and hydrogen. **Unsaturated hydrocarbons** contain double bonds. Cyclic hydrocarbons have the carbon atoms arranged in a ring. They may be saturated or unsaturated.

One or more hydrogen atoms on a hydrocarbon framework can be replaced with a **functional group.** A functional group is a reactive unit such as $-OH$, $-NH_2$, $-\overset{\displaystyle O}{\underset{}{C}}\!\!{=}\!$, $-\overset{\displaystyle O}{\underset{}{C}}\!\!-\!OH$. Much of the chemistry of organic compounds can be discussed in terms of the functional groups. Common and important hydrocarbon radicals or units are methyl ($-CH_3$), ethyl ($-C_2H_5$), propyl ($-C_3H_7$), butyl ($-C_4H_9$), and phenyl ($-C_6H_5$). New compounds can be made by appropriate reactions carried out on functional groups. For example, oxidation of an alcohol ($R-OH$, where R is the hydrocarbon radical) gives an **aldehyde** ($R-\underset{\underset{H}{|}}{C}{=}O$) or a **ketone** ($R-\overset{\overset{O}{\|}}{C}-R'$). Further oxidation of an aldehyde gives a **carboxylic acid** ($R-\overset{\overset{O}{\|}}{C}-OH$). Each of these units has its own interesting chemistry, which gives rise to the many thousands of organic compounds in the world around us.

QUESTIONS AND PROBLEMS

1 What information is revealed by the empirical formula? the molecular formula? the structural formula? Demonstrate, using ethane (C_2H_6).

2 Write the balanced equation for the complete burning of methane.

3 Draw the structural formulas for all the $C_2H_3Cl_3$ compounds.

4 Draw the structures of two isomeric compounds corresponding to the empirical formula C_3H_8O.

5 Draw the structural formulas of the isomers of butyl chloride.

6 What angle would you expect to be formed by the C, O, and H nuclei in an alcohol molecule? Explain.

7 When 0.601 g of a sample having an empirical formula CH_2O was vaporized at 200°C and 1 atm pressure, the volume occupied was 388 ml.

This same volume was occupied by 0.301 g of ethane under the same conditions. What is the molecular formula of CH_2O? One mole of the sample, when allowed to react with zinc metal, liberated (rather slowly) 0.5 mole of hydrogen gas. Write the structural formula. (*Answer:* The molecular formula is $C_2H_4O_2$.)

8 A 100-mg sample of a compound containing only C, H, and O was found by analysis to give 149 mg CO_2 and 45.5 mg H_2O when burned completely. Calculate the empirical formula.

9 A compound is found to have the empirical formula C_2H_4O. The measured molecular weight is 44. The nuclear magnetic resonance spectrum (low resolution) shows *two* distinct peaks. One covers three times as much area as the second. Which one of the structures below would be indicated by these data?

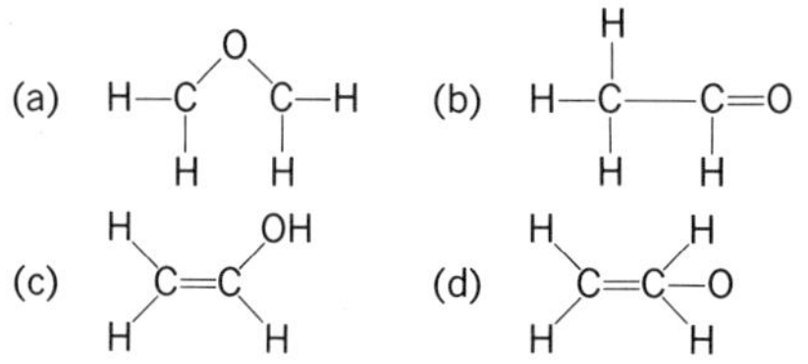

(a) H—C(=O)—C(H)(H)—H (b) H—C(H)(H)—C(H)=O

(c) H(H)C=C(H)(OH) (d) H(H)C=C(H)—O—H

10 How many separate peaks would you expect to see in the nmr (low-resolution) spectrum of methyl alcohol? Explain.

11 How much ethanol can be made from 50 g of ethyl bromide? What assumptions do you make in this calculation?

12 Write the balanced equation for the production of 2-pentanone (ketone) from 2-pentanol, using dichromate ion as the oxidizing agent.

13 One mole of an organic compound is found to react with 0.5 mole of oxygen to produce an acid. To what class of compounds does this starting material belong?

14 Using the information given in Table 9-2 on page 196, determine the reaction heat per mole of $C_2H_6(g)$ for the complete combustion of ethane.

15 An aqueous solution containing 0.10 mole/liter of chloroacetic acid (ClH_2CCOOH) is tested with indicators and the concentration of $H^+(aq)$ is found to be 1.2×10^{-2} M. Calculate the value of K_A (if necessary, refer back to Section 13-6.1). Compare this value with K_A for acetic acid—the change is caused by the substitution of a halogen atom near a carboxylic acid group.

16 Give simple structural formulas of an alcohol, an aldehyde, and an acid, each derived from methane, from ethane, from butane, and from octane.

17 Write the equations for the steps in the preparation of methylamine from methyl iodide.

18 Write equations to show the formation of the esters methyl butyrate and butyl propionate.

QUESTIONS and PROBLEMS

19 Given the structural formula

H—C(H)(H)—C(=O)—O—C(H)(H)—C(H)(H)—C(H)(H)—H

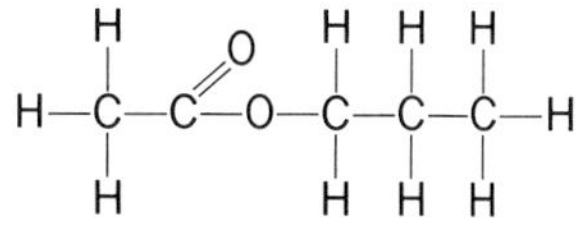

for an ester, write the formula of the acid and the alcohol from which it might be made.

20 How much acetamide can be made from 3.1 g of methyl acetate? See equation (31) on page 483. Assume the ester is completely converted. (*Answer:* 2.5 g acetamide.)

21 An ester is formed by the reaction between an acid, RCOOH, and an alcohol, R'OH, to form an ester, RCOOR', and water. The reaction is carried out in an inert solvent.

(a) Write the equilibrium relation among the concentrations, including the concentration of the product water.

(b) Calculate the equilibrium concentration of the ester if $K = 10$ and the concentrations *at equilibrium* of the other constituents are

$[RCOOH] = 0.1$ M
$[R'OH] = 0.1$ M
$[H_2O] = 1.0$ M

(c) Repeat the calculation of part (b) if the equilibrium concentrations are

$[RCOOH] = 0.3$ M
$[R'OH] = 0.3$ M
$[H_2O] = 1.0$ M

22 Give the empirical formula, the molecular formula, and draw the structural formulas of the isomers of butene.

23 There are three isomers of dichlorobenzene (empirical formula C_3H_2Cl). Draw the structural formulas of the isomers.

24 Consider the compound phenol,

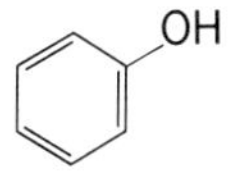

(benzene ring)—OH

(a) Predict the angle formed by the nuclei of C, O, and H. Explain your choice in terms of the valence-shell electron-pair repulsion theory.

(b) Predict qualitatively the boiling point of phenol. (The boiling point of benzene is 80°C.) Explain your answer.

(c) Write an equation for the reaction of phenol as a proton donor in water.

(d) In a 1.0 M aqueous solution of phenol, $[H^+] = 1.1 \times 10^{-5}$ mole/liter. Calculate K_A.

Nature seems to operate always according to an original general plan, from which she departs with regret and whose traces we come across everywhere.

VIC D'AZYR (1784)

CHAPTER 19 The Third Row of the Periodic Table

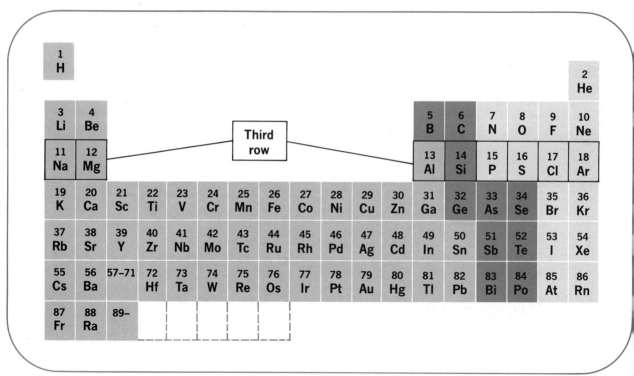

As a result of our investigations into bonding, we have a lot of information about the properties of the second row elements. We also know that elements in the same vertical column have the same configuration of valence electrons and, generally, the same chemical reactions. Yet the elements of the third row *are* different from their counterparts in row two—silicon is metallic in appearance while carbon is black in graphite or transparent and crystalline in the diamond; aluminum is widely used as a structural metal while boron is not; oxygen is a colorless gas while sulfur is a yellow solid. Obviously the addition of eight protons to the nucleus and an extra filled energy level to the exterior of the atom *does* make some difference. By comparing the two rows we can find trends that can aid our memories in understanding the oxidation-reduction and acid-base behavior of these elements.

In Chapters 8, 16, and 17 we followed changes in the properties of the elements as we moved across the second row of the periodic table. Lithium (atomic number 3) on the left-hand side of the table was distinctly metallic in character, while fluorine (no. 9) on the right-hand side was distinctly nonmetallic. Lithium is a shiny metal, while fluorine exists as distinct and discrete gaseous molecules having the formula F_2. We interpreted this gradual change in properties in terms of both the ionization energies of the valence electrons and the orbitals available for covalent bond formation.

In this chapter we shall examine the trends in physical and chemical properties as we go across the third row of the periodic table. The ionization energies and available orbitals for elements from Na (no. 11) to Ar (no. 18) suggest that we should see the same trend from metal to nonmetal that we saw in the second row. A study of this trend will expand our knowledge of the distinctive chemistry of metals and nonmetals as groups. Specifically, we shall consider the physical properties and molecular form of the elements. We shall then try to relate this information to the oxidation-reduction and acid-base processes of the third-row elements.

19-1 THE ELEMENTS OF THE THIRD ROW—
THEIR PHYSICAL PROPERTIES

All elements in row three are commercially available or can be easily prepared in the laboratory. Examine as many of these elements as possible in the laboratory as you study this chapter. If you have all the elements on hand, arrange them in order of atomic number and compare them. You can hardly imagine a more varied set. At one extreme are the elements sodium and magnesium, the very prototypes of metallic behavior. They have a bright luster, are malleable and ductile, and are excellent conductors of heat and electricity. At the other extreme is chlorine, a classic example of a gaseous nonmetal. Let us examine this trend in detail and compare the elements of the second and third rows.

19-1.1 The Metallic Region of the Third Row

Like lithium in the second row, sodium in the third row has one valence electron and three empty valence orbitals. Furthermore, the ionization energy of the sodium valence electron is slightly lower than that for the lithium valence electron (see Table 19-1). One of the criteria we developed in Chapter 17 for the formation of a metallic bond was a low ionization energy for the valence electron. Sodium, then, should be *slightly more* metallic than lithium. Experiment shows that both elements are metals. Although it is difficult to say on the basis of qualitative observation that sodium is more metallic, we do notice that sodium has a somewhat higher electrical conductance and is also somewhat more workable.

What about magnesium (third row) and beryllium (second row)? The ionization energies for magnesium are well below the values for beryllium (Table 19-1), and both magnesium and beryllium have three unfilled valence orbitals. The lower ionization energies for the

Chapter 19 / The THIRD ROW of the PERIODIC TABLE

Fig. 19-1.1 **The crystal structure of diamond (C—C distance = 1.54 Å).**

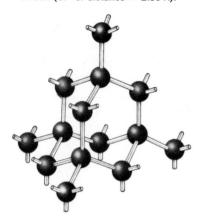

Fig. 19-1.2 **The crystal structure of silicon (Si—Si distance = 2.35 Å).**

magnesium electrons suggest that magnesium should be *more* metallic than beryllium. This is exactly what we find. Although beryllium has a metallic luster and appearance, it is hard and brittle. Its conductance is only about one-fourth that of magnesium. Both in appearance and properties, magnesium is much more metallic than beryllium. Magnesium is an often used structural metal.

The difference between the second and third rows is even more marked when aluminum (third row) is compared with boron (second row). Boron is a hard, nonmetallic material classed as a semiconductor because of its low conductance. It has something of a metallic luster but does not exhibit metallic behavior. In contrast, aluminum is clearly metallic. It is metallic in appearance and has such a high luster that it can be used as the metallic reflecting layer of mirrors. (Aluminum mirrors are frequently used for reflecting telescopes and other scientific instruments.) Aluminum is a very good electrical conductor; its conductance is about 60 percent that of copper. Aluminum can be drawn into wires which, because of their light weight, have been useful in some special areas for electrical transmission. Aluminum is rolled into thin flexible sheets so easily that aluminum foil is a common household article.

The observations on sodium, magnesium, and aluminum suggest that as we go down a column of the periodic table, the metallic characteristics of the elements increase. Properties of the third-row elements are summarized in Tables 19-1 and 19-2.

EXERCISE 19-1

From your own experience, which is the more metallic element of Group IV (the fourth column of the periodic table), tin or carbon? bismuth or nitrogen in Group V?

EXERCISE 19-2

The heats of vaporization of Na, Mg, and Al are 23.1, 31.5, and 67.9 kcal/mole. Look up the boiling points of these elements in Table 19-2. Account for the trends in heats of vaporization and boiling points of these metals in terms of the nature of the metallic bond for each solid.

19-1.2 Polymeric or Linked Solids of the Third Row— Network Solids

The electronic structure of the silicon atom, after electron promotion, is

$$\begin{array}{ccccc} 1s & 2s & 2p & 3s & 3p \\ & & p_x\ p_y\ p_z & & p_x\ p_y\ p_z \end{array}$$

[Silicon Nucleus] ◐ ◐ ◐◐◐ ◐ ◐◐◯

An sp^3 bonding arrangement with a tetrahedral geometry like that of carbon is indicated. Indeed, silicon has the geometry of diamond (see Figure 19-1). A perfect silicon crystal would be one giant molecule.

Almost all valence electrons in the silicon crystal are localized in the covalent bonds and are therefore not free to conduct heat or electricity by moving throughout the solid. On the other hand, experiments on the solid show that a few valence electrons have acquired enough energy to be nonlocalized, perhaps into $3d$ orbitals. These few electrons account for the very small, but noticeable, electrical conductivity that gives silicon one of its most important

POLYMERIC or LINKED SOLIDS

TABLE 19-1 IONIZATION ENERGIES OF THE FIRST FOUR ELEMENTS IN ROWS TWO AND THREE OF THE PERIODIC TABLE

Element	Ionization Energy (kcal/mole)			
	E_1	E_2	E_3	E_4
Li	1.25×10^2	1.75×10^3	2.82×10^3	—
Be	2.14×10^2	4.18×10^2	3.53×10^3	5.00×10^3
B	1.90×10^2	5.7×10^2	8.69×10^2	5.96×10^3
C	2.61×10^2	5.62×10^2	1.10×10^3	1.48×10^3
Na	1.18×10^2	1.09×10^3	1.65×10^3	—
Mg	1.76×10^2	3.46×10^2	1.84×10^3	2.51×10^3
Al	1.37×10^2	4.31×10^2	6.52×10^2	2.76×10^3
Si	1.87×10^2	3.76×10^2	7.7×10^2	1.04×10^3

TABLE 19-2 PHYSICAL PROPERTIES OF ELEMENTS OF ROW THREE OF THE PERIODIC TABLE

Description	Sodium	Magnesium	Aluminum	Silicon	Phosphorus		Sulfur	Chlorine	Argon
					White	Red			
Form and appearance	silvery metal	silvery metal	silvery metal	gray shiny solid	waxy, white solid	red solid	yellow solid	green gas	colorless gas
Hardness (mohs scale)	very soft	moderate 2.6	moderate 2.6–2.9	hard 7.0	very soft 0.5	very soft 0.5	soft 1.5–2.5	gas (Cl_2)	gas (Ar)
Melting point (°C)	98	650	660	≈ 1400	44	590	119*	-102	-189
Boiling point (°C)	883	1100	2270	—	280	high	445*	-34	-186
Electrical conductance (mhos)	excellent 20×10^4	excellent 23×10^4	very good 14×10^4	poor conductor	no	no	no	no	no
Ductile and malleable	yes— very	yes— very	yes— very	no	waxy	no	no	no	no
Metallic luster	yes	yes	excellent yes	some	no	no	no	no	no
Density at 25°C (g/cm³)	0.97	1.74	2.70	2.33	1.82	2.2	2.06	1.57†	0.4‡
Interesting properties	highly reactive	very strong for its low density	high luster; excellent conductor of heat	good semi-conductor	P_4 molecules dissociate above 800°C to P_2 molecules	chains of P atoms	—	—	—

* Monoclinic sulfur.
† Liquid at boiling point.
‡ Density at -120°C and 50 atm.

uses in the modern world. Very pure silicon is used to make transistors and other devices for radios, stereos, and television sets.

If you compare Table 8-7 (page 181) with Table 19-2, you will see a marked similarity between the properties of boron and silicon. Both are high-melting materials with some residual metallic character, and both are semiconductors. This similarity in the properties of elements lying on the diagonal line from the top left-hand to the bottom right-hand corner of the table is an expected consequence of the fact that the metallic character *decreases* from left to right across the periodic table and *increases* from top to bottom. This similarity along the diagonal line is sometimes called the **diagonal relationship** in the periodic table.

You may wonder whether silicon also exists in a form having the planar structure of graphite. *No graphite form for silicon has ever been isolated.* In fact, *no structure containing a true silicon-silicon double bond ($\text{Si}=\text{Si}$) has ever been found.* This has never been satisfactorily explained. Furthermore, the observation is even more general: None of the elements in the third row participate in conventional multiple bonds such as are found in $C=C$ and $N \equiv N$.

The properties of white and red phosphorus listed in Table 19-2 will remind you of the differences between the two forms of carbon—diamond and graphite. These properties suggest the existence of two different structural forms of phosphorus. Experiment shows that white phosphorus consists of P_4 molecules in liquid, solid, and vapor phases. (This applies to vapor below 800°C.) Each phosphorus atom in a P_4 molecule is located at the corner of a regular tetrahedron. Because in P_4 molecules the $3s$ and $3p$ levels of each atom are filled, as shown in Figure 19-2, the P_4 molecules are held together only by van der Waals forces, which explains the low melting and boiling points of white phosphorus. The electronic structure of the phosphorus atom before bonding is

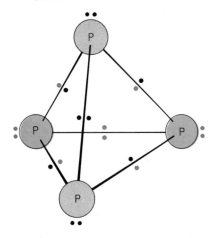

Fig. 19-2 **An electron dot representation of a molecule of white phosphorus.**

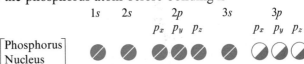

This suggests the existence of three covalent bonds per phosphorus atom—a fact consistent with the tetrahedral P_4 molecule. Repulsion beween electrons in each bond and repulsion between the free electron pair and all bonding electron pairs would be consistent, in a qualitative way, with the geometry seen in the tetrahedron.

EXERCISE 19-3

Does the theoretical electron model for P_4 suggest high or low electrical conductance for phosphorus? Explain. Would white phosphorus be a very good or relatively poor conductor of heat? Explain. Can you justify the fact that white phosphorus is soft and waxy?

If we think further about the electronic structure of a phosphorus atom, we wonder whether phosphorus atoms can link together in

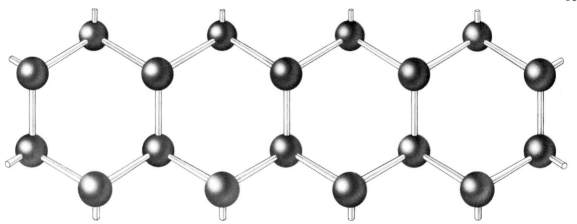

Fig. 19-3 **One possible structure for red phosphorus.**

structures other than the regular tetrahedron. Consider a structure of the general form shown in Figure 19-3, in which phosphorus atoms can link together to form long chains, double chains, or sheets. Such a polymeric structure would account for the relatively high melting and boiling points of red phosphorus. Although we do not yet know the geometry of red phosphorus in detail, black phosphorus, a form of phosphorus very much like red phosphorus, has been studied. Its geometry is shown in Figure 19-4. Double corrugated sheets have been recognized, giving a flaky, crystalline form somewhat like graphite for black phosphorus. As you may well imagine, there are many ways of linking these atoms together into solids. It is not surprising, then, that a number of somewhat different solid modifications are thought to be components of polymeric red phosphorus.

It is legitimate to ask: Why does phosphorus not form P_2 molecules, which would be inert and stable at low temperatures as are N_2 molecules. The answer seems to be that the stability of the N_2 molecule is a result of the formation of a *triple bond* between nitrogen atoms. Because the atoms of the third row have very little tendency to form regular triple bonds with each other, phosphorus forms the P_4 molecule and chains at low temperatures instead of a stable P_2 molecule. Why the elements of the third row do not form multiple bonds, as do the second-row elements, is a question which might entertain puzzled chemists on a rainy Saturday afternoon.

Fig. 19-4 **One layer of the crystalline structure of black phosphorus.**

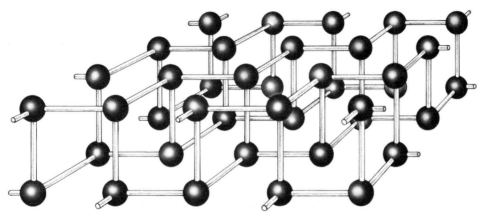

We have concluded that third-row elements do not form conventional multiple bonds with each other. Therefore, a stable, low-temperature S_2 molecule, comparable to a double-bonded O_2 molecule, should not be expected. Indeed, stable S_2 molecules are not found until very high temperatures. The electronic structure of the sulfur atom is

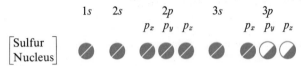

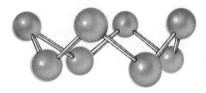

Fig. 19-5 **A structural model of an S_8 molecule.**

Two covalent bonds are expected. Two bonds per atom—each bond directed roughly toward the corners of a tetrahedron—will give the cyclic S_8 molecule; its structure is shown in Figure 19-5. This is the geometric arrangement of atoms present in solid sulfur. The molecules are held together by van der Waals forces only; but because of the extensive contact between atoms in the relatively large, puckered rings, the van der Waals forces are reasonably strong and a melting point around 100°C is observed. At higher temperatures one of the bonds in the ring can break open to form chains that intertwine and grow to give a viscous, almost rubbery liquid. These properties appear in sulfur as the temperature is raised and the chains grow in length.

EXERCISE 19-4

Draw a planar electron dot representation of an S_8 molecule. Show how this ring can break open to give chains. Show by the electron dot representation that such chains should bond to each other to give longer and longer units.

19-1.3 The Simple Gaseous Molecules in the Third Row

The electron dot representation shows that when we come to chlorine in row three, all *s* and *p* valence orbitals can be filled by the formation of a *single* covalent bond:

$$:\ddot{\text{Cl}}:\ddot{\text{Cl}}:$$

Chlorine needs no double bond to stabilize the simple diatomic structure; hence, there is no opportunity for polymerization. The net result is that the Cl_2 molecule, like the F_2 molecule above it, is stable. Chlorine exists as a gas containing distinct Cl_2 molecules.

Argon, of course, appears as a monatomic gas with the lowest condensation point in the third row ($-186°C$).

In comparing the second and third rows of the periodic table, we see that:

(1) In the third row the tendency to form metals extends farther to the right in the row because lower ionization energies make formation of the metallic structure easier.
(2) The tendency to form network solids extends farther to the right in the third row because atoms of this row show no tendency to form conventional multiple bonds.

(3) At chlorine a stable diatomic molecule can be formed without multiple bonding; hence, chlorine is the first element in this row to appear as a gas at room temperature.

TABLE 19-3 TRENDS IN PROPERTIES OF SECOND- AND THIRD-ROW ELEMENTS

Element	Heat of Vaporization (kcal/mole)	Boiling Point (°C)	Element	Heat of Vaporization (kcal/mole)	Boiling Point (°C)
Li	32.2	1326	Na	23.1	889
Be	53.5	2970	Mg	31.5	1120
B	128.8	~3900	Al	67.9	2327
C	170	~4000	Si	(105)	2355
N	0.67	−196	P	3.0	280
O	0.81	−183	S	2.5	445
F	0.78	−188	Cl	4.9	−34.1
Ne	0.42	−246	Ar	1.6	−186

19-2 THE ELEMENTS OF THE THIRD ROW AS OXIDIZING AND REDUCING AGENTS

We can interpret the physical properties of the elements of the third row fairly easily in terms of our bonding concepts. Can the chemistry of this row also be related to models for electronic and molecular structure? Let us approach this question by examining oxidation and reduction reactions.

You will recall from Chapter 14 that a reducing agent is a substance that can *give* electrons to another substance; it is a material that can bring about a reduction reaction (*gain of electrons*) in another chemical species. Let us consider this definition. What kind of substance should give up electrons most easily to another chemical species? A material with relatively mobile and loosely held electrons would make a good reducing agent because its electrons would be readily available to another species. The elements of the third row having the most mobile and loosely held electrons are on the left-hand side of the table, in the region of the metals. *The metals, then, should be good reducing agents.*

On the other hand, an oxidizing agent is a substance that can *accept* electrons from another chemical species. An oxidizing agent oxidizes something else—it makes that something else lose electrons. Atoms having a tendency to pick up or share electrons to form more stable structures should be good oxidizing agents. The electron model suggests that chlorine should be the strongest oxidizing agent of the third row. Let us examine these predictions.

19-2.1 Sodium, Magnesium, and Aluminum— Strong Reducing Agents

The elements Na, Mg, and Al are such strong reducing agents that they are found in nature only in compounds—never as free elements. This is because as free elements they give up electrons to other chemical species with great ease. For example, all three will react readily with the hydrogen ions in ordinary water to give hydrogen gas and a metal ion. For magnesium the equation is

$$2H^+(aq) \, (10^{-7} \, M) + Mg(s) \longrightarrow Mg^{2+}(aq) + H_2(g) \qquad (1)$$

Consider this statement for a minute. It suggests that water in a magnesium or aluminum pan should be decomposed to give hydrogen gas! The very thought is ridiculous. Aluminum pans that have been used for more than 50 years are still in good condition. Airplanes with magnesium parts on their surface do not dissolve in a rainstorm! Our experience does not tally with the above statement based on reducing ability. Why?

As you will remember from Chapter 14, predictions based on E^0 values (thermodynamics) do not necessarily tell us that a reaction will go at a *measurable* rate. The rate and mechanism are just as important in determining the chemistry of a species as is its oxidizing or reducing ability. The rates at which these metals react with $H^+(aq)$ vary considerably. In the case of sodium, this reaction is fast, and all the sodium is consumed rapidly and spectacularly if sufficient water is present. For magnesium and aluminum, the reaction produces a thin layer of oxide on their metallic surface. This oxide layer has low solubility in water, and the oxide adheres strongly to the metal. Hence, it forms a protective layer that prevents further contact between water (or air) and the metal. This protection accounts for the notable resistance of aluminum to weathering. If it were not for this oxide layer, neither aluminum nor magnesium would have any value as a structural metal. (If either magnesium or aluminum comes in contact with mercury, the protective layer is removed and rapid reaction takes place.)

TABLE 19-4 HALF-CELL POTENTIALS FOR REACTIONS SHOWING SODIUM, MAGNESIUM, AND ALUMINUM METALS AS REDUCING AGENTS

Half-Reaction	Half-Cell Potential
$Na(s) \longrightarrow Na^+(aq) + e^-$	$E^0 = 2.71$ volts
$Mg(s) \longrightarrow Mg^{2+}(aq) + 2e^-$	$E^0 = 2.37$ volts
$Al(s) \longrightarrow Al^{3+}(aq) + 3e^-$	$E^0 = 1.66$ volts

A glance at the table of E^0's for half-reactions in Appendix 7 will convince you that sodium, magnesium, and aluminum are among the strongest reducing agents available. Their E^0's, written to show electron loss, are given in Table 19-4. Metals have a tendency to lose electrons and become positive ions in aqueous solutions because their valence electrons are not very strongly bound. The low ionization energies for Na, Mg, and Al metal atoms indicate that the separate gaseous atoms lose electrons easily. Apparently, atoms in the metal will also readily give up electrons.*

---

* Actually several steps beyond electron loss must be considered in analysing a reduction process. A more complete analysis will be given in more advanced courses.

The third-row metals also show strong reducing properties in reactions that do not take place in aqueous solution. Magnesium metal ignited in air, for example, will react with carbon dioxide, reducing it to elemental carbon:

$$2Mg(s) + CO_2(g) \longrightarrow 2MgO(s) + C(s) \qquad (2)$$

If aluminum metal is mixed with a metal oxide such as ferric oxide (Fe_2O_3) and ignited, the oxide is reduced and large amounts of heat are evolved:

$$2Al(s) + Fe_2O_3(s) \longrightarrow 2Fe(s) + Al_2O_3(s) \qquad \Delta H = -203 \text{ kcal/mole} \qquad (3)$$

These reactions are possible because of the great stability, or low energy, of the oxides of magnesium and aluminum. These oxides are ionic compounds whose great stability can be attributed to strong electrostatic forces between small, positively charged Mg^{2+} (or Al^{3+}) ions and negatively charged O^{2-} ions. The fact that the Mg^{2+} and Al^{3+} ions are very small allows the centers of these positive ions to approach closely the center of the negative O^{2-} ions, resulting in strong attractive forces.

The ease with which magnesium oxidizes is important in the commercial manufacture of titanium metal. Titanium, when quite pure, shows great promise as a structural metal. One of the processes currently used for its preparation is the **Kroll process.** The procedure involves the reduction of liquid titanium tetrachloride with molten metallic magnesium:

$$TiCl_4(l) + 2Mg(l) \longrightarrow Ti(s) + 2MgCl_2(l) \qquad (4)$$

Titanium has a very high melting point (1812°C), so the magnesium chloride can be vaporized and distilled from the solid titanium. The gaseous magnesium chloride is condensed and then electrolyzed to regenerate magnesium and chlorine:

$$MgCl_2(l) \xrightarrow{\text{electrolysis}} Mg(s) + Cl_2(g) \qquad (5)$$

The magnesium metal is thus recovered for repeated use in the reduction of $TiCl_4$. Chlorine produced in the electrolysis of $MgCl_2$ is used to prepare $TiCl_4$, the other reactant in the preparation of titanium.

19-2.2 Silicon, Phosphorus, and Sulfur—Oxidizing and Reducing Agents of Intermediate Strength

Silicon, like the metals, can also act as a reducing agent. It reacts with molecular oxygen to form silicon dioxide (SiO_2). This network solid is held together by *very* strong bonds. However, because of the rather high ionization energy of the silicon atom and the great *stability* of the silicon crystal, its reducing properties are considerably less than those of the typical metals.

Phosphorus continues the trend away from elements of strong reducing properties. Elemental phosphorus will react rapidly with strong oxidizing agents such as oxygen and the halogens:

$$P_4(s) + 5O_2(g) \longrightarrow P_4O_{10}(s) \qquad (6)$$

$$P_4(s) + 6Cl_2(g) \longrightarrow 4PCl_3(l) \qquad (7)$$

Fig. 19-6 **Structural models of (a) P_4, (b) P_4O_6, and (c) P_4O_{10}.**

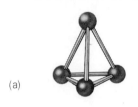

(a)

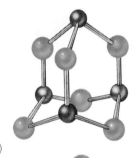

(b)

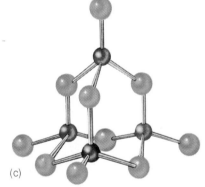

(c)

but it will also react with strong reducing agents such as magnesium to form phosphides:

$$P_4(s) + 6Mg(s) \longrightarrow 2Mg_3P_2(s) \qquad (8)$$

Elemental phosphorus can, then, show either reducing or oxidizing properties.

In the reactions of phosphorus, electrons must be released (or shared) from atoms in the molecular form—not from single atoms in the gaseous state. As a consequence, we find that different forms of phosphorus differ in reactivity. Molecular white phosphorus (P_4) is the most reactive form and may inflame spontaneously in air. Reaction with oxygen tends to insert an oxygen atom *between* each pair of phosphorus atoms on each edge of the elemental tetrahedron. A compound P_4O_6 is first formed. Its structure is seen in Figure 19-6. Further oxidation of P_4O_6 adds an oxygen to each of the free electron pairs on each phosphorus atom. The resulting compound is P_4O_{10}. Its structure is also seen in Figure 19-6. The exact mechanism for the reaction of P_4 and O_2 is not known. The reaction seems to have a low activation energy because it proceeds rapidly at low temperature.

In a rather general way it is possible to relate the behavior of phosphorus as an oxidizing or reducing agent to the electronic structure of a phosphorus atom:

$$\begin{bmatrix}\text{Phosphorus}\\\text{Nucleus}\end{bmatrix}\quad \underset{1s}{●} \quad \underset{2s}{●} \quad \underset{\underset{p_x\ p_y\ p_z}{2p}}{●●●} \quad \underset{3s}{●} \quad \underset{}{◐◐◐}$$

The three electrons of the $3p$ orbitals can be donated or shared to make phosphorus a reducing agent with reagents such as oxygen or chlorine. On the other hand, these three half-filled orbitals can accept electrons from a strong electron donor such as sodium making phosphorus an oxidizing agent.

The ionization energy of the sulfur atom shows that it is even more reluctant than phosphorus to lose electrons. Common compounds of sulfur include the sulfides, which may be formed by reactions of elemental sulfur with a large number of metals. Typical reactions are

$$16Na(s) + S_8(s) \longrightarrow 8Na_2S(s) \qquad (9)$$

$$16Ag(s) + S_8(s) \longrightarrow 8Ag_2S(s) \qquad (10)$$

These reactions show sulfur in the role of an oxidizing agent. The properties of compounds such as Na_2S suggest they contain the simple sulfide ion S^{2-}.

EXERCISE 19-5

Draw an orbital representation of the sulfur atom. Show on this diagram how sulfur should act as an oxidizing agent (sulfur itself accepts electrons). Suggest an element that might react with sulfur as you have indicated.

EXERCISE 19-6

The reaction between zinc metal and sulfur has been used as a propellant reaction for small toy rockets. Write the equation for the process if zinc loses two electrons easily. Is sulfur an oxidizing or reducing agent in this process?

As we noted in Chapter 14, sulfur reacts with molecular oxygen to form compounds in which sulfur is assigned oxidation numbers of 4+ and 6+. The reactions are

$$\tfrac{1}{8}S_8(s) + O_2(g) \longrightarrow SO_2(g) \qquad (11)$$

$$SO_2(g) + \tfrac{1}{2}O_2(g) \longrightarrow SO_3(g) \qquad (12)$$

Of these two oxides, SO_2 (sulfur dioxide) is gaseous at ordinary temperatures and pressures, while SO_3 (sulfur trioxide) is a solid with a rather high vapor pressure. Gaseous sulfur trioxide consists of discrete SO_3 molecules. Even though sulfur forms these and other compounds in which it is assigned a positive oxidation number, it does so only by reacting with the strongest of oxidizing agents. Sulfur is somewhat like phosphorus—its strength as a reducing agent or as an oxidizing agent is intermediate.

EXERCISE 19-7

Water saturated with SO_2 gas is a relatively mild but quite useful reducing agent. Which of the following aqueous ions might be reduced by it: Fe^{3+} to Fe^{2+}, Cu^{2+} to Cu^+, Sn^{4+} to Sn^{2+}, or Hg^{2+} to $Hg(l)$? Check the E^0 values in Appendix 7 and review Chapter 14.

19-2.3 Chlorine—A Strong Oxidizing Agent

The formation of several oxidation states is typical of the elements on the right side of the periodic table. Chlorine can exist in the 1+, 3+, 5+, and 7+ oxidation states as well as in the 1− state. In its compounds, chlorine is most often found in the 1− state. This preponderance of 1− compounds shows that elemental chlorine behaves as an oxidizing agent in most of its reactions.

When we review the oxidation-reduction properties of the third-row elements, we do see a rather smooth trend in behavior. Metals on the left tend to be *reducing* agents—they have easily available electrons. Nonmetals on the right tend to be good oxidizing agents—they can accept electrons. Some elements, particularly those such as P and S toward the middle of row three, are either oxidizing or reducing nonmetals. Strong reducing ability is characteristic of metals.

EXERCISE 19-8

Water containing dissolved Cl_2 is a useful oxidizing agent. Which of the following aqueous ions might be oxidized by it: (a) Fe^{2+} to Fe^{3+}, (b) Cu^+ to Cu^{2+}, (c) Sn^{2+} to Sn^{4+}, (d) Mn^{2+} to $MnO_2(s)$, or (e) Mn^{2+} to MnO_4^-? See E^0 values in Appendix 7.

19-3 THE ACIDIC AND BASIC CHARACTER OF THE HYDROXIDES OF THE ELEMENTS OF ROW THREE

We have considered the third-row elements as oxidizing or reducing agents. Let us now examine the acid-base behavior of the hydroxides of the third-row elements. We shall again use the simplest possible ideas of atomic structure to rationalize our observations. We shall deal with a series of compounds which contain the group

$$M \!:\! \ddot{O} \!:\! H$$

Here M can be any third-row element.*

Compounds of this structure may act as bases by releasing hydroxide ions, breaking the M—OH bond. The hydroxide ion, then, can accept a proton from an acid, HB:

$$M\text{—O—H} \rightleftharpoons M^+(aq) + OH^-(aq) \qquad (13)$$

$$OH^-(aq) + HB \rightleftharpoons H_2O + B^-(aq) \qquad (14)$$

Chemists recognize another way in which M—O—H compounds can act as bases. The base M—O—H can react directly with HB:

$$M\text{—O—H} + HB \rightleftharpoons M(OH_2)^+(aq) + B^-(aq) \qquad (15)$$

In reacting with HB, one of the unshared pairs of electrons on the oxygen atom of the MOH group has accepted a proton, so MOH can act as a base without actually releasing hydroxide ions. In addition, MOH compounds can act as acids, breaking the MO—H bond:

$$M\text{—O—H} \rightleftharpoons M\text{—O}^-(aq) + H^+(aq) \qquad (16)$$

or

$$M\text{—O—H}(s) + H_2O \rightleftharpoons M\text{—O}^-(aq) + H_3O^+(aq) \qquad (17)$$

We can attempt to predict whether a compound containing the MOH group will behave as an acid or base by considering the strength with which the element M binds electrons. If the atom M holds electrons strongly, we would not expect the M—O bond to break and form M^+ and OH^- ions. Also, if M holds electrons strongly, we would expect it to draw electrons away from the oxygen atom in the MOH group, with the result that the oxygen atom would be less able to acquire and bind a proton by the reaction

$$MOH(aq) + HB(aq) \longrightarrow M(OH_2)^+(aq) + B^-(aq) \qquad (15)$$

Therefore, as the ability of M to attract electrons from OH increases,† we would expect MOH to be less able to act as a base, either by loss of an OH$^-$ group or by gain of a proton.

---

*In this representation M means the element with one electron missing. The cross, ×, is that electron.

† The ionization energy might be considered as one possible clue to an atom's ability to attract electrons.

An increase in the strength with which *M* binds electrons has another consequence. As *M* draws electrons toward itself, the O—H bond may be weakened and the compound may display acidic properties.

19-3.1 Sodium and Magnesium Hydroxides—Strong Bases

Let us apply these ideas to the third-row elements. On the left side of the table we have the metallic reducing agents sodium and magnesium, which we already know have a small affinity for electrons, since they have low ionization energies and are readily oxidized. It is not surprising that the hydroxides of these elements, NaOH and Mg(OH)$_2$, are solid ionic compounds made up of hydroxide ions and metal ions. Sodium hydroxide is very soluble in water, and its solutions are alkaline because of the OH$^-$ ion. Sodium hydroxide is a strong base. Magnesium hydroxide, Mg(OH)$_2$, is not very soluble in water, but it does dissolve in acid solutions because of the reaction

$$Mg(OH)_2(s) + 2H^+(aq) \longrightarrow Mg^{2+}(aq) + 2H_2O \qquad (18)$$

Magnesium hydroxide is significantly weaker as a base than sodium hydroxide, but it is still a fairly strong base.

One way to correlate this fact is to assign an "acidic-charge" number to the atom *M using the rules for oxidation number* (refer back to Section 14-3). By these rules sodium would have an "acidic-charge" number of 1+ in NaOH. The value 1+ implies a low attraction of the relatively large 1+ ion for electrons. Magnesium in Mg(OH)$_2$ would have an "acidic charge" or oxidation number of 2+. Since Mg in Mg(OH)$_2$ would be both smaller than Na in NaOH and have a greater "acidic charge," Mg would have a stronger attraction for electrons than Na; hence, Mg(OH)$_2$ would be a weaker base than NaOH.

19-3.2 Aluminum Hydroxide—Either an Acid or a Base

We can tell from the second and third ionization energies of aluminum that this atom holds its second and third electrons rather firmly. This fact would again be implied rather neatly by the "acidic-charge" argument, since Al in Al(OH)$_3$ has an "acidic charge" of 3+ and the Al^{3+} ion would be significantly smaller than Mg^{2+}. The small 3+ ion should attract electrons more strongly than Mg^{2+} or Na$^+$.

Bearing this fact in mind, we can see why aluminum hydroxide, Al(OH$_3$), would not be as strongly basic as the hydroxides NaOH and Mg(OH)$_2$. Aluminum hydroxide has extremely low solubility in neutral aqueous solutions, but does react with strong acids according to the reaction

$$Al(OH)_3(s) + 3H^+(aq) \longrightarrow Al^{3+}(aq) + 3H_2O \qquad (19)$$

or

$$Al(OH)_3(s) + 3H_3O^+(aq) \longrightarrow Al(OH_2)_6^{3+}(aq) \qquad (20)$$

These reactions say the same thing, but the second reaction emphasizes the fact that the Al^{3+} ion is hydrated in aqueous solutions. In any case, the fact that Al(OH)$_3$ reacts with acids shows that it has the properties of a base.

Aluminum hydroxide will also react with hydroxide ion to dissolve according to the equation

$$\text{Al(OH)}_3(s) + \text{OH}^-(aq) \longrightarrow \text{Al(OH)}_4^-(aq) \qquad (21)$$

This reaction shows that Al(OH)$_3$ has the properties of an acid, since it reacts with the base OH$^-$. *A substance that acts as an acid under some conditions and as a base under other conditions is said to be* **amphoteric.** The electron arrangement in Al(OH)$_3$ is such that it can either accept a proton (act as a base) or react with OH$^-$ (act as an acid). Several other hydroxides also show amphoteric behavior.

The increasing acidity and decreasing basicity in the series NaOH, Mg(OH)$_2$, and Al(OH)$_3$ is neatly summarized by recalling the trend in electron-attracting ability of the ion M in the configuration

$$M \!\stackrel{..}{_\times}\!\! \stackrel{..}{\text{O}}\!\stackrel{..}{_\times}\! \text{H}$$

The nature of the trend is implied by the "acidic-charge" values, or oxidation numbers, for Na$^+$, Mg^{2+}, and Al^{3+}.

19-3.3 Silicon, Phosphorus, Sulfur, and Chlorine Hydroxides (or Oxyacids)

An "acidic charge" of 4+ for silicon in Si(OH)$_4$ suggests weakly acidic behavior for hydrated silicon oxides. This hydrated oxide does react with concentrated hydroxide ion to form soluble silicate ions:

$$\text{Si(OH)}_4 \cdot x\text{H}_2\text{O}(solid\ hydrate) + 2\text{OH}^-(aq) \longrightarrow$$
$$\text{H}_2\text{SiO}_4{}^{2-} + (2+x)\text{H}_2\text{O} \qquad (22)$$

Si(OH)$_4 \cdot x$H$_2$O is very weakly acidic in water solution.

Phosphorus forms the compound H$_3$PO$_4$, in which three OH groups and one oxygen atom are attached to a central phosphorus atom. The structure projected onto a plane can be represented as

$$\begin{array}{c} :\!\stackrel{..}{\text{O}}\!: \\ \text{H}\!\stackrel{..}{_\times}\!\stackrel{..}{\text{O}}\!:\!\text{P}\!:\!\stackrel{..}{\text{O}}\!\stackrel{..}{_\times}\!\text{H} \\ :\!\stackrel{..}{\text{O}}\!: \\ \stackrel{\times}{\text{H}} \end{array}$$

According to the "acidic-charge" representation, the central phosphorus, having an oxidation number of 5+, should attract electrons rather strongly. Fairly strong acidic characteristics should be displayed. As the name "phosphor*ic* acid" implies, acidic characteristics are the only characteristics apparent in water solution. The equilibrium constant for the acid indicates that phosphoric acid is an acid of moderate strength.

$$\text{H}_3\text{PO}_4(aq) \rightleftharpoons \text{H}^+(aq) + \text{H}_2\text{PO}_4{}^-(aq) \qquad K = 0.71 \times 10^{-2} \qquad (23)$$

Phosphorus forms two other oxygen acids. One has the formula H$_3$PO$_3$ and is called phosphor*ous* acid. The second has the formula H$_3$PO$_2$ and is called *hypo*phosphor*ous* acid. The equilibrium constant for the first ionization of each acid is about the same as that for H_3PO_4, approximately 10^{-2}:

Hypophosphorous acid

$$H_3PO_2 \rightleftharpoons H^+(aq) + H_2PO_2^-(aq)$$
$$K = 1 \times 10^{-2} \quad (24)$$

Phosphorous acid

$$H_3PO_3 \rightleftharpoons H^+(aq) + H_2PO_3^-(aq)$$
$$K = 1.6 \times 10^{-2} \quad (25)$$

Phosphoric acid

$$H_3PO_4 \rightleftharpoons H^+(aq) + H_2PO_4^-(aq)$$
$$K = 0.71 \times 10^{-2} \quad (26)$$

The acids H_3PO_3 and H_3PO_2 violate the earlier generalizations regarding acidic strength. The formal oxidation number of phosphorus in H_3PO_3 is $3+$, and in H_3PO_2 it is $1+$. These numbers suggest acids of relatively low strength. But the ionization constants are comparable to H_3PO_4. The reason for this violation of the generalizations based on the "acidic-charge" rule can be seen by looking at the structures of the molecules. For phosphorous acid the electron dot formulation is

$$\begin{array}{c} \ddot{\mathrm{O}}\!: \\ \mathrm{H}\!:\!\ddot{\mathrm{O}}\!:\!\mathrm{P}\!:\!\ddot{\mathrm{O}}\!:\!\mathrm{H} \\ \mathrm{H} \end{array}$$

Notice that one hydrogen atom has moved so as to become attached directly to the phosphorus atom instead of to an oxygen atom. In hypophosphorous acid the electron dot formulation is

$$\begin{array}{c} \mathrm{H} \\ :\!\ddot{\mathrm{O}}\!:\!\mathrm{P}\!:\!\ddot{\mathrm{O}}\!:\!\mathrm{H} \\ \mathrm{H} \end{array}$$

Two hydrogen atoms are bonded directly to the phosphorus atom.

The formal oxidation number rules used to obtain the "acidic charge" in previous cases cannot account for these structural features. We can only guess from the available data that each hydrogen atom attached to phosphorus in H_3PO_3 or H_3PO_2 is as effective as an —OH unit in pulling charge away from the central phosphorus. One thing becomes clear—the idea of electron attraction based on an "acidic-charge" number for the central atom, M, in MOH can be seriously altered by unexpected structural changes. No simple model for acidic strength could predict the behavior seen with H_3PO_3 and H_3PO_2. Detailed structural information was needed before after-the-fact explanations could be advanced. With knowledge of these structural changes, the theoretical rules for phosphorus in H_3PO_3 and H_3PO_2 could be altered to reproduce the results obtained (*i.e.*, H attached to P is as effective as an OH in building up "acidic charge" on P).* Chemistry remains an experimental science!

---

*One useful rule of thumb is that the approximate ionization constant can be estimated from the number of O atoms (not OH groups) attached to the central atom. An ionization constant of 10^{-2} is expected for one oxygen; a value of 10^{-7} for zero oxygens; a value of 10^3 for two; and a value of 10^8 for three.

The most common oxyacid of sulfur is sulfuric acid (H_2SO_4). In dilute aqueous solutions this substance is almost completely dissociated into ions according to the equation

$$H_2SO_4 \longrightarrow H^+(aq) + HSO_4^-(aq) \qquad (27)$$

Sulfuric acid is classified as a strong acid. The bisulfate ion, HSO_4^-, (also called the hydrogen sulfate ion), is also an acid, since the equilibrium constant for the reaction

$$HSO_4^- \longrightarrow H^+(aq) + SO_4^{2-}(aq) \qquad (28)$$

is approximately 10^{-2}. From these equilibrium constants it is clear that sulfur in the 6+ oxidation state is even more acidic than silicon and phosphorus. Work Exercise 19-9 before reading on.

EXERCISE 19-9

What prediction about the acidic strength of H_2SO_4 would be made based on the normal "acidic-charge" rules? Would you expect H_2SO_3 to be as strong an acid as H_2SO_4? H_2SO_3 has two OH groups attached to sulfur. What about $HClO_4$?

Sulfur in the 4+ oxidation state also forms an oxyacid, sulfur*ous* acid (H_2SO_3). This compound is not as strong an acid as H_2SO_4. The equilibrium constant for the reaction

$$H_2SO_3 \longrightarrow H^+(aq) + HSO_3^-(aq) \qquad (29)$$

is *approximately* 10^{-2}. The ion HSO_3^- is called the bisulfite ion or the hydrogen sulfite ion. It too is a weak acid and dissociates in water to form the sulfite ion, SO_3^{2-}

$$HSO_3^- \xrightarrow{H_2O(l)} H^+(aq) + SO_3^{2-}(aq) \qquad (30)$$

Chlorine forms four oxyacids. Their structures are shown in Figure 19-7. The acid HClO is called *hypo*chlor*ous* acid. The "acidic-charge" number of chlorine in this compound is the oxidation number of chlorine, 1+. The second member of the oxyacid family is chlor*ous* acid ($HClO_2$), with an oxidation number and an "acidic-charge" number of 3+ for chlorine. The third member of the family is chlor*ic* acid ($HClO_3$), with an oxidation number and "acidic-charge" number of 5+ for chlorine. The final chlorine oxyacid is *per*chlor*ic* acid ($HClO_4$), with an oxidation number and "acidic-charge" number of 7+ for chlorine.

Although it is not easy to handle the pure substances, aqueous solutions of these acids have been examined to determine how strong they are as proton donors. HClO is a weak proton donor, $HClO_2$ is somewhat stronger, $HClO_3$ is quite strong, and $HClO_4$ is the strongest of all. [Perchloric acid ($HClO_4$) is actually one of the strongest acids known.]

How well do the arguments on acidic strength which we developed interpret the data on the oxyacids of chlorine? Let us review the arguments before arriving at a conclusion.

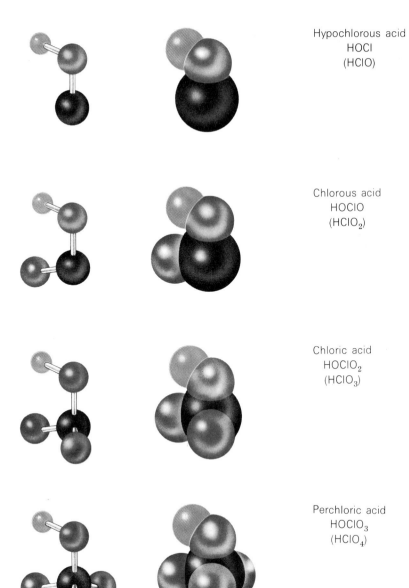

Fig. 19-7 **Structures of the four chlorine oxyacids.**

When an oxyacid serves as an acid, a proton must be broken from an OH group. A regular *decrease in the strength of the hydrogen-oxygen bond* as we proceed from hypochlorous to perchloric acid would explain the trend in acidity. How do we account for the fact that the strength of this bond varies as we go through the sequence? Formally, we say that the oxidation number of chlorine ranges from $1+$ to $3+$ to $5+$ to $7+$ across the set, but actually there are really no Cl^{1+}, Cl^{3+}, Cl^{5+}, or Cl^{7+} ions that could be recognized in, or isolated from, the oxyacids. Rather, these "acidic-charge" numbers or oxidation numbers give us an indication of how strongly the central chlorine atom attracts electrons. We must be careful not to think of the "acidic-charge" number as the actual charge on the central ion. Rather, it is an arbitrary number that gives us some clue as to the extent to which electrons are shifted inward from the O—H to the O—Cl bond. Using this interpretation, the increase in acidic strength from HClO to HClO₄ is easily understood.

In general, we can say that as the electron-attracting ability of the central atom in an oxyacid increases, the strength of the acid increases. As we move to the right in a row of the periodic table, the ability of an atom to attract electrons of the M—O bond rises. (Note that this is suggested by the increase in ionization potential across the row from left to right.) As electrons move toward the M—O bond, the O—H bond is weakened and acidic strength rises.

19-4 HIGHLIGHTS

In considering the physical properties of elements of the third row and contrasting these properties with those of the elements of the second row, we noted that in row three metallic character is stronger and is observed in elements located even farther to the right in the table. This behavior is understood in terms of the fact that metallic character increases as we go down a column of the table. The increased metallic character is associated with a decrease in ionization energies down a column. The elements of the third row show no tendency to form multiple bonds with each other using p-orbitals. This accounts for the fact that stable P_2 and S_2 molecules, comparable to N_2 and O_2, do not appear at room temperature.

The trend from metal to nonmetal parallels a trend from strong reducing to strong oxidizing ability. Metals are reducing agents. Nonmetals are both oxidizing and reducing agents, but oxidizing character becomes more important as nonmetallic character increases.

Metals tend to form hydroxides that are bases; nonmetals tend to form hydroxides or oxyhydroxides (oxyacids) that are acids. Increasing acidic strength for a hydroxide can be correlated with an increasing "acidic-charge" or oxidation number on the central atom of an oxyacid or hydroxide.

QUESTIONS AND PROBLEMS

1 Make a graph with an energy scale extending on the ordinate from zero to 3,000 kcal/mole and with the abscissa labeled at equal intervals Na, Mg, and Al. Now plot and connect with a solid line the first ionization energies, E_1, of these three elements (see Table 19-1). Plot E_2 and connect with a dashed line, E_3 with a dotted line, and E_4 with a solid line. Draw a circle around each ionization energy that identifies a valence electron.

2 Plot the ionization energy of the first electron removed from the atoms of both the second- and third-row elements (ordinate) against their atomic number (abscissa). What regularity do you observe? (See Table 8-1, page 162.)

3 Silicon melts at 1400°C and phosphorus (white) at 44°C. Explain this very great difference in terms of the structures of the solids.

4 Justify the experimental fact that white phosphorus is more reactive than red phosphorus by considering the structures of these elements.

5 Recalling the chemistry of nitrogen, write formulas for phosphorus compounds corresponding to (a) ammonia, (b) hydrazine (N_2H_4), and (c) ammonium iodide.

6 Write the formula for the fluoride you expect to be most stable for each of the third-row elements.

7 The heat of reaction for the formation of MgO(s) from the elements is -144 kcal/mole MgO(s). How much heat is liberated when magnesium reduces the carbon in CO_2 to free carbon? (See Table 9-2, page 196.) [*Answer:* $\Delta H = -97$ kcal/mole MgO(s).]

8 Magnesium oxide is an ionic solid that crystallizes in the sodium-chloride type lattice. (a) Explain why MgO is an ionic substance. (b) How many calories would be required to decompose 8.06 g of MgO? (Use the data in question 7.) (c) Draw a diagram of a crystal of MgO.

9 Aluminum oxide (Al_2O_3) is thought to dissociate at a high temperature (1950°C) according to the equation

$$2Al_2O_3(s) \longrightarrow 4AlO(g) + O_2(g)$$

The total vapor pressure at 1950°C is about 1×10^{-6} atm. (a) Which element is oxidized and which is reduced in this reaction? (b) Write the equation for the equilibrium constant. (c) Calculate its value using partial pressure as the unit of "concentration" for the gases.

10 Explain the observation that phosphorus can act either as a weak reducing or a weak oxidizing agent.

11 (a) What are the oxidation numbers of phosphorus in the two compounds phosphorous acid (H_3PO_3) and phosphoric acid (H_3PO_4)? (b) From the E^0 values in Appendix 7 and the half-reaction for phosphorous acid given below, decide which of the following substances might be reduced by phosphorous acid: Fe^{2+}, Sn^{4+}, I_2, or Cr^{3+}?

$$H_2O + H_3PO_3 \longrightarrow H_3PO_4 + 2H^+ + 2e^-$$
$$E^0 = 0.276 \text{ volt}$$

(c) Balance the equation for the reaction between phosphorous acid and Fe^{3+}. Calculate E^0 for the reaction.

12 Answer the following in terms of electron configuration and ionization energy: (a) Which elements in the second and third rows are strong oxidizing agents? Which are strong reducing agents? (b) What are the properties of strong oxidizing agents? (c) What are the properties of strong reducing agents?

13 Of the elements Na, Mg, and Al, which would you expect to form (a) a molecular solid with chlorine? (b) an ionic solid with chlorine?

14 One kilogram of seawater contains 0.052 mole of magnesium ion. What is the minimum number of kilograms of seawater that must be processed to obtain 1 kg of $Mg(OH)_2$? (*Answer:* 3.3×10^2 kg of seawater.)

15 Why is aluminum hydroxide classed as an amphoteric compound?

16 Some of the following common compounds of the third-row elements are named as hydroxides and some as acids:

NaOH	sodium hydroxide
$Mg(OH)_2$	magnesium hydroxide
$Al(OH)_3$	aluminum hydroxide
$Si(OH)_4$	silicic acid (usually written H_4SiO_4)
$P(OH)_3$	phosphorous acid (usually written H_3PO_3)
$S(OH)_2$	not known
$Cl(OH)$	hypochlorous acid (usually written HOCl)

(a) Explain why these compounds vary systematically in their acid-base behavior. (b) Write equations that show the reactions of each of these substances either as acids or bases, or both.

17 A solution containing 0.20 M H_3PO_3 (phosphorous acid) is tested with indicators, and $H^+(aq)$ concentration is found to be 5.0×10^{-2} M. Calculate the dissociation constant of H_3PO_3, assuming that a second proton cannot be removed.

18 Elemental phosphorus is prepared by the reduction of calcium phosphate, $Ca_3(PO_4)_2$, with coke in the presence of sand, SiO_2. The products are phosphorus, calcium silicate, $CaSiO_3$, and carbon monoxide. (a) Write the equation for the reaction. (b) Using 75.0 kilograms of the ore calcium phosphate, calculate how many grams of P_4 can be obtained and how many grams of coke (assumed to be pure carbon) will be used.

19 Suggest a possible reaction upon which the manufacture of Br_2 from NaBr might be based using elemental Cl_2.

20 Hydrogen is available from the water-gas reaction

$$C + H_2O \longrightarrow H_2 + CO$$

Chlorine is available from the manufacture of magnesium. Suggest a possible way to convert $MgCO_3$ to $MgCl_2$ for the electrolysis process.

Science and art have that in common that everyday things seem to them new and attractive . . . .

FRIEDRICH NIETZSCHE (1844–1900)

CHAPTER 20
Vertical Groups of the Periodic Table

Let's follow the differences discovered between the second and third row elements to its logical conclusion and see what happens to the properties of elements as one goes down an entire vertical column of the periodic chart. We will investigate one family which is clearly metallic and one which is clearly non-metallic. In trying to correlate trends in chemical properties with increasing atomic size, we find that the size of a boundary-less atom varies with the bonding situation in which the atom exists.

In the preceding chapter we looked at the third *row* of the periodic table to see what systematic changes occur in properties of elements as the number of electrons in the outer orbitals of the atom increases. These trends were conveniently discussed in terms of ionization energies and orbital occupancies.

There are similar, but less marked, trends in the properties of elements in a *column* or *family* of the periodic table. Many of the differences in properties within a family are explainable in terms of atomic size. How can such differences be investigated?

PROPERTIES of the
ALKALINE EARTH METALS

20-1 SOME IMPORTANT PROPERTIES IN TWO COLUMNS OF THE PERIODIC TABLE—THE ALKALINE EARTH METALS* AND THE HALOGENS

Let us begin our study by considering some of the differences in physical properties which are apparent as we proceed down a column of the periodic table. Since the changes observed within a family of metals may not be the same as those observed within a family of nonmetals, we shall consider two columns of the table—the metallic alkaline earth elements in the second column and the strongly nonmetallic halogens in the seventh column.

20-1.1 Some Properties of the Alkaline Earth Metals

The electron configurations for the elements of the second column are of immediate interest to us. (Data are summarized in Table 20-1.) Notice that each element has *two more electrons* than the nearest noble gas.

---

* The second-column metals—particularly Ca, Sr, and Ba—are called the **alkaline earth metals.** For convenience, Be and Mg are also considered alkaline earth metals.

TABLE 20-1 ELECTRON CONFIGURATIONS OF THE ALKALINE EARTH ELEMENTS

Element	Symbol	Nuclear Charge	Electron Configuration*
Beryllium	Be	4+	$1s^2$ $\;2s^2$ †
Magnesium	Mg	12+	$1s^2$ $\;2s^22p^6$ $\;3s^2$
Calcium	Ca	20+	$1s^2$ $\;2s^22p^6$ $\;3s^23p^6$ $\;4s^2$
Strontium	Sr	38+	$1s^2$ $\;2s^22p^6$ $\;3s^23p^63d^{10}$ $\;4s^24p^6$ $\;5s^2$
Barium	Ba	56+	$1s^2$ $\;2s^22p^6$ $\;3s^23p^63d^{10}$ $\;4s^24p^64d^{10}$ $\;5s^25p^6$ $\;6s^2$
Radium	Ra	88+	$1s^2$ $\;2s^22p^6$ $\;3s^23p^63d^{10}$ $\;4s^24p^64d^{10}4f^{14}$ $\;5s^25p^65d^{10}$ $\;6s^26p^6$ $\;7s^2$

* Remember: In a designation such as $3s^2$, 3 defines the value of the principal quantum number, s defines the sublevel and indicates the shape of the orbital in the hydrogen atom, and superscript 2 indicates that two electrons are present in the orbital (or orbitals) of that class.

† ☐ indicates valence electrons.

EXERCISE 20-1

On the basis of electron configuration and position in the periodic table, answer the following questions:

(1) Is calcium likely to be a metal or a nonmetal?
(2) Is calcium likely to resemble magnesium or potassium in its chemistry?
(3) Is calcium likely to have a higher or lower boiling point than potassium? than scandium?

EXERCISE 20-2

Predict the chemical formula and physical state at room temperature of the most stable compound formed by each alkaline earth element with chlorine, oxygen, and sulfur.

Exercises 20-1 and 20-2 pose some of the simplest questions we can ask about the alkaline earth elements. More searching questions require more detailed data. Let us examine the first three ionization energies of the alkaline earths. (Data are summarized in Table 20-2.)

TABLE 20-2 IONIZATION ENERGIES OF THE ALKALINE EARTH ELEMENTS

Element	Ionization Energy (kcal/mole)		
	E_1	E_2	E_3
Beryllium	214	420	3,533
Magnesium	175	345	1,838
Calcium	140	274	1,173
Strontium	132	253	986
Barium	120	230	811

EXERCISE 20-3

For each alkaline earth metal, calculate the ratio E_2/E_1. Account for the results in terms of the charges on the ions formed in the two ionization steps.

EXERCISE 20-4

If the ionization energy E_1 is regarded as a measure of the distance between the electron and the nuclear charge, what do the ionization energies of Be and Ba indicate about the relative sizes of the two atoms?

From Exercise 20-4 we see that the decreasing ionization energies observed as we move down the second column of the periodic table are explained in terms of increasing atomic size. Some other properties of this family are summarized in Table 20-3.

EXERCISE 20-5

From the ionization energies, predict which solid substance involves bonds having the most ionic character: $BeCl_2$, $MgCl_2$, $CaCl_2$, $SrCl_2$, or $BaCl_2$? Which substance would you expect to have the most covalent character in its bonds?

TABLE 20-3 PROPERTIES OF THE ALKALINE EARTH ELEMENTS IN THE METALLIC STATE

Element	Crystal Structure	Density (g/cm³)	Melting Point (°C)	Heat of Vaporization (kcal/mole)	Electrical Conductance (ohm⁻¹ cm⁻¹)
Beryllium	hexagonal close-packed	1.85	1283	54	1.69×10^5
Magnesium	hexagonal close-packed	1.75	650	32	2.24×10^5
Calcium	face-centered cubic	1.55	850	42	2.92×10^5
Strontium	face-centered cubic	2.6	770	39	0.43×10^5
Barium	body-centered cubic	3.5	710	42	0.16×10^5

20-1.2 Some Properties of the Halogens

We considered some aspects of halogen chemistry in Chapter 8. You will recall that the halogens are reactive elements; under standard conditions they exist as diatomic molecules held together by single covalent bonds. All the molecules have color: gaseous fluorine is pale yellow; gaseous chlorine is yellow-green; gaseous bromine is orange-red (remember the film "Equilibrium"); and gaseous iodine is violet. All halogens are toxic and dangerous substances; fluorine is the most hazardous and iodine the least, but even iodine must be handled with caution.

TABLE 20-4 ELECTRON CONFIGURATIONS AND IONIZATION ENERGIES OF THE HALOGENS

Element	Symbol	Nuclear Charge	Electron Configuration					Ionization Energy (kcal/mole)
Fluorine	F	9+	$1s^2$	$2s^22p^5$ *				402
Chlorine	Cl	17+	$1s^2$	$2s^22p^6$	$3s^23p^5$			300
Bromine	Br	35+	$1s^2$	$2s^22p^6$	$3s^23p^63d^{10}$	$4s^24p^5$		273
Iodine	I	53+	$1s^2$	$2s^22p^6$	$3s^23p^63d^{10}$	$4s^24p^64d^{10}$	$5s^25p^5$	241

* ⌐ ¬ indicates valence electrons.

Electron configurations and ionization energies for the halogen atoms are shown in Table 20-4. In every case the element lacks one electron to complete the s and p levels of its valence shell; thus, the electron dot designation for all halogens is

where X is any halogen atom. As you will recall, proper use of this notation will explain the existence of F₂, Cl₂, Br₂, and I₂ molecules and of halide ions, X⁻. Note that the halogen ionization energies are very large compared to those of comparable alkaline earth elements from column two (Table 20-2). Note further that the ionization energies decrease going down the column, a fact which suggests that electrons are farther and farther from the nucleus as the atomic number rises. (Refer also to Tables 20-1 and 20-4.) Let us investigate this effect of atomic size in both groups more carefully.

20-2 SIZES OF THE ATOMS AND IONS IN THE SECOND AND SEVENTH COLUMNS

As you will recall from Section 7-1.3 and from the description of the atom in Chapter 15, *atoms have no definite size.* To assign a size, we must first decide where an atom "stops." Any such decision will be purely arbitrary, since we know from quantum mechanics that an atom has no sharp boundaries or surfaces. What kinds of arbitrary decisions have been made?

20-2.1 Covalent Radii of Atoms

In Section 7-1.3 we used an *operational definition* to define the size of a chlorine atom. First, we determined experimentally the distance between two chlorine nuclei in Cl_2. We then assigned half this distance to each chlorine as the radius of a chlorine atom. Because the chlorine molecule we used to determine this distance is held together by a covalent bond, the radius is called the **covalent radius** of the chlorine atom. It gives an estimate of the size of a chlorine atom in a covalently bonded situation. (See Figure 20-1.) As we might well expect, the covalent radii of the halogen atoms becomes greater as more electrons are added to the electron cloud outside the atomic nucleus. The values for the covalent radii are 0.72 Å for F, 0.99 Å for Cl, 1.14 Å for Br, and 1.33 Å for I. Note that fluorine is *smaller* than we would have expected from a linear extrapolation of the sizes of the other atoms seen in Figure 20-1.

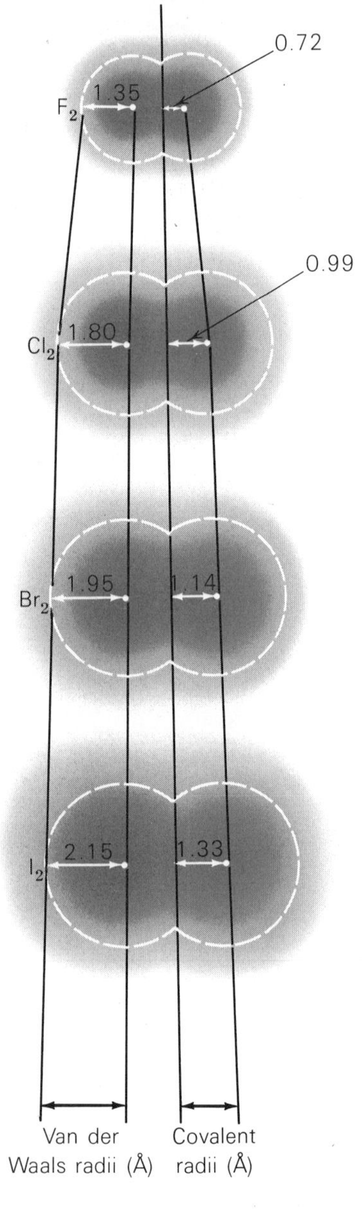

EXERCISE 20-6

Using the carbon atom covalent radius 0.77 Å and the covalent radii given above, predict the C—X bond length in the following molecules: CF_4, CBr_4, and CI_4. Compare your calculated bond lengths with the experimental values C—F in CF_4 = 1.32 Å, C—Br in CBr_4 = 1.94 Å, and C—I in CI_4 = 2.15 Å.

20-2.2 Van der Waals Radii

Another way of establishing a value for the atomic size of each halogen is to determine how closely the nuclei of the atoms *in two different molecules* approach each other during a molecular collision (see Figure 20-2). Half this distance could then be assigned as an atomic radius for each atom. These collision distances can be obtained by comparing pressure-volume data for the real gas with pressure-volume data for an ideal gas ($PV = nRT$). Because the real-gas values are related to gas imperfections and to attractions and repulsions between molecules, radii obtained by this method are called **van der Waals radii,** after the Dutch physicist who studied gas imperfections in 1873. Figure 20-1 compares the van der Waals and covalent radii for the halogens. Note specifically that van der Waals radii are always larger than covalent radii and that both the van der Waals and the covalent radii increase as the atomic number of the halogens becomes greater.

Fig. 20-1 Covalent radii and van der Waals radii of the halogens.

F_2 1.35 0.72
Cl_2 1.80 0.99
Br_2 1.95 1.14
I_2 2.15 1.33

Van der Waals radii (Å)   Covalent radii (Å)

EXERCISE 20-7

Suppose you wanted to assign a radius to the argon atom. What kind of radius could be obtained? Explain.

EXERCISE 20-8

The xenon-fluorine distance in XeF_2 is 2.00 Å. Using the data in Section 20-2.1, calculate a radius for xenon. The value usually given in textbooks and charts is 1.90 Å. Explain any discrepancy. Suppose you wanted to estimate the distance between xenon atoms in solid xenon. Which radius would be more appropriate? Explain.

20-2.3 Metallic Radii

What kind of decisions do we make in assigning a radius value to an atom of column two? The distance between magnesium nuclei in solid metallic magnesium can be determined by X-ray diffraction to be 3.20 Å. The radius of magnesium can be taken as half this value, or 1.60 Å. But we must designate this value as a **metallic radius** rather than as a covalent radius since metallic bonds are involved. Interatomic distances in the metals and values for metallic radii of the elements of column two are summarized in Table 20-5. Note that there is an increase in the size of metal atoms as we go down a column.

TABLE 20-5 METALLIC RADII FOR METALS OF COLUMN TWO

Element	Distance Between Atoms in Metal (Å)	Metallic Radius (Å)
Beryllium	2.23	1.11
Magnesium	3.20	1.60
Calcium	3.95	1.97
Strontium	4.30	2.15
Barium	4.35	2.17

20-2.4 Ionic Radii

An examination of ionic solids such as MgO and NaCl reveals another bonding situation which must be considered in assigning atomic radii. The distance between the magnesium nucleus and the oxygen nucleus in solid MgO can be measured directly by X rays. (Data are presented in Table 20-6.) To apportion this distance between oxygen and magnesium, an assumption regarding relative sizes of Mg^{2+} and O^{2-} must be made. The basis for making such an assumption is somewhat involved and cannot be given here; instead, we shall accept a value of 1.32 Å as appropriate for the oxide ion, O^{2-}. By using this value and the interatomic distances of Table 20-6, we can obtain **ionic radii** for the column-two cations. Similarly, by using an accepted radius for potassium and the distance between adjacent nuclei in KF, KCl, KBr, and KI, ionic radii for the halide ions can be obtained. (Refer to Table 20-7 and Figure 20-3.) Again, we see in both columns two and seven an increase in ionic radii as we move down the columns.

IONIC RADII

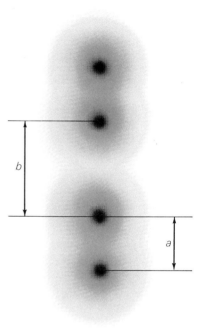

Fig. 20-2 **A collision of gaseous halogen molecules.**

a = covalent bond length

Covalent radius = $\frac{a}{2}$

b = distance of closest approach of nuclei in molecular collision

Van der Waals radius = $\frac{b}{2}$

TABLE 20-6 IONIC RADII FOR COLUMN TWO CATIONS

Element	Ionic Solid	Internuclear Distance (Å)	Cation Radius* (Å)	Cation
Beryllium	BeO	1.64	0.32	Be^{2+}
Magnesium	MgO	2.10	0.78	Mg^{2+}
Calcium	CaO	2.40	1.08	Ca^{2+}
Strontium	SrO	2.57	1.25	Sr^{2+}
Barium	BaO	2.76	1.44	Ba^{2+}

* Based on an assigned radius of 1.32 Å for O^{2-}.

TABLE 20-7 IONIC RADII FOR HALIDE ANIONS

Element	Ionic Solid	Internuclear Distance (Å)	Anion Radius† (Å)	Anion
Fluorine	KF	2.69	1.36	F^-
Chlorine	KCl	3.14	1.81	Cl^-
Bromine	KBr	3.28	1.95	Br^-
Iodine	KI	3.49	2.16	I^-

† Based on an assigned radius of 1.33 Å for K^+.

One other trend in the size of ions is important. Let us write electron dot formulas for the fluoride *anion,* the neon *atom,* and the sodium *cation:*

$$:\ddot{F}:^- \qquad :\ddot{N}e: \qquad :\ddot{N}a:^+$$

Nuclear charge F = 9+   Nuclear charge Ne = 10+   Nuclear charge Na = 11+

In this series all atoms and ions have exactly the same *number of electrons. The difference between F^-, Ne, and Na^+ is found in the number of protons in the nucleus.* Neon has one more proton in the nucleus than does fluorine. Sodium has one more proton than does neon. Increasing the number of positive charges on the nucleus should pull the electrons in more tightly toward the nucleus and the electron cloud should become smaller as we go from F^- to Na^+. Values of the ionic radii confirm this prediction. The appropriate values of the ionic or atomic radii are as follows:

$$\begin{array}{ccc} F^- & Ne & Na^+ \\ 1.36 \text{ Å} & 1.12 \text{ Å} & 0.95 \text{ Å} \end{array}$$

We see that *negative ions are larger than positive ions or neutral atoms having the same number of electrons.*

20-2.5 A Summary of Observations About Atomic Size

Because of the very nature of atoms, the actual size of an atom is a very indefinite quantity. In trying to work around this fact, chemists have set values for radii appropriate to various bonding situations.

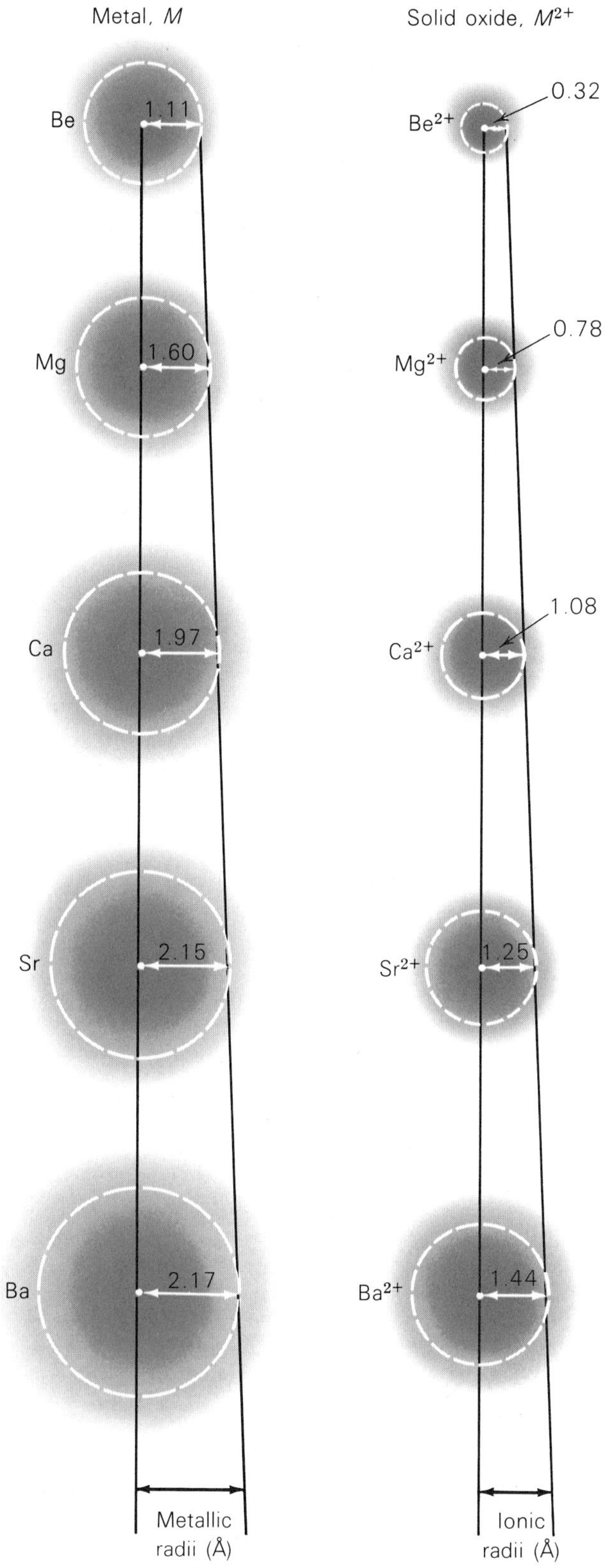

OBSERVATIONS about ATOMIC SIZE

Fig. 20-3.1 **(Left)** Metallic and ionic radii of alkaline earth atoms and ions.

Fig. 20-3.2 **(Below)** Ionic radii of halide ions.

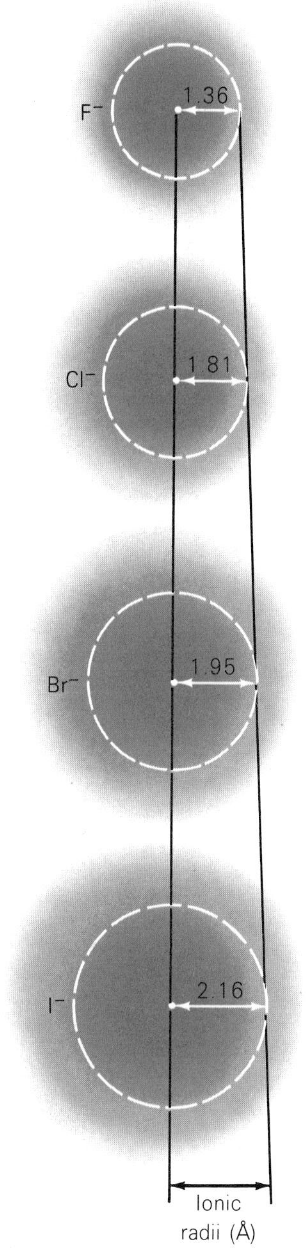

Metal, M

- Be 1.11
- Mg 1.60
- Ca 1.97
- Sr 2.15
- Ba 2.17

Metallic radii (Å)

Solid oxide, M^{2+}

- Be^{2+} 0.32
- Mg^{2+} 0.78
- Ca^{2+} 1.08
- Sr^{2+} 1.25
- Ba^{2+} 1.44

Ionic radii (Å)

- F^- 1.36
- Cl^- 1.81
- Br^- 1.95
- I^- 2.16

Ionic radii (Å)

The pertinent points we have discussed are:

(1) The radius assigned to an atom is determined by the bonding situation in which an atom exists. We have considered covalent radii, van der Waals radii, metallic radii, and ionic radii.
(2) *All* these radii increase as we move down a column of the periodic table. This is a logical result of the increasing number of electrons in higher energy levels as we move down a column.
(3) Negative ions are larger than neutral atoms or positive ions having the same number of electrons. This is a logical result of the increasing nuclear charge from negative ion to neutral atom to positive ion.

These changes in atomic size have a significant effect upon the chemical properties of the elements.

20-3 HIGHLIGHTS

As we move down any column of the periodic table, we find atoms with more electrons at higher energy levels. More electrons results in increased atomic size going down any column. In discussing atomic size, we must use rather arbitrary operational definitions, since size of an atom depends upon the bonding situation in which the element exists. We defined **covalent radii, van der Waals radii, metallic radii,** and **ionic radii.** Many properties of the elements reflect changes in atomic size within a column.

QUESTIONS AND PROBLEMS

1 Give the electron configuration for each of the following: F^-, Ne, Na^+, and Mg^{2+}. How do Cl^-, Ar, K^+, and Ca^{2+}, and Br^-, Kr, Rb^+, and Sr^{2+} differ from the above?

2 Which has the highest ionization energy in each of the following pairs? (a) Na, Mg (b) Na^+, Mg^+ (c) F^-, Ne (d) F, Na (e) Na, Rb (f) F, I (g) I, Cs.

3 Predict the molecular structures and bond lengths for SiF_4, $SiCl_4$, $SiBr_4$, and SiI_4, assuming the covalent radius of silicon is 1.16 Å.

4 Explain in terms of nuclear charge why the K^+ ion is smaller than the Cl^- ion, though they are iso-electronic (that is, they have the same number of electrons).

5 Can aqueous bromine, $Br_2(aq)$, be used to oxidize ferrous ion, $Fe^{2+}(aq)$, to ferric ion, $Fe^{3+}(aq)$ (see Appendix 7)? Can aqueous iodine (I_2) be used?

6 What will happen if F_2 is bubbled into 1 M NaBr solution? Justify your answer using E^0 values.

7 Write formulas for (a) an oxide of barium, (b) a carbonate of barium, (c) a chromate of barium, and (d) a chloride of beryllium.

8 Use solubility data to outline a scheme for separating (a) Mg^{2+} and Ba^{2+}, (b) Ca^{2+} and Ba^{2+}. (See Chapter 12 for solubility data.)

9 Using E^0 values, predict what will happen if, in turn, each halogen beginning with chlorine is added to a 1 M solution of ions of the next lower halogen: Cl_2 to Br^-, Br_2 to I^-. Which halogen is oxidized and which is reduced in each case?

10 Write a balanced equation for the reaction of dichromate and iodide ions in acid solution. Determine E^0 for the reaction

$Cr_2O_7^{2-}(aq) + I^-(aq) + H^+(aq)$ gives
$\qquad Cr^{3+}(aq) + I_2 + H_2O$

(Answer: $E^0 = +0.80$ volt.)

11 Balance the equation for the reaction of iodine with thiosulfate ion:

$I_2 + S_2O_3^{2-}(aq)$ gives $S_4O_6^{2-}(aq) + I^-(aq)$
THIOSULFATE   TETRATHIONATE
ION   ION

What is the oxidation number of sulfur in the tetrathionate ion?

12 How many grams of iodine can be formed from 20.0 g of KI by oxidizing it with ferric chloride (FeCl$_3$)? Determine E^0. (*Answer:* 15.3 g of I$_2$.)

13 Balance the equation for the reaction between SO$_2$ and I$_2$ to produce SO$_4^{2-}$ and I$^-$ in acid solution. Calculate E^0. From Le Chatelier's principle, predict the effect on the E^0 in this reaction if H$^+$ = 10^{-7} M is used instead of H$^+$ = 1 M.

14 What is the oxidation number of the halogen in each of the following: HF, HBrO$_2$, HIO$_3$, ClO$_3^-$, F$_2$, and ClO$_4^-$?

15 Comparable half-reactions for the oxidation of iodine and chlorine are shown below.

½I$_2$ + 3H$_2$O $\longrightarrow$ IO$_3^-$ + 6H$^+$ + 5e$^-$
$E^0 = -1.195$ volts

½Cl$_2$ + 3H$_2$O $\longrightarrow$ ClO$_3^-$ + 6H$^+$ + 5e$^-$
$E^0 = -1.47$ volts

(a) Which is the stronger oxidizing agent, iodate, IO$_3^-$, or chlorate, ClO$_3^-$? (b) Balance the equation for the reaction between chlorate ion and I$^-$ to produce I$_2$ and Cl$_2$.

16 Two half-reactions involving chlorine are

2e$^-$ + Cl$_2$ $\longrightarrow$ 2Cl$^-$
$E^0 = 1.36$ volts

2e$^-$ + 2HOCl + 2H$^+$ $\longrightarrow$ Cl$_2$ + 2H$_2$O
$E^0 = 1.63$ volts

(a) Balance the reaction in which self-oxidation-reduction of Cl$_2$ occurs to produce chloride ion and chlorous acid (HOCl). (b) What is the oxidation number of chlorine in each species containing chlorine? (c) What is E^0 for the reaction? (d) Explain, using Le Chatelier's principle, why the self-oxidation-reduction reaction occurs in 1 M OH$^-$ instead of 1 M H$^+$ solution.

17 How many grams of SiO$_2$ will react with 5.00 × 10^2 ml of 1.00 M HF to produce SiF$_4$? (Water is the other product.)

18 A water solution that contains 0.10 M HF is 8 percent dissociated. What is the value of its K_A? (*Answer:* 6.9 × 10^{-4}.)

19 Select the substance which best fits the requirement specified:

(a) Strongest acid — HOCl, HOClO, HOClO$_2$, Be(OH)$_2$

(b) Biggest atom — F, Cl, Br, I, Ba, Sr, Mg

(c) Smallest ionization energy — F, Cl, Br, I, Ba, Sr, Mg

(d) Best reducing agent — F$^-$, Cl$^-$, Br$^-$, I$^-$, Mg^{2+}

(e) Weakest acid — HF, HCl, HBr, HI

(f) Best hydrogen bonding — HF, HCl, HBr, HI

20 Describe two properties that the halogens have in common and explain why they have these properties in common.

The best and safest way of doing scientific work seems to be, first to enquire diligently into the properties of things, and of establishing these properties by experiment, and then to proceed slowly to theories for the explanation of them.

SIR ISAAC NEWTON (1642–1727)

CHAPTER 21
The Transition Elements: The Fourth Row of the Periodic Table

In Chapter 15 we noted briefly that the transition elements—the ones which cause the second and third rows of the periodic table to be separated into two parts—correspond to the filling of the d orbitals with electrons. Since these elements are among the most well-known, useful, and interesting of all of the 104, they deserve more attention than we have given them.

All of these elements are metals, and most of them have many common uses. Because they have several oxidation states, some of their ions were met frequently when we studied oxidation-reduction reactions. Surely you'll never forget chromium and manganese? Many of the ions are colored, and a change of color with change in oxidation state has often been used by us in the laboratory as visible evidence that a reaction has taken place. Let's see what other personality traits these elements have!

In the preceding chapters we have studied the chemistry of the elements across the top rows of the periodic table and down the sides. Now we shall consider the elements in the middle. These are usually referred to as the **transition elements** because chemists once believed that some elements behaved in a way intermediate between the extremes represented by the opposite sides of the periodic table.

ELECTRON CONFIGURATION

21-1 IDENTIFICATION AND ELECTRONIC STRUCTURE OF TRANSITION ELEMENTS

There is some disagreement among chemists as to just which elements should be called transition elements. For our purposes, it will be convenient to include all the elements in the columns of the periodic table headed by scandium through zinc.

Across the first row of the transition region, we have the elements scandium (Sc), titanium (Ti), vanadium (V), chromium (Cr), manganese (Mn), iron (Fe), cobalt (Co), nickel (Ni), copper (Cu), and zinc (Zn). On the left, we have the scandium column, which also includes yttrium (Y), lanthanum (La), and actinium (Ac). For reasons that we shall mention later, the elements that follow lanthanum (no. 58 to no. 71) are not considered transition elements but are placed in a separate group, the **lanthanides.** This is also true of the elements following actinium (no. 90 to no. 103), the **actinides.**

On the right, the transition elements end with the zinc column. Besides zinc, this includes cadmium (Cd) and mercury (Hg). It is strongly advisable during the discussion that follows to refer to the periodic table often to see where each particular element is placed.

21-1.1 Electron Configuration

Now that we know where the transition elements are in the periodic table, we must ask two questions:

(1) Why do we consider these elements together?
(2) What is special about their properties?

These questions are closely related because they both depend upon the electron configurations of the atoms. What, then, is the electron configuration we might expect for these elements?

To answer this question, we need to review some basic ideas on the electronic construction of atoms. We saw in Chapter 15 that as we progressively add electrons to build up an atom, each added electron goes into the lowest energy level not already fully occupied. With this principle as a guide, let us consider the electron configurations of the first row of transition elements from scandium to zinc.

Looking at the periodic table, we see that calcium (no. 20) comes just before scandium. The 20 electrons in a calcium atom are distributed as in the following:

In element number 21, we must accommodate one more electron. At first sight we might predict that the twenty-first electron will go into the 4p orbital, the next higher energy level after 4s. The 4p orbital *is* of higher energy than the 4s orbital, but, more important, there is a set of five 3d orbitals *between* the 4p and 4s orbitals. The twenty-first electron goes into a 3d orbital, the level of next higher energy. (See Figure 21-1, which is actually Figure 15-24 reproduced here for convenient reference.)

EXERCISE 21-1

Draw on one line a set of orbitals from 1s through 4d. Under this give the orbital occupancy for Al, Sc, and Y. Account for the fact that yttrium is much more like scandium than is aluminum.

Five 3d orbitals are available, all more or less of the same energy for an isolated atom in space. Putting a pair of electrons in each of these five orbitals means that a total of ten electrons can be accommodated before it is necessary to go to a higher energy level. Not only scandium but the nine following elements can be built up similarly by placing additional electrons in 3d orbitals. Not until we come to gallium (no. 31) do we proceed to another set of orbitals.

EXERCISE 21-2

Again using Figure 21-1, decide which orbital will be used after the five 4d orbitals have been filled. To what element does this correspond?

With the help of Figure 21-1, or with an atomic orbital chart, you should now be able to work out the electron configurations of most of the transition elements. You will not be able to deduce all of them exactly because there are some exceptions resulting from special stabilities caused by a set of orbitals being filled or half-filled. The fourth-row transition elements have the set of electron configurations shown in Table 21-1. Notice that chromium (no. 24) and copper (no. 29) provide interruptions to the continuous buildup. In the

TABLE 21-1 THE ELECTRON CONFIGURATIONS OF THE FOURTH-ROW TRANSITION ELEMENTS

Element	Symbol	Atomic Number	Electron Configuration	
Scandium	Sc	21	$1s^2\ 2s^2\ 2p^6\ 3s^2\ 3p^6$*	$3d^1\ 4s^2$
Titanium	Ti	22		$3d^2\ 4s^2$
Vanadium	V	23		$3d^3\ 4s^2$
Chromium	Cr	24		$3d^5\ 4s^1$
Manganese	Mn	25		$3d^5\ 4s^2$
Iron	Fe	26		$3d^6\ 4s^2$
Cobalt	Co	27		$3d^7\ 4s^2$
Nickel	Ni	28		$3d^8\ 4s^2$
Copper	Cu	29		$3d^{10}4s^1$
Zinc	Zn	30		$3d^{10}4s^2$

* Each fourth-row transition element has these levels filled.

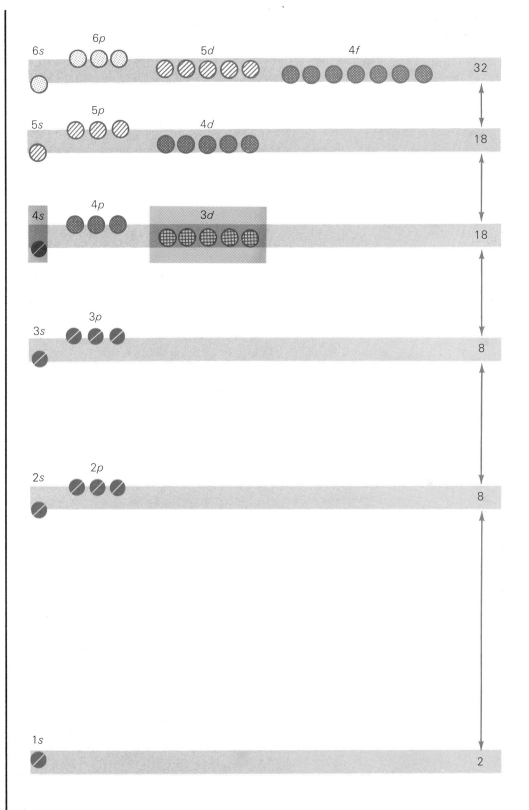

Fig. 21-1 The energy-level scheme of a many-electron atom.

case of chromium, the whole atom has lower energy if one of the 4s electrons moves into the 3d set to give a half-filled set of 3d orbitals and a half-filled 4s orbital. In the case of copper, the atom has lower energy if the 3d set is completely filled by ten electrons and the 4s orbital is half-filled, instead of the 3d orbitals having nine electrons and the 4s orbital two.

EXERCISE 21-3

Make an electron configuration table like Table 21-1 for the fifth-row transition elements—yttrium (no. 39) through cadmium (no. 48). In elements 41 through 45, one of the 5s electrons moves over to a 4d orbital. In element 46, two electrons do this.

In the sixth-row transition elements—lanthanum (no. 57) through mercury (no. 80)—there is a further complication. There are seven 4f orbitals which are very close in energy to the 5d orbitals. Putting electrons into these 4f orbitals means there will be fourteen additional elements in this row. These fourteen elements are almost identical in many chemical properties. These closely related elements (from no. 58 to no. 71), in which the 4f orbitals are filling up, are called the **inner transition elements.** Sometimes the term **lanthanides** or the older term **rare earth elements** is used. The elements following actinium, in which the 5f orbitals are being filled, are also inner transition elements. Sometimes this group is called the **actinides.**

21-1.2 General Properties

What properties do we actually find for the transition elements? What kinds of compounds do they form? How can the properties be interpreted in terms of the electron populations of the atoms?

Looking at a sample of each transition element in the fourth row, we see that they are all metallic. When clean, they are shiny and lustrous. They are good conductors of electricity and also of heat. Some—copper, silver, and gold, for example—are quite outstanding in these respects. One of them—mercury—is a liquid at room temperature; all others are solids at room temperature.

We find a tremendous range of chemical reactivity. Some of the transition elements are extremely unreactive. For example, gold and platinum can be exposed to air or water for long periods without any change. Others, such as iron, can be polished so they are brightly metallic for a while; but on exposure to air and water they slowly corrode. Still others are vigorously reactive, producing a shower of sparks when exposed to air. Lanthanum and cerium, for example, especially when finely divided, oxidize immediately when exposed to air. (Some cigarette lighters have flints containing these metals.)

It is difficult to generalize about the chemical reactivities of a group of elements because reactivities depend upon two factors:

(1) the relative stability of the specific compounds formed compared with the reactants used up and
(2) the rate at which the reaction occurs.

In special cases there are other complications. For example, chromium metal (familiar in the form of chrome plate) is highly reactive toward oxygen. Still, a highly polished piece of chromium holds its luster almost indefinitely when exposed to air. The explanation is that a very thin, invisible coat of oxide quickly forms on the surface and protects the underlying metal from contact with the oxygen in the air. We have already encountered this phenomenon in the cases of aluminum and magnesium. In other words, bulk chromium is unstable when exposed to free oxygen, but the protective layer of oxide cuts the rate of conversion so much that no reaction is observed.

What about compounds of the transition elements? Suppose we go into the chemical stockroom and see what kinds of compounds are on the shelf for a particular element, say chromium. First, we might find a bottle of green powder labeled Cr_2O_3, chromic or chromium(III) oxide. Next to it there probably would be a bottle of a reddish powder, CrO_3, chromium(VI) oxide. On an amply stocked chemical shelf, we might also find some black powder marked CrO, chromous or chromium(II) oxide. There would probably also be some other simple compounds such as $CrCl_3$, chromic or chromium(III) chloride, a flaky, violet solid, and perhaps some $Cr(CH_3CO_2)_2$, red chromium(II) acetate. Elsewhere in the stockroom we would find K_2CrO_4, potassium chromate, a bright yellow powder, probably next to a bottle of $K_2Cr_2O_7$, orange potassium dichromate. We would soon come to the conclusion that the compounds of chromium, at least the common ones, correspond to oxidation numbers of $2+$ [CrO and $Cr(CH_3CO_2)_2$], $3+$ (Cr_2O_3 and $CrCl_3$), and $6+$ (CrO_3, K_2CrO_4, and $K_2Cr_2O_7$).

EXERCISE 21-4

What is the oxidation number of chromium in each of the following: $Cr_2O_7^{2-}$, CrO_4^{2-}, $Cr(OH)_3$, and CrO_2Cl_2?

Along with these simple compounds, we might also find some more complex substances. For example, we might find next to $CrCl_3$ vials of several brightly colored solids labeled $CrCl_3 \cdot 6NH_3$ (yellow), $CrCl_3 \cdot 5NH_3$ (purple-red), $CrCl_3 \cdot 4NH_3$ (green or violet), and $CrCl_3 \cdot 3NH_3$ (violet). If you will recall that the dot in these formulas simply indicates that a certain number of moles of NH_3 are bound to 1 mole of $CrCl_3$, you will conclude that here also the oxidation number of chromium is $3+$. Looking further, we might find other complex compounds such as $Na_3Cr(CN)_6$ (bright yellow) and $KCr(SO_4)_2 \cdot 12H_2O$ (violet). In all these the chromium has a $3+$ oxidation number. As a result of our stockroom search, we would form three conclusions:

(1) Chromium forms both simple and complex compounds.
(2) Chromium forms a number of stable solids, most of them colored.
(3) Chromium may have different oxidation numbers, including $2+$, $3+$, and $6+$.

TABLE 21-2 TYPICAL OXIDATION NUMBERS FOUND FOR FOURTH-ROW TRANSITION ELEMENTS

Symbol	Representative Compounds	Common Oxidation Numbers*	Number of Valence 3d and 4s Electrons
Sc	Sc_2O_3	3+	3
Ti	TiO Ti_2O_3 TiO_2	2+ **3+** **4+**	4
V	VO V_2O_3 VO_2 V_2O_5	2+ 3+ 4+ **5+**	5
Cr	CrO Cr_2O_3 CrO_3	2+ **3+** **6+**	6
Mn	MnO Mn_2O_3 MnO_2 K_2MnO_4 $KMnO_4$	**2+** 3+ **4+** 6+ **7+**	7
Fe	FeO Fe_2O_3	**2+** **3+**	8
Co	CoO Co_2O_3	**2+** 3+	9
Ni	NiO (Ni_2O_3) (NiO_2)	**2+** (3+) (4+)	10
Cu	Cu_2O CuO	1+ **2+**	11
Zn	ZnO	**2+**	12

*The most common oxidation numbers are in bold type. Parentheses indicate uncertainty.

Similar conclusions would have resulted from a study of most of the other transition elements.

Is there any regularity to the kind of compounds the fourth-row transition elements form? Table 21-2 shows what has been found.

EXERCISE 21-5

Look through a handbook of chemistry and find one other compound for each oxidation number given for the elements in Table 21-2.

We can make the following generalizations from this table:

(1) For most of the transition elements, *several* oxidation numbers are possible.
(2) When several oxidation numbers are found for the same element, they often differ from each other by increases of one unit. For example, in the case of vanadium the common oxidation numbers form a continuous series from 2+ to 3+ to 4+ to 5+. In contrast, the halogens differ from each other by increments of two units in a series, such as Cl^-, ClO^-, ClO_2^-, ClO_3^-, ClO_4^-.
(3) The *maximum* oxidation state observed for the elements first increases and then decreases as we go across the row of the transition elements. We have 3+ for scandium, 4+ for titanium, 5+ for vanadium, 6+ for chromium, and 7+ for manganese. The 7+ represents the highest value observed for this transition row. After manganese, the maximum value diminishes as we continue toward the end of the transition row.

What explanation can we give for these observations? Why does the combining capacity vary from one transition element to another in such a way that the above pattern of oxidation numbers develops?

The combining capacity of an atom depends upon how many electrons the atom uses for bonding to other atoms. The unique feature of the transition elements is that they have several electrons in the outermost *d* and *s* orbitals, and the ionization energies of all of these electrons are relatively low. It is therefore possible for an element like vanadium to form a series of compounds in which from two to five of its electrons are either lost to or shared with other elements. Consider, for example, the oxides VO and V_2O_3, which contain the V^{2+} and the V^{3+} ions, respectively. Although more energy is needed to form V^{3+} than V^{2+}, the V^{3+} has, because of its higher charge, a greater attraction for the O^{2-} ion than does V^{2+}. This extra attraction in V_2O_3 compensates for the energy needed to form the V^{3+} ion, and both oxides (as well as VO_2 and V_2O_5) are stable compounds. Notice, moreover, that the maximum oxidation number of the transition elements never exceeds the total number of s and d valence electrons. The higher oxidation states become increasingly more difficult to form as we proceed along a row, because the ionization energies of the d and s electrons increase with the atomic number.

21-2 COMPLEX IONS

The remaining general point to be made about the transition elements is that they form a great variety of complex ions in which other molecules or ions are bonded to the central transition-element ion to form more complex units. These are called **complex ions.** Consider the series already mentioned: $CrCl_3 \cdot 6NH_3$, $CrCl_3 \cdot 5NH_3$, $CrCl_3 \cdot 4NH_3$, and $CrCl_3 \cdot 3NH_3$. How can we account for the existence of such a series? To answer this question we must consider some of the observed facts about these complex compounds. For example, if we dissolve 1 mole of each in water and add a solution of silver nitrate in an attempt to precipitate the chloride as AgCl,

$$Ag^+ + Cl^- \rightleftharpoons AgCl(s) \qquad (1)$$

we find that sometimes much of the chloride cannot be precipitated. The observed results are:

Compound	Moles of Cl^- Precipitated	Moles of Cl^- Not Precipitated
$CrCl_3 \cdot 6NH_3$	3 of 3	0
$CrCl_3 \cdot 5NH_3$	2 of 3	1 of 3
$CrCl_3 \cdot 4NH_3$	1 of 3	2 of 3
$CrCl_3 \cdot 3NH_3$	0	3 of 3

Evidently, there are two ways in which chlorine is bound in these compounds, one which allows the Cl^- to be precipitated by Ag^+ and another which does not. In $CrCl_3 \cdot 6NH_3$, all the chloride can be precipitated; in $CrCl_3 \cdot 3NH_3$, none can be precipitated. Other data also indicate different types of bonding. For example, $CrCl_3 \cdot 6NH_3$ forms a highly conductive solution. On the other hand, $CrCl_3 \cdot 3NH_3$ solution does not conduct at all. The explanation of this behavior was provided in the early 1900's by Alfred Werner (1866–1919), who suggested that complex compounds of Cr^{3+} can be accounted for by assuming that each chromium is bonded to six neighbors. In $CrCl_3 \cdot 6NH_3$, the cation consists of a central Cr^{3+} surrounded by six

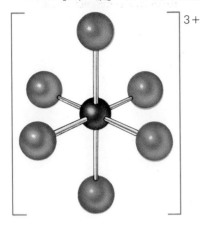

Fig. 21-2 A structural model of the cation [Cr(NH$_3$)$_6$]$^{3+}$ in CrCl$_3 \cdot$ 6NH$_3$.

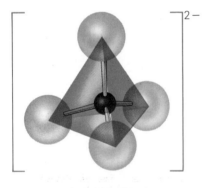

Fig. 21-3.1 The cobalt complex, [CoCl$_4$]$^{2-}$, is tetrahedral and colors its solutions blue. The coordination number of the cobalt in it is 4.

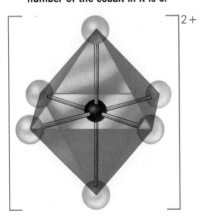

Fig. 21-3.2 The cobalt complex, [Co(H$_2$O)$_6$]$^{2+}$, is octahedral and colors its solutions red. The coordination number of the cobalt in it is 6.

NH$_3$ molecules at the corners of an octahedron; the three chlorine atoms exist as anions, Cl$^-$. In CrCl$_3 \cdot$ 5NH$_3$, the cation consists of the central chromium surrounded by the five NH$_3$ and one of the chlorines; the other two chlorines are anions. In CrCl$_3 \cdot$ 4NH$_3$, the chromium is bound to four NH$_3$ and two Cl, leaving one chloride anion. In CrCl$_3 \cdot$ 3NH$_3$, all three Cl atoms and all three NH$_3$ molecules are bonded to the central chromium. The formulas can be written [Cr(NH$_3$)$_6$]Cl$_3$, [Cr(NH$_3$)$_5$Cl]Cl$_2$, [Cr(NH$_3$)$_4$Cl$_2$]Cl, and [Cr(NH$_3$)$_3$Cl$_3$], respectively.

21-2.1 Geometry of Complex Ions

The way in which atoms or molecules are arranged in space around a central atom has a great influence on the properties of the substance. What kinds of arrangements are found in complex ions? What shapes do these complex ions show? Can we find any regularity in the transition elements that will enable us to predict what complex ions will form?

First, let us introduce a concept useful in giving spacial descriptions: *the **coordination number** is the number of near neighbors (ions or molecules) that an atom has.* For example, in the complex ion CrCl$_3 \cdot$ 6NH$_3$, structural studies show that each Cr^{3+} *ion is surrounded by six NH$_3$ molecules arranged at the corners of a regular octahedron*, as shown in Figure 21-2. The electron pair on each NH$_3$ points in toward the Cr^{3+} ion. We say that chromium has a coordination number of 6 in CrCl$_3 \cdot$ 6NH$_3$.

Some of you have probably seen humidity indicators which have a piece of colored cloth or paper as the moisture-sensitive unit. When the humidity is low, the color of the cloth or paper is blue; when the humidity is high, the color is pink or red. Many different, ingenious, and even humorous devices, purportedly to forecast the weather, have been developed to utilize commercially this change in color. The colored material absorbed by the paper or cloth is cobalt(II) chloride solution (CoCl$_2$). Some dissolved NaCl may also be present in it. The blue color is attributed to a complex ion of cobalt, presumably the well-known blue ion [CoCl$_4$]$^{2-}$. In this ion four chloride ions are arranged tetrahedrally around a central cobalt(II) cation, Co^{2+} (see Figure 21-3.1). The coordination number of cobalt in this complex is 4. The pink color is due to [Co(H$_2$O)$_6$]$^{2+}$, in which six water molecules surround the central Co^{2+} ion. The negative ends of the water molecules (electron pairs) point inward (Figure 21-3.2). The Co^{2+} ion has a coordination number of 6 in this species. The basis for the color change is found in the nature of the complex ions in the solution:

$$\underset{\substack{\text{tetrahedral}\\\text{blue}\\\text{coordination}\\\text{number}=4}}{[\text{CoCl}_4]^{2-}} + 6\text{H}_2\text{O} \longrightarrow \underset{\substack{\text{octahedral}\\\text{pink}\\\text{coordination}\\\text{number}=6}}{[\text{Co(H}_2\text{O})_6]^{2+}} + 4\text{Cl}^- \qquad (2)$$

It should be clear to us from Le Chatelier's principle that when water evaporates into the room because of low relative humidity, the equilibrium will shift to give the blue chloro- complex. On the other hand, if water is picked up from very moist air, the equilibrium shifts to give the pink water complex. Cobalt(II) shows a coordination number of 4 for chloride and 6 for water under the conditions used in the humidity indicator. This same $CoCl_2$ solution is sometimes used as an invisible ink since the pink color cannot be seen in dilute solution, but a color develops if the paper is dried.

GEOMETRY of COMPLEX IONS

EXERCISE 21-6

The mineral cryolite ($AlF_3 \cdot 3NaF$) contains a complex ion. Draw a structural formula for the complex ion. What is the coordination number of aluminum in the complex ion? What is the charge on the complex ion? What other ions are present in cryolite?

Water and NH_3 show rather striking similarities in their ability to coordinate around a metal cation. We have already mentioned $[Cr(NH_3)_6]^{3+}$ and $[Co(H_2O)_6]^{2+}$. As you might expect, we also see $[Cr(NH_3)_5H_2O]^{3+}$. One of the ammonia molecules in $[Cr(NH_3)_6]^{3+}$ has been replaced by a water molecule. Additional replacements of this type give $[Cr(NH_3)_4(H_2O)_2]^{3+}$, and so on.

Not only water molecules but many anions such as chloride ions can replace ammonia molecules. For example, the purple compound with formula $CrCl_3 \cdot 5NH_3$ found on the shelf in our hunt through the stockroom has a structure in which one of the three chloride ions of $CrCl_3$ is attached to the chromium. One of the six NH_3 molecules of yellow $CrCl_3 \cdot 6NH_3$ is replaced. Notice that this complex ion would have a charge of only 2+, $[Cr(NH_3)_5Cl]^{2+}$, since one charge on Cr^{3+} is neutralized by a Cl^-.

In the compound $CrCl_3 \cdot 4NH_3$, two species of distinctly different color were reported, one violet and the other green. Can we have isomers of $CrCl_3 \cdot 4NH_3$? If so, each color might well be identified with one distinct isomer. The octahedral models shown in Figure 21-4 suggest that two isomers should be possible. As we see, the two chlorine atoms may occupy octahedral positions which are next to each other on the *same* side of the metal atom (Figure 21-4.1) or on *opposite* sides of the metal atom (Figure 21-4.2). The isomer in which the two similar groups are located on the same side

Fig. 21-4.1 **(Left)** The violet-hued *cis*-isomer of $[Cr(NH_3)_4Cl_2]^+$.

Fig. 21-4.2 **(Right)** The green-hued *trans*-isomer of $[Cr(NH_3)_4Cl_2]^+$.

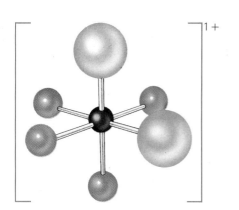
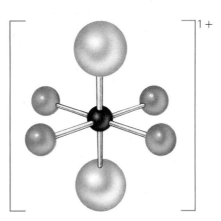

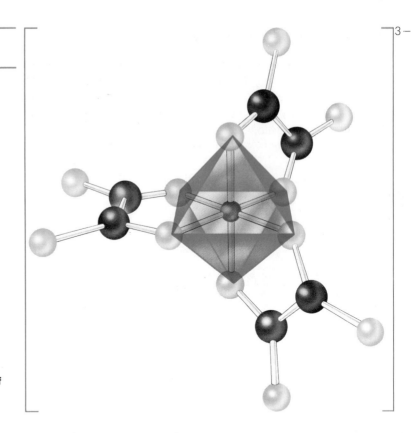

Fig. 21-5 **A structural model of** $[Fe(C_2O_4)_3]^{3-}$.

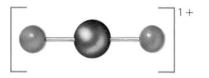

Fig. 21-6 **A linear complex,** $[Ag(NH_3)_2]^+$.

of the metal atom is called the *cis*-isomer and is violet; the other is called the *trans*-isomer and is green (see Section 16-6.2).

The complex ion $[Fe(C_2O_4)_3]^{3-}$ is formed when rust stains are bleached out with oxalic acid solution. It also has a transition element showing a coordination number of 6, even though there are only three groups ($[C_2O_4]^{2-}$ groups) around each iron ion. Figure 21-5 shows the arrangement. Each $[C_2O_4]^{2-}$, the oxalate group uses two of its oxygen atoms to bond with the central iron atom. The number of near neighbors, *as viewed from the iron atom,* is six oxygen atoms at the corners of an octahedron. Picturesquely, a *group* such as oxalate, *which can furnish simultaneously two atoms for coordination,* is said to be **bidentate,** which literally means "double-toothed."

In addition to these tetrahedral and octahedral complexes, there are two other types commonly found—the square planar and the linear. In a **square planar complex,** the central atom has four near neighbors at the corners of a square. The coordination number is 4, the same number as in the tetrahedral complexes. An example of a square planar complex is the complex nickel cyanide anion, $[Ni(CN)_4]^{2-}$.

In a **linear complex,** the coordination number is 2, corresponding to one group on each side of the central atom. An example is the silver-ammonia complex, which generally forms when a very slightly soluble silver salt such as silver chloride dissolves in aqueous ammonia (see Figure 21-6). Another example of a linear complex is $[Ag(CN)_2]^-$, which is formed during the leaching of silver ores with NaCN solution.

21-2.2 Bonding in Complex Ions

What holds the atoms of a complex ion together? There are two possibilities. In some complexes, such as $[AlF_6]^{3-}$, the major contribution to the bonding comes from the attraction between a positive ion (Al^{3+}) and a negative ion (F^-). The bonding is ionic. In other complexes, such as $[Fe(CN)_6]^{3-}$, there is thought to be substantial sharing of electrons between the central atom and the attached groups. The bonding is mainly covalent. When there is such sharing, an electron or an electron pair from the attached group spends part of its time in an orbital furnished by the central atom. In either type, as emphasized in Chapter 16, *both* atoms attract the electron.

For transition elements there are usually empty d orbitals to accommodate electrons from attached groups. This is by no means always necessary; witness the case of Zn^{2+}, a good complex-former even though all its $3d$ orbitals are already occupied. Any vacant orbital, low enough in energy to be populated, can be used to form the coordinate covalent bond of a complex.

The geometry of a complex ion often can be correlated with the orbitals of the central atom that are used in the bonding process. For example, if only one s and one p orbital are used, the bonding is called *sp* bonding. We have already seen in Section 16-4.8 that this bonding situation is associated with a linear arrangement. Some complexes are linear; $[Ag(NH_3)_2]^+$ is a good example. If one s and three p orbitals are used, the bond is called an sp^3 bond. In Section 16-4.8 we associated sp^3 bonding with tetrahedral geometry. $[Zn(NH_3)_4]^{2+}$ is a good example. When d orbitals are involved, other geometries are found. For example, dsp^2 has square planar geometry; d^2sp^3, octahedral geometry.

21-2.3 Significance of Complex Ions

In addition to their occurrence in solid compounds, complex ions such as those we have mentioned are important for two other reasons:

(1) They may determine what species are present in aqueous solutions.
(2) Some of them are very important in biological processes.

As an example of the problem of species in solution, consider the case of a solution made by dissolving some potassium chrome alum $[KCr(SO_4)_2 \cdot 12H_2O]$ in water. On testing, the solution is found to be distinctly acidic. A currently accepted explanation of the observed acidity is based upon the assumption that the chromic ion, $[Cr(H_2O)_6]^{3+}$, in water solution can act as a weak acid, dissociating to give a proton (or hydronium ion). Schematically, the dissociation can be represented as the transfer of a proton from one water molecule in the $[Cr(H_2O)_6]^{3+}$ complex to a neighboring H_2O to form a hydronium ion, H_3O^+. Note that removal of a proton from an H_2O bound to a Cr^{3+} leaves an OH^- group at that position. The reaction is reversible and comes to equilibrium:

$$[Cr(H_2O)_6]^{3+} + H_2O \rightleftharpoons [Cr(H_2O)_5OH]^{2+} + H_3O^+ \quad (3)$$

We see that $[Cr(H_2O)_6]^{3+}$ acts as a proton-donor—that is, as an acid.

21-2.4 Amphoteric Complexes

Another reason chemists find the above complex-ion picture of aqueous solutions useful is that it is easily extended to explain amphoteric behavior. Consider chromium hydroxide [$Cr(OH)_3$], a good example of an amphoteric hydroxide. It dissolves very little in water, but is quite soluble both in acid and in excessively strong base. Presumably it can react with either. How can this behavior be explained in terms of the complex-ion model?

First, consider the equilibrium represented by equation (3). If NaOH is added to the solution, the OH^- ion combines with the H_3O^+ ion to form H_2O. This removes one of the species on the right side of the equation (H_3O^+), so formation of the other species, $[Cr(H_2O)_5OH]^{2+}$, is favored. In other words, as OH^- is added to $[Cr(H_2O)_6]^{3+}$, the reaction which is favored is that which will pull a proton from $[Cr(H_2O)_6]^{3+}$. What will happen when enough NaOH has been added to remove *three* protons from each $[Cr(H_2O)_6]^{3+}$? Loss of three protons leaves the neutral species $Cr(H_2O)_3(OH)_3$, or $Cr(OH)_3 \cdot 3H_2O$. Because this neutral species has no charges to repel other molecules of its own kind, it precipitates. But as more NaOH is added to this solid phase, one more proton can be removed to produce $[Cr(H_2O)_2(OH)_4]^-$, and the $Cr(OH)_3 \cdot 3H_2O$ dissolves. [In principle, more protons could be removed, perhaps eventually to form $[Cr(OH)_6]^{3-}$, but there is as yet no evidence for this.]

The following equations summarize the steps believed to occur when NaOH is slowly added to a solution of $[Cr(H_2O)_6]^{3+}$. Step (*3c*) corresponds to formation of solid hydrated chromium hydroxide; step (*3d*) corresponds to its dissolving in excess NaOH.

$$[Cr(H_2O)_6]^{3+} + OH^- \rightleftharpoons [Cr(H_2O)_5OH]^{2+} + H_2O \quad (3a)$$

$$[Cr(H_2O)_5OH]^{2+} + OH^- \rightleftharpoons [Cr(H_2O)_4(OH)_2]^+ + H_2O \quad (3b)$$

$$[Cr(H_2O)_4(OH)_2]^+ + OH^- \rightleftharpoons [Cr(H_2O)_3(OH)_3](s) + H_2O \quad (3c)$$

$$[Cr(H_2O)_3(OH)_3](s) + OH^- \rightleftharpoons [Cr(H_2O)_2(OH)_4]^- + H_2O \quad (3d)$$

When an acid is added to a solution, as in equation (*3d*), the above set of reactions is progressively reversed: precipitation of chromium hydroxide, $Cr(OH)_3$, is first observed; and this then dissolves to give $[Cr(H_2O)_6]^{3+}$.

21-2.5 Complexes Found in Nature

Complex ions have important roles in certain physiological processes of plant and animal growth. Two such complexes are **hemin**—a part of **hemoglobin,** the red pigment in the red corpuscles of the blood—and **chlorophyll**—the green pigment in plants. Hemoglobin contains iron (Fe, no. 26), so it belongs in a discussion of complex compounds of the transition elements. Chlorophyll, however, is a complex compound of magnesium (Mg, no. 12), which is not a transition element. We shall discuss chlorophyll here both because it has some features in common with hemoglobin and because considering it here will help us avoid the misconception that only transition elements form complexes.

As extracted from plants, chlorophyll is actually made up of two closely related compounds, chlorophyll A and chlorophyll B. These differ slightly in molecular structure and can be separated because they have different tendencies to be adsorbed on a finely divided solid, such as powdered sugar.

COMPLEXES FOUND in NATURE

EXERCISE 21-7

If you wish to prepare some chlorophyll, grind up some fresh leaves and extract with alcohol. The alcohol dissolves the chlorophyll, as the solution color shows.

To show the complexity of this biologically important material, the structural formula of chlorophyll A is shown in Figure 21-7(a). The formula need not be memorized. Note simply that it is a large organic molecule with a magnesium atom in the center. Around the magnesium atom are four near-neighbor nitrogen atoms, each of which is a part of a five-membered ring. Also consider the vast amount of knowledge and experimentation that was necessary to learn the structure of this complicated molecule.

EXERCISE 21-8

If a typical plant leaf yields 40.0 mg of chlorophyll A, how many milligrams of this will be magnesium? The molecular weight of chlorophyll A is 893.

Figure 21-7(b) shows the structure of hemin. It appears next to the model of chlorophyll A to emphasize their astonishing similarity. The portions within dotted lines identify the differences. Except for the central metal atom, the differences are all on the

Fig. 21-7 **The structure of (a) chlorophyll and (b) hemin.**

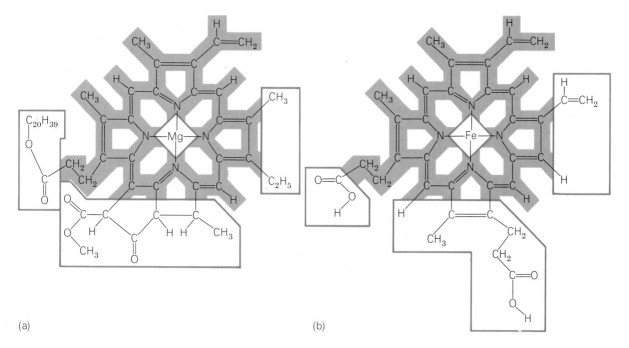

(a)   (b)

periphery of these cumbersome molecules. We cannot help wondering how nature managed to standardize this molecular skeleton for molecules with such different functions.

The most important function of hemoglobin in the blood is to carry oxygen from the lungs to the tissue cells. This is done through a complex between the iron atom of the hemin part and an oxygen molecule. Just how the O_2 molecule is bound to the hemin is not yet clear, but it must be a rather loose combination since O_2 is readily released to the cells. The complex is bright red, the characteristic color of arterial blood. When the O_2 is stripped from the hemin group, it becomes purplish red, the color of venous blood.

Groups in addition to oxygen molecules can be bound to the iron atom of hemoglobin. Specifically, carbon monoxide molecules can be so attached and, in fact, CO is more firmly bound to hemoglobin than is O_2. This is one detail of the carbon monoxide poisoning mechanism. If we breathe a mixture of CO and O_2 molecules, the CO molecules are preferentially picked up by the red blood cells. Since the sites normally used to carry O_2 molecules are thus filled by the CO molecules, the tissue cells starve for lack of O_2. If caught in time, carbon monoxide poisoning can be treated by raising the ratio of O_2 to CO in the lungs (in other words, by administering fresh air or oxygen). Both reactions,

$$O_2(g) + \text{hemoglobin} \rightleftharpoons \text{complex}_1 \qquad (4)$$

$$CO(g) + \text{hemoglobin} \rightleftharpoons \text{complex}_2 \qquad (5)$$

have tendencies to go to the right. If the concentration of O_2 is much higher than that of CO, O_2 will be used in complex formation instead of CO. Another remedial measure is to inject methylene blue directly into the bloodstream. CO bonds more strongly to methylene blue than to hemin. Equilibrium conditions then favor the transfer of CO to the methylene blue, thus freeing hemoglobin for its normal oxygen-transport function.

Another biologically significant material, cobalmin—the coenzyme of vitamin B_{12}—has been identified recently as a cobalt complex. Indeed, many catalysts, both industrial and biological, are known to function as a result of complex formation. Chemists are now very excited about complexes of cobalt and other metals which contain a *nitrogen molecule* in one coordination position. The interest in these complexes partially stems from the fact that they may help us understand how a colony of bacteria on the root of a clover plant in a field at 20°C can take nitrogen from the air (partial pressure = 0.8 atm) and convert it to protein. To do this in a laboratory or factory would require temperatures of 450°C to 500°C and pressures of hundreds of atmospheres. We hope to learn from the bacteria; their chemistry is better than ours for this particular process.

21-3 SPECIFIC PROPERTIES OF FOURTH-ROW TRANSITION ELEMENTS

The preceding discussion of the transition elements has been quite general. Now we shall consider their specific properties and discuss two in detail.

TABLE 21-3 SOME PROPERTIES OF FOURTH-ROW TRANSITION ELEMENTS

Description	Sc	Ti	V	Cr	Mn	Fe	Co	Ni	Cu	Zn
Atomic number	21	22	23	24	25	26	27	28	29	30
Atomic wt	45.0	47.9	51.0	52.0	54.9	55.9	58.9	58.7	63.5	65.4
Abundance* (% by wt)	0.005	0.44	0.015	0.020	0.10	5.0	0.0023	0.008	0.0007	0.01
Melting point (°C)	1400	1812	1730	1900	1244	1535	1493	1455	1083	419
Boiling point (°C)	3900	3130†	3530†	2480†	2087	2800	3520	2800	2582	907
Density (g/cm³)	2.4	4.5	6.0	7.1	7.2	7.9	8.9	8.9	8.9	7.1
First ionization energy (kcal/mole)	154	157	155	155	171	180	180	175	176	216
2+ ion radius (Å)	—	0.90	0.88	0.84	0.80	0.76	0.74	0.72	0.72	0.74
3+ ion radius (Å)	0.81	0.76	0.74	0.69	0.66	0.64	0.63	0.62	—	—
E^0 $M(aq)^{2+} + 2e^- \longrightarrow M(s)$ (volt)	−2.1‡	−1.6	−1.2	−0.90	−1.18	−0.44	−0.28	−0.25	+0.34	−0.76

* In the Earth's crust.
† Estimated.
‡ $M(aq)^{3+} + 3e^- \longrightarrow M(s)$.

Table 21-3 organizes some of the data ordinarily found useful for the transition elements of the fourth row of the periodic table. The following are some notes on regularities observed.

ATOMIC WEIGHT

Atomic weight *increases regularly* across the row except for the inversion of cobalt and nickel. The reason for this inversion lies in the distribution of naturally occurring isotopes. Natural Co consists entirely of the isotope $^{59}_{27}\text{Co}$; natural Ni consists primarily of the isotopes $^{58}_{28}\text{Ni}$ and $^{60}_{28}\text{Ni}$, the 58-isotope being about three times as abundant as the 60-isotope.

ABUNDANCE IN THE EARTH'S CRUST

With the exception of iron, which is very abundant, and titanium, which is moderately abundant, all the other elements of the first transition row are relatively scarce. However, some of them, such as copper, are quite familiar. Copper is one of the few metallic elements found free in nature. The existence of deposits of metallic copper undoubtedly accounts for the fact that the Bronze Age preceded the Iron Age in man's evolution. Copper, the essential ingredient of bronze, did not require the difficult smelting process needed for iron.

MELTING POINT

Except for zinc at the end of the row, the melting points are quite high. This is appropriate, since these elements have a large number of valence electrons and also a large number of vacant valence orbitals. Toward the end of the row, the 3d orbitals of zinc become filled. Since d electrons and orbitals are no longer involved in metal bonding, the melting point drops.

DENSITY

There is a steady increase in density through this row, with some leveling off toward the right. But because the atoms are of almost constant size, the main cause of density change is the increasing nuclear mass.

IONIZATION ENERGY

In contrast to what we might expect, the values of ionization energies for the transition elements are neither very high nor very low. They are all rather similar in magnitude. The sequential increase in nuclear charge, which would tend to increase the ionization energy, seems to be almost offset by the extra screening of the nucleus provided by the added electrons.

IONIC RADIUS

Ionic radii do not change very much in going across a transition row. The reason for this is essentially a balance of the following two effects:

(1) As nuclear charge increases across the row, the electrons are pulled in, so the ions ought to shrink.
(2) But as more $3d$ electrons are added, these electrons repel each other, so the ions ought to swell.

These effects just about cancel each other. As expected, the size of the $3+$ ion is smaller than the size of the $2+$ ion of that same element. Holding nuclear charge constant, if one $3d$ electron is removed, the repulsion between the remaining $3d$ electrons is reduced, allowing them all to be pulled closer to the nucleus.

COLOR

Many solid compounds of the transition metals and their aqueous solutions are colored. This color indicates light is absorbed from the visible part of the spectrum. The energy levels that account for this absorption are relatively close together and involve unoccupied d orbitals. The environment of the ion changes the spacing of these levels, thereby influencing the color. A familiar example is the $Cu^{2+}(aq)$ ion, which changes from a light blue to a deep blue when NH_3 is added. The formation of the ammonia complex alters the energy-level spacing of the central Cu^{2+} ion to produce the color observed.

E^0

The last row of Table 21-3 gives the values of the reduction tendencies for these metallic ions. Except for Sc^{3+}, these values correspond to the reaction

$$M^{2+}(aq) + 2e^- \longrightarrow M(s) \tag{6}$$

As can be seen from the table, all the ions except Cu have negative values, which means the metals are more easily oxidized than is hydrogen gas, for which E^0 is zero. Thus, manganese metal should dissolve in acid to liberate hydrogen gas. The E^0 for the overall reaction

$$Mn(s) + 2H^+(aq) \rightleftharpoons Mn^{2+}(aq) + H_2(g) \tag{7}$$

is $+1.18$ volts, so the reaction should proceed spontaneously to the right. (Note that this is an equilibrium consideration; it does not give the rate, which may be slow.) For Cu the reaction

$$Cu(s) + 2H^+(aq) \rightleftharpoons Cu^{2+}(aq) + H_2(g) \tag{8}$$

has a negative E^0 (-0.34 volt), so we do not expect the reaction to go to the right.

At the end of the row, Zn^{2+} has an E^0 value of -0.76 volt, which is intermediate between the values at the beginning of the row and those toward the end. In view of this E^0 value, we can predict that Zn metal will reduce Fe^{2+}, Co^{2+}, Ni^{2+}, and Cu^{2+} to their respective metals but will not be able to reduce Sc^{3+}, Ti^{2+}, V^{2+}, Cr^{2+}, or Mn^{2+} to their metals.

SPECIFIC PROPERTIES of FOURTH-ROW TRANSITION ELEMENTS

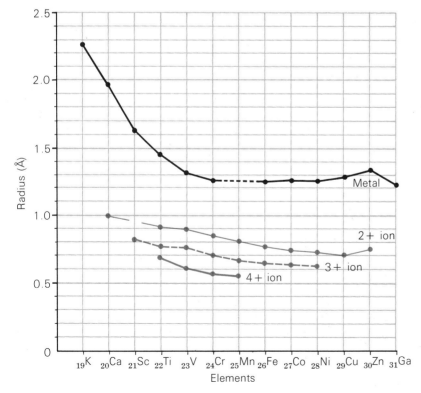

Fig. 21-8 **The atomic sizes of the transition elements.**

21-4 SOME CHEMISTRY OF CHROMIUM— A TYPICAL TRANSITION METAL

Most of us are familiar with chromium as the shiny bright metal on household appliances and auto parts. We recognize chromium as an inert metal which is not attacked by the oxygen in the air. It then comes as something of a surprise to learn that E^0 for the half-reaction

$$Cr(s) \longrightarrow Cr^{2+}(aq) + 2e^- \qquad (9)$$

is +0.90 volt. *This value suggests that chromium should be a very reactive element, that it should be significantly more reactive than zinc.* We have already mentioned an impervious, closely adhering oxide coat which forms over the surface and protects the chromium metal underneath from further chemical attack. Acids such as HCl that dissolve the oxide coating attack chromium readily, as do reagents such as chlorine that destroy the oxide layer.

Most chromium we see today is in the form of a thin plate over a nickel or copper coating on steel. The coat is applied by making the object the cathode in an electrolysis cell containing chromic acid (H_2CrO_4) and sulfuric acid (H_2SO_4) in a ratio of about 100 to 1. By controlling conditions chromium can also be deposited from salts of lower oxidation number, such as Cr^{3+} and Cr^{2+} compounds. Chromic oxide (Cr_2O_3) can also be reduced by metallic aluminum.

EXERCISE 21-9

Write the equation for the reduction of Cr_2O_3 by Al. If it takes 399 kcal/mole to decompose Al_2O_3 into the elements and 270 kcal/mole to decompose Cr_2O_3, what will be the net heat liberated in the reaction you have just written?

Chromium exhibits those properties most characteristic of a transition element. Chromium is a relatively heavy metal which alloys well with many other transition elements; it is an important component in stainless steel. It has a number of oxidation states—2+,

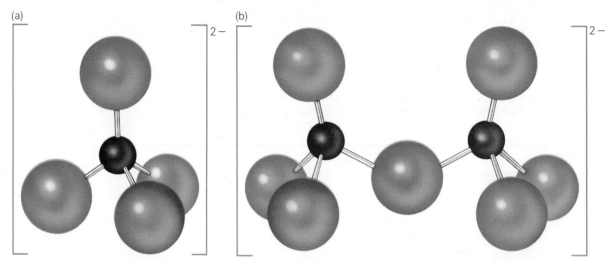

Fig. 21-9 **Structural models of (a) chromate, $[CrO_4]^{2-}$, and (b) dichromate, $[Cr_2O_7]^{2-}$, ions.**

3+, and 6+. Its compounds are usually colored and its ions are always colored in water solution. The chromium(II) salts are blue in water; the chromium(III) salts are green; potassium chromate (K_2CrO_4) is yellow; and potassium dichromate ($K_2Cr_2O_7$) is orange. The chromium(II) salts are such strong *reducing* agents that they will liberate hydrogen from water solution:

$$2Cr^{2+} + 2HOH \longrightarrow H_2 + 2Cr^{3+} + 2OH^- \qquad (10)$$

This reaction is usually rather slow, but it is strongly catalyzed by the presence of platinum metal. On the other hand, the chromium(VI) salts are very strong oxidizing agents:

$$[Cr_2O_7]^{2-} + 14H^+ + 6e^- \longrightarrow 2Cr^{3+} + 7H_2O \quad E^0 = 1.33 \text{ volts} \qquad (11)$$

and are used extensively for this purpose in all branches of chemistry. For example, iron can be determined quantitatively in a solution through titration of Fe^{2+} with standard potassium dichromate ($K_2Cr_2O_7$) solution.

Chromium can form an extremely large number of complex ions. The yellow $[Cr(NH_3)_6]^{3+}$ and the red $[Cr(NH_3)_5H_2O]^{3+}$ ions are two members of a very large family of brightly colored complexes. Chromium frequently shows up in pigments such as chrome-yellow and chrome-green. The very name chromium indicates the frequent incidence of color in chromium compounds.

Like many transition elements, chromium is frequently used in the preparation of catalysts. Its ability to pass through a variety of oxidation states is probably responsible for its effectiveness as a catalyst for many reactions.

One other characteristic of the transition elements is found in the low basic strength of the hydroxides. In general, transition-element hydroxides are insoluble and are very weak bases when compared to hydroxides of such elements as potassium and calcium. You will recall from Section 21-2.4 that chromium(III) hydroxide even has weakly acidic properties.

As chromium demonstrates, the transition elements are an interesting and useful group.

21-5 SOME CHEMISTRY OF IRON—OUR MOST WIDELY USED TRANSITION METAL

No discussion of metals is complete without some mention of iron, the element which gives its name to the so-called Iron Age. Much of our iron is obtained commercially from naturally occurring deposits of an iron oxide, usually Fe_2O_3 or Fe_3O_4.

Examination of the formulas Fe_2O_3 and Fe_3O_4 shows that oxygen must be removed from the iron by the process of reduction. Commercial reduction is carried out on a tremendous scale in a blast furnace. Raw materials—iron ore, limestone, and the reducing agent coke—are fed into the top and oxygen is blown in at the bottom. The limestone ($CaCO_3$) reacts with the sand (SiO_2) in the ore to give calcium silicate ($CaSiO_3$) slag, which floats on top of the iron and can be removed. The coke serves a double function. It burns in oxygen to produce heat, thus generating the high temperatures needed in the furnace. It also provides carbon monoxide, which reduces the iron oxide to metallic iron. The overall equation is

$$3CO(g) + Fe_2O_3(s) \longrightarrow 2Fe(s) + 3CO_2(g) \qquad (12)$$

The impure pig iron produced in the above process is brittle and of low tensile strength. It is made into steel by burning out most of the carbon, sulfur, and phosphorus in a furnace such as an open-hearth furnace, basic oxygen furnace, or electric arc furnace.

21-5.1 The Rusting of Iron

The rusting of iron and its consequences are so well known that it does not need to be described visually. What is the chemistry of corrosion? Let us review the known facts about corrosion.

First, H_2O and O_2 are *both* necessary; $H^+(aq)$ speeds up the reaction; some metals, such as zinc hinder corrosion, while other metals, such as copper, speed it up. It is also known that strains in the metal (like those produced when a car fender is pressed out during manufacture) accelerate corrosion.

How can these observations be interpreted? The most widely accepted mechanism involves many steps. First, the iron acts as an anode to *give up* two electrons and form Fe^{2+}. Second, the electrons are picked up by H^+ to give reactive H atoms. Third, the reactive H atoms combine with the required O_2 to form OH or H_2O. Fourth, the OH or O_2 oxidize the Fe^{2+} to hydrated $Fe(OH)_3 \cdot xH_2O$, or rust.

Various strategies for fighting corrosion interfere with key steps in the above corrosion process. For example, applying a coat of paint completely excludes O_2 and H_2O. Another scheme is to attach the iron by a conducting wire to a more readily corroded metal such as Zn or Mg. Under these circumstances the Zn or Mg instead of the iron loses electrons and corrodes.

Corrosion is a fascinating problem of great economic importance. Many treatises have been written on it, yet we still find that about one man in every four in the steel industry works to replace iron lost by corrosion. The Iron Age would be history now were it not for the steel industry.

21-6 HIGHLIGHTS

The groups of elements formed by the filling of the d electron levels occupy a position in the middle of the periodic table. These elements, known as the **transition elements,** include atomic numbers 21 to 30, 39 to 48, and 72 to 80. The elements in which the f levels are filling are known as the **inner transition elements.** These include atomic numbers 58 to 71 and 90 to 103. The similarities among the transition elements in any one row of the periodic table are fairly strong, and they are very strong among the inner transition elements.

Transition elements have metallic characteristics, relatively high density, many oxidation states, colored ions and compounds, hydroxides which are weak bases, and metal ions with the ability to form **complex ions.** Many of the transition elements have marked activity as catalysts.

Some of our most useful metals (*e.g.*, copper, chromium, and iron) are transition elements.

QUESTIONS AND PROBLEMS

1 Why are the elements with atomic numbers 21 to 30 placed in a group and considered together in this chapter?

2 Write the orbital representation for (a) chromium, (b) molybdenum, and (c) tungsten.

3 What properties of the transition elements are consistent with their being classified as metals?

4 Ferrous ion, iron(II), forms a complex with six cyanide ions, CN^-; the octahedral complex is called ferrocyanide. Ferric ion, iron(III), forms a complex with six cyanide ions; the octahedral complex is called ferricyanide. Write the structural formulas for the ferrocyanide and the ferricyanide complex ions.

5 Draw all possible structures for an octahedral cobalt complex containing four NH_3 and two NO_2^- groups.

6 Draw the structures of the compounds $Cr(NH_3)_6(SCN)_3$ and $Cr(NH_3)_3(SCN)_3$ (SCN^- is the thiocyanate ion). Consider the oxidation number of chromium to be $3+$ and the coordination number to be 6 in both compounds. Estimate (a) the solubility of these compounds in water, (b) their relative melting points, and (c) the relative conductivity of the liquid phases.

7 Why does NH_3 readily form complexes while NH_4^+ does not?

8 Chromic oxide (Cr_2O_3) is used as a green pigment and is often made by the reaction between $Na_2Cr_2O_7(s)$ and $NH_4Cl(s)$ to give $Cr_2O_3(s)$, $NaCl(s)$, $N_2(g)$, and $H_2O(g)$. Write a balanced equation and calculate how much pigment can be made from 1.0×10^2 kg of sodium dichromate.

9 Chromic hydroxide [$Cr(OH)_3$] is a compound with low solubility in water. It is usually hydrated and does not have the definite composition represented by the formula. It is quite soluble either in strong acid or strong base. (a) Write an equation showing the ions produced by the small amount of $Cr(OH)_3$ that dissolves. (b) Explain, using Le Chatelier's principle, why $Cr(OH)_3$ is more soluble in strong acid than in water. (c) What is the significance of the fact that $Cr(OH)_3$ dissolves in base as well as in acid?

10 What is the oxidation number of manganese in each of the following: $MnO_4^-(aq)$, $Mn^{2+}(aq)$, $Mn_3O_4(s)$, $MnO_2(s)$, $Mn(OH)_2(s)$, $MnCl_2(s)$, and $MnF_3(s)$?

11 Manganese(III), $Mn^{3+}(aq)$, spontaneously disproportionates to $Mn^{2+}(aq)$ and $MnO_2(s)$. Balance the equation for the reaction.

12 Use the E^0 values in Table 21-3 to predict what might happen if a piece of iron is placed in a 1 M solution of Mn^{2+} and if a piece of manganese is placed in a 1 M solution of Fe^{2+}. Balance the equation for any reaction that you feel would occur to an appreciable extent.

13 One of the important cobalt ores is $Co_3(AsO_4)_2 \cdot 8H_2O$. How much of this ore is needed to make 1.0 kg of Co?

14 Nickel carbonyl [$Ni(CO)_4$] boils at 43°C and uses the sp^3 orbitals of Ni for bonding. Give reasons to justify the following: (a) It forms a molecular solid. (b) The molecule is tetrahedral. (c) Bonding to other molecules is of the van der Waals type. (d) The liquid is a nonconductor of electricity. (e) It is not soluble in water.

15 Write balanced equations to show the dissolving of $Cu(OH)_2(s)$ when $NH_3(aq)$ is added and the reprecipitation caused by the addition of an acid. Cu^{2+} forms a complex $[Cu(NH_3)_4]^{2+}$ ion.

16 Copper(II) sulfide, reacts with hot nitric acid to produce nitric oxide gas (NO) and elemental sulfur. Only the oxidation numbers of S and N change. Write the balanced equation for the reaction.

17 The solubility of copper(I) iodide (CuI) is 0.008 g/liter. Determine the value of the solubility product.

18 Iron exists in one cubic crystalline form at 20°C (body-centered cubic, with a cube-edge length of 2.86 Å) and in another form at 1100°C (face-centered cubic, with a cube-edge length of 3.63 Å). (a) Draw a picture of each unit cell, showing the nine atoms involved in a body-centered cubic cell and the fourteen atoms involved in a face-centered cubic cell (see Figure 17-10.2). (b) Decide the number of unit cells with which each atom is involved in each structure. (c) How many atoms are in each unit cell if we take into account that some atoms are shared by two or more adjoining unit cells? (d) Calculate the volume of the unit cell and, with your answer to (c), the volume per atom for each structure. (e) What conclusion can be drawn about the "effective size" of an iron atom?

"... Science seldom proceeds in the straightforward logical manner imagined by outsiders."

From the Preface to *The Double Helix*

JAMES D. WATSON (1928–   )

CHAPTER 22 Some Aspects of Biochemistry

Biochemistry is the mechanism of both animal and plant life.

The chemistry of living things is of special interest to us because, of course, we *are* living things. Is the chemistry of life processes unique, or can we apply our knowledge of the chemistry of non-living things to the chemistry of life?

In Chapter 18 we were introduced to the great number and complexity of organic compounds and also to some of the techniques which enable chemists to study them. The compounds which make up living things are among the most complex known, but recent discovery of the structure of DNA and the synthesis of both a virus and an enzyme from non-living materials confirm our belief that investigation of biochemical reactions is within man's ability—and control of undesirable reactions, such as those causing disease, perhaps not too far off.

One and a half centuries ago men regarded the chemistry of living organisms as something quite distinct from the chemistry of rocks, minerals, and other nonliving things. Indeed, men were inclined to believe that living things were imbued with some mysterious "vital force" that was beyond their power to define and understand. But as time went on, it became apparent that the mystery in the chemistry of living things was due to ignorance. As men's understanding of chemical principles increased, the mystery began to disappear. Compounds that were obtained only from plants and animals were produced in the laboratory from ordinary inorganic substances. By the middle of the nineteenth century the superstitious belief in a "vital force" had disappeared; most chemists now believe that man can understand the chemistry of living organisms.

This point of view received dramatic support in the fall of 1967 when L. A. Kornberg and M. Goulian of Stanford University reported the laboratory synthesis of a biologically active virus from nonliving compounds called **nucleotides** (see Section 22-3). The synthetic virus of Drs. Kornberg and Goulian could attack and destroy bacterial cells just like the natural living virus.* Their work suggests that biological chemistry is not fundamentally different from the chemistry we have been discussing.

We still, however, describe a large area of chemical study by the term **biochemistry.** This is not because biochemistry is fundamentally different from chemistry in general. It is because a chemist working effectively in a *particular* area of science must devote special (but not exclusive) attention to what is known about that field of knowledge. Biochemists are concerned fundamentally with the chemical processes that go on in living organisms, but they must use information from all branches of chemistry to answer the questions they ask. Some of these questions are:

(1) What kinds of molecules make up living systems?
(2) What structures do biologically significant materials have?
(3) How is biologically significant information passed from parents to offspring?

22-1 MOLECULAR COMPOSITION OF LIVING SYSTEMS

The chemical system of even the smallest plant or animal is extremely complex. This system has a multitude of compounds, many of polymeric nature, existing in hundreds of interlocking equilibrium reactions whose rates are influenced by a number of specific catalysts. We shall not try to study such a complex system. Instead, we shall focus on some parts of it that serve as examples of material that has been well studied and which illustrate the applicability of chemical principles. *All our knowledge of biochemistry has*

---

* These experiments raise a fascinating philosophical question. Have Drs. Kornberg and Goulian synthesized life in a test tube? Certainly the synthesized virus, produced from nonliving compounds by the action of a biological catalyst (an enzyme) and a template-pattern molecule, has all the characteristics of a natural virus, which is usually defined as a living body. The identity of natural and synthetic viruses was established by R. Sinsheimer of the California Institute of Technology in 1967. (See Section 22-3.)

MOLECULAR COMPOSITION of LIVING SYSTEMS

**MARSHALL W. NIRENBERG
(1927-   )**

A native of New York City, Marshall W. Nirenberg received the Ph.D. in biochemistry from the University of Michigan in 1957. He is now affiliated with the National Institutes of Health, Bethesda, Maryland.

Nirenberg's research has centered on the chemical mechanism by which information carried in the DNA molecule is transferred and used in the synthesis of proteins. He devised a series of experiments to identify the combinations of nucleotides which determine the specific amino acids used in the formation of protein. Previous work by others had suggested that a three-nucleotide "code" exists for each amino acid. Nirenberg was able to "break the code"—to identify the sequence which carries information for each of the twenty amino acids. For his work, Marshall Nirenberg was one of the recipients of the 1968 Nobel Prize in Medicine or Physiology.

come through use of the same basic ideas and the same experimental method you have learned in this course. Specifically, we shall consider in this chapter four classes of compounds that have great importance in biochemistry. **Sugars, fats,** and **proteins** occur in most animals and plants, while **cellulose** is more common in plants.

22-1.1 Sugars

SIMPLE SUGARS

The word *sugar* brings to mind the sweet, white, crystalline grains found on any dinner table. The chemist calls this substance **sucrose** and knows it as just one of many "sugars" which are classed together because they have a related composition and undergo similar reactions. Sugars are part of the larger family of **carbohydrates,** a name given because many such compounds have the empirical formula CH_2O.

EXERCISE 22-1

Glucose, a sugar that is simpler than sucrose, has a molecular weight of 180 and the empirical formula CH_2O. What is its molecular formula?

$C_6H_{12}O_6$

The structure of the **glucose** molecule was deduced by a series of steps similar to those described in Chapter 18 for ethanol. Glucose was found to contain one aldehyde group

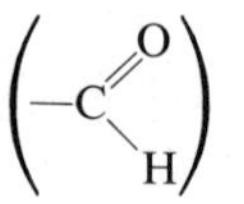

$$\left(-C\underset{H}{\overset{O}{\diagup\!\!\!\!\diagdown}}\right)$$

and five hydroxyl groups (—OH). These functional groups exhibit their typical chemistry. The aldehyde part can be oxidized to an acid group. This reaction is like equation (*22*) for aldehyde oxidation in Section 18-3.5. If a mild oxidizing agent (such as the hypobromite ion in bromine water) is used, the aldehyde group can be oxidized without oxidizing the hydroxyl groups.

EXERCISE 22-2

Write the equation for the oxidation of

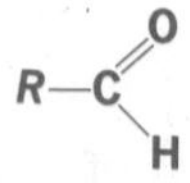

$$R-C\underset{H}{\overset{O}{\diagup\!\!\!\!\diagdown}}$$

by the hypobromite ion, BrO^-, to produce Br^-. What are the oxidation numbers of carbon and bromine before and after reaction? Which element is oxidized? Which is reduced?

If all the oxygen-containing groups are reduced, *n*-hexane,

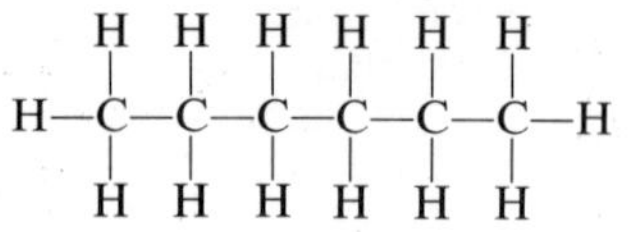

$$H-\underset{H}{\overset{H}{\underset{|}{C}}}-\underset{H}{\overset{H}{\underset{|}{C}}}-\underset{H}{\overset{H}{\underset{|}{C}}}-\underset{H}{\overset{H}{\underset{|}{C}}}-\underset{H}{\overset{H}{\underset{|}{C}}}-\underset{H}{\overset{H}{\underset{|}{C}}}-H$$

results. This test helps establish that the glucose molecule has a chain structure. One representation of the structural formula of glucose ($C_6H_{12}O_6$) is

$$\begin{array}{c} H\diagdown \!\!\!\!\! \diagup O \\ C \\ | \\ CHOH \\ | \\ CHOH \\ | \\ CHOH \\ | \\ CHOH \\ | \\ CH_2OH \end{array}$$

Using a delicate reduction method, the aldehyde group can be converted to a sixth hydroxyl group, giving the substance called **sorbitol**. This compound shows the typical behavior of an alcohol. For example, it forms esters with acids:

$$\underset{\text{SORBITOL}}{\begin{array}{c} CH_2OH \\ | \\ CHOH \\ | \\ CHOH \\ | \\ CHOH \\ | \\ CHOH \\ | \\ CH_2OH \end{array}} + \underset{\text{ACETIC ACID}}{6CH_3COOH} \longrightarrow \underset{\text{ESTER}}{\begin{array}{c} H \quad O \\ \| \\ H-C-O-C-CH_3 \\ | \\ O \\ \| \\ H-C-O-C-CH_3 \\ | \end{array}} + 6H_2O \quad (1)$$

and so on for all six carbons. This is a hexa-acetate.

The sugar **fructose** also occurs naturally and also has the molecular formula $C_6H_{12}O_6$. It is an isomer of glucose with the carbon of the $C=O$ group at the second position in the carbon chain instead of at the end. This makes fructose a ketone (see Section 18-3.4).

EXERCISE 22-3

Draw a structural formula for the fructose molecule. (Remember that fructose is an isomer of glucose.) Explain why fructose cannot be oxidized to a six-carbon acid.

Another aspect of the structure of glucose and fructose is that they, like other simple sugars, can exist as a straight chain in equilibrium with a cyclic structure. In solutions the cyclic structure prevails. Equation (2) shows both forms of glucose:

$$\begin{array}{c} H\diagdown \!\!\!\!\! \diagup O \\ C \\ | \\ CHOH \\ | \\ CHOH \\ | \\ CHOH \\ | \\ CHOH \\ | \\ CH_2OH \end{array} \rightleftharpoons \text{[cyclic structure]} \quad (2)$$

The ring form can be written in a simpler way, showing the hydrogen atoms attached to carbon atoms by lines only and omitting the symbols for the ring carbons:

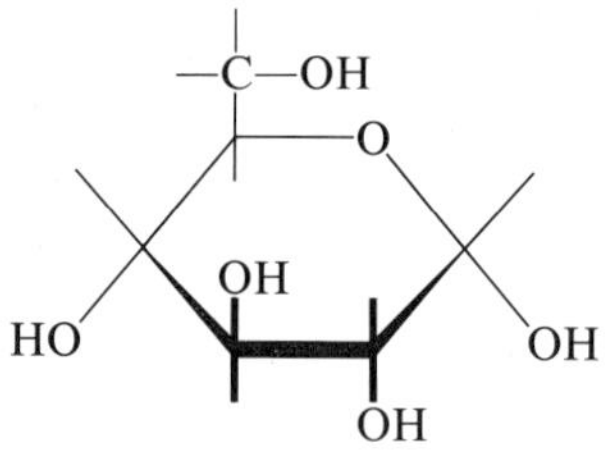

The sugar structure will be important in our discussion of DNA in Section 22-3.

EXERCISE 22-4

At equilibrium in a 0.1 M solution of glucose in water, only 1 percent of the glucose is in the straight chain form. What is K for the cyclization of glucose (the formation of a cyclic structure)?

The two sugars we have discussed are **monosaccharides**—they have a single, simple sugar unit in each molecule. The sugar on your table is a **disaccharide**—it has two units. Each molecule of sucrose contains one molecule of glucose and one of fructose hooked together (losing a molecule of water in the joining reaction). Fructose has a slightly different ring structure because the $\diagdown\!\!C\!\!=\!\!O$ group is not on the end carbon. The formation of sucrose is shown in equation (3):

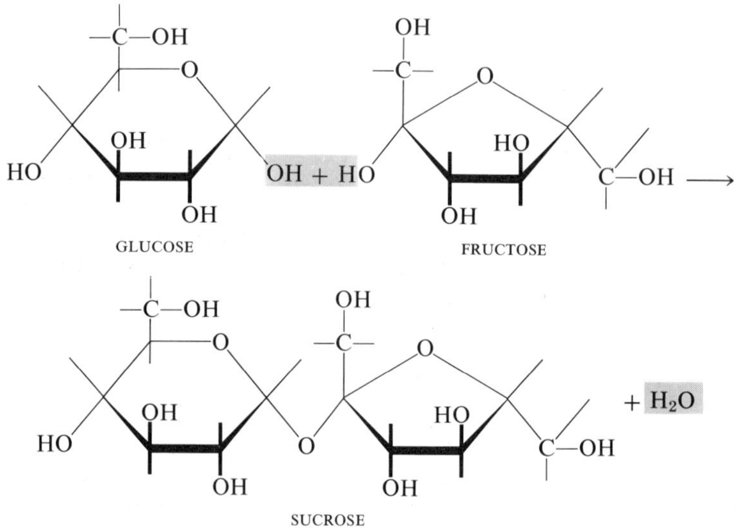

GLUCOSE + FRUCTOSE →

SUCROSE + H₂O   (3)

PROPERTIES OF SUGARS

Sugars occur in many plants. The major commercial sources are sugar cane (a large, specialized grass which stores sucrose in the stem) and sugar beet (as much as 15 percent of the root is sucrose).

In addition, fruits, some vegetables, and honey contain sugars. On the average, every American eats almost 100 pounds of sugar per year. The nation requires about 2×10^{10} pounds per year, about one fourth of which is grown in the U.S. Sugar, retailing at about ten cents a pound, is one of the cheapest pure chemicals produced.

Sugars are fairly soluble in water; about 5 moles dissolve per liter (solubility varies somewhat with the sugar). High solubility in water is readily explained because sugars have many functional groups that can form hydrogen bonds. From your study of hydrogen bonding in Section 17-6, you will recall that from 3 to 10 kcal of energy are released for each hydrogen bond formed. This energy can then be used as part of the energy needed to disrupt the structure of the crystal.

Sugars are easily oxidized, as in the oxidation reaction that involves cupric hydroxide:

$$R-\underset{H}{\overset{O}{C}} + 2Cu(OH)_2 \longrightarrow R-\underset{OH}{\overset{O}{C}} + Cu_2O(s) + 2H_2O \qquad (4)$$

The reaction is more complicated than shown. The $Cu(OH)_2$ is not very soluble in the basic solution used, so tartaric acid is added to form a complex ion. The $Cu_2O(s)$ is a red solid which precipitates from solution because Cu^+ does not form such a complex ion. The reaction is characteristic and is used as a qualitative test for simple aldehydes and sugars.

An important metabolic reaction of disaccharides is the reverse of sucrose formation—the **hydrolysis** of sucrose to give glucose and fructose. Water in the presence of $H^+(aq)$ reacts with sucrose to give glucose and fructose. The term hydrolysis means "reaction with water."

22-1.2 Cellulose and Starch

Cellulose is an important part of woody plants, occurring in cell walls and making up part of the structural material of stems and trunks. Cotton and flax are almost pure cellulose. Chemically cellulose is a **polysaccharide**—a polymer made by successive reaction of many glucose molecules giving a high molecular weight ($\sim 600,000$) structure. This polymer is not basically different from the polymers that were discussed in Section 18-6.

Starch is a mixture of glucose polymers, part of which is water-soluble. This soluble portion consists of chains (molecular weight $\sim 4,000$). The polymer insoluble portion consists of much longer chains with frequent branches.

EXERCISE 22-5

The monomer unit in starch and cellulose has the empirical formula $C_6H_{10}O_5$. These units are about 5.0 Å long. Approximately how many units occur and how long are the molecules of cellulose and soluble starch?

22-1.3 Fats

Fats, as well as animal and plant oils, are esters. Actually they are triple esters of glycerol (1,2,3-propanetriol or $C_3H_8O_3$):

$$\begin{array}{c}HHH\\|||\\H-C-C-C-H\\|||\\OOO\\|||\\HHH\end{array}$$

When carboxylic acids, similar to those you studied in Section 18-3.3, react with glycerol OH groups, a fat is formed. In natural fats the acids usually have 12 to 20 carbon atoms, C_{16} or C_{18} acids being most common.

EXERCISE 22-6

Write the formula for glycerol tributyrate, and then write the formula of the fat made from glycerol and one molecule each of stearic ($C_{17}H_{35}COOH$), palmitic ($C_{15}H_{31}COOH$), and myristic ($C_{13}H_{27}COOH$) acids. How many isomers are possible for the last fat? How many would be possible if all possible combinations of the three acids were used? Compare your answer with that for Exercise 18-18.

Common fats (*e.g.*, butter, tallow) and oils (*e.g.*, olive, palm, and peanut) are mixed esters: each molecule has either (most often) three, (sometimes) two, or (rarely) one kind of acid combined with a single glycerol. There are so many such combinations in a given sample that fats and oils do not have sharp melting or boiling points. Melting and boiling occur over a range of temperatures instead.

An important reaction of fats is the reverse of ester formation. They hydrolyze, or react with water, just as disaccharides do. Usually hydrolysis is carried out in aqueous $Ca(OH)_2$, NaOH, or KOH solution. The metal salts of natural carboxylic acids, such as sodium stearate, are called **soaps**. The products of hydrolysis are, then, soap (*e.g.*, sodium stearate) and glycerol. Because soap is formed, the alkaline hydrolysis is called **saponification**:

$$\underset{\text{A FAT}}{\begin{array}{c}O\\|\|\\-C-O-C-C_{17}H_{35}\\|O\\|\|\\-C-O-C-C_{17}H_{35}\\|O\\|\|\\-C-O-C-C_{13}H_{27}\\|\end{array}} + 3Na^+(aq) + 3OH^-(aq) \longrightarrow \underset{\text{GLYCEROL}}{\begin{array}{c}|\\-C-OH\\|\\-C-OH\\|\\-C-OH\\|\end{array}} + \begin{array}{l}O\\\|\\2Na-O-C-C_{17}H_{35}\\\text{SODIUM STEARATE}\\O\\\|\\Na-O-C-C_{13}H_{27}\\\text{SODIUM MYRISTATE}\end{array} \quad (5)$$

Fats make up as much as half the diet of many people. Fats are a good source of energy because when they are completely "burned" in the body, they supply twice as much energy per gram as do proteins and carbohydrates. As many people know, this is often a mixed blessing, particularly when weight is a problem.

22-2 MOLECULAR STRUCTURES IN BIOCHEMISTRY

Some of the most exciting recent advances in biochemistry have come from recognition of the importance of the structural arrangement of molecular parts. You saw in Chapter 18 that the chemistry of a C_2H_6O compound depends upon structure. Thus, an ether (CH_3—O—CH_3) behaves quite differently from an isomeric alcohol (CH_3CH_2OH). You also learned how interactions between molecules can influence the properties of water (Section 17-6), and how attractions between ions and water can arrange the water molecules in preferred positions around an ion (Figure 17-13). Structure influences the observed properties of systems. Both covalent bonds and intermolecular interactions are involved in fixing the structure of biochemical substances. We shall consider a few examples.

22-2.1 The Structure of Starch and Cellulose

A striking example of the effect of structure is the difference between cellulose and water-soluble starch. Both contain the same monomer since hydrolysis gives only glucose in each case. But the glucose ring differs slightly in the arrangement of the OH groups, resulting in two different polymers. Let us represent the ring structure of glucose by

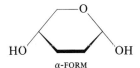
α-FORM

There is another isomer, identical in all parts, except for the placement of the right-hand OH group. Its symbol is

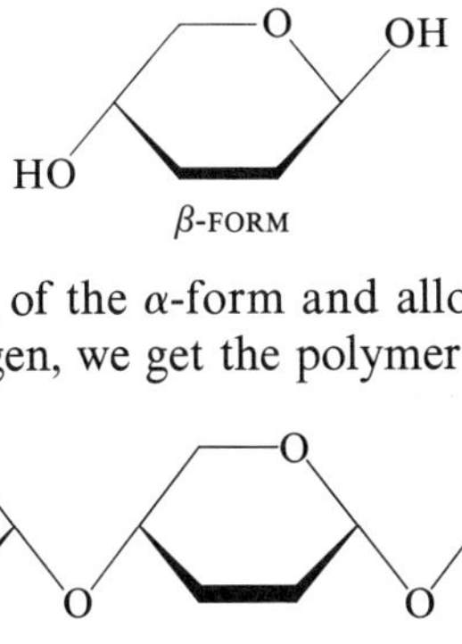

β-FORM

If we connect a string of the α-form and allow for the normal 105° angle of bonds to oxygen, we get the polymer called starch:

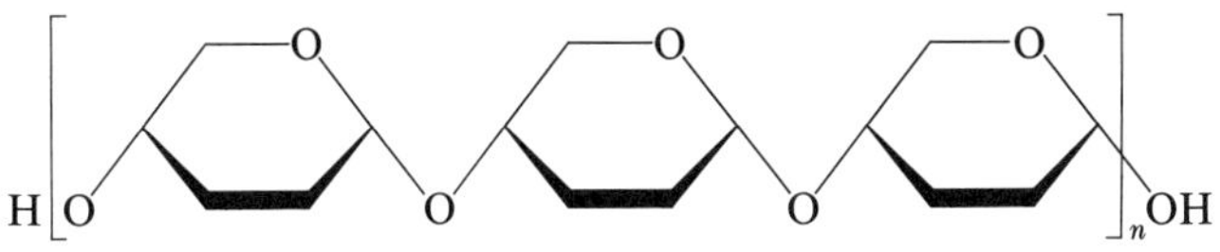

STARCH

On the other hand, a chain of the β-form of glucose gives the polymer called cellulose:

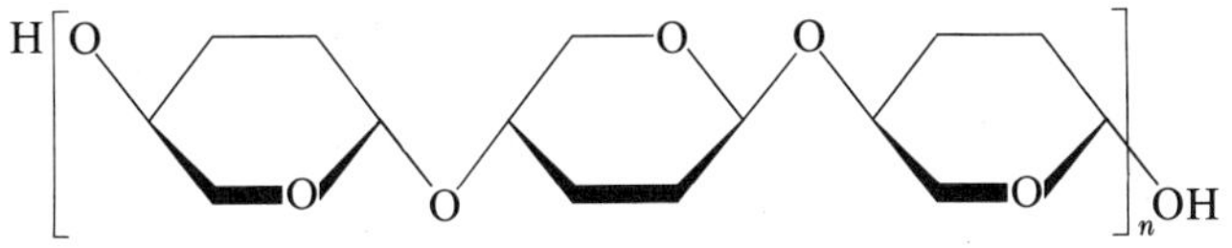

CELLULOSE

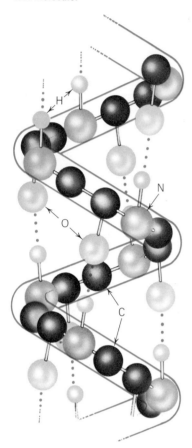

Fig. 22-1 The coiled form of a protein molecule.

The very different geometry of the ether linkage in starch and cellulose causes these two polymers to have different chemical properties.

22-2.2 Proteins

In Section 18-6.3 the composition of proteins was given. They are large, amide-linked polymers of amino acids. However, the long chain formula (Figure 18-14) does not represent all that is known about the structure of proteins. It properly shows the covalent structure but does not indicate the relative positions of the atoms in space.

The use of X-ray diffraction (Section 16-7.2) and the principles that describe hydrogen bonding (Section 17-6) have led to the recognition of a coiled form of the chain in natural proteins. This model is consistent with other tests also and has received general acceptance. It is shown in Figure 22-1. This form has a great deal of regularity —it is not at all a random shape. Order must have some energy factor sustaining it; in the protein molecule, this energy is provided by the hydrogen bonds. These are represented in Figure 22-1 by dotted lines, just as they were in Section 17-6. When the hydrogen bonds are broken (by heating or putting the protein in alcohol), the order disappears and the coiled form loses its shape. Often this damage cannot be repaired and the coil is permanently deformed. Cooking an egg destroys the coiled form of the proteins it contains. A few moments of thought concerning the profound differences of an egg before and after cooking will suggest the very great importance of molecular structure in biochemistry.

22-2.3 Enzymes

Nearly all biochemical reactions proceed at ordinary temperatures and pressures. Most biochemical reactions (especially those in the human body) take place at about 37°C (98°F) and proceed at a rate adequate for the role they play. Man lives, grows, and reproduces himself under very mild conditions. Most reactions would not proceed at a measurable rate at this temperature outside living organisms. Glucose, starches, and fats are very stable compounds and can remain in contact with oxygen without apparent change. This is so, even though their oxidation to carbon dioxide and water releases large amounts of energy. To make these reactions proceed, nature uses catalysts to provide new paths with lower activation energies. Measurable reaction rates are then achieved.

Biological catalysts are called **enzymes.** Nearly every step in the breakdown of a complex molecule to a series of smaller ones is catalyzed by a specific enzyme within living cells. For instance, when acetaldehyde is reduced to ethanol in yeast cells, the reaction takes place in the presence of a specific enzyme called alcohol dehydrogenase:

$$CH_3-\underset{H}{\overset{O}{C}} + 2H^+ + 2e^- \xrightarrow[\text{dehydrogenase}]{\text{alcohol}} CH_3-CH_2\overset{OH}{} \quad (6)$$

You can see that the hydrogenation of acetaldehyde is the reverse of the dehydrogenation of ethanol. The enzyme is named for the latter reaction, but of course it catalyzes the reaction in either direction. Conditions at equilibrium are not affected by the enzyme, but the rate at which the reacting substances reach the equilibrium state is affected by the enzyme (as with any catalyst).

The synthesis of enzymes remained an impossible problem until early in 1969. At that time Drs. Robert B. Merrifield and Bernd Gutte of Rockefeller University and Drs. Robert G. Denkewalter and Ralph F. Hirschmann of the Merck, Sharp & Dohme Research Laboratories reported the synthesis of ribonuclease, an enzyme which helps to break down ribonucleic acid (RNA). (See Section 22-3.6.)

REACTION MECHANISM

Enzymes are protein molecules. Although all enzymes are proteins, not all proteins can act as enzymes. The protein molecules of enzymes are very large, with molecular weights of the order of 100,000. In comparison the substance upon which the enzyme acts (called a **substrate**) is made up of small molecules. Thus, the reaction involves a small substrate molecule which has become attached to the surface of a large protein molecule, where the reaction occurs. [Refer to equations (6), (10), and (11).] The products of the reaction then dissociate from the enzyme surface; a new substrate molecule is attached to the enzyme and the reaction is repeated. We can write the following sequence:

$$\text{enzyme} + \text{substrate} \longrightarrow \text{enzyme-substrate complex} \qquad (7)$$

$$\text{enzyme-substrate complex} \longrightarrow \text{enzyme} + \text{reaction products} \qquad (8)$$

Adding these equations and canceling gives

$$\text{substrate} \longrightarrow \text{reaction products} \qquad (9)$$

Despite the large size of an enzyme molecule, there is reason to believe that there is only one or a few places on its surface at which reaction can occur. These are usually referred to as **active centers.** The evidence supporting this view comes from many kinds of observations. One such observation is that enzyme reactions can often be stopped or slowed down by adding only a small amount of a "false" substrate. A false substrate is a molecule so similar to the real substrate that it can attach itself to the active center, but sufficiently different that no reaction (and consequently no release) can occur. Thus, the active center is "blocked" by the false substrate. (See the CHEMS film "Biochemistry and Molecular Structure.")

SPECIFICITY OF ENZYMES

Most enzymes are quite specific for a given substrate. For example, the enzyme urease that catalyzes the reaction

$$\underset{\text{UREA}}{O=C{\overset{NH_2}{\underset{NH_2}{\diagup\!\!\!\diagdown}}}} + H_2O \underset{}{\overset{\text{urease}}{\rightleftharpoons}} CO_2 + 2NH_3 \qquad (10)$$

is specific for urea. If we use urease to try to catalyze the reaction of a very similar molecule, N-methyl urea, no catalysis is observed:

$$O=C\begin{matrix}NHCH_3\\NH_2\end{matrix} + H_2O \underset{}{\overset{urease}{\rightleftarrows}} \text{no reaction occurs} \quad (11)$$

N-METHYL UREA

This suggests that on the surface of the enzyme there is a special arrangement of atoms (belonging to the amino acids of which the protein is constructed) that is just right for attachment of the urea molecule but upon which the methyl urea will not "fit."

Specificity is not always perfect. Sometimes an enzyme will work with any member of a *class* of compounds. For example, some esterases (enzymes that catalyze the reaction of esters with water) will work with numerous esters of similar, but different, structures. Usually, in cases of this kind, one of the members of the substrate class will react faster than the others, so the rates will vary from one substrate to another.

A PRACTICAL APPLICATION OF ENZYME INHIBITION BY A FALSE SUBSTRATE

It is now believed that many of our useful drugs exert their beneficial action by the inhibition of enzyme activity in bacteria. Bacteria such as *staphylococcus* require for their growth the simple organic compound *para*-aminobenzoic acid and can grow and multiply in the human body because sufficient amounts of this compound occur in blood and the tissues. The control of many diseases caused by these (and other) bacteria was one of the first triumphs of **chemotherapy**,* and the first compound found to be an effective drug of this type was sulfanilamide:

PARA-AMINOBENZOIC ACID   SULFANILAMIDE

It seems reasonable that an enzyme which uses *para*-aminobenzoic acid as a substrate might be "deceived" by sulfanilamide—the two compounds are very similar in size and shape and in many chemical properties. To explain the success of sulfanilamide, it is proposed that the amide can form an enzyme-substrate complex that uses up the active centers normally occupied by the natural substrate.

Usually fairly high concentrations of such a drug are needed for effective control of an infection because the inhibitor (the false substrate) should occupy as many active centers as possible, and also because the natural substrate will probably have a greater affinity for

---

* Chemotherapy is the control and treatment of disease by synthetic drugs. Most of these are organic compounds, often of remarkably simple structure. Sulfanilamide is one example of an organic compound synthesized by chemists for the treatment of bacterial infections.

the enzyme. Thus, a high concentration of the false substrate must be used so that the false substrate-enzyme complex will predominate. The bacteria, deprived of a normal metabolic process, cannot grow and multiply. Now the body's defense mechanisms can take over and destroy them.

22-3 THE "NUCLEIC ACIDS," DNA AND RNA, NATURE'S MESSENGERS

So far we have discussed the *composition* of living systems, molecular changes in living systems, and certain *structural features* associated with living systems. But we have not considered one of the most characteristic and still mysterious properties of a living body—its ability to reproduce an organism which is like itself, yet different enough to be a new individual.

An egg cell from a female and a sperm cell from a male meet and immediately growth begins. The "nonliving" chemicals surrounding the now fertilized egg are *organized* and a new *living body* begins to grow. The living body which grows may be a mouse, an elephant, a red-headed human male, or a dark-haired human female. How are the "directions" for the synthesis of an elephant or a mouse or a man or a woman carried in the egg and sperm cells? Where is the biological pattern which seems to provide a means of organizing nonliving raw materials into a living body? This is a profound and even frightening question.

Men have made real progress toward answering it during the last 15 years, progress such as the already mentioned dramatic synthesis of a virus from nucleotides by Drs. Kornberg and Goulian. Let us see if we can use the ideas we have learned to understand what happened in this challenging synthesis of a "living" virus. First we must learn something about the structure of the **"nucleic acids"**—**DNA** (deoxyribonucleic acid) and **RNA** (ribonucleic acid).

22-3.1 DNA—Its Early History and Significance

DNA is the active component in almost all **genes,** the small units which transmit the hereditary characteristics of living organisms. The existence of DNA has been known for nearly a century. It was first isolated in rather impure form as a gelatinous material in 1869 by the German biochemists F. Hoppe-Seyler and F. Miescher.* The name *nucleic acid* is from the name *nuclein,* the substance they first used. DNA's role as a carrier of genetic information was suggested in 1884, but it was not until 1943 that Oswald Avery and his co-workers at Rockefeller Institute provided sound evidence to resolve questions in a confused field. DNA is now recognized as the gene-carrying molecule. DNA is also the living component of a virus. A virus is known to be an active DNA thread covered with a nonactive protective sheath, usually of protein.

The study of DNA was approached like that of any other interesting organic chemical; that is, its structural formula had to be

---

* See E. A. Mirsky, "The Discovery of DNA," *Scientific American,* **218,** 78 (June 1968).

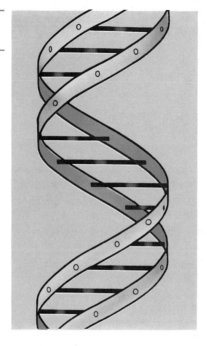
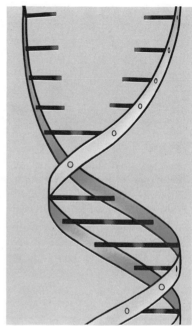

Fig. 22-2 **The double helix of DNA: (a) coiled, (b) partially uncoiled.**

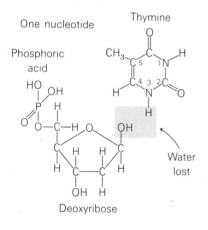

Fig. 22-3 **The formation of a nucleotide of DNA.**

determined. The same principles used to study simpler molecules were used to determine its structure, but the DNA "molecule," because of its tremendous polymeric size, was much more complex and difficult to determine. Many scientists struggled to obtain the structure; a major breakthrough was made in 1953 when J. D. Watson, Maurice Wilkins, and Francis Crick established with reasonable certainty the now accepted structure of DNA. Their work was recognized by a Nobel Prize in 1962.

22-3.2 The Structure of DNA—Nucleotides

The most important structural feature of DNA is that it consists of two very long thin polymeric chains twisted about each other in the form of a regular **double helix** (see Figure 22-2). These chains are made up of nucleotide units fastened together. The nucleotide units themselves are not unduly complex; in fact, four different nucleotide units appear to provide the basis for transmitting all necessary hereditary information. A nucleotide unit includes a phosphoric acid molecule to which a sugar molecule, deoxyribose, is linked by loss of water:

To the resulting structure one of four nitrogen bases is linked also by loss of water. The base thymine can serve as an illustration (see Figure 22-3). Four different nucleotides are found in DNA. Notice in Figure 22-4 that different nucleotides differ only in the identity of the base linked to the deoxyribose phosphate structure. The phosphate and sugar molecules in each nucleotide are *identical* to those found in every other nucleotide of DNA. The bases attached to these groups differentiate the four distinct nucleotides.

22-3.3 The Single Helix

Nucleotides can also link together under the influence of a catalyst by losing one water molecule. A polymeric unit results (see Figure 22-5). The resulting polymeric chain, made by removing a water molecule from between nucleotide units, is shown in Figure 22-5. Notice that each chain is composed of a *backbone* consisting of sugar molecules linked together by phosphate groups in a very regular fashion. The backbone of the chain is seen as the shaded portion of Figure 22-6 (page 564). Attached to each sugar molecule is *one* of the nitrogen-containing bases—adenine, cytosine, guanine, or thymine. The method of fastening these bases to the chain is seen in Figure 22-6. Thymine (abbreviated T) and cytosine (C) are *single, six-membered ring structures containing two nitrogens per ring.* They are of a chemical family called **pyrimidines.** Adenine (A) and guanine (G) are *double ring structures, each having one five- and one six-membered ring, and each ring containing two nitrogen atoms.* These larger bases are of a family called **purines.**

These four base molecules identify the four nucleotides used as starting materials for DNA synthesis. These long polymeric chains formed by linking nucleotides assume a helical form with different nucleotides making up the chain (*i.e.*, different bases attached to the

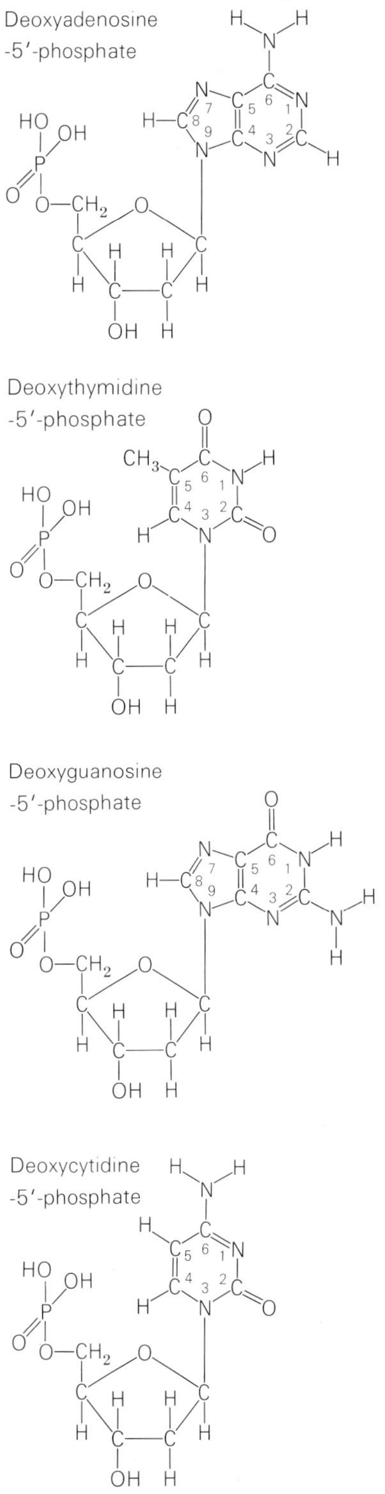

Fig. 22-4 **The four nucleotide building blocks of DNA.**

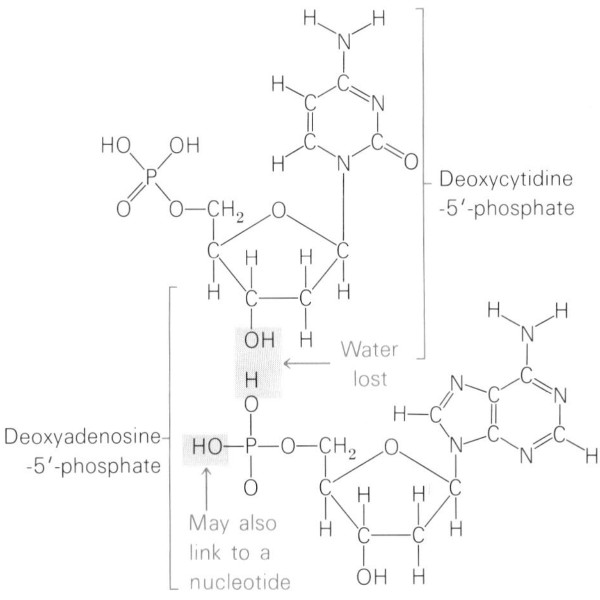

Fig. 22-5 **Linkage of nucleotide units in a DNA chain.**

backbone). The order in which nucleotides are linked provides the means of communicating hereditary information. Or, *the hereditary information is transmitted by the order in which the bases are attached to the backbone of deoxyribose-phosphate.* The purine and pyrimidine bases are flat, relatively water-insoluble molecules which tend to stack above each other perpendicular to the direction of the helical axis of the backbone (see Figure 22-7). The order of the purine and pyrimidine bases along the backbone chain varies greatly from one DNA molecule to another. Each carries a definite message.

22-3.4 The Double Helix—Template Action

Two such chains joined by hydrogen bonds between base pairs give a double helix. The convincing experimental data for this structure is from X-ray diffraction studies. A recent electron micrograph (see Figure 22-8) provides a visual image which appears to be in agree-

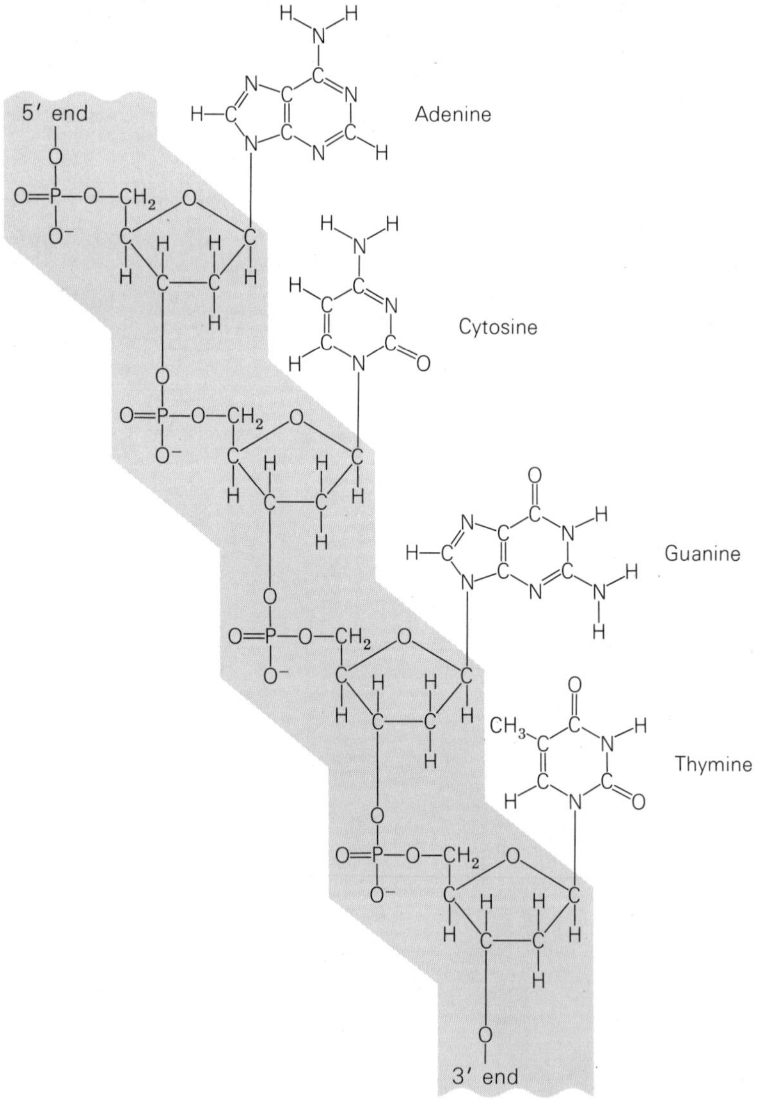

Fig. 22-6 **A section of the DNA chain.**

Fig. 22-7 **A space-filling model of DNA.**

ment. Because of the physical sizes of the bases, adenine (A) is always bonded to thymine (T) and cytosine (C) to guanine (G) (see Figure 22-9). These pairing rules restrict the possible sequence of bases on two intertwined chains. If we have a sequence of ATGTC on one chain, the complementary chain must be TACAG. The two polynucleotide chains must be similar but running in reverse order; one is a **template** or mold for the synthesis of the other.

Genetic information is carried by the sequence of the four bases A, G, C, and T. Perhaps you will worry that having only four nitrogen bases on the DNA molecule will limit the amount of information carried, but such fears are groundless. The number of possible arrangements is 4^n, where n is the number of nucleotide units (*i.e.,* number of separate base units) in a molecule of DNA. Since each molecule is very long, many arrangements are possible. DNA molecules have an average molecular weight in excess of 10^6. That means there are at least 1,500 nucleotide units in the chain; hence, *the possible number of different molecules of DNA is $4^{1,500}$, more than the number of different molecules of DNA that have existed in all of the chromosomes present since the origin of life.* The **genetic "code"** is the sequence of bases on the backbone chain. Hereditary information is passed on in this way.

The next question of significance is: How does this order within the DNA molecule pass on biological information? The current view of this process is that the hydrogen bonds holding the DNA chains together are weak enough to break under certain conditions, permitting the chains to uncoil [see Figure 22-2(b)]. Each helix can then serve as a template for combining and organizing smaller nucleotide units into the complementary helix for that chain. One chain determines the order of new chains made from it. The nucleotide units are the raw material.

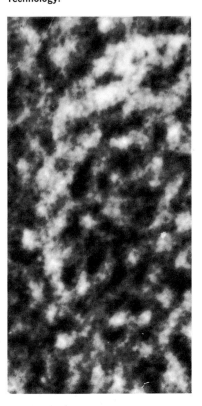

Fig. 22-8 **An electron micrograph of a short section of a DNA molecule.** This electron micrograph is consistent with the established model. Note double helix in picture. Courtesy J. Griffith, California Institute of Technology.

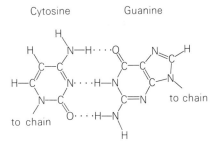

Fig. 22-9 **Hydrogen bonding to link bases of two helical DNA chains.**

Fig. 22-10 **Natural DNA (top) and synthetic DNA (bottom).**

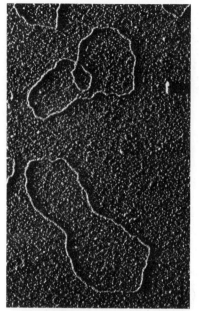

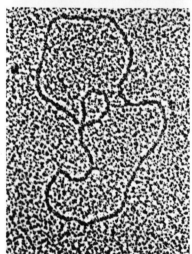

22-3.5 The Synthesis of a Virus

We are now in a position to understand the experiments of Kornberg and Goulian described at the beginning of this chapter. These chemists took from the shelf the four essential, nonliving nucleotides (Figure 22-4). They then added DNA polymerase, a *catalyst* which *accelerates* the linking of nucleotide units by removing water. They then added some natural DNA to serve as a template. The result was a strand consisting of about 6,000 nucleotide units, the *mirror image* of the strand in the original helix. Then, using the mirror-image molecule as a template, the process was repeated to produce a precise but *synthetic* duplicate of the natural DNA molecule. The result was a virus which could attack bacteria and kill them as effectively as the natural DNA (see Figure 22-10).

22-3.6 RNA

A very close relative of DNA is the compound RNA, ribonucleic acid. RNA differs from DNA in two important ways. First, the sugar in the chain is ribose,

not deoxyribose,

(*deoxy-* means "oxygen removed"). Second, the thymine on the chain,

is replaced by uracil,

Most RNA does *not* exist as double-helical strands, but rather as non-hydrogen-bonded, single *polyribonucleotide* strands. The role of RNA in biology is still not well understood.

As you will see in the CHEMS film "Biochemistry and Molecular Structure," these ideas about template structure have been used successfully against certain forms of cancer. By putting a fluorine in the 5 position on uracil,

5-FLUORO-URACIL

the molecule is changed just enough to block the use of uracil in cancer growth and destroy enzyme catalysts.

The modern approach to disease is closely related to structural biochemistry.

22-4 HIGHLIGHTS

Living organisms are complex chemical factories employing and synthesizing the same types of molecules studied in Chapter 18. Biologically significant molecules are frequently large polymeric units made by joining simpler units. **Starches, cellulose, fats,** and **proteins** are typical of these large molecules.

The material of which genes are made, **DNA,** is the material by which hereditary information is passed from generation to generation. DNA has a "backbone" of alternating sugar and phosphate units to which is attached at each sugar unit one of four nitrogen bases—thymine, cytosine, adenine, and guanine. The hereditary information is transmitted by the order in which bases are attached to the backbone of the DNA molecule. This is known as the **genetic code.**

We still have a long way to go in answering the questions concerning heredity and life, but progress is being made. Perhaps man will someday understand the chemistry of life itself.

Where the telescope ends, the microscope begins. Which of the two has the grander view?

VICTOR HUGO (1802–1885)

CHAPTER 23 Astrochemistry

Earth, as photographed from Apollo 8. The horizon of the moon appears in the foreground.

The success of Apollo 8 in orbiting the moon with astronauts Borman, Lovell, and Anders has renewed man's thirst for knowledge about the universe he lives in. Obviously the moon is unfit for habitation by any form of life which we can imagine. But what of other planets—in our solar system and other solar systems? What can we learn about our neighbors in space—and how? Where can life exist —and what kind of life? This is one of the most exciting frontiers of chemistry.

A population growing at an exponential rate on Earth has focused man's attention on problems associated with his activities. Air and water pollution and depletion of fuel and mineral reserves are attracting the attention of thoughtful people. Further, the age-old problem of food supply for an expanding population is a real and painful part of life for many parts of our world. We live in a challenging age, an age when an understanding of our surroundings becomes more and more important to our survival. As a study of our surroundings, chemistry, based on those same principles we developed in the earlier chapters of this book, will continue to play an important part in the development of man's future.

As biological pressures on planet Earth have multiplied, imaginative men have looked to the sky and wondered about the other planets and the stars. Many have dreamed of the planets as centers for future expansion of man's activities; all have wondered about the laws governing the heavenly bodies. People ask: Are the laws of chemistry unchanged throughout the universe? Are the planets similar in composition to the Earth? Can life as we know it survive outside the Earth's protective atmosphere? We are just entering an era when the answers to these questions are becoming available. Let us see where we stand as we enter the eighth decade of the twentieth century.

To enable us to compare the chemistry and composition of the moon, planets, and stars with the Earth's chemistry and composition, we shall briefly review the planet Earth.

23-1 THE CHEMISTRY OF OUR PLANET EARTH

The Earth is conveniently broken into three parts: the **lithosphere,** the solid portion of the Earth; the **hydrosphere,** the liquid portion of the Earth; and the **atmosphere,** the gaseous envelope surrounding the Earth. The lithosphere is a sphere of solid material about 4,000 miles in radius. About 80 percent of the surface of this sphere is covered with water having an average depth of 3 miles. Man sees only a thin *crust* less than 20 miles thick on the outside of the Earth's surface. The exact nature of the inner lithosphere and the Earth's core can

Fig. 23-1 **The parts of the Earth.**

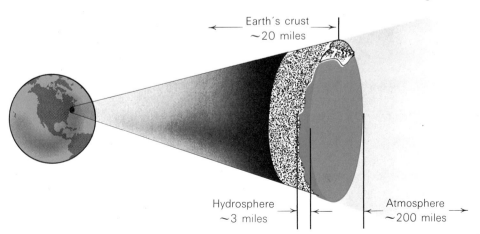

only be inferred from indirect observations of its density, magnetic properties, and ability to transmit sound waves.

23-1.1 The Atmosphere of the Earth

The atmosphere surrounding our sphere forms a blanket in which 98 percent of the air is less than 40 miles above the Earth's surface. The composition of air varies rather widely depending upon local circumstances. For example, the composition of the air above a crowded freeway is different than that over an isolated section of Greenland. Nevertheless, a representative sample of dry air (air from which all water vapor has been removed) would have the approximate composition shown in Table 23-1. Water vapor in normal air would range from about 0.1 percent to 1.6 percent of the total molecules in the air.

TABLE 23-1 COMPOSITION OF A SAMPLE OF DRY AIR*

Substance	Formula	Molecules (%)
Nitrogen	N_2	78.08
Oxygen	O_2	20.95
Argon	Ar	0.93
Carbon dioxide	CO_2	0.03
Neon	Ne	0.0018
Helium	He	0.00052
Krypton	Kr	0.0001
Hydrogen	H_2	0.00005
Xenon	Xe	0.000008

* Traces of other compounds (less than 0.0002 percent of the molecules) are also known to be present.

Some interesting chemical processes occur in the upper atmosphere where ultraviolet light impinges upon oxygen molecules of the air. The oxygen molecules absorb energy present in ultraviolet light to give two oxygen atoms:

$$O_2(g) + \text{ultraviolet light} \longrightarrow 2O(g) \qquad (1)$$

Many of these highly reactive oxygen atoms combine with oxygen molecules to give **ozone** (O_3):

$$O(g) + O_2(g) \longrightarrow O_3(g) \qquad (2)$$

This highly reactive form of oxygen seems to reach a maximum concentration at a height of about 15 miles. Ozone is also an excellent absorber of ultraviolet radiation. Thus, O_2 and O_3 together make the atmosphere opaque to most ultraviolet light. This is indeed fortunate for man as he is now constituted, since ultraviolet radiation has many highly undesirable physiological effects. Presumably the chemistry of life on this planet would have evolved quite differently if all the Sun's ultraviolet radiation reached the Earth's surface.

At the opposite end of the spectrum, the infrared, the atmosphere again becomes virtually opaque, mainly because of absorption by gaseous water and carbon dioxide. Thus, air, normally regarded as transparent, actually serves to filter the Sun's rays striking the Earth. The very high energy photons (in the ultraviolet) and the very low energy photons (in the infrared) are removed and the spectral region between is transmitted.

EXERCISE 23-1

On this planet, of what value would be an eye that is sensitive only to light in the ultraviolet spectral region? Discuss the evolutionary significance of the facts that the human eye and the photosynthesis process are both dependent upon light in the part of the spectrum called the "visible" region.

23-1.2 Composition of the Hydrosphere

Water dissolves some of the air to which it is exposed. Oxygen is highly soluble in water (twice as soluble as nitrogen at the same pressure). It is this dissolved elementary oxygen that is used by living organisms for their oxidative processes. The concentration of dissolved carbon dioxide is low because its concentration in air is low. Dissolved carbon dioxide is necessary for photosynthesis in marine plants and is in part responsible for the pleasant taste of water. Boiled water has lost almost all dissolved gases; it tastes "flat."

Ocean water contains dissolved solids in addition to the gases of the air. About 3.3 percent of the weight of the ocean water is from dissolved salts. Actually, more than 40 elements have been found to be present in ocean water; but half of these are present in very small concentrations, less than 1 g per billion grams (10^9 g) of water. Table 23-2 shows the concentrations of the most abundant ions in ocean water.

TABLE 23-2 AVERAGE COMPOSITION OF OCEAN WATER (DISREGARDING DISSOLVED GASES)

Element	Symbol	Predominant Species	Ocean Water (moles/kg)
Hydrogen	H	H_2O	53.7
Oxygen	O		
Chlorine	Cl	$Cl^-(aq)$	0.535
Sodium	Na	$Na^+(aq)$	0.460
Magnesium	Mg	$Mg^{2+}(aq)$	0.052
Sulfur	S	$SO_4^{2-}(aq)$	0.028
Calcium	Ca	$Ca^{2+}(aq)$	0.010
Potassium	K	$K^+(aq)$	0.010
Bromine	Br	$Br^-(aq)$	0.008

23-1.3 Composition and Properties of the Lithosphere

We know very much about the outermost portion of the lithosphere because we can study it directly. In contrast, we know almost nothing about the inner lithosphere, even though it constitutes over 99.5 percent of the mass of the Earth. The temperature at the center of the Earth is thought to be a few thousand degrees. Although at the Earth's surface rock would melt at this temperature, solids can remain stable at the Earth's core because of the exceedingly high pressure thought to exist there. Very much remains to be learned about the chemistry of the inner lithosphere. It is a high-temperature and high-pressure laboratory whose door has not yet been opened.

Oxygen and silicon are the most abundant elements in the Earth's crust. Table 23-3 shows that 60 percent of the atoms of the crust are oxygen atoms and 20 percent are silicon atoms. If we were to include the hydrosphere as well as the crust, hydrogen would displace aluminum as the third most abundant element (remember that water contains two hydrogen atoms for each oxygen atom). If we were to consider the whole lithosphere, iron would move into second place (ahead of silicon) and magnesium would be fourth. Thus, the exact order is determined by the sample or part of the Earth chosen. In any listing of elements, *the most abundant elements are those having low atomic numbers* (26 or less). All elements beyond iron (no. 26) account for less than 0.2 percent of the weight of the Earth's crust.

In daily life most of us are more concerned with the availability of the elements than with their general abundance in the Earth. The air is all around us and equally accessible to all. Water is somewhat more restricted. Some regions have a surplus of water, whereas other regions are deficient. Even in regions of abundant rainfall, the use of water may be so great that the reserves become depleted gradually. Thus, as population increases and more and more water is used, water threatens to become a scarce natural resource. Many metals used by ancient man—copper (or *cuprum*, Cu), silver (or *argentum*, Ag),

TABLE 23-3 ABUNDANCE OF ELEMENTS IN THE EARTH'S CRUST*

Rank	Element	Symbol	Atomic Number	Number of Atoms (per 10,000)
1	Oxygen	O	8	6050
2	Silicon	Si	14	2045
3	Aluminum	Al	13	625
4	Hydrogen	H	1	270
5	Sodium	Na	11	258
6	Calcium	Ca	20	189
7	Iron	Fe	26	187
8	Magnesium	Mg	12	179
9	Potassium	K	19	138
10	Titanium	Ti	22	27
11	Phosphorus	P	15	8.6
12	Carbon	C	6	5.5
13	Manganese	Mn	25	3.8
14	Sulfur	S	16	3.4
15	Fluorine	F	9	3.3
16	Chlorine	Cl	17	2.8
17	Chromium	Cr	24	1.5
18	Barium	Ba	56	0.75

* From calculations of I. Asimov, *J. Chem. Ed.*, **31**, 70 (1954), based on the data of B. Gutenberg, Ed., *Internal Constitution of the Earth*, 2nd ed, Dover Publications, New York, 1951, p. 87. Reprinted with permission of the publisher.

gold (or *aurum,* Au), tin (or *stannum,* Sn), and lead (or *plumbum,* Pb) —are now in relatively short supply.

Ancient man found deposits of copper, silver, and gold occurring as elementary metals. These three may also be separated from their ores by relatively simple chemical processes. On the other hand, aluminum and titanium, although abundant, are much more difficult to prepare from their ores. Fluorine is more abundant in the Earth than chlorine, but chlorine and its compounds are much more common because they are easier to prepare and easier to handle.

As the best sources of the elements now common to us become depleted, we shall have to turn to the elements that are now little used.

During geological time, a number of separating and sorting processes—melting, crystallization, solution, precipitation—have concentrated various elements in local deposits. In these, the elements tend to be grouped together in rather stable compounds. These are called **minerals.** Many minerals have compositions similar to compounds we can make in the laboratory but most are not so pure. *Minerals concentrated sufficiently to act as commercial sources of desired elements are called* **ores.**

EXERCISE 23-2

Explain in terms of energy the significance of the fact that a piece of wood is stable in air at room temperature; but if the temperature is raised, it will burn and release heat.

Replacing depleted rich ores such as the iron ore Fe_3O_4 will require imaginative chemistry if we are to extract metals from poor ores and guarantee to future generations a continued supply of our natural resources.

One of the fascinating questions about the Earth is: How old is the world? Answers to this question have been obtained by using radioactivity as a clock. Uranium-238 has a known half-life (see Section 7-4.2) and decays to give lead as the end product. Assuming that all lead in uranium deposits arises from the radioactive decay of uranium-238, and knowing its half-life, scientists have estimated the age of the Earth to be about five billion years. Analyses of other radioactive decay processes support this figure. Thus, the best available evidence suggests that the Earth was formed about five billion years ago. Its origin is anybody's guess.

23-2 THE CHEMISTRY OF THE MOON

The moon, our nearest neighbor in space, has already received man's first mechanical envoys. In the short period since June 2, 1966, four space probes of the United States and the Union of Soviet Socialist Republics have orbited the moon and sent back pictures of the moon's surface, while six more space vehicles* have made soft landings on the moon. In December, 1968, three U.S. astronauts,

---

* Soft landings were made by Surveyor I (June, 1966), Luna 13 (December, 1966), Surveyor III (April, 1967), Surveyor V (September, 1967), Surveyor VI (November, 1967), and Surveyor VII (January, 1968).

TABLE 23-4 COMPOSITION OF THE LUNAR SURFACE AS MEASURED BY α-PARTICLE SCATTERING DEVICE (PERCENTAGE BY WEIGHT)

Element	Location		
	Mare Tranquillitatis (Surveyor V)	Sinus Medii (Surveyor VI)	Near Tycho (Surveyor VII)
Carbon	less than 3%	less than 2%	less than 2%
Oxygen	$58 \pm 5\%$	$57 \pm 5\%$	$58 \pm 5\%$
Sodium	less than 2%	less than 2%	less than 3%
Magnesium	$3 \pm 3\%$	$3 \pm 3\%$	$4 \pm 3\%$
Aluminum	$6.5 \pm 2\%$	$6.5 \pm 2\%$	$8 \pm 3\%$
Silicon	$18.5 \pm 3\%$	$22 \pm 4\%$	$18 \pm 4\%$
Calcium	$13 \pm 3\%$	$6 \pm 2\%$	$6 \pm 2\%$
Iron		$5 \pm 2\%$	$2 \pm 1\%$

Colonel Frank Borman, Captain James Lovell, and Major William Anders, orbited the moon at an elevation of about 70 miles. They took many colored photographs in their historic flight. Moon traffic is becoming fairly heavy! These flying laboratories have examined the moon in some detail. Table 23-4 gives an analysis of the moon's surface as determined at three different locations by the Surveyor moon probes V, VI, and VII. Figure 23-2 shows photographs of the Surveyor in operation. This photograph was radioed back from the moon's surface, as was Figure 23-3.

Probably you are wondering how analytical data for three separate locations on the moon's surface were obtained without someone having landed on the moon to do an analysis. The answer to this question was provided by Dr. Anthony Turkevich, who designed a small and ingenious device based on the scattering of

Fig. 23-2 **Surveyor VII sampling the soil of the moon.**

The CHEMISTRY of the MOON

Fig. 23-3 **A mosaic panorama of the lunar surface, prepared from Surveyor VII pictures.**

α-particles* which can estimate the composition of the surface layer. A beam of α-particles, generated by a curium source, was shot at the moon's surface. A few of the α-particles hit nuclei of atoms in the lunar surface and bounced back at almost 180° after a head-on collision. The particles which bounced back directly were caught and their energy was measured. *By determining the energy of the bouncing particle, the mass of the nucleus responsible for the scattering could be estimated.* Particles hitting light nuclei bounced back with only a fraction of their original energy [see Figure 23-4(a)]; e.g., particles hitting a helium nucleus or hydrogen nucleus would not bounce back enough to measure. Particles hitting heavy nuclei bounced back with almost their original velocity [Figure 23-4(c)]. The energy of the rebounding particle identified the element whose nucleus had been hit. The number of rebounding α-particles having a given energy told the number of atoms of that element present where the α-particle struck. The data on α-particle number and energy were

---

* The device reminds us of Rutherford's classic experiment in α-particle scattering. Rutherford determined the scattering angles using a single element, gold. In the moon probe, scattering at a fixed angle (almost 180°) by a target of varying composition was studied.

Fig. 23-4 **(a) Ball striking lighter ball, (b) ball striking heavier ball, (c) ball striking brick wall.**

radioed back to earth and interpreted. The analysis was obtained while the closest man was some 239,000 miles away!

The interesting conclusion to be drawn from the data in Table 23-4 is that the moon's surface is very much like the Earth's. The samples analyzed seem to be typical of the fine-grained rock known as basalt.

One of the misconceptions that troubled possible space voyagers was the idea that the moon's surface was so soft and dusty that a spacecraft landing on it would sink out of sight. All current evidence indicates that this hypothesis is not true. We know it is not true of the area covered by the Surveyor and Luna probes. The lunar materials tested by the Surveyors resembled a light garden soil; when claws from a spacecraft lifted a piece of lunar soil above the surface and dropped it, the lunar material crumbled like soil from your flower garden. Even more encouraging is the fact that its strength increased with depth.

A lunar day lasts about 15 of the Earth's 24-hour days. During the day the temperature on the moon's surface climbs to 120°C. During the night, lasting 360 hours, the surface temperature drops to −170°C. One of the interesting consequences of the long days and nights on the moon is the slow movement of the twilight zone. On Earth the frontier between light and dark moves at 1,000 miles per hour. Even our fastest commercial jets lose out in their race with the sun. On the moon this frontier moves at about 10 miles per hour. An ambitious man, jogging at a good clip, could easily keep up with the setting sun on the moon.

The Surveyor and Luna series have enabled us to find out the relative abundance of elements on the moon's surface. They and the Orbiter series also have enabled us to refine earlier Earth-bound observations. We now know that the mass of the moon is 1/81.29 times the mass of the Earth, that its density is 3.33 g/cm^3, and that the moon's gravity at its surface is 0.165 times that of the Earth; we also know that its diameter is 2,160 miles or 0.2723 times that of the Earth. In addition these probes have confirmed what we always believed, that the moon's surface has no water or atmosphere.

23-3 THE CHEMISTRY OF THE PLANETS

Our solar system consists of the Sun, the planets and their moon satellites, asteroids (small planets), comets, and meteorites. The planets are generally divided into two categories: **Earth-like (terrestrial) planets**—Mercury, Venus, Earth, and Mars; and **giant planets**—Jupiter, Saturn, Uranus, and Neptune. Little is known about Pluto, the most remote planet.

Space-probe data on the terrestrial planets are now becoming available. The close approach of Venus to Earth in June, 1967, inspired both the U.S.S.R. and the U.S. to launch space probes to Venus. The Soviet probe, Venera 4, landed a small instrument package on the surface of Venus. The U.S. probe, Mariner V, flew by the planet at a distance of 2,540 miles from the surface. Mariner II first flew by the surface in 1962. Valuable data were radioed back to

Earth by all of these flying laboratories. *The data received indicated quite clearly that the Venusian environment will not support life as we know it.*

In designing and using space probes, man is coming face to face with the vastness of space. The voyage to Venus took some four months, the spaceships reaching the planet in mid-October, 1967. The trip to Mars was even longer. On November 28, 1964, Mariner IV was launched toward Mars. It passed by Mars at a distance of 6,118 miles near mid-July, 1965. Twenty-one high-quality photographs were sent back over the 135,000,000 miles to Earth (see Figure 23-5). These photographs established that Mars has craters closely resembling those found on areas of the moon. The very good state of these craters indicates the absence of erosion of the type associated with liquid water. Since water is a necessary constituent of living things, it seems fairly certain that complex forms of life have not existed on Mars during the last few hundred million years.

Despite our prying probes and picture-taking satellites, we still know very little about our neighboring planets. Much of what we now believe to be true may in time turn out to be incorrect. Still, we can only advance by taking steps, even though some of these steps may take us backward a short distance.

The CHEMISTRY of the PLANETS

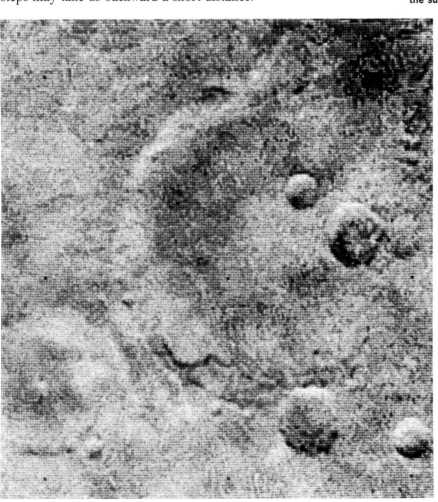

Fig. 23-5 **Mariner IV photograph of the surface of Mars.**

Table 23-5 presents a factual survey of the planets with a comparison of their masses, radii, and densities. These facts are probably the most reliable known because they are deduced from the planets' orbital movements in the solar system.

TABLE 23-5 DATA ON THE SOLAR SYSTEM

Body	Relative Mass	Radius (kilometers)	Density (g/ml)
Sun	3.32×10^5	695×10^3	1.41
Mercury	0.05	2.5×10^3	5.1
Venus	0.81	6.2×10^3	5.0
Earth	(1.00)	6.371×10^3	5.52
Mars	0.11	3.4×10^3	3.9
Jupiter	3.18×10^2	71×10^3	1.33
Saturn	95	57×10^3	0.71
Uranus	14.6	25.8×10^3	1.27
Neptune	17.3	22.3×10^3	2.22
Pluto	uncertain; est. 0.03	2.9×10^3	uncertain; est. 2

23-3.1 Meteorites

Even before Surveyors V, VI, and VII, with their miniature analytical laboratories, landed on the moon, we had some basis for believing that the material in space is not very different from that found on Earth. Occasionally a piece of debris wandering in space is caught in the Earth's gravitational field and pulled in. Such **meteorites,** as they are called, become white hot as they pass through the Earth's atmosphere and most often burn up before they hit the Earth. Occasionally, however, a large fragment will survive the atmospheric screen and we can analyze it. Such analysis has shown that meteorites are of two kinds: **stony meteorites** are like the rocks of the Earth; **metallic meteorites** are pieces of metal such as iron and nickel. One theory suggests that these meteorites are fragments from exploded planets that may have resembled the Earth. Another hypothesis is that meteorites and the Earth share a common origin and that they became separated after the elements were created.

23-3.2 The Planetary Atmospheres

Much information about planetary atmospheres has been inferred from the light which reaches us on Earth. But firm conclusions are difficult because the light reaching us has always passed through the atmosphere of the Earth. For example, if absorption frequencies characteristic of water are observed in a planetary spectrum, the question arises: Is it Earth's atmosphere or the planetary atmosphere we are seeing? Since water is present in our atmosphere, conclusions are tenuous at best. These questions are finally being resolved as automated observatories are sent *outside* the Earth's atmosphere. In fact, an observatory has been placed right in the atmosphere of the planet Venus.

Table 23-6 summarizes our present information on the atmospheres of the planets. Data on Venus and Mars have been obtained by space probes as well as by spectral observation. There is some disagreement over both the temperature and pressure at the surface of Venus. The results shown in Table 23-6 are derived from the latest U.S. probe, Mariner V. The analytical results for the atmosphere of Venus are taken in large part from results sent back by the Soviet flying laboratory, Venera 4. The temperature and pressure recorded by the Soviet craft are lower than those recorded by the Mariners and by other observations. This has given rise to speculation that the Soviet laboratory may have landed on a 15-mile high mountain on the cloudy surface of Venus. In any event, all scientists agree that the atmospheric pressure of Venus at the surface ranges from 20 to 100 Earth atmospheres. The higher value is probably correct. The temperature on the surface of Venus is probably above the melting point of lead and is certainly above the boiling point of mercury. The cloud cover and dense atmosphere make it likely that the surface of Venus is in total darkness, and pressures on Venus would be comparable to those found half a mile below the surface of the sea. It is believed that winds of unimaginable fury blow across the face of Venus and help to equalize the temperatures on the dark and light side of the planet. These extremes of climate indicate that Venus is a most inhospitable environment. Any exploration of Venus will require not only great technical skill, but almost superhuman courage.

TABLE 23-6 THE PLANETARY ATMOSPHERES

Planet	Maximum Surface Temperature (°C)	Minimum Surface Temperature (°C)	Gases Present	Atmospheric Pressure (atm)
Mercury	530 (day = 59 ± 5 Earth days)	−240	—	less than 0.01
Venus	450–500 (day = 247 ± 5 Earth days)	uncertain; perhaps below 100	more than 90% CO_2 N_2, Ar less than 1% O_2 trace H_2O (less than 1%)	20–100
Earth	60 (day = 1 Earth day)	−95	78% N_2 21% O_2 1% Ar trace H_2O	1
Mars	20–30 (day = 24.6 Earth hours)	−50 to −60	50–100% CO_2 less than 1% O_2 N_2, Ar trace H_2O	0.01
Jupiter	−138	—	H_2, CH_4, NH_3, H_2O	nature of surface unknown
Saturn	−153	—	CH_4, NH_3	nature of surface unknown
Uranus	−184	—	CH_4	nature of surface unknown
Neptune	−200	—	CH_4	nature of surface unknown
Pluto	—	—	—	

The conditions on Mars, while certainly not congenial to life, suggest that Mars could be explored with less technical difficulty than any of the other planets.

The giant planets—Jupiter, Saturn, Uranus, and Neptune—have an amazing constancy in the composition of their atmospheres. Notice that all contain CH_4; NH_3 is believed to be present in the atmospheres of Jupiter and Saturn. It is difficult to understand the presence of NH_3 in the atmospheres of Jupiter and Saturn since NH_3 freezes at $-78°C$ and the surface temperatures of these planets are never above $-138°C$. Perhaps clouds of solid NH_3 crystals swirl over the planets' surfaces. These giant planets have atmospheres that extend several thousand miles out from the surface.

--- EXERCISE 23-3 ---

Examine the components believed to be present in Jupiter's atmosphere. Would you expect oxygen to be a likely component of the atmosphere?

Information on the composition of the gaseous mantle surrounding these planets comes from a study of their infrared spectra. Absorptions observed in the light reaching Earth are identical to the absorptions of NH_3 and CH_4. Estimates of the average molecular weight of the atmosphere of Jupiter are around the value 3. If these values are correct, we would expect hydrogen also to be an important constituent of Jupiter's atmosphere.

23-3.3 The Planetary Lithospheres

The extreme difficulty we have probing the composition of the Earth beneath us should suggest that little is known about the inner composition of the planets. The evidence available is indirect (*e.g.*, average density and surface composition) and the interpretations are conflicting.

23-4 THE SUN

The surface temperature of the Sun is about 5500°C. Moving inward from the surface, the temperature rises, probably to above

Fig. 23-6 **Some molecules detected in the solar atmosphere.**

one million degrees. At these high temperatures the tendency toward greatest randomness (favoring atoms) is stronger than the tendency toward lowest energy (favoring molecules). As a result, only the simplest molecules are to be expected.

Some data from Earth-based observatories and orbiting probes have been obtained. At the present time, the spectra of many diatomic molecules have been detected. These are not the familiar, chemically stable molecules we find on the stockroom shelf, but the molecules that would be stable on a solar stockroom shelf. Figure 23-6 shows some of these and their location in the periodic table.

Inside the Sun, thermal energies are sufficient to destroy all molecules and to ionize the atoms. These ions emit their characteristic line spectra and tens of thousands of lines are observed. The lines that have been analyzed show the existence of atoms ionized as far as O^{5+}, Mn^{12+}, and Fe^{13+}. At this time, over sixty of the elements have been detected in the Sun through their spectral emissions and absorptions.

23-5 STELLAR ATMOSPHERES

Our Sun is the nearest **star.** It is a relatively cool star and, as such, contains a number of diatomic molecules, as we saw in Figure 23-6. There are many stars, however, with still lower surface temperatures and these contain chemical species whose presence can be understood in terms of the temperatures and the usual chemical equilibrium principles. For example, as the temperature of the star drops, the spectral lines attributed to CN and CH become more prominent. At lower temperatures, TiO becomes an important species along with the hydrides MgH, SiH, and AlH, and the oxides ZrO, ScO, YO, CrO, AlO, and BO.

Detailed consideration of the chemical equilibria among these species provides evidence of the presence of other molecules that cannot be observed directly. The chemical properties of the molecules mentioned provide a firm basis for predicting the presence and concentrations of such important molecules as H_2, CO, O_2, N_2, and NO. Thus, the faint light by which we view the distant stars is rich with information. We need but learn how to read it.

23-6 INTERSTELLAR SPACE

In addition to the stars, the space between them and the Earth, **interstellar space,** is part of the astronomical spectroscopy laboratory. Light from a distant star must traverse fantastic distances to reach our telescopes and the absorption of this light by the most minute concentrations of atoms and molecules in space becomes important and detectable. Absorption spectra have shown that diatomic molecules such as CH, CN, and CH^+ are present in "empty" space at an average concentration of about one molecule per 1,000 liters. These molecules are probably concentrated in "clouds" with one molecule per 100 liters.

EXERCISE 23-4

Calculate the volume in liters of a sphere of radius 6,400 kilometers (the radius of the Earth). How many grams of oxygen are needed to fill this volume to a concentration of one molecule per 1,000 liters?

In addition to these molecules, atoms are present, as shown by absorption lines of Ca, Na, K, Fe, and other atoms. There are some absorption lines that have not been identified, but these may be due to small, solid particles. How these particular molecules and atoms came to be present in these almost nonexistent "clouds," and what other, yet undetected molecules and atoms are there, are matters to wonder about. But wondering is at once the pleasure and the driving force of science.

23-7 EPILOGUE

For centuries men dreamed of flying. Magic carpets became the hallmark of the fairy tale. Today we ride in "encased" magic carpets at speeds far in excess of 500 miles per hour, a speed much greater than the wildest dreams of the ancient storyteller.

More recently men have dreamed of space travel and of fantastic voyages to other planets. We stand on the threshold of such voyages. But today we are as ignorant of the planets as the fifteenth century geographers were of Africa and America. We can, however, be sure that the unexplored lands in space will contain bigger surprises than Victoria Falls or the Grand Canyon. In undertaking this exploration we depend upon the truth of the assumption that the laws of chemistry and physics remain the same throughout the universe. So far we feel this is a safe assumption. But if this assumption is incorrect, our technical skills cannot be applied to distant space exploration. All our technology and apparently life itself are founded on the laws of physics and chemistry. We believe today that these laws when properly understood are eternal truths, remaining applicable throughout time and space.

Appendices

Appendix 1

A DESCRIPTION OF A BURNING CANDLE

A photograph of a burning candle is shown[1]* in Figure A1-1. The candle is cylindrical in shape[2] and has a diameter[3] of about ¾ inch. The length of the candle was initially about 8 inches[4] and it changed slowly[5] during observation, decreasing about ½ inch per hour[6]. The candle is made of a translucent[7], white[8] solid[9] which has a slight odor[10] and no taste[11]. It is soft enough to be scratched with the fingernail[12]. There is a wick[13] which extends from the top to the bottom[14] of the candle along its central axis[15] and which protrudes above the top of the candle about ½ inch[16]. The wick is made of three strands of string braided together[17].

The candle is lighted by holding a source of flame close to the wick for a few seconds. The source of the flame can then be removed and the flame will sustain itself at the wick[18]. The burning candle makes no sound[19]. While burning, the body of the candle remains cool to the touch[20] except near the top. Within about ½ inch from the top the candle is warm (but not hot)[21] and sufficiently soft to mold easily[22].

The flame flickers in response to air currents[23] and tends to become quite smoky while flickering[24]. In the absence of air currents, the flame is of the form shown in Figure A1-1, although it exhibits some movement at all times[25]. The flame begins about ⅛-inch above the top of the candle[26] and at its base the flame has a blue tint[27]. Immediately around the wick in a region about ¼-inch wide and extending about ½-inch above the top of the wick[28], the flame is dark[29]. This dark region is roughly conical in shape[30]. Around this zone and extending about ½-inch above the dark zone is a region which emits yellow light[31], which is bright but not blinding[32]. The flame has rather sharply defined sides[33] but a ragged top[34].

The wick is white where it emerges from the candle[35]; from the base of the flame to the end of the wick, it is black, appearing burnt[36], except for the last 1/16 inch, where it glows red[37]. The wick curls over about ¼-inch from its end[38]. As the candle becomes shorter, the wick becomes shorter too, so as to extend roughly a constant length above the top of the candle[39].

Heat is emitted by the flame[40], enough so that it becomes uncomfortable in 10 or 20 seconds if a finger is held ¼-inch to the side of the flame[41] or 3 or 4 inches above the flame[42].

The top of a quietly burning candle becomes wet with a colorless liquid[43] and becomes bowl-shaped[44]. If the flame is blown, one side of this bowl-shaped top may liquefy, and the liquid trapped in the bowl may drain down the side of the candle[45]. As it runs down, the colorless liquid cools[46], becomes translucent[47], and gradually solidifies from the outside[48] and attaches itself to the side of the candle[49]. When there is no draft, the candle can burn for hours without such drippings[50]. Under these conditions, a stable pool of clear liquid remains in the bowl-shaped top of the candle[51]. The liquid rises slightly around the wick[52], wetting the base of the wick as high as the base of the flame[53].

Fig. A1-1 A burning candle.

* The numbers refer to distinct observations made by the observer. Note the careful differentiation between observation and interpretation.

Several aspects of this description deserve specific mention. Compare your own description with this one in each of the following characteristics:

DESCRIPTION of a BURNING CANDLE

(1) The description is comprehensive in *qualitative* terms. Did *you* mention appearance? smell? taste? feel? sound? (Note: A chemist quickly becomes reluctant to taste or smell an unknown chemical. A chemical should be considered poisonous unless it is *known* not to be!)
(2) Wherever possible, the description is stated *quantitatively*. This means the question "How much?" is answered, the quantity is specified. The remark that the flame emits yellow light is made more meaningful by the "how much" expression "bright but not blinding." Any statement to the effect that heat is emitted might lead a cautious investigator who is lighting a candle for the first time to stand in a concrete blockhouse 100 yards away. A few words telling him "how much" heat would save him this precaution.
(3) The description does not make assumptions regarding the relative importance of observations. Thus the observation that a burning candle does not emit sound deserves to be mentioned just as much as the observation that it does emit light.
(4) The description does not confuse observation and interpretation. To say that the top of the burning candle is wet with a colorless liquid is to make an observation. To suggest a possible composition for this liquid is to offer an interpretation.

Appendix 2

THE PRESSURE DUE TO THE ATMOSPHERE

A2-1 Determining the Pressure Due to the Atmosphere from a Plot of P Versus 1/V

A plot of the number of bricks versus the volume will give a curve (see Figure A2-1). A plot of the number of bricks versus one divided by the volume (1/V) will give a straight line (see Figure A2-2). The algebraic expression for a straight line is

$$y = mx + b \qquad (1)$$

where m is the slope. This can be written as

$$y + (-b) = mx \qquad (1a)$$

If we let the pressure due to the bricks be P_B and the pressure due to the atmosphere be P_A, we can write

$$\underbrace{P_B + P_A}_{y + (-b)} = C\left(\frac{1}{V}\right) \quad \text{This is an experimental equation from our plot (Figure A2-2).} \qquad (2)$$

$$y + (-b) = mx \quad \text{This is the equation for a straight line from algebra.} \qquad (1a)$$

These two equations are identical if $P_B = y$, $P_A = -b$, $C = m$, and $1/V = x$.

We have values for P_B, $1/V$, and C, which is the slope of the line. We can determine P_A in the following way:

$$P_A = C\left(\frac{1}{V}\right) - P_B \qquad (2a)$$

Fig. A2-1 **A plot of pressure vs. volume.**

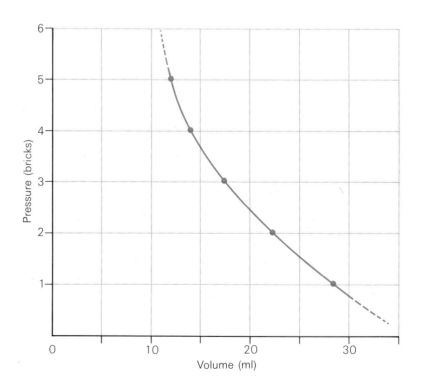

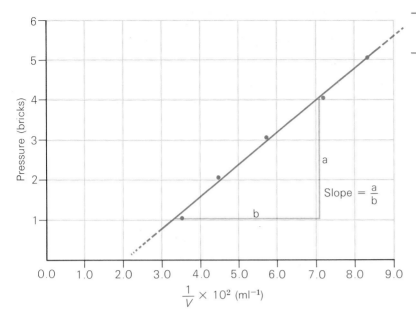

Fig. A2-2 **A plot of pressure vs. 1/volume.**

Let us extend the straight line to $P_B = 0$; then, since $P_B = 0$ (corresponding to $1/V$ for zero bricks), we can write

$$P_A = C\left(\frac{1}{V}\right) \qquad (3)$$

The value of $1/V$ can be read from the line at $P_B = 0$: $1/V = 2.1$ (1/ml). The slope, C, is obtained from the equation

$$\text{slope} = C = \frac{a}{b} = \frac{4.00 - 1.00}{7.10 - 3.35} = \frac{3.00}{3.75} = 0.80 \qquad (4)$$

Then

$$P_A = C\left(\frac{1}{V}\right) = 2.1 \times 0.80 = 1.7 \text{ bricks} \qquad (5)$$

A2-2 Determining the Pressure Due to the Atmosphere in "Bricks" from Weight of Brick and Size of Plunger

The value for P_A obtained in Section A2-1 can also be estimated using the size of the syringe plunger, the weight of the brick used, and the known value of 14.7 lb/in² for the atmospheric pressure. The calculations are as follows.

The diameter of the piston is $15/16$ in; hence, its area is $(15/16 \text{ in})^2 \times (3.14/4)$. (Remember that the area of a circle is $\pi d^2/4$.) The area of the plunger is 0.69 in². Each brick weighed 5.2 lb when placed on a balance. The known pressure of the air at sea level is 14.7 lb/in².

These numbers can be combined as follows:

$$\frac{\text{pounds per brick} \times \text{number of bricks}}{\text{area of plunger where } V \text{ acts}} = P_A$$

Substituting numbers, we write

$$\frac{(5.2 \text{ lb/brick}) \times (\text{number of bricks})}{0.69 \text{ in}^2} = 14.7 \text{ lb/in}^2 \quad (6)$$

Then

$$\text{number of bricks} = \frac{14.7 \frac{\text{lb}}{\text{in}^2} \times 0.69 \text{ in}^2}{5.2 \frac{\text{lb}}{\text{brick}}} \quad (6a)$$

number of bricks = 1.9 bricks

The value is in acceptable agreement with the value of 1.7 bricks obtained in Section A2-1. More precise measurements should bring the results of the two procedures into agreement.

DETERMINING THE RATIO OF CHARGE TO MASS FOR AN ELECTRON

Appendix 3

One of the key steps in obtaining the weight of an electron is determining the ratio of charge to mass for the particle—that is, e/m. Most experimental procedures used are based upon the behavior of the electron in electric and magnetic fields. Let us begin this study by considering the behavior of electrons as they move in the space between the poles of a magnet. Here is the experiment.

Rapidly moving electrons are generated in the cathode ray tube shown in Figure A3-1. A narrow beam of electrons can be obtained in the tube by using the metal plate Z with a small slit in it. When this narrow beam hits the glass plate coated with zinc sulfide that is at the end of the tube, a glowing line appears. If the tube is now placed between the pole faces of a very large electromagnet, the electron beam will be moving in a magnetic field. The strength of the field is determined by the amount of current flowing through the electromagnet. Under the influence of the magnetic field created by the electromagnet, the electron beam bends into a circular path of radius r. The circular path is parallel to the two pole faces (see Figure (A3-1). Because the beam is bent, it will now hit the zinc-sulfide coated glass plate at a higher point. By measuring the distance which the beam has moved upward on the screen and by applying a little geometry, it is possible to calculate the radius r of the circle in which the electrons moved between the poles of the magnet.

What is the force acting on an electron in the beam? The force acting on an electron moving within and perpendicular to a magnetic field is dependent upon the electron's velocity and charge and upon the strength of the perpendicular magnetic field. The equation for this force is

$$\begin{Bmatrix} \text{force acting on electron} \\ \text{moving in magnetic field} \end{Bmatrix} = \begin{Bmatrix} \text{velocity of} \\ \text{electron} \end{Bmatrix} \times \begin{Bmatrix} \text{charge on} \\ \text{electron} \end{Bmatrix} \times \begin{Bmatrix} \text{strength of} \\ \text{magnetic field} \end{Bmatrix}$$

or

$$\text{force}_1 = v \times e \times B \qquad (1)$$

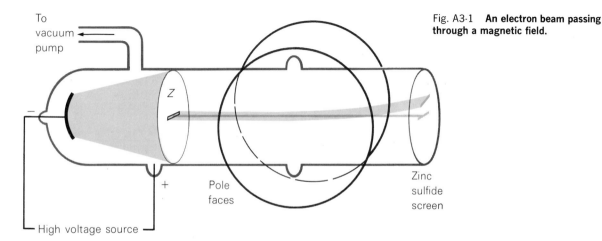

Fig. A3-1 **An electron beam passing through a magnetic field.**

where v = the velocity of the electron, e = the charge on the electron, and B = the strength of the magnetic field.

The force required to make an electron move in a circular path of a certain radius is dependent upon the mass of the electron and the square of the velocity of the electron. The equation for this force is

$$\left\{\begin{array}{l}\text{force required}\\ \text{to make electron}\\ \text{move in a}\\ \text{circular path}\end{array}\right\} = \frac{\{\text{mass of electron}\} \times \{\text{velocity of electron}\}^2}{\{\text{radius of circular path}\}}$$

or

$$\text{force}_2 = \frac{m \times v^2}{r} \qquad (2)$$

where m = the mass of the electron, v = the velocity of the electron, and r = the radius of the circular path described by the electron.

If the electron is actually to follow a circular path, then force$_1$ must be equal to force$_2$, or

$$v \times e \times B = \frac{m \times v^2}{r} \qquad (3)$$

Solving this equation for e/m, we have

$$\frac{e}{m} = \frac{v}{r \times B} \qquad (3a)$$

or

$$\frac{\text{charge on electron}}{\text{mass of electron}} = \frac{\text{velocity of electron}}{\text{strength on magnetic field} \times \text{radius of circular path}}$$

The radius of the circular path, r, can be calculated from the geometry of the tube and the upward shift of the line on the screen as the magnetic field is applied. The strength of the magnetic field, B, can be determined from the construction of the electromagnet and from the current flowing through its coils.

The quantity e/m can now be calculated if the velocity of the electron is known. This quantity has been determined in several ways. One way is by using both an electrostatic and a magnetic field. As you will recall from Section 6-3, the force of attraction between an electron and a proton, which are separated by distance d, measured in centimeters, is given by the expression

$$\left\{\begin{array}{l}\text{force of attraction}\\ \text{between electron}\\ \text{and proton}\end{array}\right\} = \frac{\text{charge on electron} \times \text{charge on proton}}{\text{distance between electron and proton}^2}$$

$$\text{force} = \frac{e_1 \times e_2}{d^2} \qquad (4)$$

where e_1 = charge on the electron, e_2 = charge on the proton, and d = distance between them.

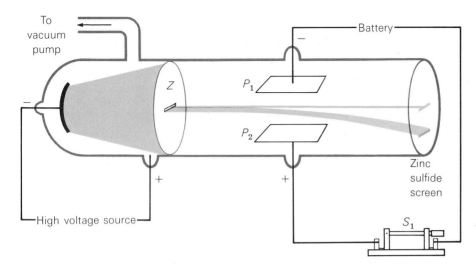

Fig. A3-2 An electron beam passing through an electrostatic field.

If an electron is placed between two plates, one negatively and the other positively charged, the quantity e_2/d^2 is replaced by a quantity called the **electrostatic field strength.** The electrostatic field strength is determined by the charge on the plates and by the distance between them. If we use the symbol Q to represent the electrostatic field strength, the earlier equation for force on the electron becomes

$$\left\{\begin{array}{l}\text{electrostatic force}\\ \text{on electron}\end{array}\right\} = e \times Q \qquad (5)$$

where Q = electrostatic field strength.

If a beam of electrons is now generated in the cathode ray tube and passed between the two charged plates, P_1 and P_2, the electrostatic field acting between the plates makes the beam bend (Figure A3-2).

If a tube is now made with plates P_1 and P_2 on the top and bottom of the tube, respectively, and with magnetic-pole faces on each side of the tube (see Figure A3-3), the moving electron can be subjected to both electrostatic and electromagnetic forces at the same

Fig. A3-3 An electron beam passing through magnetic and electrostatic fields.

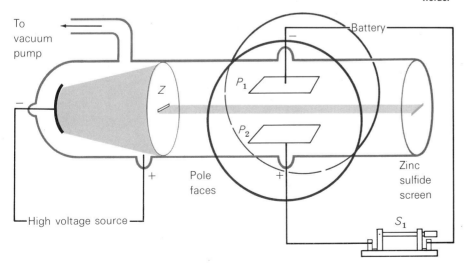

time. If the electrostatic force tending to bend the beam downward is just balanced by the magnetic force tending to bend the beam upward, the electrons will pass through both fields without deflection. Under these conditions

$$\left\{\begin{array}{l}\text{force acting on electron}\\ \text{moving in magnetic field}\end{array}\right\} = \left\{\begin{array}{l}\text{electrostatic force}\\ \text{on electron}\end{array}\right\}$$

or

$$v \times e \times B = e \times Q \tag{6}$$

$$v = \frac{Q}{B} \tag{6a}$$

$$\left\{\begin{array}{l}\text{velocity}\\ \text{of electron}\end{array}\right\} = \frac{\text{electrostatic field strength}}{\text{magnetic field strength}}$$

If this value for v is now put into our earlier expression for e/m, (3a), the following is obtained:

$$\frac{e}{m} = \frac{v}{r \times B} = \frac{Q}{B \times r \times B} = \frac{Q}{r \times B^2} \tag{3b}$$

All quantities on the right can be measured directly. Q is the electrostatic field strength and can be determined for the tube in Figure A3-3 by reading the voltmeter attached to the plates; B is the magnetic field strength and is caused by the current flowing in the coils of the electromagnet; r is the radius of the circle described by the path of the electron in the magnetic field and can be obtained from the shift in the bright line appearing on the zinc sulfide screen when the magnet is turned on. The quantity e/m can thus be obtained.

THE OPERATION OF THE MASS SPECTROGRAPH OR THE MASS SPECTROMETER

Appendix 4

In Appendix 3 a procedure was described for determining the ratio of charge to mass for the electron. This is written as e/m. The value obtained can be coupled with experiments giving the charge on the electron to yield a precise determination of the mass of the electron. The same procedure can be modified slightly to determine the mass of positive ions. The instrument used to carry out this procedure is known as a **mass spectrograph** or a **mass spectrometer**.* It permits the measurement of the masses of individual atomic or molecular ions.

One type of mass spectrograph is shown in Figure A4-1. Positive ions are generated from atoms in the tube by bombardment with an electron beam. The rapidly moving electrons of the beam knock electrons from the atoms in the tube. The positive ions produced are accelerated to a known velocity by attraction for the highly charged negative electrode, which has a slit in it. This rapidly moving beam of positive ions, emerging from the slit in the negative electrode, is made into a narrower beam by passage through a slit in another charged metal disk. This thin beam then passes through a uniform magnetic field.

Figure A4-1 shows neon gas entering at the bottom. The gas passes through the electron beam; some atoms collide with the electrons of the beam to form neon ions. Both Ne^+ and Ne^{2+} ions are formed and they are accelerated by passing through the slits in the negative electrodes.

After the beam of positive ions is formed, it enters the magnetic field, where the charged ions move in a circular path. The circular path has a large radius if the *mass* of the particle is high and a small

---

* A *mass spectrograph* records the beam intensities on a photographic plate; a *mass spectrometer* records the beam intensity with a pen on a piece of chart paper.

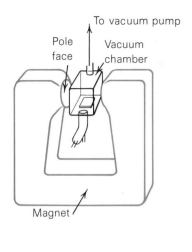 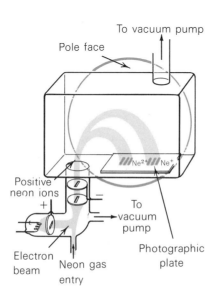

Fig. A4-1 **A mass spectrograph.**

APPENDIX 4

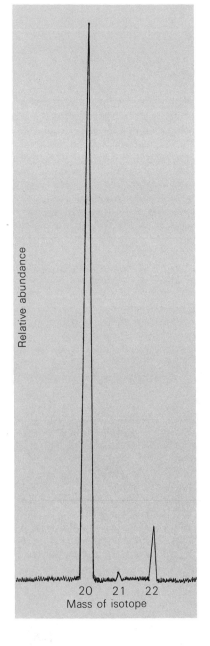

Fig. A4-2 **A mass spectrometer tracing for a sample of neon gas containing 90.5% ^{20}Ne$^+$, 0.3% ^{21}Ne$^+$, and 9.2% ^{22}Ne$^+$.**

radius if the *charge* is high. Each positive ion follows a distinctive circular path fixed by its mass and charge. After traveling half a circle (an arc of 180°), the ions hit a photographic plate. The impact of ions on the plate causes a chemical reaction which darkens the plate, just as exposure to light darkens the plate. The plate, then, will have a dark spot for each ion at a position fixed by the ratio of charge to mass for each ion. The record is called a **mass spectrum.**

Ions can be detected by methods other than the use of the photographic plate. After being sorted according to mass and charge, the ions can be "counted" by a charge-measuring device. The advantage of such a detector is that the result can be presented continuously on a paper chart, thus eliminating a cumbersome photographic process. The mass spectrum obtained in this way consists of peaks on a chart.

Closer examination of the mass spectrum for neon (see Figure A4-2) shows in more detail how the data are interpreted. The record consists of two widely separated groups of three peaks each. The three peaks corresponding to ions moving in a circle with large radii are caused by neon ions with a single positive charge. The other set of three peaks corresponding to ions moving in circles of smaller radii are caused by ions carrying two positive charges. For each ionic charge there are three slightly separated peaks. This second group of peaks is not visible in Figure A4-2. These peaks indicate that neon has three different isotopes. The relative abundance of each isotope can be determined by measuring the relative heights of the peaks caused by each ion beam.

The mass of the ion causing a given spot can be calculated from the following equation:

$$m = \frac{n \times e \times r^2 \times B^2}{2V}$$

where n = an integer indicating number of charges on the ion, e = the electron charge, r = the radius of the circular path of the ions (a value which is obtained from the position of the line on the plate), B = the strength of the magnetic field causing the ions to move in a circular path, and V = the voltage between positive and negative electrodes, which accelerates ions.

Atomic weights are today determined by the mass spectrometer. It is possible to determine the precise mass of all isotopes making up an element. It is also possible to determine the percent of each isotope present in a normal sample of the element. (See Figure A4-2.) These numbers permit calculation of the atomic weight of that element.

MOSLEY'S EXPERIMENT: THE RELATIONSHIP BETWEEN ATOMIC NUMBER AND THE FREQUENCY OF LINES IN X-RAY SPECTRA

Appendix 5

If a beam of cathode rays (a stream of electrons) is allowed to impinge upon a piece of metal in a vacuum tube, X rays or light of very short wavelength is produced (see Chapter 15). The metal hit by the electrons is called the target. A device of this type which generates X rays is known as an X-ray tube (see Figure A5-1). In 1912, H. G. J. Mosley studied the nature of the X radiation given off

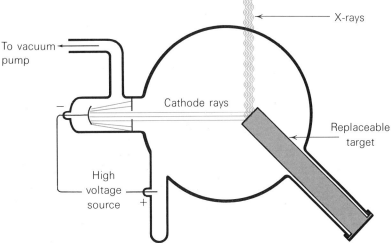

Fig. A5-1 **A schematic drawing of an X-ray tube.**

when X-ray tubes were made which contained different elements as the target. For example, he measured the frequency (see Chapter 15) of the X radiation when elemental cobalt was the target. The same measurements were made using nickel, copper, zinc, and so on, as targets. He then found that a plot of the square root of the wavelength against the atomic number of the target gave a straight line (see Figure A5-2). This result was of great significance since it

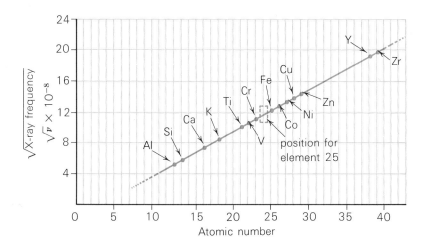

Fig. A5-2 **The relationship of X-ray spectra to atomic number.**

clearly identified the atomic number mentioned by Bohr and Rutherford as a fundamental characteristic of the atom. The relationship also was important in showing the atomic numbers for undiscovered elements. For example, if in the graph of Figure A5-2 the element of atomic number 25 were still undiscovered, a blank space would appear in the Mosley plot at this atomic number. If chemists knew the atomic number of a missing element, guesses as to its properties could be made and procedures for its separation could be devised. Furthermore, the expected characteristic X-ray line (predicted by the plot) would provide an easy and certain method for identifying the new element when it was found.

It was also possible to derive the Mosley relationship using Planck's equation $E = h\nu$ (see Section 15-2.1) and the Bohr model of the atom as well as the quantum-theory model of the atom (see Section 15-2 and Section 15-3). This derivation will be delayed until your college course in chemistry.

RELATIVE STRENGTHS OF ACIDS IN AQUEOUS SOLUTION AT ROOM TEMPERATURE

Appendix 6

The equation for the ionization is

$$HB(aq) \rightleftharpoons H^+(aq) + B^-(aq)$$

Since all ions and molecules in water solution are aquated, the (aq) is assumed in the notation. We then write for K_A

$$K_A = \frac{[H^+][B^-]}{[HB]}$$

Acid	Strength	Reaction	K_A
Perchloric acid	very strong	$HClO_4 \longrightarrow H^+ + ClO_4^-$	very large
Hydriodic acid	↓	$HI \longrightarrow H^+ + I^-$	very large
Hydrobromic acid		$HBr \longrightarrow H^+ + Br^-$	very large
Hydrochloric acid	↓	$HCl \longrightarrow H^+ + Cl^-$	very large
Nitric acid		$HNO_3 \longrightarrow H^+ + NO_3^-$	very large
Sulfuric acid	very strong	$H_2SO_4 \longrightarrow H^+ + HSO_4^-$	very large
Oxalic acid		$HOOCCOOH \longrightarrow H^+ + HOOCCOO^-$	5.4×10^{-2}
Sulfurous acid ($SO_2 + H_2O$)	↓	$H_2SO_3 \longrightarrow H^+ + HSO_3^-$	1.7×10^{-2}
Hydrogen sulfate ion	strong	$HSO_4^- \longrightarrow H^+ + SO_4^{2-}$	1.3×10^{-2}
Phosphoric acid		$H_3PO_4 \longrightarrow H^+ + H_2PO_4^-$	7.1×10^{-3}
Ferric ion		$Fe(H_2O)_6^{3+} \longrightarrow H^+ + Fe(H_2O)_5(OH)^{2+}$	6.0×10^{-3}
Hydrogen telluride	↓	$H_2Te \longrightarrow H^+ + HTe^-$	2.3×10^{-3}
Hydrofluoric acid	weak	$HF \longrightarrow H^+ + F^-$	6.7×10^{-4}
Nitrous acid		$HNO_2 \longrightarrow H^+ + NO_2^-$	5.1×10^{-4}
Hydrogen selenide		$H_2Se \longrightarrow H^+ + HSe^-$	1.7×10^{-4}
Chromic ion		$Cr(H_2O)_6^{3+} \longrightarrow H^+ + Cr(H_2O)_5(OH)^{2+}$	1.5×10^{-4}
Benzoic acid		$C_6H_5COOH \longrightarrow H^+ + C_6H_5COO^-$	6.6×10^{-5}
Hydrogen oxalate ion	↓	$HOOCCOO^- \longrightarrow H^+ + OOCCOO^{2-}$	5.4×10^{-5}
Acetic acid	weak	$CH_3COOH \longrightarrow H^+ + CH_3COO^-$	1.8×10^{-5}
Aluminum ion		$Al(H_2O)_6^{3+} \longrightarrow H^+ + Al(H_2O)_5(OH)^{2+}$	1.4×10^{-5}
Carbonic acid ($CO_2 + H_2O$)		$H_2CO_3 \longrightarrow H^+ + HCO_3^-$	4.4×10^{-7}
Hydrogen sulfide		$H_2S \longrightarrow H^+ + HS^-$	1.0×10^{-7}
Dihydrogen phosphate ion		$H_2PO_4^- \longrightarrow H^+ + HPO_4^{2-}$	6.3×10^{-8}
Hydrogen sulfite ion	↓	$HSO_3^- \longrightarrow H^+ + SO_3^{2-}$	6.2×10^{-8}
Ammonium ion	weak	$NH_4^+ \longrightarrow H^+ + NH_3$	5.7×10^{-10}
Hydrogen carbonate ion		$HCO_3^- \longrightarrow H^+ + CO_3^{2-}$	4.7×10^{-11}
Hydrogen telluride ion	↓	$HTe^- \longrightarrow H^+ + Te^{2-}$	1.0×10^{-11}
Hydrogen peroxide	very weak	$H_2O_2 \longrightarrow H^+ + HO_2^-$	2.4×10^{-12}
Monohydrogen phosphate ion		$HPO_4^{2-} \longrightarrow H^+ + PO_4^{3-}$	4.4×10^{-13}
Hydrogen sulfide ion		$HS^- \longrightarrow H^+ + S^{2-}$	1.2×10^{-15}
Water		$H_2O \longrightarrow H^+ + OH^-$	1.8×10^{-16}*
Hydroxide ion	↓	$OH^- \longrightarrow H^+ + O^{2-}$	$<10^{-36}$
Ammonia	very weak	$NH_3 \longrightarrow H^+ + NH_2^-$	very small

* The equilibrium constant, K_A, for water equals

$$\frac{K_w}{H_2O} = \frac{1.00 \times 10^{-14}}{55.5}$$

Appendix 7

STANDARD REDUCTION POTENTIALS FOR HALF-REACTIONS: IONIC CONCENTRATIONS = 1 MOLAR IN WATER, TEMPERATURE = 25°C

All ions are in water.

Strength As Oxidizing Agent	Half-Reaction	E^0 (volts)	Strength As Reducing Agent
very strong oxidizing agents	$F_2(g) + 2e^- \longrightarrow 2F^-$	+2.87	very weak reducing agents
	$H_2O_2 + 2H^+ + 2e^- \longrightarrow 2H_2O$	+1.77	
	$MnO_4^- + 8H^+ + 5e^- \longrightarrow Mn^{2+} + 4H_2O$	+1.52	
↑	$Au^{3+} + 3e^- \longrightarrow Au(s)$	+1.50	↓
	$Cl_2(g) + 2e^- \longrightarrow 2Cl^-$	+1.36	increasing strength as reducing agent
	$Cr_2O_7^{2-} + 14H^+ + 6e^- \longrightarrow 2Cr^{3+} + 7H_2O$	+1.33	
	$MnO_2(s) + 4H^+ + 2e^- \longrightarrow Mn^{2+} + 2H_2O$	+1.28	
	$\tfrac{1}{2}O_2(g) + 2H^+ + 2e^- \longrightarrow H_2O$	+1.23	
	$Br_2(l) + 2e^- \longrightarrow 2Br^-$	+1.06	
	$AuCl_4^- + 3e^- \longrightarrow Au(s) + 4Cl^-$	+1.00	
	$NO_3^- + 4H^+ + 3e^- \longrightarrow NO(g) + 2H_2O$	+0.96	
	$\tfrac{1}{2}O_2(g) + 2H^+ (10^{-7} M) + 2e^- \longrightarrow H_2O$	+0.82	
	$Ag^+ + e^- \longrightarrow Ag(s)$	+0.80	
	$\tfrac{1}{2}Hg_2^{2+} + e^- \longrightarrow Hg(l)$	+0.79	
	$Hg^{2+} + 2e^- \longrightarrow Hg(l)$	+0.78	
	$NO_3^- + 2H^+ + e^- \longrightarrow NO_2(g) + H_2O$	+0.78	
	$Fe^{3+} + e^- \longrightarrow Fe^{2+}$	+0.77	
	$O_2(g) + 2H^+ + 2e^- \longrightarrow H_2O_2$	+0.68	
	$I_2(s) + 2e^- \longrightarrow 2I^-$	+0.53	
	$Cu^+ + e^- \longrightarrow Cu(s)$	+0.52	
	$Cu^{2+} + 2e^- \longrightarrow Cu(s)$	+0.34	
increasing strength as oxidizing agent	$SO_4^{2-} + 4H^+ + 2e^- \longrightarrow SO_2(g) + 2H_2O$	+0.17	
	$Cu^{2+} + e^- \longrightarrow Cu^+$	+0.15	
	$Sn^{4+} + 2e^- \longrightarrow Sn^{2+}$	+0.15	
	$S + 2H^+ + 2e^- \longrightarrow H_2S(g)$	+0.14	
↑	$2H^+ + 2e^- \longrightarrow H_2(g)$	0.00	
	$Pb^{2+} + 2e^- \longrightarrow Pb(s)$	−0.13	
	$Sn^{2+} + 2e^- \longrightarrow Sn(s)$	−0.14	
	$Ni^{2+} + 2e^- \longrightarrow Ni(s)$	−0.25	
	$Co^{2+} + 2e^- \longrightarrow Co(s)$	−0.28	
	$Se + 2H^+ + 2e^- \longrightarrow H_2Se(g)$	−0.40	
	$Cr^{3+} + e^- \longrightarrow Cr^{2+}$	−0.41	
	$2H^+ (10^{-7} M) + 2e^- \longrightarrow H_2(g)$	−0.41	
weak oxidizing agents	$Fe^{2+} + 2e^- \longrightarrow Fe(s)$	−0.44	
	$Ag_2S + 2e^- \longrightarrow 2Ag(s) + S^{2-}$	−0.69	
↑	$Te + 2H^+ + 2e^- \longrightarrow H_2Te(g)$	−0.72	
	$Cr^{3+} + 3e^- \longrightarrow Cr(s)$	−0.74	
	$Zn^{2+} + 2e^- \longrightarrow Zn(s)$	−0.76	
	$2H_2O + 2e^- \longrightarrow 2OH^- + H_2(g)$	−0.83	
	$Mn^{2+} + 2e^- \longrightarrow Mn(s)$	−1.18	
	$Al^{3+} + 3e^- \longrightarrow Al(s)$	−1.66	
	$Mg^{2+} + 2e^- \longrightarrow Mg(s)$	−2.37	
	$Na^+ + e^- \longrightarrow Na(s)$	−2.71	
	$Ca^{2+} + 2e^- \longrightarrow Ca(s)$	−2.87	
	$Sr^{2+} + 2e^- \longrightarrow Sr(s)$	−2.89	
	$Ba^{2+} + 2e^- \longrightarrow Ba(s)$	−2.90	↓
	$Cs^+ + e^- \longrightarrow Cs(s)$	−2.92	
very weak oxidizing agents	$K^+ + e^- \longrightarrow K(s)$	−2.92	very strong reducing agents
	$Rb^+ + e^- \longrightarrow Rb(s)$	−2.92	
	$Li^+ + e^- \longrightarrow Li(s)$	−3.00	

NAMES, FORMULAS, AND CHARGES OF SOME COMMON IONS

POSITIVE IONS (CATIONS)

Name	Symbol	Name	Symbol
Aluminum	Al^{3+}	Lithium	Li^+
Ammonium	NH_4^+	Magnesium	Mg^{2+}
Barium	Ba^{2+}	Manganese(II), manganous	Mn^{2+}
Calcium	Ca^{2+}	Mercury(I),* mercurous	Hg_2^{2+}
Chromium(II), chromous	Cr^{2+}	Mercury(II), mercuric	Hg^{2+}
Chromium(III), chromic	Cr^{3+}	Potassium	K^+
Copper(I),* cuprous	Cu^+	Silver	Ag^+
Copper(II), cupric	Cu^{2+}	Sodium	Na^+
Hydrogen, hydronium	H^+, H_3O^+	Tin(II),* stannous	Sn^{2+}
Iron(II),* ferrous	Fe^{2+}	Tin(IV), stannic	Sn^{4+}
Iron(III), ferric	Fe^{3+}	Zinc	Zn^{2+}
Lead	Pb^{2+}		

NEGATIVE IONS (ANIONS)

Name	Symbol	Name	Symbol
Acetate	CH_3COO^-	Hydrogen oxalate ion, binoxalate	$HC_2O_4^-$
Bromide	Br^-	Perchlorate	ClO_4^-
Carbonate	CO_3^{2-}	Permanganate	MnO_4^-
Hydrogen carbonate ion, bicarbonate	HCO_3^-	Phosphate	PO_4^{3-}
Chlorate	ClO_3^-	Monohydrogen phosphate	HPO_4^{2-}
Chloride	Cl^-	Dihydrogen phosphate	$H_2PO_4^-$
Chlorite	ClO_2^-	Sulfate	SO_4^{2-}
Chromate	CrO_4^{2-}	Hydrogen sulfate ion, bisulfate	HSO_4^-
Dichromate	$Cr_2O_7^{2-}$	Sulfide	S^{2-}
Fluoride	F^-	Hydrogen sulfide ion, bisulfide	HS^-
Hydroxide	OH^-	Sulfite	SO_3^{2-}
Hypochlorite	ClO^-	Hydrogen sulfite ion, bisulfite	HSO_3^-
Iodide	I^-		
Nitrate	NO_3^-		
Nitrite	NO_2^-		
Oxalate	$C_2O_4^{2-}$		

* Aqueous solutions are readily oxidized by air.

NOTE: In ionic compounds the relative number of positive and negative ions is such that the sum of their electric charges is zero.

Appendix 8

Appendix 9

ATOMIC WEIGHTS ACCEPTED BY THE INTERNATIONAL UNION OF PURE AND APPLIED CHEMISTRY

Name	Symbol	Atomic Number	Atomic Weight
Actinium	Ac	89	—
Aluminium	Al	13	26.9815
Americium	Am	95	—
Antimony	Sb	51	121.75
Argon	Ar	18	39.948
Arsenic	As	33	74.9216
Astatine	At	85	—
Barium	Ba	56	137.34
Berkelium	Bk	97	—
Beryllium	Be	4	9.0122
Bismuth	Bi	83	208.980
Boron	B	5	10.811
Bromine	Br	35	79.904
Cadmium	Cd	48	112.40
Calcium	Ca	20	40.08
Californium	Cf	98	—
Carbon	C	6	12.01115
Cerium	Ce	58	140.12
Cesium	Cs	55	132.905
Chlorine	Cl	17	35.453
Chromium	Cr	24	51.996
Cobalt	Co	27	58.9332
Copper	Cu	29	63.546
Curium	Cm	96	—
Dysprosium	Dy	66	162.50
Einsteinium	Es	99	—
Erbium	Er	68	167.26
Europium	Eu	63	151.96
Fermium	Fm	100	—
Fluorine	F	9	18.9984
Francium	Fr	87	—
Gadolinium	Gd	64	157.25
Gallium	Ga	31	69.72
Germanium	Ge	32	72.59
Gold	Au	79	196.967
Hafnium	Hf	72	178.49
Helium	He	2	4.0026
Holmium	Ho	67	164.930
Hydrogen	H	1	1.00797
Indium	In	49	114.82
Iodine	I	53	126.9044
Iridium	Ir	77	192.2
Iron	Fe	26	55.847
Krypton	Kr	36	83.80
Lanthanum	La	57	138.91
Lawrencium	Lr	103	—
Lead	Pb	82	207.19
Lithium	Li	3	6.939

ATOMIC WEIGHTS

Name	Symbol	Atomic Number	Atomic Weight
Lutetium	Lu	71	174.97
Magnesium	Mg	12	24.305
Manganese	Mn	25	54.9380
Mendelevium	Md	101	—
Mercury	Hg	80	200.59
Molybdenum	Mo	42	95.94
Neodymium	Nd	60	144.24
Neon	Ne	10	20.179
Neptunium	Np	93	—
Nickel	Ni	28	58.71
Niobium	Nb	41	92.906
Nitrogen	N	7	14.0067
Nobelium	No	102	—
Osmium	Os	76	190.2
Oxygen	O	8	15.9994
Palladium	Pd	46	106.4
Phosphorus	P	15	30.9738
Platinum	Pt	78	195.09
Plutonium	Pu	94	—
Polonium	Po	84	—
Potassium	K	19	39.102
Praseodymium	Pr	59	140.907
Promethium	Pm	61	—
Protactinium	Pa	91	—
Radium	Ra	88	—
Radon	Rn	86	—
Rhenium	Re	75	186.2
Rhodium	Rh	45	102.905
Rubidium	Rb	37	85.47
Ruthenium	Ru	44	101.07
Samarium	Sm	62	150.35
Scandium	Sc	21	44.956
Selenium	Se	34	78.96
Silicon	Si	14	28.086
Silver	Ag	47	107.868
Sodium	Na	11	22.9898
Strontium	Sr	38	87.62
Sulfur	S	16	32.064
Tantalum	Ta	73	180.948
Technetium	Tc	43	—
Tellurium	Te	52	127.60
Terbium	Tb	65	158.924
Thallium	Tl	81	204.37
Thorium	Th	90	232.038
Thulium	Tm	69	168.934
Tin	Sn	50	118.69
Titanium	Ti	22	47.90
Tungsten	W	74	183.85
Uranium	U	92	238.03
Vanadium	V	23	50.942
Xenon	Xe	54	131.30
Ytterbium	Yb	70	173.04
Yttrium	Y	39	88.905
Zinc	Zn	30	65.37
Zirconium	Zr	40	91.22

Appendix 10

REACTION RATES AND THE EQUILIBRIUM CONSTANT

The kinetic model for equilibrium demands that the rate of the forward reaction be equal to the rate of the reverse reaction. For the process

$$2A + B \rightleftarrows C + 3D \tag{1}$$

we write the equilibrium constant as

$$K = \frac{[C][D]^3}{[A]^2[B]} \tag{2}$$

The kinetic model for the equilibrium constant would suggest that if we measured the rate of the reaction between A and B and made it equal to the measured rate of the reverse reaction between C and D we could obtain the equilibrium constant. This would be true if the measured rates were

$$\text{rate}_{forward} = k_1[A]^2[B] \tag{3}$$

$$\text{rate}_{reverse} = k_2[C][D]^3 \tag{4}$$

Then

$$k_1[A]^2[B] = k_2[C][D]^3 \tag{5}$$

or

$$k = \frac{k_1}{k_2} = \frac{[C][D]^3}{[A]^2[B]} \tag{6}$$

Suppose, however, that measurements of rate gave expressions which were quite different from those shown above. Suppose the rate of the forward reaction was expressed as

$$\text{rate}_{forward} = k_1[A][B] \tag{7}$$

Suppose also that measurements showed that the rate of the reverse reaction was expressed as

$$\text{rate}_{reverse} = k_2[C][D] \tag{8}$$

Equating these two rate expressions would *not* give the proper expression for the equilibrium constant. Indeed it is usually found that a proper equilibrium constant is *not* obtained when *measured* rates of forward and reverse reactions are made equal to each other. How is this apparent paradox explained? Let us consider a specific example.
Consider the process

$$H_2 + 2ICl \rightleftarrows I_2 + 2HCl \tag{9}$$

The *experimentally measured* rate for the forward process is

$$\text{rate}_{forward} = k_1[H_2][ICl] \tag{10}$$

The *experimentally measured* rate for the reverse process is

$$\text{rate}_{reverse} = k_2[\text{I}_2][\text{HCl}] \qquad (11)$$

If we equate these two rates, we obtain the expression

$$k_1[\text{H}_2][\text{ICl}] = k_2[\text{I}_2][\text{HCl}] \qquad (12)$$

This leads to an equilibrium constant expression of

$$K = \frac{[\text{I}_2][\text{HCl}]}{[\text{H}_2][\text{ICl}]} \qquad (13)$$

The above expression is clearly wrong; it should be

$$K = \frac{[\text{I}_2][\text{HCl}]^2}{[\text{H}_2][\text{ICl}]^2} \qquad (14)$$

The explanation for the incorrect expression, (13), is found when we consider the mechanism for the process in some detail.

Assume that the process proceeds by a two-step mechanism:

Step 1	$\text{H}_2 + \text{ICl} \rightleftharpoons \text{HI} + \text{HCl}$	SLOW	(15)
Step 2	$\text{HI} + \text{ICl} \rightleftharpoons \text{I}_2 + \text{HCl}$	FAST	(16)
Overall	$\text{H}_2 + 2\text{ICl} \rightleftharpoons \text{I}_2 + 2\text{HCl}$		(9)

Let us first consider the equation for Step 1. The expected and *measured* rates for the forward and reverse reactions are

$$\text{rate}_{forward} = k_1[\text{H}_2][\text{ICl}] \qquad (10)$$

$$\text{rate}_{reverse} = k_2[\text{HI}][\text{HCl}] \qquad (11)$$

If these two equations are combined to obtain the equilibrium constant for reaction (15), Step 1, we obtain

$$K_{Step\ 1} = \frac{k_1}{k_2} = \frac{[\text{HI}][\text{HCl}]}{[\text{H}_2][\text{ICl}]} \qquad (17)$$

In like manner the rates and equilibrium constant for reaction (16), Step 2,

$$\text{HI} + \text{ICl} \rightleftharpoons \text{I}_2 + \text{HCl} \qquad (16)$$

can be written

$$\text{rate}_{forward} = k_3[\text{HI}][\text{ICl}] \qquad (18)$$

$$\text{rate}_{reverse} = k_4[\text{HI}][\text{ICl}] \qquad (19)$$

$$K_{Step\ 2} = \frac{k_3}{k_4} = \frac{[\text{I}_2][\text{HCl}]}{[\text{HI}][\text{ICl}]} \qquad (20)$$

Since we have a valid constant for Step 1 and another for Step 2, a third constant should be obtained if these two are multiplied. We can write

$$K_{Step\ 1} \times K_{Step\ 2} = \frac{[\cancel{\text{HI}}][\text{HCl}]}{[\text{H}_2][\text{ICl}]} \times \frac{[\text{I}_2][\text{HCl}]}{[\cancel{\text{HI}}][\text{ICl}]} \qquad (21)$$

The values for [HI] reduce to 1 and we obtain

$$K_{Step\ 1} \times K_{Step\ 2} = \frac{[I_2][HCl]^2}{[H_2][ICl]^2} \qquad (21a)$$

You will now note that the above expression is the correct expression for the equilibrium constant of the overall process

$$H_2 + 2ICl \rightleftharpoons I_2 + 2HCl \qquad (9)$$

$$K_{Overall} = \frac{[I_2][HCl]^2}{[H_2][ICl]^2} \qquad (14)$$

Thus, by applying the ideas of kinetics to *each step* of the process, the correct overall equilibrium expression can be obtained.

Appendix 11

AN EXAMPLE OF A LEWIS ACID-BASE INTERACTION NOT INVOLVING A PROTON

The fluoroborate ion is an example of a Lewis acid-base interaction not involving a proton. We start with fluoride ion which can be written as

$$:\!\ddot{F}\!:^{-}$$

where F represents the fluorine nucleus plus two electrons in the first electron level. Each dot represents an electron. Remember that this gives an *ion* with the same electron configuration as neon. The compound boron trifluoride (BF_3) can be represented as

$$\begin{array}{c} :\ddot{F}: \\ B\!:\!\ddot{F}: \\ :\ddot{F}: \end{array}$$

where B represents the boron nucleus plus the two electrons of the first electron level and F again represents the fluorine nucleus plus two electrons. We see immediately that boron in BF_3 has only six electrons around it rather than the eight electrons characteristic of a stable configuration for elements in the second row of the periodic table. We wonder: Can the fluoride *ion* share two of its electrons with the boron of BF_3 to give a more stable structure? The resulting structure would be

$$\left[\begin{array}{c} :\ddot{F}: \\ :\ddot{F}\!:\!B\!:\!\ddot{F}: \\ :\ddot{F}: \end{array}\right]^{-}$$

This is just what happens; this complex negative ion known as the fluoroborate ion forms. Of course, once the bond between the fluoride ion and boron is formed, all fluorines become equivalent. There is no way to tell which fluorine originated in the fluoride ion.

The fluoroborate ion can be pictured as a tetrahedral arrangement of four F^- ions around a central B^{3+} ion. The written representation is

$$\left[\begin{array}{c} :\ddot{F}: \\ :\ddot{F}\!:\!B\!:\!\ddot{F}: \\ :\ddot{F}: \end{array}\right]^{-}$$

In this process the fluoride ion is an electron-pair donor. By the Lewis definitions, F^- is a base. The boron of BF_3 *accepts* the electron pair. We say that BF_3 is an acid; it is an electron-pair acceptor. Using this definition, the concept of acids and base can be extended to systems not involving protons.

Acknowledgments

FOR PHOTOGRAPHS

COVER AND TITLE PAGE   Drs. Cagnet, Françon, and Mallick, Laboratoire d'Optique, Faculté des Sciences de Paris, France.

CHAPTER ONE   Page 2 Swiss National Tourist Office.   Page 5 (top) Stephen W. Mulvey.   Page 5 (bottom) Esso Research and Engineering Co.   Page 9 The Department of the Interior, National Park Service. Photos by John M. Kauffmann.   Page 17 (left) General Mills.   Page 17 (right) Zimble from Monkmeyer.

CHAPTER TWO   Page 20 C. L. Martonyi, Photographic Services, The University of Michigan.   Pages 22–23 Jerrold J. Stefl.   Page 45 National Bureau of Standards.

CHAPTER THREE   Page 50 Basil King Photo.

CHAPTER FOUR   Page 66 Charles E. Miller.   Page 70 C. L. Martonyi, Photographic Services, The University of Michigan.

CHAPTER FIVE   Page 92 Kenneth Hill.   Page 110 Corning Glass Works.

CHAPTER SIX   Page 114 "Ascending and Descending" and "Waterfall," M. C. Escher, Escher Association.

CHAPTER SEVEN   Page 138 Brookhaven National Laboratory.   Page 146 U. S. Atomic Energy Commission.

CHAPTER EIGHT   Page 160 Sovfoto.   Pages 170–171 Neil Bartlett.   Page 186 The Bettmann Archive Inc.

CHAPTER NINE   Page 190 Rapho Guillumette Pictures (by Gerry Granham).   Page 198 The Bancroft Library, University of California, Berkeley.

CHAPTER TEN   Page 210 E. R. Degginger.   Page 217 Henry Eyring.

CHAPTER ELEVEN   Page 234 Jerrold J. Stefl. Automobiles courtesy of Eldon Industries, Inc.   Pages 236–237, 239–240, 244 C. L. Martonyi, Photographic Services, The University of Michigan.   Page 246 Alburtus, Yale News Bureau.   Page 257 (left) Aspen Corporation.   Page 257 (right) Oregon State Highway Department.   Pages 259–261 C. L. Martonyi, Photographic Services, The University of Michigan.

CHAPTER TWELVE   Page 266 Stockpile, New York City.   Page 267 New Mexico Department of Development.

ACKNOWLEDGEMENTS

CHAPTER THIRTEEN Page 288 Jerrold J. Stefl. Pages 294, 306 C. L. Martonyi, Photographic Services, The University of Michigan. Page 311 American Swedish News Exchange.

CHAPTER FOURTEEN Page 318 C. L. Martonyi, Photographic Services, The University of Michigan. Page 333 Cornell University.

CHAPTER FIFTEEN Page 356 Rapho Guillumette Pictures (by Hanna Schreiber). Page 359 C. L. Martonyi, Photographic Services, The University of Michigan. Page 360 Lockwood, Kessler, Bartlett. Page 367 Charles L. Finance. Page 371 C. L. Martonyi, Photographic Services, The University of Michigan. Pages 384–385 Don T. Cromer, University of California, Los Alamos Scientific Laboratory, and W. T. Lippincott, Editor, *Journal of Chemical Education*. Reproduced from *J. Chem. Ed.*, 45, 626 (1968). Page 399 Luis W. Alvarez.

CHAPTER SIXTEEN Page 402 Jerrold J. Stefl. Page 417 The California Institute of Technology. Page 418 C. L. Martonyi, Photographic Services, The University of Michigan.

CHAPTER SEVENTEEN Page 434 Rapho Guillumette Pictures (by Bruce Roberts). Page 448 The American Chemical Society.

CHAPTER EIGHTEEN Page 460 C. L. Martonyi, Photographic Services, The University of Michigan. Page 493 Fabian Bachrach.

CHAPTER TWENTY-TWO Page 550 Rapho Guillumette Pictures (by Marc & Evelyne Bernheim). Page 551 M. W. Nirenberg. Page 564 M. H. C. Wilkins. Page 565 J. Griffith, The California Institute of Technology. Page 566 H. Fernandez-Moran. Page 566 United Press International and L. A. Kornberg.

CHAPTER TWENTY-THREE Pages 568, 574–575, 577 National Aeronautics and Space Administration.

APPENDIX ONE Page 584 Jerrold J. Stefl.

Index

A

Absolute temperature, 81
  absolute zero, 82
  Kelvin scale, 81
Accuracy and precision, 16
Acetaldehyde, 477, 496
Acetamide, 483, 484
Acetanilide, 490
Acetate group, 491
Acetic acid:
  K_A for, 306
  structural formula for, 478
Acid-base:
  Brønsted-Lowry theory of, 314
  definition of, 300, 315
  indicators, 304
  Lewis theory of, 315, 605
  proton transfer concept in, 312
  titrations, 302
Acidic character of third-row elements, 510
Acids:
  carboxylic, 477, 481, 482
  charge rule for, 513
  complex ions as, 539–540
  definitions of, 300
  derivatives of, 481, 482
  equilibrium constants of, K_A, 307
  experimental identification of, 297
  model for properties of, 298
  organic, 484
  reactions with water, 275
  relative strengths of, table, 597
  solutions of, 176
  strengths of, 305
Actinides, 529, 532
Activated complex, 224, 474, 475
Activation energy, 222, 225, 231
Active centers, enzyme, 559
Addition polymerization, 493
Additivity of heats of reaction, 194
Adenine, 564–565
Adipic acid, 494
Affinity, electron, 407
Air:
  composition of, 570
  pressure volume behavior of, 22–26
Alanine, 495
Alcohol dehydrogenase, 558
Alcohols:
  chemistry of, 473
  definition, 473
  ethanol, 462, 465, 467
  ethyl, 462
  nomenclature of, 484
  oxidation of, 475, 480
Alkali metals:
  chemical reactions of, 171–173
  chemistry, summary, 174
  compounds of, solubility, 276
  crystal structure of, 445
  ionization energies of, 388
  physical properties of, 171
Alkaline earth metals:
  electron configurations of, 519
  ionic radii, 524, 525
  ionization energies of, 520
  metallic radii, 523, 525
  properties of, 521
Alkanes, 485
  nomenclature of, 484
  properties of some, 486
  structure of, 485
Alkyl group, 481
Alpha particle, 152
Aluminum:
  abundance in Earth's crust, 572
  abundance on lunar surface, 574
  crystal structure of, 447
  electron arrangement, 167
  electron configuration, 392, 447
  first ionization energy, 162
  heat of vaporization, 444, 505
  ionization energies, 501
  manufacture of, 131
  properties of, 500
  as reducing agent, 505, 507
  X-ray spectrum of, 595
Aluminum hydroxide, 511
Alvarez, Luis W., 399
American Chemical Society, 18
Amides:
  chain of, 495
  nomenclature of, 484
  polymeric, 494
Amines, 481
  aromatic, 490
  nomenclature of, 484
  preparation of, 481
Amino acids, 494
Ammonia, 183
  boiling point, 96
  burning of, 59, 60
  combining volumes of, 29, 38
  in complex ions, 537
  Haber process for, 249–251
  heat of combustion, 197
  heat of formation, 196
  liquefaction of, 94
  manufacture of, 249–251
  molar volume of, 69–70, 93
  molecular geometry of, 422
  molecular models of, 41
  in planetary atmospheres, 579–580
  pressure-volume data of, 27, 69, 93
  reaction with water, 299
Ammonium ion, model for, 294
Ampère, André, 30
Ampere, 133
Amphoteric complexes, 540
Amphoteric substance, 512
Amplitude, 360
Analogy; see also Model and Theory:
  balloon, 21
  billiard ball, 198
  bird feeder, 383
  blender, 261
  bookcase, 374
  dancing couple, 243–245, 294, 308
  dozen eggs, 51
  earthworm, 3
  garbage collector, 115
  Martin the Martian, 6
  pole-vaulting, 222
  plum pudding, 141
  Sherlock Holmes, 4
  "super-rubber" ball, 21, 93, 100, 218, 365
  traffic flow, 234
Anders, William, 568, 574
Angstrom, 142
Aniline, 490
Anions:
  and aqueous solubilities, 276
  in electrochemical cells, 324
  names, formulas, and charges of, 599
Anode:
  definition, 276, 323–324
  in electrolysis, 349
Apollo 8, 568
Aq notation, 124
Aqueous, 124
Aqueous solutions, 123, 274–280
  of acids, 297
  of bases, 298
  electrical conductivity of, 125
  nature of, 291
  precipitation reactions in, 282
Argon, 168
  abundance in dry air, 570
  boiling point, 436
  electron arrangement, 167
  electron configuration, 391
  first ionization energy, 162
  heat of vaporization, 188, 505
  ionization energy, 169
  melting point, 436
  properties of, 169
Aromatic compounds, 487
Arrhenius, Svante A., 311
Aspirin, 491
Astatine, 175
Astrochemistry, 568
Atmosphere, 569
  composition of:
    earth, 570
    planetary, 578–580
  infrared light in, 570
  ozone in, 570
  standard, 75
  stellar, 581
  ultraviolet light in, 570
Atmospheric pressure:
  measurement of, 72–73
  of other planets, 578–580
Atomic mass unit, 34
Atom, 32–47, 117–154, 369–399
  chemical evidence for existence of, 117
  components of, 139
  definition, 37
  energy levels of, 387
  hydrogen; see Hydrogen, atom
  ionization energy of, 161
  many electron, 386

Atom (cont'd.)
  model of;
    Bohr, 379–380
    classical, 377
    electrical, 125
    mechanical, 376, 379
    nuclear, 140
    quantum, 374, 379
    Rutherford, 152–154, 377
    Thomson, 140, 153
    wave, 380–382
  radius of, 142, 522–526
  size of, 142, 522–526
  velocity of; see Atomic velocity
Atomic nucleus; see Nucleus
Atomic number, 141, 146, 154
  determination of, 595
  of elements, 184–185, 600–601
  and energy levels, 397
  and ionization energy, 162, 165
Atomic size, 142, 522–526
  alkaline earth metals, 523, 525
  halogens, 522
  transition elements, 545
Atomic theory, 20; see also Atom:
  chemical evidence for, 117
  as fundamental postulate, 41
  and quantitative relationships, 51
Atomic velocity:
  effect of temperature change on, 221
  and kinetic energy, 221
  measurement of, 219
Atomic volume:
  alkali metals, 171
  halogens, 175
Atomic weight:
  definition, 43
  table of, 44, 600–601
  trend in transition elements, 543
Atomic weight scale, 140
Atoms:
  belief in, 115
  chemical implications of, 40
  interaction between, 404
Avery, Oswald, 561
Avogadro, Amedeo, 30
Avogadro's hypothesis, 30, 47, 115, 119, 132
  and kinetic theory, 86
  reliability of, 76
Avogadro's number, 34, 51, 53
Azo dyes, 490

B

Balanced equations, 54
  and mass balance, 56
  and the mole, 55
  by use of half-reactions, 340
  by use of oxidation numbers, 347
  writing, 54–55
Barium:
  abundance in Earth's crust, 572
  electron configuration, 519
  heat of vaporization, 444, 521
Barium chloride, flame test for, 358
Barium chromate, K_{sp}, 280
Barium sulfate, K_{sp}, 280

Barometer, 73
Bartell, L. S., 418, 456
Bartlett, Neil, 8, 170
Bases:
  definition, 300
  experimental identification, 298
  model for properties, 298
Basic character of third-row elements, 510
Basic oxygen furnace, 548
Bent, H. S., 418
Benzene, 10, 488
  derivatives, 490–492
  structure, 488
  substitution reactions, 489, 490
  vapor pressure, 100, 101
Benzene ring, 488
Benzoic acid:
  determination of K_A, 307
  solubility of, 309
Beryllium, 180
  bonding capacity of, 414
  chemistry of, 182
  compared with magnesium, 499
  electron arrangement, 167
  electron configuration, 388, 414, 519
  heat of vaporization, 444, 505, 521
  ionization energies, 162, 164, 398
  properties of, 181
Beryllium difluoride, molecular dipole in, 424
Berzelius, J. J., 30
Beta decay, 156
Beta particle, 156
Bidentate complex ions, 538
Binary compound, 46
Binding energy, 206
Biochemistry, 551
Bird feeder analogy, 383
Bisulfate ion, 514
Bisulfite ion, 514
Blender analogy, 261
Body-centered cubic packing, 445
Bohr, Niels, 167, 379–381, 434, 596
Bohr atom, 379
  and Mosley's experiment, 596
  orbits of, 379
  shortcomings of, 380
Boiling point:
  of alkali metals, 171
  of CX_4 molecules, 437
  of common gases, 96
  of fourth-row transition elements, 543
  of halogens, 175
  of hydrides, 454
  and molar heat of vaporization, 98
  of noble gases, 169
  of noble gases and halogens, compared, 436
  normal, defined, 104
  of pure substances, 98
  of saturated hydrocarbons, 486
  of second and third-row elements, compared, 505

  and vapor pressure, 103
Bond, chemical:
  angle;
    in BeF_2, 422
    in BF_3, 422
    in CF_4, 422
    in CH_4, 422
    in F_2O, 422
    in H_2O, 420
    in NF_3, 422
    in NH_3, 422
  covalent; see Covalent bond
  covalent and ionic, compared, 416
  dipole, 433
  double, 426, 433
  effect on reaction rate, 212
  energy of, 202
  hydrogen; see Hydrogen bond
  ionic; see Ionic bond
  representation of;
    by electron dot formula, 407
    by orbital notation, 406
    by structural formula, 38–40
  single, 426, 433
  stability, 403
  symbol for, 38–40
  three-center, 450
  trends of, in second-row fluorides, 415
  triple, 429, 433
  types of, in flourine compounds, 417
Bonding:
  in complex ions, 539
  in boron, 449
  dsp^2, 539
  d^2sp^3, 539
  electron dot representation of, 406
  in fluorine molecules, 407
  in gaseous lithium fluoride, 415
  in hydrogen flouride, 408
  in hydrogen peroxide, 410
  in hydroxyl ion, 410
  metallic, 441, 443
  in molecular hydrogen, 406
  orbital representation of, 406
  sp, 423
  sp^2, 423, 439
  sp^3, 423, 438
  in water, 410, 419–421
Bonding capacity, 409
  of beryllium atoms, 414
  of boron atoms, 413
  of carbon atoms, 412
  of lithium atoms, 414
  of nitrogen atoms, 411
  of oxygen atoms, 409
Bonding patterns in second-row elements, 186
Bookcase analogy, 374
Borman, Frank, 568, 574
Boron, 180–182, 448–451
  bonding capacity, 413
  electron arrangement, 167
  electron configuration, 388, 413
  first ionization energy, 162
  heat of vaporization, 444, 505
  hydrides, 450, 451

INDEX

ionization energies of, 398
isotopes, composition of, 144
physical properties compared to aluminum, 500
properties of, 181
structure of, 448
Boron triflouride:
  as example of Lewis acid-base interaction, 605
  molecular dipole of, 425
  molecular geometry of, 422
Brackets, [ ], meaning of, 251
Bromination, 489
Bromine:
  abundance in ocean water, 571
  atomic radius, 142
  boiling point, 436
  color, 521
  covalent radius, 522
  electrolysis of, 352
  electron configuration, 521
  internuclear distance, 142
  ionic radius, 524, 525
  ionization energy, 521
  melting point, 436
  properties of, 175
  reaction with hydrogen, 176
  reaction with sodium, 175
  van der Waals radius, 522
Bromobenzene, 489
Bromoethane; see Ethyl bromide
Bromthymol blue, 305
Brønsted, J. N., 314
Brønsted-Lowry theory of acids and bases, 314
Bronze Age, 543
Bubble formation, 103
Butane, 484
iso-Butane, properties, 486
n-Butane, properties, 486
1-Butanol, 484
Butyl alcohol, 484
1-Butylamine, 484
Butyric acid, 484
Butyramide, 484

C

c, velocity of light, 361, 369
Cady, George, 170
Calcium:
  abundance in Earth's crust, 572
  abundance on lunar surface, 574
  electron arrangement, 167
  electron configuration, 391, 519, 529
  first ionization energy, 162
  heat of vaporization, 444, 521
  ionic radius, 523–525
  ionization energies, 520
  metallic radius, 523
  orbital notation, 529
  physical properties of, 521
  X-ray spectrum of, 595
Calcium carbonate:
  in cave formations, 267, 283
  in coral reefs, 283
  equilibrium with calcium oxide, 241
  in oyster shells, 283
Calcium chloride:
  flame test for, 358
  naming, 46
Calcium hydroxide dissociation in water, 299
Calcium oxide, equilibrium with calcium carbonate, 241
Calcium sulfate:
  K_{sp}, 280
  solubility of, 283
Calculations:
  based on chemical equations, 59
  of equilibrium constant, 252–254
  of H^+ concentration in an acid solution, 309–311
  of heat of reaction, 196
  with moles, 51
  of solubility of CuCl in H_2O, 281
  of solubility product, 280
Calorie, 61
Cancer, chemical treatment of, 567
Candle, description of burning, 584
Caprylamide, 484
Caprylic acid, 484
Carbohydrate, definition and origin of word, 46
Carbohydrates, 552
Carbon, 181–182
  abundance in Earth's crust, 572
  abundance on lunar surface, 574
  boiling point, 181
  bonding capacity of, 412
  carbon-carbon double bond, 426
  compounds, 461–497
    composition of, 462
    sources of, 461
    structure of, 462
  covalent radius, 500, 522
  electron arrangement, 167
  electron configuration, 388, 412
  first ionization energy, 162
  heat of combustion, 192, 195, 196
  heat of vaporization, 505
  ionization energies, 501
  isotopes, composition of, 144
  melting point, 181
  properties of, 181
  structure of, 438
    diamond, 187, 438, 500
    graphite, 438–440
Carbon dioxide:
  abundance in dry air, 570
  at equilibrium with $CaCO_3$ and CO, 241
  formula, 46
  heat of formation, 196
  molar volume, 35, 71
  molecular motions, 201
  pressure-volume data, 27
  reaction with nitric oxide, 224
  relative weight, 31
  solubility in water, 571
Carbon disulfide, infrared spectrum, 431
Carbon monoxide:
  at equilibrium with $CaCO_3$ and CO_2, 241
  formula, 46
  heat of combustion, 192, 197
  heat of formation, 196
  molar volume, 35, 71
  poisoning, 542
  reaction with nitrogen dioxide, 211, 224
Carbon tetrabromide:
  boiling point, 437
  internuclear distance, 142, 522
  melting point, 437
Carbon tetrachloride:
  boiling point, 104, 437
  infrared spectrum, 431
  internuclear distance, 142, 522
  melting point, 437
  poison, 113
  solubility of iodine in, 271
  use in fire extinguishers, 113
  vapor pressure, 101
Carbon tetrafluoride:
  boiling point, 437
  dipole in, 425
  geometry of, 422
  internuclear distances, 522
  melting point, 437
Carbon tetraiodide:
  boiling point, 437
  internuclear distance, 522
  melting point, 437
Carbonates, solubility of, 278; *see also* Calcium carbonate
Carbonyl group, 479
Carboxyl group:
  bonding in, 481
  structure of, 478
Carboxylic acids:
  acid behavior, 481–482
  definition, 478, 496
  derivatives of, 482
  in fat formation, 556
  reaction with glycerol, 556
Catalysis, 227
Catalysts:
  acids as, 483
  action of, 227
  biological; *see* Enzymes
  chromium as, 547
  and complex formation, 542
  and decomposition of formic acid, 228
  definition, 227, 231
  enzymes, 231, 558; *see also* Enzymes
  and equilibrium, 245
  examples of, 229
  platinum as, 547
  transition elements as, 549
Cathode, 323, 324, 349
Cations:
  definition, 324

Cations (cont'd.)
  names, formulas, charges, 599
Cell:
  electrochemical, 319
  electrolytic, 129
  voltaic, 348
Cellulose, 552
  function in plants, 555
  structure, 557
Celsius temperature, 31
  definition, 75
  relation to °K, 81
Cerium:
  oxidation of, 532
  reaction rate of ion, 213
Cesium:
  crystal structure of, 445
  heat of vaporization, 444
  physical properties, 171
  reaction with chlorine, 172
Chain reaction, 203
Change, chemical, 36
Characteristics of chemical process, 62
Charge:
  conservation of;
    in chemical reactions, 135
    in nuclear reactions, 156
  electric, 120
  on electron, 149
  of fundamental particles, 140
  on nucleus, 146
  unit of, definition, 132
Charge/mass ratio of electron, 150, 589
Charles' law, 80
Charles, Jacques, 80
Chemical bonding; see Bonding and Bonding capacity
Chemical bonds; see Bond, chemical
Chemical change:
  definition, 36–37
  energy change in, 63
Chemical equation; see Equation, chemical
Chemical equilibrium; see Equilibrium
Chemical formula; see Formula, chemical
Chemical kinetics, 211
Chemical process, defining characteristic of, 62
Chemical reaction; see Reactions, chemical
Chemical symbols:
  origin of, 45
  periodic table of, 184–185
Chemical systems, equilibrium in, 238
Chemistry, definition, 2
Chemotherapy, 560
Chloric acid:
  strength, 514
  structure, 515
Chloride ion, hydration of, 243
Chlorides:
  of alkali metals, 172
  chemistry of, 176

  of second-row elements, 182
  solubilities of, 277
Chlorine:
  abundance in Earth's crust, 572
  abundance in ocean water, 571
  atomic volume, 175
  boiling point, 105, 175
  bonding in, 174
  color, 521
  combining volume of, 36
  compounded with chromium, 535
  covalent radius, 522
  dissociation of molecules, 174
  electron arrangement, 167
  electron configuration, 521
  electron dot representation, 521
  first ionization energy, 162, 521
  heat of melting, 105
  heat of solution in water, 274
  heat of vaporization, 98, 505
  hydroxides, 512, 514
  internuclear distance, 142
  ionic radius, 529
  melting point, 105, 175
  molar volume, 96
  occurrence in ocean water, 571
  oxidation states, 509
  as oxidizing agent, 509
  oxyacids, 512
  properties of, 175
  reaction with alkali metals, 172
  reaction with hydrogen, 36, 176
  reaction with sodium, 175
  van der Waals radius, 522
Chloroacetic acid, 497
Chlorobenzene, 491
Chlorophyll:
  as complex ion, 540
  structure, 541
Chlorous acid:
  strength, 514
  structure, 515
Chromate-dichromate equilibrium, 297
Chromate ion:
  oxidation number of chromium in, 533
  structure of, 546
Chrome plate, 546
Chromium:
  abundance in Earth's crust, 572
  amphoteric complexes, 540
  atomic size, 543, 545
  chemistry of, 546
  complex compounds, 533, 539, 540
  complex ions of, 547
  electron configuration of, 530
  oxidation states, 533, 534
  physical properties of, 543
  X-ray spectrum of, 595
Cis-isomers, 429
Coal, 461
Coal tar, 461
Cobalmin, 542
Cobalt:
  atomic size, 545
  complex ion of in humidity indicator, 536

  complex ions of, 536, 542
  electron configuration, 530
  oxidation numbers, 535
  physical properties of, 543
  X-ray spectrum of, 595
Coenzymes, 542
Coke, 191, 461
Collision theory:
  an application of, 218
  and concentration, 213
Collisions, elasticity of, 12
Color, 357–358
  of acid-base indicators, 304–306
  of chromium compounds, 533
  of diffraction jewelry, 359
  effect of concentration on, 216, 244, 246, 247
  of $FeSCN^{2+}$ solutions, 244
  of flame tests, 359
  of iodine in solution, 236
  of mercuric oxide, 32
  of methyl violet in water, 259
  of nitrogen dioxide, 211
  of nitrogen oxide bulb, 238–240
  of oil film, 359
  of platinum hexafluoride, 170
  of soap bubble, 359
  of solutions, 359
  of transition elements, 544
  and wavelength of light, 365–368
  of xenon hexafluoroplatinate, 170
Color television tube, 358
Colorimetric analysis, 251
Combining volumes, 36–38
Combining volumes, law of, 119
Combustion:
  of coke, 192
  of hydrogen, 54
  of magnesium, 55
  of methane, 57
  molar heat of, 61
    for hydrogen, 62
    for magnesium, 61
  of nitric oxide, 196–197
  of paraffin, 58
  of water gas, 192
  of wood, 214
Communication in science, 4, 17
Complex ions, 535
  acid-base behavior, 539
  ammonia in, 537
  amphoteric, 540
  bidentate, 538
  in biological processes, 540
  bonding in, 539
  chlorophyll, 540
  of chromium, 535, 547
  cobalmin, 542
  of cobalt, 536
  coordination number of, 536
  geometry of, 536
  hemin, 540
  hemoglobin, 540
  isomers of, 537
  linear, 538
  significance of, 539
  square planar, 538
Components of solution, 167

613

INDEX

Composition of dry air, 570
Composition of Earth's crust, 572
Composition of isotopes, 144
Composition of lunar surface, 574
Composition of ocean water, average, 571
Composition of solutions, 109
Compound, 42, 46
Concentration:
  in acid-base equilibria, 295–297
  effect on cell voltage, 331–332, 337–338
  effect on conductivity, 125
  effect on equilibrium, 244, 247
  effect on reaction rate, 213, 216
  and Le Chatelier's principle, 247
  molar, 110
  pH, 305
  and pressure, 248
  of solids in the equilibrium constant, 253
  of solutions, 109
  of standard state, 333
  symbol for, 251
  of water in the equilibrium constant, 255
Conceptual definition, 301, 315
Conclusions:
  and accuracy of observations, 13
  and observation, 115
Condensation polymerization, 493
Condensed phases:
  electrical properties, 123–125
  nature of, 104, 112
Conductivity, electrical:
  of alkali metals, 171
  of alkaline earth elements, 521
  of aqueous solutions, 123–125
  of graphite, 181
  of ionic crystals, 452
  and the metallic bond, 444
  of metals, 441
  of second-row elements, 181
  of solutions, 112, 123–125
  of strong and weak electrolytes, 289–290
  of third-row elements, 501
  of transition elements, 532
  of water, 290
Conservation of atoms:
  in chemical equation, 54
  in oxidation-reduction equations, 340–342
Conservation of charge, 135
  in nuclear equations, 156
  in oxidation-reduction equations, 340–342
Conservation of energy, 61, 198
  in billiard ball collisions, 198
  in a chemical reaction, 198–200
  law of, 63, 64, 200
  in storage of billiard balls, 198
Conservation of mass, 54
  in nuclear reactions, 156
  law of, 43, 63, 64
Constant heat summation, law of, 195

Coordination number, 536
Copper:
  atomic size, 545
  boiling point, 98
  electron configuration, 530
  heat of fusion, 105
  heat of vaporization, 98
  melting point, 105
  natural occurrence, 543, 572–573
  oxidation numbers, 534
  in oxidation-reduction reactions, 324–332, 336–337, 349–353
  properties, 543
  reference cell, 330–331
  standard reduction potential, 333
  X-ray spectrum of, 595
Copper(II)bromide, K_{sp}, 280
Copper(II)chloride:
  electrolysis, 127–129
  K_{sp}, 280
  solubility in water, 281
Copper(II)iodide, K_{sp}, 280
Copper(II)ion:
  color, 544
  in flame test, 358
Corrosion of iron, 548
Coulomb, 133, 148
Coulomb's Law, 158
Covalent bonds, 175, 178, 432
  in boron, 449
  in carbon tetrachloride, 185
  in complex ions, 539
  elements forming solids using, 441
  in fluorine, 407–408
  in fluorine compounds, 417
  in halogens, 174, 521
  in hydrogen, 403–405
  in hydrogen fluoride, 408–409
  ionic character in, 415–416
  in network solids, 187, 438
  in sulfur, 504
Covalent radius, 522, 526
  definition, 522
  of halogens, 522
Crawford, Bryce, Jr., 402
Crick, Francis H. C., 496, 562
Critical mass, 204
Cryolite, 131, 537
Crystal:
  ionic, 452
    properties of, 452
    stability, 452
  metal, 442
  molecular, 271, 435
Crystal packing; see Packing, crystal
Crystallization; see also Precipitation:
  definition, 107
  and equilibrium, 238
  rate of, 238
Cyclohexane:
  boiling point, 486
  heat of combustion, 486
  melting point, 486
Cyclopentane, 487
Cyclopentene, 487
Cytosine, 562, 564–565

D

d orbital, 385, 530
d^2sp^3 orbital, 539
dsp^2 orbital, 539
"Dacron," 494
Dalton, John:
  and atomic theory, 118
  and Avogadro's hypothesis, 30
  and conservation of mass, 54
  and formula for water, 136
  and scientific communication, 17
Dampier, Sir William Cecil, 50
"Dancing couple" analogy, 243–245, 294, 308
Data:
  extrapolation from, 165
  interpolation between points of, 165
  presentation of, 24
Davisson, C. J., 381
Davy, Humphrey, 18
D'Azyr, Vic, 498
De Broglie, Louis, 356, 380
De Broglie's matter-waves, 380
Debye, Peter J., 93, 333
Decomposition:
  of formic acid, 228
  of water, 63, 199
Definite composition, law of, 117
Definition; see also item of interest:
  conceptual, 301, 315
  operational, 301, 315
Delocalized electrons:
  in benzene rings, 389
  definition, 440
Delta H, 194, 208
Delta H‡, 226
Denkewalter, Robert G., 559
Dense, definition, 146
Density:
  of alkali earth metals, 521
  of alkali metals, 171
  of second-row elements, 181
  of third-row elements, 501
  of transition elements, 43
Deoxyadenosine-5'-phosphate, 563
Deoxycytidine-5'-phosphate, 563
Deoxyguanosine-5'-phosphate, 563
Deoxyribonucleic acid; see DNA
Deoxyribose, 562, 566
Deoxythymidine-5'-phosphate, 563
Deuterium, 143
Diagonal relationship, 502
1,6-Diaminohexane, 494
Diamond, 181, 183
  heat of combustion, 209
  properties of, 181
  structure, 438, 500
Diatomic molecules, 38
Diborane, 414

Diborane (cont'd.)
  determination of molecular weight, 33
  structure, 451
Dichloroethylene, isomers, 428
Dichromate ion, 546
Differential thermometer, 16
Diffraction:
  jewelry, 359
  of light, 359
  patterns, 430
  X-ray, 430
Dimer, 450
Dimethyl ether:
  nmr spectrum, 472
  structure, 469
Dinitrogen difluoride, 46
Dipole, bond, 433
Dipole:
  of BeF_2, 425
  of BF_3, 425
  of CF_4, 425
  of $CHCl_3$, 453
  of CO_2, 426
  electric, 416
  of F_2O, 426
  geometrical sum of, 424
  of HF, 417
  of H_2O, 426, 453
  of ionic bond, 416
  of LiF, 416
  of LiH, 417
  molecular, 423
  and molecular shape, 423
  of N—H bond in ammonia, 417
  of O—H bond in water, 417
  and solvent action, 453
Disaccharide, 554–555
Discharge tube, 147, 148
Dissolving:
  of LiCl in water, 291–294
  of NaCl in water, 453
  rate of, 268–269
Distance, effect of, on electrical attraction, 122
Dissociation:
  of electrolytes, 289
  of water, 290
Distillation, 107
Divalent atoms, 410, 426
DNA, 495, 554, 567
  definition, 554
  electron micrograph of, 565
  history, 561
  model, 564
  natural and synthetic, 566
  structure, 562–563
  template action, 564
Döbereiner, J. W., 186
Double arrow, use of, 240
Double bond, 426
Double helix, 562; see also DNA
Double Helix, The, 550
Doyle, Sir Arthur Conan, 4, 5, 92
Dyes, azo, 490
Dynamic nature of equilibrium, 268

E

e/m, 150, 589
$E = mc^2$, 205
E^0:
  definition, 333
  and equilibrium, 339
  of hydrogen half-cell, 333
  predicting reactions from, 336
  table, 598
Earth:
  age of, 573
  atmosphere of, 570
  chemistry of, 569
  crust of;
    composition of, 571
    transition elements in, 543
  parts of, 569
Einstein, Albert:
  relationship of mass to energy, 205, 208
  special theory of relativity, 205, 362
Elastic collisions, 12
Electric arc furnace, 548
Electric charges:
  detection of, 120
  and electrolysis, 130
  of electron, 150
  and Faraday's experiment, 130
  interaction of like, 121
  interaction of unlike, 121
  negative, 126
  net, 126
  neutral, 126
  in nuclear reactions, 156
  particulate nature of, 129
  positive, 126
  production of, 121
Electric current, 120
Electric dipole, 416, 433; see also Dipole
Electric discharge tube, 148
Electric field:
  in electromagnetic radiation, 362–364, 368, 399
  in oil drop experiment, 149
Electric force:
  effect of distance on, 122
  in electrometer, 122
  in oil drop experiment, 149
Electric motor, 363
Electrical conductivity; see Conductivity, electrical
Electrical nature of matter, 119
Electrical phenomena, 119
Electrical properties of condensed phases, 123
Electrochemical cell, 319
  chemistry of, 319
  examples, common, 319
  operation of, 323
  in opposition, 351, 354
  terminology of, 323
Electrode:
  defined, 127
  in electrochemical cell, 323, 324
  negative, 127
  positive, 127
Electrode processes, 127
Electrolysis, 348
  of copper chloride, 128, 322
  model for, 128
  and particulate nature of electric charge, 129
  of water, 63, 199
Electrolytes, 289
  in aqueous solutions, 275
  dissociation of, 290
  strong, 290
  weak, 290
Electrolytic cell, 348, 352
Electrolytic solutions, 275
Electromagnetic radiation, 364, 399
Electromagnetic spectrum, 365, 368
Electromagnetic waves, 368
Electrometer, 120
Electron:
  affinity, 407
  affinity of fluorine atom, 407
  arrangements;
    of first 20 elements, 167
    stable, 164
  charge on, 149
  charge/mass ratio, 589
  competition for, 326, 328
  configuration, 388
    alkaline earth metals, 519
    of elements; see individual elements
    halogens, 521
    and periodic table, 392
    orbital chart of, 531
    transition metals, 391, 530
  definition, 126
  delocalized, 440
  diffraction, 430
  energy of, 161
  excitation, 380
  extranuclear, 141, 144
  flow, 349
  kinetic energy, 161
  location, 384
  mass of, 150
  measurement of a mole of, 133
  and metallic properties, 443
  naming of, 127
  in nuclear reactions, 156
  promotion of, 412, 413, 414
  "sea" of, 187, 442
  "seeing," 147
  spin, 388
  trajectory, 384
  transfer, 328
  transition, 386
  valence, 396
  van der Waals forces and, 436
  volt, 162
  wave nature of, 142, 381
Electron dot notation, 406, 432
Electron micrograph of DNA, 565
Electronic structure and periodic table, 386
Electrostatic field strength, 591
Elements:

INDEX

availability of, 573
abundance in Earth's crust, 572
abundance in ocean water, 571
abundance on lunar surface, 574
definition, 42
discovery of, 42
names, 43, 114, 600
symbols, 43, 114, 600
Empirical formula:
   compared to molecular formula, 463
   definition, 173
Empty orbital, 413
Endothermic reaction, 63, 194, 208, 225
Energy:
   absorption of, 191
   activation, 222
   binding, 206
   changes;
      in chemical reactions, 61, 63, 191
      in phase changes, 63, 97
      on warming, 202
   content; *see* Enthalpy
   of covalent bonds, 455
   and electromagnetic radiation, 369
   and equilibrium, 256, 262
   of fats, 556
   of hydration, 293
   of hydrogen bond, 455
   and hydrogen spectrum, 369
   in ion formation, 145
   ionization; *see* Ionization energy
   kinetic; *see* Kinetic energy
   law of conservation of, 61, 63, 198, 200
   levels; *see* Levels, energy
   liberation of, 191
   of motion; *see* Kinetic energy
   of position; *see* Potential energy
   potential; *see* Potential energy
   quantitative relationships, 192
   and reaction rates, 222
   of rotation, 201
   and solubility, 271, 273
   stored in a molecule, 192, 201
   stored in a nucleus, 203
   of sugars, 555
   thermal, 79
   threshold, 221
   of translation, 201
   of van der Waals forces, 455
   of vibration, 201
Energy levels:
   and atomic number, 397
   of bookcase analogy, 374
   of hydrogen atom, 376, 385, 394
   and ionization energies, 393
   of many-electron atom, 387
   and the periodic table, 393
Enthalpy:
   change in, 194
   defined, 193
   and entropy, 262
   molar, 202
   molecular, 202
   tendency toward minimum, 262, 272, 285
   symbol for, 194
Entropy:
   and chemical equilibrium, 262
   defined, 259
   and solubility, 271
   and temperature, 260
Enzymes:
   active centers, 559
   as catalysts, 230
   and false substrates, 560
   importance of, 558
   inhibition of, 560
   reaction mechanism, 559
   specificity of, 559
   substrate, 559
   synthesis of, 559
Equations:
   balanced, and mass balance, 56
   balanced, defined, 54
   balancing ionic, 135
   calculations based on, 59
   chemical, 38
   conservation of charge in balanced, 135
   definition, 54
   writing balanced, 54
Equilibrium:
   altering, 244
   application of, 249
   in chemical systems, 238
   dynamic nature of, 269–270
   effect of catalyst on, 245
   effect of concentration on, 244
   effect of temperature on, 245, 246
   effect of volume changes on, 248
   and energy, 256, 262
   and entropy, 258–262
   E^0 and, 339
   factors determining, 256
   Haber process and, 249
   law of chemical, 252
   and Le Chatelier's principle, 246
   liquid-vapor, 98, 235
   and meaning of equations, 242
   in phase changes, 235
   quantitative aspects of, 250
   rate of approach to, 245
   and reaction rate, 236, 238, 262
   recognizing, 238
   solid-solutions, 237
   solubility, 267
   solubility in ionic solutions, 280
   sugar structure, 553
   thermal, 74, 80
Equilibrium constant:
   definition, 253
   meaning, 255
   and reaction rate, 602
   table of, 254
Erg, 206
Error, experimental; *see* Uncertainty
Esters, 481
   definition, 482
   naming, 482
   nomenclature, 484
   occurrence and use, 483
   of sorbitol, 553
Estimation, 15
Ethane:
   compared to ethanol, 462
   heat of formation, 196
   model of, 428
   properties of, 486
Ethanol, 462, 482
   boiling point, 104
   chemical reactions, 469, 470, 473
   determination of molecular formula, 464
   determination of structural formula, 467
   empirical formula, 465, 469
   molecular formula, 466
   nmr spectrum, 472
   physical properties, 471
   structural formula, 467, 469
   vapor pressure, 101
N-Ethyl acetamide, 483
Ethyl alcohol; *see* Ethanol
Ethylamine, 481, 483, 484
Ethyl bromide:
   chemistry of, 473
   structural formula, 470
Ethyl group, 472, 473
Ethyl iodide, 481
Ethylene:
   double bond in, 488
   structure, 428
   as unsaturated hydrocarbon, 487
Ethylene glycol, 466, 467, 472, 493
Exothermic reaction:
   definition, 62
   and ΔH, 194
   and reversibility, 225
Experiment, 4, 116
Experimental error; *see* Uncertainty
Extranuclear electron, 141, 144
Eyring, Henry, 210, 217

F

f orbital, 386, 532
Fable, Martin the Martian, 6
Face-centered cubic packing, 447
Family, chemical, 168
Faraday, Michael, 129
Fats, 556
Ferric ion, equilibrium with thiocyanate ion, 244
Ferricthiocyanate ion, equilibrium of, 244, 247, 252
Ferrous ion, reaction rate of, 212–213, 215
First ionization energy, 162, 393; *see also* Ionization energy
Fission, nuclear, 203
5-Fluoro-uracil, 567
Flame tests, 358, 359
Fleming, Sir Alexander, 114

"Floppy ring" species, 456
Fluorides:
　bonding orbitals, 424
　bond types, 417
　dipoles in, 424–425
　ionization energies, 417
　molecular shape, 424
　second-row, bonding capacity, 415
Fluorine:
　abundance in Earth's crust, 572
　boiling point, 181
　color, 521
　covalent radius, 522
　electron affinity, 407
　electron arrangement, 167
　electron configuration, 388, 407, 521
　electron dot representation, 407, 408
　energy in molecular formation, 407
　first ionization energy, 162, 521
　heat of vaporization, 505
　ionic character in bonds, 416
　ionic radius, 524, 525
　melting point, 181
　orbital representation, 407, 408
　properties of, 175, 181
　reaction with hydrogen, 176
　reaction with sodium, 175
　reactivity of, 184
　van der Waals radius, 522
Fluorine compounds:
　bonding in and geometry of, 424
　bond types in, 417
Fluorine oxide:
　geometry, 421
　molecular dipole in, 426
Fluorine molecule, bonding in, 407, 408
Fluoroborate ion, 605
Force, electric, 122
Force field, 363
Formaldehyde, 475
　formation of, 476
　oxidation of, 478
　structure of, 476
Formic acid:
　catalyzed decomposition, 228
　structural formula, 477
Formula:
　chemical, 39
　empirical, 173, 463, 496
　molecular, 463, 464, 496
　structural, 40, 47, 463, 469, 496
Fourth row of the periodic table, 528
Free radical, 410
Frequency:
　of colors, different, 366, 368
　of electromagnetic radiation, 368
　of hydrogen spectrum, 372
　meaning, 360
　relation to wavelength, 361
Fructose, 553
Fucson, R. C., 460
Fumaric acid, 459
Functional groups:
　definition, 472
　examples, 480
Fundamental property, 122
Furnace:
　basic oxygen, 548
　electric arc, 548
　open hearth, 548
Fusion, molar heat of:
　definition, 105
　pure substances, 105
Fusion, nuclear, 205

G

Gallium:
　atomic radius, 545
　electron configuration, 391
　halides of, 392
　oxides of, 392
Garbage collector analogy, 115
Gas:
　bulb, 70
　combining volumes of, 29, 36
　distance between particles of, 67
　effect of temperature change on, 69
　elements found as, 96
　ideal, 83, 94
　ideal gas constant, 84
　ideal gas law, 84
　　calculations, examples, 84, 85
　　from kinetic theory, 88
　inert, 8, 168
　kinetic theory of, 76
　liquid-gas phase change, 97
　liquid-vapor equilibrium, 98
　measuring pressure of, 71
　model for, 10
　molar volume of, 34, 67
　molecular structures of, 402
　natural gas, 461
　noble, 8, 168
　non-ideal, 93
　pressure, 11, 71
　pressure-volume data for, 27
　properties of, 28
　relative weights of, 31, 47
　solubility in a liquid, 273
Gasoline:
　composition of, 486
　and van der Waals forces, 436
Geiger, H., 141, 151–154
General gas equation, 82
　development of, 82, 83
　values of R in, 84
Generalization, 7
　melting points, 9
　reliability of, 7
Genetic code, 565, 567
Geometry; *see also individual elements and compounds:*
　of complex ions, 536
　determination of, 430
　of fluorine compounds, 424
　of metals, 445
　molecular, 417; *see also* Molecular geometry
　of molecular collisions, 218, 223
　of water, 419
Germer, L. H., 381
Gillespie, R. J., 418
Glucose, 552
Glutamic acid, 495
Glycerol, 556
Glyceryl tributyrate, 556
Glycine, 495
Gold, components of atom, 140
Goulian, M., 551, 566
Graphing, value of, 26
Graphite, 181–182
　delocalized electrons in, 440
　double bonds in, 440
　electrical conductivity, 440
　heat of combustion, 209
　properties of, 181
　structure of, 438
Guanine, 564–565
Gutte, Bernd, 559

H

h (Planck's constant), 370
H and ΔH, 194
$\Delta H^{\ddagger}$, 226
[H^+]; *see also* Hydrogen ion:
　in acidic solution, 305
　in basic solution, 305
　in neutral solution, 294
Haber, Fritz, 250
Haber process, 249
Half-cell, 330
　selection of standard, 330
　standard, 331
Half-cell potential; *see* Half-cell voltage *and* E^0
Half-cell reaction; *see* Half-reactions
Half-cell voltage, 330
　against copper reference cell, 330, 331
　against nickel reference cell, 331
　effect of concentration on, 331
　and Le Chatelier's principle, 331
Half-life, 158, 231, 573
Half-reactions, 321, 353
　balancing, 341
　definition, 321
　standard reduction potentials of, 598
　use of in balancing oxidation-reduction equations, 340
Halides:
　of hydrogen, 176
　metal, solubility of, 177
　solubility of, 277, 279
　variety, 177
Halogens:
　boiling points, 436
　chemistry of, 175
　color, 521
　and covalent bond, 174
　covalent radii, 522
　electron configuration, 521
　ionic radii, 524, 525
　ionization energies, 521
　melting points, 436
　properties of, 175

reactions with hydrogen, 176
reactions with sodium, 175
summary of, 178
toxicity, 521
van der Waals radii, 522
Hardness of metals and alloys, 443
Heat, definition, 191
Heat changes and chemical reactions; see Enthalpy and Energy
Heat conductivity of metals, 444
Heat content, 193, 208; see also Enthalpy and Energy
Heat of combustion, 61
  n-butane, 486
  cyclohexane, 486
  of diamond, 209
  of ethane, 209, 486
  of graphite, 209
  n-hexane, 486
  of hydrazine, 65
  of hydrogen, 62
  isobutane, 486
  of magnesium, 61
  of methane, 209, 486
  molar, 61
  n-octadecane, 486
  n-octane, 486
  propane, 486
Heat of formation, 198; see also Heat of reaction
Heat of fusion, 105
Heat of melting; see Fusion, molar heat of
Heat of reaction, 194, 225
  between elements, 196
  Law of additivity of, 194, 195, 200
  measurement of, 195
  predicting, 196
Heat of reaction to form:
  ammonia, 196
  CO and H_2, 192
  carbon dioxide, 196
  carbon monoxide, 196
  Fe_2O_3, 265
  ethane, 196
  fluorine molecules, F_2, 408
  hydrogen atoms, 403
  hydrogen bromide, 208
  hydrogen chloride, 263
  hydrogen iodide, 196
  magnesium oxide, 516
  nitric oxide, 196
  nitrogen atoms, 429
  nitrogen dioxide, 196
  P_4O_{10}, 232
  propane, 196
  sulfur dioxide, 196
  sulfur trioxide, 264
  sulfuric acid, 196
  water, 196
  water vapor, 196
Heat of solution, 272
  of chlorine in water, 273
  of chloroform in acetone, 286
  of iodine in alcohol, 272
  of iodine in benzene, 273
  of iodine in carbon tetrachloride, 272
  of nitrous oxide in water, 273
  of oxygen in water, 273
Heat of vaporization, 97
  of aluminum, 444, 505
  of argon, 188, 505
  of barium, 444, 521
  of beryllium, 444, 505, 521
  of boron, 444, 505
  of calcium, 444, 521
  of carbon, 505
  of cesium, 444
  of chlorine, 98, 505
  of copper, 98
  of fluorine, 505
  of helium, 188
  and kinetic theory, 102
  of krypton, 188
  of lanthanum, 444
  of lithium, 444, 505
  of magnesium, 444, 505, 521
  of neon, 98, 188, 505
  of nitrogen, 505
  of oxygen, 505
  of phosphorus, 505
  of potassium, 444
  of radon, 188
  of rubidium, 444
  of scandium, 444
  of silicon, 505
  of sodium, 98, 444, 505
  of sodium chloride, 98
  of strontium, 444, 521
  of sulfur, 505
  of water, 97, 98
  of xenon, 188
  of yttrium, 444
Helium:
  abundance in dry air, 570
  boiling point, 82, 169, 436
  electron arrangement, 167
  electron configuration, 388, 406
  first ionization energy, 162, 169
  formation from radium decay, 231
  heat of vaporization, 188
  interaction between atoms, 405
  isotopes, composition of, 144
  molar volume, 96
  orbital representation of, 406
  properties of, 169
  stability of, 169
Helix:
  double, 562
  single, 562
Hemin, 540, 541
Hemoglobin, 540–542
Hertz, Heinrich, 364
Hess, G. H., 195
Hess' Law of Constant Heat Summation, 195
Heterogeneous system, reaction rate of, 214
Hexagonal close-packing, 446
n-Hexane:
  properties of, 486
  structural formula, 552
Higgins, William, 17, 117

Hildebrand, Joel H., 266
Hirschmann, Ralph F., 559
Homogeneous systems, reaction rate, 214
Hoppe-Seyler, F., 561
Hugo, Victor, 568
Hybrid orbital, 420; see also Bonding
Hydrate, 46
Hydration:
  energy of, 293
  of hydrogen ion, 293
  of lithium chloride, 292
  of lithium ion, 293
  and randomness, 454
  of sodium chloride, 453
Hydrazine:
  dissociation of, 429
  heat of combustion, 65
  stability of, 411
Hydrides:
  boiling points, table, 454
  boron, 450, 451
  of second-row elements, 182
  sodium, 179
Hydrocarbons, 461, 496
  saturated, 485
  unsaturated, 485, 487
Hydrogen:
  abundance in dry air, 570
  abundance in Earth's crust, 572
  abundance in ocean water, 571
  apparent ionization energy, 417
  Bohr model for, 380
  boiling point, 82, 90
  bomb, 207
  bonding in, 403
  boron compounds, 450
  burning of, 62
  chemistry of, 178
  combining volumes, 36–38
  electron arrangement, 161
  electron configuration, 388, 406
  electron dot representation of, 406
  energy level scheme, 376, 394
  first ionization energy, 181
  freezing point, 90
  halides of, 176, 454
  heat of combustion of, 192, 196
  ion; see Hydrogen ion
  ionic character in bonds to, 417
  isotopes, composition of, 143–144
  in Jupiter's atmosphere, 580
  molar heat of combustion, 62
  molecular model, 41
  molecule, covalent bond in, 403
  nuclear fusion, 205
  orbital representation, 406
  physical properties of, 178
  production, 90
  quantum model of, 374
  reactions with halogens, 176

Hydrogen (cont'd.)
  relative weight, 31
  spectrum of, 370–374
  uniqueness of, 178
  wave model for, 380
Hydrogen atom:
  Bohr model for, 380
  electron dot representation, 406
  energy and interpretation of spectrum of, 369
  energy changes and spectrum of, 370
  energy changes in, 374
  energy level scheme for, 376, 394
  heat of reaction to form, 403
  orbital representation, 406
  quantum model for, 374
  spectrum of, 371, 376
  wave model for, 380
  zero energy of, 382
Hydrogen bonds, 454
  in DNA, 565
  energy of, 455
  nature of, 456
  occurrence, 455
  in proteins, 558
  representation of, 455
  significance of, 457
  in sugars, 555
Hydrogen bromide:
  boiling point, 454
  combining volumes, 29
  dissociation in water, 176
  formation, 176
  pressure-volume data, 27
  reaction mechanism for oxidation of, 215
Hydrogen chloride:
  acid equilibrium constant for, 308
  boiling point, 454
  dissociation in water, 176, 290
  equation for formation of, 39
  formation of, 176
  formula, 46
  heat of formation, 263
  heat of oxidation, 263
  molar volume of, 96
  molecular nature of, 174
  and NaOH in the same solution, 302–305
  oxidation, 263
Hydrogen difluoride ion, 456
Hydrogen fluoride, 176–177
  boiling point, 454
  bonding in, 408
  "floppy ring," species of, 456
  hydrogen bond in, 455
  liquid form, 455–456
Hydrogen halides, 176, 454
Hydrogen iodide:
  boiling point, 454
  dissociation in water, 176
  equilibrium constant, 251–253
  formation of, 176
  heat of formation, 196
  rate of formation, 216
Hydrogen ion:
  concentration of in solution, 295
  finding concentration of from K_A values, 310
  hydration of, 293
  importance of concentration of, 296, 315
  nature of in solution, 291, 293
  and pH scale, 305
  reduction of, 325
  transfer of in acid-base reactions, 313
Hydrogen peroxide:
  bonding in, 410
  and multiple proportions, 118
Hydrogen selenide, boiling point, 454
Hydrogen sulfide, boiling point, 454
Hydrogen spectrum:
  energy and, 369–370
  frequencies of, 372
Hydrogen telluride, boiling point, 454
Hydrogenation, 558
Hydrolysis:
  of cellulose, 557
  of fats, 556
  of starch, 557
  of sucrose, 555
Hydronium ion:
  model for, 293
  in proton-transfer theory, 314
Hydroquinone, 492
Hydrosphere:
  composition of, 571
  definition, 569
Hydroxide ion:
  concentration of in solutions, 295
  interaction of proton with, 421
Hydroxides:
  solubility of, 278
  of third-row elements, 510
Hydroxybenzene, 491; see also Benzene
Hydroxyl group, 472
Hydroxyl radical, 410
Hydroxylamine, 159
Hypobromite ion, 552
Hypochlorous acid:
  as proton donor, 514
  structure, 515
Hypophosphorous acid, 512
  electron dot formula, 513
  equilibrium constant, 513
Hypothesis, 7
Hypothesis, Avogadro's:
  and kinetic theory, 86
  reliability of, 76

I

Ice, 9, 96
  hydrogen bonding in, 455
  melting of, 104
  melting point, 9, 97
  molar heat of melting, 105
Icosahedral structure, 449–450
Icosahedron, 448
Ideal gas, 92, 96
Ideal gas constant, 84
Ideal gas law, 84, 96
  calculations using, 86
  from kinetic theory, 88
Identification:
  of acids, 297
  of bases, 298
Indicators:
  acid-base, 304–306
  litmus, 297–298, 300–301, 482
  methyl orange, 308, 490
Inert gases, 8, 170; see also Noble gases
Information:
  accumulations, 4, 18
  communicating, 4, 17–18
  organizing, 4, 18
Infrared radiation:
  energy, 370
  frequency, 369
  reactions with atmosphere, 570–571
  wavelength, 368
Infrared region, 367
Infrared spectroscopy, 430
Infrared spectrum:
  of carbon disulfide, 431
  of carbon tetrachloride, 431
  of dimethyl ether, 471
  of ethanol, 471
Inhibition of enzymes, 560
Inner transition elements, 392, 532, 548
Interference, 359
Intermolecular distance, 68
Intermolecular forces, 95, 112, 100–104
International Union of Pure and Applied Chemistry, 45, 600
Interstellar space, 581
Iodides:
  reaction of alkyl iodides, 481
  solubility, 277, 279
  solubility products, 280
Iodine:
  atomic volume, 175
  boiling point, 175
  covalent radius, 522
  crystal, unit cell, 271
  electron configuration, 521
  heat of solution;
    in alcohol, 272
    in benzene, 273
    in water, 272
  ionic radius, 524–525
  ionization energy, 521
  melting point, 175
  properties of, 175
  reaction with hydrogen, 176; see also Hydrogen iodide
  reaction with sodium, 175
  solubility in alcohol, 267–269
  solubility in carbon tetrachloride, 271
  solution in alcohol-water, 237–238
  van der Waals radius, 522
Ionic bond, 172, 415–416, 432, 452
  consequences of, 452
  electric dipole of, 416

Ionic character, 415–416
  in bonds to fluorine, 416–417
  in bonds to hydrogen, 417
Ionic crystals:
  properties of, 452
  stability of, 452
  structure of, 173, 175
Ionic radii, 523, 526
  of column two cations, 524–525
  of halides, 524–525
  of transition elements, 543–545
Ionic solids, 125, 275
  conductivity of, 173
  hydration of, 453–454
  solubility in water, 453
Ionic solutions, 275
  conductivity of, 124, 127–129, 289
  equilibrium in, 280
Ionization constants, 315
Ionization energy, 161, 393, 399
  of alkali metals, 388
  of alkaline earth elements, 520
  and atomic number, 396
  chemical significance of, 168
  determination of, 161
  and electron distance, 167
  and electron levels, 167
  of elements, 162, 167, 395
  of elements bonded to fluorine, 417
  and energy of electron, 202
  first, 187, 393
  of halogens, 521
  of hydrogen atom, 393–395
  and ionic character of bonds, 417
  maximum and minimum in, 165–166
  and the metallic bond, 444
  of noble gases, 169, 170
  and nuclear change, 167
  second, 163, 187, 395
  and stable electron configurations, 162
  successive:
    of Li, Be, B, 398
    of second-row elements, 501
    of third-row elements, 501
  of transition elements, 543–544
  trends in, 162, 396
  and valence electrons, 396
Ionization of water, 290–291
Ions, 123
  balancing equations involving, 135
  complex; see Complex ions
  electron attracting ability of, 328
  electron configuration of, 168
  formation of, 124, 144, 154
    energy relations in, 145
    positive and negative, 145
  hydration of, 291–294
  ionization energy of, 395
  movement of, 128
  names, formulas, and charges of common, 599
  in precipitation reactions, 134, 282
  "seeing" positive, 150
Iron:
  abundance in Earth's crust, 572
  abundance on lunar surface, 574
  atomic radii, 545
  chemistry of, 547
  commercial preparation of, 547
  in complex ion with cyanide, 539, 549
  in complex ion with oxalate, 538
  crystalline forms, 549
  electron configuration, 530
  in hemoglobin, 540
  in meteorites, 578
  oxidation numbers, 534
  properties of, 543
  rusting of, 548
  X-ray spectrum of, 595
Iron age, 543, 547, 548
Iso-, 485
Isomers:
  cis- and *trans*-, 429
  of C_2H_6O, 468–471
  of complex ions, 537
  dichloroethylene, 428
  propanol, 479
  structural, 428, 429
Isotopes:
  composition of selected, 144
  definition, 143
  determination of mass, 594
  and mass spectrograph, 151, 594
  of neon, 151

J

Janzen, Jay, 456
John of Salisbury, 138
Journals, research, 18
Jupiter, 576
  atmosphere, 579–580
  density, 578
  radius, 578
  relative mass, 578

K

K:
  definition, 253
  relation between value and equilibrium, 255
  table of selected values, 254
K_A:
  calculation of for benzoic acid, 308
  definition, 307
  determination of, 307
  tables of values, 308, 597
  use of in finding [H⁺], 310
K_{sp}:
  definition, 280
  use of, 280–283
K_w:
  determination of, 291
  and Le Chatelier's principle, 295
  size and significance of, 294
  variation with temperature change, 295
Kelvin, Lord (William Thomson), 66
Kelvin temperature scale, 81, 89

Ketones, 478, 479, 496
Kilocalorie, definition, 192
Kerosene, 486
Kinetic energy:
  of billiard balls, 198
  and chemical bonds, 202–203
  definition, 76
  determination of, 79, 87
  distribution, 221
  of electrons, 161
  in liquids, 95
  and molecular mass, 79
  and molecular motion, 201
  and molecular velocity, 79
  and pressure, 86
  and temperature of gases, 80
  three kinds in molecules, 201–202
  in vaporization, 102
Kinetic theory, 67
  and Avogadro's hypothesis, 86
  of gases, 76
  and ideal gas law, 88
  summary of, 89
  and temperature, 79
Kinetics, chemical, 211
Knudsen cell, 91
Kornberg, L. A., 551, 566
Kroll process, 507
Krypton:
  abundance in dry air, 570
  boiling point, 436
  electron configuration, 392
  heat of vaporization, 188
  ionization energy, 169
  melting point, 436
  properties of, 169

L

Lambda, λ, 361
Lanthanides; see Inner transition elements
Lanthanum:
  heat of vaporization, 444
  oxidation of, 532
Latimer, Wendell M., 318
Law:
  of additivity of heats of reaction, 195, 200
  Avogadro's hypothesis, 30
  Charles', 80
  of chemical equilibrium, 252
  of chemical equilibrium, derivation of, 256
  of combining volumes, 119
  of conservation of atoms, 54
  of conservation of charge, 135
  of conservation of energy, 63, 200
  of conservation of mass, 44, 54, 63
  of definite composition, 117
  definition, 21

Law (cont'd.)
  Hess's, of constant heat summation, 195
  ideal gas, 84
  of octaves, 186
  of simple multiple proportions, 117
Lead chromate, K_{sp}, 280
Lead sulfate, K_{sp}, 280
Le Chatelier, Henry Louis, 246
Le Chatelier's principle:
  and cobalt complex ions, 536
  concentration changes and, 247
  and equilibrium, 247
  and the Haber process, 249
  and methyl acetate-methyl alcohol equilibrium, 482
  and prediction of reactions, 337
  and pressure changes, 248
  temperature and, 246
  and values of K_{ic}, 295
  and voltage in electrochemical cells, 331
Levels, energy:
  of hydrogen atom, 376, 385
  of many-electron atom, 387
  model of, 374
Lewis, G. N., 198, 234, 314
Lewis acid-base interaction:
  defined, 315
  example of, 605
Light, 357
  and color, 357
  diffraction of, 359
  nature of, 358
  particle model of, 358
  spectrum of infrared, 367
  spectrum of ultraviolet, 366
  spectrum of visible, 365, 366
  wave nature of, 360
  wavelengths of visible, 366
Lime, manufacture of, 241
Limestone:
  in manufacture of lime, 241
  solubility equilibrium in cases of, 267
Linear complex, 538
Linear molecule, 423
Linear orbitals, 423
Linked solids; see Polymeric solids
Liquid, 9, 92, 434
  atoms and molecules in, 41
  disordered structure of, 104
  elements found as, 96
  intermolecular attraction in, 95
  kinetic energy in, 95
  and partial pressure, 100
  and phase change, 96
  solid phase changes, 104
  solutions, 105, 108
  and van der Waals forces, 436
  vapor equilibrium, 98
  vapor pressure of, 101
Lister, Joseph, 491
Lithium, 180
  boiling point, 181, 505
  bonding capacity of, 414
  bonding with fluorine, 415
  electron arrangement, 167
  electron configuration, 388
  first ionization energy, 162
  heat of vaporization, 444, 505
  ionization energies of, 398, 501
  isotopes, composition of, 144
  melting point, 181
  molecule of, bonding in, 442
  properties of, 171, 181
  physical properties, 499
  second ionization energy, 163
  structure of, 445
Lithium chloride, flame test for, 358
Lithium fluoride:
  bond type in, 417
  bonding in, 415, 424
  electric dipole in, 416
  molecular dipole of, 424
  molecular geometry of, 424
Lithium ion:
  hydration of, 292, 293
  nature of in solution, 291
Lithosphere, 569
  composition of, 571, 572
  of moon, 574
  of planets, 580
  properties of, 571
Litmus, 304–305
  color in acid solutions, 297, 300
  color in basic solutions, 298, 300
Lovell, James, 568, 574
Lowry, T. M., 313
Lucite, 493
Luna spacecraft, 573–576

M

M, 110
Macroscopic properties and equilibrium, 236
Magnesium:
  abundance in Earth's crust, 572
  abundance in ocean water, 571
  abundance on lunar surface, 574
  "acidic charge" of in $Mg(OH)_2$, 511
  boiling point, 501
  burning of, 55, 59
  in chlorophyll, 540
  complex ions of, 540
  compared to beryllium, 499
  concentration in seawater, 517
  crystal structure of, 446
  electron arrangement, 167
  electron configuration, 446, 519
  first ionization energy, 162
  heat of vaporization, 444, 505, 521
  ionic radius, 524
  ionization energies of, 501
  ionization of, half-cell potential of, 506
  melting point, 501
  metallic radius, 523
  molar heat of combustion, 61
  as reducing agent, 505
Magnesium hydroxide:
  "acidic charge" of magnesium in, 511
  basic strength, 511
  solubility in water, 299
Magnesium oxide, 55–57
  crystal type, 517
  heat of formation, 516
Magnesium sulfite hexahydrate, solubility in water, 287
Magnetic fields, 362
  in electromagnetic radiation, 364
  oscillating, 363
Magnetic wave, 363
Maleic acid, 459
Malm, John, 170
Manganese:
  abundance in Earth's crust, 572
  atomic sizes, 545
  electron configuration, 530
  oxidation numbers, 534
  properties of, 543
Manometer, 72
Manufacture of:
  aluminum, 131
  ammonia, 232, 249
  aspirin, 491
  calcium oxide, 241
  chromium plate, 546
  hydrogen, 63, 90
  iron, 547
  methanol, 263
  nitric acid, 64
  nylon, 494
  phenol, 491
  phosphorus, 517
  polyethylene, 493
  polymers, 493
  quicklime, 241
  sodium, 131
  titanium, 507
  "water gas," 191–192
Many-electron atoms, energy level scheme, 387
Mariner space probes, 576
Mars, 576–579
  atmosphere of, 579
  density, 578
  radius, 578
  relative mass, 578
  surface of, 577
Marsden, E., 141, 151–154
Mass:
  of atom, 154
  conservation of, 54
  conservation of in nuclear reactions, 156
  of electron, 140, 150
  of helium atom, 205
  of hydrogen atom, 205
  of neutron, 205
  of proton, 140
  and weight, 32
Mass balance, in balanced equations, 56
Mass, critical, 204
Mass-energy relationship, 205
Mass numbers:
  conservation of, 156
  definition, 146
  and isotopes, 143

Mass spectrograph, 140, 151, 155, 162, 593
Mass spectrometer; see Mass spectrograph
Mass spectrum, 594
Mass spectrum of neon, 594
Matter:
  electrical nature of, 119
  fundamental property of, 122
Matter-waves, 380
Maxwell, James Clerk, 364, 377
Measurement:
  precision of, 14
  uncertainty in, 14
Mechanism, reaction:
  for $CH_3Br + OH^-$, 475
  for decomposition of formic acid, 228
  definition, 215
  of $H_2 + I_2$, 217
  for oxidation of HBr, 215
  problems of, 217
  rate determining step of, 216
Melting:
  of fats and oils, 556
  of ice, 104
  of solids, regularities in, 8–9
Melting, heat of; see Fusion, molar heat of
Melting point; see also *individual substances*:
  of alkaline earth elements, 521
  of alkali metals, 171
  of CX_4 molecules, 437
  as characteristic property, 9
  definition, 97
  of halogens, 175
  of noble gases, 169
  of noble gases and halogens, compared, 437
  of pure substances, 105
  of saturated hydrocarbons, 486
  of second-row elements, 181
  of third-row elements, 501
  of transition elements, 543
  uncertainty in, 14
Melting temperature, 9
Mendel, Gregor Johann, 288
Mendeleev, Dmitri, 183, 186
Mercuric nitrate, electrolysis of, 130
Mercuric perchlorate, electrolysis of, 130
Mercurous perchlorate, electrolysis of, 130
Mercury (element), spectral lines, frequency of, 367
Mercury (planet):
  atmosphere, 579
  density, 578
  radius, 578
  relative mass, 578
Mercury vapor lamp, 357, 367
Merrifield, Robert B., 559
Metallic bond
  properties of, 443
  strength of, 444
Metallic crystals; see Packing, crystal

Metallic elements, 442
Metallic meteorites, 578
Metallic radius:
  of alkaline earth metals, 523, 525
  definition, 523
  of transition elements, 545
Metallic state, 440
Metals:
  alkali, 171
  alkaline earth, 519
  alloys of, 443
  bonding in, 441, 443
  electrons in, 442
  geometry of, 445
  heat conductivity of, 444
  heats of vaporization of selected, 444
  on periodic table, 442
  physical properties of, 186, 441, 443
  recognition of, 440
  slippage of planes in, 443
  structure of, 445
Meteorites, 578
Methane:
  boiling point, 486
  bond angle in, 422
  burning of, 57
  geometry of, 422
  heat of combustion, 486
  melting point, 486
  molar volume of, 96
  molecular geometry, 418
  reaction rate with oxygen, 212
  relative weight, 32
  space-filling model, 418
  structural formula, 40, 476
Methanol:
  equilibrium with methyl acetate, 482
  oxidation of, 475, 477
  structural formula of, 476
Method, scientific, 4
Methyl:
  acetate, 482–483
  alcohol; see Methanol
  amine, 484
  ammonium ion, 495
  bromide, 473
  butyrate, 484
  caprylate, 484
  formate, 484
  group, 473
  octanoate, 484
  orange;
    as acid-base indicator, 308
    structure of, 490
  propionate, 484
  salicylate;
    odor of, 491
    vapor pressure, 101
  violet, 259
N-Methyl acetamide, 495
n-Methyl urea, 560
Methylene blue, in carbon monoxide poisoning, 542
Meyer, Lothar, 186

Microscopic processes and equilibrium, 236
Microwave spectroscopy, 431
Miescher, F., 561
Millikan, Robert, 149
Millikan's oil drop experiment, 149
Minerals, 573
Mirsky, E. A., 561
Miscible liquids, 471
Model:
  atomic theory, 20
  characteristics of "good," 13
  difference from system, 11
  electrical, for atoms, 125
  for electrode processes, 127
  for electron arrangement, 166
  for gases, 10
    growth of, 21
    PV product, 26
  kinetic theory of gases, 76
  of metals, 186
  nuclear atom, 140
  quantum, for hydrogen, 374
  Rutherford, 153, 377
  Thomson, 153
  wave, for hydrogen atom, 380
Model, molecular:
  acetaldehyde, 477
  acetic acid, 478
  acetone, 479
  $[Ag(NH_3)_2]^+$, 538
  ammonia, 41
  boron, elemental, 449, 450
  carbon;
    diamond, 438, 500
    graphite, 438
  carbon dioxide, 201
  carbon monoxide, 229
  chloric acid, 515
  chlorous acid, 515
  chromate ion, 546
  $CoCl_4^{2-}$, 536
  $[Co(H_2O)_6]^{2+}$, 536
  $[Cr(NH_3)_4Cl_2]^+$, 537
  $[Cr(NH_3)_6]^{3+}$, 536
  cyclopentane, 485
  cyclopentene, 487
  diborane, 451
  dichloroethylene, isomers, 428
  dichromate ion, 546
  dimethyl ether, 469
  DNA, 564
  ethane, 428
  ethanol, 469
  ethyl bromide, 470
  ethylene, 428, 487
  $[Fe(C_2O_4)_3]^{3-}$, 538
  formaldehyde, 476
  formic acid, 228, 477
  hydrogen, 41, 118, 198
  hydrogen peroxide, 118
  hydronium ion, 293

Model, molecular (cont'd.)
  hypochlorous acid, 515
  lithium chloride, 292
  methane, 418, 476
  methanol, 475, 476
  methyl bromide, 475
  neopentane, 438
  oxygen, 118, 199
  P_4O_6, 508
  P_4O_{10}, 508
  iso-pentane, 485
  n-pentane, 438, 485
  perchloric acid, 515
  phosphorus:
    black, 503
    red, 503
    white, 508
  1-propanol, 479
  2-propanol, 479
  propylene, 487
  protein, 558
  silicon, 500
  sodium chloride, 173
  sulfur, 504
  types of, 40, 47
  water, 41, 118, 199
Molar concentration; *see* Molarity
Molar heat:
  of combustion; *see* Heat of combustion
  of fusion; *see* Fusion, heat of
  of melting; *see* Fusion, heat of
  of reaction; *see* Heat of reaction to form
  of solution; *see* Heat of solution
  of vaporization; *see* Heat of vaporization
Molar volume, 34; *see also* Atomic volume:
  of gases, common, 35, 71, 96
  of nitrogen, 68, 71
  of non-ideal gas, 96
  and temperature change, 69
Molarity, 110, 112
Mole, 30, 33–35
  in calculations, 51
  determination of, in laboratory, 51
  of electrons, measurement of, 133
  meaning, 34
  volume of; *see* Molar volume
Molecular architecture:
  of gases, 402
  of liquids, 434
  of solids, 434
Molecular dipole, 433; *see also* *individual substances*
Molecular formula:
  determination of, 464
  meaning of, 463
  relationship to empirical formula, 463
Molecular geometry, 417
  of BF_3, 422
  of BeF_2, 422
  of CF_4, 422
  of F_2O, 421
  of NF_3, 422

  of NH_3, 422
  and electric dipoles, 423
  linear, 423
  of methane, 418, 422
  planar triangular, 423
  summary, 423
  tetrahedral, 423
  of water, 419
Molecular models; *see* Models, molecular
Molecular shape:
  and boiling point, 438
  and melting point, 438
  and van der Waals forces, 437
Molecular size and boiling point, 436
Molecular solids:
  elements forming, 436
  van der Waals forces in, 435
Molecular structure:
  in biochemistry, 557
  method of determining, 430
Molecular velocity, 76, 77
  distribution of, 219
  and temperature, 79
Molecular weight:
  calculation of, 44
  definition, 32
  determination of, 466
Molecules:
  collision of, 213
  definition, 28
  determining number of, 53
  diatomic, 38
  energy stored in, 201
  formulas for, 38
  models of, 41
  polar, 416
  relative weights of, 31
Momentum, 87
Monomer, 493
"Monomeric" units, 494
Monosaccharide, 554
Moon:
  chemistry of, 573
  density of, 577
  diameter of, 577
  gravity of, 577
  Luna landings on, 573–576
  mass of, 577
  surface appearance, 575
  surface composition, 574
  Surveyor landings on, 573–576
  temperature of, 576
Mosley, H. G. J., 140, 595
Motion, types of, 201
Multiple bonds, 426
Multiple Proportions, Law of, 117
Myristic acid, 556

N

n, principal quantum number, 382
n-, in names of organic compounds, 485
Nagaoka, H., 377
Names:
  of common ions, 599

  of elements, the origin, 44–46
  of organic compounds, 483–484
Natural gas, 461
Negative charge, 126
Negative electrode, 127
Neon:
  abundance in dry air, 570
  boiling point, 436
  electron arrangement, 167
  electron configuration, 388
  first ionization energy, 162, 169
  heat of fusion, 105
  heat of vaporization, 98, 188, 505
  ionic radius, 524
  melting point, 436
  properties of, 169
  sign, 357
  spectrum of, 367, 371
Neon gas, mass spectrum of, 594
Neopentane:
  molecular shape and behavior, 437
  properties of, 438
  structure of, 438
Neptune, 576–580
  atmosphere, 579
  density, 578
  radius, 578
  relative mass, 578
Net electrical charge, 126
Network solid:
  and covalent bonds, 438
  definition, 187
  diamond and graphite, 187, 438
  silicon, 500
  of third-row elements, 500
  three-dimensional, 438
  two-dimensional, 439
Neutral:
  atom, 144
  electrically, 126
  particle, 139
  solution, 294
Neutron, 140–141
  charge of, 140
  and nuclear fission, 203
  and nuclear fusion, 205
  mass of, 140
Newlands, J. A. R., 160, 186
Newton, Sir Isaac, 20, 358, 528
Newton's laws of motion, 377
Newton's particle model, 369
Nickel:
  electron configuration, 530
  ionic radius, 545
  isotopes predominant, 543
  metallic radius, 545
  in meteorites, 578
  oxidation numbers, 534
  properties of, 543
  X-ray spectrum of, 595
Nietzsche, Friedrich, 518
Nirenberg, Marshall W., 551
Nitrates, solubility of, 276
Nitration, 490
Nitric acid:
  K_A of, 308
  reaction with water, 275
Nitric oxide:

combining volume, 37
heat of formation, 196
reaction with carbon dioxide, 224
Nitrobenzene, 490
Nitrogen:
  abundance in dry air, 570
  boiling point, 96, 181
  bonding capacity, 411
  combining volume, 38
  combustion of, 250
  density, 68, 181
  dissociation of, 113
  electron arrangement, 167
  electron configuration, 388, 411
  first ionization energy, 162
  heat of vaporization, 505
  intermolecular distance of, in gas phase, 68
  isotopes, composition of, 144
  melting point, 90, 181
  molar volume, 68, 71
  nuclear decay, 157
  pressure-volume data, 27
  properties of, 181
  reactions of, 113
  solubility in water, 571
  triple bond in, 429
  velocity of molecule, 77
Nitrogen dioxide:
  combining volume, 37
  equilibrium between two isomers of, 238–241
  heat of formation, 196
  reaction with carbon monoxide, 211, 224
Nitrogen trifluoride:
  electron dot formula, 411
  geometry of, 422
  naming, 46
Nitrous acid, K_A of, 307
Nitrous oxide:
  heat of solution in water, 273
  solubility in water, 273
nmr spectroscopy, 432
nmr spectrum:
  dimethyl ether, 471
  ethanol, 471
Nobel prize, 93, 246, 399, 417, 496, 562
Noble gases, 168, 187
  chemical reactivity of, 170
  compounds of, 170
  electron arrangements, 166
  ionization energies of, 169
  melting and boiling points, compared to halogens, 436
  properties of, 169
Nomenclature; see Names
Novocain, 492
n-pentane:
  properties, 438
  structure, 485
nu, ν, frequency of light, 360
Nuclear:
  atom, model of, 140
  charge, 146
  chemistry, 155
  decay processes, 155

decay, rate of, 158
energy, relationship to chemical energy, 203
fission, 203
fusion, 205
reactions;
  effect of temperature on, 231
  rate of, 230
reactor, 204
stability, factors in, 158
Nuclear magnetic resonance, 432, 471, 472
Nucleic acids, 561
Nucleon, 203, 207
Nucleotides, 551, 562, 563
Nucleus:
  binding energy per nucleon of, 206
  components of, 141, 154
  decay processes of, 155
  diameter of, 142
  energy stored in, 203
  properties of, 155
  "seeing," 151
  size of, 142
  stability of, 155, 158
  structure of, 357
Number; see Atomic number
Nyholm, R., 418
Nylon, 494

O

Observation:
  accuracy of, and conclusions, 13
  and belief in atoms, 117
  characteristics of, 5
  and conclusion, 115
  power of, 4
Ocean water, composition of, 571
n-Octadecane, properties of, 486
Octahedral complex, 537
Octanamide, 484
Octane, 484
n-Octane, properties of, 486
Octanoic acid, 484
1-Octanol, 484
"Octaves, Law of," 186
Octyl alcohol, 484
1-Octylamine, 484
OH group:
  behavior with ethyl group, 472
  bonding in, 410
Oil-drop experiment, Millikan's, 149
Oil film, color of, 359
Oil of wintergreen, 491
Onsager, Lars, 246
Open hearth furnace, 548
Operational definition, 301, 315
Oppenheimer, J. Robert, 2
Orbital representation of chemical bonding, 406, 432
Orbitals, 382
  arrangement of electrons in, 418
  computer plots of:
    $1s$, $2s$, $3s$, 384
    $2p$, 385
    d, 385
  empty, 413

hybrid, 420
introduction of protons into, 419
notation of, 423
p, 385
s, 384
shapes, 384
Ores, 573
Organic chemistry, 462
Organic compounds, 462
  nomenclature of, 484
  oxidation of, 475–480
Oscillating force fields, 362, 363, 364, 399
Overall reaction, 322
Oxalate ion, 212
Oxidation, 318
  of acetaldehyde, 478
  of alcohol, 480
  of ammonia, 197
  of chlorine, 527
  of copper, 322, 323
  definition, 322
  of ethanol, 477
  of formaldehyde, 480
  of hydrogen bromide, 215
  of iodine, 527
  of iron, 548
  of methanol, 475, 476, 477
  of organic compounds, 475, 477, 478
  and oxidation number, 345
  of phosphorus, 507
  of 1-propanol, 478
  of 2-propanol, 479
  of sugars, 555
  of sulfur, 509
  of zinc, 325
Oxidation numbers, 342–347
  of chromium, 533, 546
  and oxidation, 344
  and reduction, 344
  rules for assigning, 345
  of transition elements, 534
  use in balancing oxidation-reduction equations, 347
  of vanadium, 535
Oxidation-reduction reactions, 318
  balancing by half-reactions, 340
  balancing by oxidation numbers, 347
  balancing of, 339, 354
  in a beaker, 324, 325
  standard state in, 333
Oxide ion:
  proton interaction with, 420
  radius of, 524
  schematic representation, 419
Oxides:
  of aluminum, 506
  of chromium, 533
  of magnesium, 506
  of second-row elements, 182

Oxidizing agent:
 definition, 343
 third-row elements as, 507–509
Oxyacids, 512
 of chlorine, 514–516
 of phosphorus, 512–513
 of silicon, 512
 of sulfur, 514
Oxygen, 113–114
 abundance in dry air, 570
 abundance in Earth's crust, 572
 abundance in ocean water, 571
 abundance on lunar surface, 574
 boiling point, 90, 96
 bonding capacity of, 409
 combining volume, 37
 electron arrangement, 167
 electron configuration, 388, 409
 first ionization energy, 162
 fluorine compounds, 411
 freezing point, 90
 heat of solution in water, 273
 heat of vaporization, 505
 ionic radius of, 524
 isotopes, composition of, 144
 molar volume, 35, 71, 96
 molecular weight, 33
 preparation from $KClO_3$, 65
 properties of, 181
 reactions of, 113–114
 solubility in water, 273
Ozone, 570

P

p electron, 385
p orbitals, 385
Packing, crystal:
 body-centered cubic, 445
 face-centered cubic, 447
 hexagonal close-packed structure, 446
Palmitic acid, 556
Para-aminobenzoic acid, 560
Paraffin:
 burning of, 58
 composition of, 486
*Para*phthalic acid, 493
Partial pressure:
 explained, 77–79
 and vapor pressure, 100–102
Particle model for light, 358, 369
Particles, fundamental, 140
Pauli Exclusion Principle, 386
Pauling, Linus, 171, 417
Peat, 461
n-Pentane:
 properties of, 438
 structure of, 485
iso-Pentane, 485
Pepsin, 230
Perchloric acid:
 reaction with water, 275
 strength of, 514
 structure of, 515
Perfect gas, 96
Periodic table, 184–186
 and electronic structure, 386

 and elements forming metallic crystals, 442
 and elements forming network solids, 440, 441
 and elements forming van der Waals solids, 436
 and energy levels, 393
 history of, 186
 and ionization energies, 393
 number of elements in each row of, 393
 and orbitals, 386
 original of, 160
 the second column of, 519
 the seventh column of, 519, 521
 the sixth row of, 532
 the third row of, 498
 transition elements of, 529
Permanganate ion:
 reaction with ferrous ion, 212, 215
 reaction with hydrogen sulfide, 341, 347
 reaction with oxalate ion, 212
Petroleum, 461
pH scale, 305, 316
Phase:
 condensed, 104
 factors determining, 96
 gas, 9
 liquid, 9
 solid, 9
Phase changes, 9, 96
 energy change in, 63, 202
 energy requirements of, 97
 equilibrium in, 225
 liquid-gas, 96
 solid-liquid, 104
 and solutions, 106
Phases, condensed, electrical properties of, 123
Phenacetin, 492
Phenol, 491
Phosphates, solubility of, 278
Phosphides, 508
Phosphoric acid:
 in DNA, 562
 electron dot formula for, 512
 K_A, 307, 512
Phosphorous acid
 K_A, 513
 electron dot formula, 513
Phosphorus:
 abundance in Earth's crust, 572
 black, 503
 boiling points, 501, 505
 chemistry of, 507
 electron arrangement of, 167
 electron configuration, 502
 heat of vaporization, 505
 hydroxides of, 512
 ionization energies, 501
 oxides of, 508
 oxyacids of, 512
 preparation of, 517
 red, 502
  structure of, 503
  physical properties of, 501
 white;

 as oxidizing and reducing agent, 507
 physical properties of, 501
 structure, 502, 508
Phosphorus pentachloride, equilibrium:
 with phosphorus trichloride, 243, 247
 with phosphorus pentachloride, 243, 247
Photon, 381
π (pi) cloud, 489
Pig iron, 548
Pimentel, George, 171
Planar molecule, 422
Planck, Max, 369, 379
Planck's constant of action, 370
Planck's equation, 596
Planets:
 atmospheres of, 578, 579
 chemistry of, 576
 data on, 578
 giant, 576, 580
 lithospheres of, 580
 space probes of, 576
 terrestrial, 576
Plastics, 493
Platinum:
 as catalyst, 230
 reactivity in air, 532
 "Plexiglas," 493
Pluto, 576
 density, 578
 radius, 578
 relative mass, 578
Plutonium, nuclear decay, 156
Polar molecule:
 of chloroform, 453
 of lithium fluoride, 416
 of water, 293
Polar solvent, 453
Pole-vaulting analogy, 222
Polyamides, 494
Polyethylene, manufacture of, 493
Polymeric solids, of third row elements, 500
Polymerization, 493
 addition, 493
 condensation, 493–494
Polymers, 492
 cellulose, 555, 557
 DNA, 495, 562
 "nylon," 494
 phosphorus, 503
 plastics, 493
 proteins, 494, 558
 starch, 555, 557
Polyribonucleotide, 566
Polysaccharide, 555
Polystyrene, 492
Positive charge, 126
Positive electrode, 127
Positive ions, "seeing," 150
Positron, 157
Potassium:
 abundance in Earth's crust, 572
 abundance in ocean water, 571
 crystal structure of, 445

electron arrangement, 167
electron configuration, 391
first ionization energy, 162
heat of vaporization, 444
ionic radius, 524
metallic radius, 545
properties of, 171
second ionization energy, 163
spectral lines, frequency of, 367
X-ray spectrum of, 595
Potassium chrome alum, 539
Potassium hydroxide, dissociation in water, 299
Potassium permanganate, 341
Potassium vapor lamp, 367
Potential energy:
  barrier, 222
  of billiard balls, 198
  defined, 198
  diagrams, 222, 224, 226–228, 475
  of electron in hydrogen atom, 376
  of molecules, 201
  of paperweight in bookcase, 374–376
  and reaction rates, 222–225
Potentials, standard half-cell, E^0, 322
  concentration, effect of, 338
  defined, 333
  and equilibrium, 339
  and Le Chatelier's principle, 338
  table of, 598
  using to predict reactions, 336
  using to understand cells, 334
Precipitation:
  definition, 134
  in industrial processes, 283
  in natural processes, 283
  prediction of, 282
  rate of, 269–270
  reactions, 134
  separation by, 284
Precision:
  and accuracy, 16
  of measurement, 15
Prediction:
  from E^0 values, 336
  and effect of concentration, 337
  of heat of reaction, 196
  and Le Chatelier's principle, 337
  of new equilibrium condition, 246
  of precipitation, 282
  of reactions, 328
  reliability of, 338
Predominant reacting species, 134
Pressure:
  atmospheric, 24
  definition, 11, 71
  and Le Chatelier's principle, 248
  measurement, 71–73
  and number of molecules, 77
  partial, 77–79, 100–102
  and reaction rate, 214
  and relation to concentration, 248
  standard, 75
  vapor, 100, 236–237
Pressure-volume behavior:
  of ammonia, 27, 69, 93, 94
  of air, 22, 24

of carbon dioxide, 27
of hydrogen bromide, 27
of nitrogen, 27
of non-ideal gases, 93
Principle, 21
  Le Chatelier's, 246
  Pauli's Exclusion, 386
Principal quantum numbers, 382
Probability, and electron orbitals, 383–384
Procaine, 492
Promoting the electron, 412, 413
Propane:
  boiling point, 486
  heat of combustion, 486
  heat of formation, 196
  melting point, 486
  naming of, 484
1-Propanol, 478, 479
2-Propanol, 479
Property, fundamental, 122
Propionaldehyde, 478
Propionamide, 484
Propionic acid, 478, 482, 484
Propyl alcohol, 484
1-Propylamine, 484
Propylene, 487
Propyl group, 473
Proteins:
  composition of, 494
  hydrogen bonds in, 558
  model of, 558
Proton; see also Hydrogen ion
  and atomic number, 141
  definition, 126
  mass of, 140, 205
  transfer, 312
  transfer and hydronium ion, 314
  transfer concept of acids and bases, 312
Ptyalin, 230
Purines, 563
Pyrimidines, 563

Q

Qualitative analysis, 284
Qualitative presentation of data, 26, 585
Qualitative view of aqueous solubilities, 276
Quantitative aspects of equilibrium, 250
Quantitative correlation, 22
Quantitative presentation of data, 26, 585
Quantitative relationships and atomic theory, 51
Quanta, 370
Quantum mechanics:
  definition, 382
  and hydrogen 1s orbital, 403
  and orbital shape, 384
Quantum model for hydrogen atom, 374
Quantum numbers:
  definition, 382
  and energy levels, 382–384

and orbitals, 384
principal, 382
Quantum theory, 168, 358, 382
Quicklime, formation of, 241

R

R, ideal gas constant, 84
Radiation, electromagnetic, 364
Radio transmitting station, 365
Radioactivity:
  and age of the Earth, 573
  and nuclear chemistry, 155
  types of, 156–157
Radium:
  electron configuration, 519
  nuclear decay, 156, 207
  rate of decay, 230
Radium sulfate, K_{sp}, 280
Radius:
  atomic, 142
  covalent, 522, 526
  ionic, 523–526
  metallic, 523, 526
  van der Waals, 522, 526
Radon:
  boiling point, 436
  heat of vaporization, 188
  ionization energy, 169
  melting point, 436
  properties of, 169
Randall, Merle, 234
Randomness; see also Entropy:
  effect on solubility, 271–273, 285
  and equilibrium, 258
  in solar atmosphere, 581
  and solubility of ionic solids, 453
Rare earths; see Inner transition elements
Rate:
  of crystallization, 238
  determining step, 214
  of nuclear decay, 158
  of precipitation, 269
  of solution, 268
Rate of reaction:
  and catalysis, 227
  collision theory and, 213
  and concentration, 213
  and energy, 222
  and equilibrium, 236–238
  from equilibrium constant, 602
  of heterogeneous systems, 214
  measurement of, 211
  nature of reactants and, 212
  nuclear, 230
  rate determining step, 216
  and reaction mechanism, 214
  and temperature, 218
Reacting species, predominant, 134
Reaction coordinate, 223

Reaction, heat of; *see also* Heat of reaction:
  additivity of, 195
  definition, 194
  measurement of, 195
  and reaction rate, 225
Reaction mechanism, 214, 217, 231; *see also* Mechanism, reaction
Reaction rate; *see* Rate of reaction
Reactions:
  acid-base, 302
  of alkali metals with water, 173
  balancing half-cell, 341
  balancing oxidation-reduction, 339–341, 347–348
  conservation of charge in, 135
  conservation of energy in, 198
  endothermic, 63, 194, 225
  energy change in, 61, 191
  equations for, 38
  equilibrium in chemical, 238
  exothermic, 62, 194, 225
  half-cell, 321
  of halide ions, 177
  mechanisms of, 214, 217, 231
  nuclear, 155, 203
  oxidation-reduction, 318, 322, 324, 326, 339–342
  predicting, and electron transfer, 328
  predicting from E^0, 336–339
  structure determination from, 430
  substitution, of benzene, 489
  writing balanced equations for, 54
Red-Headed League, The, 4
Reducing agents:
  definition, 343
  in third-row elements, 505–509
Reduction:
  of copper ion, 326
  definition, 322, 353
  in electrolytic cells, 349
  of hydrogen ion, 325
  and oxidation numbers, 344
Reduction potentials, standard, table of, 333, 598
Reduction tendencies of transition metals, 545
Regularities, 3
  in melting of solids, 8
  search for, 6
Relativity, special theory of, 205, 362
Reproducibility of measurement, 15
Repulsions, of two atoms, 404
Resonance absorption frequency, 432
Ribonucleic acid; *see* RNA
Ribose, 566
RNA, 559, 561, 566
Rotational motion, 201
Rubidium:
  crystal structure of, 445
  heat of vaporization, 444
  properties of, 171
  reaction with chlorine, 172
Rule, 21
Rusting of iron, 548
Rutherford atom, 357, 377–379

Rutherford, Ernest, 140, 141, 151–154, 596
Rutherford experiment, 151–154, 575
Rydberg constant, 372, 380
Rydberg equation, 370
  predictions from, 373
  validity of, 372
Rydberg, J. R., 370

S

s electron, 384
s orbitals, 384
sp bonding, 423, 424
sp^2 bonding, 423, 424
sp^3 bonding, 423, 424
Salicylic acid, 491
Salt bridge, 319
Saponification, 556
"Saran," 493
Saturated compound, 467, 496
Saturated hydrocarbon, 485, 496
  pentanes, 485
  properties of, 486
Saturated solution, 110
Saturn, 576
  atmosphere, 579
  density, 578
  radius, 578
  relative mass, 578
Scandium:
  electron configuration, 391, 530
  ionic radii, 545
  metallic radius, 545
  oxidation numbers, 534
  properties of, 543
Scattering of alpha particles, 152–153
Science:
  activities of, 3, 4
  communication in, 4, 17
  uncertainty in, 14
Scientific method, 4
"Sea" of electrons, 187, 441, 442
Seaborg, Glenn T., 146
Second-column metals; *see* Alkaline earth metals
Second-row elements, 179
  boiling points, 505
  bonding capacity of, 409
  bonding patterns in, 186
  compared to third row, 504
  formulas of compounds, 182
  heats of vaporization, 505
  metals, 186–187
  network solids, 187
  properties of, 181
  trends in chemical properties, 182
  trends in physical properties, 180, 505
  trends in, summary, 187
Second-row fluorides, bonding in, 415
Self-oxidation-reduction, 527
Separations:
  by crystallization, 107
  by distillation, 107
  by precipitation, 284

Seventh-column elements; *see* Halogens
Significant figures, 15; *see also* Precision
Silica, 439
Silicic acid, 512, 517
Silicon:
  abundance in Earth's crust, 572
  abundance on lunar surface, 574
  chemistry of, 507
  covalent radius, 526
  crystal structure, 187, 500
  electron arrangement, 167
  electron configuration, 500
  electronics, use in, 501
  first ionization energy, 162
  geometry of, 500
  heat of vaporization, 505
  hydroxides of, 512
  ionization energies of, 501
  oxyacids of, 512
  properties of, 501
  X-ray spectrum of, 595
Silver bromate, K_{sp}, 280
Silver bromide, K_{sp}, 280
Silver chloride:
  K_{sp}, 280
  solubility in water, 281
Silver, complex with cyanide, 538
Silver iodate, K_{sp}, 280
Silver iodide, K_{sp}, 280
Silver ion, reduction of, 322, 323
Simple multiple proportions, law of, 117
Simple unit analysis, 51
Single bond, 426, 433
Single helix, 562
Sinsheimer, R., 551
Sixth-row elements, 532
Size:
  of atoms, 142
  compared, of atoms and ions, 524, 526
  of nucleus, 142
Slightly soluble, 111
Smog, 212
Snyder, H. R., 460
Soap bubble, color of, 359
Soaps, 556
Sodium:
  abundance in Earth's crust, 572
  abundance in ocean water, 571
  abundance on lunar surface, 574
  boiling point, 98
  as coolant in nuclear power plant, 113
  crystal structure of, 445
  electron arrangement, 167
  electron configuration, 389
  flame test for, 358–359
  heat of melting, 105
  heat of vaporization, 98, 444, 505
  ionic radius, 524
  ionization energies, 162, 163, 501
  manufacture of, 131
  melting point, 105
  metallic properties of, 499
  properties of, 171, 501

reaction with;
  chlorine, 109
  ethanol, 469
  halogens, 175
  hydrogen, 109
  water, 173
  as reducing agent, 505
  storage of, 486
Sodium benzoate, 311
Sodium carbonate, dissociation in water, 299
Sodium chloride:
  boiling point, 98, 452
  crystal;
    empirical formula for, 173
    ionic bond in, 172
    nature of, 172
  dissociation in water, 453
  electrolysis of, 130
  flame test for, 358
  heat of melting, 105
  heat of vaporization, 98
  hydration energy, 453
  melting point, 105, 452
  naming of, 46
  solubility in water, 274
Sodium ethoxide, 469
Sodium hydride, 109
Sodium hydroxide:
  basic strength, 511
  dissociation in water, 289
Sodium myristate, 556
Sodium stearate, 556
Sodium uranyl acetate, 276
Sodium vapor lamp, 357
Solids, 9, 92
  atoms and molecules in, 41
  concentration of in equilibrium constant, 253
  ionic, 125, 173, 452
  melting of, 8
  metallic, 186–187, 440–448
  molecular, 435–436
  molecular architecture of, 434
  network, 187, 438–440
  ordered structure of, 104
  phase changes of, 104–105
  polymeric, 492, 500
  solid-solution equilibrium, 237, 267
  solubility of, 271–272
  solutions, 108
Solubility, 110, 237, 266–287
  of acetates, 276
  of alkali metal compounds, 276
  of alkaline earth compounds, 277, 278
  aqueous, 276
  of ammonium compounds, 276
  of bromides, 277
  calculations, 281
  of carbonates, 278
  of chlorides, 277
  of common compounds in water, 279
  dynamic nature of, 268, 269
  of electrolytes in water, 275
  equilibrium, 237
  of ethyl alcohol in water, 267
  factors fixing, 271–273
  of gases, 273
  of hydrogen compounds, 276
  of hydroxides, 278
  of iodides, 277
  of ionic solids in water, 453
  of nitrates, 276
  of phosphates, 278
  product constant, K_{sp}, 280–285
  product constants, K_{sp}, selected, 280
  qualitative, 276
  quantitative, 280–281
  range of, 111
  and rate of solution, 268
  of sugars, 555
  of sulfates, 277
  of sulfides, 277
  of sulfites, 278
  tables of, 276–279
Soluble, 111
Solute, 109
Solutions, 105
  acid, 300
  aqueous, 123–125, 134, 274
  basic, 300
  boiling behavior of, 107
  components of, 107
  concentrations of, 109
  conductivity of, 112, 125, 127
  conductivity of, model for, 128
  electrolytic, 129–132, 275
  expressing composition of, 109
  gaseous, 108
  heat of, 272
  ionic, solubility equilibrium in, 280
  liquid, 108
  molarity of, 110, 112
  neutral, 294
  and phase changes, 106
  properties of, 111
  rate of, 268
  saturated, 111
  solid, 108
  solid-solution equilibrium, 237
Solvent, 109
Solvent, polar, 453
Sorbitol, 553
Space, interstellar, 581
Spectrograph, light, 566–567
Spectrograph, mass, 151, 593
Spectrometer; see Spectrograph
Spectroscopy:
  infrared, 430–431
  mass, 151, 593
  microwave, 431
  nmr, 432
Spectrum:
  electromagnetic, 365, 368
  of hydrogen atom;
    ultraviolet, 371, 372
    visible, 371
  of neon;
    mass, 594
    visible, 371
  tungsten, hot, 367
  visible, 365
Spontaneous reactions, 257, 336

endothermic, 258, 260
exothermic, 257
predicting, rules for, 337
Square planar complex, 538
Stability:
  of chemical bond, 403, 416
  of nucleus, 155
Stable electron arrangements, 166
Standard half-cell, 330
Standard half-cell potentials; see Potentials, standard half-cell
Standard pressure, 75
Standard state, 333
  for gases, 333
  for ions, 333
  for pure substances, 333
Standard temperature, 75
Staphylococcus bacteria, 560
Starch:
  composition, 555
  structure, 557
Stars, 581
State:
  standard, 333
  steady, 241
"Stationary states," 379
Steady-state system, 241
Stearic acid, 556
Steel:
  composition of, 108
  manufacture of, 548
Stock, Alfred E., 448
Stony meteorites, 578
Stored energy, 193, 200
STP, 75
Strontium:
  beta decay of, 156
  ionic radius, 524, 525
  metallic radius, 523, 525
  properties of, 521
  ionization energies, 52
  electron configuration, 519
  heat of vaporization, 444, 521
Strontium chromate, K_{sp}, 280
Strontium sulfate, K_{sp}, 280
Structural formula, 40, 463; see also Model, molecular
  alanine, 495
  α-amino acid, 495
  benzene, 488
  cellulose, 555
  chlorophyll, 541
  "Dacron," 494
  deoxyribose, 566
  determining chemically, 467
  determining physically, 471
  DNA, 564
  fructose, 554
  glucose, 553, 554
  glutamic acid, 495
  glycerol, 556
  glycine, 495

Structural formula (cont'd.)
  graphite, 439–440
  hemin, 541
  n-hexane, 552
  "Nylon," 494
  paraphthalic acid, 493
  polyethylene, 493
  a protein, 495
  ribose, 566
  starch, 557
  sucrose, 554
  uracil, 566
Structural isomers, 463, 468
Structure determination, methods of, 430
Styrene, 492
Sublimation, 285
Substance, pure, 107
Substitution reactions, 489
Substrate:
  false, 560
  function, 559
Sucrose, 552
  hydrolysis of, 555
  structure, 554
Sugars, 552
  burning of, 58
  conductivity of aqueous solution, 112
  consumption of, 555
  dissolving of, 111, 274
  hydrogen bonding in, 555
  hydrolysis of, 555
  oxidation of, 555
  properties of, 554, 555
  simple, 552
Sulfanilamide, 560
Sulfates, solubility of, 277
Sulfides, solubility of, 277
Sulfites, solubility of, 278
Sulfur:
  abundance in Earth's crust, 572
  abundance in ocean water, 571
  boiling point, 505
  electron arrangement, 167
  electron configuration, 504
  first ionization energy, 162
  heat of vaporization, 505
  hydroxides of, 512, 514
  oxidation numbers of, 343–347
  oxides of, 509
  oxidizing and reducing properties, 507
  oxyacids of, 512, 514
  properties of, 501
  structures of, 504
Sulfur dioxide:
  boiling point, 96
  formation of, 343
  heat of formation, 196
  molar volume, 96
  oxidation of, 344
  reaction with water, 347
Sulfur trioxide:
  formation of, 344
  formula for, 46
  reaction with water, 347
  vapor pressure, 509

Sulfuric acid:
  acid equilibrium constant, 308
  formation of, 196
  heat of formation, 196
  reaction with water, 275, 308
Sulfurous acid, K_{sp}, 307
Sun:
  atmosphere of, 581
  molecules in atmosphere, 203, 580
  temperature of, 580
Superconductivity, 82
Superfluidity, 82
"Super-rubber" ball analogy, 10–13, 21, 27, 93, 218, 365, 380
Surveyor spacecraft, 573–576
Symbols, 43
  in formulas, 46
  for isotopes, 155
  origin of, 45
Synthesis:
  of enzymes, 230, 559
  of proteins, 495
  of ribonuclease, 559
  of viral DNA, 551
System:
  closed, 241
  heterogeneous, 108, 214
  homogeneous, 105, 214
  open, 241

T

T, 83
"Teflon," 493
Television transmitting station, 365
Television tube, 358
Temperature:
  and atomic velocity, 221
  boiling, 103–104
  and control of equilibrium, 237
  definition, 74
  effect on equilibrium, 245
  effect on gas volume, 80, 81
  effect on K_w, 294–295
  effect on solubility, 272, 273
  and flow of thermal energy, 79
  and kinetic energy, 80, 221
  and Le Chatelier's principle, 246
  and kinetic theory, 79
  measurement of, 74
  melting, 9, 97, 105
  of planetary atmospheres, 579
  and reaction rate, 218
  scales:
    absolute, 81
    Celsius, 75, 81
    centigrade, 75
    Fahrenheit, 75
    Kelvin, 81
    standard, 75
  of sun's surface, 580
Template, DNA, 564–565
Tetrahedral:
  arrangement of H_2O around Li^+ and H^+, 293
  complex, 536
  geometry, 423
  molecule, 422

Tetrahedron, 40, 422
Tetrathionate ion, 526
Tetravalent atom, 412, 426
Thallium bromide, K_{sp}, 280
Thallium chloride:
  K_{sp}, 280
  solubility in water, 282–283
Thallium iodide:
  K_{sp}, 280
  solubility in water, 281
Theory:
  atomic, 20, 51
  Brønsted-Lowry, 314
  collision, 213, 218
  definition, 21
  garbage collector, of garbage disappearance, 116
  kinetic, 66, 76, 79, 86, 100, 102, 139
  phlogiston, 17
  quantum, 358, 382
  relativity, 205, 362
  valence-shell electron-pair repulsion, 423
Thermal energy, 79
Thermal equilibrium, 74, 80
Thermite reaction, 209, 507
Thermometer:
  Bureau of Standards, 16
  differential, 16
Thermometers, 74
Thiocyanate ion, equilibrium with ferric ion, 244
Thiosulfate ion, 526
Third-row elements, 498
  boiling points of, 505
  compared to second-row elements, 499, 501, 502, 504, 505
  gaseous molecules of, 504
  heats of vaporization, 505
  hydroxides of, 510–516
  ionization energies of, 501
  metals, 499
  oxyacids of, 512–516
  as oxidizing agents, 505, 507
  physical properties of, 499, 501
  polymeric solids of, 500
  as reducing agents, 505, 507
  trends in properties of, 505
Thomson, J. J., 140
Thomson atom, 140, 152, 153
Three-center bond, 450
Three-dimensional network solid, 438
Threshold energy, 218, 221
Thymine, 562, 564–565
Tin, velocity of gaseous molecule, 219
Titanium:
  abundance in Earth's crust, 572
  electron configuration, 530
  ionic radii, 545
  manufacture of, 507
  metallic radius, 545
  oxidation numbers, 534
  properties of, 543
  X-ray spectrum of, 595
Titanium tetrachloride, 507

Titrations:
 acid-base, 302
 definition, 304
Toluene, 10
Trajectory of electron, 384
Trans-isomers, 429, 537–538
Transition elements, 528
 as catalysts, 547, 549
 complex ions of, 535–542
 compounds of, 533
 electron configuration of, 391, 530
 hydroxides of, 547
 ionic radii, 545
 metallic radii, 545
 oxidation numbers of, 534
 properties of, 532, 543
 reactivities, 532
Transition metals; *see* Transition elements
Translational motion, 201
Trinitrotoluene, melting point, 97
Triple bond, 429, 433
Trivalent atom, 411
Tungsten:
 carbides of, 136
 spectrum of hot, 367
Turkevich, Anthony, 574
Two-dimensional network solid, 438, 439

U

Ultraviolet radiation:
 in formation of HCl, 36
 frequency of, 369
 reactions in atmosphere, 571
Uncertainty, 14–16
Uncertainty principle, 380
Unit analysis, simple, 51
Univalent atom, 409
Unsaturated hydrocarbons, 487
 definition, 485
 reactions of, 488
 structural formulas for, 487
Uracil, 566
5-fluoro-Uracil, 567
Uranium, compounds of, 355
Uranium-235, fission of, 203, 204, 208
Uranium-238:
 and age of earth, 573
 nuclear decay of, 158
Uranium hexafluoride, 48
Uranus, 576
 atmosphere, 579–580
 density, 578
 radius, 578
 relative mass, 578
Urea, 559
Urease, 559

V

Valence electrons:
 definition, 396
 representation of, 406
Valence-shell electron-pair repulsion theory, 423

Vanadium:
 electron configuration, 530
 ionic radii of, 545
 metallic radius of, 545
 oxidation numbers, 534
 oxides of, 535
 properties of, 543
 X-ray spectrum of, 595
van der Waals forces:
 compared to hydrogen bond, 455
 definition, 435
 and molecular shape, 437
 and molecular solids, 435
 and number of electrons, 436
van der Waals, J. D., 435, 522
van der Waals liquids, 436
van der Waals radius:
 compared to covalent radius, 522
 definition, 522
van der Waals solids, 436
Vanillin, 492
Vapor pressure:
 of benzene, 100, 101
 of carbon tetrachloride, 101
 definition, 99
 and equilibrium, 235
 of ethyl alcohol, 101
 and kinetic theory, 100
 of methyl salicylate, 101
 and temperature change, 100
 of water, 99, 100, 101
Vaporization, heat of; *see also* Heat of vaporization:
 definition, 97
 metals, 444
 second- and third-row elements, 505
Velocity:
 atomic and molecular, distribution of, 220
 and kinetic energy, 221
 definition, 219
 measurement of atomic and molecular, 219
 of nitrogen molecule, 77
 temperature and molecular, 79
Venera space probes, 576
Venus, 576
 atmosphere of, 579
 density of, 578
 radius of, 578
 relative mass of, 578
Very slightly soluble, 111
Vibrational motion, 201
Virus, synthesis of, 551, 566
Voltage, 328, 333
Voltaic cell, 348, 350, 352, 354
Volume; *see also* Molar volume *and* Atomic volume:
 relationship to pressure, 24–28
 relationship to temperature, 80–84

W

Wall, F. T., 190
Water:
 as acid, 315

as base, 315
boiling point, 98, 454
bonding in, 419–421
conductivity of, 123
Dalton's formula for, 136
decomposition of, 63, 198
density, 255
dipole in, 426, 453
dissociation of, 289
electrolysis of, 63, 198
as electrolyte, 290
equilibrium constant for, 295
in equilibrium expressions, 255
formation of, 62, 63
gases in, solubility of, 273, 571
geometry of, 419
heat of formation, liquid, 196, 199
heat of formation, vapor, 196
heat of vaporization, 97, 98
and hydrogen bonds, 455
melting of, 104
melting point, 105
molarity of, 291
molecular model, 40, 41
molecule, polarity of, 293
as polar solvent, 453
reactions with alkali metals, 173
solubility of substances in, 276–279
solutions, conductivity of, 123
structure of, liquid, 104
structure of, solid, 104
vapor pressure of, 101
vaporization of, 97
"Water Gas":
 energy changes in manufacture of, 193
 manufacture of, 191
Water, ocean, composition of, 571
Watson, James D., 496, 550, 562
Wavelength, λ:
 of light, 361, 366, 368–369
 of ocean waves, 360–361
 relationship to frequency and velocity, 361, 369
Wave mechanics, 382
Wave nature:
 of electron, 142, 381
 of light, 360
 of matter, 381
Waves:
 characteristics of, 360
 electric, 363–364
 magnetic, 363
 ocean, 360–361
Weight and mass, 32; *see also* Atomic weight *and* Molecular weight
Weinstock, Bernard, 170
Werner, Alfred, 535
Wichers, Edward, 45
Wilkins, M. H. F., 496, 562
Wood, reaction rate of burning, 214

Woodward, Robert B., 493
Work, 145, 198, 259

X

Xenon:
   abundance in dry air, 570
   boiling point, 436
   compounds of, 170–171
   electron arrangement of, 166
   heat of vaporization, 188
   ionic radius, 523
   ionization energy, 169
   melting point, 436
   properties of, 169
Xenonhexafluoroplatinate, 170
X rays:
   diffraction of, 430
   frequency and wavelength of, 368
   spectra of, and atomic number, 595
Xylene, 10

Y

Yttrium:
   heat of vaporization, 444
   X-ray spectrum of, 595

Z

Zinc:
   complex with ammonia, 539
   in corrosion prevention, 548
   electron configuration, 391, 530
   ionic radius, 545
   metallic radius, 545
   oxidation numbers, 534
   oxidation of, 325
   properties of, 543
   X-ray spectrum of, 595
Zirconium, X-ray spectrum of, 595

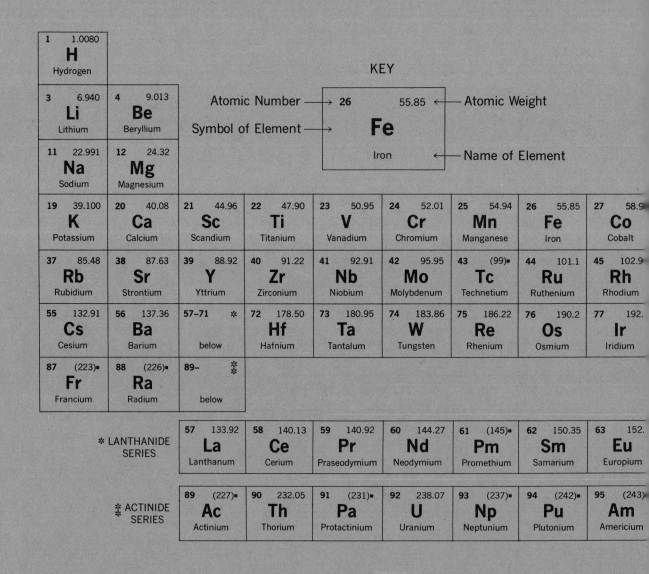